Lecture Notes in Electrical Engineering

Volume 880

The book series *Lecture Notes in Electrical Engineering* (LNEE) publishes the latest developments in Electrical Engineering - quickly, informally and in high quality. While original research reported in proceedings and monographs has traditionally formed the core of LNEE, we also encourage authors to submit books devoted to supporting student education and professional training in the various fields and applications areas of electrical engineering. The series cover classical and emerging topics concerning:

- Communication Engineering, Information Theory and Networks
- Electronics Engineering and Microelectronics
- Signal, Image and Speech Processing
- Wireless and Mobile Communication
- Circuits and Systems
- Energy Systems, Power Electronics and Electrical Machines
- Electro-optical Engineering
- Instrumentation Engineering
- Avionics Engineering
- Control Systems
- Internet-of-Things and Cybersecurity
- Biomedical Devices, MEMS and NEMS

For general information about this book series, comments or suggestions, please contact leontina.dicecco@springer.com.

To submit a proposal or request further information, please contact the Publishing Editor in your country:

China

Jasmine Dou, Editor (jasmine.dou@springer.com)

India, Japan, Rest of Asia

Swati Meherishi, Editorial Director (Swati.Meherishi@springer.com)

Southeast Asia, Australia, New Zealand

Ramesh Nath Premnath, Editor (ramesh.premnath@springernature.com)

USA, Canada:

Michael Luby, Senior Editor (michael.luby@springer.com)

All other Countries:

Leontina Di Cecco, Senior Editor (leontina.dicecco@springer.com)

**** This series is indexed by EI Compendex and Scopus databases. ****

More information about this series at https://link.springer.com/bookseries/7818

Yi Wang · Kristian Martinsen ·
Tao Yu · Kesheng Wang
Editors

Advanced Manufacturing and Automation XI

Springer

Editors
Yi Wang
School of Business
Plymouth University
Plymouth, UK

Tao Yu
Shanghai University of Engineering Science
Shanghai, China

Kristian Martinsen
Department of Manufacturing
and Civil Engineering
Norwegian University of Science
and Technology
Gjøvik, Norway

Kesheng Wang
Department of Mechanical and Industrial
Engineering
Norwegian University of Science
and Technology
Trondheim, Norway

ISSN 1876-1100 ISSN 1876-1119 (electronic)
Lecture Notes in Electrical Engineering
ISBN 978-981-19-0571-1 ISBN 978-981-19-0572-8 (eBook)
https://doi.org/10.1007/978-981-19-0572-8

This Springer imprint is published by the registered company Springer Nature Singapore Pte Ltd.
The registered company address is: 152 Beach Road, #21-01/04 Gateway East, Singapore 189721, Singapore

Preface

IWAMA—International Workshop of Advanced Manufacturing and Automation—aims at providing a common platform for academics, researchers, practising professionals and experts from industries to interact, discuss current technology trends and advances, and share ideas and perspectives in the areas of manufacturing and automation.

IWAMA began in Shanghai University in 2010. In 2012 and 2013, it was held at the Norwegian University of Science and Technology, in 2014 at Shanghai University again, in 2015 at Shanghai Polytechnic University, in 2016 at Manchester University, in 2017 at Changshu Institute of Technology, in 2018 at Changzhou University, in 2019 at Plymouth University and in 2020 at Lingnan Normal University. The sponsors organizing the IWAMA series have expanded to many universities throughout the world, including Plymouth University, Changzhou University, Norwegian University of Science and Technology, SINTEF, Manchester University, Shanghai University, Shanghai Polytechnic University, Changshu Institute of Technology, Xiamen University of Science and Technology, Tongji University, University of Malaga, University of Firenze, Stavanger University, The Arctic University of Norway, Shandong Agricultural University, China University of Mining and Technology, Indian National Institute of Technology, Donghua University, Shanghai Jiao Tong University, Changshu Institute of Technology, Dalian University, St. Petersburg Polytechnic University, Hong Kong Polytechnic University, Lingnan Normal University, Civil Aviation University of China, China Instrument and Control Society, Henan Polytechnics, Shandong University of Science and Technology, Jimei University, etc. As IWAMA becomes an annual event, we are expecting more sponsors from universities and industries, who will participate in the international workshop as co-organizers.

Manufacturing and automation have assumed paramount importance and are vital for the economy of a nation and the quality of daily life. The field of manufacturing and automation is advancing at a rapid pace, and new technologies are also emerging. The main challenge faced by today's engineers, scientists and academics is to keep on top of the emerging trends through continuous research and development.

COVID-19 has affected almost more than 40 million people in 216 countries. Almost in all the affected countries, the number of infected and deceased patients has been enhancing at a distressing rate. It continues to have a critical impact on society, business and manufacturing. It appears that the impact of the pandemic will continue for the foreseeable future and life as we have known it has changed in a very short period of time. But what does this mean for manufacturing industries and the way we do business? In manufacturing, the new normal is accelerating towards us. How can technology and digitalization help manufacturers adapt and survive this new paradigm?

Several enabling technologies could help us to fight the pandemics, such as:

- Condition monitoring,
- Artificial intelligence,
- IIoT,
- Digital twin,
- Industrial additive manufacturing and other information technology (WMS, MES, PAM, HMI).

As the early prediction can reduce the spread of the virus, it is highly desirable to have intelligent prediction and diagnosis tools. The inculcation of efficient forecasting and prediction models may assist the government in implementing better design strategies to prevent the spread of virus. In this proceedings, some papers are using machine learning (ML) and deep learning (DL) methods in the diagnosis and prediction of COVID-19. Especially, the intelligent manufacturing systems are promising to create a safe working environment by using the automated manufacturing assets which are monitored by the networked sensors and controlled by the intelligent decision-making algorithms. These technologies can be used in the fields of manufacturing:

- Predictive maintenance and machine inspection,
- Production planning,
- Quality control and
- Others.

IWAMA 2021 takes place in Henan Polytechnics, Zhengzhou, China, 11–12 October 2021, organized by Plymouth University, Norwegian University of Science and Technology, Shanghai University and Henan Polytechnics. Because of the COVID-19 pandemic, we arrange IWAMA 2021 as a virtual conference. In this proceedings, there are some papers shown how we can use intelligent manufacturing to increase the quality and productivity of manufacturing during the period of COVID. All participants in the conference are able to present their papers and discuss their options in an online way. The programme is designed to improve manufacturing and automation technologies for the next generation through discussion of the most recent advances and future perspectives, and to engage the worldwide community in a collective effort to solve problems in manufacturing and automation.

The main focus of the workshop is the transformation of present factories, towards reusable, flexible, modular, intelligent, digital, virtual, affordable, easy-to-adapt, easy-to-operate, easy-to-maintain and highly reliable "smart factories". Therefore, IWAMA 2020 has mainly covered 5 topics in manufacturing engineering:

- Industry 4.0
- Manufacturing systems
- Manufacturing technologies
- Production management
- Design and optimization.

All papers submitted to the workshop have been subjected to strict peer-review by at least 2 expert referees. Finally, 98 papers have been selected to be included in the proceedings after revision processes, which will be published in in Lecture Notes of Electrical Engineering (LNEE) by Springer and will be indexed by EI. We hope that the proceedings will not only give the readers a broad overview of the latest advances, and a summary of the event, but also provide researchers with a valuable reference in this field.

On behalf of the organization committee and the international scientific committee of IWAMA 2021, I would like to take this opportunity to express my appreciation for all the kind support, from the contributors of high-quality keynotes and papers, and all the participants. My thanks are extended to all the workshop organizers and paper reviewers, to Henan Polytechnics, Plymouth University and NTNU for the financial support, and to co-sponsors for their generous contribution. Thanks are also given to Jian Wu, Haishu Ma, Xueping Chu, Long Xiao and Shifeng Chen, for their hard editorial work of the proceedings and arrangement of the workshop.

Yi Wang

Yi Wang
General Chair of IWAMA 2021

Organization

Organized and Sponsored by

HP (Henan Polytechnics, China)
PLYU (Plymouth University, UK)
NTNU (Norwegian University of Science and Technology, Norway)

Co-organized by

LNU (Lingnan Normal University)
SHU (Shanghai University, China)
SSPU (Shanghai Second Polytechnic University, China)
TU (Tongji University, China)
SJTU (Shanghai Jiao Tong University, China)

Honorary Chairs

Minglun Fang, China
Kesheng Wang, Norway
Jan Ola Strandhagen, Norway

General Chairs

Yi Wang, UK
Kristian Martinsen, Norway
Tao Yu, China

Local Organizing Committee

Long Xiao (Chair)
Yi Wang
Xueping Chu
Jian Wu

Haisshu Ma
Shifeng Chen
Lilan Liu

International Programme Committee

Jan Ola Strandhagen, Norway
Kesheng Wang, Norway
Odd Myklebust, Norway
Per Schjølberg, Norway
Knut Sørby, Norway
Erlend Alfnes, Norway
Kristian Martinsen, Norway
Hirpa L. Gelgele, Norway
Wei D. Solvang, Norway
Yi Wang, UK
Chris Parker, UK
Jorge M. Fajardo, Spain
Maria J. Calle, Spain
Torsten Kjellberg, Sweden
Fumihiko Kimura, Japan
Gustav J. Olling, USA
Michael Wozny, USA
Byoung K. Choi, Korea
Wladimir Bodrow, Germany
Guy Doumeingts, France
Van Houten, Netherlands
Peter Bernus, Australia
Janis Grundspenkis, Latvia
George L. Kovacs, Hungary
Rinaldo Rinaldi, Italy
Gaetano Aiello, Italy
Romeo Bandinelli, Italy
Jinhui Yang, China
Dawei Tu, China
Lilan Liu, China
Minglun Fang, China
Binheng Lu, China
Xiaoqien Tang, China
Ming Chen, China
Xinguo Ming, China
Keith C. Chan, China
Shouqi Cao, China
Guiqing Li, China
Jin Yuan, China

Chuanhong Zhou, China
Jianqing Cao, China
Yayu Huang, China
Shirong Ge, China
Guijuan Lin, China
Shanming Luo, China
Benlian Xu, China
Zumin Wang, China
Guohong Dai, China
Sarbjit Singh, India
Vishal S. Sharma, India
Bo Chen, China
Ziqiang Zhou, China
Yuan He, China
Zhengyu Hong, China

Secretariat

Jian Wu
Haishu Ma
Xueping Chu

Contents

About the Editors

Dr. Yi Wang obtained his PhD from Manufacturing Engineering Center, Cardiff University, in 2008. He is an associate professor in Business School, Plymouth University, UK. Previously, he worked in Southampton University, Nottingham Trent University and Manchester University. He holds various visiting lectureships in several universities worldwide. He has special research interests in supply chain management, logistics, operation management, culture management, neuromarketing, big data and data analytics, and Industry 4.0/5.0. He has published over 100 technical peer-reviewed papers in international journals, chapters and conferences. He has authored 3 books, for example, Operations Management for Business, Fashion Supply Chain Management and Data Mining for Zero-defect Manufacturing, edited 6 books and made 5 chapters.

Dr. Kristian Martinsen took his PhD at the Norwegian University for Science and Technology (NTNU) in 1995, with the topic “Vectorial Tolerancing in Manufacturing”. He has 15 years’ experience from manufacturing industry. He is a professor at Faculty of Engineering and Department for Manufacturing and Civil Engineering, the Norwegian University for Science and Technology (NTNU), and is the manager of the Manufacturing Engineering Research Group in this department. He is a corporate member of the international academy for production engineering and a member of the High-Level Group of the EU Technology Platform for manufacturing: MANUFUTURE. He is the manager of the Norwegian national infrastructure for manufacturing research laboratories: MANULAB, and is the international coordinator for the Norwegian Centre for Research-based Innovation SFI MANUFACTURING. He has published many papers in international journals and conference. His major research area is within the field of measurement systems, variation/quality management and tolerancing: Towards Industry 5.0: Research Challenges in Human-Machine Systems.

Dr. Tao Yu is the president of Shanghai University of Engineering Science (SUES), China, and professor of Shanghai University (SHU). He was the president of Shanghai Second Polytechnic University (SSPU), China. He received his PhD from

SHU in 1997. He is a member of the Group of Shanghai manufacturing information and a committee member of the International Federation for Information Processing IFIP /TC5. He is also an executive vice president of Shanghai Science Volunteer Association and executive director of Shanghai Science and Art Institute of Execution. He managed and performed about 20 national, Shanghai, enterprises commissioned projects. He has published hundreds of academic papers, of which about thirty were indexed by SCI, EI. His research interests are mechatronics, computer integrated manufacturing system (CIMS) and grid manufacturing.

Dr. Kesheng Wang holds a PhD in production engineering from the Norwegian University of Science and Technology (NTNU), Norway. Since 1993, he has been appointed professor at the Department of Mechanical and Industrial Engineering, NTNU. He was a director of the Knowledge Discovery Laboratory (KDL) at NTNU. He is also an active researcher and serves as a technical adviser in SINTEF. He was elected member of the Norwegian Academy of Technological Sciences (NTVA) in 2006. He has published 27 books, 10 chapters and over 300 technical peer-reviewed papers in international journals, chapters and conferences. His current areas of interest are intelligent manufacturing systems, applied computational intelligence, data mining and knowledge discovery, predictive/cognitive maintenance and Industry 4.0.

Experimental Study on the Piezoresistive Effect of Modified Multi-walled Carbon Nanotube-Modified Sensor

Yue Li[1,2](✉), Yin He[1,2](✉), and Hao Liu[1,2](✉)

[1] School of Textile Science and Engineering, Tiangong University, Tianjin 300387, China
liyue_05@sina.com, heyin@tiangong.edu.cn
[2] Institute of Smart Wearable Electronic Textiles, Tianjin 300387, China

Abstract. This study reports the use of a silane coupling agent (KH550) and sodium dodecyl benzene sulfonate (SDBS) for the modification of the MWCNTs to explore the influence of chemical modification on the piezoresistive performance of the flexible sensor film filled with multi-walled carbon nanotubes (MWCNTs). Modified MWCNTs and polyurethane (PU) were used to prepare the sensing materials. The fabric electrode was made of silver-plated yarn, and the film was assembled into a flexible piezoresistive sensor, following which it was integrated into the openings for fingers in the knitted gloves. The results revealed that the modification could significantly improve the electrical and mechanical properties of the MWCNTs. It could also be used to determine the forces generated by the hand during pressing, holding, etc. Thus, the sensor presents huge application prospects in the field of fabrication of wearable devices, medical detection, and electronic skin.

Keywords: Flexible piezoresistive sensor · Multi-walled carbon nanotube/polyurethane composite material · Health monitoring · Wearable device

1 Introduction

In recent years, the development of telemedicine has promoted the rapid development of smart wearable monitoring equipment. This monitoring equipment can be used to record data such as heart rate, ECG, rate of respiration, body temperature, and force [1, 2]. At present, the research on and development of smart wearable devices primarily starts with flexible sensors. While the flexible sensor satisfies the physiological monitoring parameters, it also has the advantages of low elastic modulus and large strain. It is also bendable, foldable, and washable [3, 4]. The piezoresistive sensors are the most widely used among all. This sensor can monitor dynamic and static pressure signals and exhibits a wide response range [5, 6].

As a one-dimensional conductive nanomaterial, carbon nanotube has a small percolation threshold due to its size (nanometer) and high aspect ratio. It exhibits excellent electrical conductivity [7, 8]. Multi-walled carbon nanotubes (MWCNTs) have extremely high strength and toughness. The Young's modulus is approximately 1.8 Tpa [9].

Y. Wang et al. (Eds.): IWAMA 2021, LNEE 880, pp. 1–11, 2022.
https://doi.org/10.1007/978-981-19-0572-8_1

The conductive composite material of MWCNTs and high molecular polymers exhibit good electronic conductivity and excellent mechanical properties [10]. These are currently the new favorites for the fabrication of flexible sensors [3]. The improved performances could be achieved by modifying the MWCNTs. Wei et al. used PDMS as the matrix and silane coupling agent-modified MWCNTs as conductive fillers to prepare stretchable, highly sensitive piezoresistive strain sensing fibers [11]. Perez et al. studied a composite film made of modified MWCNTs, multilayer graphene sheets, and polyurethane [12]. The piezoresistive response of this sensing film is nonlinear, and the highest sensitivity of 4.59 kPa can be achieved at a specific concentration of MWCNTs. Jang et al. used modified MWCNTs and PDMS to prepare a dielectric elastomer which exhibits an enhanced response compared to pure PDMS [13]. These researchers improved the sensitivity of the flexible piezoresistive sensor or expanded its response range by modifying the MWCNTs. However, its application in the field of specific wearable devices was largely ignored.

The sensor proposed in this study uses KH550 and SDBS double modifiers to modify the surface of carbon nanotubes for the first time. This can effectively improve the uniformity of nanoparticles in the polymer matrix and enhance the piezoresistive effect of the conductive film. The sensitivity of the modified MWCNTs/PU conductive film reaches 4.208 kPa^{-1}, and its nonlinear error, hysteresis error, and repeatability error are $\pm 8\%$, $\pm 8.2\%$, and $\pm 6.63\%$, respectively. These values are respectively 9%, 16.72%, and 54.95% lower than the corresponding values of the unmodified MWCNTs/PU conductive film. This paper, for the first time, reports the combination of flexible sensors based on MWCNTs/PU composite film with gloves and integrates health monitoring systems with gloves.

2 Experiment

2.1 Experimental Materials

MWCNTs (Chengdu Organic Chemistry Co. Ltd, China); thermoplastic polyurethane (PU) (Jiangbeilu Chemical Technology Co. Ltd, Shanghai, China); N,N-dimethyl formamide (DMF) (Yuyuan Technology Co. Ltd, Tianjin, China); plating Silver yarn (Qingdao Hengtong Weiye Special Fabric Technology Co. Ltd, Shandong, China); 0.4 mm single-core multi-strand silver-plated wire (Shanghai Xiangyu Wire and Cable Co. Ltd, Shanghai, China); 30 D knitted fusible interlining (Suzhou Dushi Garment Accessories Co. Ltd, Jiangsu, China); fiber Stretch cotton gloves (Suzhou Camelot Labor Insurance Products Co. Ltd, Jiangsu, China); 0.2 mm-thick copper foil (Shenzhen Hubao Jincheng Metal Products Co. Ltd, Guangdong, China); KH550 and SDBS (Beijing Chemical Reagent Company, China); and anhydrous ethanol (Tianjin Blue Crystal Company, China) were used to conduct the experiments.

2.2 Preparation of the Sensing Layer

MWCNTs and 1wt.% SDBS and/or KH550 were added to an ethanol solution, and the mixed solution was sonicated for 2 h (power: 80 W). Then the mixed solution was centrifuged at 6000 r/min for 30 min using a high-speed centrifuge. Subsequently, the suspension was vacuum filtered, following which the modified MWCNT on the filter paper were heated in a 100 °C drying oven for 3 h. The modified MWCNT powder was collected and placed in the drying oven for later use.

As shown in Fig. 1, the solution method was used to prepare the MWCNT/PU films, KH550 multiwall carbon nanotube (K-MWCNT)/PU films, SDBS multiwall carbon nanotube (S-MWCNTs)/PU films, and KH550/SDBS multiwall carbon nanotube (KS-MWCNT)/PU films. Of each MWCNTs, K-MWCNTs, S-MWCNTs, and KS-MWCNTs powders, 0.015 g was dissolved in 10 mL of DMF (dispersion solution). After dispersing the powders under conditions of ultrasonic vibration for 1 h, 3 g of the PU monomer that exhibits good solubility was added to the solution at 60 °C. Magnetic stirring was carried out in the water bath for 2 h. Following this, the mixture was ultrasonicated for another 30 min. The vacuum oven was evacuated over 15 min. Finally, the four dispersed mother liquors were uniformly cast in a glass mold and dried in air for 5 h. A film sample with a thickness of 50 μm ±0.24 was prepared.

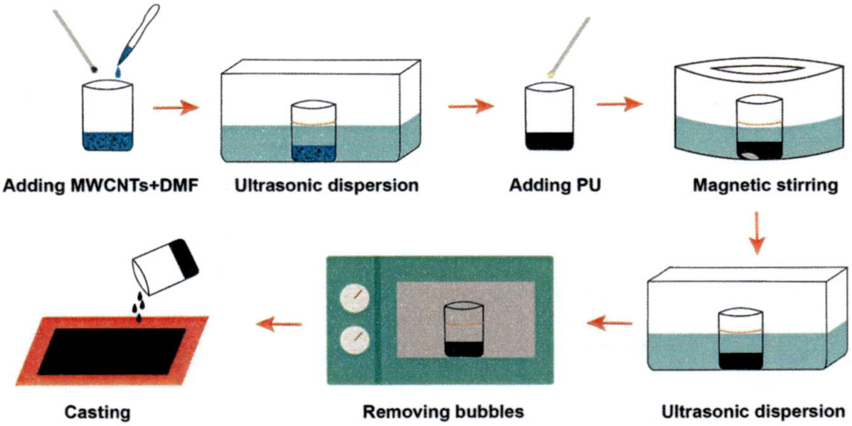

Fig. 1. Preparation of sensing film

2.3 The Mechanism of the Piezoresistance Enhancement of the Sensing Film

2.3.1 Particle Size Analysis of MWCNTs

Figure 2 presents the probability distribution density plots of the diameters of the unmodified MWCNTs and modified MWCNTs. Analysis of the results presented in figures b, c, and d reveals that the probability distribution density (for the diameter) of the modified MWCNTs was more concentrated than that of the unmodified MWCNTs. Thus, the mean value and confidence level in probability can be used to express the size of the diameter: the diameter of MWCNTs: $x_1 = 218.28$ nm $\pm$ 9.99 ($P\alpha = 0.9973$); the diameter of S-MWCNTs: $x_2 = 96.18$ nm $\pm$ 4.62 ($P\alpha = 0.9973$); the diameter of K-MWCNTs: $x_3 = 66.58$ nm $\pm$ 3.00 ($P\alpha = 0.9973$); and the diameter of KS-MWCNTs: $x_4 = 63.8$ nm $\pm$ 2.63 ($P\alpha = 0.9973$).

The test data shows that the particle size of unmodified MWCNT is approximately 218.48 nm, which is an increase of 247% from the nominal size (50 nm) of the used MWCNT. This indicates that the unmodified MWCNTs have agglomerated in the DMF solution. The particle size of the MWCNT modified with KH550 + SDBS double modifier is approximately 63.8 nm, which is close to its nominal value (36% of the diameter of unmodified MWCNTs) and smaller than the average particle diameter of MWCNTs modified with KH550 or SDBS.

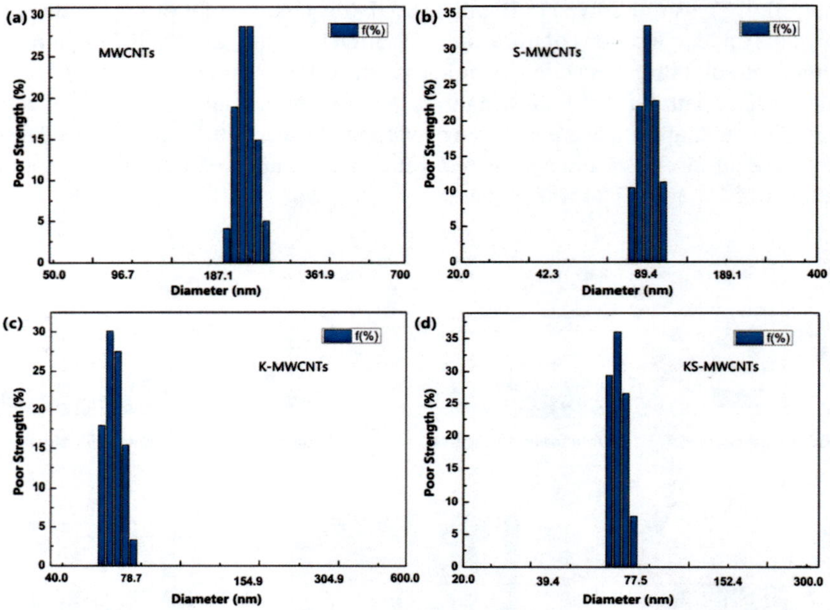

Fig. 2. Probability density distribution of particle size and diameter of unmodified and modified MWCNTs (a) MWCNTs, (b) S-MWCNTs, (c) K-MWCNTs, and (d) KS-MWCNTs

2.3.2 X-ray Photoelectron Spectroscopy (XPS) Analysis

The coexistence of C in the KS-MWCNTs/PU conductive film is evidenced by a shoulder on the main peak at 284.58 eV assigned to the C species. During the modification process, the modifier removes the amorphous carbon and impurities present on the surface of the MWCNTs. Numerous oxygen-containing groups are generated in the nozzle or defective sites of the MWCNTs, improving the dispersibility of the MWCNTs. The distance between the MWCNTs becomes more uniform, and the effect of pressure makes it easier for the particles to come in contact with each other, leading to the increase in conductivity (Fig. 3).

2.3.3 Mechanical Properties of the Sensing Film

Figure 4(a) presents the relation between the filler content and tensile strength of the MWCNTs/PU and modified MWCNTs/PU conductive films. The tensile strength of the MWCNTs/PU film decreases gradually with an increase in the MWCNT content. As the

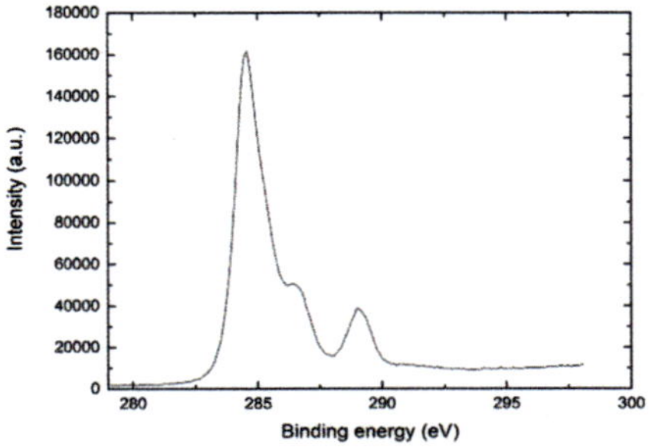

Fig. 3. C 1 s XPS spectrum of the KS-MWNTs/PU conductive film

doping content of the modified MWCNTs increases, the tensile strength of the modified MWCNT/PU conductive film first increases and then decreases. When the conductive filler content is 1wt.%, the tensile strength of the MWCNTs/PU is the maximum (1554.24 ± 14.06 MPa). The modified MWCNT/PU conductive films exhibit the maximum tensile strength value when the filling content is 10 wt.%. The value is estimated to be 1114.43 ± 29.36 MPa, which is 46.7% higher than the tensile strength recorded for the MWCNT/PU conductive films (10 wt.%). In addition, the tensile strength of the KS-MWCNT/PU conductive film is greater than that of the conductive film prepared following other modification methods under conditions of varying filler contents. When the filler content is 10 wt.%, the tensile strength of the KS-MWCNT/PU conductive films is 18.2% and 50.2% higher than that of the K-MWCNT/PU and S-MWCNTs/PU conductive films, respectively.

Figure 4(b) presents the relation between the filler content and elongation at break of the MWCNT/PU and modified MWCNT/PU conductive films. The data presented in the figure reveals that the elongation at break of the MWCNT/PU conductive film first increases and then decreases with an increase in the number of unmodified carbon nanotubes. The maximum elongation at break for the MWCNT/PU (5 wt.%) films is 355.5%. As the filler content increases, the elongation at break for the modified MWCNT/PU conductive films gradually decreases. The elongation at break of the KS-MWCNT/PU conductive films drops (from approximately 404% to 169%). When the filler content is 15 wt.%, the elongation at break of the modified MWCNT/PU conductive film is 62.5% higher than that of the MWCNT/PU conductive films.

2.3.4 Conductive Network Model and Electrical Characteristics of the Sensing Films

The data presented in Fig. 5(a) presents the relationship between the conductivity of the MWCNT/PU and MWCNT/PU conductive films and the modification methods and filler contents. The conductivity of the composite conductive film increases with the increase in filler content. Analysis of the data reveals that the conductivity of the modified MWCNT/PU conductive films is lower than that of the unmodified MWCNT/PU conductive films.

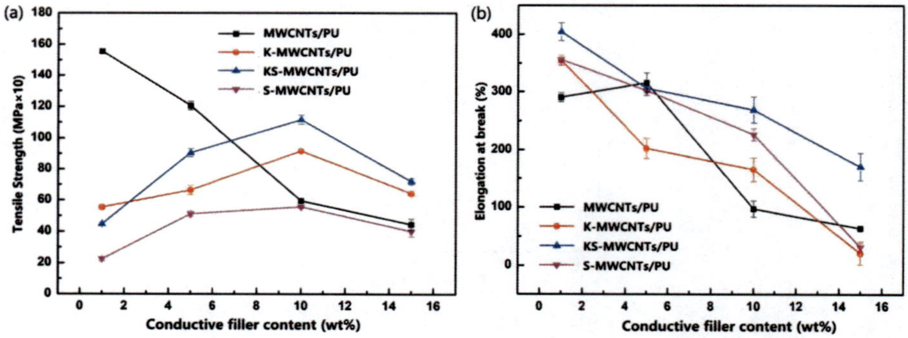

Fig. 4. (a) Tensile strength of the modified MWCNT/PU conductive films and MWCNT/PU conductive films under different filler contents and (b) elongation at break of modified MWCNT/PU conductive films and MWCNT/PU conductive films at different filler contents

The data presented in Fig. 5(b) shows the relationship between the filler content and the conductivity of the MWCNT/PU and modified MWCNT/PU conductive films under conditions of varying excitation signal frequencies. At different frequencies, the conductivity of the composite conductive film increases with the increase in the filler content. Under conditions of the same frequency, the conductivity of the modified MWCNT-doped conductive film is lower than that of the unmodified MWCNT-doped conductive film. The difference decreases with the increase in the filler content.

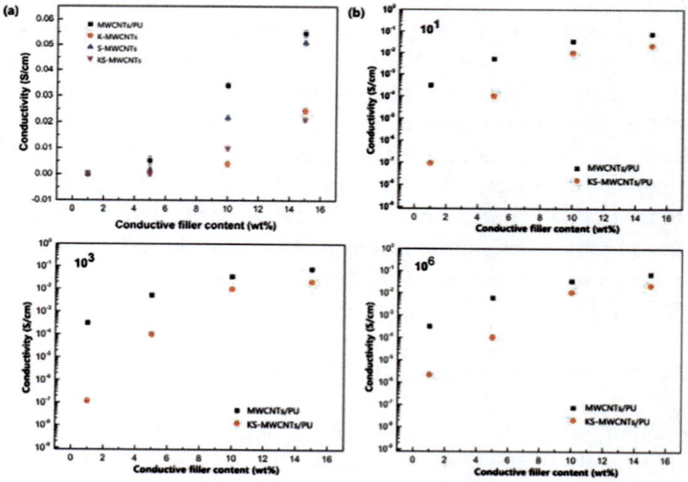

Fig. 5. (a) Conductivity of modified MWCNT/PU conductive film and MWCNT/PU conductive film under conditions of varying filler content and (b) filler content-conductivity relationship of modified MWCNT/PU conductive film and MWCNT/PU conductive film at 10^1, 10^3, and 10^6 Hz of excitation signal frequency

3 Application Verification of Sensing Films

3.1 Fabrication and Calibration of Smart Gloves

The flexible piezoresistive sensor is a key component that can be used to obtain biological information of the human body through a wearable electronic system [14]. We assembled it on the glove and determined its piezoresistive performance through several application scenarios. The single-sided patterned electrode of the glove finger, as shown in Fig. 6(a), was sewn with silver-plated yarn. The sensor used an MWCNT/PU film and is fixed with a circular cloth lining. A wire was sewn at the end of the silver wire to connect the experimental instrument.

Copper sheets were used as electrodes on a horizontal table, and weights were added to the sensor to conduct a calibration experiment. It was observed that the resistance of the sensor changed with the change in stress. Calibration experiments can help understand the conductivity of the selected sensor composite material under force [15]. The resistance value presented in Fig. 6(b) was recorded for the state of the sensor composite material stabilized under force. It was fitted and calculated the functional relationship between sensor resistance and force:

$$y = 1.88846 \times 10^5 - 9.30252 \times 10^5 \times x + 2.21034 \times 10^6 \times x^2 - 1.69076 \times 10^6 \times x^3 \quad (1)$$

where x is the force and y is the resistance.

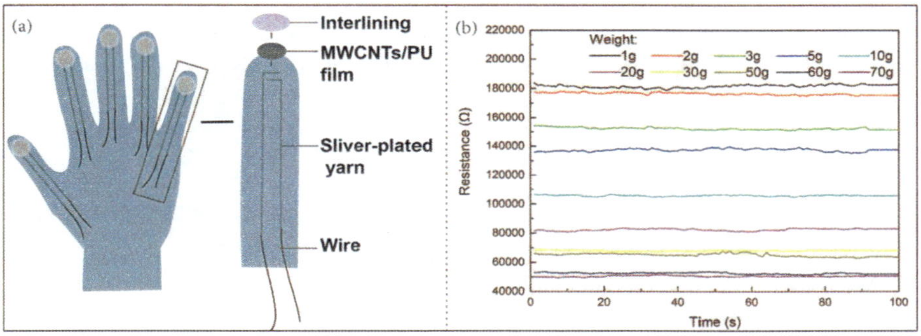

Fig. 6. (a) Integration of the flexible sensors on gloves and (b) the stable resistance value of the flexible sensor in the calibration experiment

3.2 Single Finger Pressing Test

As shown in Fig. 7(a), the index finger of the left-hand presses the table top regularly. In the first group of experiments, the index finger pressed the table for 5 s, and relaxed for 5 s (repeat operation: 1000 s). In the second group of experiments, the index finger pressed the tabletop every 20 s.

Figures 8(b)(c) show the resistance change curves of the two sets of tests. When the index finger presses the tabletop, the resistance decreases, and when the index finger rises, the resistance rises and returns to the initial value. It fails to return to the initial value. According to Eq. (1), the maximum force exerted by the index finger in the first set of tests is in the range of 4.28–4.57 N, and the maximum force exerted by the index finger in the second set of tests is in the range of 2.20–2.30 N.

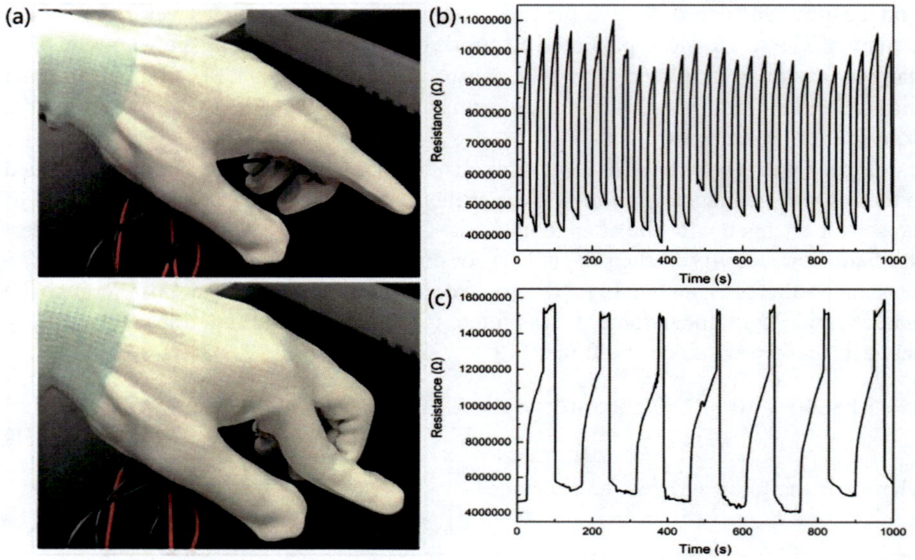

Fig. 7. (a) Schematic diagram of index finger pressing, (b) change of resistance when the sensor is pressed once every 5 s, and (c) change of resistance when the sensor is pressed once every 20 s

3.3 Five-Finger Pressing Test (Piano Experiment)

The song "Painter" with only five tones in Fig. 8(b) was selected. Five fingers imitate five different tones, and the fingers were pressed evenly on the table in rhythm. The thumb corresponded to tone 1, the rest of the fingers were similar.

Figure 8(c) shows the change in the resistance curve of the right hand with five fingers (simulating playing the piano). It can be seen that when the finger presses the key, the conductivity of the sensor on the finger corresponding to each tone changes regularly. When the key is pressed, the conductivity becomes stronger, and the resistance becomes smaller. When the key is loosened, the conductivity becomes weaker, and the resistance becomes larger. During the whole song, the conductivity of the sensor composite material increased as the key was pressed and vice versa.

3.4 Ten-Finger Grabbing Test (The Holding Ball Experiment)

As shown in Fig. 9(a), the two hands hold a basketball at a constant speed and at regular intervals. The ball was help for 10 s and relaxed for 20 s in an action cycle.

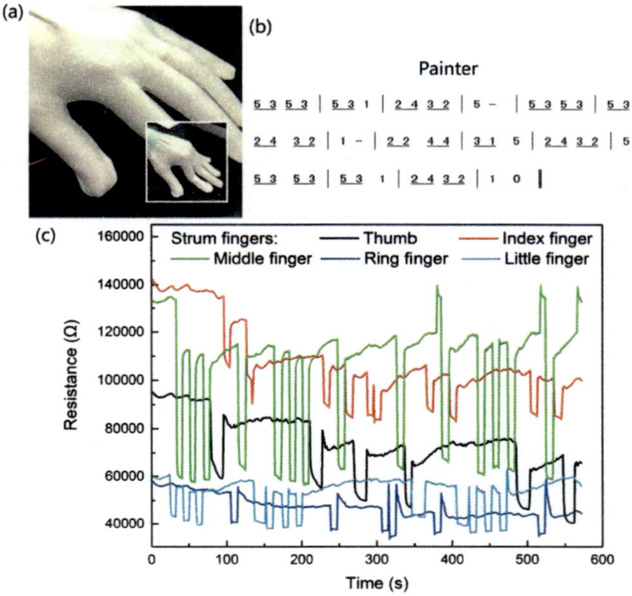

Fig. 8. (a) Schematic diagram of the piano experiment, (b) music notation for "Painter", and (c) change in resistance of the five fingers when playing eight pieces of music

When holding the ball with both hands, the resistance change of the sensor composite material connected by ten fingers can be recorded at the same time. When holding the ball, the resistance value of the ten fingers decreases at the same time, and the conductivity of the sensor composite material becomes stronger. Under conditions of relaxation, the resistance value of the ten fingers increases at the same time, and the conductivity of the sensor composite material becomes weaker. The average applied force for the fingers during the time the basketball held was calculated to be approximately 0.7 N based (Eq. (1)).

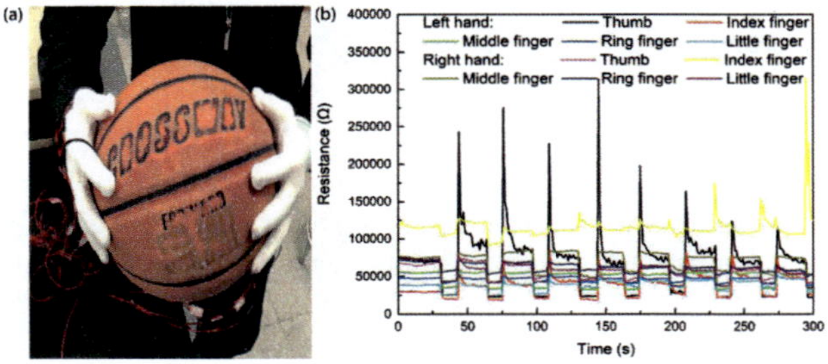

Fig. 9. (a) Schematic diagram of the holding ball experiment and (b) change in resistance of the ten fingers while holding the ball with both hands

3.5 Strain Test at Joint

In order to monitor the changes in the stress and resistance of the sensing layer at the joint, as shown in Fig. 10(a), the long strip MWCNT/PU conductive layer at the joint was pressed at a constant speed (press time: 10 s and relaxation time: 20 s).

Figure 10(b) shows the resistance change curve during the process of pressing the index finger joint. When the joint was pressed, the resistance of the sensor composite exhibited a periodic change.

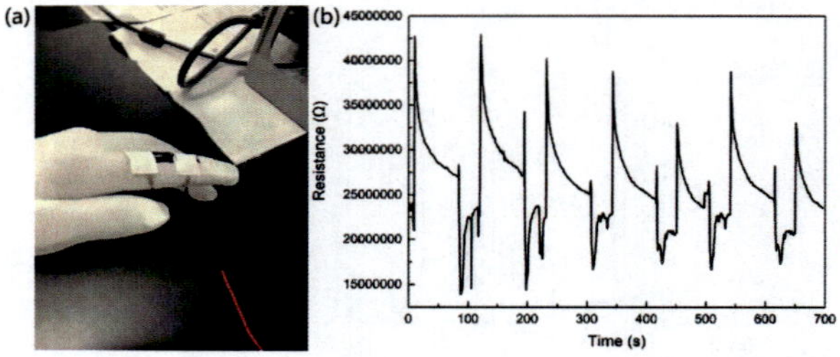

Fig. 10. (a) Schematic diagram of pressing joint and (b) the resistance change of the sensor during the process of single finger joint compression.

4 Conclusion

The KH550 and SDBS double-modified MWCNTs proposed in this study can enhance the piezoresistive performance of the MWCNT/PU sensing films. The particle size of the modified MWCNTs was 36% of that of the unmodified MWCNTs. This could be attributed to the higher degree of dispersion. The polymer composite membrane prepared using the modified carbon nanotubes exhibited better tensile strength, elongation at break, and better electrical conductivity. By assembling the modified MWCNT/PU films on the gloves and conducting tests under different conditions of pressing and grasping, the good piezoresistive effect was verified. This new type of sensing material will provide new ideas for the development of flexible force-sensitive sensors that have huge application prospects in the fields of health monitoring, motion control, and human-computer interaction.

References

1. Salehi, H., Burgueno, R., Chakrabartty, S.: A comprehensive review of self-powered sensors in civil infrastructure: state-of-the-art and future research trends. Eng. Struct. **234**, 111963 (2021)
2. Cheng, M., Zhu, G.T., Zhang, F.: A review of flexible force sensors for human health monitoring. J. Adv. Res. **26**, 53–68 (2020)
3. Pan, Z.Y., Ma, J.Z., Zhang, W.B.: Flexible conductive polymer composites in strain sensors. Prog. Chem. **32**(10), 1592–1601 (2020)
4. Ha, M., Lim, S., Ko, H.: Wearable and flexible sensors for user-interactive health-monitoring devices. J. Mater. Chem. B **6**(24), 4043–4064 (2018)

5. Peng, X.W., Wu, K.Z., Hu, Y.J.: A mechanically strong and sensitive CNT/rGO-CNF carbon aerogel for piezoresistive sensors. J. Mater. Chem. A **6**(46), 23550–23559 (2018)
6. Zhao, Y.M., Liu, Y., Li, Y.Q.: Development and application of resistance strain force sensors. Sensors **20**(20), 5826 (2020)
7. Natarajan, T.S., Eshwaran, S.B., St Ckelhuber, K.W.: Strong strain sensing performance of natural rubber nanocomposites. ACS Appl. Mater. Interfaces **9**(5), 4860–4872 (2017)
8. Zheng, Y., Li, Y., Dai, K.: Conductive thermoplastic polyurethane composites with tunable piezoresistivity by modulating the filler dimensionality for flexible strain sensors. Compos. A Appl. Sci. Manuf. **101**, 41–49 (2017)
9. Zhang, D., Yang, H., Pan, J.J.: Multi-functional CNT nanopaper polyurethane nanocomposite fabricated by ultrasonic infiltration and dip soaking processes. Comp. Part B Eng. **182**, 107646 (2020)
10. Yamakawa, A., Suzuki, S., Oku, T.: Nanostructure and physical properties of cellulose nanofiber-carbon nanotube composite films. Carbohyd. Polym. **171**, 129–135 (2017)
11. Wei, A.J., Li, Y.T., Ma, Z.L.: Flexible stretchable and highly sensitive silicone rubber@multiwalled carbon nanotubes/silicone rubber wearable strain sensing fibers. Acta Materiae Compositae Sinica **37**(8), 2045–2054 (2020)
12. Perez-Aranda, C., Valdez-Nava, Z., Gamboa, F.: Electro-mechanical properties of thermoplastic polyurethane films and tubes modified by hybrid carbon nanostructures for pressure sensing. Smart Mater. Struct. **29**(11), 115021 (2020)
13. Jang, H., Yoon, H., Ko, Y.: Enhanced performance in capacitive force sensors using carbon nanotube/polydimethylsiloxane nanocomposites with high dielectric properties. Nanoscale **8**(10), 5667–5675 (2016)
14. He, Y., Zhao, L.D., Wang, X.J.: Microstructured hybrid nanocomposite flexible piezoresistive sensor and its sensitivity analysis by mechanical finite-element simulation. Nanotechnology **31**(18), 185502 (2020)
15. Konin, Y.A., Bulatov, M.I., Shcherbakova, V.A., Garanin, A.I., Tokareva, Y.D., Mosheva, E.V.: investigation of the properties of an all-fiber temperature sensor created using the melting effect. Instrum. Exp. Tech. **63**(4), 511–515 (2020). https://doi.org/10.1134/S0020441220040284

Kanban System in Industry 4.0 Era: A Systematic Literature Review

Mirco Peron(✉), Erlend Alfnes, and Fabio Sgarbossa

Department of Mechanical and Industrial Engineering, Norwegian University of Science and Technology, S P Andersens V 3, 7031 Trondheim, Norway
mirco.peron@ntnu.no

Abstract. In the last years, many researchers and practitioners have investigated the possibility to integrate Lean Production (LP) and Industry 4.0 (I4.0) technologies. From what emerges from the literature, I4.0 technologies can boost LP practices. Based on an extensive systematic literature review where more than 300 papers were initially considered, this work provides a deep understanding of how I4.0 technologies can modify and support the LP practice of pull, specifically Kanban. The best practices available have been gathered together in this work, aiming to constitute a "catalogue" for practitioners who are facing the problem of improving the LP practice of pull to be able to cope with the today's market requirements.

Keywords: Kanban · Pull production · Industry 4.0 · Best practices

1 Introduction

The interest of researchers and practitioners in Lean Production (LP) has been constantly increasing over the last years [1]. LP, in fact, has proved to eliminate wastes, to reduce inventories and both set-up and lead times, and to improve performance, employees' satisfaction and decision-making attitude, leading to costs and efforts reduction and contributing to reach customer satisfaction [2, 3]. However, nowadays, in a scenario where the market is increasingly competitive and requires highly customized products, LP is facing several challenges, such as the fact that pull production must face rapid changes in scheduling [4].

Many researchers have considered the potentialities of Industry 4.0 (I4.0) technologies able to overcome the LP's limits and to cope with the market's requirements. Several studies, in fact, demonstrate that I4.0 technologies significantly improve industrial performances like flexibility, productivity, delivery time, cost, and quality [5–7]. Furthermore, I4.0 technologies are attracting many companies since they can allow an individual and customized production thanks to the integration of smart machines and components into a digital network that enables a modular and changeable production [8–13].

However, I4.0 does not mean the sunset of lean, actually. Literature in fact recognizes a strong interdependency between the two paradigms. On the one side, LP is reported to facilitate the implementation of I4.0 technologies thanks to the standardization of work

Y. Wang et al. (Eds.): IWAMA 2021, LNEE 880, pp. 12–19, 2022.
https://doi.org/10.1007/978-981-19-0572-8_2

and places and to its emphasis on visual control and transparency [14, 15], while, on the other side, I4.0 technologies can boost LP practices [15, 16].

In this work we will concentrate on the latter, demonstrating how I4.0 technologies can boost LP practices. Specifically, due to space limitation, we will focus only on pull production as LP practice, and particularly on Kanban. Thanks to a systematic literature review, in fact, we were able to identify the evolution of Kanban after the advent of I4.0, that represents the aim of the paper. The details of the systematic literature review are reported in Sect. 2, while Sect. 3 reports its results. Section 4, then, concludes the work.

2 Systematic Literature Review

The literature analysis was carried out as described in Ref. [17]. Scopus was used as database, and the keywords used were ("lean" OR "lean4.0") AND ("industry 4.0" OR "industry4.0" OR "I4.0"). It is worth mentioning that the search was not limited to Kanban, but instead the general term "lean" was used. This is because many papers embrace the more general topic of LP and I4.0, and they discuss the integration of Kanban and I4.0 together with other LP practices. The literature review was limited to journal articles and conference papers written in English, published till the end of 2020 and adherent to the subject areas of "Engineering", "Decision sciences" and "Business, Management and Accounting". The literature analysis resulted in 311 documents which, after a title and abstract screening, were reduced to 61 documents dealing with Kanban and I4.0. Finally, after full text review, the documents were reduced to 32. An overview of the distribution of the published papers over the years and between journal articles and conference papers (book chapters are included in this typology) is reported in Table 1. It is interesting to see that the interest in the topic has increased over the years and that the most common publication type has changed from conference papers to journal articles.

Table 1. Details of the papers considered.

Year	Conference papers	Journal articles	Total
2015	1	0	1
2016	0	1	1
2017	3	2	5
2018	4	0	4
2019	2	6	8
2020	4	9	13

3 Impact of I4.0 Technologies on Kanban

Kanban systems are one of the best methods of implementing pull production, in which a successive station generates Kanban cards to initiate operation for a particular station. The functioning of the Kanban systems can be divided into four areas [18]:

(i) ***Demand assessment***, i.e., the practice of assessing the demand at a particular workstation given the remaining components stored in Kanban bins;
(ii) ***Kanban signal***, i.e., the transmission of the demand once this has been assessed;
(iii) ***Disposition and production***, i.e., the triggering of production once received the Kanban signals (both within the factory and by the suppliers);
(iv) ***Collection and delivery***, i.e., the collection and delivery of materials (often following the milk run approach)

From the literature review it emerged that I4.0 technologies can improve all these four areas. Dealing with (i), it emerged that Auto-ID technologies (e.g., sensors, RFID tags, scanning devices, etc.) can improve this area since they can enable an accelerated and more precise demand assessment process by tracking fill levels in real time.

With respect to (ii), then, many studies reported how with I4.0 technologies like Cyber Physical System (CPS) and vertical and horizontal integration can improve the transmission of the demand. Specifically, the demand can be signaled digitally (i.e., the so-called Electronic Kanban systems), and this information can either be fed into the material planning system, which then sends the order to the suppliers, or directly to the suppliers.

An example of the application of I4.0 technologies in areas (i) and (ii) is the "iBin" system introduced by Würth in 2013: a camera in the module detects the charging level of the bin and then iBin sends orders automatically to suppliers.

Dealing with (iii), it emerged that CPS can be highly beneficial in this area. It was in fact suggested that if different machines of the same factory or production facilities of buyers and suppliers are able to communicate directly with each other, production might initiate production without any human interaction.

Finally, with respect to (iv), the literature review identified two main contributions of I4.0 technologies, both with respect to the milk run approach:

a) Simulations can lead to highly dynamic, demand-driven milk run by simulating the delivery processes with respect to adjacent processes.
b) Automated Guided Vehicle (AGV) and/or Autonomous Mobile Robot (AMR) can support the milk run.

These findings are summarized in Table 2, together with the corresponding documents found in the literature.

Table 2. Summary of the impact of I4.0 technologies on the four main Kanban areas and corresponding documents.

Kanban area	I4.0 technologies	Integration of I4.0 and Kanban area	Documents
Demand assessment	Auto-ID technologies (sensors, RFID tags, scanning devices, etc.)	The use of Auto-ID technologies can enable an accelerated and more precise demand assessment process by tracking fill levels in real time	[2, 4, 14, 16, 18–40]
Kanban signal	Cyber physical system Vertical integration Horizontal integration	With I4.0 technologies, demand will be signaled digitally and either fed into the material planning system or sent directly to the suppliers	[2, 4, 14, 16, 18, 20–42]
Disposition and production	Cyber physical system Connected plants/production facilities	When different machines of the same factory or production facilities of buyers and suppliers are connected to each other, cyber-physical bins can initiate production	[2, 14, 18, 20, 22–24, 29, 30, 33, 36, 39, 41, 43, 44]

(*continued*)

Table 2. (*continued*)

Kanban area	I4.0 technologies	Integration of I4.0 and Kanban area	Documents
Collection and delivery	Simulation Autonomous mobile robots	– Simulations can lead to highly dynamic, demand-driven milk run – AGV and AMR can support the milk run	[2, 16, 18, 19, 22, 23, 25, 31, 39, 40, 45]

4 Conclusions

The interdependency between the LP and I4.0 technologies has been increasingly investigated over the last few years, focusing both on the possibilities offered by LP to facilitate the implementation of I4.0 technologies and on the positive effect that I4.0 technologies can have on LP practices. In this work we have focused on the possibilities provided by I4.0 technologies to improve LP practices and particularly, due to space limitations, we have considered only pull production as LP practice. In particular, we have focused on Kanban systems.

Through a systematic literature, we have identified that I4.0 technologies can improve Kanban systems in four different areas, i.e., (i) demand assessment, (ii) Kanban signal, (iii) disposition and production, and (iv) collection and delivery. We have described in detail the different I4.0 technologies that can improve these Kanban systems, as well as how they can improve them.

This work, although limited in the analysis due to space constrictions, can be useful for researchers and practitioners busy with the integration of I4.0 technologies in Kanban systems. Specifically, it can offer (i) examples of best practices for practitioners who are seeking to improve their current Kanban systems and (ii) an overview for researchers of which are the hot topics in the field of Kanban4.0.

Moreover, from the literature analysis we were able to identify an important lack. All the works considered reported (in different ways) the potentialities associated with coupling I4.0 technologies and Kanban systems, but none of them discussed nor the critical success factors to successfully combine the two nor the criticalities that can be encountered when trying to combine I4.0 technologies and Kanban systems.

References

1. Ciano, M.P., Pozzi, R., Rossi, T., Strozzi, F.: How IJPR has addressed ‘lean’: a literature review using bibliometric tools. Int. J. Prod. Res. **57**(15–16), 5284–5317 (2021)

2. Ciano, M.P., Dallasega, P., Orzes, G., Rossi, T.: One-to-one relationships between Industry 4.0 technologies and Lean Production techniques: a multiple case study. Int. J. Prod. Res. **59**(5), 1386–1410 (2021)
3. Kovács, G.: Combination of Lean value-oriented conception and facility layout design for even more significant efficiency improvement and cost reduction. Int. J. Prod. Res. **58**(10), 2916–2936 (2020)
4. Sanders, A., Elangeswaran, C., Wulfsberg, J.: Industry 4.0 implies lean manufacturing: research activities in industry 4.0 function as enablers for lean manufacturing. J. Ind. Eng. Manag. **9**(3), 811–833 (2016)
5. Sgarbossa, F., Peron, M., Fragapane, G.: Cloud material handling systems: conceptual model and cloud-based scheduling of handling activities. In: Sokolov, B., Ivanov, D., Dolgui, A. (eds.) Scheduling in Industry 4.0 and Cloud Manufacturing. ISORMS, vol. 289, pp. 87–101. Springer, Cham (2020). https://doi.org/10.1007/978-3-030-43177-8_5
6. Fragapane, G., Ivanov, D., Peron, M., Sgarbossa, F., Strandhagen, J.: Increasing flexibility and productivity in Industry 4.0 production networks with autonomous mobile robots and smart intralogistics. Ann. Oper. Res., 1–19 (2020). https://doi.org/10.1007/s10479-020-03526-7
7. Peron, M., Fragapane, G., Sgarbossa, F., Kay, M.: Digital facility layout planning. Sustainability **12**(8), 3349 (2020)
8. Sgarbossa, F., Peron, M., Lolli, F., Balugani, E.: Conventional or additive manufacturing for spare parts management: an extensive comparison for Poisson demand. Int. J. Prod. Econ. **233**, 107993 (2021)
9. Simonetto, M., Peron, M., Fragapane, G., Sgarbossa, F.: Digital assembly assistance system in Industry 4.0 era: a case study with projected augmented reality. In: Wang, Y., Martinsen, K., Yu, T., Wang, K. (eds.) IWAMA 2020. LNEE, vol. 737, pp. 644–651. Springer, Singapore (2021). https://doi.org/10.1007/978-981-33-6318-2_80
10. Peron, M., Sgarbossa, F., Strandhagen, J.O.: Decision support model for implementing assistive technologies in assembly activities: a case study. Int. J. Prod. Res., 1–27 (2020)
11. Roda, I., Arena, S., Macchi, M., Orrù, P. F.: Total cost of ownership driven methodology for predictive maintenance implementation in industrial plants. In: Ameri, F., Stecke, K.E., von Cieminski, G., Kiritsis, D. (eds.) APMS 2019. IAICT, vol. 566, pp. 315–322. Springer, Cham (2019). https://doi.org/10.1007/978-3-030-30000-5_40
12. Orrù, P.F., Zoccheddu, A., Sassu, L., Mattia, C., Cozza, R., Arena, S.: Machine learning approach using MLP and SVM algorithms for the fault prediction of a centrifugal pump in the oil and gas industry. Sustain **12**(11), 4776 (2020)
13. Arena, S., Roda, I., Chiacchio, F.: Integrating modelling of maintenance policies within a stochastic hybrid automaton framework of dynamic reliability. Appl. Sci. **11**(5), 1–18 (2021)
14. Kolberg, D., Zühlke, D.: Lean automation enabled by Industry 4.0 technologies. In: IFAC-PapersOnLine, pp. 1870–1875. Elsevier (2015)
15. Saxby, R., Cano-Kourouklis, M., Viza, E.: An initial assessment of Lean Management methods for Industry 4.0. TQM J. **32**(4), 587–601 (2020)
16. Peron, M., Alfnes, E., Sgarbossa, F.: Best practices of just-in-time 4.0: multi case study analysis. In: Wang, Y., Martinsen, K., Yu, T., Wang, K. (eds.) IWAMA 2020. LNEE, vol. 737, pp. 636–643. Springer, Singapore (2021). https://doi.org/10.1007/978-981-33-6318-2_79
17. Tranfield, D., Denyer, D., Smart, P.: Towards a methodology for developing evidence-informed management knowledge by means of systematic review. Br. J. Manag. **14**(3), 207–222 (2003)
18. Hofmann, E., Rüsch, M.: Industry 4.0 and the current status as well as future prospects on logistics. Comput. Ind. **89**, 23–34 (2017)
19. Mayr, A., et al.: Lean 4.0 - a conceptual conjunction of lean management and Industry 4.0. Procedia CIRP **72**, 622–628 (2018)

20. Ayabakan, M., Yilmaz, E.: Digitalization of the kanban system. In: PICMET 2019 - Portland International Conference on Management of Engineering and Technology: Technology Management in the World of Intelligent Systems, Proceedings. Institute of Electrical and Electronics Engineers Inc. (2019)
21. Tissir, S., Fezazi, S.El., Cherrafi, A.: Industry 4.0 impact on Lean Manufacturing: literature review. In: 2020 13th International Colloquium of Logistics and Supply Chain Management, LOGISTIQUA 2020. Institute of Electrical and Electronics Engineers Inc. (2020)
22. Núñez-Merino, M., Maqueira-Marín, J.M., Moyano-Fuentes, J., Martínez-Jurado, P.J.: Information and digital technologies of Industry 4.0 and Lean supply chain management: a systematic literature review. Int. J. Prod. Res. **58**(16), 5034–5061 (2020)
23. Rosin, F., Forget, P., Lamouri, S., Pellerin, R.: Impacts of Industry 4.0 technologies on Lean principles. Int. J. Prod. Res. **58**(6), 1644–1661 (2020)
24. Tortorella, G., Sawhney, R., Jurburg, D., de Paula, I.C., Tlapa, D., Thurer, M.: Towards the proposition of a Lean Automation framework: integrating Industry 4.0 into Lean Production. J. Manuf. Technol. Manag. **32**(3), 593–620 (2020)
25. Valamede, L.S., Santos Akkari, A.C.: Lean manufacturing and Industry 4.0: a holistic integration perspective in the industrial context. In: International Conference on Industrial Technology and Management. Institute of Electrical and Electronics Engineers (IEEE), pp. 63–68 (2020)
26. Ghobakhloo, M., Fathi, M.: Corporate survival in Industry 4.0 era: the enabling role of lean-digitized manufacturing. J. Manuf. Technol. Manag. **31**(1), 1–30 (2019)
27. Ejsmont, K., Gladysz, B., Corti, D., Castaño, F., Mohammed, W.M., Martinez Lastra, J.L.: Towards 'Lean Industry 4.0'–Current Trends and Future Perspectives, vol. 7, p. 1781995. Cogent Business and Management, Cogent (2020)
28. Majiwala, H., Sharma, S., Gandhi, P.: Lean and Industry 4.0 strive to create smart factory through integration of systems: an exploratory review. In: Nain, N., Vipparthi, S.K. (eds.) ICIoTCT. AISC, vol. 1122, pp. 184–195. Springer, Cham (2020). https://doi.org/10.1007/978-3-030-39875-0_20
29. Tortorella, G.L., Giglio, R., van Dun, D.H.: Industry 4.0 adoption as a moderator of the impact of lean production practices on operational performance improvement. Int. J. Oper. Prod. Manag. **39**(6/7/8), 860–886 (2019)
30. Bittencourt, V., Saldanha, F., Alves, A., Leão, C.: Contributions of lean thinking principles to foster Industry 4.0 and sustainable development goals. In: Alves, A.C., Kahlen, F.-J., Flumerfelt, S., Siriban-Manalang, A.B. (eds.) Lean Engineering for Global Development, pp. 129–159. Springer, Cham (2019). https://doi.org/10.1007/978-3-030-13515-7_5
31. Pereira, A.C., Dinis-Carvalho, J., Alves, A.C., Arezes, P.: How Industry 4.0 can enhance lean practices. FME Trans. **47**(4), 810–822 (2019)
32. Longo, F., Nicoletti, L., Padovano, A.: Smart operators in industry 4.0: a human-centered approach to enhance operators' capabilities and competencies within the new smart factory context. Comput. Ind. Eng. **113**, 144–159 (2017)
33. Romero, D., Gaiardelli, P., Powell, D., Wuest, T., Thürer, M.: Digital lean cyber-physical production systems: the emergence of digital lean manufacturing and the significance of digital waste. In: Moon, I., Lee, G.M., Park, J., Kiritsis, D., von Cieminski, G. (eds.) APMS 2018. IAICT, vol. 535, pp. 11–20. Springer, Cham (2018). https://doi.org/10.1007/978-3-319-99704-9_2
34. Kolberg, D., Knobloch, J., Zühlke, D.: Towards a lean automation interface for workstations. Int. J. Prod. Res. **55**(10), 2845–2856 (2017)
35. Wagner, T., Herrmann, C., Thiede, S.: Industry 4.0 impacts on lean production systems. Procedia CIRP **63**, 125–131 (2017)

36. Sanders, A., Subramanian, K.K., Redlich, T., Wulfsberg, J.: Industry 4.0 and lean management – synergy or contradiction? In: Lödding, H., Riedel, R., Thoben, K.-D., von Cieminski, G., Kiritsis, D. (eds.) APMS 2017. IAICT, vol. 514, pp. 341–349. Springer, Cham (2017). https://doi.org/10.1007/978-3-319-66926-7_39
37. Buer, S.V., Fragapane, G.I., Strandhagen, J.O.: The data-driven process improvement cycle: using digitalization for continuous improvement. IFAC-PapersOnLine **51**(11), 1035–1040 (2018)
38. Shahin, M., Chen, F.F., Bouzary, H., Krishnaiyer, K.: Integration of Lean practices and Industry 4.0 technologies: smart manufacturing for next-generation enterprises. Int. J. Adv. Manuf. Technol. **107**(5–6), 2927–2936 (2020). https://doi.org/10.1007/s00170-020-05124-0
39. Pekarcikova, M., Trebuna, P., Kliment, M., Rosocha, L.: Material flow optimization through e-Kanban system simulation. Int. J. Simul. Model. **19**, 243–254 (2020)
40. Valamede, L.S., Cristina, A., Akkari, S.: Lean 4.0: a new holistic approach for the integration of lean manufacturing tools and digital technologies. Int. J. Math. Eng. Manag. Sci. **5**(5), 851–868 (2020)
41. Chiarini, A., Kumar, M.: Lean six sigma and Industry 4.0 integration for operational excellence: evidence from Italian manufacturing companies. Prod. Plan. Control **32**, 1084–1101 (2020)
42. Pekarčíková, M., Trebuňa, P., Kliment, M.: Digitalization effects on the usability of lean tools. Acta Logist. **6**(1), 9–13 (2019)
43. Slim, R., Rémy, H., Amadou, C.: Convergence and contradiction between lean and industry 4.0 for inventive design of smart production systems. In: Cavallucci, D., De Guio, R., Koziołek, S. (eds.) TFC 2018. IAICT, vol. 541, pp. 141–153. Springer, Cham (2018). https://doi.org/10.1007/978-3-030-02456-7_12
44. Mrugalska, B., Wyrwicka, M.K.: Towards lean production in Industry 4.0. In: Procedia Engineering, pp. 466–473. Elsevier Ltd. (2017)
45. Goienetxea Uriarte, A., Ng, A.H.C., Urenda, M.M.: Bringing together Lean and simulation: a comprehensive review. Int. J. Prod. Res. **58**, 87–117 (2020)

Research on Digital Twin System of Intelligent Workshop and Application of Historical Data

Muchen Yang(✉), Lilan Liu, Zenggui Gao, and Wentao Wei

Shanghai Key Laboratory of Intelligent Manufacturing and Robotics, Shanghai University, Shanghai, China
{michael_ymc,lancy}@shu.edu.cn

Abstract. Digital twin technology has attracted much attention in the manufacturing industry in recent years. It can provide practical functions such as real-time monitoring, status parameter monitoring, fault diagnosis, fault prediction, and offline debugging. With the development of information technology, more and more data are collected in manufacturing, and these data have great application value. However, in the current research, the integration of the digital twin system and historical data is not close. This article combines historical data with the digital twin system. It introduces the application of historical data in the digital twin system by reproducing historical scenes as an example, which has specific application value. This article first introduces the basic framework of the digital twin system, then introduces the method of historical accessing data in the digital twin system, and also proposes some application scenarios of historical data in the digital twin system, and finally puts forward the development direction of the digital twin system Some opinions.

Keywords: Digital twin · Historical data · Smart manufacturing

1 Introduction

In an environment where many countries, led by Germany, call for intelligent manufacturing, the transformation and upgrading of the manufacturing industry have become an urgent need to strengthen the competitiveness of the manufacturing industry. To make it possible, traditional manufacturing companies need to introduce intelligent manufacturing systems made of intelligent end-to-end processes throughout the supply chain by using technological solutions such as smart decision-making and cyber-physical systems [1]. Digital twin technology can well meet the needs of manufacturing intelligent transformation.

The digital twin is to create a virtual model of a physical entity digitally and simulate the behaviour of the physical entity in the natural environment with the help of data. The digital twin system can expand new functions for the physical entity through virtual and real interactive feedback, data fusion analysis, decision-making iterative optimization, and other means [2]. Digital twin technology has been widely used in aerospace, traffic management, manufacturing, and other fields. The intelligent workshop digital twin

Y. Wang et al. (Eds.): IWAMA 2021, LNEE 880, pp. 20–27, 2022.
https://doi.org/10.1007/978-981-19-0572-8_3

system is a virtual-real mapping system based on digital twin technology in the manufacturing field. It monitors the product's production process through accurate process state tracking and complete real-time data obtaining and further realizes the interconnection between the physical world and the information world [3].

A large amount of production data is necessary to support the operation of the digital twin system. In the past applications of digital twin systems of the intelligent workshop, Yang Jinho et al. proposed an integrated cyber-physical system-based platform that can be used to utilize information from various distributed manufacturing sites in real-time [4]. Miao Qiang et al. use digital twin technology to supervise the various operating parameters and indicators of the product and realize the system health management of the intelligent workshop [5]. These studies focus on data collection and the fusion of virtual and reality and have not explored the value of historical data in the digital twin system. This paper will focus on applying historical data in the digital twin system in the intelligent workshop. First, it introduces the basic framework of the digital twin system in the intelligent workshop. Second, it clarifies the storage and reading of historical production data based on InfluxDB. Then, it describes the application of historical data. Finally, it analyzes the development direction of the digital twin system of the intelligent workshop.

2 The Basic Framework of the Digital Twin System of the Intelligent Workshop

In 2003, Professor Michael Grieves of the University of Michigan proposed three parts of the digital twin in the PLM course for the first time: real space, virtual space, and the information interaction between the two [6]. According to the traditional theory of digital twins, as shown in Fig. 1, the intelligent workshop's digital twin system includes a physical entity layer, a data transmission layer, and a simulation application layer. The historical data is processed in the data transmission layer.

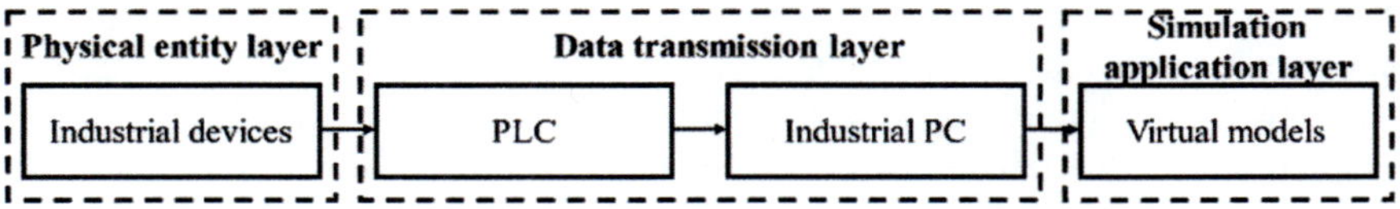

Fig. 1. Three-tier framework of digital twin system

2.1 Physical Entity Layer

The physical entity layer includes all physical elements in the intelligent workshop, including robots, conveyor belts, AGVs, guardrails, rolling doors, and other devices. According to the importance, these devices are divided into devices with a large amount of data and devices with a medium amount of data, devices with a small amount of data, and devices without data. Among them, the robot, as a device with a large amount of data, requires high-frequency, low-latency posture data and terminal work data to reflect the operating state of the robot. Suppose the data collection frequency is too low. In that case, the movement of the virtual robot model will be distorted, and it will be hard to reflect the interaction process between the robot and other devices in time. AGVs, conveyor belts, etc., belong to devices with a medium amount of data because these devices follow a fixed law of motion. When the data collection accuracy is insufficient, the virtual model animation is used to compensate for the missing data area, allocating more data transmission channel resources to devices with a large amount of data. Rolling shutters and other devices are devices with a small amount of data. These devices only need a few Boolean variables to reflect their movement. Fixed facilities such as guardrails and trusses do not require data.

The data generated from the intelligent workshop presents multi-source, multi-modal, and heterogeneous characteristics. There are multiple types of devices in the intelligent workshop, multiple data types, and different data volume requirements. Multiple PLCs, industrial computers, and lines are necessary to build the data transmission channel of the digital twin system.

2.2 Data Transmission Layer

The data transmission layer is used to exchange and process data between the physical entity and simulation application layers. The data from the physical entity layer to the simulation application layer needs to pass through the PLC and the industrial computer. The data acquisition program is deployed in the industrial computer and communicates with the PLC through the Modbus protocol or the S7 protocol. The signal value is read according to the pre-stored PLC address and signal address, and the read production data correspondingly stored in the database. Limited by the PLC communication channel bandwidth, the data acquisition frequency, related to the amount of data, is constant. According to the user's requirements, it is also necessary to install a data push program on the industrial computer, package the data in the database into the format required by the user, and send it to the network port through Internet Information Service for the users.

To deal with the diversified data generated by the intelligent workshop, the OPC unified architecture, which defines the primary data type, the address space, the access mechanism, and the session mechanism, is usually used in the research of digital twin technology. In addition, it is necessary to name the signals of various devices and make a detailed comparison table between the signals and PLC addresses and points. It is essential to determine the sign bit of the sign and combine the high and low bits of the number to obtain accurate data in some cases.

The core indicators of the data transmission layer are data refresh frequency and data delay. The data refresh frequency refers to the time that elapses between two consecutive updates of a single signal. The data delay refers to the time that elapses from the action of the physical device to when the simulation application layer receives the corresponding signal. These two indicators are essential for the stability of the digital twin system. Therefore, there are some requirements for the hardware and software in the data transmission process.

In intelligent workshops, various types of devices are often connected with PLC to achieve unified control. The connection between PLC and industrial computers can be achieved through Modbus protocol or S7 protocol. It should be noted that the bandwidth of the communication channel between the PLC and the industrial computer is limited, so it has an impact on the data transmission speed. According to the previous section, different devices have different requirements of data transmission speed. Taking the assembly workshop as an example, as shown in Fig. 2, the assembly robot needs to collect data such as posture, tightening torque, tightening angle, and tightening force, which is a device with a large amount of data. The real-time nature of these data is essential for assembly quality monitoring. Therefore, it is recommended that the data collection cycle be less than 200 ms, and each robot is connected to one PLC. Devices with a medium amount of data, such as AGVs and conveyor belts, need to collect information such as operating status and target location. For this kind of device, multiple are connected to one PLC. A data collection cycle within 500 ms is recommended. Rolling shutters, rotating tables, and other devices only need in-position, in-situ signals, which are devices with a small amount of data and can use one PLC to control similar devices.

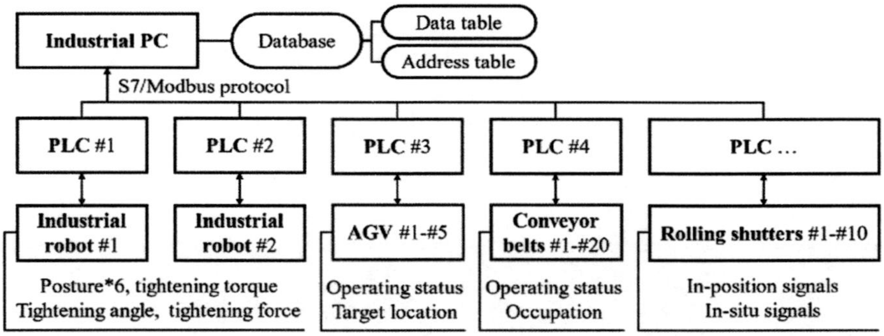

Fig. 2. Data transmission layer hardware composition

2.3 Simulation Application Layer

After recorded, the production data can be applied to real-time simulation, fault alarm, fault prediction, quality prediction, etc., which is highly valuable. In the digital twin system, the foundation of the simulation application is the virtual model. The virtual model, which is mapping all physical production factors in the virtual environment, runs synchronously with the physical entity. Construct a virtual model of the production line and apply it to the digital twin system to feedback the production status. Then, add an information bar to the virtual model of the equipment to display general production data or high-dimensional data extracted by algorithms to represent production quality accurately.

The simulation application layer should establish a perfect virtual model operation law based on the production logic. In the case of incomplete production data, make up for the missing data according to the virtual model operation law to improve the synchronization of the system mapping.

3 Historical Production Data Access Based on InfluxDB

It is a very classic problem to store historical data, and there are already very mature solutions in the Internet industry. When the manufacturing industry enters the informatization, production data also needs to be stored. The production data has high frequency and high precision and has higher requirements on the reading and writing ability of the database. Since production data is closely related to time, time-series databases have emerged and have been widely used in the industrial Internet field. Di Martino Sergio et al. made an empirical comparison of three NoSQL Databases Management Systems, including Cassandra, MongoDB, and InfluxDB, in maintaining and retrieving gigabytes of real Industrial Internet of Things data, and the results show that InfluxDB is the best one [7]. Although time-series databases have been widely applied in the Industrial Internet of Things, few people apply them to digital twin systems.

The historical data access function is added to the data transmission layer of the digital twin system, and the time-series databases InfluxDB is deployed in the industrial computer. When storing production data, each piece of data needs to add a Unix timestamp and make the data match the field name. Depending on the device, the data is stored in a table named after the device name. The automatic storage of data is realized by a program written in C# language. Every time data is collected, the data is formed into a data storage statement and sent to the time series database through the local network port. The data stored in InfluxDB is shown in Fig. 3.

name: Display_Rob

time	Date	J1	J2	J3	J4	J5	J6	J7	Robname	XL_E2J	XL_E2S	XL_J	XL_S
1604558240002551	201105	-28.7	15.1	-20.3	-0.5	-69.8	76.6	0	ROB03	false	false	false	false
1604558240003548	201105	29.7	-2.4	-36.9	-179.9	53.8	88.5	0	ROB06	false	false	false	false
1604558240019092	201105	5.9	-23.9	-27.3	85.6	-97.3	152.6	0	ROB01	false	false	false	false
1604558240065692	201105	35.3	16.2	-51.6	-31.3	57.2	17.9	0	ROB05	false	false	false	false
1604558240080720	201105	25.6	-39.5	-22.5	-47.7	31.1	-44.7	0	ROB09	false	false	false	false
1604558240097075	201105	-7.5	-42	9.7	0.1	-99.2	5.3	0	ROBTD	false	false	false	false
1604558240111090	201105	16.5	-42	-27.7	132.5	-32	310.4	0	ROB08	false	false	false	false
1604558240115079	201105	4.3	-55.4	-8.1	0.5	-80.3	-4.3	0	ROB13	false	false	false	false
1604558240139689	201105	60.2	-5.9	-35.5	0.5	-54.6	-31.2	0	ROB11	false	false	false	false
1604558240173760	201105	0.8	-6.8	8.8	-1	-99.9	-1.5	0	ROB04	false	false	false	false
1604558240177787	201105	23.9	2.5	-67.7	-207.5	-68.9	-199.4	0	ROB12	false	false	false	false

Fig. 3. The data stored in InfluxDB

According to different historical data applications, the reading methods are also different. This paper will introduce a data read method with time sequence alignment. The usual digital twin system uses real data to simulate real-time scenarios and drives the virtual model according to real-time changes in signals. Using historical data to simulate historical scenes needs to read historical data in sequence and time intervals according to the actual occurrence time. Otherwise, the operating speed will be unstable. The reading method proposed in this paper combines the digital twin system to use historical data to simulate historical scenes, and the operating speed is consistent will reality.

As shown in Fig. 4, the data change cycle of different devices is different. Therefore, the historical data is read through interval scanning. The length of a single query interval is T_{read}, and the offset of the interval is ΔT. T_{read} must be equal to ΔT to ensure complete coverage of historical data. The key to keeping the running speed consistent with reality is to make the period for the program to query historical data equal to the length of the interval for querying historical data, that is, to query data within 0.1 s every 0.1 s. When the amount of historical data is not large, such as 15 pieces of data per second, the historical data can be read at a realistic speed. When the amount of historical data is large, such as 300 pieces of data per second, the actual speed cannot be reached. In this case, the actual speed should be reduced. Moreover, make the period for the program to query historical data slightly longer than the interval length for querying historical data, that is, for querying data within 0.1 s every 1 s. Then, historical scenes can be reproduced at 0.1 times speed.

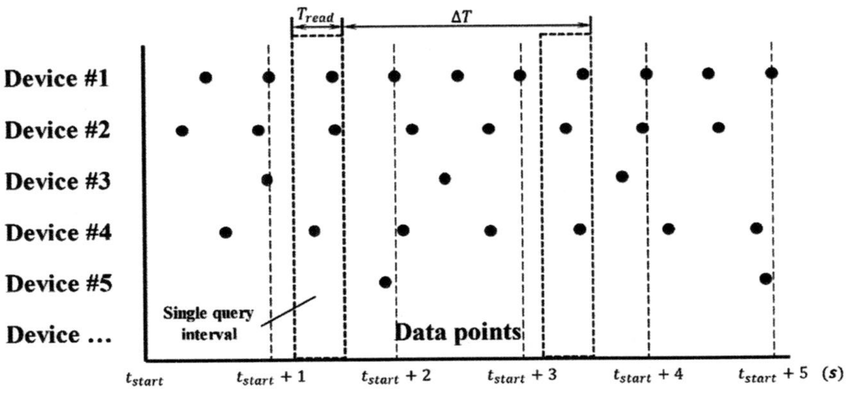

Fig. 4. Schematic diagram of historical data reading

The specific steps of historical data query are as follows:

Step 1. Set the starting time T_{start} of historical data according to user needs, determining the starting point of historical scenes. In addition, the simulation platform data refresh period $T_{determined}$ is set, which represents the accuracy with which the simulation platform will read historical data. Let T_{read} and ΔT be equal to $T_{determined}$.
Step 2. Calculate the start time T^n_{start} and end time T^n_{over} of each query interval, where $T^n_{start} = T_{start} + n \times \Delta T$, $T^n_{over} = T_{start} + (n+1) \times \Delta T$, and $\boldsymbol{n}$ is the number of queries. Set the running period of the query program of the simulation platform to $T_{determined}$.
Step 3. Cyclically execute the simulation platform query program, package and send the queried data to the virtual model driver to complete the historical scene reproduction.

4 Application Scenarios of Historical Data

The value of mining historical data belongs to the category of big data technology. The value of historical data often lies in extracting experience and providing guidance for future work. In the field of manufacturing, many scholars apply deep learning to analyze historical production data and play the value of the data.

For equipment, the voltage, current, temperature, force, speed and other data of the equipment can reflect the operational status of the equipment. Through long-term data comparison and analysis, the equipment's health status and development trend can be obtained, and the life expectancy of the equipment can be further realized. For products, the surface quality, assembly quality and other data of the product can reflect the production quality of the product. Through a large amount of data comparison and analysis, the stability and durability of the product can be obtained. There are more types of data and more complex logic rules for a more extensive range of production lines. Analysis of this type of historical data can optimize the layout of production lines and improve production efficiency.

In addition, historical data based on the digital twin system can reproduce historical production scenarios. Restore historical production scenarios through complete production data, high-precision models and drive logic modules, which facilitates tracing the causes of production line failures and discovering production process defects. Provide workshop managers and fault maintenance engineers with monitoring of the historical operation of the workshop, which can provide more intuitive and comprehensive auxiliary analysis methods for workshop production scheduling optimization, process optimization, and equipment failure cause analysis.

5 Discussion and Conclusion

Digital twin technology is an emerging technology formed by combining cutting-edge technologies such as the Industrial Internet of Things and cyber-physical integration. It deeply embodies the combination of computer technology, communication technology and manufacturing. It has developed rapidly in the manufacturing industry in recent years. This article focuses on applying historical production data in the digital twin system, introduces the basic framework of the digital twin system, proposes a data storage method based on InfluxDB, and illustrates some historical data application scenarios. Among them, the application of historical data to the reproduction of historical scenes is relatively novel and valuable in use. However, at the same time, this application is still immature, which is reflected in its extremely high requirements for data comprehensiveness. In particular, due to the lack of data support, it is difficult to reflect the impact of human behaviour on the production line, resulting in distortion of historical scenes.

This problem is also a problem faced by digital twin technology. Whether it is a smart city or a smart factory, the application of digital twins in all walks of life requires the support of a large amount of data. The more data, the higher the authenticity of the digital twin, and the more similar it is to the real world, the higher the value it can generate. Thus, the question becomes whether more data can be collected. However, when the amount of data required increases, the cost of data collection is also rising rapidly. Therefore, it is necessary to consider which data collected can generate more excellent value.

In the future, digital twin technology should also focus on increasing the physical laws of the virtual world so that a large amount of data collection is no longer necessary. Instead, more state parameters can be calculated and derived from a small amount of data to realize a digital twin in a broader sense.

Acknowledgment. The support of Shanghai University and Huayu Intelligent Equipment Technology Co., Ltd for author's research is greatly appreciated. The work described in this article has been conducted as part of the research project Development and Application of Key Technologies for Car Intelligent Chassis Assembly Line (No. 19511105200), which is supported by Shanghai Science and Technology Committee of China.

References

1. Schioenning, L., Maria, S., Lassen, A.H.: Design parameters for smart manufacturing innovation processes. Procedia CIRP **93**, 365–370 (2020)
2. Tao, F., et al.: Digital twin and its potential application exploration. Comput. Integr. Manuf. Syst. **24**(1), 1–18 (2018)
3. Guo, L., et al.: Data collection and transmission method for workshop production in intelligent manufacturing terminal. Mach. Electron. **37**(8), 21–24 (2019)
4. Yang, J., et al.: Integrated platform and digital twin application for global automotive part suppliers. In: Lalic, B., Majstorovic, V., Marjanovic, U., von Cieminski, G., Romero, D. (eds.) APMS 2020. IAICT, vol. 592, pp. 230–237. Springer, Cham (2020). https://doi.org/10.1007/978-3-030-57997-5_27
5. Miao, Q., Zou, W., Liu, L., Wan, X., Wu, P.: Intelligent workshop digital twin virtual reality fusion and application. In: Wang, Y., Martinsen, K., Yu, T., Wang, K. (eds.) IWAMA 2019. LNEE, vol. 634, pp. 585–592. Springer, Singapore (2020). https://doi.org/10.1007/978-981-15-2341-0_73
6. Grieves, M., Vickers, J.: Digital twin: mitigating unpredictable, undesirable emergent behavior in complex systems. In: Kahlen, F.-J., Flumerfelt, S., Alves, A. (eds.) Transdisciplinary Perspectives on Complex Systems, pp. 85–113. Springer, Cham (2017). https://doi.org/10.1007/978-3-319-38756-7_4
7. Di Marino, S., Fiadone, L., Peron, A., Vitale, V.N., Riccabone, A.: Industrial internet of things: persistence for time series with NoSQL databases. In: Proceedings – 2019 IEEE 28th Internetional Conference on Enabling Technologies: Infrastructure for Collaborative Enterprises, pp. 340–345 (2019)

Research on UWB Driving Positioning Technology in Smart Warehouse

Kuiliang Liu[1], Guiqin Li[1](✉), Tiancai Li[2], Yicong Shen[1], and Peter Mitrouchev[3]

[1] Shanghai Key Laboratory of Intelligent Manufacturing and Robotics, Shanghai University, Shanghai 200072, China
leeching@shu.edu.cn
[2] Shanghai Tigers CO., Ltd., Shanghai, China
[3] G-SCOP, Grenoble Alpes University, 38031 Grenoble, France

Abstract. A driving positioning system based on Ultra Wide Band (UWB) indoor wireless positioning technology is designed to overcome the difficulty in supervising the movable property of commercial banks. The system includes crane, crane hook controller, database server, positioning server, system server, cargo information identification module and high precision positioning module based on UWB. The UWB positioning module includes UWB positioning Base Station, the UWB positioning TAG and System Controller. Through the UWB positioning module, the accurate positioning of the crane is realized. The abnormality can be triggered under the situation that the position of the crane is consistent with the pledged goods, ensuring the effective supervision of movable property.

Keywords: UWB indoor positioning · Chattel supervision · Train operation

1 Introduction

Chattel pledge supervision can become the main value-added service of warehouse storage. In recent years, some enterprises and researchers began to study chattel pledge supervision. Huang Jiajie [1] and others put forward a logistics enterprise pledge supervision scheme, which proposed a matrix supervision system, standardized process, improved internal control measures, strict operation and management, etc. which can effectively avoid some mistakes in warehouse management. Xi Jun [2] proposed to combine video technology, FRID technology and intelligent access control technology into the field of chattel pledge to build a chattel pledge supervision system based on the Internet of things. After the chattel pledge formed a certain scale, the corresponding standardization standards also appeared one after another. Feng Ying [3] and others gave the corresponding opinions to standardize the domestic pledge supervision service by analyzing the current situation of enterprise pledge supervision business and the standardization demand of enterprise pledge supervision service. In the field of technology, UWB technology is becoming more and more mature. Jiang Chunsheng [4] and others put forward the positioning system of warehouse logistics automatic guided vehicle based on UWB, which has higher accuracy than the traditional positioning method. A supervision system for

Y. Wang et al. (Eds.): IWAMA 2021, LNEE 880, pp. 28–34, 2022.
https://doi.org/10.1007/978-981-19-0572-8_4

the movable property of bulk goods and precious metal goods is designed. A driving positioning device based on UWB technology is developed, and the supervision of goods is completed through the information identification module.

2 Design of Chattel Pledge Supervision System

2.1 Supervision Mode of Movable Property

The supervision of chattel pledge can be divided into three roles: pledgor (loan enterprise or individual), pledgee (commercial bank or other commercial institution that can provide loan), and Supervisor (logistics or storage enterprise entrusted by the pledgee to keep the pledged chattel). The relationship between them is: the pledgor mortgages the mortgagee with his own movable property, the pledgee grants the pledgor credit financing guarantee, and the supervisor is entrusted by the pledgee to supervise the mortgaged property according to the pledgee's instructions during the pledge period (Fig. 1).

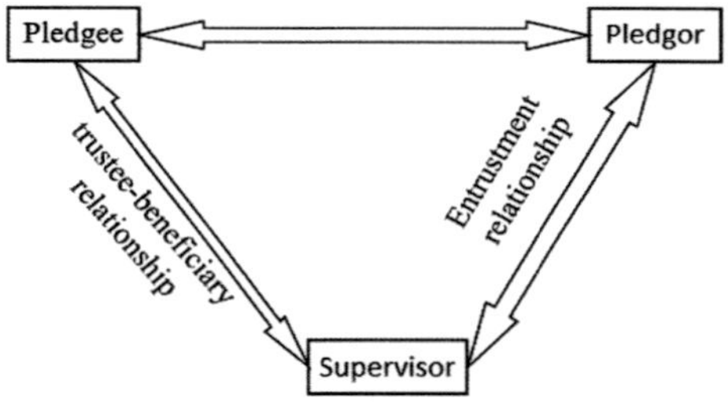

Fig. 1. Tripartite relationship of pledge supervision

This model is innovative in solving the difficulties of small and medium-sized enterprises' chattel financing. It is a new model of bank's chattel credit, and can also improve the bank's risk control ability in the credit process.

In the pledge supervision business, warehousing enterprises play an important role. As a bridge between the bank and the customer enterprise, it should supervise and keep the movable property pledged by the enterprise to the bank, and protect the interests of the customer enterprise and both sides. It is more convenient and professional for warehousing enterprises to help banks supervise the goods than for banks themselves. It also saves the bank's business and reduces the bank's credit risk. Warehousing enterprises can help customers get better bank loans through pledge supervision services, reducing financing costs.

2.2 The UWB Positioning

The UWB wireless positioning system includes three roles: Base Station, TAG and Console. Base Station and TAG are used for positioning design. The layout mode of the Base Station is cellular structure. The UWB positioning Base Station installed in the warehouse forms an equilateral triangle to form a cellular structure. In order to avoid reflection interference, the distance between UWB positioning Base Station and wall is at least 30 cm. The UWB positioning Base Station and its antenna are connected by a radio frequency line. At the same time, the UWB positioning Base Station shall be installed face down, and the cable shall be routed along the wall or column through the conduit to the installation position of the UWB positioning Base Station. The positioning TAG is placed on the vehicle, and the position of the TAG is calculated by the positioning algorithm. Set up an electronic fence, which cannot be reached by driving or when some supervised goods are not released. The location data is uploaded to the location server directly through the TAG.

The UWB positioning system we studied adopts TOF (time of flight) positioning method, which is essentially the same as TOA. The improvement is that TOF ranging does not rely on the time synchronization between the Base Station and the TAG. Besides, there is no error caused by clock synchronization deviation, so the accuracy is better. The basic principle is shown in Fig. 2.

The time of the TAG transmitting signal to different Base Stations is measured. The distance from the transmitting point to the receiving point is obtained by simple calculation. Take the location of the Base Station as the center of the circle and the measured distance as the radius to draw a circle. The intersection of the three circles is the location of the TAG. The relationship between distance and coordinates is as follows.

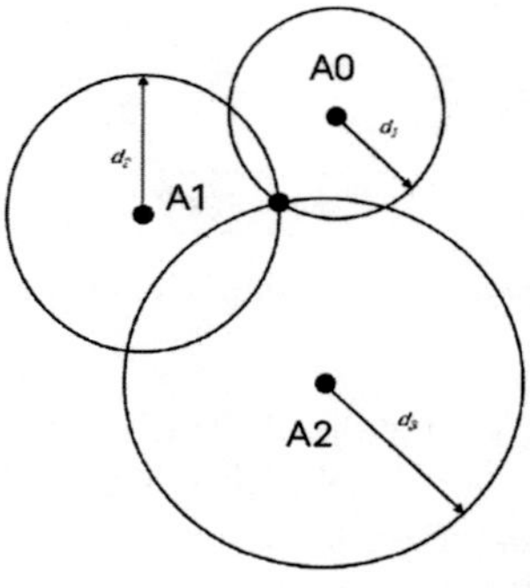

Fig. 2. Positioning principle

$$\begin{cases} d_1 = \sqrt{(x - x_1)^2 + (y - y_1)^2} \\ d_2 = \sqrt{(x - x_2)^2 + (y - y_2)^2} \\ d_3 = \sqrt{(x - x_3)^2 + (y - y_3)^2} \end{cases} \tag{1}$$

The TAG coordinates can be obtained by solving (1):

$$\begin{bmatrix} x \\ y \end{bmatrix} = \begin{bmatrix} \frac{(y_2-y_1)(x_2^2-x_3^2+y_2^2-y_3^2+d_3^2-d_2^2)+(y_2-y_3)(x_1^2-x_2^2+y_1^2-y_2^2+d_2^2-d_1^2)}{2[(x_2-x_3)(y_2-y_1)+(x_1-x_2)(y_2-y_3)]} \\ \frac{(x_2-x_1)(x_2^2-x_3^2+y_2^2-y_3^2+d_3^2-d_2^2)+(y_2-y_3)(x_1^2-x_2^2+y_1^2-y_2^2+d_2^2-d_1^2)}{2[(x_2-x_1)(y_2-y_3)+(x_2-x_3)(y_1-y_2)]} \end{bmatrix} \tag{2}$$

3 System Design and Implementation

The QR code and RFID tags are installed on the pledged goods to store the product information. The RFID reader used to identify the RFID tag information of goods and the camera used to identify the two-dimensional code information of goods are installed on the crane hook. The data is transmitted to the system server, which analyzes the data and controls the crane hook controller according to the analysis results. The crane hook controller is used to control the lifting and falling of crane hook. When the system server triggers an abnormal alarm, the crane hook controller forbids the crane hook to drop and protects the pledged goods that are not allowed to move from being lifted by the crane.

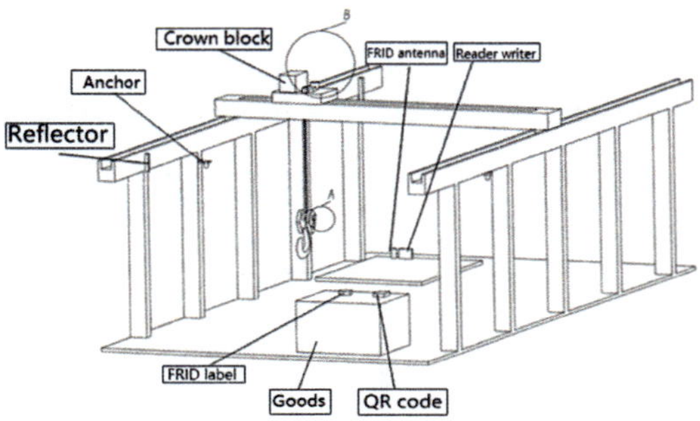

Fig. 3. Schematic diagram of crane monitoring system

The design of vehicle monitoring system for chattel supervision is shown in Fig. 3 and Fig. 4, and its system framework is shown in Fig. 5.

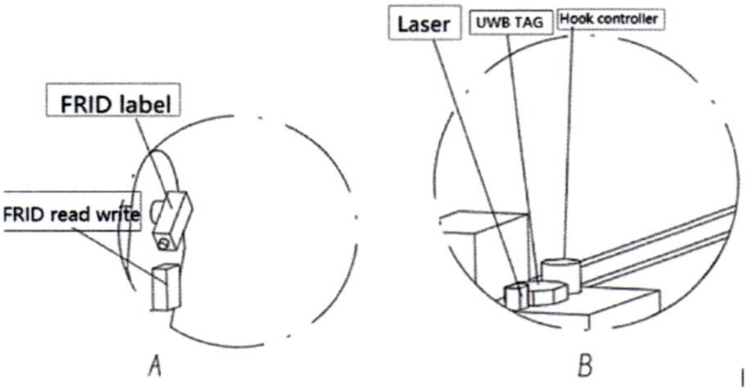

Fig. 4. System detail

The system abnormalities can be caused by the following conditions: when the position of the crane is inconsistent with that of the goods to be lifted, when the driving position is consistent with the inventory position, when the goods are under pledge supervision, when the two-dimensional code data information read by the camera is inconsistent with the information, when the goods are to be lifted allocated by the system server, when the camera reads the label information when the goods are not allowed to move.

The RFID antenna and RFID reader are installed in each storage location for goods in the storage area. The database server is the key hub of the whole network environment architecture. It is mainly responsible for the data interaction between the upper management system and the cargo transportation system. As the server of the digital workshop map, it realizes the data interaction with RFID reader and driver through the wireless communication network, and stores the data in the real-time database, provide real-time data support for the upper application system. The system server has the following functions: it is responsible for setting the perception area in the workshop environment, completing the event triggered accordingly, user network authority management and information interaction with other servers through setting communication interface.

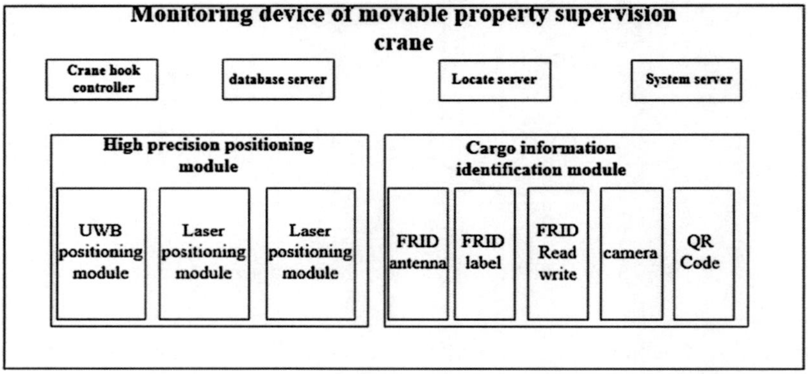

Fig. 5. System framework

When the goods need to be supervised enter the warehouse, the goods information is stored in the two-dimensional code and RFID tag. The RFID tag and two-dimensional code are attached to the goods. After the back-end system server has allocated the storage location of the goods, it will start the instruction to the crane. The crane will identify the goods to be lifted through the camera head above the hook and the RFID reader, and then the crane will be lifted. When the location information of the UWB tag of the crane is consistent with the location information of the warehouse provided by the system, the crane hook lowers the goods and stores them in the warehouse designated by the system. At the same time, the RFID reader on the warehouse reads the RFID tag on the goods and sends the goods information stored in the warehouse to the background database server for storage.

When the goods need to be removed or moved, the crane will search for the goods to be transported according to the instructions of the system server. When the goods need to be removed or moved, the crane shall search for the goods to be transported according to the system server instruction. When the driving position is consistent with the position of the goods to be handled, the crane hook controller can allow the crane

Table 1. Data error table

Error table					
Group	Error	Group	Error	Group	Error
1	(0.551, 0.455)	14	(−0.086, −0.015)	27	(−0.552, −0.299)
2	(0.436, −0.631)	15	(−0.089, 0.390)	28	(−0.893, 0.979)
3	(0.487, −0.344)	16	(−0.228, 0.089)	29	(−0.178, 0.770)
4	(0.202, 0.288)	17	(−0.024, 0.014)	30	(−0.447, 0.568)
5	(0.302, −0.091)	18	(−0.483, 0.166)	31	(−0.552, 0.342)
6	(0.428, 0.049)	19	(−0.842, −0.261)	32	(0.555, −0.141)
7	(−0.424, −0.350)	20	(0.573, −0.178)	33	(−0.582, 0.082)
8	(0.203, 0.298)	21	(−0.406, −0.052)	34	(−0.358, −0.175)
9	(−0.507, −0.463)	22	(−0.826, −0.583)	35	(−0.647, 0.152)
10	(−0.865, 0.585)	23	(0.382, 0.583)	36	(−0.353, −0.299)
11	(0.460, 0.772)	24	(−0.798, −0.271)	37	(0.106, 0.643)
12	(−0.319, 0.569)	25	(−0.596, −0.188)	38	(0.471, 0.078)
13	(−0.178, 0.166)	26	(−0.517, 0.253)		

hook to descend. The tracking and positioning of the traffic can help to supervise the pledged bulk goods and precious metal goods accurately.

Fig. 6. Installation drawing of base station

The positioning measurement error is the actual two-dimensional coordinate minus the measurement coordinate, as shown in Table 1. The measurement results are shown in Fig. 6. According to the table, the maximum error between the measured value and the actual value is about 0.8 m, and the minimum error is about 0.01 m.

4 Conclusion

A kind of traffic monitoring system is designed for the supervision of movable property of bulk goods and precious metal goods. High precision positioning module and cargo information identification module are developed by using UWB technology. The test verified that the system can track and supervise the quantity change of movable property in real time.

References

1. Jiajie, H., Jian, W.: Scheme design of logistics enterprise pledge supervision. Mod. Commer. **24**, 203–204 (2007)
2. Xi, J., Ren, J.: Research and design of chattel pledge supervision platform based on Internet of Things. Comput. Program. Skills Maintenance **12**, 14–15 + 18 (2020)
3. Zhan, J.: An analysis of the pledge supervision business of China reserve shares. Huazhong University of Science and Technology (2008)
4. Shuyi, C., Li Jianhui, W., Mingjuan, X.T.: Application and standardization of chattel pledge supervision system of Internet of Things. Inf. Technol. Stand. **07**, 20–22 (2017)
5. Chai, F., Zhubing, H.: Design of crane positioning and automatic weighing system. J. Chengde Pet. Coll. **23**(02), 51–54 (2021)
6. Wang, M., Zhou, A., Chen, X., Shen, Y., Li, Z.: A novel asynchronous UWB positioning system for autonomous trucks in an automated container terminal. SAE Int. J. Adv. Curr. Pract. Mob. **2**(6), 3413–3422 (2020)
7. Li Jieren, X., Qingshan, L.X., Jianmin, W.: Application of wireless positioning technology in mobile crane. China Mech. Eng. **30**(06), 716–721 (2019)
8. Fangmin, J.: Application of Internet of things technology in dynamic supervision system of chattel pledge. Enterp. Technol. Dev. **29**(19), 13–15 (2010)
9. Ying, F., Jinyuan, L., Peiwu, Z., Zhewen, L.: Analysis and Research on the status quo and standardization demand of domestic enterprise pledge supervision service. China Storage Transp. **11**, 105–109 (2016)
10. Li, J., Liu, H., Lan, Z.: Internet of things international standard for chattel pledge supervision. Inf. Technol. Stand. **12**, 52–55 (2018)
11. Jiang, C., Liao, Y., Cai, B.: Research on positioning system of logistics and warehousing AGV based on UWB. Comput. Age **05** (2019)

Trajectory Planning of a Six-Degree-of-Freedom Robot for Spraying Automobile Roof Beams

Guiqin Li[1](✉), NanShan Yan[1], and Peter Mitrouchev[2]

[1] Shanghai Key Laboratory of Intelligent Manufacturing and Robotics, Shanghai University, Shanghai 200072, China
leeching@shu.edu.cn

[2] University Grenoble Alpes, G-SCOP, 38031 Grenoble, France
peter.mitrouchev@g-scop.inpg.fr

Abstract. In order to improve the surface spraying quality of the car roof beam, the spraying trajectory of the car roof beam is planned in this paper. Firstly, the parabolic β distribution model is fitted through a one-way plane spraying experiment, and then the distance between adjacent spraying tracks is calculated through the paint deposition rate model. Secondly, the Z-shaped reciprocating spraying path of the robot is determined according to the stroke mode of the spray gun trajectory and the actual application scenario of the robot. The automatic spraying experiment results show that the proposed scheme effectively improves the spraying quality and spraying efficiency of the surface of the car roof beam.

Keywords: Trajectory planning · Robot · β distribution model

1 Introduction

With the development of industrial robots, robot automatic spraying technology has gradually become a research focus. A. Klein proposed an offline programming system for spraying robots based on CAD models to realize visual programming and motion simulation of spraying trajectories [1]. Balkan et al. established a coating deposition β distribution model, and analyzed the different conditions of the coating distribution model curve when different β values were taken, and finally verified the accuracy of the β distribution model optimization method on the plane through experiments [2]. G. Duelen et al. proposed the electrostatic field distribution model on the surface of the sculpture and began to apply offline programming technology in electrostatic spraying [3]. W. Sheng et al. used the grid topology and normal vector direction to divide the complex surface into several plane areas, and then analyzed and controlled the trajectory motion of each area to optimize the spraying trajectory and coating thickness uniformity, and verified through experiments [4, 5]. Chen et al. proposed a method that can be used to optimize paint distribution errors due to changes in spray width [6, 7].

A six-degree-of-freedom spray robot trajectory planning method is proposed by this paper based on the β distribution model, which effectively improves the spraying quality and efficiency of the surface of the car roof beam.

Y. Wang et al. (Eds.): IWAMA 2021, LNEE 880, pp. 35–43, 2022.
https://doi.org/10.1007/978-981-19-0572-8_5

2 Establishment of Paint Deposition Rate Model

2.1 Analysis of the β Distribution Model

The β distribution model is a better distribution model obtained through experimental data. Therefore, the β distribution model is selected as the prototype of the mathematical model of paint deposition rate. Assuming that the spraying space is a cone as shown in Fig. 1, the finite range model expression is as follows:

$$T = T_{\max}\left(1 - \frac{4m^2}{\omega^2}\right)^{\beta}, \ 0 \le m \le {}^{\omega}/_{2}, \ \omega = 2h\tan\emptyset \tag{1}$$

Where ω is the width of the spraying area, m is the distance from any point in the spraying area to the projection point of the spray gun, h represents the vertical distance from the spray gun to the surface of the workpiece, $\emptyset$ represents half of the spray cone opening angle, and T_{max} is the maximum spraying thickness, which is related to the parameters β, diameter ω and paint flow rate α, expressed as $(4\alpha\omega)/(\pi\omega^2)$.

The characteristic of the β distribution model is that different coating deposition models can be determined by different β values. For example, when β is 1.5, it is an ellipse, and when β is 2, it is a parabola. Figure 2 shows the model thickness curve for different β.

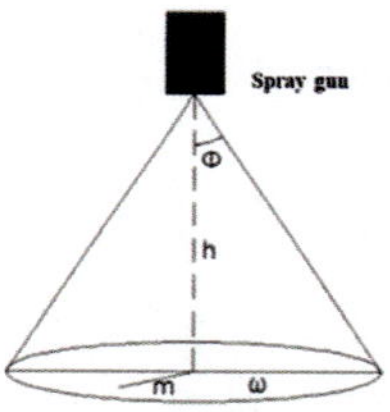

Fig. 1. β distribution model

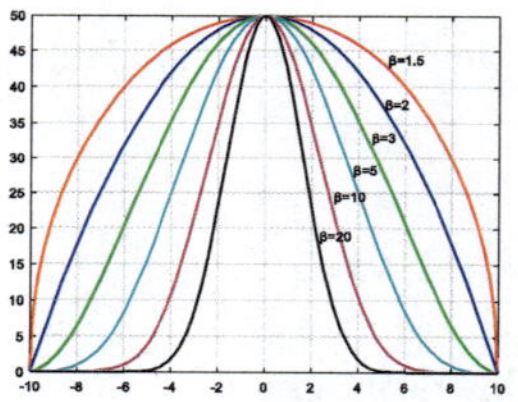

Fig. 2. Distribution model of different β values

2.2 Fitting of the Paint Deposition Rate Model

The coating thickness data in different sprayed areas can be obtained through spraying test, and the coating deposition rate model can be fitted by the obtained coating thickness data. The linear trajectory of the spray gun axis is marked on the workpiece and it is used as the X axis. The thickness of the paint film at different positions along the vertical direction of the X axis (set as the Y_i axis) are measured, and then 5 sets of experimental data at the same distance points on different cross-sections along the spraying direction are recorded to obtain the average value. The schematic diagram of the spraying method is shown in Fig. 3.

The thickness of the paint film at each point is measured by the paint film thickness gauge, and the recorded data is shown in Table 1.

Fig. 3. Schematic diagram of spraying method

Table 1. Average value of cross-section paint film thickness at different positions

Distance (cm)	−20	−18	−16	−14	−12	−10	−8	−6	−4	−2	0
Thickness (μm)	0	0	14.4	33.8	40.6	54.8	66.2	74.4	78.4	78	81.2
Distance (cm)	20	18	16	14	12	10	8	6	4	2	
Thickness (μm)	12.6	16.2	35.4	43.8	57	60.2	67.8	76.2	78.4	81.6	

The analysis results of Table 1 show that the data distribution in Table 1 is the closest to the β distribution model when $\beta = 2$. According to the expression of the β distribution model, combined with the data in Table 1, the function of the coating deposition model can be obtained by numerical fitting.

When $\beta = 2$, the β distribution model can be expressed as:

$$T = \frac{T_{\max}}{R^2}\left[R^2 - m^2\right],\ R = \frac{\omega}{2} \tag{2}$$

Assuming that the paint film thickness model conforms to the above-mentioned β distribution model, the relationship between the paint film thickness and the measuring point radius y can be obtained by fitting the data in the above table through the MATLAB fitting tool:

$$T = -0.1501y^2 + 0.1456y + 59.24 \tag{3}$$

The following formula can be obtained by deforming the above formula:

$$T = 0.1501\left[19.87^2 - (y - 0.4850)^2\right] \tag{4}$$

The fitting result is shown in Fig. 4.

By comparing with the β distribution model, it can be seen that the radius of the spraying area is $R = 19.87$, and y is the distance from any point in the spraying area to the projection point of the spray gun. It is also found that the center origin is not the place where the paint film thickness is the largest, but is shifted to the right by 0.4850 cm, which is in line with the actual situation.

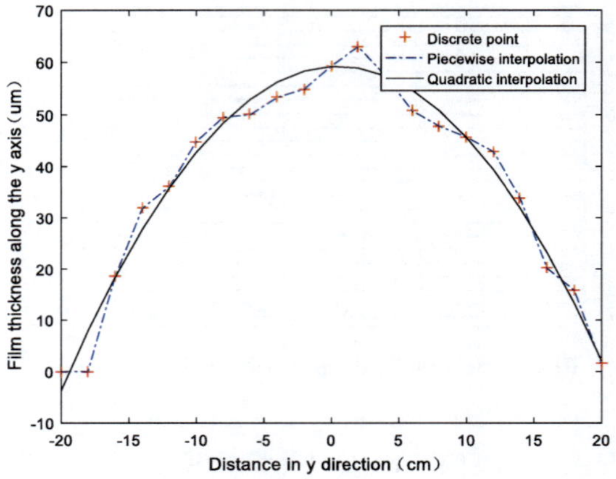

Fig. 4. The β distribution model fitting

For the convenience of further theoretical expression, simplified to the algebraic formula about x:

$$T(x) = k\left(R^2 - x^2\right) \tag{5}$$

$k = 0.1501, R = 19.87, x = y - 0.4850.$

3 Calculation of Adjacent Spraying Distance and Spraying Overlap Rate

The distance between adjacent sprays and the coincidence rate of sprays can be obtained by the paint deposition rate model. In the plane spraying process, in order to determine the optimal overlap distance d, it is assumed that at a certain moment, the adjacent spraying distance is shown in Fig. 5.

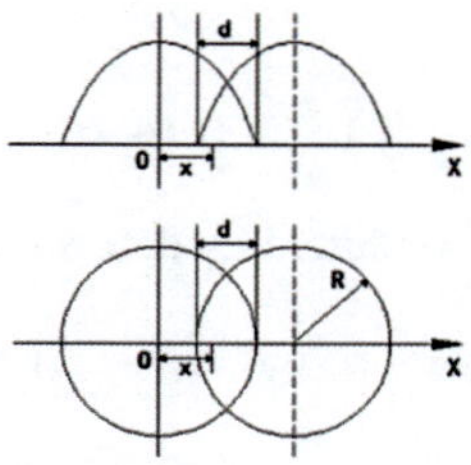

Fig. 5. Schematic diagram of spraying trajectory overlap

Point O is the center point of the area, x represents the distance from a point in the spraying area to the center of the first path, R is the radius of the spraying area, and d is the width of the overlapping area of the two spray gun paths, It can be known from the fitting experiment that the algebraic equation of paint thickness is as follows in the case of no overlap:

$$\mathrm{T(x)} = \mathrm{k}\left(\mathrm{R}^2 - \mathrm{x}^2\right)$$

Then when x belongs to $(-R, 3R - d)$, the expression of paint film thickness is:

$$T = \begin{cases} T(x) & x \in (-R,\ R - d) \\ T(x) + T(x - (2R - d)) & x \in (R - d,\ R) \\ T(x - (2R - d)) & x \in (R,\ 3R - d) \end{cases} \tag{6}$$

It can be seen that T(x) is a piecewise quadratic function with parameter d. When d takes 5, 10, 15 respectively, the coating distribution graph is shown in Fig. 6.

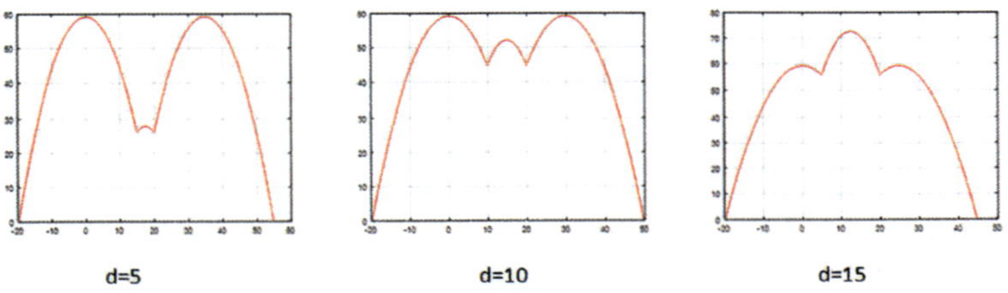

Fig. 6. Distribution of coating film when d takes different values

The Fig. 6 shows that when d is set to 10, the coating distribution is relatively uniform and stable than the other two spacing cases, but obviously this is not the optimal result. Let $\mathrm{T(R - d/2) = T(0)}$, the optimal distance $\mathrm{d} = \left(2 - \sqrt{2}\right)\mathrm{R} = 11.64$ can be obtained [8]. The coating distribution graph is shown in Fig. 7.

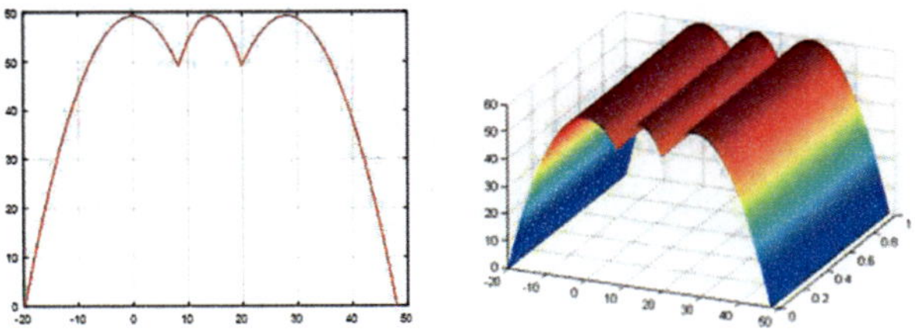

Fig. 7. Distribution of coating film at d = 11.64

By comparing Fig. 6 and Fig. 7, we can see that the distribution effect when d is 11.64 is more stable than when d is 10. At same time, $T_{min} = 49.10\ \mu m$, $T_{max} = 59.26\ \mu m$, $\Delta T = 8.63\ \mu m$, overlap rat $w = d/R = 58.58\%$, And this coincidence rate is general. When $\beta = 2$, by adjusting the spray gun parameters, it is applicable to different adjustment radii. At same time, the distance between two adjacent spray gun paths is $S = \sqrt{2}\,R$.

4 Selection of Spray Gun Path

There are generally two kinds of stroke paths of spraying robots: Z-shaped path and spiral-shaped path. As shown in Fig. 8.

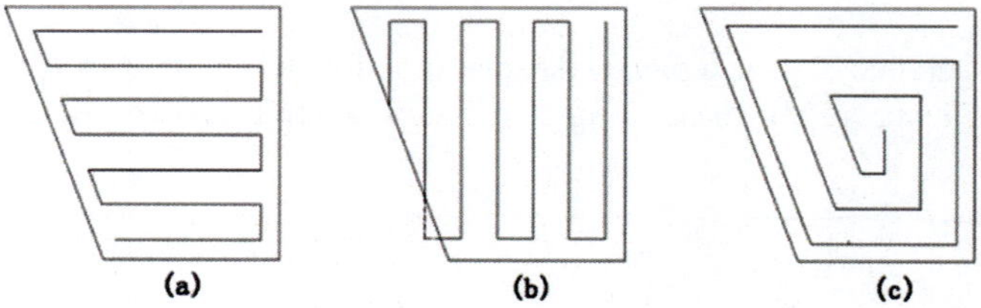

Fig. 8. Z-type and spiral-type spray gun tracks

Usually, spraying robots use Z-shaped paths when spraying, and the motion mode is generally horizontal reciprocating or vertical reciprocating. The difference between these two motion modes is that for workpieces with different aspect ratios, when reciprocating spraying in different directions, the inflection point of the spray gun during the spraying process is different, and the absolute spraying time is also different. In order to achieve a wider spray coverage, and based on the style of the sprayed workpiece, the Z-shaped path with the horizontal reciprocating motion mode is finally adopted.

This paper aims to plan the spraying trajectory for the car roof beam, and the final spraying path is shown in Fig. 9.

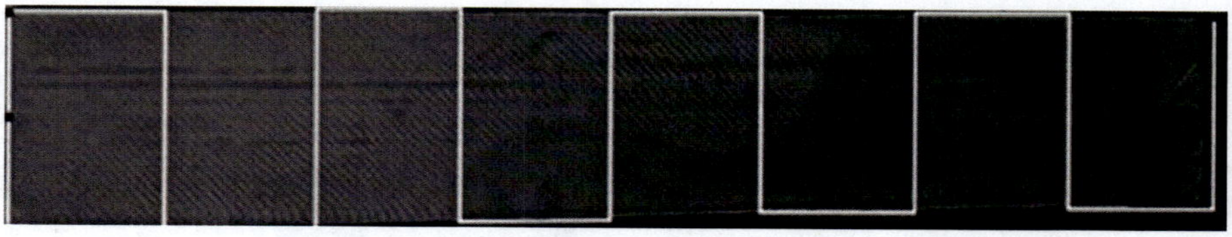

Fig. 9. Spraying path of car roof beam

5 Automatic Spraying Experiment

According to the Z-type motion trajectory, the offline programs are written, and then import it into the on-site robot. And the parameters are set as follows: spraying distance is set to 250 mm, coincidence rate is set to 58.58%, robot speed is set to 20% (At same time, the spraying speed is 400 mm/s) and the speed of the governor is set to 667 r/min (At same time, the flow rate is 400 ml/min). And then start the system and set it to automatic mode to spray the top cover beam sample, the experimental site is shown in Fig. 10. The spraying task is completed through a series of spraying processes such as automatic spraying, polishing, cleaning, drying, and spraying. The effect diagram of each stage during the period is shown in Fig. 11. The total process completion time is 7.8 min, which is far below the 18 min requirement. The robot spraying time is only 53 s, which meets the production requirements. After the spraying is completed, the workpiece is removed and cured in a large oven for 90 min, and the curing temperature is set to 80 °C. Finally, the sprayed samples are tested. The test results show that the gloss of the paint film reaches 90+, and the average thickness of the paint film is about 40 μm. The cross-cut test shows that the paint film does not fall off, which meets the requirements.

Fig. 10. Robot spraying experiment site

(a) To be sprayed (b) Finish spraying to be cured (c) Curing is complete

Fig. 11. Appearance of each stage of automatic spraying

Through experimental test comparison, Table 2 is obtained.

Table 2. Comparison of experimental results of manual spraying and automatic spraying

Performance		Before spraying	Manual spraying	Automatic spraying
Appearance quality		Rough and dull	Smooth and bright surface	Smooth and bright surface
Thickness	Max	—	78 μm	43 μm
	Minimum	—	30 μm	38 μm
	Average	—	—	40.73 μm
Gloss		Less than 5	80–95	80–95
Adhesion		—	Level 0	Level 0
Hardness		—	4H	4H

6 Conclusions

The robot spraying trajectory to improve the surface spraying quality and spraying efficiency of the car roof beam is planned in this paper. By establishing the coating deposition rate model, the optimal spraying distance $S = \sqrt{2}R$ is obtained. At same time, the overlap ratio of the adjacent spraying spacing is 58.58%, and the coating distribution is the most even and stable. According to the actual application scenario, the Z-shaped path with the horizontal reciprocating motion mode is selected as the spray gun motion path. The results of automated spraying experiments show that the surface glossiness of the car roof beam has increased from 83.2 for manual spraying to 90.5. In addition, compared with manual spraying, the thickness of automatic spraying has a smaller change, and it is more ideal to maintain it at about 40 μm.

References

1. Klein, A.: CAD-based off-line programming of painting robots. Robotica **5**(4), 267–271 (1987)
2. Balkan, T., Arikan, M.A.S.: Modeling of paint flow rate flux for circular paint sprays by using experimental paint thickness distribution. Mech. Res. Commun. **26**(5), 609–617 (1999)
3. Duelen, G., Stahlinann, H.D., Liu, X.: An off-line planning and simulation system for the programming of coating robots. CIRP Ann. Manuf. Technol. **38**(1), 369–372 (1989)
4. Sheng, W.H., Chen, H.P., Xi, N., Tan, J.D.: Optimal tool path planning for compound surfaces in spray forming processes. In: Proceedings of IEEE International Conference on Robotics and Automation, pp. 45–50, New Orleans (2004)
5. Sheng, W.H., Chen, H.P., Xi, N.: Tool path planning for compound surfaces in spray forming processes. Autom. Sci. Eng. **2**(6), 240–249 (2005)
6. Chen, H.P., Xi, N., Masood, S.K.: Development of automated chopper gun trajectory planning for spray forming. Ind. Robot. **31**(3), 297–307 (2004)

7. Chen, H.P., Xi, N., Sheng, W.H.: Optimizing material distribution for tool trajectory generation in surface manufacturing. In: Proceedings of IEEE/ASME International Conference on Advanced Intelligent Mechatronics. Monterey, pp. 1389–1394 (2005)
8. Arikan, M.A., Sahir, T.B.: Process modeling, simulation, and paint thickness measurement for robotic spray painting. J. Robot. Syst. **17**(9), 479–494 (2000)
9. Diao, X.D., Zeng, S.X., Tam, V.W.Y.: Development of an optimal trajectory model for spray painting on a free surface. Comput. Ind. Eng. **8**, 209–216 (2009)

Design of the Breathing Exerciser Integrated the Functions of Flutter and Expectoration

Zhen Li[1], Guiqin Li[1](✉), and Zhenwen Liang[2]

[1] Shanghai Key Laboratory of Intelligent Manufacturing and Robotics, Shanghai University, Shanghai 200072, China
leeching@shu.edu.cn
[2] College of Rehabilitation Science, Shanghai, China

Abstract. A breathing exerciser is designed to assist patients with COPD for discharging respiratory secretions in this paper. This breathing exerciser has the training lung function and can visually display expiratory parameters. Based on vibratory sputum extraction method, it promotes the patient's lungs to cough up secretions. By using the sensor and control module, the expiratory parameters are collected and visualized on the screen as well. The results of the final breath parameter collection experiment show that the breathing exerciser can remove excessive or retained secretions in the airway and improve the lung function of patients through parameterized training.

Keywords: Expectoration · Breathing training · Expiratory parameters

1 Introduction

Chronic obstructive pulmonary disease (COPD) is a preventable and treatable disease characterized by airflow limitation [1]. The disease have a high disability fatality rate [2]; an experimental research by the medical school of Shanghai Jiaotong University has proved that the application of breathing exerciser-assisted training sessions can effectively improve patients' dyspnea symptoms and enhance their respiratory function [3], but the Chinese Academy of Sciences investigates and studies It shows that there are few domestic companies in related fields, and the COPD regulatory market is almost blank in my country [4], while the promotion of respiratory rehabilitation systems in foreign countries is earlier and more professional [5].

Flutter and Acapella are two types of breathing training equipment abroad [6, 7]. In this paper, the mechanical part of the breathing exerciser combines the principles of the above two equipment, which can realize the comprehensive functions of expectoration and breathing training. There is also a visualization function of exhalation parameters, statistics of the change trend of exhalation parameters and analysis of the reasons for the changes. This plays an important guiding value for the subjects' follow-up training [8]. The purpose of this study is to explore a new physical expectoration and expiratory parameter collection method for the clinic, and to verify its feasibility through experiments.

Y. Wang et al. (Eds.): IWAMA 2021, LNEE 880, pp. 44–50, 2022.
https://doi.org/10.1007/978-981-19-0572-8_6

2 Theoretical Analysis of Vibration Sputum Discharge

2.1 Vibration Sputum Expulsion Method

The powerful flapping vibrations in the airway can loosen the sputum and promote sputum expulsion. This sputum expulsion method is called the vibratory sputum expulsion.

The mechanism is designed based on aerodynamic principles. Using the rapid opening and closing action of the mechanical device during the exhalation process to produce pressure oscillations and short-accelerated airflow. On the one hand, it greatly increases the pressure oscillation amplitude in the airway and the swing amplitude of the airway wall. On the other hand, due to the rapid external thrust of the airway under the condition of short and frequent opening and closing, the external thrust of the secretions is obviously strengthened. The combination of the violent oscillation of the airway wall and the rapid external airflow is extremely conducive to the secretion adhered to the inner wall of the airway to escape from the wall (Fig. 1).

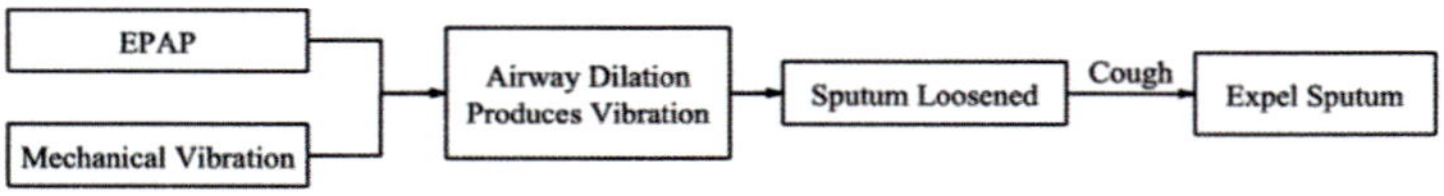

Fig. 1. Flow chart of vibration sputum discharge

The main indicators of pulmonary function and the important criteria for determining COPD are forced expiratory volume in one second (FEV1), forced vital capacity (FVC), and expiratory volume in one second (FEV1).

In addition, peak expiratory flow (PEF), the ratio of forced expiratory volume in the first second to forced expiratory capacity (FEV1/FVC) and instantaneous expiratory flow also play an important reference significance in the evaluation of lung function.

3 Mechanical Structure Design

The breathing exerciser is designed to assisting patients in expectoration and strengthening the lung function of patients through breathing training. The function of the breathing exerciser is divided into two parts: loosen sputum part and move sputum part (Fig. 2).

The structure of the breathing exerciser includes sputum loosening device, sputum moving device, sensing device and other devices, as is shown in Fig. 3. The sputum loosening device and the sputum moving device are installed in parallel inside the box. The sensor and the display screen are connected with the information processing module to cooperate to complete the collection and visualization of exhalation parameters.

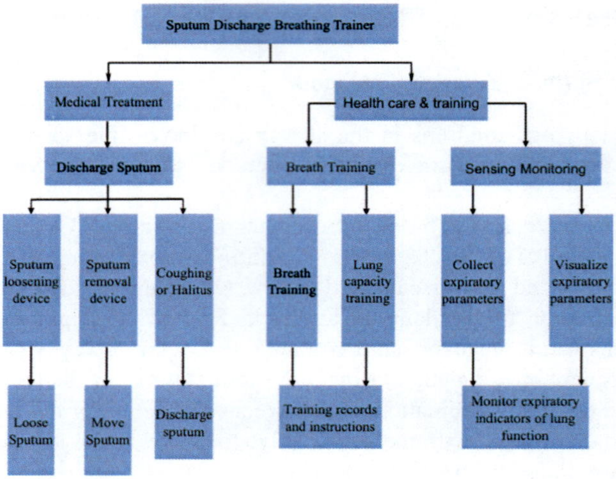

Fig. 2. Functional design

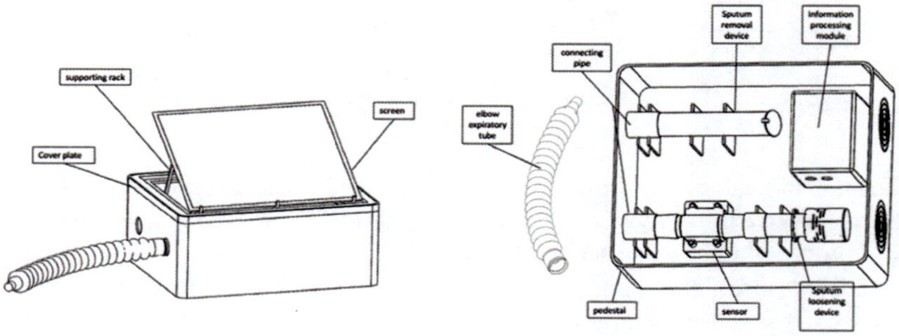

Fig. 3. General structure of sputum discharge breathing training apparatus

Exhalation parameters are collected using air flow sensor. The CAFS4000-1501V-A gas mass flowmeter sensor (as shown in Fig. 4 and Fig. 5) produced by Consensic, ICN Company in the United States is selected, and its flow range is 0-150SLM. The maximum allowable flow rate is 6.580 m/s, and its output is analog output. The linear output of the sensor is unidirectional air flow mode, where:

$$\text{flow} = [\frac{\text{VOUT} - 1\,\text{V}}{4\,\text{V}}] \tag{3-1}$$

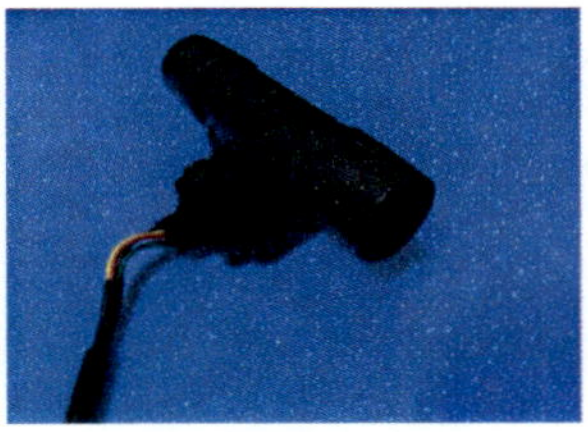

Fig. 4. Airflow sensor

Fig. 5. Mega 2560 development board

4 System Function Design

4.1 Data Acquisition and Parameter Calculation

Respiration-related parameters of subjects are collected by sensors to realize the system parameter visualization and data analysis. These parameters need to be calculated by corresponding formulas or directly measured.

Currently, body mass index (BMI) is believed that there is a positive correlation between BMI and FEV1\FVC in the population.

$$\mathrm{BMI} = \frac{\mathrm{M}}{\mathrm{H}^2} \tag{4-1}$$

M—Weight (Kg).

H—Height (m).

Forced vital capacity (FVC) is the maximum amount of air that can be exhaled after inhaling as much as possible and exhaling as soon as possible.

$$\mathrm{FVC} = \sum\nolimits_{i=0}^{n-1} [\frac{1}{2} \times \frac{t}{60} \times (F_i + F_{i+1})] \tag{4-2}$$

t—The time interval between two instantaneous flow measurement points, t = 0.05 s.

Fi—Instantaneous flow measuring point (before).

Fi + 1—Instantaneous flow measuring point (after).

FEV1 is the expiratory volume of air exhaled for 1 s at the total volume of the lung, which is the most commonly used parameter to determine the type of ventilation dysfunction and the degree of damage.

$$\mathrm{FVC} = \sum\nolimits_{i=0}^{19} [\frac{1}{2} \times \frac{t}{60} (F_i + F_{i+1})] \tag{4-3}$$

Peak expiratory flow (PEF) is the fastest expiratory rate of forced exhalation, which mainly reflects the subject's expiratory limitation [8]. In the system, the peak velocity can be measured directly by the sensor.

Inspiration and expiration ratio (I/E):

$$\frac{\mathrm{I}}{\mathrm{E}} = \frac{I_t}{E_t} \tag{4-4}$$

It—Inhalation time.
Et—Exhalation time.
Respiratory rate (RR) indicates the number of breaths completed in 1 min.

$$RR = \frac{60}{I_t + E_t} \tag{4-5}$$

4.2 User Interaction Design

The user interface is designed to visualize the expiratory parameters of subject, including FVC, FEV1, PEF, etc., in which the continuous value is displayed in the way of line graph, and the discrete value is displayed in the way of number. The user interface is shown in Fig. 6.

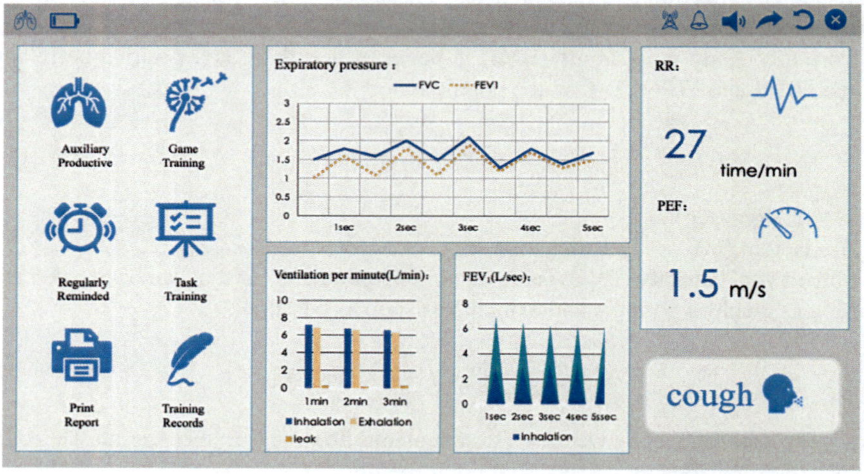

Fig. 6. Host computer interface

Arduino is used to read serial port data systematically, and python is used to read and call serial port data, so that the collected data can be processed and graphs of related parameters can be drawn.

5 Experiment and Test

A total of six subjects participated in the experiment. The experimental data are shown in Table 1, [BMI (Kg/m^2), RR (time/min), FEV1 (ml), PEF (SLM)]:

After repeated trials for several times, FEV1, FVC, FEV1/FVC% and PEF values of all subjects before and after the experiment is slightly increased. The long-term and regular use of breathing exerciser for training will be helpful to effectively improve

Table 1. Exhalation parameters test results

	BMI	RR	FEV_1	FVC	PEF	FEV_1/FVC
Subject 1	18.76	15.79	858.75	1835.83	72	46.78%
Subject 2	20.45	17.93	1121.25	1510.83	81	74.21%
Subject 3	24.98	14.43	797.08	2220.00	60	35.90%
Subject 4	16.56	22.32	922.92	1386.67	78	66.56%
Subject 5	24.30	20.45	802.08	2019.58	79	39.72%
Subject 6	19.09	15.36	797.50	1648.33	76	48.38%

clinical ventilation. For patients with chronic respiratory inflammation or high-risk respiratory diseases, the effective removal of excessive or retained secretions in the airway and reduction of airflow resistance will significantly reduce the incidence of bronchial infection, or the prevention and treatment of atelectasis caused by mucus obstruction of the airway.

6 Conclusion

The breathing exerciser collecting and visualizing exhalation parameters is designed based on the principle of vibration expectoration method. Compared with the traditional expectoration device or breathing training device, this expectoration and breathing trainer combines the exhalation channel and the combination switch to combine the expectation and breathing training. The weight of the breathing trainer has been reduced from the traditional 3–5 kg to 1.5 kg, which improves the portability of the breathing trainer. In addition, this breathing trainer increases the error range of the expiratory flow to 1.5% FS. The expiratory parameters are visually displayed on the display screen, so that patients and their families can clearly read the relevant expiratory parameters and understand the lungs at any time. The experimental data verifies its effective effect.

References

1. Li, Y., Jing, Z., Shaobin, J.: Observation on clinical efficacy of patients with acute exacerbation of chronic obstructive pulmonary disease with different body mass indexes. Ningxia Med. J. **42**(11), 1014–1017 (2020)
2. ARDS Definition Task Force, Ranieri, V.M., Rubenfeld, G.D., et al.: Acute respiratory distress syndrome: the Berlin Definition. JAMA **307**(23), 2526 (2012)
3. Pan, H., et al.: Clinical efficacy of respiratory rehabilitation assisted by respiratory training apparatus in patients with COVID-19. J. Second Mil. Med. Univ. **42**(03), 255–260 (2012)
4. Zhao, R., et al.: Research of chronic obstructive pulmonary disease monitoring system based on four-line turbine-type. J. Electron. Inf. Technol. **41**(02), 469–476 (2019). (in Chinese)
5. Tang, S., Li, Y.: Design of respiratory rehabilitation training device for children. Sci. Technol. Innov. Appl. **2020**(36), 28–29+32
6. Yilong, W., Xiangyu, Z., Qixing, W., Jianfeng, R.: Clinical application of portable oscillating sputum discharge device in free position. Shanghai Nurs. **16**(03), 63–64 (2016). (in Chinese)

7. Wang, R., Zhao, Y., Jing Li, H., Liu, G.Z., Jianying, X.: Risk factors of first-second forced expiratory volume decline in community residents. China Materia Medica Clinica **20**(11), 1775–1777 (2020)
8. The Acute Respiratory Distress Syndrome Network: Ventilation with lower tidal volumes as compared with traditional tidal volumes for acute lung injury and the acute respiratory distress syndrome. N. Engl. J. Med. **342**(18), 1301 (2000)

Design of Warehouse Chattel Supervision System Based on AI Video

Yicong Shen[1], Guiqin Li[1](✉), Tiancai Li[2], Kuiliang Liu[1], and Peter Mitrouchev[3]

[1] Shanghai Key Laboratory of Intelligent Manufacturing and Robotics, Shanghai University, Shanghai 200072, China
leeching@shu.edu.cn
[2] Shanghai Tigers Co., Ltd., Shanghai, China
[3] Grenoble Alpes University, G-SCOP, 38031 Grenoble, France

Abstract. In this paper, the AI video supervision system of chattel is designed to monitor and manage chattel. The video is logged to AI processors by means of the image or video collecting hardware in this system. Once the system finding difference between the video and original setting, the data is output to alarms. The function of automatic, all-weather, intelligent and no dead angle is realized in the supervision of chattel system, and the system provides an effective and feasible way for the high efficiency and low cost of intelligent warehouse operation.

Keywords: AI processors · Chattel supervision · Virtual fence · Machine vision · Image acquisition

1 Introduction

AI system can be used to read the status of goods in real time and improve the safety of goods. Based on this, Maarten van Geest [1] designs a complete reference architecture of intelligent warehouse under the background of industry 4.0. Furthermore, according to a method which is proposed by N. Irino [2], status of goods and surroundings can be read by vision camera. KalleKärhä [3] applies vision camera to the measurement of piles of woods in the large wharf yard, and the very accurate data are obtained this time. Therefore, warehouses can also try to utilize vision camera to read the data of goods itself status and its surrounding environment. Further, the system can set the effective range of monitoring through the electronic virtual fence technology which has been popular in recent years.

Electronic virtual fence technology originated in Australia, which is originally used in animal husbandry fence, and then gradually used in other fields. Through the methods [4] about electronic fence improvement and the application in detection, recognition [5], this system combines electronic virtual fence and depth camera as a goods monitoring system in intelligent warehouse. This paper will introduce the basic scheme of this design.

Y. Wang et al. (Eds.): IWAMA 2021, LNEE 880, pp. 51–57, 2022.
https://doi.org/10.1007/978-981-19-0572-8_7

2 Intelligent Warehouse Supervision Design Scheme

The AI video chattel supervision includes AI processor (1), image and video acquisition module (2), virtual fence setting module (3), alarm module (4), information display and processing module (5) as well as image and video storage module (6). The output port of image and video acquisition module (2) and the virtual fence setting module (3) are connected with the input port of AI processors (1). And the output port of the AI processor is connected with the input port of the information display and processing module (5), the alarm module (4), and image and video storage module (6). structural representation of an AI video is as shown in the Fig. 1.

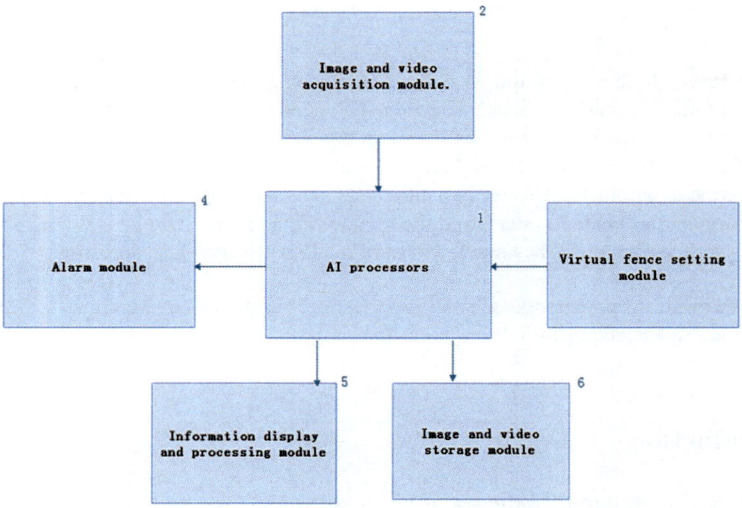

Fig. 1. The structural representation of an AI video chattel supervision

The image and video acquisition module consists of infrared night vision HD camera as well as 3D vision depth camera. In order to improve the accuracy and robustness of the current input signal while the machine completes the tasks of scene recognition and self-localization in the process of environment interaction, the camera must take advantage of relocation technology. Compared with other kinds of sensors with relocation function, RGB [6] camera is cheaper and more informative. The RGB camera usually uses SFM (Structure From Motion) technology to locate. SFM seeks the corresponding relationship between local features and 3D points in 2D image, meanwhile matches the 2D points and 3D points in shared high-dimensional feature descriptor space. These features points can be traditional manual feature points (such as sift, hog), and also be the output of deep neural network, combined with 2D-3D matching, then utilize RANSAC algorithm to select the best pose estimation, finally getting the appropriate position.

The virtual fence setting module (3) can set the virtual fence for storing the movable property according to the need of storage location, occupied area and specification of the movable property to be supervised. The fence can be regular rectangular, circular and triangular area, or a arbitrarily closed area surrounding the supervised chattel. Refer to Fig. 2, where (2) is the image and video acquisition module, (4) is the alarm module (5) is the system information display and processing module, and (7) represents different movable property.

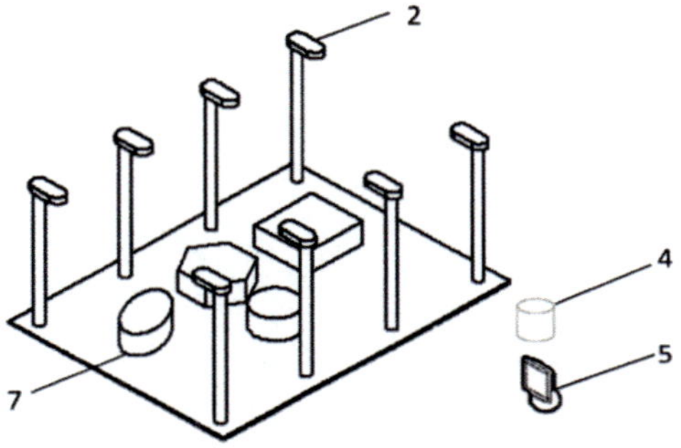

Fig. 2. The sketch diagram of module placing

3 The Design of Module Function

In the intelligent warehouse, the supervision of goods is mainly through a way of AI video chattel supervision, which is principally based on machine vision technology [6], and through six modules to establish a closed-loop system of goods supervision.

3.1 AI Processor and Mechanical Vision Module

The AI processor (1) is mainly used to decode and calculate the image and video transmitted by the module (6), and monitors the volume, shape, color, position, status. It can authorize operator of the movable property through this method, and fill the intermediate device to receive the input data stream. Then the calculated data is transmitted to the alarm and display module through the output port for entity display.

The image and video acquisition module (2) in the system is used to obtain video stream and depth information, including infrared night vision HD camera and 3D vision depth camera [7]. The function of image and video acquisition module (2) is reflected in the supervision area without dead space. According to the property of the supervised chattel, arranging conventional infrared night vision HD cameras and adopting 3D visual depth cameras, for special, small and valuable movable property pledge, can cover all the monitoring area [8]. No matter whether the environment of the goods is night or day, the system can always accurately collect the real-time image as well as video data of the goods to guarantee the safety of those goods.

3.2 Virtual Fence Module

The virtual fence setting module (3) is used to set the virtual fence for storing the movable property according to the storage location, occupied area and some else conditions to be supervised [9]; The virtual fence is a virtual boundary line that separates the security area from the surrounding area along the physical protection line on the screen obtained from the imaging device.

The virtual fence is set with a spline curve [10], which is marked with control points, and the operators can input the coordinates by operating the control point through GUI (graphical user interface). It can also extend a certain distance from the virtual fence as an early warning range (Fig. 3).

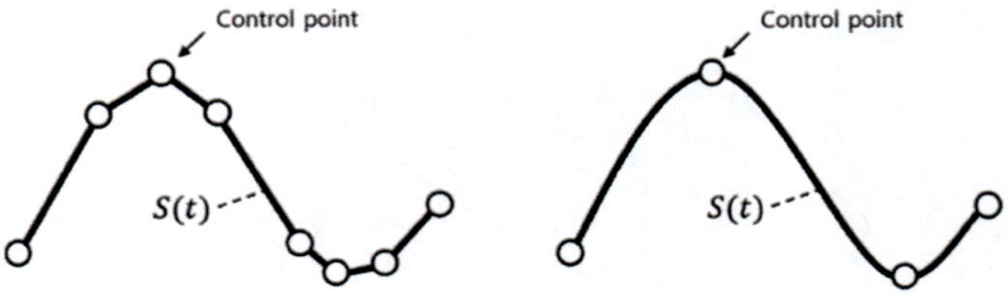

Fig. 3. Spline curve with linear interpolation

Spline interpolation is a form of interpolation. Interpolation is a special piecewise polynomial, which is called spline. In the closed interval [1, b], there are (n−1) intervals, the leading vertex is a, then x1, x2... and the final is B. The minimum order of the qualified spline function is the third order, and the part of the function expressed by the trinomial polynomial is called the cubic spline function. The general form of the cubic spline function is given by the following equation:

$$S(x) = \alpha_i(x - x_i)^3 + b_i(x - x_i)^2 + c_i(x - x_i)^1 + d_i \tag{1}$$

When $x_i < x < x_{(i+1)}$, the coefficient a_i, b_i, c_i, needs to be calculated to get the specific value. Firstly, the system needs to consider the second derivative of cubic spline. And when it chooses a spline as cubic polynomial, the system should know its second derivative is a linear polynomial in the interval $\left[x_{j-1}, x_j\right]$, and it means

$$S(x)'' = \frac{(x_{j+1} - x)Q_j + (x - x_j)Q_{j+1}}{\Delta x}, j = 0, 1, \ldots, N - 1 \tag{2}$$

The Δx is the interval. The spacing between nodes, however, usually is not equal, thus we must replace the denominator with Δx_j, $\Delta x_j = x_{j+1} + x_j$. According to the formula of the second derivative, the second derivative is continuous in the whole process. Therefore, through this continuity, we can integrate it twice to obtain the

$$S(x) = \frac{(x_{j+1} - x)^3 Q_j + (x - x_j)^3 Q_{j+1}}{6\Delta x} + A(x_{j+1} - x) + B(x - x_j) \tag{3}$$

The A and B are arbitrary constants which are need to be determined. In order to confirm them, it needs to use certain conditions, brings these conditions, rearranges the calculation, and then we can get the expression of two constants as follows:

$$A = \frac{y_j}{\Delta x} - \frac{\Delta x Q_j}{6}, \tag{4}$$

$$\mathrm{B} = \frac{\mathrm{y_{j+1}}}{\Delta \mathrm{x}} - \frac{\Delta \mathrm{x} \mathrm{Q_{j+1}}}{6} \tag{5}$$

After substituting constants into the calculation,

$$S(x) = \frac{(x_{j+1} - x)^3 Q_j + (x - x_j)^3 Q_{j+1}}{6\Delta x} + \frac{(x_{j+1} - x)y_j + (x - x_j)y_{j+1}}{\Delta x} - \frac{\Delta x((x_{j+1} - x)Q_j + (x - x_j)Q_{j+1})}{6} \tag{6}$$

The results obtained from the above formula are shown in the following equation:

$$h_{i-1}S_{i-1} + (2h_{i-1} + 2h_i)S_i + h_i S_{i+1} = 6\left(\frac{y_{i+1} - y_i}{h_i} - \frac{y_i - y_{i-1}}{h_{i-1}}\right) \tag{7}$$

Through the formula, we can infer the coordinates of each point on the formula.

Then, in the monitoring area, operators should set up an early warning area, which is the part between the virtual fence and the alarm curve. This part of the distance is set through the distance value input by the GUI. Assuming that each control point of the spline of the virtual fence is p, then each control point of the new spine is P′. The dista-nce between two splines is r, and the distance between each control point is r too. The system can get three boundary lines f(x), $g_1(x)$, $g_2(x)$ by getting the sign of the tangent of P, the slope of S and the position of quadrant, meanwhile, divide them into two regions. You can select the early warning area from two space through GUI to make f(x) be the corresponding virtual fence, and make $g_1(x)$, $g_2(x)$ be two boundary lines in the same time. As shown in the Fig. 4.

3.3 Alarm and Image Storage Module

The system information display and processing module (5) is used to display the status information of the supervision area, as well as some abnormal information, image and video data fed back by the processing system. After being decoded and transmitted by the AI processor, the data is sent to the system information display and processing module for further processing, and the data is displayed on the screen in a certain specification and format according to the design. The intelligent warehouse staff can directly observe the real-time situation of the supervised goods, and can deal with the abnormal goods in time. Part of the data decoded by AI processor is sent to the system information display and processing module, and the other part is transmitted to the image and video storage module for encoding into video format, and then the video is backed up, which is also the last guarantee to ensure the safety of goods in the system.

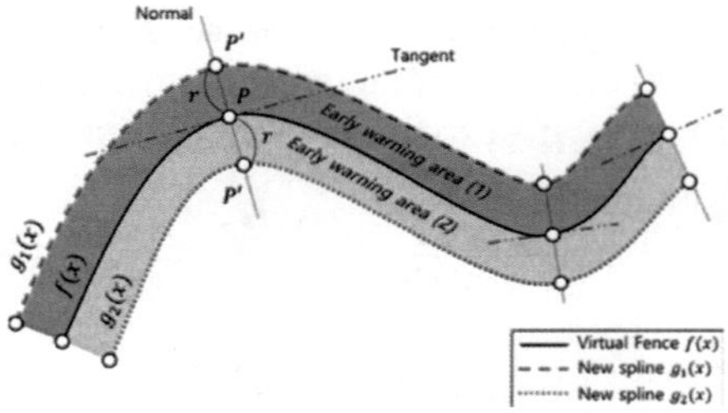

Fig. 4. Virtual fence and alarm curve

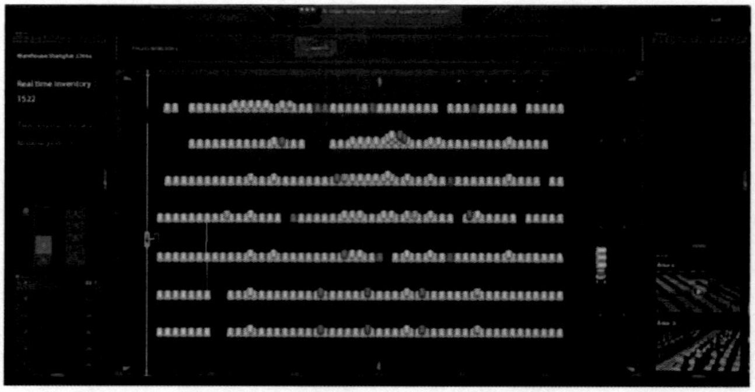

Fig. 5. Operation interface on Mobile terminal

3.4 Operation Interface Display

The GUI interface is shown as Fig. 5.

The area included in the preset virtual fence will be displayed on the display interface of APP end and PC end. When goods move out of the fence without lifting the ban or people enter the fence, the data will be transmitted to the alarm end and the video will be saved.

4 Conclusion

This paper designs an AI video chattel supervision system in modern intelligent warehouse, including AI processor, image and video acquisition module, virtual fence setting module, alarm module, system information display and processing module, image and video storage module. The video calculation and monitoring is performed by the AI processor while the operation of each module is coordinated by it too. The image and video acquisition module is designed to acquire video stream and depth information. The virtual fence setting module is designed to set the virtual fence to store the movable property according to the storage location, occupied area and area of the movable property to be supervised. The AI video chattel supervision system can overcome the shortcomings of the existing physical fence protection method of chattel storage, which requires a lot of manpower and material resources to perform monitoring. It can also realize the intelligent unattended of chattel in the storage area all day and effectively improve the storage security.

References

1. van Geest, M., Tekinerdogan, B., Catal, C.: Design of a reference architecture for developing smart warehouses in industry 4.0. Comput. Ind. **123**, 103343 (2021)
2. Irino, N., Shimoike, M., Mori, K., Yamaji, I., Mori, M.: A vision-based machine accuracy measurement method. CIRP Ann. **69**(1), 445–448 (2020)
3. KalleKärhä, S.N., Karvonen, H., Kivinen, V.-P., Melkas, T., Nieminen, M.: Comput. Electron. Agric.**158**, 167–182 (2019)
4. Umstatter, C.: The evolution of virtual fences: a review. Comput. Electron. Agric. **75**(1), 10–22 (2011)
5. Kim, S.H., Lim, S.C., Kim, D.Y.: Intelligent intrusion detection system featuring a virtual fence, active intruder detection, classification, tracking, and action recognition. Ann. Nucl. Energy **112**, 845–855 (2019)
6. Pei, H., Wang, J., Chen, Z.: A review of camera relocation methods based on deep learning. In: The 21st China Conference on System Simulation Technology and Its Applications
7. Karanam, S.R., Srinivas, Y., Krishna, M.V.: Study on image processing using deep learning techniques. Mater. Today Proc. (2020)
8. Khan, A.I., Al-Habsi, S.: Machine learning in computer vision. Procedia Comput. Sci. **167**, 1444–1451 (2020)
9. Kashi, A.: Electronic fence using wireless mesh network (2009)
10. Goldman, R.: B-Spline approximation and the de Boor algorithm. Pyramid Algorithms **20**, 347–443 (2003)

Fault Diagnosis of Massage Chair Movement Based on Attention-GRU-MLP

Lixin Lu[1], Xianhua Cai[1], Guiqin Li[1](✉), and Peter Mitrouchev[2]

[1] Shanghai Key Laboratory of Intelligent Manufacturing and Robotics, Shanghai University, Shanghai 200072, China
leeching@shu.edu.cn
[2] G-SCOP, University Grenoble Alpes, 38031 Grenoble, France

Abstract. An optimized recurrent neural network fault diagnosis model for the failure of the massage chair movement is proposed in this paper. An optimized Attention-RNN-MLP, Attention-LSTM-MLP and Attention-GRU-MLP fault diagnosis models are constructed by adding the Attention mechanism and MLP network to RNN, LSTM and GRU. The Attention-GRU-MLP fault diagnosis model is proved to be more rational by comparing the differences among these 3 models.

Keywords: Fault detection · Massage chair · Recurrent neural network

1 Introduction

Massage is one of the effective techniques used to relieve pain, reduce stress, and increase relaxation [1]. As early as 2020, RNN was used for fault diagnosis of air conditioning systems by some scholars [2]. However, RNN cannot be relied on for a long time, and it is prone to gradient disappearance or gradient explosion problems because of its structural defects. These problems are solved by scholars by improving the RNN model. The most commonly used ones are Long Short-Term Memory (LSTM) networks and Gated Recurrent Unit (GRU) networks that add internal gates control mechanism to maintain long-term dependence [3].

Pan et al. [4] achieved a high accuracy rate for fault diagnosis of bearings by combining CNN and LSTM. F. Zhang et al. [5] combined Attention mechanism and double structure GRU for the location of power system fault. R. Fu et al. [6] used LSTM and GRU neural network methods to predict short-term traffic flow and Obtained a better optimized network model. Wang Z et al. [7] combined the DLSTM and CNN to establish a diagnostic model for the fault diagnosis of rolling bearings, the model proposed by this method has higher performance.

2 Fault Diagnosis Method Based on RNN

2.1 Optimized RNN Based on Attention Mechanism and MLP

The two output forms of RNN are as follows.

Y. Wang et al. (Eds.): IWAMA 2021, LNEE 880, pp. 58–64, 2022.
https://doi.org/10.1007/978-981-19-0572-8_8

1. Output the final result which is a single sequence output.
2. Output all state values in the process which is an equal sequence output.

In general, the RNN is used to build models for related research and the output result is a single sequence output, which can achieve higher prediction results, but ignores the value of each sequence state in the historical moment. The MLP that is the multi-layer Dense layer is added to the original RNN to make full use of the historical state values of the RNN in this paper. The Attention mechanism is added to RNN is used as the feature extraction network, and the output result is used as a feature. The MLP performs non-linear changes to extract the association between these features, and maps to the output space to improve the Diagnostic effect. Dropout is added between the RNN and the MLP to prevent over fitting. Figure 1 shows the optimized network that is Attention-RNN-MLP, Attention-LSTM-MLP, Attention-GRU-MLP.

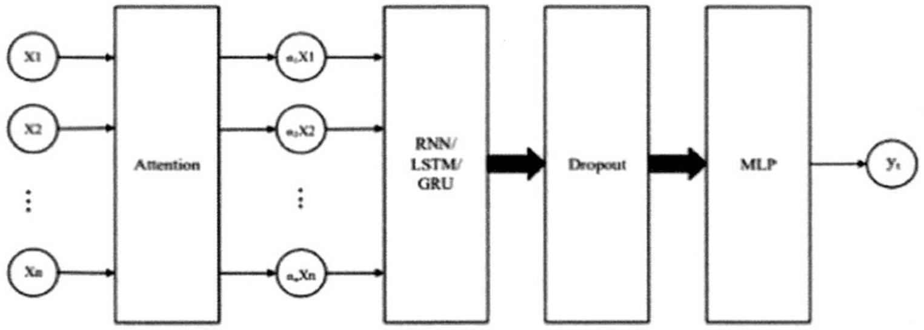

Fig. 1. Optimize recurrent neural network

Table 1. Attention-GRU-MLP diagnosis model parameter table

Network layer	Number of neurons/hyper-parameter value	Number of parameters
Dense1	1400	2800
Multiply	1400	/
GRU1	64	12864
GRU2	64	24960
GRU3	64	24960
Dropout	0.5	/
Dense2	64	4160
Dense2	128	8320
Dense3	128	8320
Soft-max	6	774

The Attention-GRU-MLP diagnostic model includes 1-layer of Dense and Multiply that is used to achieve Attention weight distribution, the 3-layer GRU that is used to extract current signal characteristics, the 3-layer perceptron that is used for mining and processing data information, the 1-layer Dropout layer that is used to reduce over fitting and the 1-layer Soft-Max layer that is used for fault classification. Table 1 shows the specific parameters of each network of the model.

2.2 Fault Diagnosis Model Architecture Based on Optimized Network

Figure 2 shows the fault diagnosis framework of movement that includes the input layer, the hidden layer, the output layer and the network training layer. The specific fault diagnosis process is as follows.

1) The input layer divides the collected original movement current signal into training set and test set.
2) The number of layers of RNN, LSTM and GRU, and the hyper-parameters, such as random seed, learning rate, number of iterations, etc. are determined.
3) The features of movement current data are extracted through RNN, LSTM and GRU based on Attention mechanism.
4) Dropout layer is used to suppress over fitting, and MLP for weighted calculation to achieve feature space conversion.
5) The classification results of the model and the actual results are compared to calculate loss value, and optimizes the parameters of the neural network of the system through the optimization function to achieve a better fault diagnosis effect.
6) Movement failures are classified by the Soft-Max multi-classifier of the output layer.
7) The accuracy and loss value of each model are evaluated.

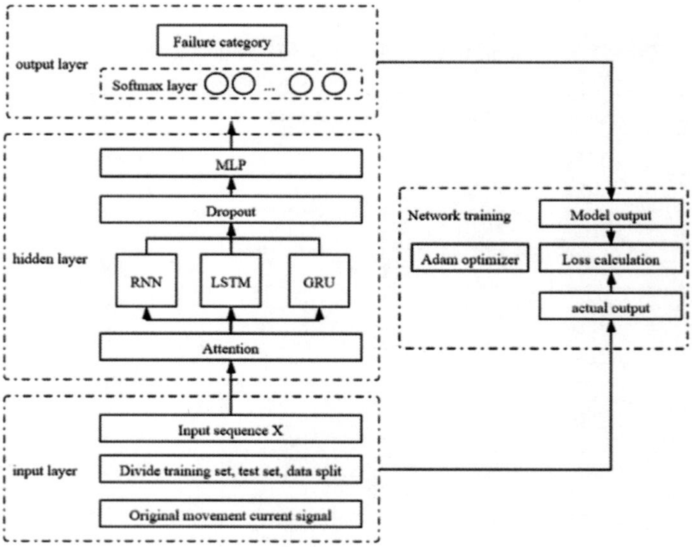

Fig. 2. Frame diagram of movement fault diagnosis

3 Testing and Verification of Fault Diagnosis System

3.1 Movement Data Set Loading

The experiment takes the 18N movement as the object, the frequency of the acquisition card is 1 kHz. It collected 10 times a total of 14000 continuous data points f_t in normal state, kneading motor turns to short circuit state, percussion motor armature winding element to open circuit state, 3D motor fails to run state, 3D percussion motor failure state and 3D motor drive mechanism failure state. The data set under each state is processed for noise reduction in the perception layer is transmitted to the application layer is $F_0 = \{f_1 f_2 \ldots f_{140000}\}$. The data used includes a total of 840,000 current signal sampling points. The data set F_0 is divided into training set $F_1 = \{f_1 f_2 \ldots f_{100000}\}$ and test set $F_2 = \{f_{100001} f_{100002} \ldots f_{140000}\}$ by the ratio of 5:2. The collected data is divided into 600 groups and 1,400 continuous data points in each group to adapt to the input characteristics of the input layer. The training set contains a total of 420 groups, and the test set contains 180 groups. The model input in each state is cut into $X = \{X_1 X_2 \ldots X_L\}$, $X_P = \{f_{1400P} f_{1400P+1} \ldots f_{1400P+1400}\}$, where $1 \leq P \leq L$. They X are inputted into the hidden layer network and output P.

3.2 Comparative Analysis of Different Models

The experiment is carried out with the same experimental conditions and data.

Figure 3 shows the diagnosis results of the un-optimized RNN. The evaluation indicators of the RNN and LSTM models are basically similar, and there is no obvious difference in the trend during the training process. The overall loss of GRU is lower, the accuracy, recall rate and AUC values are higher, but the recall rate is still insufficient compared to the precision and accuracy rate. Moreover, its various indicators fluctuate greatly, and the training results are unstable for each model.

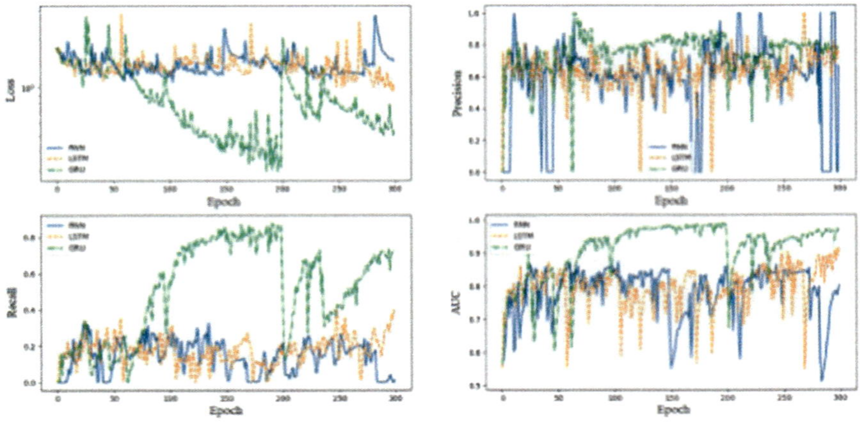

Fig. 3. Changes in training indicators of different recurrent neural networks

Figure 4 shows the diagnosis results of the optimized RNN.

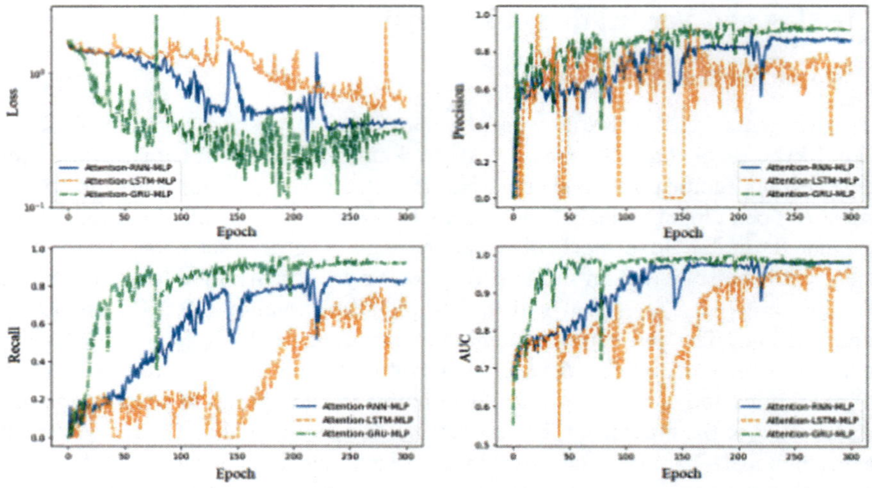

Fig. 4. Different optimized recurrent neural network training index changes

Table 2 shows the best evaluation data obtained by the Attention-RNN-MLP, Attention-LSTM-MLP, and Attention-GRU-MLP models during the training process after adding the Attention mechanism and MLP to the RNN. Overall, the stability of each optimized model has been improved. The precision and accuracy of the Attention-RNN-MLP and Attention-LSTM-MLP models are high, but the corresponding recall level is still too low, no more than 80%. It means that a large amount of data will be mistakenly diagnosed as faults in the actual diagnosis process which will affect the production volume of the movement and increase the manufacturing cost. Attention-GRU-MLP is easier to train, and all indicators can reach higher standards compared with Attention-RNN-MLP and Attention-LSTM-MLP.

Table 2. Comparison table of evaluation indicators for each model

Network model	Data sources	Loss	Accuracy	Precision	Recall	AUC
Attention-RNN-MLP	Train	1.2431	0.8583	0.7603	0.7290	0.8281
	Test	1.2634	0.8500	0.6800	0.7089	0.7822
Attention-LSTM-MLP	Train	1.1077	0.8583	0.6981	0.6643	0.8684
	Test	1.1858	0.8407	0.5556	0.6222	0.8482
Attention-GRU-MLP	Train	0.0392	0.9960	0.9881	0.9881	0.9998
	Test	0.2858	0.9741	0.9222	0.9222	0.9833

Figure 5 shows the confusion matrix of each mode to compare the test results of each model for different fault classifications. The 0 represents the normal state of the movement, and 1–5 represent the rest of the fault classification results.

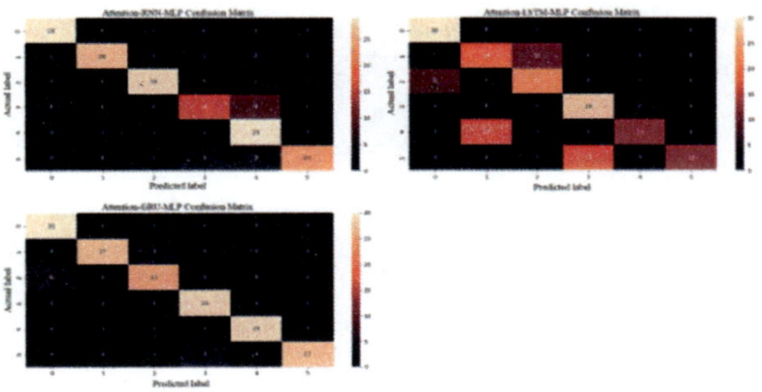

Fig. 5. Prediction confusion matrices of different recurrent neural networks

Obviously, Attention-LSTM-MLP only has high accuracy for the diagnosis of categories 0 and 4, Attention-RNN-MLP has a large diagnostic error for category 4, while Attention-GRU-MLP can achieve high accuracy for each classification fault.

Table 3. Noise reduction effect index value comparison table

Network model	Data sources	Loss	Accuracy	Precision	Recall	AUC
Attention-GRU-MLP	Train	0.0301	0.9982	0.9905	0.9917	0.9998
	Test	0.2264	0.9838	0.9437	0.9319	0.9902

In summary, the Attention-GRU-MLP model has a better fault diagnosis effect. Under the optimal noise reduction ratio, the indicators of the Attention-GRU-MLP model are shown in Table 3, and the diagnostic accuracy rate can reach 98%.

4 Conclusion

An Attention-GRU-MLP fault diagnosis model is proposed for the failure of the massage chair movement in this paper. It is compared with the Attention-LSTM-MLP and Attention-RNN-MLP models before and after optimization. Experiments show that the Attention-GRU-MLP model can achieve high accuracy for each classification fault and the diagnostic accuracy rate can reach 98% with the best noise reduction ratio.

References

1. Jaafar, H., Fariz, A., Ahmad, S.A., Md. Yunus, N.A.: Intelligent massage chair based on blood pressure and heart rate. In: 2012 IEEE-EMBS Conference on Biomedical Engineering and Sciences, 2012, pp. 514–518 (2012), https://doi.org/10.1109/IECBES.2012.6498196
2. Samarasinghe, H.K.U., Hashimoto, S.: Automated trend diagnosis using neural networks. In: SMC 2000 Conference Proceedings 2000 IEEE International Conference on Systems, Man and Cybernetics. 'Cybernetics Evolving to Systems, Humans, Organizations, and Their Complex Interactions' (cat. no.0, 2000), vol. 2, pp. 1186–1191. https://doi.org/10.1109/ICSMC.2000.886013

3. Le, P., Zuidema, W.: Quantifying the vanishing gradient and long distance dependency problem in recursive neural networks and recursive LSTMs. In: Workshop on Representation Learning for NLP (2016)
4. Pan, H., et al.: An improved bearing fault diagnosis method using one-dimensional CNN and LSTM. J. Mech. Eng. **64**, 1–10 (2018)
5. Zhang, F., Liu, Q., Liu, Y., Tong, N., Chen, S., Zhang, C.: Novel fault location method for power systems based on attention mechanism and double structure GRU neural network. IEEE Access **8**, 75237–75248 (2020). https://doi.org/10.1109/ACCESS.2020.2988909
6. Fu, R., Zhang, Z., Li, L.: Using LSTM and GRU neural network methods for traffic flow prediction. In: 2016 31st Youth Academic Annual Conference of Chinese Association of Automation (YAC), pp. 324–328 (2016), https://doi.org/10.1109/YAC.2016.7804912
7. Wang, Z., Liu, Q., Chen, H., et al.: A deformable CNN-DLSTM based transfer learning method for fault diagnosis of rolling bearing under multiple working conditions. Int. J. Prod. Res. **59**, 1–15 (2020)

Research on Automatic Cupping Device

Yulin Jiang[1], Guiqin Li[1(✉)], Zhengwei Li[2], and Zhenwen Liang[3(✉)]

[1] Shanghai Key Laboratory of Intelligent Manufacturing and Robotics, Shanghai 200072, China
leeching@shu.edu.cn
[2] Shanghai Wemade Electromechanical Equipment Co., Ltd., Shanghai 201108, China
[3] Shanghai University of Medicine and Health Sciences, Shanghai 201318, China
Liangzw@sumhs.edu.cn

Abstract. An automatic cupping device is put forward in this paper. It is mainly used for cupping treatment on the back. Data acquisition module and treatment module are designed in this device. The air pressure and temperature in the cups can be shown on PC and can be automatically controlled to stabilize within the set range by control module during cupping. That multiple cups are integrated into the support makes the cupping process convenient.

Keywords: Medical equipment · Data collection · Automatic control · Cupping device

1 Introduction

Cupping therapy is an important part of traditional Chinese medicine. It has the characteristics of easy operation, wide application, remarkable curative effect, economy and safety, etc. It has been loved by the majority of patients and medical workers. Li Dandan's team [1] mentioned that cupping therapy acts on the skin to dredge collaterals and promote blood circulation. As a traditional folk remedy, cupping therapy can be used to treat a variety of diseases, such as neck, shoulder, waist and leg pain, joint disease, etc. But its mechanism has not been fully understood. Many researchers at home and abroad have studied it. Now here are some representative theories, like Blood circulation promotion theory that Moncada and others mentioned [2], Neural reflex theory that Abele mentioned [3], Theories of pain that Moayedi and Davis mentioned [4], Tension change theory that Muanjai P's team mentioned [5], and Immune inflammatory response theory that Guo Y's team mentioned [6].

After long-term clinical practice, it has been proved that cupping therapy is effective and feasible in curing, preventing and maintaining health. Of course, there are still shortcomings in cupping. On one hand, its treatment cannot be explained by science so that many people doubt it, on the other hand, it cannot be used in clinical first aid because of long treatment time.

Y. Wang et al. (Eds.): IWAMA 2021, LNEE 880, pp. 65–70, 2022.
https://doi.org/10.1007/978-981-19-0572-8_9

2 Design of Automatic Cupping Device

2.1 Overall Design

The negative pressure in the cup is the main factor to achieve the purpose of cupping treatment. An automatic cupping device that multiple cups are integrated on a support is put forward in this paper. Its negative pressure is generated by a vacuum pump rather than open flame. It has the characteristics of high efficiency of cupping, controllable pressure and controllable temperature. The automatic cupping device, as shown in Fig. 1, is mainly composed of power supply module, control module, treatment module, and data acquisition module.

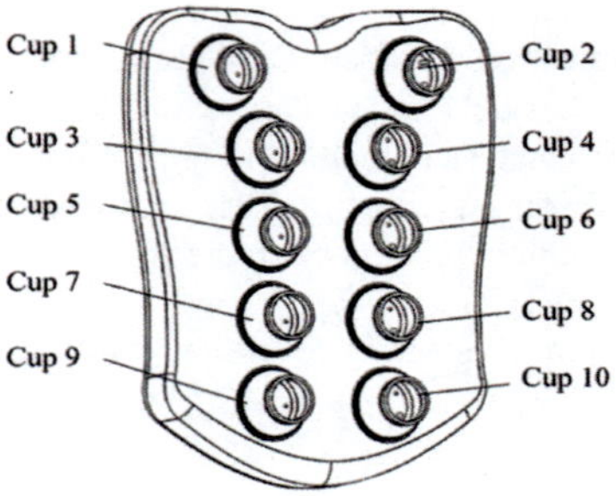

Fig. 1. Automatic cupping device

In this automatic cupping device, the cups are made of bio-ceramic materials with more stable physical structure, higher mechanical strength, more comfortable cupping feel, excellent affinity with human tissues, and good sterilization. 3D printed cups are shown in Fig. 2.

The ten cups are installed on the support, so there is no need to put them on the patient's back one by one. The position of each cup is relatively fixed. They are connected to the support through a rigid corrugated tube, so they can be adjusted within a certain range, avoiding the situation that one automatic cupping device is only suitable for one person. Each cup has an independent start-stop switch so that they can be used arbitrarily.

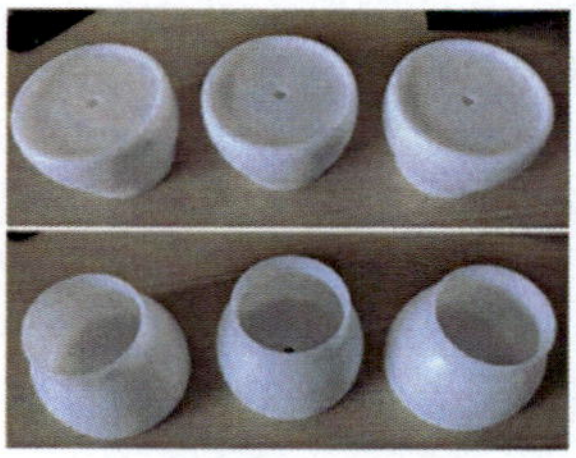

Fig. 2. 3D printed cups

2.2 Module Design

(1) Data acquisition module

The data acquisition module mainly collects data through sensors, including temperature sensors and pressure sensors. In the process of cupping, the temperature and pressure in the cups are two important data indicators.

The selected temperature sensor DS18B20 is shown in Fig. 3. Its applicable voltage range is wider that from 3.0 to 5.5 V, and the temperature measurement range is −55 °C to +125 °C with the accuracy is ±0.5 °C at −10 to +85 °C. The measurement result is directly converted into the digital temperature signal. The measuring range of pressure sensor that shown in Fig. 4 is −0.1 MPa–0 MPa. Its power supply voltage is 5 V or 3.3 V or 3.0 V.

Fig. 3. Temperature sensor DS18B20

Fig. 4. Pressure sensors

(2) Treatment module

When a patient is treated by the automatic cupping device, the negative pressure in the cup is formed by the vacuum pump to remove the air. On and off of the air flow is controlled by solenoid valves. Each cup has two solenoid valves. The configuration diagram is shown in Fig. 5. When the solenoid valve is powered on, port 1 and port 2 are in communication, and when is powered off, port 1 and port 3 are in communication. While the cupping device is connected to the power supply, the solenoid valve 1 and 2 is in the de-energized state. At this time, the air in the cups is pumped out by the air pump. As the negative pressure in the cups reaches a certain value, the solenoid valve 1 is energized and the pumping stops, the negative pressure in the cups is maintained a constant. At the end of the cupping time, the solenoid valve 2 is energized so that the air pressure in the cups gradually rises and returns to the ambient air pressure.

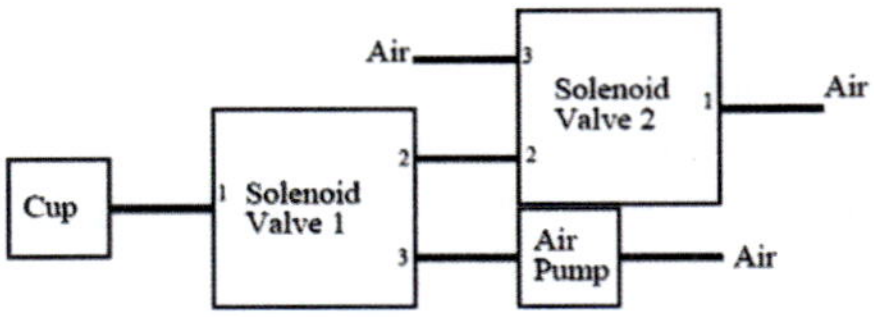

Fig. 5. Airflow on-off configuration diagram

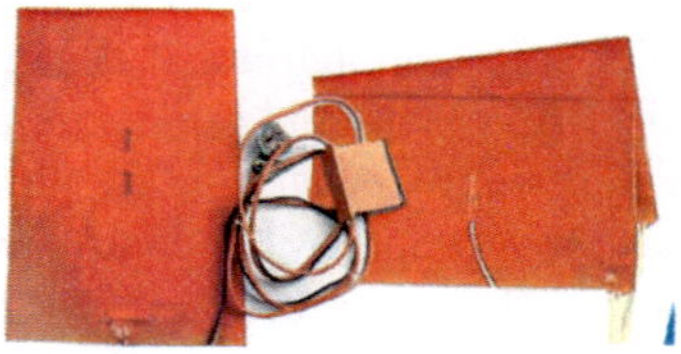

Fig. 6. Silicone rubber heating plate

In addition, each cup is equipped with a silicone rubber heating plate, as shown in Fig. 6. The size of heating plate can be customized. The temperature can be heated to 220 °C. When the temperature in the cup reaches the maximum value that set, the heating plate is off, and when the temperature drops to the minimum value, it is turned on.

(3) Control module

The parameters that need to be controlled are the negative pressure and temperature in the cups in the process of cupping treatment. Each cup is designed with a small circuit board as shown in Fig. 7 that is used to receive the data collected by the pressure sensor and the temperature sensor, and control on and off of the heating plate and cup. A main control board is designed to summarize the information of small circuit boards and control on and off of the solenoid valves.

Fig. 7. Cup and its small circuit board

The small circuit board is communicated with the pressure sensor by the IIC interface protocol to collect pressure data, communicated with temperature sensor by single bus to collect temperature data and switch signals. The heating plates are controlled by it through PWM method. The small circuit boards are communicated with the main control board by the 485 protocol. The main control board is communicated with the upper computer by the network port W5500 to send the collected pressure data, temperature data and switch signals, and receive the return data from upper computer to control on and off of the heating plates and solenoid valves. The overall block diagram of the hardware is shown in Fig. 8.

The software interface is designed, as shown in Fig. 9. The upper and lower limits of the air pressure and temperature in the cup and the cupping time can be preset in the interface. When starting cupping, the pressure value and temperature value are kept within the set value by the control module. When the set time is reached, the air pressure in the cup is increased, and the heating plate is turned off. This completes a cupping.

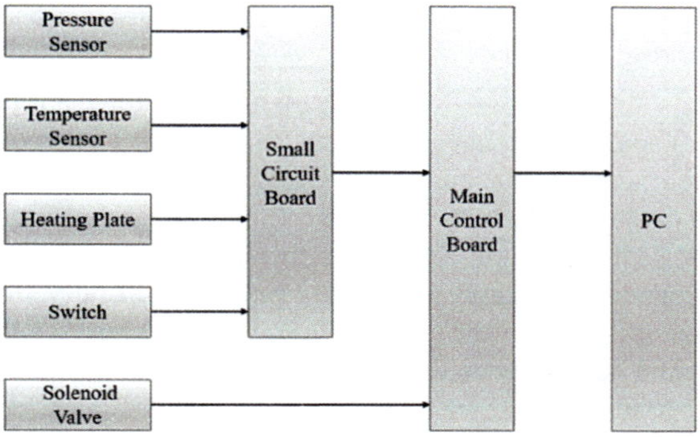

Fig. 8. Hardware overall block diagram

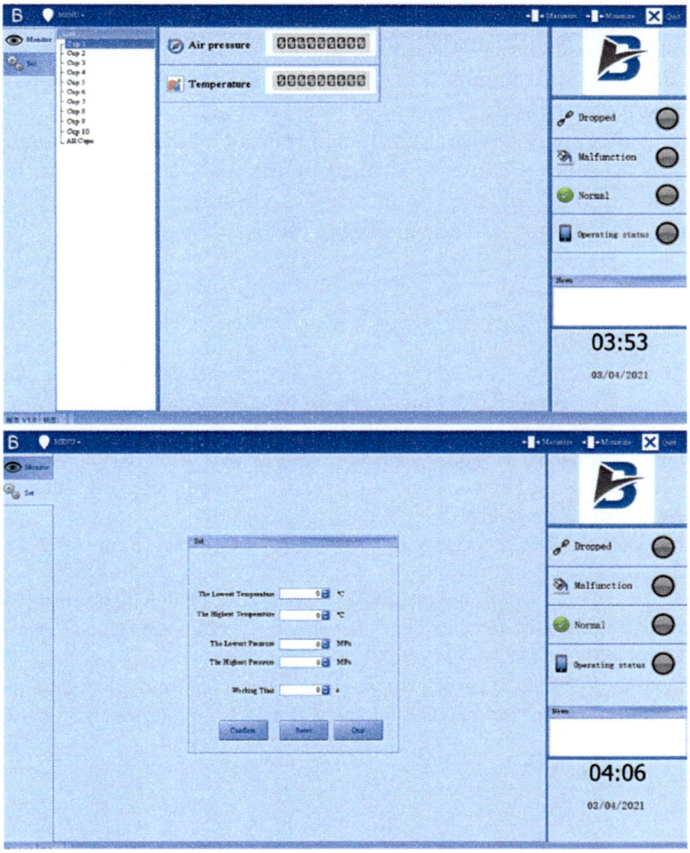

Fig. 9. Software interface

3 Application

When the automatic cupping device is used for cupping treatment, it is placed on the back of the patient first. The positions of the cups can be adjusted. The power supply module and the independent switch on the cup that needed cupping are turned on. The negative pressure value, temperature value in the cups and cupping time are determined according to the patient's physical condition, and set on the software interface. If confirm, each module starts to work immediately. If device is working normally, the indicator lights of Normal and Operating status are green. Otherwise, they will be turned red and the device should be turned off immediately.

4 Conclusion

Support of the automatic cupping device is equipped with ten cups to realize the purpose of simultaneous treatment of multiple cups and greatly improves the efficiency of cupping. Compared with traditional cupping, the method of using vacuum pump to generate negative pressure is safer and more reliable, and avoids the shortcomings of traditional cupping that burns human skin if the operation is improper. The physiotherapy of temperature on human body is ensured by the heating plate.

The automatic cupping device is combined sensors to achieve quantitative control. On or off of the vacuum pump is controlled by the main control board to adjust the negative pressure in the cups. And on or off of the heating plate is controlled by each small circuit board to adjust the temperature in the cups.

References

1. Li, D.D., Meng, X.W., Liu, H.P., Zhu, C.H., Pu, S.A.: Research overview on mechanism of cupping therapy. Liaoning J. Trad. Chin. Med. **41**(11), 2506–2508 (2014)
2. Moncada, S., Palmer, R.M., Higgs, E.A.: Nitric oxide: physiology, pathophysiology, and pharmacology. Pharmacol. Rev. **43**(2), 109–142 (1991)
3. Abele, J.: Das Schroepfen: einebewaehrte alternative methode, cupping: reliable alternative healing method. Munchen Urban Fischer **2003**, 48–53 (2003)
4. Moayedi, M., Davis, K.D.: Theories of pain: from specificity to gate control. J. Neurophysiol. **109**(1), 5–12 (2013)
5. Muanjai, P., Mickevicius, M., Kamndulis, S., et al.: The relationship between stiffness and pain following unaccustomed eccentric exercise: the effects of gentle stretch and repeated bout. Eur. J. Appl .Physiol. **119**(5), 1183–1194 (2019)
6. Guo, Y., Chen, B., Wang, D.Q., et al.: Cupping regulates local immunomodulation to activate neural-endocrine-immune worknet. Complement Ther. Clin. Pract. **28**, 1–3 (2017)

Intelligent Recognition of Automatic Production Line of Metal Sodium Rod

Rongrong Pan[1], Guiqin Li[1](✉), and Peter Mitrouchev[2]

[1] Shanghai Key Laboratory of Intelligent Manufacturing and Robotics, Shanghai University, Shanghai 200072, China

[2] University of Grenoble Alpes, G-SCOP, 38031 Grenoble, France

Abstract. A scheme and algorithm for identifying and locating metal sodium bars and placing them into barrels intelligently were designed. In the detection and positioning of sodium bars, deep learning is mainly used to quickly identify sodium bars based on Open CV and YOLO convolutional neural network, and then morphological processing is carried out on each recognition frame of sodium bars to obtain the central location of sodium bars. When determining where the remaining sodium bars can be placed, it is assumed that the long axis of the sodium bar is tangent to the barrel wall and a cluster of circles is drawn to predict all possible positions of the sodium bars. Then, whether each circle has the intersection with other sodium bars is calculated one by one, so as to output the positions of the sodium bars that can be placed.

Keywords: Intelligent recognition · Deep learning · YOLO

1 Introduction

Intelligent recognition is to perceive and locate the external environment and objects through machines. With the development of the Internet economy, intelligent manufacture gradually become mainstream, edging towards industrial production, manufacturing in the realization of intelligent manufacturing, automation, at the same time is also looking for a kind of new technology, in order to realize the automatic detection and assembly of parts, people turn to the intelligent recognition technology, using the technology to complete the workpiece measurement and identification, saves a lot of tedious work. In recent years, intelligent identification technology has become increasingly mature with the research of the majority of scholars. Alberto Pretto et al. [1] proposed a new recognition technique, which is mainly based on Hough voting algorithm. In the image edge detection, it is easy to be disturbed by the noise in the image. In order to solve this problem, Alberto Pretto et al. identified the features of the object effectively through the Hough voting algorithm. Keiji Yanai et al. [2] proposed a method for food recognition based on deep convolution features. Firstly, an appropriate model is trained, and then the model features are brought into the recognition system, which can achieve a classification accuracy of up to 72.26%. Zhu Jianxiong et al. [3] proposed a global adaptation model for Synthetic Aperture Radar (SAR) automatic recognition, established a

Y. Wang et al. (Eds.): IWAMA 2021, LNEE 880, pp. 71–78, 2022.
https://doi.org/10.1007/978-981-19-0572-8_10

complete image template with multiple perspectives and raw data volume, and predicted different data, so that the regional features at different levels were preserved. Pluim et al. [4] proposed a method to improve image matching performance by combining mutual information and image gradient information. Cui et al. [5] proposed a method combining depth cues with image features, which extracted depth and motion cues through background subtraction and histogram segmentation. Under the guidance of depth cues extraction, color image features were extracted by skin color region segmentation. Karpushin et al. [6] proposed a feature extraction method for local vision of texture and depth content based on depth image analysis. The depth information is used to derive the local projection transform, and the descriptor facet can be calculated from the texture image. The geometric information provided by the depth information is used to improve the stability of the local visual features, and the experiments show that the feature extraction of geometric perception can bring advantages.

Intelligent recognition system of workflow is: first picked by cameras and other facilities to deal with image, and then the obtained image in image processing, image analysis and calculation, the final will be processed by image processing system and image characteristics of the output image information to the control system, control system based on the system of instruction in operation [7].

2 Sodium Bar Image Recognition Model Construction

Based on deep learning theory, a set of intelligent recognition algorithm of sodium bar is designed. The images collected by the camera are processed with data, and then the YOLO target recognition algorithm is used for deep matching calculation and recognition, and the sodium bar is selected by the box. Then the sodium bar in each target box was taken as the research object, and the center of the sodium bar was located through morphological processing to get the specific location of the center of the sodium bar in the barrel.

2.1 Principles of Deep Learning

The essence of deep learning is to learn target features through deep neural network and establish the relationship between feature data and labels, while "learning" is a process of extracting features through a large number of data training, and its depth is the number of hidden layers. The number of hidden layers is the number of layers from the input layer to the output layer, which is the depth. The number of hidden layers is the basis to judge whether deep learning or shallow learning. In general, deep learning refers to more than 8 layers of neural networks.

There are many neurons in each layer of the neural network. They are the basic components of the neural network, and it is the information in the network that they transmit that constitutes information transmission. The output of each neuron will become the input of the next neuron, passing layer by layer. The structure of a single neuron is shown in Fig. 1:

The number of inputs to each neuron is determined by the previous layer, and each input produces a different number of inputs to the neuron Stimulus, but only one output

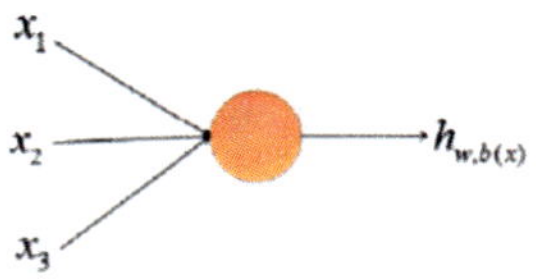

Fig. 1. Single neuron structure diagram

number. The diagram has three inputs, x_1, x_2, and x_3, and $h_{w,b\,(x)}$ Output, that is:

$$h_{w,b(x)} = f(W^T x) = f\left(\sum\nolimits_{i=1}^{3} W_i x_i + b\right) \tag{1}$$

Where W is referred to as weight, and is the stimulus degree of the input to the neuron. A large positive weight represents a strong excitement, while a small negative weight represents a weak inhibition. The offset B is used to define the minimum value before the processing of the activated function. When used for classification, B is equivalent to the threshold value. Those greater than B are classified into one class, and those less than or equal to B are classified into another class, as follows:

$$y = \begin{cases} 1, w^T x + b > 0 \\ 0, w^T x + b \le 0 \end{cases} \tag{2}$$

Neural convolution of the core is essentially a linear filter for 2 d image processing, often used in signal processing and image information processing, specific operation is to calculate the pixels in the image and the corresponding filter product of matrix elements, pixel point is the product and, this is equivalent to perform pixel values of the filtering operation (Fig. 2).

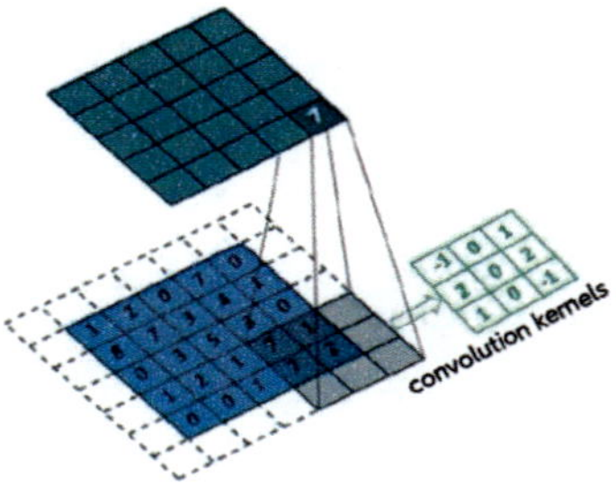

Fig. 2. Image convolution flow

Convolutional layer is an important part of complex neural network. The deeper the network, the richer the target attributes. CNN (Convolutional Neural Networks) is a method to extract target attributes through multi-layer convolution, while shallow Convolutional layers can only extract low-level features such as the target edge contour, while deep Convolutional Networks extract more complex and abstract functions by adding Convolutional layers.

If the input image is a two-dimensional RGB three-channel color image, since the image is composed of each pixel, its convolution operation is expressed as follows:

$$z(x, y) = f(x, y)^{*}g(x, y) = \sum_{t}\sum_{h} f(t, h)g(x - t, y - h) \quad (3)$$

Its integral form is as follows:

$$z(x, y) = f(x, y)^{*}g(x, y) = \int\int f(t, h)g(x - t, y - h)dtdh \quad (4)$$

A good deep learning framework is not only easy to understand and code, but also can automatically calculate gradients and execute parallel processes. Commonly used deep learning frameworks are shown in Table 1.

Table 1. Common deep learning frameworks

Framework	Language	Development institutions
Caffe	C++	University of California, Berkeley
Keras	Python	Fchollet
Tensorflow	C++, Python	Google
Pytorch	Python, C	Facebook
PaddlePaddle	C++, Python	Baidu

Tensorflow has the advantages of wide application range, convenient operation, portability and so on. It can support languages such as C++, Python, and contains a variety of deep models. Tensorflow is a commonly used open source deep learning framework. Keras, on the other hand, allows users to build simple and fast models with minimal code. It supports convolutional neural networks, recursive neural networks, and both approaches. Keras typically runs on either TensorFlow or Theano as a backend port, simplifying the programming complexity and making it very easy to get started.

2.2 Sodium Bar Identification Algorithm Design

This algorithm is based on the unique stage model of the regression method of YOLO (You Only Look Once). It uses the regression method to detect and classify the coordinate frame. It is the fastest and most widely used detection model at present.

YOLO algorithm has many advantages: first, the algorithm process is simple and fast, which can be used for real-time detection; Second, YOLO is based on the global feature information of the image for prediction. Compared with other detection methods, the error rate of the background detection as the object is greatly reduced, which is practical. Third, strong induction ability, can summarize the target, and has a high accuracy.

When running the algorithm of this project, the first thing to do is to configure the environment and get the data set. The algorithm consists of three parts: data set part, training part and test part.

1. Data set

   A. Partition of data sets. The data set needs to be divided into a training part and a testing part. The training part is divided into training set and verification set.
   B. Data set format conversion. Since YOLO source code is based on VOC data set, the data set format needs to be converted. Into VOC data set format.

2. Training

   A. Processing data. Nine priori box size parameters were generated by KMeans clustering and imported into the model.
   B. Execute the training function. The first step is to further configure the environment by configuring missing or incorrect packages based on the error message at execution time. Then set the parameters, adjust the batchsize and num_worker parameters according to their own CPU performance, so as to improve training efficiency.

3. Test section

   Select the trained weight file or model, and determine the category to be identified, and execute the prediction program.

   The sodium bar image was put into the above target detection model for testing. From the image results, it can be seen that the algorithm can effectively detect and identify the location of the sodium bar in the barrel, and almost every sodium bar is tangent to the target box, which completes the detection and recognition of the sodium bar (Fig. 3).

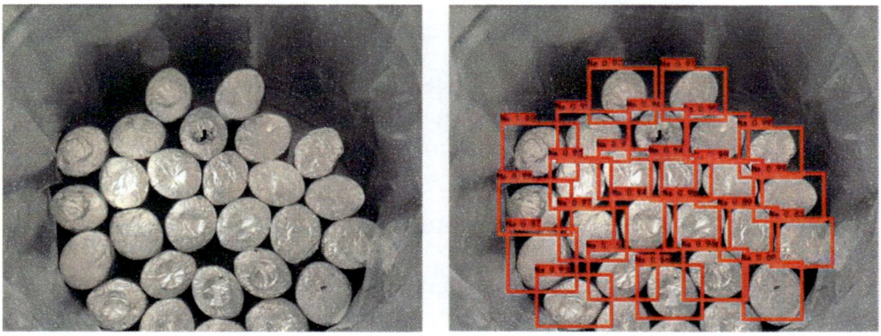

Fig. 3. Pictures before and after intelligent recognition of sodium rod

Sodium bars have been identified through the above process, and on this basis, sodium bars in each target box are centrally located. The idea of the sodium bar localization algorithm is to take the sodium bar in each target box as the research object. The location has been detected. Through basic morphological operation, the small noise points in the sodium bar are removed by expansion first, and then the sodium bar is corroded until only the long axis is left. After that, the excess edges and corners in the target box were removed, and the optimal ellipse fitting was used to find the center point of the sodium bar.

Morphological processing is the most popular method to study the image features. It is often used in image processing, which can effectively remove the noise in the image and improve the image quality.

The basic idea of morphological processing is to measure and extract the corresponding image shape by using the specific image structure elements to achieve the purpose of image analysis and image recognition. The main operation is expansion and corrosion of two methods. The expansion can expand or connect the narrow area. The pixel points on the image surface can be repeatedly removed while the corrosion of small holes is removed. The point graphics can be removed. All the sodium bars in the barrel were positioned above, and the center positions were marked with blue dots. The experimental results were shown in Fig. 4.

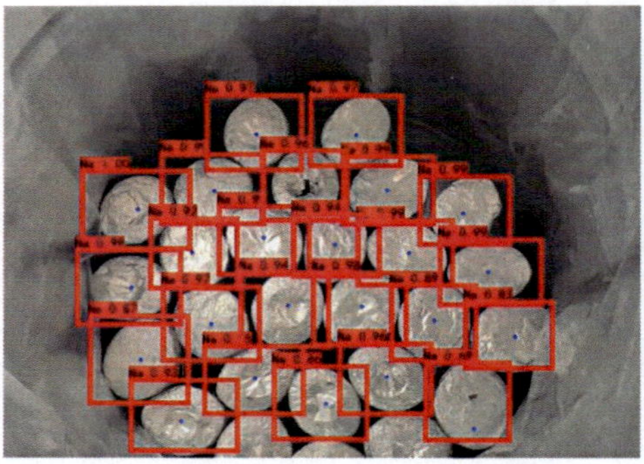

Fig. 4. Location of sodium bar center

It can be seen from the experimental results that the accuracy of the algorithm is high, and the sodium bar in the image can be accurately identified, and the center position of the sodium bar can be given, which has a strong practicability.

3 Intelligent Arrangement Scheme Implementation

In order to predict the remaining space in the bucket and arrange the sodium bars intelligently, the position of the bucket wall is firstly determined in the image. In practical application, since the position and Angle of the bucket are fixed and unchanged every time, the position of the bucket wall can be assumed to be unchanged. The center is determined according to the size of the image and a circle is drawn as the bucket wall in the image, as shown in Fig. 5.

In practice, the bucket is tilted, and the sodium bar will automatically move down the bucket due to gravity when it is loaded into the bucket, so the gap usually appears on the top of the bucket. When there is not much room left in the top of the barrel, you need to consider where to put the sodium bar next. After the location of the bucket wall

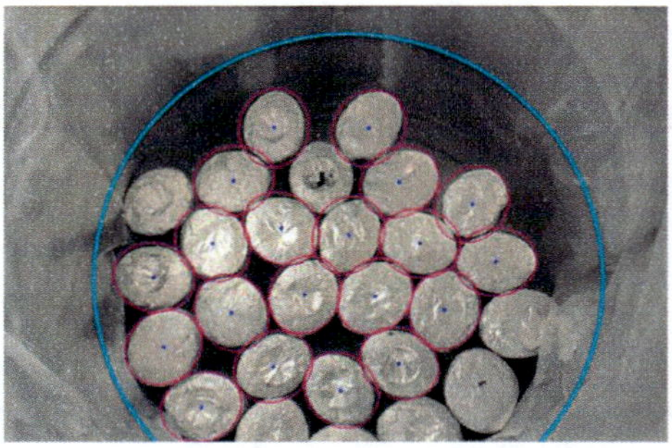

Fig. 5. Barrel wall location drawing

is determined, a cluster of circles is made tangent to the bucket wall with the long axis of the ellipse as the radius, which is approximate to the sodium bar. Whether there is an intersection between each circle and its adjacent sodium bar is judged. If there is an intersection, it will be removed. If there is no intersection, it will be the location where the sodium bar can be placed. The effect diagram is shown in Fig. 6, and the white circle in the diagram is the location where the sodium bar can be placed.

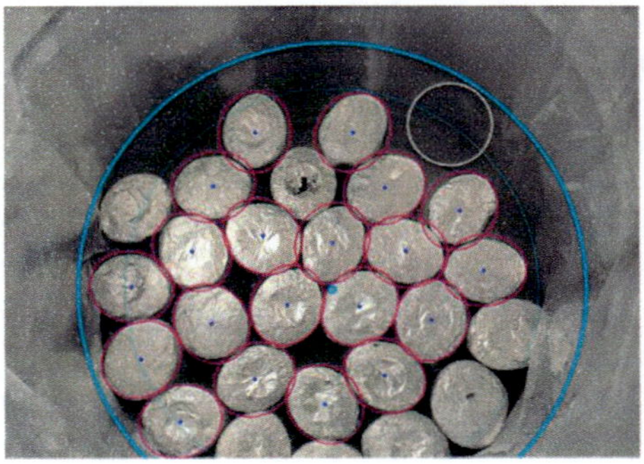

Fig. 6. Map of sodium rod placement

The position of the barrel wall is fixed according to the posture of the barrel, thus assuming that the sodium rod can be placed in an area. The feasibility of the proposed position is judged by the intersection of the assumed position and the existing sodium Rod. The method has certain error, but it can solve some practical problems.

4 Conclusions

Aiming at the automatic production line of visual inspection, this paper introduces the application of intelligent recognition technology based on the combination of theory and practice. In this paper, Yolo Algorithm is used to identify and detect the sodium Rod, and the basic frame of deep learning is expounded, which can accurately detect the sodium rod in the barrel, the morphology treatment and the best ellipse fitting were used to get the sodium Rod Center. The possible position of the sodium rod remaining in the barrel is analyzed, and the possible position of the sodium rod is selected by using the intersection of geometric operation.

References

1. Pretto, A., Tonello, S., Menegatti, E.: Flexible 3d localization of planar objects for industrial bin-picking with mono camera vision system. In: 2013 IEEE International Conference on Automation Science and Engineering (CASE), IEEE (2013)
2. Yanai, K., Kawano, Y.: Food image recognition using deep convolutional network with pre-training and fine-tuning. IEEE International Conference on Multimedia and Expo Workshops, pp. 1–6. IEEE (2015)
3. Jianxiong, Z., Zhiguang, S., et al.: Automatic target recognition of SAR images based on global scattering center model. IEEE Trans. Geosci. Remote Sens. **49**(10), 3713–3729 (2011)
4. Josien, P.W., Pluim, J.B., Maintz, A., Viergever, M.: Image registration by maximization of combined mutual information and gradient information. In: Delp, S.L., DiGoia, A.M., Jaramaz, B. (eds.) Medical Image Computing and Computer-Assisted Intervention – MICCAI 2000, pp. 452–461. Springer Berlin Heidelberg, Berlin, Heidelberg (2000). https://doi.org/10.1007/978-3-540-40899-4_46
5. Cui, W., Wang, W., Liu, H.: Robust hand tracking with refined CAM Shift based on combination of depth and image features. In: 2012 IEEE International Conference on Robotics and Biomimetics (ROBIO), IEEE (2012)
6. Karpushin, M., Valenzise, G., Dufaux, F.: Local visual features extraction from texture+depth content based on depth image analysis. IEEE (2014)
7. Cheng, W.B., Li, C.P.: Design of image acquisition system based on machine vision. J. Guangxi. Teach. College. (Natural Science Edition) **23**(002), 42–45 (2006)

Signal Denoising Algorithm of Massage Chair Movement Based on iForest-EEMD

Lixin Lu[1], Dongcai Wu[1], Guiqin Li[1](✉), and Peter Mitrouchev[2]

[1] Shanghai Key Laboratory of Intelligent Manufacturing and Robotics, Shanghai University, Shanghai 200072, China
leeching@shu.edu.cn

[2] University Grenoble Alpes, G-SCOP, 38031 Grenoble, France

Abstract. Aiming at the problem of massage chair movement signal detection, a signal denoising algorithm based on iForest-EEMD is proposed. Wavelet threshold is used to improve the denoising effect of EEMD algorithm on high frequency signals, and the iForest algorithm is used to eliminate the local noise in the signal. The experimental results show that compared with the EMD and CEEMD noise reduction algorithm, this method has higher accuracy and noise reduction efficiency.

Keywords: Massage chair movement · EEMD · iForest · Noise reduction treatment

1 Introduction

The quality of the signal has an important influence on the accuracy of massage chair fault diagnosis [1]. The signal of the massage chair movement in the working state usually contains global high-frequency noise and local noise, so it is necessary to reduce the noise of the collected data.

Isolated forest (iForest) algorithm is an unsupervised anomaly detection algorithm proposed by Liu Fei et al. This algorithm can effectively deal with a large number of data [2]. In order to suppress the mode aliasing problem, Huang and Wu proposed an integrated empirical mode decomposition method (EEMD) [3]. The proposal of EEMD effectively suppresses the phenomenon of mode aliasing in EMD decomposition. Wavelet transform is proposed by Haar, and its advantage is to better represent the frequency characteristics of time-domain signals [4]. The shape of the time and frequency window of the wavelet transform can change adaptively with the signal, and the resolution of the signal is higher.

2 Signal Denoising Algorithm Based on iForest-EEMD

The noise in the massage chair movement signal includes the global high frequency noise and the local noise. The signal denoising algorithm described first uses the iForest algorithm to eliminate the local noise, and then uses the improved EEMD algorithm to deal with the global high frequency noise. The process of noise reduction algorithm for iForest-EEMD is shown in Figs. 1.

Y. Wang et al. (Eds.): IWAMA 2021, LNEE 880, pp. 79–84, 2022.
https://doi.org/10.1007/978-981-19-0572-8_11

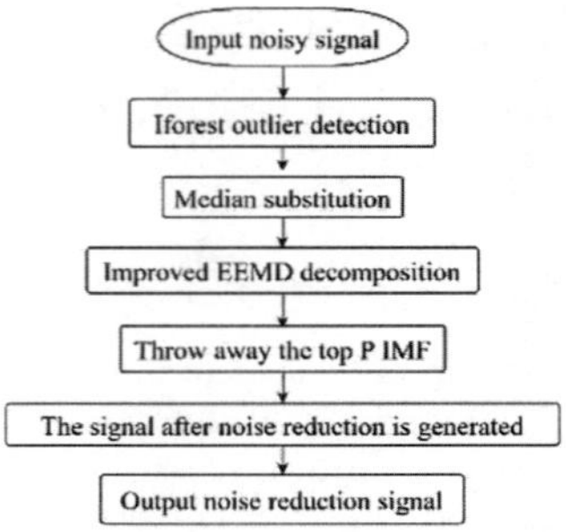

Fig. 1. The flow of iForest-EEMD signal denoising algorithm

2.1 Outlier Detection and Correction Algorithm Based on iForest

EEMD algorithm is not effective in dealing with local noise. In order to enhance the reliability of data, it is necessary to combine other algorithms to correct local noise.

Isolated forest (iForest) is selected to detect outliers in the data. In isolated forests, outliers are defined as "easily isolated points". In the feature space, if the distribution of data points in a region is sparse, it can be considered that the probability of data falling is very low. IForest is based on this theory to detect outliers in data sets.

IForest consists of t isolated trees. For each isolated tree in an isolated forest, the training process is as follows:

1. φ data are randomly selected from the data as subsamples, and one data is selected as the root node.
2. Randomly select a data as the child node and partition point p.
3. The data less than p is placed on the left branch of the current node, and the data greater than p is placed on the right branch of the current node.
4. Repeat steps 2 and 3 in the left and right branches until the segmentation cannot continue or the specified height has been reached.

After t isolated trees are obtained, iForest training is completed and outliers can be detected. In general, the path between outliers and roots is much shorter than normal data points because they are more easily isolated. Therefore, the length of the path can be used to determine whether the data is an exception. For each isolated tree, the path length is calculated as follows:

$$h(x) = e + c(T.size)$$

For a training data x, it is easy to go through all the isolated trees, the anomaly index of iForest is obtained as follows:

$$\mathrm{s}(\mathrm{x}, \varphi) = 2^{\left(\frac{-\mathrm{E}(\mathrm{h}(\mathrm{x}))}{\mathrm{c}(\varphi)}\right)}$$

The $E(h(x))$ is the average path length of x in multiple isolated trees, the shorter the length, the closer the anomaly index is, which indicating that the data is more likely to be an outlier.

When the iForest is trained, it is necessary to determine the appropriate expected ratio of outliers l. In this paper, the initial value of lump 0.05 is taken. The median is more robust because it is not affected by extremes. The detected outliers are replaced with the median of this group of data.

2.2 Denoising Algorithm of High Frequency Signal Based on EEMD

The signal collected by the system usually contains global high-frequency noise. This part of the noise contains redundant information, which will adversely affect the accuracy of movement fault diagnosis. The EEMD algorithm is used to reduce the noise of the movement signal.

This method can decompose the original signal directly. After EEMD decomposition, N intrinsic mode function (IMF) components and final residual terms can be obtained, as follows:

$$x(t) = \sum_{i=1} imf_i(t) + r_n(t)$$

In the formula, $x(t)$ is the original signal, $imf_i(t)$ is the IMF component, and $r_n(t)$ is the remainder.

For each IMF, decomposed, there are two conditions:

1. In the data sequence, the number of extreme points and zero-crossing points must be equal or only one difference.
2. At any point, the average value of the envelope formed by the maximum and minimum points in this coordinate is zero.

The EEMD method not only reduces noise, but also removes the effective signal in the high-frequency IMF component [5]. The wavelet optimized EEMD denoising method is used to Denoise the IMF components obtained after decomposition. The wavelet threshold is used to Denoise the IMF components. Then the processed IMF component is reconstructed to remove the noise signal while retaining the effective signal.

The steps of denoising massage chair movement signal based on wavelet improved EEMD denoising method are as follows:

1. The massage chair movement signal is decomposed by EEMD to get a set of IMF components.
2. Each IMF component is denoised by wavelet threshold, and the denoised IMF component is obtained.
3. The IMF components after wavelet threshold denoising are reconstructed to get the de-noised massage chair signal.

3 Experiment and Results Analysis

3.1 Comparison of iFores Local Noise Processing Algorithms

The original core fault current data is used as input signal, which to verify the noise reduction effect of the iForest algorithm. The results are shown in Figs. 2. As can be seen from Figs. 2, there is a set of obvious outliers in the original core current signal. After the iForest local noise processing, the outliers in the signal are effectively eliminated.

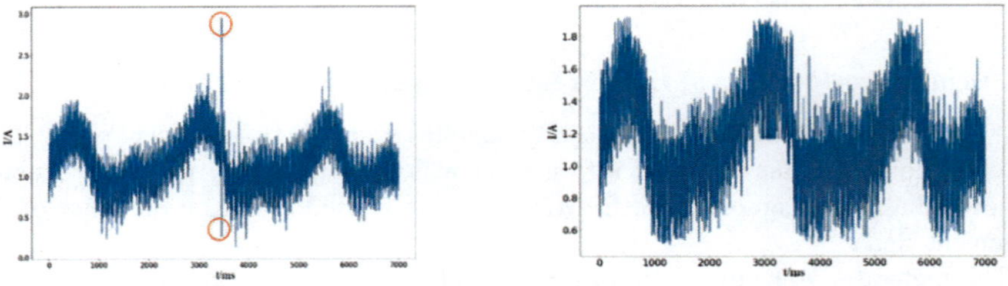

Fig. 2. Comparison of original movement signal and iForest correction signal

When the current signal of the movement is processed under the normal working condition, the iForest method will correct the intermittent extreme point of the signal and make the signal distortion. In order to solve this problem, the outlier correction ratio L needs to be adjusted. The experiment shows that when L = 0.015, the iForest algorithm can effectively eliminate the outliers. The modified waveform is shown in Figs. 3.

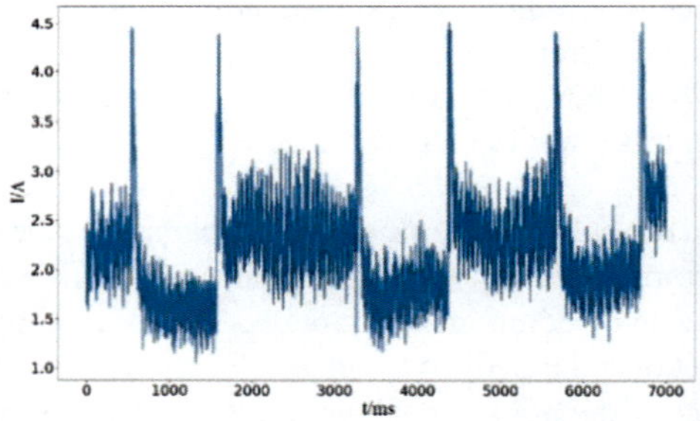

Fig. 3. Abnormal value correction waveform under normal condition when L = 0.015

3.2 Comparison of Global High-Frequency Noise Processing Algorithms Based on iForest-EEMD

The mean square error (RMSE), signal-to-noise ratio (SNR) and smoothness (R) are used to evaluate the noise reduction effect. The current signal of the movement in normal state is used as the experimental data (Fig. 4). EMD, CEEMD and iForce-EEMD algorithms are used to reduce noise in this paper, and the decomposed IMF is shown in Figs. 4.

As can be seen from Figs. 5, compared with CEEMD, iForest-EEMD and CEEMD have a certain degree of reduction in mode aliasing and can decompose more IMF components. Compared with different algorithms, under the same Noise Reduction Ratio, EEMD has higher mean square error and signal-to-noise ratio, smaller smoothness, and the best processing effect.

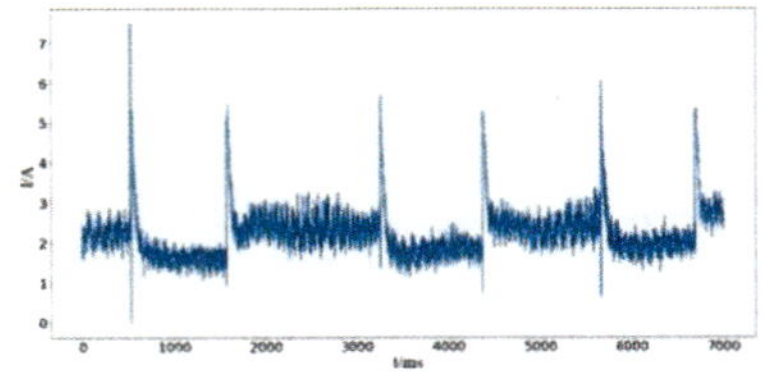

Fig. 4. Normal movement signal waveform

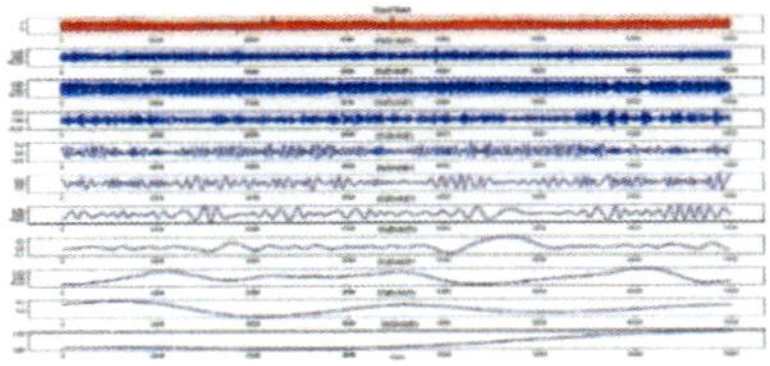

(a) IMF decomposition of waveforms by EMD algorithm

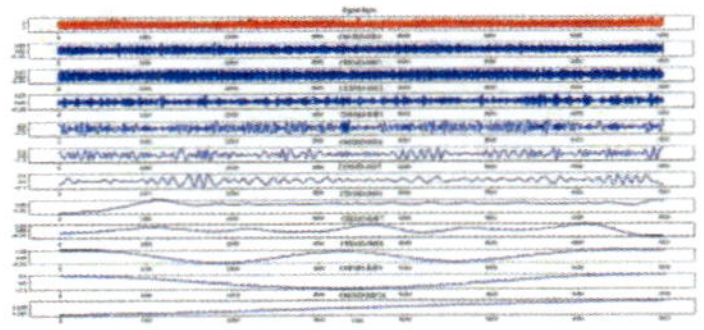

(b) IMF decomposition waveform of iForce-EEMD algorithm

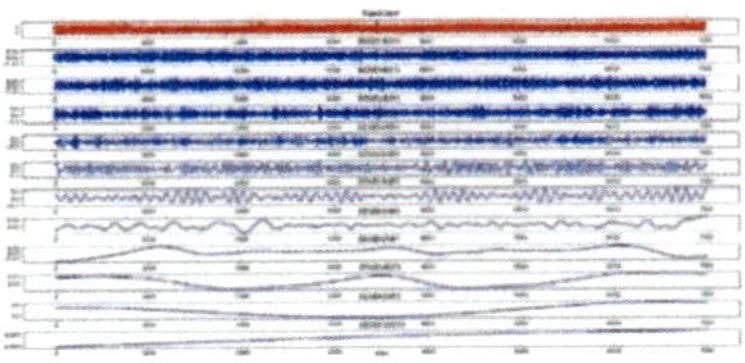

(c) IMF decomposition waveform of CEEMD algorithm

Fig. 5. IMF decomposition waveforms with different algorithms

4 Conclusions

A signal denoising algorithm based on iForce-EEMD is proposed for the noise reduction of massage chair movement signal. This algorithm uses IFOREST to eliminate the local noise of the motor current signal, improves the signal quality, and uses EEMD algorithm to reduce the influence of high frequency noise. The experimental results show that the algorithm based on iForest-EEMD can effectively detect and correct the abnormal points in the movement signal, and it has higher accuracy and efficiency than EMD and CEEMD algorithm.

References

1. Long, D., Wang, X., Tian, M., Mao, Y., He, Y.: Estimation of fatigue status by sEMG signal using SVM algorithm in massage assessment. In: IEEE International Conference on Mechatronics and Automation, pp. 1316–1320 IEEE (2019)
2. Zhao, X., et al.: iForest: interpreting random forests via visual analytics. IEEE Trans. Visualizat. Comput. Graph. **25**(1), 407–416 (2018)
3. Jiang, H., Li, C., Li, H.: An improved EEMD with multiwavelet packet for rotating machinery multi-fault diagnosis. Mech. Syst. Signal Process. **36**(2), 225–239 (2013)
4. Gilles, J.: Empirical wavelet transform. IEEE Trans. Signal Process. **61**(16), 3999–4010 (2013)
5. Zhigang, L.I.U., Cui Yan, L.I., et al.: A classification method for complex power quality disturbances using EEMD and rank wavelet SVM. IEEE Trans. Smart Grid **6**(4), 1678–1685 (2015)

Prediction of Remaining Life of Massage Chair Movement Based on ARIMA-BP Model

Lixin Lu[1], Yuanzhe Li[1], Guiqin Li[1](✉), and Peter Mitrouchev[2]

[1] Shanghai Key Laboratory of Intelligent Manufacturing and Robotics, Shanghai University, Shanghai 200072, China
leeching@shu.edu.cn
[2] University Grenoble Alpes, G-SCOP, 38031 Grenoble, France

Abstract. A detection method for predicting the remaining service life of the movement is proposed in this paper. This optimized model is constructed by combining ARIMA and BP neural network models. Based on method of signal feature fusion, a single evaluation index is constructed that can reflect the changing law of the movement performance. The ARIMA model, BP model and combined model are compared in experiments to prove the combined model which has the smallest prediction error among three methods and has a good predictive effect.

Keywords: Massage chair movement · ARIMA model · BP neural network

1 Introduction

The massage chair movement is a device installed on the back of the massage chair. The basic functions are realized through the mechanical structure of the movement and the control module. Using the massage chair movement life detection model can improve the reliability of the factory movement products and ensure the quality of the massage chair products.

Based on experience and indirect monitoring data, Wang and Zhang predicted the remaining service life of the bearing based on random filtering [1]. Shi Hui et al. took the wind turbine generator gearbox as the research object, and proposed a life prediction method based on random filtering theory and kernel density estimation to predict the remaining life of the gearbox in real time [2]. Peng Y et al. proposed an age-related HSMM-based equipment health prediction method for the HSMM-based prediction algorithm that assumes that the transition probability is only related to the state [3].

The drive current signal of the movement is selected as the original data reflecting the performance state of the movement. After the signal noise reduction and feature extraction are completed, analysis model is used to predict the life of the movement based on the performance state evaluation index of the movement [4]. Good model is aim to achieve the purpose of accurately and quickly predicting the remaining service life of the movement under the condition of a small amount of known data.

Y. Wang et al. (Eds.): IWAMA 2021, LNEE 880, pp. 85–92, 2022.
https://doi.org/10.1007/978-981-19-0572-8_12

2 Movement Life Prediction Process of ARIMA Model

In order to quickly predict the life of the massage chair movement with fewer samples, Autoregressive Integrated Moving Average Model (ARIMA) is used in this paper. It is a mixed model composed of Auto Regressive model (AR) and moving average model (MA). The method is simple to model and has strong adaptability to different data. The basic form is *ARIMA (p,d,q)*, where d is the difference Order, p is the order of autocorrelation, and q is the order of moving average [5]. The prediction principle is to obtain the trend of change by observing the correlation fitting function of the time series of the movement itself, and accurately predict the sequence information of the unknown time.

The life prediction steps based on the ARIMA model are as follows:

(1) Preprocess the known movement time series including sequence centralization and difference operation smoothing. The purpose is to obtain a stationary and zero mean time series.
(2) Choose an appropriate model to fit the observation sequence, estimate the autocorrelation order p and the moving average order q, and realize the order determination of the model. The selection of the autoregressive model and the moving average model is determined according to the tailing characteristics of the autocorrelation coefficient sequence and the partial correlation coefficient sequence. This paper determines the specific order of the model based on the Akaike Information criterion (AIC) and the Bayesian Information criterion (BIC). The functional forms of the AIC and BIC criteria are as follows:

$$AIC = N\ln(\sigma^2) + 2(p + q + 1) \tag{2-1}$$

$$BIC(p) = N\ln(\sigma^2) + p\ln N \tag{2-2}$$

Where σ^2 represents the residual variance of the model, and N is the number of samples collected. When the value of BIC(p) is the smallest, then the value of P is the autocorrelation order applicable to the model.
(3) By using the known sample results and models to inversely infer the maximum probability based on the maximum likelihood estimation method, the model parameter values of the known results are derived to realize the estimation of the ARIMA model parameters. The log-likelihood function and the basic form of parameter estimation are shown in Eqs. (2-3) and (2-4) respectively. By setting $\partial_L(\theta|X)/\partial_\theta = 0$ the corresponding value θ is the parameter of the ARIMA model. This article realizes the parameter estimation of the model by calling the estimate function.

$$L(\theta|X) = p(\boldsymbol{X};\theta) = \sum_{k=1}^{N}\log(p((x_k;\theta))) \tag{2-3}$$

$$\hat{\theta}_{ML} = \underset{\theta}{arg\max}\prod_{k=1}^{N}p(x_k;\ \theta) = \underset{\theta}{arg\ \max}\sum_{k=1}^{N}\log(p(x_k;\ \theta)) \tag{2-4}$$

Where $p(\boldsymbol{X};\theta)$ represents the probability density function; $\boldsymbol{X} = x_1, x_2, \cdots, x_N\}$ is the set of samples randomly selected from the probability density function; θ represents the model parameters.

(4) Verify the effectiveness of the fitted model by performing white noise test on the residual term of the fitted model.
(5) Fit the movement's future performance changes. By setting the movement performance failure threshold, the first failure time of the movement is obtained and the remaining service life of the movement is predicted.

3 Movement Life Prediction Process of BP Model

The BP neural network structure used in this paper is mainly composed of an input layer, a hidden layer and an output layer [6], which can well solve the highly non-linear and time series-related prediction problems presented by the life prediction of the movement.

Three layers are fully connected to form the neural network in this paper. The starting point of the movement performance state prediction is defined as ST. The sliding time window length is L and time is t. The actual value and predicted value of movement at a certain time is HI(t) and HIe(t). Output value of the performance state at a certain time is HI0(t). The transfer functions of all neurons are Sigmoid functions. The process is as follows.

(1) Initialize the network, determine the input layer, hidden layer and output layer of the network. This paper selects the data before ST as the training data set, and divides it into $X = [x^1\ x^2\ \cdots\ x^n]$ based on the length of the sliding time window, X is an n * L dimensional vector, n = ST-L and X is selected as the input of the network, and the next time point of each neuron in the input layer corresponds to the vector composed of the performance state of the movement is used as the output of the network, as shown in Eq. (3-1).

$$\begin{cases} \{HI(1),\ HI(2),\ \cdots,HI(L)\} \Rightarrow HI_o(L+1) \\ \{HI(2),\ HI(3),\ \cdots,HI(L+1)\} \Rightarrow HI_o(L+2) \\ \vdots \\ \{HI(ST-L),\ HI(ST-L+1),\ \cdots,HI(ST-1)\} \Rightarrow HI_o(ST) \end{cases} \tag{3-1}$$

(2) Calculate the output of the hidden layer. Calculate the hidden layer output value P according to the input X and the set weight and threshold through the transfer function between the input layer and the hidden layer.
(3) Calculate the output value OK of each neuron in the output layer based on the transfer function between the hidden layer and the output layer, that is, calculate the output value of the performance state of the movement at the corresponding time. The calculation is as follows:

$$O_k = f\left(\sum_{j=1}^{l} \omega_{jk} P_j + b_k\right) \tag{3-2}$$

(4) Calculate the total error of all neurons in the output layer of the network, which is calculated as follows:

$$E = \frac{1}{2}\sum_{k=1}^{m}(HI_k - O_k)^2 = \frac{1}{2}\sum_{k=1}^{r}(e_k)^2 \tag{3-3}$$

Where HIK represents the expected output value of each output layer neuron, that is, the actual performance state of the movement; e_k represents the error between the output value of each output layer neuron and the expected value.

(5) Update the weights and thresholds according to the rules of the gradient descent momentum method.

(6) Determine whether the training iteration is over. If the error is within the required range, the network training stops; if the error does not reach the required range, return to the second step to continue training.

Finally, based on the trained BP neural network, the performance state of the movement is predicted point by point in an iterative manner. Each time, L continuous points before the predicted time point is used as the input of the model, and the performance state of the movement at the predicted time point is output. Iterate continuously until the failure threshold is reached.

4 Combination Forecasting Model Combining ARIMA and BP

A combination prediction model for the life of the movement that integrates ARIMA and BP neural network is proposed in this paper. ARIMA model is suitable for short-term prediction. BP neural network has strong fitting ability and Non-linear prediction ability, but there are problems such as long training time and slow calculation speed. Combined prediction model in this article is compatible with the advantages of the above two models, which has strong fitting and approximation capabilities and is simple to calculate. The combined prediction idea is shown in Fig. 1. The specific process of combined prediction is as follows:

(1) By using the known history data of the movement life as the input of the ARIMA model, the movement life is predicted according to the above model prediction process. A linear sequence is fitted. The part of the linear sequence in the known data period is recorded as L(t), and the ARIMA model predicting result is $L'(t)$. Historical data of the movement life is I(t) and you can get:

$$I(t) = L(t) + E(t) \tag{4-1}$$

Where E(t) represents the error between the partial linear sequence of the known data time period and the actual data of the movement, which is a non-linear sequence.

(2) Use E(t) as the input of the BP neural network, and predict the subsequent error according to the above-mentioned network prediction process. Record the prediction result of the BP neural network as $E'(t)$.

(3) The two prediction results are organically combined to realize the prediction of the remaining service life of the movement by the combined model. Record the prediction result of the combined model as $I'(t)$, then:

$$I'(t) = L'(t) + E'(t) \tag{4-2}$$

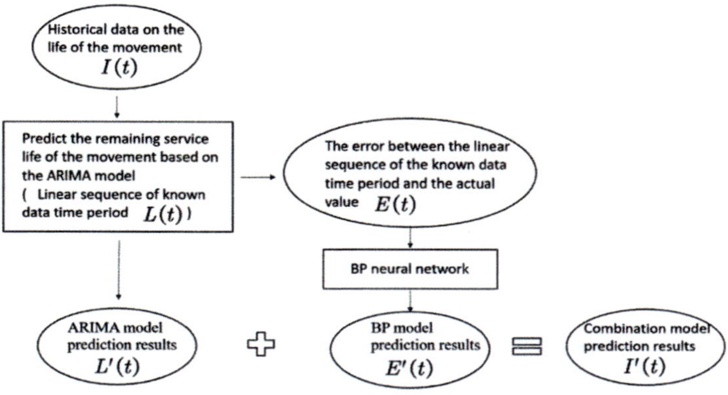

Fig. 1. Combination model forecast flow chart

5 Comparison Results of Different Models

The experiment randomly selects an 18 N model movement to extract performance indicators. The frequency of the data acquisition card is 1000 Hz and acquisition time is 12 noon every day. The duration of each acquisition is 3 s to obtain the actual performance changes from the beginning of the data acquisition to the failure period.

Parameters such as maximum value, minimum value, peak-to-peak value, standard deviation, root mean square, form factor, center of gravity frequency, and root mean square frequency are all strongly related to the life of the movement. They are used as the characteristic parameters of the movement life prediction in this paper. The paper uses the kernel principal component analysis method (KPCA) to reduce the dimension of the selected feature indicators, and the main characteristics and changes of the extracted movement current signal are shown in Fig. 2.

Feature fusion method based on the Mahalanobis distance (MD) index [7] is used to realize the fusion of the main features of the movement current signal, and generate a single evaluation index H1 that can reflect the changing law of the movement performance.

The performance degradation failure threshold of the 18N model movement is set to 0.28. As shown in Fig. 3, the result proves that the movement was in a failed state during the 218th data collection, that is, the remaining service life of the movement was 218 days.

Three models are used the paper predicts the remaining service life of the movement. In order to significantly reflect the prediction performance of each model, this paper

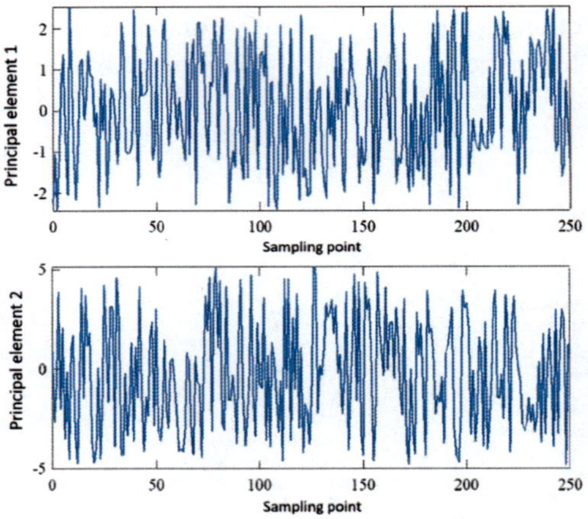

Fig. 2. KPCA optimization characteristic curve

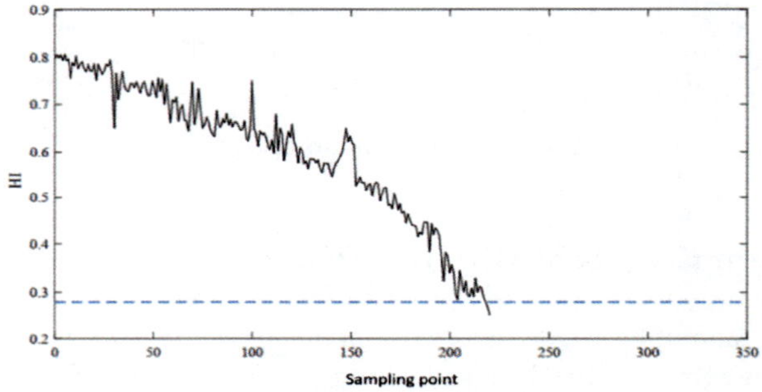

Fig. 3. Changes in the actual performance of the movement

selects the first 130 points and the first 198 points of known data as the training data of the model. Figure 4 shows the prediction results of the three models with different amounts of training data. The top-down distribution is the ARIMA model, the BP neural network model and the combined model corresponding result graphs. The solid line represents actual data. The double dotted line represents 95% confidence interval. The dotted line represents the predicted data and the horizontal dotted line represents the core performance failure threshold.

Prediction results and prediction error statistics of the three models are shown in Table 1. The comparative analysis results are as follows: ARIMA model is approaching linear prediction and the prediction speed is faster. BP neural network has a stronger fit but the prediction speed is slower, and the combined model has a better prediction effect than a single model. As the amount of training data increasing, the greatest impact is on the ARIMA model. The prediction accuracies of the BP neural network and combined model are both improved rapidly.

In summary, the combined model is compatible with the advantages of the ARIMA model and the BP neural network, weakening the shortcomings of the single model,

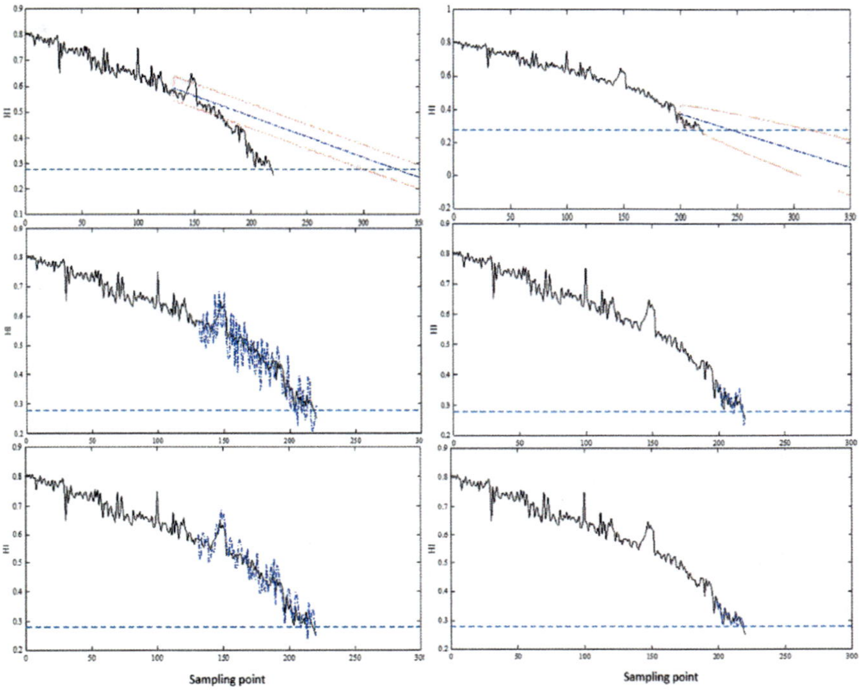

(a) 130 points result (b) 198 points result

Fig. 4. Prediction results based on the first 198 data

Table 1. Error analysis table

Models	Training set	Original life	Predicted life	Prediction error
ARIMA	130	218	328	50.46%
BP	130	218	203	6.88%
Combination	130	218	208	4.59%
ARIMA	198	218	244	11.93%
BP	198	218	210	3.67%
Combination	198	218	214	1.83%

which can achieve a more accurate and rapid prediction of the remaining service life of the movement with a small amount of training data.

6 Conclusion

In order to achieve accurate and effective prediction of the remaining service life of the movement with a small amount of data, this paper proposes a combination prediction model that integrates ARIMA and BP neural network. The experiment uses three models for comparison, taking the single evaluation index of the movement as the test object, and uses the error between predicted result and actual result as the basis. It proves that the combined model has a higher prediction accuracy than the single model, which enables the prediction of the remaining service life of the movement more reliable and efficient.

References

1. Wang, W., Zhang, W.: An asset residual life prediction model based on expert judgments. Eur. J. Oper. Res. **188**(2), 496–505 (2008)
2. Shi, H., Songren, W., Zhang, Y.: The remaining life prediction method of gearbox based on kernel density estimation and random filtering theory. Comput. Integr. Manuf. Syst. **2020**(3), 632–640 (2020)
3. Peng, Y., Dong, M.: A prognosis method using age-dependent hidden semi-Markov model for equipment health prediction. Mech. Syst. Signal Process. **25**(1), 237–252 (2011)
4. Ming, D., He, D.: Hidden semi-Markov model-based methodology for multi-sensor equipment health diagnosis and prognosis. Eur. J. Oper. Res. **178**(3), 858–878 (2007)
5. Kozitsin, V., Katser, I., Lakontsev, D.: Online forecasting and anomaly detection based on the ARIMA model. Appl. Sci. **11**(7), 3194 (2021). https://doi.org/10.3390/app11073194
6. Zhouxi, Y., Qin, L.: Stock price forecasting based on LLE-BP neural network model. Phys. Statist. Mech. App. (2020)
7. Parthiban, T., Ravi, R., Kalaiselvi, N.: Exploration of artificial neural network [ANN] to predict the electrochemical characteristics of lithium-ion cells. Electrochim. Acta **53**(4), 1877–1882 (2008)

Real-Time Data-Driven Digital Twin Workshop Web Interactive Application

Heng Cao(✉), Lilan Liu, Shuaichang Zhou, Jiaying Li, Yuxing Chang, and Wentao Wei

Shanghai Key Laboratory of Intelligent Manufacturing and Robotics, Shanghai University, Shanghai, China

Abstract. With the increasing development of industrial intelligence, the application of digital twin workshops has become popular, but there are still humanization issues such as convenience, safety, and visualization. Therefore, based on the digital twin system of the actual workshop production line, using Web development technology to realize real-time alarm, process monitoring, data display and interaction with the digital twin system in response to problems that be encountered in the actual production process, such as failures, monitoring, and data analysis in this article. The real-time production situation of the workshop is functionally presented on the digital twin page, which can observe the on-site production line more clearly, effectively improving the overall intuition, failure and safety. At the same time, the realization of the interactive function with the digital twin system also enhances the closeness of the connection with reality.

Keywords: Digital twin · Web interactive · Real-time alarm · Process monitoring

1 Introduction

Digital Twin (DT) is one of the key enabling technologies for intelligent manufacturing. It is usually defined as the construction of a virtual model equivalent to a physical entity, through the virtual-real interaction feedback and data fusion between the physical entity and the virtual model Analysis, decision-making iterative optimization, etc., to add or expand new capabilities for physical entities [1].

With the continuous maturity and improvement of digital twin-related technologies, its functional development is on the rise and is moving towards artificial and intelligent development. For example, Tao Fei and others have studied the theory and technology of digital twin workshop cyber-physical fusion [2]. Wei Yixiong and others studied the realization of a digital twin workshop based on real-time data-driven [3]. Web interaction design has also developed in various fields to varying degrees. For example, Chen Yanhong and others have studied the 3D dynamic display technology of web pages from the perspective of digital interaction based on 3D models [4]. In the research on the interaction between the digital twin workshop and the web page, Liu Aihua and others studied the 3D visualization of the network based on WebGL technology, combining 3D

Y. Wang et al. (Eds.): IWAMA 2021, LNEE 880, pp. 93–100, 2022.
https://doi.org/10.1007/978-981-19-0572-8_13

technology with visualization [5]. However, the 3D visualization application of the digital twin workshop based on WebGL technology is not perfect enough. In summary, this article combines the intelligent workshop digital twin system based on Unity3D software with the web design to design a digital twin system Platform that can be observed, monitored, and interacted, and realizes data display, real-time alarm, process monitoring and other features.

2 Web Interaction Implementation Plan

Realizing real-time data-driven interaction between the digital twin workshop and the front-end requires a digital twin workshop system and a digital twin UI interface, and then collect real-time data according to the needs of the production line, design the layout of the web page, and publish the data to users through the web, as shown in the following framework Fig. 1:

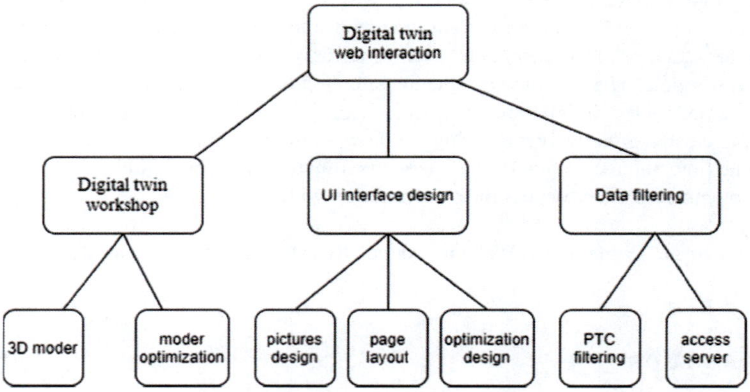

Fig. 1. Web interaction implementation plan

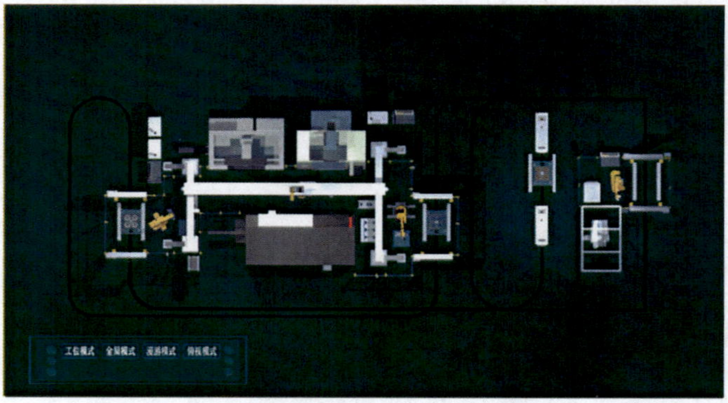

Fig. 2. Digital twin workshop

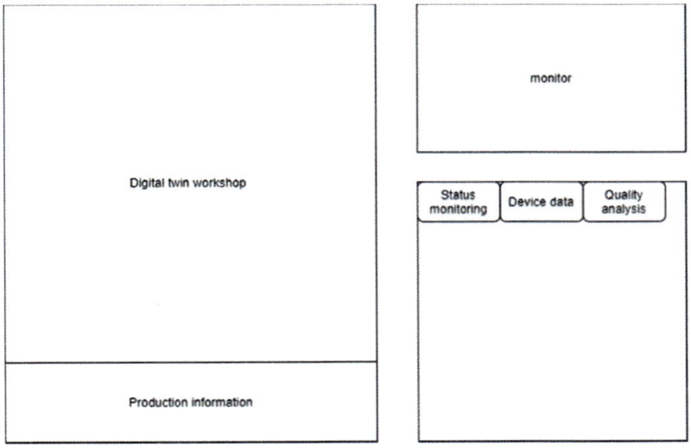

Fig. 3. UI surface design

As shown in Fig. 1, there are three main levels of Web interaction implementation plan listed below.

1) Digital twin workshop: The workshop production line digital twin system combines the automotive production lines, aerospace vehicles, steel heating furnaces and other industries to build technical systems such as industrial interconnection, digital twins, and intelligent predictive diagnostic analysis to realize in-depth perception based on the Industrial Internet of Things (IIoT)-Data and material integration-process monitoring-predictive maintenance-intelligent decision-making and other functions. This system is built using 3D modeling and simulation software and Unity3D software based on the overview of the on-site workshop. Figure 2 shows the entire digital twin model constructed based on Unity3D software according to the actual workshop production line. Its main structure is composed of three industrial robots, three turning-milling compound machine tools, a measuring instrument, and an AGV smart car responsible for handling. It is divided into flexible loading and unloading System, automatic stacking system, intelligent vision destacking system. The entire processing process is handled by the AGV, grabbed by the robots, and processed by the compound machine tools to complete the loading and unloading and stacking of the brake disc.
2) UI surface design: After the digital twin platform environment of the production line is built, the interactive function of the web needs to be realized by designing the UI interface layout. First of all, the design of the on-site monitoring screen corresponds to the DT (Digital twin) to better reflect the virtual and real combination of the digital twin. Secondly, in order to present status monitoring, data representation, quality analysis and other functions, design three sections corresponding to the functional areas, the first lower right column is the status monitoring, design a schematic diagram of the field equipment with indicator lights in this section; the second lower right column is the equipment data, which displays the robot joint data; the third lower right columns is the quality analysis, which is used to display

the quality analysis of the processed brake disc. Finally, optimizing the design to improve the visual effect. The design UI interface is shown in Fig. 3.

3) Data filtering: The realization of the interactive function of the web is through the filtering of real data on the spot, and disposing the data obtained through the signal processing that needs feedback, and finally provide users with real convenience. This filtering process first filter the record data of the production line through PTC, and then accesses the interface data stored in the server through the web corresponding to the site, and uses Ajax to access the internal server URL to get the Json format string, and convert it to string format, since the data is driven in real time, so we need to set the refresh frequency.Codes are showed in the following below:

```
setInterval(function () {
  $.ajax({
    url:"http://10.0.0.24/Thingworx/Things/shu_KepServerThing/Ser-
        vices/queyProperties?userid=test&pass-
        word=shu12345678&method=post&Accept=application/json",
    dataType: "json",
    success: function (res)}, 500);
```

3 Web Interaction Functions Realization

3.1 Real-Time Alarm Function

Real-time alarm is a key function based on the actual production line that encounters faults or problems in the production process and requires timely feedback to ensure safety. For the digital twin system, this function can truly reflect the on-site situation and provides a key role for maintenance, is a necessary decision-making function realizing real-time alarm need filter on-site signal sources. Through careful investigation of the actual work situation on the site, it is found that the main signal processing source comes from the data drive of the robot gripper and the turning-milling compound machine tool, with a total of 6 signal sources. The filtered alarm signals are bool type alarm signals, which are the robot alarm signal R1_alarm at the beginning of the line, the robot alarm signal R2_alarm between the lines, the robot alarm signal R3_alarm at the end of the line, and the alarm signals of three compound machine tools Machine1_alarm, Machine2_alarm, Machine3_alarm.The alarm signals are false in the default state. At this time, the production line is running normally, and these signals are bound to the status monitoring signal lights. When an alarm occurs at a station in the field, the corresponding signal value is true. At this time, this station will represent a red light flashing, which means that it is malfunctioning and presents specific alarm information. At the same time, an interface is also added to Unity3D. When an alarm occurs on the site, the corresponding jump to the faulty station. At this time, the UI interface and the digital twin both display the alarm status, which is consistent with the on-site correspondence. Data-driven web interactive alarm function has achieved.

3.2 Process Monitoring Function

Since the processing flow of brake discs cannot be directly observed from the site, the addition of the process monitoring function can observe the on-site operation in real time. It show the specific processing technology, and have a more intuitive and clear understanding of the process flow of each process. The entire brake disc processing process goes through 6 processing steps. How to determine the specific process and processing technology of a certain processing step? filtering three robot start signals R1_startL, R2_startL, R3_startL, and three machine tool feeding request signals Machine1_feed, Machine2_feed, Machine3_feed, Determine whether a specific station starts operation by judging its change value. When a certain signal changes, its corresponding operation position starts to work, then receiving this signal and publishing it to the web. There are a total of 6 devices, and the web corresponds to 6 ports, which receive different signals respectively and are displayed in the status monitoring column. At this time, the web page will show the device name, device status, and its production process. At the same time, corresponding equipment stations also appear in the twin, which realizes the combination of virtual and reality, and the web interacts with the digital twin.

3.3 Data Display and Quality Analysis Function

In the process of brake disc processing and transportation, in order to reflect the working status of the equipment, the most intuitive way is to display the equipment data. Since the production process is mainly handled by three robots, the working state of the robot can reflect whether the state of the production line is normal. At the same time, when the digital twin is synchronized with the on-site production line, the parameter of the robot joint can also be real-time. Changes can reflect its authenticity. Then obtain the coordinate values of the 6 joints and find the corresponding interface signals through the interface document, which are the signal values of the first 6 joints of the line R1_J1, R1_J2, R1_J3, R1_J4, R1_J5, R1_J6, the signal values of the 6 joints between lines R2_J1, R2_J2, R2_J3, R2_J4, R2_J5, R2_J6, and the signal values of the 6 joints at the end of the line R3_J1, R3_J2, R3_J3, R3_J4, R3_J5, R3_J6. According to the on-site robot's motion state, set obtaining data every 0.5s, which basically meets the actual working state of the robot. After obtaining the signal, it is published to the web and displayed in the device data column, which can respond to users concisely and clearly, and realize data display Function.

After the brake disc is processed, it is necessary to use a measuring instrument to check the integrity of the product to judge whether conforms to the standard. In order to complete this part of the work, it is necessary to obtain the data in the on-site inspection equipment. By accessing the internal server, the PDF files obtained from the measuring instrument include the diameter, parallelism, flatness, and roundness of the brake disc, as the Fig. 4 shows:

	Actual	Nominal	Upper Tol.	Lower Tol.	Deviation	
True Position1^4					⊢	
	0.1053	0.0000	0.5000			0.1053
R	49.9429	49.9000				0.0429
Angle	180.0351	180.0000				0.0351
Flatness_Plane_1					⊢	
	0.0507	0.0000	0.3000			0.0507
Flatness_Plane_2					⊢	
	0.0111	0.0000	0.3000			0.0111
Roundness_D65					⊢	
	0.0230	0.0000	0.3000			0.0230

Fig. 4. Quality analysis PDF file

4 Result of Web Interaction Function

After achieving the implementation plan and functional requirements, completing the production of the web page and the realization of the interactive function with the digital twin through VS Code. The first step is the alarm function. After accessing the server address to obtain the data, the alarm information content is bound to the obtained alarm signal. Figure 5 shows the actual situation at the scene and the situation on web page. At this time, the line head, line middle, line end all have alarms, the alarm signal changes, and the alarm message appears on the UI page, as shown by the successful realization of the production line early warning function.

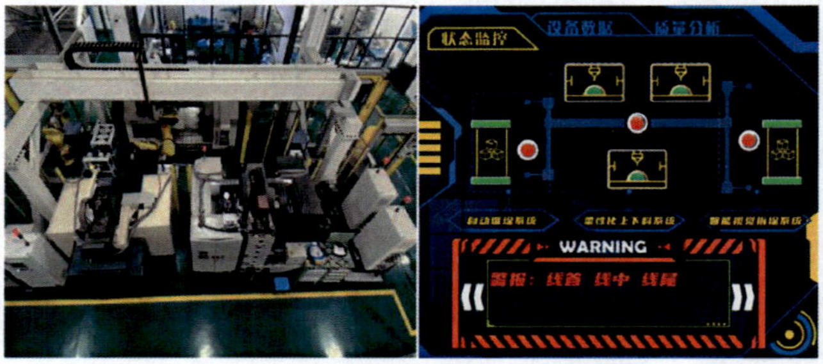

Fig. 5. Real-time alarm function

The second function is process monitoring. As shown in the Fig. 6, it can be seen that at this time the line robot is transporting the brake disc to one of the turning and milling machine tools, and the machine tool starts to work. At this time, the working status of the machine tool appears on the page, and the production line processing process is realized Monitoring function.

The Fig. 7 shows the joint data display of the three robots and the quality analysis of the brake discs, which effectively reflect the working status of the real robots and the quality of processed products.

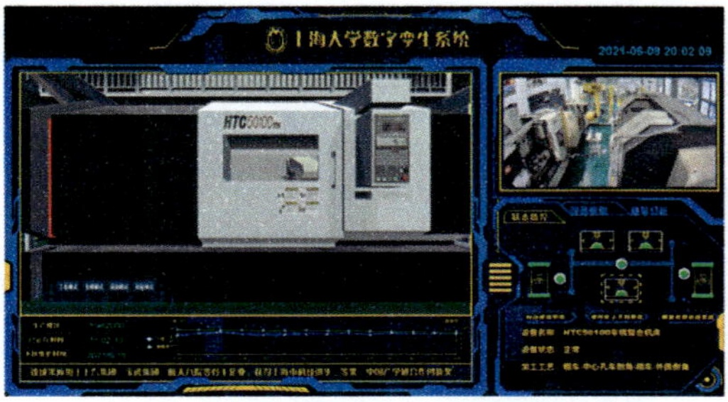

Fig. 6. Process monitoring function

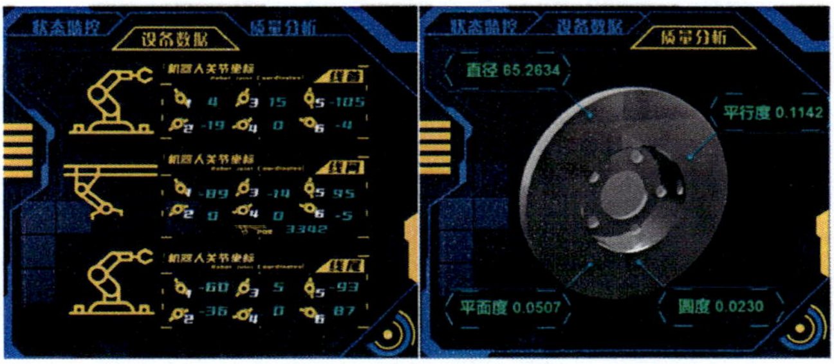

Fig. 7. Data display and quality analysis function

5 Conclusions

As more and more digital twins are beginning to move towards the development prospects of convenience, functionality, and safety, the web development based on digital twins is particularly important. It can better display the digital twin platform and make the combination of virtual and real visible. At the same time, it can also meet the functional needs of users. Based on the basic functions of the digital twin, a web page that can realize interactive functions is constructed in this paper, and the construction of a multifunctional platform for real-time alarming of fault information, real-time monitoring of on-site processes, real-time analysis and display of data and quality is completed. The twin system has been optimized, and the function of interacting with the twin has also been realized, which can better meet the needs of users.

Acknowledgements. The authors would like to express appreciation to mentors in Shanghai University and Huayu Intelligent Equipment Technology Co., Ltd for their valuable comments and other helps. Thanks for the pillar program supported by Shanghai Economic and Information Committee of China (No. 19511105200).

References

1. Fei, T., Qi, Q.: Make more digital twins. Nature. **573**(7775), 490–491 (2019)

2. Fei, T., Ying, C., Jiangfeng, C., Zhang Meng, X., Wen-jun, Q.Q.: The theory and technology of cyber-physical fusion in digital twin workshop. Comput. Integr. Manuf. Syst. **23**(08), 1603–1611 (2017)
3. Yixiong, W., Lei, G., Liangxi, C., Zhang Hongqi, H., Xiangtao, Z.H., Guang, L.: Research and implementation of digital twin workshop based on real-time data drive. Comput. Integr. Manuf. Syst. **27**(02), 352–363 (2021)
4. Chen, Y., Gulimila, K., Xie, W., Lu, Y.: Research and application of three-dimensional dynamic display technology on Web pages. Mod. Electron. Technol. **41**(20), 24-27+32 (2018)
5. Liu, A., Yong, H., Zhang, X., Chen, G.: Research and implementation of network 3D visualization based on WebGL technology. Geospatial. Inf. **10**(05), 79-81+7 (2012)

Digital Twin Construction Method for Docking Mechanism Test-Bed

Yuxing Chang(✉), Lilan Liu, Shuaichang Zhou, Tao Xu, and Shibo Yuan

Shanghai Key Laboratory of Intelligent Manufacturing and Robotics, Shanghai University, Shanghai, China

Abstract. In order to solve the problems of model visualization, control logic consistency and data interaction in the process of docking mechanism test-bed, this paper proposes a digital twin construction method for docking mechanism test. Through the construction of visual model, logical model and data model of the docking mechanism test-bed, the digital twin of realize the mapping of the physical test-bed in digital space and the health data management of the test process. Taking a docking mechanism test-bed as an example, the docking mechanism of a certain type was tested and verified, through repeated tests, the test data and key information consistent with the physical test bed were obtained, the test cycle and data management time were shortened, and the test cost was saved.

Keywords: Digital twin · Visual model · Logical model · Data model · Space docking mechanism test-bed

1 Introduction

With the development of astronautical technology and space station, space rendezvous and docking technology has been rapidly developed and widely used. In order to realize the safe and reliable docking of two spacecraft, space-docking mechanism plays a vital role. However, to ensure the adaptability of the docking mechanism in the face of complex working conditions in the future, and to ensure the high success rate and reliability of space docking, the simulation test of the docking mechanism has become an essential and important link. The docking mechanism test-bed is mainly used to simulate the dynamic process of the docking of two space vehicles on the ground, and to verify and evaluate the rationality of the mechanism design and the stability of the docking process [1].

At present, the semi-physical simulation platform was used for the test of the docking mechanism in China. A part of the simulated object system is introduced into the simulation loop in the form of physical object (or physical model), and the rest of the simulated object system is described by mathematical model and transformed into the simulation calculation model. By means of the physical effect model, the real time mathematical simulation and physical simulation are combined. However, there are a large number of sensor data and the interaction of various test platform's parameters in the test process, a large number of data and tedious test links greatly aggravate the test platform burden.

Y. Wang et al. (Eds.): IWAMA 2021, LNEE 880, pp. 101–108, 2022.
https://doi.org/10.1007/978-981-19-0572-8_14

The docking test-bed is a high precision and dynamic response system with multiple inputs and outputs. In order to improve the effectiveness of the ground simulation test, this paper proposes a construction method of the digital twin of the docking mechanism test-bed. With the characteristics of obvious data and strong visualization ability, the docking mechanism test-bed will become more intelligent and digital.

2 Design of Digital Twin Construction Scheme

Digital twin can realize the fusion and iterative optimization of the physical test-bed and the digital test-bed, and partially replace the physical test to a certain extent, shorten the test period and reduce the test cost [2].

One of the typical elements of digital twin is the high degree of simulation of the physical device by the simulation model. The collected state data of the physical test bed can drive the model, which enables the visual model to restore the state of the physical test-bed in the process of data collection. In this paper, the general framework of the digital twin of the docking test-bed is proposed.

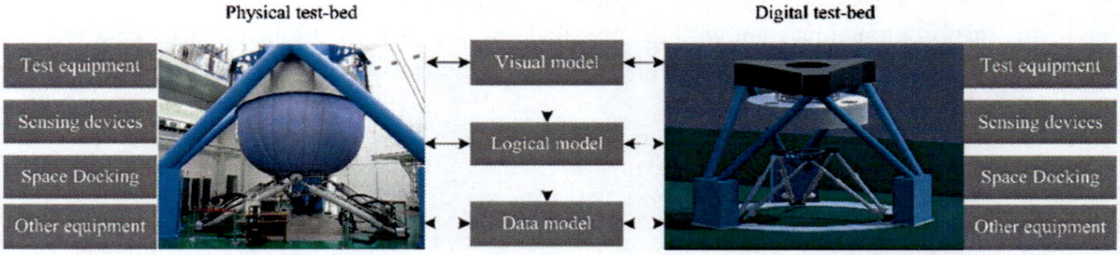

Fig. 1. The general frame diagram of the digital twin of the docking mechanism test-bed

As shown in Fig. 1, there are five main levels of general frame diagram of digital twin listed below.

1. Physical test-bed, also called semi-physical simulation test-bed, including six degrees of freedom motion simulator, test equipment, six dimensional force sensor sensing devices, such as docking mechanism, test environment, in addition to have the product of traditional test function, also have the awareness of multi-source heterogeneous data access and integration ability.
2. Digital test-bed is a set of twin models, which refers to the simulation equipment model and environment model corresponding to the physical test-bed. It mainly describes the test elements such as products, test-bed and environment, and can realize the test task analysis, deduction and simulation of the physical test-bed in digital space.
3. Visual model. It mainly uses 3D modeling software (SolidWorks or NX UG) to model and import it into the Unity3D. To achieve the same static appearance of physical test-bed and digital test-bed and dynamic running higher similarity.
4. Logic model, used to further debug and optimize the logic control program of the digital test-bed base on the visual model until it meets the requirements of the digital-test.

5. Data model used for collecting and analyzing physical system information, with upper and lower computer communication and data processing functions.

Figure 2 shows the modeling process of the digital twin of the docking test-bed.

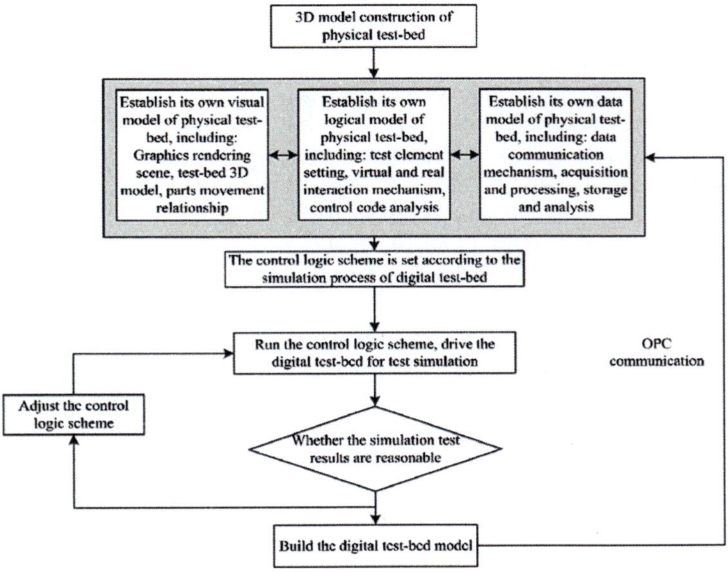

Fig. 2. Digital twin modeling process of docking mechanism test-bed

Firstly, the 3D model of the docking physical test-bed was established by using the 3D modeling software, and the simulation software was imported to render the scene model, define the motion relationship, and establish the visual model of the docking comprehensive test bed. Then, the logic model and data model of the docking synthesis test bed are established [3]. According to the interaction among the visual model, logic model and data model, the testable digital twin with high consistency between the physical test bed and the digital test bed is established. Finally, the simulation process, control logic and data are tested repeatedly in the testable digital twin, and the running data and key information of the digital test bed are obtained, which are similar to the physical test bed in appearance, consistent in test control logic and smooth in data extraction. Thus, the test cycle and debugging cost can be shortened to a certain extent.

3 Visual Model Construction

In order to realize the visualization of the docking test-bed, 3D model of the docking test-bed and docking mechanism must be established. The visualization model aims to build a digital test system with highly similar appearance to the physical test bed and completely consistent movement of equipment, and to provide a platform for subsequent data exchange. The digital twin can be built through three steps: digital test-bed model construction, motion planning and experimental simulation.

(1) Construction of digital test-bed. The docking test-bed consists of test equipment, sensing equipment, docking mechanism model and other equipment. Test equipment consists of 6-DOF motion platform and temperature simulation equipment. Sensing device contains six dimensional force sensor, displacement sensor, grating sensor and acceleration sensor movement data acquisition sensors, thin film temperature sensor, air point temperature sensor, duct type temperature sensor, temperature sensor, gravity sensor, etc. The butt mechanism model is mainly composed of butt cone (active collision device), butt ball (passive collision device), spring mechanism, damping mechanism, capture lock (butt pin) and so on.
(2) Motion planning. Firstly, the established device model was imported into the Unity3D to complete the preliminary layout of the digital test-bed. Then set the physical attributes of the model (such as material and colour, etc.), and give the motion attributes to the hydraulic driven platform model (including the space attitude Angle, the motion relationship between the moving platform and the hydraulic cylinder, etc.). Finally, the path planning of the motion equipment is completed (including the path of the motion platform, the extension length of each hydraulic cylinder, etc.). When planning the path of the moving platform, it is necessary to simulate and traversal all accessible position points to ensure that no interference and collision will occur during the movement of each hydraulic cylinder. At the same time, the extension length of each hydraulic cylinder needs to be given in the path editor.
(3) Experimental simulation. Chain all action in accordance with the "contact-capture-buffer-correction-closer-lock tight-seal-separation" of the time sequence together [4]. The obtained data are simulated with the digital test-bed, and the simulation is compared with the real one, and the path is iterated and optimized continuously. The digital test-bed can replace the physical test-bed to complete the simulation test and obtain the test data and conclusion, to replace the physical test to a certain extent.

4 Logical Model Construction

The logical model can be built based on the dynamic simulation of the visual model. The basic framework is shown in Fig. 3. The control logic between physical test stand and digital test stand is connected with PLC by OPC server. The iterative optimization of the digital twin logic program requires repeated tests in the digital test bed to verify the accuracy of its control logic. So that the test function of the physical test-bed can be truly restored in the digital test bed, and the control logic model of the physical test bed is consistent with that of the physical test bed.

The logic model is mainly based on the physical test bench to test the control system. The control program can be written in the digital test bench control system and the logic program can be debugged.

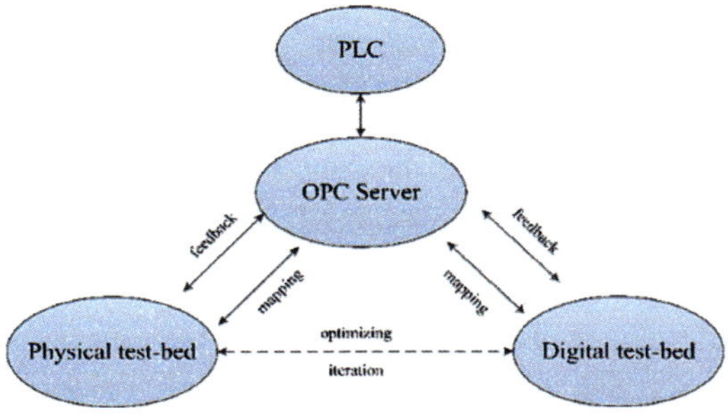

Fig. 3. Logical model framework

(1) OPC information exchange. In view of the data needed in digital twin modeling, the signal points needed in PLC are selected respectively [5]. By adopting a multi-device information correlator based on OPC server, which encapsulates nearly 150 kinds of communication protocols, the multi-node real-time acquisition can be accomplished by using this kind of aggregator. Because OPC protocol can be compatible with most of the industrial Internet sensor protocol and PLC protocol, so the use of OPC protocol can greatly reduce the amount of programming. In the process of transmission can reduce the number of threads, improve the efficiency of the system.
(2) Scheduling strategy programming. Control program based on event logic program simulation, the execution of all actions by the received signals driven. The scheduling strategy realizes the virtual debugging of the digital test-bed, introduces the mapping mechanism into the platform, debuts the control program of the digital test-bed, verifies the correctness of the program logic, and iteratively adjusts the program by controlling the input and output operations until the logic of the digital test-bed is completely consistent with the logic of the physical test-bed.
(3) Virtual debugging based on Unity software. After importing the written control program, debug the control program of the digital test bench, check the correctness of the program logic, and adjust the program iteratively until the logic program is completely consistent with the test logic of the physical test bench.

5 Data Model Construction

Different devices and sensors have different interface protocols. In order to solve the compatibility problem of heterogeneous data, the communication network architecture of digital twin system based on OPC should be established (see Fig. 4).

In the communication network architecture of digital twin system, OPC server of physical test bed control system, PLC signal frame and field equipment are connected through fieldbus, and I/0 port data of device control components (PLC, sensor, etc.) are obtained to realize data acquisition of physical test bed. Integrate the data of the physical test bed and convert it into data compatible with OPC protocol to provide data support for OPC client [6]. As the receiving end of OPC, the digital test-bed twin can obtain the test data of the corresponding physical test-bed from the OPC server. By driving the

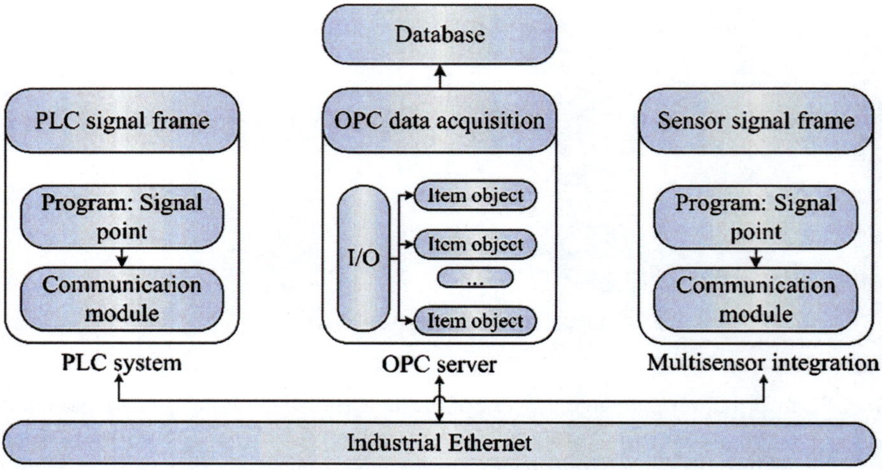

Fig. 4. Data collection process

digital twin element model, it can update all kinds of test data and make analysis and intelligent decision.

The data model includes data acquisition model and data mapping mechanism. In order to acquire the test data of the physical test bench efficiently, it is necessary to build the data acquisition model of the physical test bench. Data mapping includes mapping of grid data between different physical field computing models and grid management after grid deformation.

(1) Data acquisition model. Taking the six-degree-of-freedom motion simulator of a physical test-bed as an example, the data acquisition model of the test-bed as shown in Fig. 5 is constructed. The six-degree-of-freedom motion simulator consists of six hydraulic actuators, hydraulic cylinders, hydraulic motion platform and base. To build the data model, data such as the length of the hydraulic cylinder of the motion simulator, the position and attitude of the control point of the motion simulator, the velocity and angular velocity of the motion simulator, and the action signal, position and attitude of the motion platform need to be collected. The simulated quantities of all the entities in the physical test bed are organized in the same hierarchical relationship, and all the data of the entities in the test bed are obtained.

(2) Data mapping mechanism. The signal logic of the digital test bed is different from that of the physical test bed. In order to make use of the high degree of simulacrization of the data-driven model, the data and signals are processed in various ways. Model data or signals generally exist in the form of Boolean, integer, real and string, in which Boolean can be connected to IO signal of entity mechanism, opening and closing signal of cylinder and other high and low level signals; Integer type can connect no precision requirements of floating point, state data, etc.

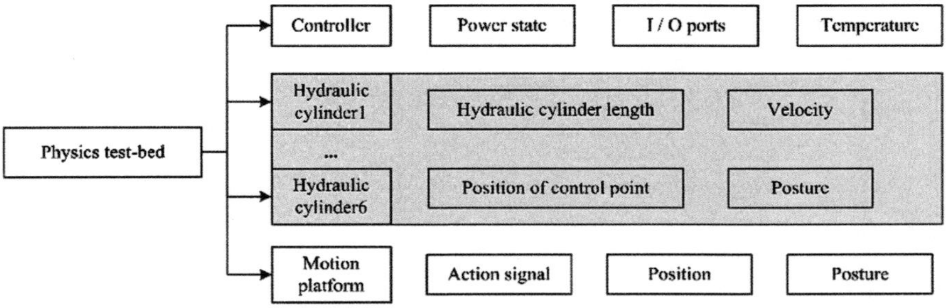

Fig. 5. Comparison before and after optimization

The real number corresponds to the high precision requirements of the joint value of the docking mechanism, cylinder value, monitoring data, etc. The string type can receive flexible data according to a custom format. The driving data of the digital test bed can be divided into four categories:

Motion-driven data, such as joint value, coordinate position and cylinder value of the docking mechanism, are directly driven by the corresponding data in production [7].

Action signal. Action signal triggers the corresponding response of the virtual world when the data changes to a specific value. The Boolean signal here generally takes two forms in the physical world: a variation of the pulse form; For the high and low level state signal, it is necessary to use the virtual service program to capture the changes of the semaphore and trigger the corresponding actions at the corresponding time.

State data and state signals correspond to state information of docking mechanism and space environment, etc.

Instruction data, motion control instructions of each subsystem and module of the test-bed control system. Digital test bed needs to be interpreted and transformed according to the meaning of the instructions to control the operation of the digital test bed.

By processing the driven data of physical space and then acting on the twin model, the virtual service programs of the digital test-bed can map the mechanism, environment and system in the digital test-bed in the way of multi-thread parallel.

6 Conclusion

(1) Based on problems of model visualization, control logic and data consistency in the process of interaction, this paper put forward the docking test-bed digital twin simulation model, data model and logic model construction method, implements the physical test-bed in the digital space mapping and test process of health data management.

(2) This paper studies the data acquisition and transmission methods of docking comprehensive test bed in the digital twin environment, as well as the existing data sources and characteristics. And we divide the data types according to the data acquisition methods, and establish the data acquisition model. The OPC client is connected with the programmable control equipment of the docking test bed, and the data acquisition and application of the whole process are realized.

Acknowledgements. The authors would like to express appreciation to mentors in Shanghai University for their valuable comments and other helps. Thanks for the pillar program supported by Shanghai National Defense Science, Technology and Industry Office of China (JKCY2020413C002).

References

1. Jin, F.W.: Analysis and Experimental research of Docking Mechanism Test-bed System. Harbin Institute of Technology (2006)
2. Li, H., et al.: The modeling method of the testable digital twins for automatic white body welding production line. J. Zhongyuan Univ. Technol. **32**(01), 1–7 (2021)
3. Wang, J., Xiang, Y., He, Z.: Models and implementation of digital twin based spacecraft system engineering. Comput. Integr. Manuf. Syst. **25**(06), 1348–1360 (2019)
4. Schleich, B., et al.: Shaping the digital twin for design and production engineering. CIRP Ann. Manuf. Technol. **66**(1), 141–144 (2017)
5. Tao, F., et al.: Theories and technologies for cyber-physical fusion in digital twin shop-floor. Comput. Integr. Manuf. Syst. **23**(8), 1603–1611 (2017)
6. Zhang, Q., Liu, J.: Spacecraft system design. Beijing Institute of Technology Press, Beijing, 515–523 (2018)
7. Gehrmann, C., Gunnarsson, M.: A digital twin based industrial automation and control system security architecture. IEEE Trans. Indust. Inf. **16**(1), 669–680 (2020)

Visualized Interaction Method of Mechanical Arm Based on Augmented Reality

Shuaichang Zhou(✉), Lilan Liu, Heng Cao, Yuxing Chang, Jiaying Li, and Wentao Wei

Shanghai Key Laboratory of Intelligent Manufacturing and Robotics, Shanghai University, Shanghai, China
z1s1c1@shu.edu.cn

Abstract. A visualized interactive application of a mechanical arm based on augmented reality is designed and developed by using Unity 3D software and Vuforia development kit. The visual application design scheme based on augmented reality is proposed, and the technical process is specifically explained. The key technology of the visual interaction of the mechanical arm is achieved by constructing the three-dimensional modeling of the mechanical arm, realizing the augmented reality function of this application and setting the virtual button for human-computer interaction. The feasibility of realizing the visual interaction of the mechanical arm by using Unity 3D and Vuforia SDK and the interaction method based on virtual buttons is verified through the display of augmented reality effects and functional testing of the application on the mobile terminal.

Keywords: Augmented reality · Mechanical arm · Virtual button · Unity 3D

1 Introduction

Augmented reality (AR) is an emerging technology developed on the basis of virtual reality (VR), which generates three-dimensional virtual information through a computer system, including virtual scenes, virtual objects, etc., and then superimposes these information into the real scene to realize the function of enhancing the real world and enhance the user's perception of the real world. There are wide range of applications of AR, including engineering, military and other fields [1].

With the continuous maturity and improvement of mechanical arm related technologies, the function of mechanical arm is becoming more and more powerful It is developing towards modularization and intelligence, and is gradually applied to various fields. There are many researchers having some achievement in the field of augmented reality related to mechanical arm. Wang Zhiqiang et al. designed a virtual manipulator system based on mobile augmented reality technology, and demonstrated the action characteristics of the virtual manipulator on the mobile terminal [2]. Zhang Yu et al. realized the design of the mechanical arm simulation platform based on augmented reality technology, and researched the unmarked three-dimensional registration algorithm and the point cloud-based collision detection algorithm [3].

Y. Wang et al. (Eds.): IWAMA 2021, LNEE 880, pp. 109–116, 2022.
https://doi.org/10.1007/978-981-19-0572-8_15

The visualization interactive application of the mechanical arm based on Unity 3D and Vuforia platform is designed and implemented in this article, And it is published to the mobile terminal for functional testing, and successfully achieved the interactive operation of the mechanical arm using virtual buttons. The human-computer interaction function is expended and the feasibility of the interactive method based on virtual buttons is verified comparing the existing works.

2 Visualization Scheme Design Based on Augmented Reality

2.1 Development Platform

Unity 3D is a game development tool released by the Unity company in Danish. It can be operated on multiple platforms, including Windows, Mac, Android, etc. And it can also run the developed system directly in the form of a web page. Unity 3D uses an advanced 3D engine, which has the advantages of high development quality, strong interaction, easy operation, and excellent picture quality. It is widely used in the creation of real-time rendered 3D animations, visualized buildings and other interactive fields [4].

Voforia is an augmented reality development platform launched by Qualcomm, which is mainly used in the development of Android or IOS smart device application software. Vuforia can recognize and capture images or three-dimensional targets in realtime by using machine vision and other technologies. Developers can place virtual objects through the camera and adjust the position of the virtual objects on the physical background. Its structure mainly includes cloud target database, Vuforia cloud engine, target management system, Vuforia SDK and other parts [5].

2.2 Overall Framework

The overall design framework of this application can be divided into two parts, three-dimensional modeling and realization of AR function, as shown in Fig. 1.

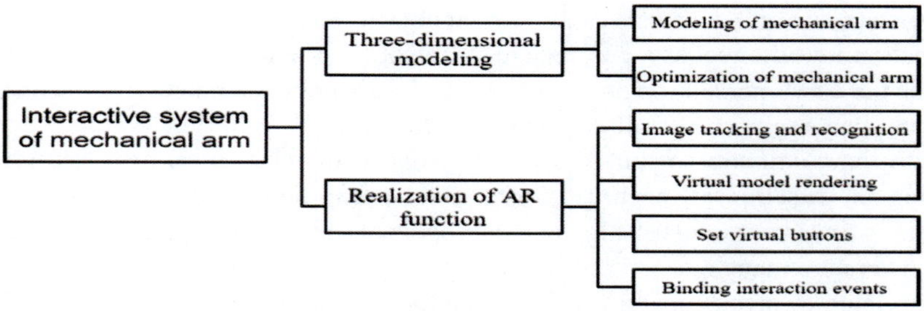

Fig. 1. Overall design framework of the system

It is desired for three-dimensional modeling to replicate the real mechanical arm device and optimize the generated model of the mechanical arm to form the equipment model used by the interactive system.

Realization of AR function is the core module, including core steps such as recognition image production, virtual model rendering, and virtual button setting. The visual interaction function of the mechanical arm is realized in this model.

2.3 Technical Process

The technical process of this application is mainly divided into t image tracking registration, modeling and import model, and human-computer interaction. The specific process is shown in Fig. 2.

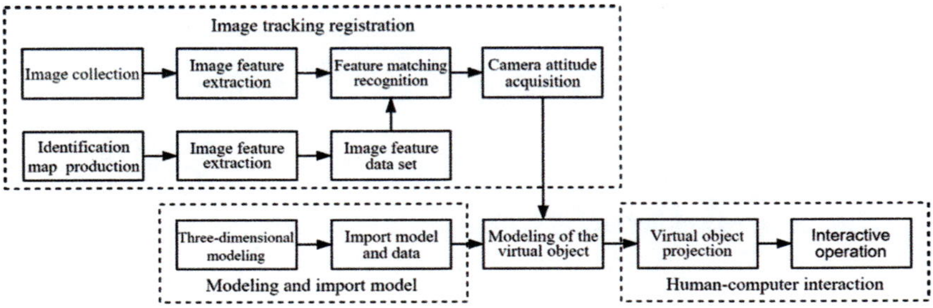

Fig. 2. System technical process

The feature points of the image is detected and saved to the image feature data set by using Vuforia. Vuforia SDK captures the image using the camera module and converts the captured image to make it suitable for OpenGL ES rendering. Then use the tracking module to call the database with the data obtained after converting the image, identify the target content inside and track it. The previously set models or scenes are then superimposed through the application code module. The images collected by the camera and the superimposed images are finally rendered, and the final effect image is presented to the mobile device. And the staff can control the device Model for interactive operation [6].

3 Realization of Key Technologies for Visualized Interaction of Mechanical Arm

3.1 Modeling of Mechanical Arm

Three-dimensional modeling of mechanical arm is the most basic part. The final realization effect is directly affected by the accuracy of the 3D model. The more realistic the restoration equipment, the better the effect will be. The mechanical arm is modeled using 3D Max software. 3D Max is a three-dimensional animation rendering and production software developed by Autodesk, which has very powerful functions, good scalability, a wealth of plug-ins and realistic effect [7]. More lines are needed to depict the entire model due to the complexity of the mechanical arm. Since the system is to be displayed on the mobile terminal, the model with a large number of lines has high requirements for the current mobile equipment. Therefore, in order to improve the display effect of the system as much as possible, it is necessary to optimize the initially built model of the mechanical arm. The optimized model of the mechanical arm is shown in Fig. 3.

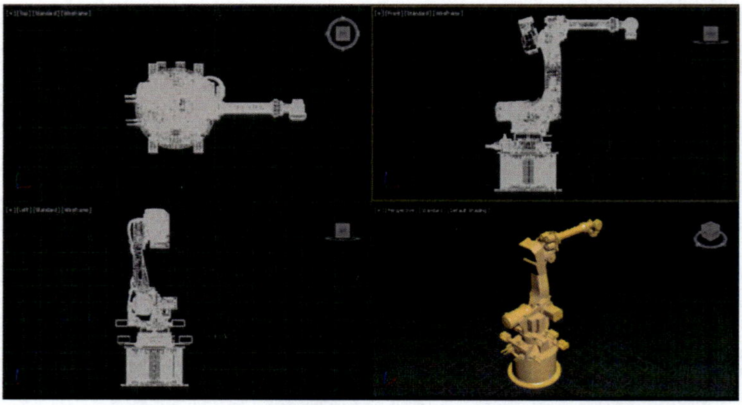

Fig. 3. Model of the mechanical arm

3.2 Realization of AR Function

The production of the identification map is the first step to realize the AR function. The AR device generates and renders the model by recognizing the feature points of the identification map, and anchors the model on the surface of the map. The identification map is made by using the Target Manager online tool provided by Qualcomm, which can greatly reduce the difficulty and workload of the identification map production [8].

Firstly, create a database file on the Vuforia official website, fill in the name of the database you have set and select "Device" as the type of database. And then, enter the database and uploading the identification map. Select the local image to upload to complete the addition of the machine identifying map of the arm. Select the added identification map, you need to pay attention to the "Rating" attribute of the added identification map. It represents the quality of the identification map. The indicator is the star rating. The more stars, the better the quality of the identification map. Finally, click the "Download Dateset" button to complete the creation and download of the identification map.

After the identification map is created, use Unity 3D and Vuforia to integrate and interact with it. Create a new Unity3D project and install the Vuforia SDK development kit, and import the 3D model and identification map made in the previous article into the project.

There are two important core components that need to be adjusted, namely AR Camera and ImageTarget. The camera is the core tool to realize the AR function, and the ARCamera module is used to realize the parameter setting of the camera function. It should be noted that the App License Key is needed to obtained when setting the AR

Camera module, which can be obtained on the Vuforia official website. The ImageTarget module is an important module for the AR function. The binding of models, scripts, and virtual button settings are all based on the ImageTarget module.

3.3 Set Virtual Buttons

The setting of the virtual button is about to start after above works. Virtual buttons allow us to touch virtual buttons in reality to trigger events. User's sense of experience can be increased by using virtual buttons for interaction. In layman's terms, people feel like they can touch virtual objects.

A virtual button can be added after clicking the "Add Virtual Button" button in the "Advanced" option of the ImageTarget module. Multiple virtual buttons can be added, adjusted size and position as needed. It should be noted that when adding a virtual button, the size of the virtual button should be about 10% of the target recognition map, and there should be sufficient space between the virtual button and the boundary to avoid the inability to track the identification map after the button is pressed.

You can add Scripts can be added to bind events that occur when the virtual button is pressed or released In the ImageTarget module. Button is pressed judged by Vuforia according to the "name" attribute of the virtual button. Therefore, the "name" attribute should be set to different virtual buttons. The system is equipped with four virtual buttons, the functions are "Show/Hide Model", "Rotate", "Replace Model Material", and "Reset Model". The system preview is shown in Fig. 4.

Fig. 4. Mechanical arm interaction system

4 Test of Visual Interactive Applications of Mechanical Arm

Applications can be published to multiple platforms by Unity 3D. A smartphone based on the Android operating system is used as a testing tool due to the design and development of this application based on the Android platform.

The SDK package of the Android platform must be installed before publishing the apk file on the PC. And then start Unity3D, then click the "Build" button, select the save path of the apk, and complete the release of the apk after confirming [9].

The released apk should be installed on the Android phone for testing. The augmented reality effect of the mechanical arm is shown in Fig. 5.

Fig. 5. Augmented reality effect of mechanical arm

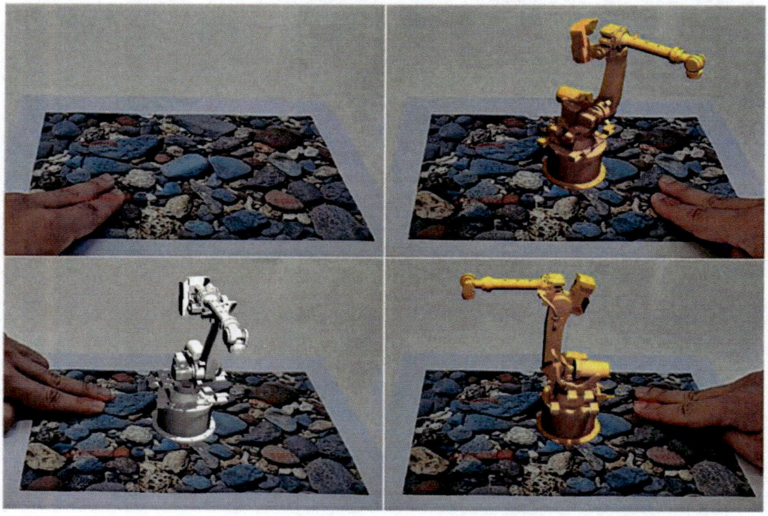

Fig. 6. Visualized interactive operation of the mechanical arm

After running the App, point the phone camera at the map, and the App will display the 3D simulation model of the mechanical arm in the video. By clicking the virtual button with fingers, different human-computer interaction functions can be completed. As shown in Fig. 6, the functions are "shown/hiding the model", "rotating the model", "changing the model material", and "resetting the model".

It can be seen from the above results that this application can complete the expected core functions, and the model of the virtual mechanical arm is well displayed on the mobile phone. The AR function also realizes the function of model rendering and the visual interaction of the mechanical arm based on virtual buttons.

5 Conclusions

Augmented reality has been gradually applied in people's lives. Augmented reality based on Unity3D and Vuforia can be more convenient to develop augmented reality applications with interactive functions.

An application for visual interaction of a mechanical arm based on augmented reality technology is developed, and a human-computer interaction method using virtual buttons is implemented. The expected core function can be achieved after releasing and testing the system on the mobile terminal. The model of the virtual mechanical arm can be well displayed and the human-computer interaction function can be completed.

Acknowledgment. The authors would like to express appreciation to mentors in Shanghai University and Huayu Intelligent Equipment Technology Co., Ltd for their valuable comments and other helps. Thanks for the pillar program supported by Shanghai Science and Technology Committee of China (No. 19511105200).

References

1. Ruan, W.H.: Design and implementation of AR vehicle display model based on unity3D. In: Proceedings of 2018 3rd International Conference on Mechatronics and Information Technology (ICMIT2018). International Information and Engineering Association: Computer Science and Electronic Technology International Society, vol. 8 (2018)
2. Zhiqiang, W.: Research and Implementation of Virtual Mechanical Arm System Based on Mobile Augmented Reality. Hebei University of Engineering, Handan (2015)
3. Zhang, Y.: Development of a mechanical arm simulation platform based on augmented reality. Beijing University of Posts and Telecommunications (2018)
4. Tang, H.: Design and implementation of a training system for virtual collaboration and augmented reality of electric vehicles based on Unity3D. East China Jiaotong University (2020)
5. Liu, W.: Research on cloud-based object recognition and grabbing in smart space. Shandong University (2016)
6. Haiyong, W., Jindong, Z., Yangping, W., Wenrun, W.: Research and application of railway equipment visualization based on augmented reality. J. Railway Sci. Eng. **15**(08), 2092–2098 (2018)
7. Shatte, A., Holdsworth, J., Lee, I.: Mobile augmented reality based context-aware library management system. Expert Syst. Appl. **41**(5), 2174–2185 (2014)

8. Sarosa, M., et al.: Developing augmented reality based application for character education using unity with Vuforia SDK. J. Phys. Conf. Ser. **1375**(1), 012035 (2019)
9. Min, X., Qiang, T.: Development of an augmented reality interactive app based on unity3D+Vuforia. Mod. Comput. **12**, 71–75 (2016)

Research on Image Acquisition Preprocessing and Data Augmentation Algorithm Based on Strip Surface Defect Detection Technology

Xiang Wan(✉), Lilan Liu, Zenggui Gao, and Xiangyu Zhang

Shanghai Key Laboratory of Intelligent Manufacturing and Robotics, Shanghai University, Shanghai, China

Abstract. The identification of strip steel surface defect is an important index to test its quality. With the development of deep learning technology, strip surface detection has developed from traditional machine learning algorithm recognition to deep neural network detection. Existing deep learning strip detection algorithms mainly focus on the tuning and identification of standard defect data sets, while few focus on the preliminary preparation processes such as the collection and pretreatment of the original data of the production line. To address this problem, this paper proposes a series of image data acquisition and pre-processing algorithms for strip surface defect detection technology, such as out-of-area background shielding, noise filtering, suspected defect image acquisition, light equalization and image enhancement processing. Light equalization and image enhancement processing and a series of image data acquisition and pre-processing algorithms to quickly eliminate invalid data from the collected image data and retain valid data. And for the image ROI region data broadening, effectively solve the problem of insufficient defect samples, for the existing deep learning strip inspection algorithm to provide a standard training and testing data set. The image acquisition pre-processing and data augmentation algorithm in this paper is also of practical value when extended to other products in the field of surface inspection.

Keywords: Strip surface defects · Image acquisition preprocessing · Data augmentation

1 Introduction

Strip steel is one of the main products of iron and steel industry, its quality will directly affect the quality and performance of the final product. Due to many factors such as raw materials, rolling equipment and processing technology in the production process, strip steel surface will appear various defects. In order to accelerate the integration of information technology and industrialization, iron and steel manufacturing enterprises gradually begin to use the non-contact nondestructive testing technology represented by machine vision. Through the online deployment of image acquisition equipment, various sensors are installed on the existing production line to collect the image data of strip steel surface in the production process, and the real-time collected images are

Y. Wang et al. (Eds.): IWAMA 2021, LNEE 880, pp. 117–125, 2022.
https://doi.org/10.1007/978-981-19-0572-8_16

transmitted to the monitoring equipment, so that the surface state of the product can be monitored in real time.

Through the deep neural network algorithm, the on-line detection system can automatically detect the surface of the strip steel in real time and classify and recognize the defects. However, there will be a large number of pictures and the mismatch between the speed of image acquisition and the speed of image data transmission during its use. This not only presents a high challenge to the real-time transmission efficiency of the data acquisition system, but also presents a severe test to the real-time storage and processing efficiency of subsequent data [1]. This is also the traditional manufacturing industry to smart manufacturing must overcome a big obstacle. Although the larger the size and quality of the data, the better the generalization capability of the model. However, in the actual steel rolling production line, the effective data collected is often too small (compared with the model), the samples are not balanced, and it is difficult to cover all the required application scenarios. Among them, the amount of data in some specific states is very rare, and small sample data sets may occur. If there is a large gap between sample categories, there can also be a *long-tail Distribution* that severely affects the classifier's classification performance [2].

Aiming at the problem of data collection and preprocessing in strip steel manufacturing process, this paper quickly eliminates invalid data from the collected data and retains effective data. Because a large number of invalid data does not need to be processed, the effective number of operations needed in the subsequent processing links is greatly reduced, which not only saves the time of data transmission, but also improves the real-time data processing efficiency of the algorithm, which is very suitable for solving the transmission and subsequent processing problems of massive redundant data. And for the small sample data, data enlargement is carried out to solve the learning problem of *small sample long-tail distributed data set* in the industrial field, so as to remove the obstacles for the deployment and application of various algorithms in industrial production.

2 Data Acquisition and Preprocessing Algorithm of Strip Steel Manufacturing Process

2.1 Real-Time Image Data Acquisition Algorithms for Suspected Defect of Strip Steel

In recent years, a large number of data preprocessing and feature extraction algorithms have been proposed to solve the problem of manufacturing process data collection, preprocessing and feature extraction. However, in the actual industrial production, due to the strict production environment restrictions and large-scale promotion and application of the budget restrictions, this requires not only the data preprocessing and feature extraction algorithm real-time good, but also requires the algorithm simple and compact and easy to deploy on low-end cheap hardware equipment. In view of this kind of problem, this section proposes algorithms such as out-of-area background shielding, noise filtering and suspected defect image acquisition to process the collected image data, and quickly remove invalid data from the collected image data and retain effective data.

(1) Out-of-area Background Shielding

In the actual production line, the angle between the CCD camera and the light source and the distance between the surface of the product to be tested is carefully designed. In order to collect the surface image of the product to be detected more clearly, light sources (such as bright field and dark field light source) will be set in scenes with different illumination intensity to ensure uniform illumination on the surface of the product to be detected in the image collection area. However, the background area outside the image collection area is usually outside the light source, which will lead to abnormal brightness (dark, highlighting) or uneven brightness. Some of them only have narrow conveyor belts, and the color is often relatively simple, and the average brightness is far lower than the gray value of the product surface image.

To the problem of image background screening this section through the average gray level to solve this problem, namely the image grayscale average in horizontal or vertical projection, if a horizontal or vertical projection of the image grayscale average is too low and below a specific threshold, argues that the image line or as a background region outside the territory, and shielding removed. The threshold value can be selected according to the lighting situation and the gray value of the surface of the product without defects for the preliminary adjustment test.

(2) Noise Filtering

Image signal in the process of acquisition, transmission, conversion and processing, due to the light and the CCD camera and workshop of dust particles, photoelectric signal conversion sensor sensitivity of different precision, in the process of digital image quantization noise and error, the platform of the acquisition object factors such as vibration, will cause the collected images contained different levels of noise, The main noise is Gaussian noise. Gaussian noise, also known as normally distributed noise, is produced by the superposition of all kinds of interference factors in the imaging system. When the acquired image data is mixed with Gaussian noise, there will be some problems, such as contrast reduction, poor hierarchical sense and edge blurring, which will seriously affect the subsequent pretreatment and feature extraction of the data. Therefore, the noise filtering method in this section is denoising by Gaussian filtering.

(3) Suspected Defects Image Acquisition

At present, the most widely used algorithm for rapid screening is the *Background Subtraction Method*, whose principle is to perform algebraic operation on the input image and the filtered image, and Subtraction operation on the gray value of the corresponding storage rectangular dot column of the two images. In the figure obtained after subtraction of difference, the part with the same gray value becomes a black dot, and no defects exist. However, due to the influence of light refraction, the gray value of the defect area is large, and the gray value after subtracting is greater than the preset threshold, then the pixel is considered to be a suspicious defect, and the system will cache the image and wait for further screening.

2.2 Illumination Equalization and Image Enhancement Processing Algorithms

Following the suspected defect image acquisition in the previous section, this section performs surface illumination equalization and image enhancement on the acquired suspected defect images, which can greatly reduce the size and time of the data to be processed, and also make the non-formatted data more standardized and uniform to facilitate the subsequent algorithm processing.

(1) Retinex Algorithm for Surface Light Equalization
In actual production, due to the lighting environment, camera hardware platform vibration properties and the image acquisition process, lead to the surface appear to be detected image blur, appear uneven illumination problems, not only cover up the real defect on the surface of the object to be detected, can also lead to subsequent algorithm error identification, and extract a large number of pseudo defects. But in practice there is no defect in this part. To solve this problem, Retinex algorithm is used in this section to balance illumination on the detected object surface, so that it is not affected by non-uniform illumination.

Retinex algorithm is an image adaptive enhancement algorithm proposed by Edwin H. Land in 1963. It can achieve balance in three aspects of dynamic range compression, edge enhancement and color constance, and is suitable for balancing various types of illumination [3]. As for the reflective problem on the surface of the object to be detected, it is mainly removed by hardware devices such as polarizing lens and light source, which is not considered in this section. The Retinex algorithm only solves the problem of uneven surface illumination. As shown in Fig. 1, it is the illumination balance processing of the image with uneven surface brightness by Retinex algorithm. It can be seen from the enhanced gray histogram that after processing, the peak value of the brightness concentration of the original image with unbalanced high brightness is reduced, and the gray distribution is more uniform than the original image, but it can be distinguished from the background. Therefore, the processed image can more easily highlight the details of defects in the current background, which is conducive to subsequent algorithm detection.

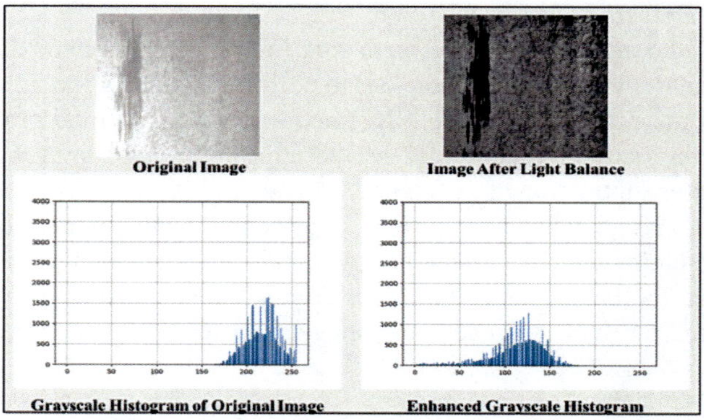

Fig. 1. Retinex algorithm for light equalization of bright surface

However, such low-pass filtering algorithms as Retinex algorithm are prone to halo interference in the processing of small details such as high-frequency textures and the transition area of strong light shadow, leading to image blurring after processing, covering up small details of the original image and producing false defects [4]. As shown in Fig. 2, Retinex algorithm is used to equalize the illumination of the uneven image on the dark light surface. As you can see, in the dark on the surface of the image after processing, gray peak down further, although the grayscale distribution more uniform than the original, but because of the original image brightness is darker and the transition area between light and shade after processing the halo disturbance, defects of the original image details is completely the halo cover, appeared serious falsified distortion image.

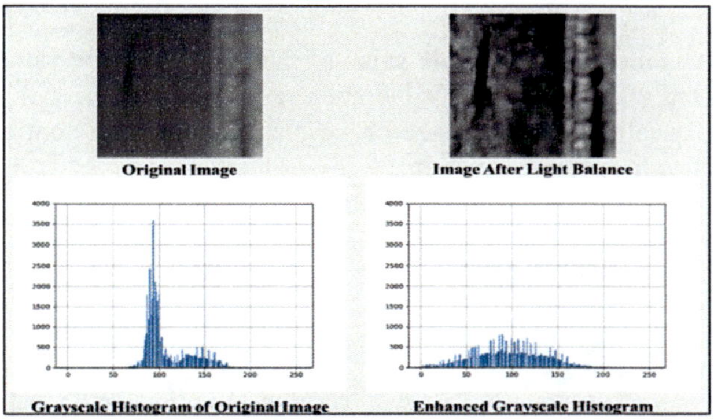

Fig. 2. Retinex algorithm for light equalization of dark light surface

The above experiments show that Retinex algorithm is not suitable for image processing of uneven illumination in the shadow transition region. Therefore, in the following sections, this paper will use gamma correction algorithm to enhance the image in the transition region of strong light shadow.

(2) Gamma-corrected Image Enhancement of Dark-light Surface
Image enhancement is an image distortion process, the purpose of which is to improve the visual effect of the image. Due to the low gray value of the whole image collected in the dim light, it is difficult to identify the features and small details of defects effectively. However, due to the high reflectivity of the surface, there are strong light shadow transition regions around some defects. The image enhancement of Retinex and other light equalization algorithms is easy to generate halo interference. However, the gray transformation enhancement of spatial domain processing on the original image can effectively avoid this problem.

The enhancement of gray transformation in spatial domain includes linear transformation and nonlinear transformation. Linear transformation enhancement is sensitive to threshold value and the enhancement effect is not ideal. The nonlinear transformation includes: logarithmic transformation, exponential transformation. The logarithmic transformation can extend the low gray value of the image, which is mainly used to improve the dark pixel value. The exponential transformation of the picture is used to improve the pixel value of the highlight part of the image. However, since the background gray value of the collected images is between the high brightness gray value defect and the dark light low gray value defect, the logarithmic transformation of the collected images is not conducive to the effective separation of the defects with different gray values, so this section adopts exponential transformation to enhance the collected images.

Gamma correction algorithm is one of the representatives of exponential transformation enhancement. Gamma correction is mainly used to correct image gray scale and enhance contrast. Through different γ values, it can achieve the effect of enhancing the details of low or high gray scale, as shown in Eq. 1:

$$s = cr^{\gamma} \tag{1}$$

Where r is the input grayscale value of the image, and the value range of r is [0, 1]; s is the grayscale output value after gamma correction; c is the gray scale coefficient, usually 1; γ is the size of gamma factor, which controls the scaling degree of the whole transformation.

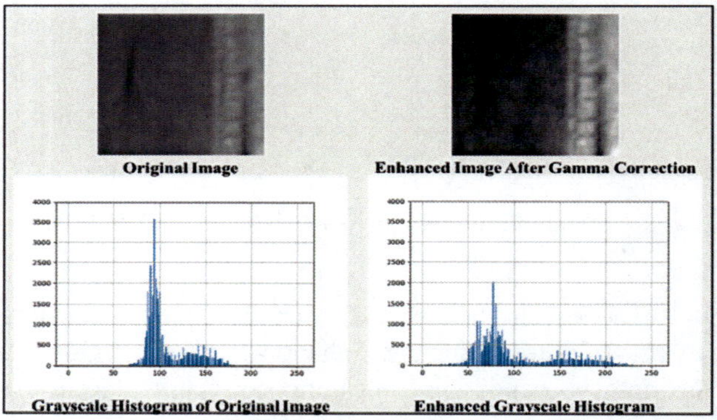

Fig. 3. Image detail enhancement based on gamma correction algorithm

As shown in Fig. 3, the image details are enhanced by the gamma correction algorithm. It can be seen from the enhanced histogram that after gamma correction, most of the pixel gray value of the original image is migrated to a higher range, and the contrast and pixel gray value distribution are improved compared with the original image, and there is no halo interference and false defects, which is suitable for the image enhancement of the dark light surface in this paper.

3 Data Augmentation Algorithm for Strip Defect Image

3.1 Small Samples of Data and the Long-Tailed Distribution Problem

Standard machine learning algorithms are based on the assumption that there are enough samples and the categories are evenly distributed. However, in the actual data collected by sensors, there is a certain degree of class imbalance, that is, each class has a different number of samples. This kind of data set is a typical long-tail Distribution. When the sample data has a long-tail distribution, it will lead to the bias of the classifier. The classifier is more inclined to identify the categories with sufficient sample size and feature diversity. Therefore, the key to this kind of problem is to deal with the distribution of small sample data. The methods to solve the long-tail distribution problem with small samples can be divided into two categories: (1) data augmentation with small samples; (2) learning with small samples based on modeling data characteristics. This section mainly solves the "long-tail Distribution" problem of the data set by enlarging the small sample data.

3.2 Data Augmentation Algorithm for Image ROI Region

Data Augmentation, or Data Augmentation, is a strategy to increase the quantity and diversity of limited Data, aiming to extract more useful information from it. For industrial image acquisition, due to the high-speed operation of the production line and many interference factors, the image background is complex, and the image captured by CCD camera has low resolution and few qualified images. Traditional image data augmentation techniques, such as horizontal or vertical roll-over, random scaling, adding various kinds of noise and other methods, are mainly based on the assumption that the ROI region of the sampled image has local geometric correlation and semantic correlation. However, various defects of industrial images do not satisfy the above assumptions [5]. For this kind of data, the data amplification method is mainly based on data deformation. The geometric transformation method is used for industrial image data amplification, which is represented by image local Random Erasing.

Inspired by the normalization of dropout, Random Erasing is a simple regularization technique. Its principle is to erase part of the image area randomly in the training process, forcing the model to learn more descriptive features about the image, so as to prevent over-fitting of a specific visual feature. However, such methods also have obvious disadvantages. Local random erasure may tamper with the semantic information of the original image, thus causing the model to lose targets and features (especially small targets and local features). After important local information is erased, the model may be unable to recognize the image [6].

Aiming at the limitations of the existing image data augmentation technology of local random erasure, this section proposes an image data augmentation algorithm of ROI Region random local erasure. Its principle is to randomly select a rectangular Region ROC (Region of Classification) from the ROI Region of the input image. The change range of ROC is controlled by scaling and scaling scaling parameters, and scaling to a fixed size is used to realize the geometric transformation image enlargement algorithm. As shown in Fig. 4, the product surface defect picture captured by the CCD camera, the

ROI area marks the position of the defect, and the rectangular side lengths are W and H respectively. A rectangular ROC region with gray value 0 is randomly generated in the ROI region, and the rectangular side lengths are W and H, respectively. The rectangular ROC region shall meet the requirement that the length of each side is between 1/3 and 1/2 of the length of the corresponding ROI rectangular side, that is:

$$\frac{1}{3}(W, H) \leqq (w, h) \leqq \frac{1}{2}(W, H) \tag{2}$$

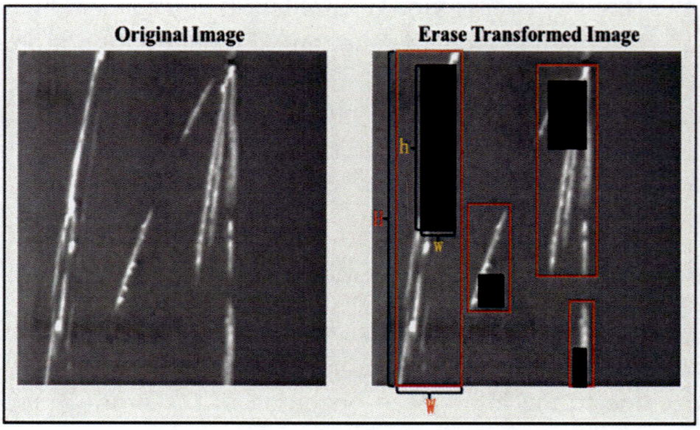

Fig. 4. ROI region random local erasure image data augmentation

This method can not only avoid the problem that the ROC area is too large to erase the important features of the desired target image, but also effectively ensure that the local features of the image can be erased randomly to complete data enlargement. There is no loss of small targets or the erasure of areas outside the target image, resulting in ineffective enhancement of the image data.

4 Conclusions

In this paper, we propose a series of image data acquisition and pre-processing algorithms such as background shielding outside the area, noise filtering, suspected defect image acquisition, light equalization and image enhancement processing for the pre-processing process of strip production line raw data. Light equalization and image enhancement processing" and a series of image data acquisition and pre-processing algorithms. And through the ROI region random local erasure of the defective data data to expand the data, to achieve the original data from data acquisition, pre-processing, data expansion of a complete set of data processing process. It not only solves the problem of class disequilibrium in the data set from the data level algorithm level, but also solves the problem of traditional geometric transformation image data augmentation technology (easy to introduce noise enhancement image). This also provides multi-scale feature data for various defect recognition algorithms, alleviates the problem of small sample data sets and model overfitting, and facilitates the training and deployment of deep neural networks. The algorithm of image acquisition preprocessing and data augmentation in this paper also has some practical value to be extended to the surface detection of other products.

Acknowledgements. The authors would like to express appreciation to Shanghai Key Laboratory of Intelligent Manufacturing & Robotics and all members of the CIMS laboratory for their support. Thanks for the funding from Shanghai Science and Technology Committee of China (No. 19511105200).

References

1. Srinivasan, K., Dastoor, P.H., Radhakrishnaiah, P., et al.: FDAS: a knowledge-based framework for analysis of defects in woven textile structures. J. Text. Inst. **83**(3), 431–448 (1992)
2. Cui, Y., Jia, M., Lin, T.Y., et al.: Class-balanced loss based on effective number of samples. In: Proceedings of the 2019 IEEE Conference on Computer Vision and Pattern Recognition (CVPR), Long Beach, CA, USA, 15 June–20 June 2019, pp. 9260–9269 (2019)
3. Land, E.H.: The retinex theory of color vision. Sci. Am. **237**(6), 108–128 (1978)
4. Wei, W., Shan, S., Wen, G., et al.: An improved active shape model for face alignment. In: Proceedings of the 2002 IEEE International Conference on Multimodal Interfaces (ICMI), Pittsburgh, PA, USA, 16 October 2002, pp. 523–528 (2002)
5. Singh, K.K., Lee, Y.J.: Hide-and-seek: forcing a network to be meticulous for weakly-supervised object and action localization. In: Proceedings of the 16th IEEE International Conference on Computer Vision (ICCV), Venice, Italy, 22 October–29 October 2017, pp. 3544–3553 (2017)
6. Chawla, N.V., Bowyer, K.W., Hall, L.O., et al.: SMOTE: synthetic minority over-sampling technique. J. Artif. Intell. Res. **16**(1), 321–357 (2002)

Experimental Investigation of the Effect of Ultrasonic Wave on the Saturated Hydrocarbons in Castilla Crude Oil

Shichun Zhu[1,2(✉)], Xuedong Liu[1,2], and Zhihong Zhang[3]

[1] School of Mechanical Engineering, Changzhou University, Changzhou, China
zhushichun@cczu.edu.cn
[2] Jiangsu Key Laboratory of Green Process Equipment, Changzhou, China
[3] School of Petrochemical Engineering, Changzhou University, Changzhou, China

Abstract. Saturated hydrocarbons in heavy crude oil are rarely investigated under ultrasonic treatment since the transformation of asphaltenes is a key part in heavy oil upgrading. To gain a more complete understanding of the cracking of compounds in heavy oil under ultrasonic wave, the effects of ultrasonic power and irradiation time on the saturated hydrocarbons in Castilla crude oil (Colombia) have been investigated. Functional groups and compound identification of saturated hydrocarbons in the crude oil before and after the ultrasonic treatment have also been analyzed by Fourier transform infrared spectroscopy (FTIR) and gas chromatography-mass spectrometry (GC-MS) respectively. Experimental results show that at least 13 kinds of long-chain or multi-branched chain saturated compounds are cracked and disappeared in crude oil by ultrasonic wave. The irradiation time has more effect on the kind of saturated compounds than does the ultrasonic power. The content of saturated hydrocarbons declines by 32.5% at 40 min, but increasing 38.4% dramatically at 50 min. It seems that the cracking of heavy aromatic hydrocarbons contributes more to the increase of the content than that of asphaltenes or resins. The ratio value of infrared absorbance of methylene (=CH_2) to methyl (–CH_3) increases by 40.0% when the irradiation time is 40 min, but declining 22.6% at 50 min. The optimum value of the irradiation time or ultrasonic power in forming methyl compounds should be more than 50 min or 80 W respectively.

Keywords: Heavy oil · Ultrasound · Cavitation · Saturated hydrocarbons

1 Introduction

The resources of heavy crude oil are more than twice than those of conventional light oil worldwide. However, a difficult issue arises during the gathering and transportation of heavy crude oil due to its high viscosity. In the process of viscosity reduction of heavy crude oil, various existing conventional methods have not been recognized as profitable methods because of high pressure, low production efficiency and low product quality [1]. In recent years, heavy crude oil upgrading issue and converting it into lighter products using ultrasonic cavitation technique has attracted the attention of researchers.

Y. Wang et al. (Eds.): IWAMA 2021, LNEE 880, pp. 126–134, 2022.
https://doi.org/10.1007/978-981-19-0572-8_17

Ultrasonic cavitation can produce conditions of very high localized temperature (up to 20,000 K) and pressure (several thousand bar) at room temperature in very short time [2]. Many researchers reported that the long-chains and heavy molecules in crude oil can be cracked by this cavitation energy, which leads to a reduction in viscosity [3–6]. Under optimum operation conditions, the heavy oil even can be greatly reduced in viscosity by 63.95% [7]. However, Taheri et al. [8] and Najafi et al. [9] observed that the viscosity of heavy crude oil increases as the irradiation time is prolonged. They reported that there are two reasons for the increase in viscosity of crude oil. One is the evaporation of light compounds by radiant heat due to the ultrasonic wave, and the other is the formation of heavy molecules.

Based on previous studies of ultrasonic viscosity reduction technique [3–14], it is clear that most of the researchers focused on the transformation of asphaltenes in crude oil exposed by ultrasonic wave. Although the cracking of asphaltenes molecular chains is a dominant factor of heavy oil upgrading, the disintegration and reintegration of saturated hydrocarbons should be investigated further to gain a more complete understanding of the viscosity reduction in heavy crude oil under ultrasonic cavitation treatment.

In this paper, the effect of ultrasonic power and irradiation time on the content of the saturated hydrocarbons in Castilla crude oil (Colombia) is investigated. Functional groups and compounds of saturated hydrocarbons under different ultrasonic conditions are also analyzed via Fourier transform infrared spectroscopy (FTIR) and gas chromatography-mass spectrometry (GC-MS). The purpose of this work is to provide the experimental basis for the application of ultrasonic viscosity reduction technique in heavy oil upgrading.

2 Experiment

2.1 Experimental Setup and Properties of Castilla Crude Oil

The schematic diagram of ultrasonic device is shown in Fig. 1, which mainly consists of two parts, one is the ultrasonic generator and controller, and the other is ultrasonic transducer. The frequency of the ultrasonic transducer is 28 kHz.

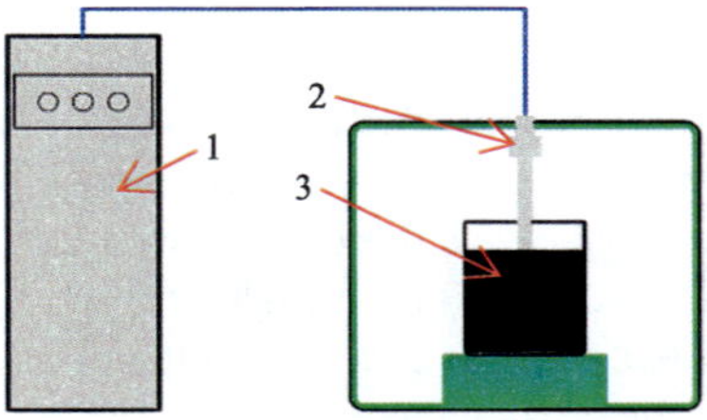

Fig. 1. Schematic diagram of ultrasonic device. 1, ultrasonic generator and controller; 2, ultrasonic transducer; 3, oil sample.

In this study, Castilla crude oil from Colombia was used as the heavy oil sample, the main properties are shown in Table 1.

Table 1. Properties of Castilla crude oil (Colombia).

Parameter	Value
API	20.2
Density at 20 °C	0.9287 g/cm^3
Viscosity at 20 °C	212.45 mm^2/s
Saturates	24.24 wt.%
Aromatics	27.23 wt.%
Asphaltene and resin	48.53 wt.%

2.2 Procedure

A 30 g sample of Castilla crude oil was exposed by ultrasonic wave according to the arrangement of experiments shown in Table 2. In each experiment, the initial oil temperature was approximately 20 to 25 °C due to environment temperature. After the experiments were finished, oil samples were collected and kept for 24 h at room temperature to reach stability.

Table 2. The arrangement of experiments.

No.	Ultrasonic power (W)	Irradiation time (min)
Spec. 1	80	20
Spec. 2	80	30
Spec. 3	80	40
Spec. 4	80	50
Spec. 5	50	5
Spec. 6	80	5
Spec. 7	110	5

Saturated hydrocarbons before and after the ultrasonic treatment were separated by a chromatography column according to the Industrial Specification of China Petroleum Standard (SY/T 5119). Oil sample was dissolved into n-hexane and stood for 12 h to precipitate asphaltenes. After filtering out asphaltenes, the filtrate was separated by a chromatography column filled with the active alumina. Then, saturated hydrocarbons was prepared and ready for the next analysis.

The FTIR spectra of saturated hydrocarbons was recorded using FTIR-8400s (Shimadzu, Japan). GC-MS analysis of saturated hydrocarbons was carried out using GCMS-QP2010 Ultra (Shimadzu, Japan). In the GC-MS analysis, separations were accomplished using a 30 m × 0.25 mm × 0.25 μm Rxi-5ms column. The flow rate of Helium carrier was 0.92 mL/min. Saturates was injected in the GC through the injection port at

250 °C. The GC oven was kept at the temperature of 50 °C for 5 min and then kept the incremental rate of 3 °C/min until the temperature rose to 280 °C, at which the reaction was kept for 20 min.

3 Results and Discussion

3.1 Effect of Irradiation Time

Figure 2 shows the effect of irradiation time on the content of saturated hydrocarbons in the crude oil. As the irradiation time is prolonged, the content of saturated hydrocarbons does not decline or increase continuously. At 40 min, saturates content declines greatly by 32.5%. However, at 50 min, saturates content increases dramatically by 38.4% relative to that at 40 min. When saturates content is declining, it means that long-chain saturated compounds may be cracked, and new compounds may be formed. On the contrary, new saturated compounds would be formed when the irradiation time is 50 min.

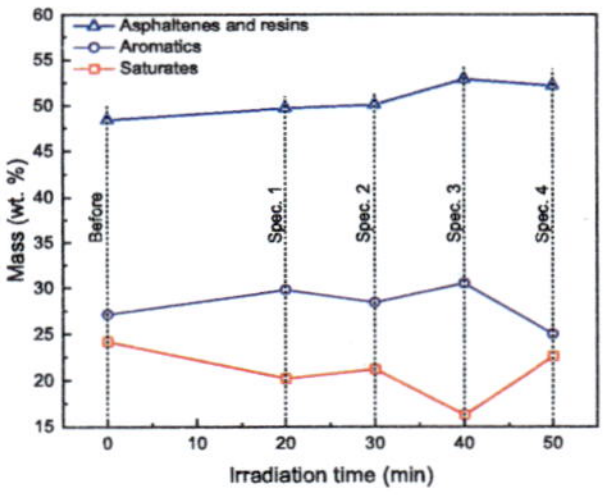

Fig. 2. The effect of irradiation time on the content of saturated hydrocarbons in crude oil.

There is an interesting phenomenon shown in Fig. 2 that the total content of asphaltenes and resins does not decline as the irradiation time is prolonged. At 40 min, the total content of asphltenes and resins increases by 9.2%. By comparing three curves in Fig. 2, it seems that the irradiation time has a little effect on the change of the total content of asphaltenes and resins, and the cracking and forming of compounds in crude oil would mainly occur at saturated and aromatic hydrocarbons in current experiments. It can be inferred that there should be a critical value of irradiation time for the cracking of asphaltenes and resins molecules.

Figure 3 shows the FTIR transmittance spectra of saturated hydrocarbons in oil samples under different irradiation time. On the transmittance spectra, the range 2500–3700 cm^{-1} is called the hydrogen-stretching zone, because the vibrations of C-H, N-H, and O-H appear in the frequencies in this region. The range 1000–1600 cm^{-1} is referred to as the fingerprint region, because various bonds such as C-O, C-N, C-C (single bonds), the C-H bending bond, and some bonds related to the benzene ring are located in this region, which is used to determine the type of functional group. The range 400–1000 cm^{-1} is aromatic region, it shows the aromatic bonds. In this study, peaks of the samples under different irradiation time are in the same position, but the shapes are different.

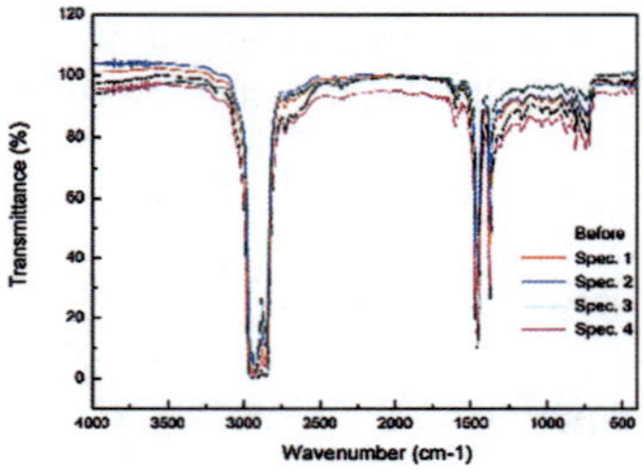

Fig. 3. FTIR transmittance spectra of saturated hydrocarbons in crude oil under different irradiation time.

Peaks at the frequencies of 2960 cm^{-1} and 2920 cm^{-1} represent the asymmetrical stretching of C-H bonds in methyl and methylene respectively. When the bonds are cracked during the treatment, one group of $-CH_2-CH_2-$ can be divided into two groups of $-CH_3$. Thus the change of chain length in heavy oil can be determined by the ratio value of infrared absorbance of methylene to methyl.

Table 4 shows the infrared absorbance of methylene and methyl. It is clear that the ratio value of infrared absorbance increases, and reaching up to 40.0% at 40 min (Spec. 3). This phenomenon implies that long-chain compounds may be formed during the treatment. However, at 50 min (Spec. 4), the ratio value declines by 22.6% relative to that at 40 min, which means that long-chain compounds may be cracked dramatically.

Table 3 shows the compound identification of saturated hydrocarbons under different irradiation time using GC-MS. It is clear that there are 36 kinds of saturated compounds in the crude oil before the treatment. These compounds include long-chain molecules and multi-branched molecules, such as n-tetracontane, hexatriacontane, 4,9,13-trimethyl octadecane and so on.

By comparing the treated samples to the untreated shown in Table 3, it is quite obvious that some long-chain molecules are cracked under ultrasonic wave, such as $C_{24}H_{50}$, $C_{29}H_{60}$, $C_{34}H_{70}$, $C_{36}H_{74}$ and $C_{40}H_{82}$. Some multi-branched molecules are also cracked, such as 7-hexylpentadecan, 4,9,13-trimethyl octadecane, 4,8,12-trimethyl heptadecane, 3,7,11-trimethyl hexadecane and so on. However, the irradiation time of ultrasonic wave seems to have little effect on some light compounds with branched chain, such as trans-3,4-dimethyl-2-pentene, 2,4-dimethyloctane and 4,6-dimethylnonane.

At 30 min, there are 15 kinds of saturated compounds. However, as the irradiation time is prolonged, the kind of compounds increases, and some new long-chain compounds are formed. By taking the change of the contents shown in Fig. 2 into consideration, the cracking of some aromatic compounds with long branched chain or radical molecules should have greatly effect on the forming of new saturated compounds.

3.2 Effect of Ultrasonic Power

When the irradiation time is 5 min, the effect of ultrasonic power on the contents in crude oil is shown in Fig. 4. As the ultrasonic power is improved, the content of saturated hydrocarbons declines, and reaching up to 18.4% at 110 W. However, at 80 W, there is a slight increase by 9.5% relative to that at 50 W. It can be inferred that there should be a

Table 3. Identification of saturated hydrocarbons in crude oil under different irradiation time.

Before		30 min (Spec. 2)	40 min (Spec. 3)	50 min (Spec. 4)
Formula	Compound			
C_7H_{14}	Methylcyclohexane	Methylcyclohexane	Methylcyclohexane	Methylcyclohexane
	Trans-3,4-Dimethyl-2-Pentene	Trans-3,4-Dimethyl-2-Pentene	Trans-3,4-Dimethyl-2-Pentene	Trans-3,4-Dimethyl-2-Pentene
C_7H_{16}	3-Methyl hexane			
	Heptane			
C_8H_{18}	2-Methylheptane	2-Methylheptane	2-Methylheptane	2-Methylheptane
	n-Octane	n-Octane		n-Octane
	2,4-Dimethylhexane	2,4-Dimethylhexane		2,4-Dimethylhexane
		4-Methylheptane	4-MethylHeptane	4-Methylheptane
C_9H_{20}	4-Methyloctane	4-Methyloctane		4-Methyloctane
$C_{10}H_{22}$	2,4-Dimethyloctane	2,4-Dimethyloctane	2,4-Dimethyloctane	2,4-Dimethyloctane
	3,4,5-Trimethyl heptane	3,4,5-Trimethyl Heptane	3,4,5-Trimethyl Heptane	
	3,5-Dimethyloctane	3,5-Dimethyloctane	3,5-Dimethyloctane	
	Decane	Decane	Decane	Decane
	2,5-Dimethyloctane			
	3-Methylnonane			
$C_{11}H_{24}$	4,6-Dimethylnonane	4,6-dimethylnonane	4,6-Dimethylnonane	4,6-Dimethylnonane
	n-Hendecane	n-Hendecane	n-Hendecane	n-Hendecane
$C_{12}H_{26}$	Dodecane	Dodecane	Dodecane	Dodecane
$C_{13}H_{28}$	4-Methyldodecane	4-Methyldodecane		4-Methyldodecane
	n-Tridecane		n-Tridecane	n-Tridecane
$C_{14}H_{30}$	Tetradecane		Tetradecane	Tetradecane
$C_{15}H_{32}$	n-Pentadecane		n-Pentadecane	n-Pentadecane
$C_{16}H_{34}$	n-Hexadecane-D34		n-Hexadecane-D34	n-Hexadecane-D34

critical value of ultrasonic power for the forming of saturated compounds. The optimum ultrasonic power should be 80 W in current experiments.

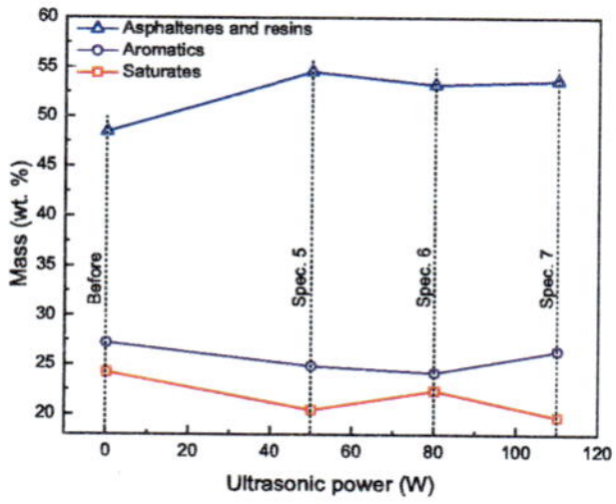

Fig. 4. The effect of ultrasonic power on the content of saturated hydrocarbons in crude oil.

By comparing three curves in Fig. 4, the contents of saturates and aromatics decline by 15.6% and 8.6% respectively when the ultrasonic power is 50 W, but the total content of asphaltenes and resins increases by 12.6%. The increase of the total content of asphaltenes and resins most likely results from the forming of new resin compounds by cracked saturates and aromatics. At 50, 80, and 110 W, variation value of the total content of asphaltenes and resins is below 1.3%, the reorganization of molecules seems to mainly occur between saturates and aromatics according to the shape of other two curves.

Figure 5 shows the FTIR spectra of saturated hydrocarbons in crude oil under different ultrasonic power, the infrared absorbance of methylene and methyl is shown in Table 4. At 50 W (Spec. 5), the ratio value of infrared absorbance of methylene to methyl increases by 7.6%, which implies that long-chain compounds may be formed during the treatment.

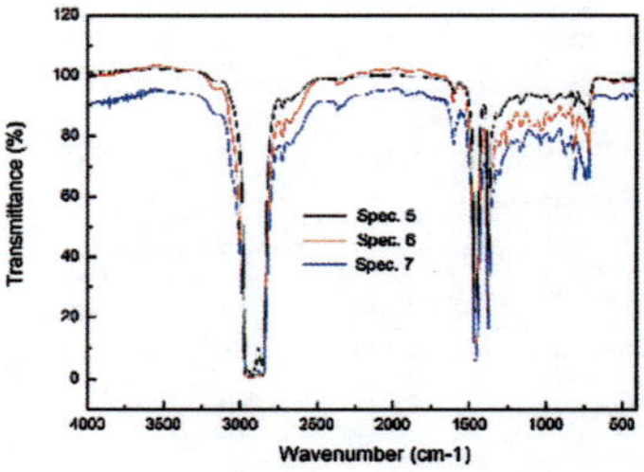

Fig. 5. FTIR transmittance spectra of saturated hydrocarbons in crude oil under different ultrasonic power.

However, the ratio value declines by 10.4% at 80 W (Spec. 6) relative to that at 50 W shown in Table 4, which means that long-chain compounds may be cracked by the ultrasonic wave. As the ultrasonic power is improved to 110 W (Spec. 7), the ratio value almost keeps unchanged, but the content of saturates declines by 11.7% and the content of aromatic hydrocarbons increases by 9.2% shown in Fig. 4.

Table 4. Infrared absorbance of methylene and methyl. Saturated hydrocarbons in crude oil under different ultrasonic power.

No.	Methylene at 2920 cm^{-1}	Methyl at 2960 cm^{-1}	Ratio value (Methylene/Methyl)
Before	2.604640	2.175643	1.1971817
Spec. 5	1.876900	1.457323	1.2879094
Spec. 6	2.515184	2.180242	1.1536261
Spec. 7	2.155365	1.870967	1.1520059

By using GC-MS analysis, compound identification of saturated hydrocarbons under different ultrasonic power is shown in compounds in Table 5. It is clear that there are two obvious features of the saturated compounds before and after the treatment. One is that long-chain compounds and some multi-branched chain compounds are cracked when the oil samples are exposed by ultrasonic wave, such as $C_{24}H_{50}$, $C_{29}H_{60}$, 4,9,13-trimethyl octadecane and so on. The other is that the kind of saturated compounds seems to be stable when the ultrasonic power is improved. Especially when the ultrasonic power is improved from 80 to 110 W, the kind of saturated compounds does not change.

Table 5. Identification of saturated hydrocarbons in crude oil under different ultrasonic power.

Before		50 W (Spec. 5)	80 W (Spec. 6)	110 W (Spec. 7)
Formula	Compound			
C_7H_{14}	Methylcyclohexane	Methylcyclohexane	Methylcyclohexane	Methylcyclohexane
	Trans-3,4-Dimethyl-2-Pentene	Trans-3,4-Dimethyl-2-Pentene	Trans-3,4-Dimethyl-2-Pentene	Trans-3,4-Dimethyl-2-Pentene
C_7H_{16}	3-Methyl hexane	3-Methylhexane	3-Methylhexane	3-Methylhexane
	Heptane	Heptane	Heptane	Heptane
C_8H_{18}	2-Methylheptane	2-Methylheptane	2-Methylheptane	2-Methylheptane
	n-Octane	n-Octane	n-Octane	n-Octane
	2,4-Dimethylhexane	2,4-Dimethylhexane	2,4-Dimethylhexane	2,4-Dimethylhexane
		4-Methylheptane	4-Methylheptane	4-Methylheptane
C_9H_{20}	4-Methyloctane	4-Methyloctane	4-Methyloctane	4-Methyloctane
$C_{10}H_{22}$	2,4-Dimethyloctane	2,4-Dimethyloctane	2,4-Dimethyloctane	2,4-Dimethyloctane
	3,4,5-Trimethyl heptane			
	3,5-Dimethyloctane			
	Decane	Decane	Decane	Decane
	2,5-Dimethyloctane			
	3-Methylnonane			
$C_{11}H_{24}$	4,6-Dimethylnonane	4,6-Dimethylnonane	4,6-Dimethylnonane	4,6-Dimethylnonane
	n-Hendecane	n-Hendecane	n-Hendecane	n-Hendecane
$C_{12}H_{26}$	Dodecane	Dodecane	Dodecane	Dodecane
$C_{13}H_{28}$	4-Methyldodecane	4-Methyldodecane	4-Methyldodecane	4-Methyldodecane
	n-Tridecane	n-Tridecane	n-Tridecane	n-Tridecane
$C_{14}H_{30}$	Tetradecane	Tetradecane	Tetradecane	Tetradecane
$C_{15}H_{32}$	n-Pentadecane	n-Pentadecane	n-Pentadecane	n-Pentadecane
$C_{16}H_{34}$	n-Hexadecane-D34	n-Hexadecane-D34	n-Hexadecane-D34	n-Hexadecane-D34
$C_{17}H_{36}$	n-Heptadecane	n-Heptadecane	n-Heptadecane	n-Heptadecane
$C_{18}H_{38}$	Octadecane			
$C_{19}H_{40}$	n-Nonadecane	n-Nonadecane	n-Nonadecane	n-Nonadecane
	3,7,11-Trimethyl hexadecane			
$C_{20}H_{42}$	n-Eicosane		n-Eicosane	n-Eicosane
	4,8,12-Trimethyl heptadecane			
$C_{21}H_{44}$	n-Heneicosane	n-Heneicosane	n-Heneicosane	n-Heneicosane
	7-Hexylpentadecane			
	4,9,13-Trimethyl octadecane			
$C_{24}H_{50}$	n-Tetracosane			
$C_{29}H_{60}$	n-Nonacosane			
$C_{34}H_{70}$	n-Tetratriacontane			
$C_{36}H_{74}$	Hexatriacontane			
$C_{40}H_{82}$	n-Tetracontane			

4 Conclusions

Saturated hydrocarbons with long-chain or multi-branched chain in heavy crude oil can be cracked by ultrasonic wave. Cracked molecules can form light saturates or aromatics. There should be a critical value of ultrasonic power or irradiation time for the cracking of asphaltenes and resins in crude oil. When the ultrasonic power is no more than 110 W or the irradiation time does not exceed 50 min, the cracking and structural reorganization of molecules mainly occur at heavy saturates and aromatics.

As irradiation time is increasing or the ultrasonic power is improving, long-chain compounds can be formed. However, after the critical values, long-chain compounds are cracked dramatically.

Saturated compounds in crude oil is sensitive to ultrasonic wave significantly, and the irradiation time seems to play a larger role than the ultrasonic power in changing the kind of saturated compounds.

References

1. Nasri, Z., Mozafari, M.: Multivariable statistical analysis and optimization of Iranian heavy crude oil upgrading using microwave technology by response surface methodology (RSM). J. Pet. Sci. Eng. **161**, 427–444 (2018)
2. Gogate, P.R.: Application of cavitational reactors for water disinfection: current status and path forward. J. Environ. Manag. **85**, 801–815 (2007)
3. Mehdi, R., Jafar, Q.: Experimental investigation of the ultrasonic wave effects on the viscosity and thermal behavior of an asphaltenic crude oil. Chem. Eng. Process. **153**, 1–11 (2020)
4. Gopinath, R., Dalai, A., Adjaye, J.: Effects of ultrasound treatment on the upgradation of heavy gas oil. Energy Fuels **20**, 271–277 (2005)
5. Abramov, V.O., Abramova, A., Bayazitov, V., Mullakaev, M.S., Marnosov, A.V., Ildiyakov, A.: Acoustic and sonochemical methods for altering the viscosity of oil during recovery and pipeline transportation. Ultrason. Sonochem. **35**, 389–396 (2017)
6. Shi, C., et al.: Application and mechanism of ultrasonic static mixer in heavy oil viscosity reduction. Ultrason. Sonochem. **37**, 648–653 (2017)
7. Huang, X., Zhou, C., Suo, Q., Zhang, L., Wang, S.: Experimental study on viscosity reduction for residual oil by ultrasonic. Ultrason. Sonochem. **41**, 661–669 (2018)
8. Taheri, J., Shekarifard, A., Naderi, H.: Analysis of the asphaltene properties of heavy crude oil under ultrasonic and microwave irradiation. J. Anal. Appl. Pyrolysis. **129**, 171–180 (2018)
9. Najafi, I., Amani, M.: Asphaltene flocculation inhibition with ultrasonic wave radiation: a detailed experimental study of the governing mechanisms. Adv. Pet. Explor. Dev. **2**, 32–36 (2011)
10. Taheri, J., Shekaifard, A., Naderi, H.: Experimental investigation of the asphaltene deposition in porous media: accounting for the microwave and ultrasonic effect. J. Pet. Sci. Eng. **163**, 453–462 (2018)
11. Mohapatra, D.P., Kirpalani, D.M.: Bitumen heavy oil upgrading by cavitation processing: effect on asphaltene separation, rheology, and metal content. Appl. Petrochem. Res. **6**(2), 107–115 (2016). https://doi.org/10.1007/s13203-016-0146-1
12. Rad, M.H., Tavakolian, M., Najafi, I., Ghazanfari, M.H., Taghikhani, V., Amani, M.: Modeling the kinetics of asphaltene flocculation in toluene–pentane systems for the case of sonicated crude oils. Sci. Iran. **20**, 611–616 (2013)
13. Avvaru, B., Venkateswaran, N., Uppara, P., Iyengar, S.B., Katti, S.S.: Current knowledge and potential applications of cavitation technologies for the petroleum industry. Ultrason. Sonochem. **42**, 493–507 (2018)
14. Kaushik, P., Kumar, A., Bhaskar, T., Sharma, Y.K., Tandon, D., Goyal, H.B.: Ultrasound cavitation technique for up-gradation of vacuum residue. Fuel Process. Technol. **93**, 73–77 (2012)

Random Assembly Task Evaluation Based on Human-Robot Collaboration

Jiaying Li[1], Haiyun Wang[2], Zenggui Gao[1](✉), Lilan Liu[1], and Changru Wang[2]

[1] Shanghai Key Laboratory of Intelligent Manufacturing and Robotics, Shanghai University, Shanghai, China
gaozg@shu.edu.cn

[2] Shanghai Robot Industrial Technology Research Institute, Shanghai, China

Abstract. Based on human-robot collaboration, this paper discusses the human factor assessment in independent assembly and cooperative assembly in random assembly task, and provides a theoretical basis for the extensive application of human-robot collaboration. An experimental method was proposed to obtain real-time eye movement data of eight participants by using an eye tracker. Then relevant eye movement parameters are analyzed. And human factors are comprehensively evaluated in both subjective and objective aspects by combining NASA-TLX Subjective Rating Scale. Compared with independent operation, human-robot collaboration has more advantages in random task and lower consumption in all aspects of humans.

Keywords: Human-robot collaboration · Eye tracking · Fixation time · Pupil diameter · NASA-TLX

1 Introduction

In the traditional production mode of complete separation of man and machine, robots are generally used to completely replace manual work. This kind of work is basically relatively simple and repetitive. However, for some complex or flexible tasks, robots still can't do it alone. With the development of artificial intelligence technology, it is emphasized in manufacturing that robots can cooperate and adapt to their environment to assist people to complete their work. "Intelligent" production is gradually replacing "automatic" production, robots and humans work together in workplaces without guardrail isolation, and human-robot collaboration is gradually possible. If you choose to use human-robot collaboration instead of pure labor or pure machine to complete tasks, the first premise is that the working mode of human-robot collaboration can be more efficient and accurate [1]. At the same time, as an important factor in the task process, people's comfort, fatigue and work efficiency should also be taken into consideration.

In this paper, an experimental method is proposed to test whether people face random tasks in a certain difficulty range, and compare and evaluate with NASA-TLX subjective rating scale and related eye movement behavior data in two scenarios with or without robot cooperation. Based on subjective and objective data, the influence of robot's participation on people's work performance and physiological load when they complete the assembly task is obtained through comprehensive analysis.

Y. Wang et al. (Eds.): IWAMA 2021, LNEE 880, pp. 135–140, 2022.
https://doi.org/10.1007/978-981-19-0572-8_18

2 Task Evaluation Based on Eye Tracking

2.1 Eye Movement Tracking Data

People's basic eye movements can be roughly divided into three types: fixation, saccade and smooth follow-up movement [2], This paper mainly focuses on the changes of total fixation duration, fixation times and pupil diameters of both eyes. In human factor analysis, eye tracking is a quantifiable measurement technology based on physiology, and eye movement data can objectively reflect people's concentration and fatigue during tasks.

The fixation duration refers to the sum of the residence time of all fixation points in an area of interest, which reflects the user's attention to the area of interest [3]. In the experiment, the fixation duration also reflects the difficulty of extracting task information. The more difficult it is to obtain and process information when performing tasks, the longer the corresponding gaze duration will be. Attention times refer to the total number of people's attention points in the region of interest corresponding to an object. Similar to the duration of gaze, the number of fixations can also reflect the user's attention to the object.

Pupil diameter is the size of the pupil diameter of the left and right eyes of a person. The change of pupil size is biologically related to the intensity of light and the distance of observing objects. When facing strong light or observing nearby objects, the pupil will shrink accordingly to reduce the light entering the eyeball, thus helping to distinguish objects, and vice versa. From a psychological point of view, pupil diameter can reflect the degree of psychological load in the cognitive process, that is, pupil enlargement often thinks that the degree of psychological load increases [4]. In addition, fatigue will also affect the pupil diameter, which is inversely related, that is, fatigue will cause the pupil diameter to shrink.

2.2 Experimental Methods

In this experiment, Tobii Glasses 2 eye tracker with 50 Hz sampling was used to collect the eye movement data of the subjects in real time. A total of 8 participants, with an average age of 22 years, all with bachelor degree or above. The experimental tasks are divided into cooperative tasks and independent tasks. Finally, the subjects are required to complete two sets of jigsaw puzzles with the same pat-tern at the same time. One set of jigsaw puzzles will be completed with the help of mechanical arm. The jigsaw puzzles will be randomly selected by the subjects be-fore the experiment begins. The task scene is shown in Fig. 1. In the experiment, 8 subjects wore eye movement instruments, cooperated with mechanical arm to complete random jigsaw puzzles and collected their eye movement data, with an average recording time of about 5 min.

2.3 Experimental Results

2.3.1 Fixation Duration and Fixation Times

The original data collected by the eye tracker includes many eye movement recognition parameters, such as saccade, fixation and blinking. Pre-process the original data, remove

Fig. 1. Desktop layout for collaborative task (red) and single task (blue)

the invalid recording time and invalid data, and finally keep the effective experimental time for about 2 min on average. The effective experimental time is distinguished and marked according to tasks, and the interest area is drawn. The differentiated data are processed and exported by "ErgoLAB".

The chart of fixation duration and fixation times is shown in Fig. 2 and 3 by analyzing the data of all subjects. In Fig. 2, the total fixation duration of 8 subjects on independent tasks is higher than that on cooperative tasks. In Fig. 3, in the process of independent task operation, the attention frequency of the ROI is obviously lower than that of the cooperative task, but the related data of other subjects are basically higher than that of the cooperative task. According to the chart analysis, when two tasks are parallel, the subjects tend to pay more attention to the corresponding interest areas of independent tasks. With the increase of attention time, the number of fixations also increases, that is, the task switching rate is higher. For the subjects, after randomly selecting tasks, their cognition only stays on the size, color and shape of the puzzle. In order to complete the puzzle task, we must analyze the extracted task patterns, form the cognition of tile splicing sequence and placement position in our mind, and then take the corresponding tiles for puzzle. Compared with collaborative tasks, independent tasks require the subjects to complete the cognitive process and hand operation independently, and the need for attention increases with the increase of task difficulty.

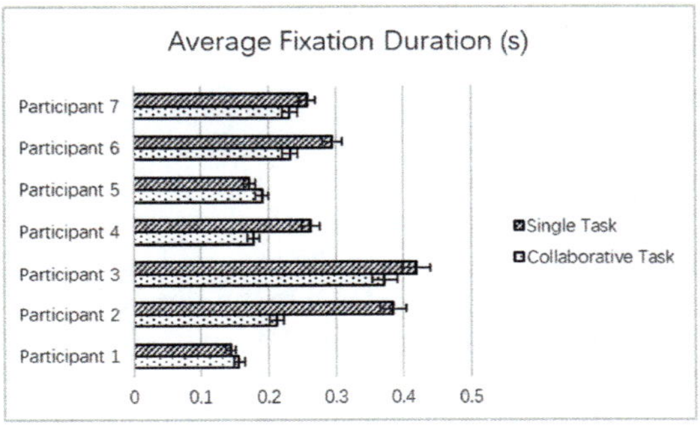

Fig. 2. Total fixation duration

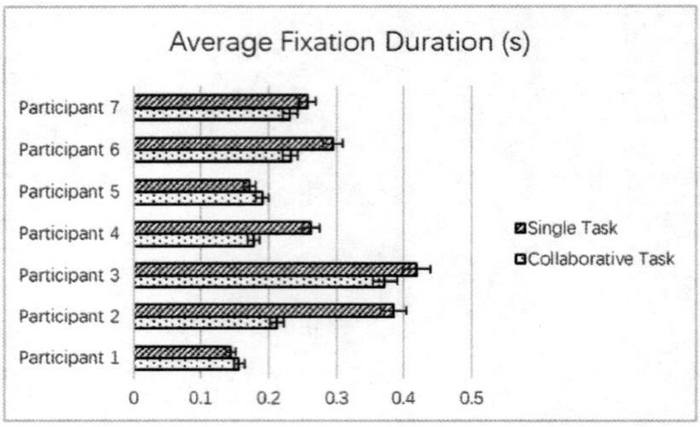

Fig. 3. Fixation count

3 Subjective Task Evaluation of NASA-TLX

NASA-TLX is used in this paper, which has a wide range of applications and good results. It is widely used in human performance level research experiments and can effectively evaluate subjective psychological load. This scale regards people's psychological load as a multidimensional evaluation scale, which is divided into six dimensions: psychological demand, physical strength demand, time demand, performance level, effort level and frustration level. After completing the above experiments, the subjects scored their performance and psychological feelings in the two tasks from six dimensions. Because the subjective assessment scale is filled by the subjects, it is necessary to express the meaning of scores of each dimension accurately in the scale to avoid the result error caused by the wrong interpretation.

3.1 Reliability of Scale

Through reliability analysis, the internal consistency among six factors of NASA-TLX, such as psychological demand, physical strength demand, time demand, performance level, effort degree and frustration degree, was calculated, and Cronbach α coefficient was 0.915. The result is greater than 0.9, which indicates that the data reliability of the scale is of high quality and can be used for further analysis (Table 1).

3.2 Analysis of Evaluation Results

After the experiment, the subjects filled out NASA-TLX to evaluate the psychological load degree of each dimension after the task was completed. The analysis results are shown in Fig. 4. It can be seen from the figure that the psychological needs of the subjects in performing independent tasks are significantly higher than those in cooperative tasks, that is, the mental level spent in completing the tasks is higher. The physical needs of these two tasks are very close, which is related to the simple jigsaw puzzle with upper limbs. According to the average scores of time demand, effort level and performance level, the participants all think that collaborative tasks can achieve higher performance level with less time and effort. With assistance, the degree of frustration felt during the

Table 1. Cronbach reliability analysis

Name	CITC	Alpha coefficient of deleted item	Cronbach α coefficient
Psychological need	0.600	0.920	0.915
Physical strength demand	0.976	0.867	
Time requirement	0.318	0.945	
Performance level	0.876	0.888	
Effort	0.883	0.884	
Degree of frustration	0.974	0.866	

task is relatively low. Combined with the data obtained from eye movement recognition, it is found that the evaluation results of NASA-TLX are consistent with the objective experiments.

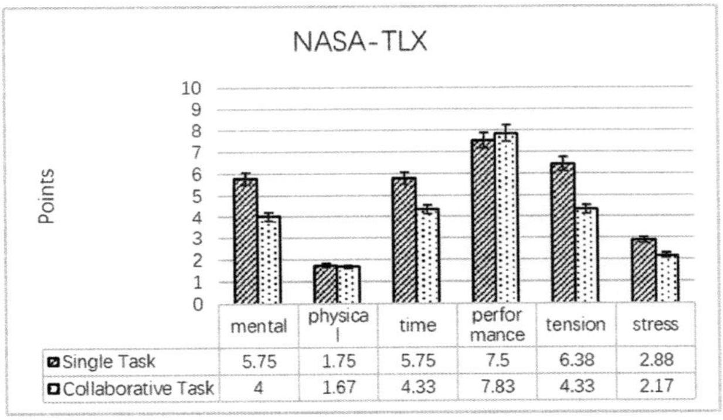

Fig. 4. Analysis of NASA-TLX results

4 Conclusion

In this paper, aiming at man-machine cooperation tasks, the human factors differences between independent tasks and man-machine cooperation tasks un-der random conditions are comprehensively compared and analyzed from objective and subjective aspects. By collecting eye movement data, the attention distribution of task-related interest areas is fully quantified by analyzing the fixation duration, fixation times and pupil diameter. Combined with NASA-TLX, the results of human factors evaluation in two task situations are finally obtained. It is proved that when people are faced with random tasks, man-machine cooperation can give better auxiliary work, which provides a theoretical basis for the wider application of man-machine cooperation in the future.

Acknowledgements. Project of Shanghai Robot R&D and Transformation Functional Platform and the 2nd batch of production-learning cooperation and collaborative education project by the Ministry of Education in 2020.

References

1. Wang, Q., Fan, X., He, Q., et al.: AR-based simulation interaction and human factor assessment for human robot cooperation assembly planning. J. Syst. Simul. **33**(02), 389–400 (2021)
2. Haber, R.N., Hershenson, M.: The psychology of visual perception. Leonardo **9**(1) (1973)
3. Hu, W.: Research on semantic representation of eye movement behaviour and application in image retrieval. Nanjing University (2016)
4. Yan, G.: Application of Eye Movement Analysis in Psychological Research, pp, 340–355. Tianjin Publishing House, Tianjin (2004)
5. Li, Y., Yin, G., Chen, Y.: The regulating effect of fatigue and mental load on pupil size in text reading. Stud. Psychol. Behav. (03), 545–548+560 (2004)

A Review on Application of Eddy Current Separation for the Recycling of Scraped Vehicles

Youdong Jia, Jianxiong Liu(✉), Hongshen Zhang, Jiaxing Zeng, and Mingjiang Jiang

Faculty of Mechanical and Electrical Engineering, Kunming University of Science and Technology, 727 Jingming South Road, Chenggong District, Kunming, Yunnan, China

Abstract. Since the 1970s, eddy current separation has experienced five development stages. Currently, it has been widely used for non-ferrous metal of end-of-life vehicles. This paper briefly summarizes the current development of scrapped vehicle recycling in China, meanwhile the research work of eddy current separation at home and abroad is comprehensively described, in the end, the application of eddy current separation is summarized and prospected.

Keywords: End-of-life vehicles · Eddy current separation · Non-ferrous metals

1 Introduction

With the growth of automobile production and consumption, the number of automobiles in China has reached 217 million by the end of 2017 [1]. At the same time, according to statistics, the number of scrapped cars will exceed 7 million in 2018 and 13 million in 2019, with a growth rate of more than 10%. At present, the scrapped rate of old cars in China is about 3%, and the recovery and dismantling rate is about 20%. The dismantling process is mainly based on the sale of scrapped metal [2]. The life of cars is generally about 8 to 10 years, so the recycling of scrapped cars is urgent.

Although China has many kinds of non-ferrous metal mineral resources and abundant reserves, the reserves of some important pillar minerals such as aluminum and copper account for a relatively low proportion of the world's total, only 3.9% and 2.3%, At the same time, due to the huge consumption, the external dependence of some non-ferrous metals is high. Aluminum and copper respectively reach 47% and 59% [3]. So effective non-ferrous metal recycling will greatly contribute to the sustainable use of resources.

At present, eddy current separator (ECS) is an important means for non-ferrous metals separation in China and most European and American countries. Although many industrial separators, such as metal separators and alloy separators have been developed, eddy current separation is still widely used as the main method of sorting non-ferrous metals. This paper provides a comprehensive overview of the state of the study of ECS both at home and abroad. This review mainly suggests future research directions in the field of ECS for recycling of end-of-life vehicles.

Y. Wang et al. (Eds.): IWAMA 2021, LNEE 880, pp. 141–147, 2022.
https://doi.org/10.1007/978-981-19-0572-8_19

2 The Study Status of ECS

When non-magnetic metal particles pass through a changing magnetic field, alternating eddy current will be generated inside the particle. The alternating eddy current will generate a new magnetic field changing in a new direction around the particle. The two magnetic fields are opposite in direction and repel each other. The process of separating mixed materials by eddy current force is called eddy current separation.

Eddy current separation is a kind of separation technology using material conductivity different, eddy current separation device is particularly suited to high conductivity and low density material with low conductivity high density material and the separation between the conductor and non-conductor, is now mainly used for wood, heavy plastic, rubber, paper and other separated from the hybrid mixture of non-ferrous metals [4].

2.1 The Current Study Situation Abroad

At present, the research of ECS abroad (mainly in European countries) mainly focus on the separation of fine granular materials, some universities and research institutes have designed a lot of novel structures, and mainly take experimental methods to observe the separation effect, exploring the potential of its industrial application. In addition, the research on the basic type of ECS, namely the drum type ECS, has not stopped, the main content of the research is the mechanical model of magnetic partial repulsion force of particles in ECS, and the separation efficiency under the composite influence of multiple factors. Recently, how to use computer to control product quality online in ECS has also aroused people's interest, and some scholars have started the preliminary research.

Based on the ECS principle, various eddy current separators are designed, and the process of ECS are put forward by the scholars abroad.

To reduce the cost of separators and increase the rate of recovery, following separators designed by Lungu, M, Lungu, M.; Schlett, Schlett, Z.; Claici, F.; Mihalca, I., Lungu, M, Meier-Staude, R.; Schlett, Z.; Lungu, M.; Baltateanu, D, Kohnlechner, R.; Schlett, Z.; Lungu, M.; Caizer, C and so on. Several articles also have been reported on the study of theory related to separating for recycling on ECS and are as follows:

Rem et al. [4] described the trajectory model of non-magnetic particles in a changing magnetic field through a first-order linear differential equation. Through the model, it is concluded that particle size, shape and electrical conductivity can affect the trajectory of particles. At the same time, the calculation equation of magnetic field intensity in the cylindrical coordinate (r, φ, z) of the vortex separator is given.

$$\mathrm{Br} = \sum_{n=0}^{2} b_n (r/R_{drum})^{-(2\mathrm{n}+1)\mathrm{k}-1} \sin(2n+1)k(\Phi - \omega_{drum} t) \quad (1)$$

$$\mathrm{B}\phi = \sum_{n=0}^{2} -b_n (r/R_{drum})^{-(2\mathrm{n}+1)\mathrm{k}-1} \cos(2n+1)k(\Phi - \omega_{drum} t) \quad (2)$$

$$B_z = 0 \quad (3)$$

Where k is the magnet logarithm of the separator, ω_{drum} is the rotating speed of the magnetic roller, b_n is the Fourier coefficient, R_{drum} is the radius of the magnetic roller, and n is the natural constant.

Zhang and Forssberg [6] studied the novel invented two-drum eddy current separator, they proposed the calculation model of tangential and radial forces of eddy current force on particles in the process of eddy current separation.

$$\mathrm{F_t} = \frac{2\pi sV}{\mu_0 w}\frac{(k\omega_{drum}+\Omega)\tau}{1+(k\omega_{drum}+\Omega)^2\tau^2}B_a^2 \tag{4}$$

$$\mathrm{F_t} = \frac{2\pi sV}{\mu_0 w}\frac{(k\omega_{drum}+\Omega)^2\tau^2}{1+(k\omega_{drum}+\Omega)^2\tau^2}B_a^2 \tag{5}$$

Where s is the factor of particle shape, V is the particle volume, τ is the particle conductivity, μ_0 is the vacuum permeability, Ba is the magnetic field intensity of the magnetic roller, k is the logarithm of the magnetic pole of the magnetic roller, ω_{drum} is the rotational speed of the magnetic roller, Ω is the angular velocity of the particle, and w is the width of a single magnetic pole in the magnetic roller.

Maraspin and Rem [6] proposed a torque calculation model for non-magnetic metal particles in the process of eddy current separation when they were simulated to calculate the horizontal projectile distance.

$$\mathrm{F_m} = \frac{(\mathrm{k}+1)\widehat{\mathrm{B}}^2\mathrm{V}}{\mu_0\mathrm{r}}\begin{pmatrix}\mathrm{R}_{(\xi)}\\ \mathrm{I}_\xi\end{pmatrix} \tag{6}$$

$$\mathrm{T_m} = -\frac{\widehat{\mathrm{B}}^2\mathrm{V}}{\mu_0}\mathrm{I}_{(\xi)} \tag{7}$$

$$\xi = \mu_0(\mathrm{k}\omega_\mathrm{m} + \Omega)\delta\mathrm{d}^2 \tag{8}$$

Where k is the number of pole, B is magnetic field intensity of the roller, r for radial distance from which particles to the center of the roller, V is the volume of a particle, including μ_0 for vacuum magnetic permeability, for the shape factor, as the magnetic roller speed, Ω for particle motion velocity, sigma for particles electrical conductivity, d for particles in vertical direction of the magnetic roller radial magnetic field the largest cross-sectional area.

Lungu and Schlett [7] proposed a calculation model of radial and tangential forces on particles when they were separated from copper and aluminum in optical cable by a newly developed eddy current separation device.

$$F_n = s\mu_0(\omega - \Omega)\sigma R^2 F_t \tag{9}$$

$$\mathrm{F_t} = \left(\frac{2\pi\mathrm{R}}{\lambda}\right)\frac{\mathrm{T}}{\mathrm{R}} \tag{10}$$

$$\mathrm{T} = \int_\mathrm{V} r\times f\,dV \tag{11}$$

$$f = j \times BdV \tag{12}$$

Where s is the shape factor of the particle, μ_0 is the vacuum permeability, Ω is the speed of the magnetic roller, Ω is the angular velocity of the particle motion, σ is the electrical conductivity of the particle, R is the maximum cross-sectional area perpendicular to the radial direction of the magnetic field of the magnetic roller, λ is the width of a single magnetic pole of the magnetic roller, V is the volume of the particle, r is the radial distance between the particle and the center of the magnetic roller. j is the eddy current intensity generated inside the particle, and B is the magnetic field intensity of the magnetic roller.

Zhang and Forssberg et al. [8, 9] analyzed the effects of different particles (particle size, shape, electrical conductivity) and feeding speed on material separation when using eddy current separation to recover aluminum materials from waste computers. When the particle size is less than 2–3 mm, the particles cannot be separated effectively. The effect of particle shape factors on eddy current sorting is irregular shape > spherical > slice, and the sorting effect of slice particle is the best. The sorting effect of pure metal particles is better than that of metal alloy particles, and the feeding method of overlapping particles will reduce the sorting effect of particles. In addition, when the particle size is small, the separation quality can be improved by strengthening the magnetic field strength and increasing the rotation speed of magnetic roller.

R. Meier-Staude [10] shows through experiments that translational acceleration is proportional to the particle surface, while angular acceleration is independent of particle size. This causes the small particles not to move in the direction of the band, but to start jumping over it or moving in the opposite direction. In order to increase the translational motion caused by particle rotation, in a horizontal drum at the bottom of the eddy current separator designed on the basis of a feeding type non-ferrous metal eddy current separator used for from a mixture of millimeter level of Al - Cu line separating aluminum, as shown in Fig. 4, also found that by increasing the friction coefficient between particles and surface vibration makes the separation effect is better. The purpose of separation is single, and the use has great limitation.

R. Meier-Staude [11] shows through experiments that translational acceleration is proportional to the particle surface, while angular acceleration is independent of particle size. This causes the small particles not to move in the direction of the band, but to start jumping over it or moving in the opposite direction. In order to increase the translational motion caused by particle rotation, in a horizontal drum at the bottom of the eddy current separator designed on the basis of a feeding type non-ferrous metal eddy current separator used for from a mixture of millimeter level of Al - Cu line separating aluminum, as shown in Fig. 2, also found that by increasing the friction coefficient between particles and surface vibration makes the separation effect is better. The purpose of separation is single, and the use has great limitation.

The static method used by Z. Schlett [12] for the separation of metal particles (2–3 mm) from a metal-plastic mixture is shown in Fig. 1. When the conductive particles are in the magnetic field of the coil run by the capacitor's discharge current, a force is generated so that the recovery rate of the particles in the mixture can reach up to 90%.But for the feeding requirements are high, only in the single particle through the effect can reach the best, the recovery efficiency is low.

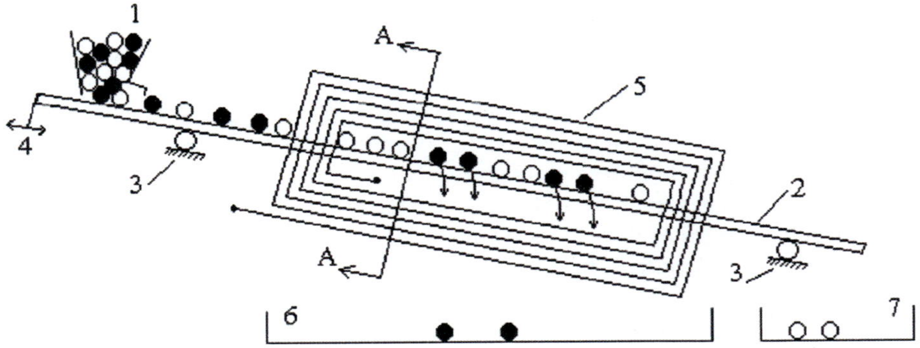

Fig. 1. Front view of static electromagnetic ECS

Magnetic top/cross belt invented by Bonifazi and Serranti [13]: Magnetic top/cross belt technology for partial separation of iron-containing wastes in mixed waste streams. The magnetic overhead zone has a magnetic field acting perpendicular to the direction of the mixed waste stream (Fig. 2). Thus, the sheet metal is attracted and removed by the mixed waste. The metal parts are then discharged into the collection box through a moving belt.

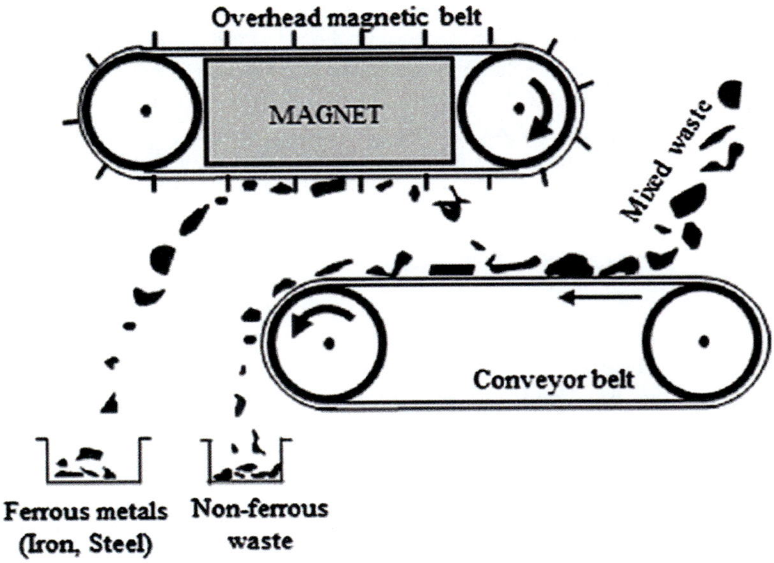

Fig. 2. Magnetic overhead belt sort

2.2 The Current Study Situation in China

When Dr. Jujun Ruan studied the recycling technology of waste toner cartridge and waste refrigerator box, he analyzed the force and movement of particles in the eddy current sorting in and out of the magnetic field. Ruan Jujun [13] made a detailed study on the calculation of eddy current force, and derived the calculation model of eddy current force on circular, rectangular and triangular waste particles in alternating magnetic field through formula. The repulsive force model based on the detachment Angle has a guiding significance for discarding the toner cartridge and the crushed refrigerator.

Zhang Dehao [14] didn't analysis the calculation model of eddy current force theoretically, but only by using the eddy current force calculation model and the test of waste separation, it is concluded that the magnetic field intensity, the waste electrical conductivity, magnetic roller speed, movement speed, the volume of waste and waste cross-sectional area will affect the waste in the process of sorting by eddy conclusion.

Jia Li et al. [74] made use of the classic usage of eddy current separation, that is, separation of light metals (aluminum) and plastics, and mainly studied the separation of plastics in printed circuit boards (PCB) and crushed mobile phones with ECS. A theoretical model of particle trajectories is established, which can be used to select separators for specific applications. On the basis of computer simulation, it was found that ECS could be used to separate PCB from plastic by specific parameters. With the help of software, using different feed belt speed (V), magnetic roller speed (U) and particle radius (RP) to analyze the optimal separation conditions and selected operating parameters, through MATLAB simulation of particle trajectory. The optimal separation conditions were obtained by model calculation ($V = 1.18$ m/s, $U = 3000$ RPM, $RP = 8.44$ mm).The experimental results show that the separation efficiency reaches 95.54%

Jianbo Wang et al. [75] made an exploratory research on the recycling of aluminum electrolytic capacitor (AEC) on printed circuit board (PCB). Under the optimal conditions, The total separation efficiency and aluminum grade are 87.9% and 98.25%, respectively. Among them, the two post-treatment links are screening, eddy current separation for >1.6 mm particle size, electrostatic separation for particle size <1.6 mm. This study provides a feasible way for waste AECS to be recycled in an efficient and environmentally friendly manner.

3 Conclusion

In China, eddy current separation mainly based on the introduction of foreign advanced equipment, and because of the difference of sorting materials, the sorting effect of the equipment is not good and single-function, currently, the independent ability of research and development of eddy current separation equipment is poor, and lacking in theoretical research, moreover, it seldom applied to the recycling for scrapped vehicles.

Eddy current separation has great advantages in the field of material recycling. With the increase of scrapped cars, electronic waste, household waste and so on in China, the recycling of non-ferrous metals and other materials is becoming more and more important.

References

1. Cui, X.: The Market is Expected to Exceed 100 Billion Yuan, 14 March 2018. [EB/OL]. http://www.chinanews.com/auto/2018/03-14/8467560.shtml
2. Liu, H.: The Regular Recovery Rate is less than 30%. Where do Millions of Cars Go Every Year. [EB/OL]. http://www.chinanews.com/cj/2018/05-02/8504395.shtml
3. Yang, B.: Analysis and discussion on dependency of overseas market and supply of China non-ferrous metals resources (2014)
4. Rem, P.C., Leest, P.A., van den Akker, A.J.: A model for Eddy current separation. Int. J. Miner. Process. **49**(3–4), 193–200 (1997)
5. Zhang, S., Rem, P., Forssberg, E.: Particle trajectory simulation of two-drum Eddy current separators. Resour. Conserv. Recycl. **26**(2), 71–90 (1999)
6. Schlett, Z., Lungu, M.: Eddy-current separator with inclined magnetic disc. Miner. Eng. **15**(5), 365–367 (2002)
7. Zhang, S., Forssberg, E.: Mechanical recycling of electronics scrap: the current status and prospects. Waste Manag. Res. **16**(2), 119–128 (1998)
8. Zhang, S., Forssberg, E., Arvidson, B., Moss, W.: Separation mechanisms and criteria of a rotating Eddy-current separator operation. Resour. Conserv. Recycl. **25**(3–4), 215–232 (1999)
9. Meier Staude, R., Schlett, Z., Lungu, M., et al.: A new possibility in Eddy-current separation. Miner. Eng. **15**(4), 287–291 (2002)
10. Schlett, Z., Claici, F., Mihalca, I., et al.: A new static separator for metallic particles from metal plastic mixtures, using Eddy currents. Miner. Eng. **15**(1–2), 111–113 (2002)
11. Bonifazi, G., Serranti, S.: Encyclopedia of Sustainability Science and Technology (2012)
12. Ruan, J.: Study on eddy current separation and its engineering application for recovery crushed waste toner cartridges and refrigerator cabinets (2012)
13. Zhang, D.: Study on the Technology of Eddy Current Separation for Scrap Non-Ferrous Metal (2016)
14. Li, J., Jiang, Y., Xu, Z.: Eddy current separation technology for recycling printed circuit boards from crushed cell phones. J. Clean. Prod. **141**, 1316–1323 (2017)

Influence of Magnetic Fluid Hydraulic Medium on Cushioning Performance of Hydraulic Cylinder

Jiajun Jiang[1], Zhangyong Wu[1](✉), Hua Li[2], Qichen Zhu[1], and Ziyong Mo[1]

[1] Faculty of Mechanical and Electrical Engineering, Kunming University of Science and Technology, Kunming, China
zhyongwu63@163.com

[2] City College, Kunming University of Science and Technology, Kunming, China

Abstract. Improving the cushioning performance of hydraulic cylinder has always been the focus of research in hydraulic field. Magnetic fluid is a kind of functional nano-fluid and has the advantages of viscosity control, good flow performance, good lubrication performance and hard to settle, it has a positive prospect to use magnetic fluid as transmission medium in hydraulic field. In this paper, magnetic fluid is used as hydraulic medium to improve the cushioning performance of hydraulic cylinder. $Ni_{0.5}Zn_{0.5}Fe_2O_4$ magnetic fluid hydraulic medium was prepared and its magneto-viscous properties were tested. The Amesim software is used to simulate the cushioning process of the hydraulic cylinder. The results show that: when the mass fraction of magnetic particles is 2wt%, the viscosity of magnetic fluid hydraulic medium can be increased by about 3 times in the range of 0–80 mT magnetic induction intensity. Magnetic fluid hydraulic medium has the ability to improve the cushioning performance of hydraulic cylinder with applied magnetic field.

Keywords: Hydraulic cylinder · Cushioning performance · Magnetic fluid · Magneto-viscosity · Hydraulic medium

1 Introduction

Hydraulic technology is an important transmission and control technology in the manufacturing industry. Therefore it is of great practical significance to develop a more reliable, efficient and intelligent hydraulic system. Hydraulic cylinder is the executive component of the hydraulic system, its performance will directly affect the quality of manufacturing and processing, so it is very important to improve the performance of hydraulic cylinder [1–3]. When the working stroke of the hydraulic cylinder reaches the terminal, it usually produces shock and makes the work unstable. To improve the cushioning performance of hydraulic cylinder, many researchers have studied the cushioning characteristics of the cylinder.

Antonio Algar et al. [4] studied 3D movement in cushioning of the hydraulic cylinder by experiments. During the experiment, the radial displacement was measured by an

Y. Wang et al. (Eds.): IWAMA 2021, LNEE 880, pp. 148–157, 2022.
https://doi.org/10.1007/978-981-19-0572-8_20

eddy current sensor assembled in the piston and the cushioning characteristics of two hydraulic cylinder cushioning forms are also discussed. The experimental result shows that the 3D displacement of piston plays a significant role as a self-adjusting element.

Li et al. [5] established a model for the cushioning of linear actuator with a cushioning system of conical geometry. The author associated the over pressure with cushioning, relating actuator's position as a function during three successive stages. The study combines an equation of the actuator to simulate its functional behavior.

Borghi et al. [6] performed a study of mechanical cushioning systems at the end of stroke of hydraulic linear actuators, analyzing various geometries (cylindrical, conical, double conical, etc.). The study established a simulation mathematical model, which had been validated through experimental results.

With the development of nanotechnology, applying nanomaterials to improve the efficiency of advanced manufacturing equipment becomes realistic. Magnetic fluid is a typical type of smart functional nano-fluid with both magnetism and fluidity. It can be regarded as Newtonian fluid without magnetic field. When a magnetic field is applied, its viscosity will increase, transforming into bingham fluid characteristics [7]. Different from magneto-rheological fluid (MR fluid), magnetic fluid consists of nanoscale magnetic particles with a diameter of 1–100 nm, which are stable and dispersed in the base fluid. It has the characteristics of good flow significant, controllable viscosity, outperforming lubrication performance and hard to settle. Accordingly, the magnetic fluid basically meets the requirement of the hydraulic system to the transmission medium. Therefore, many research scholars focus on the application of magnetic fluid to the hydraulic system in recent years:

Hu et al. [8] studied the viscosity characteristics of $NiFe_2O_4$ magnetic fluid and proposed that $NiFe_2O_4$ magnetic fluid should be applied to the hydraulic medium to solve the wear problem of vane pump to a certain extent.

Zhu et al. [9] used a two-step method to prepare nano magnetic fluid hydraulic medium. The research compared and analyzed magnetic fluid components, temperature, magnetic field, pH and other factors, and concluded that the magnetic fluid with zero magnetic field and magnetic field conditions had better sedimentation stability under different temperatures. The research demonstrated the feasibility of magnetic fluid as a hydraulic medium in viscosity.

Mohammadfam et al. [10] studied the rheological properties of hydraulic oil-based Fe_3O_4 magnetic fluid. The results revealed that the viscosity of the magnetic fluid increased as the concentration of nanoparticles increases, the magnetic fluid changed from Newtonian fluid to non-Newtonian fluid and showed expansion characteristics.

It is found that above researchers have systematically studied the cushioning characteristics of the traditional hydraulic cylinder, which used hydraulic oil as the hydraulic medium. And some scholars prepared magnetic fluid hydraulic medium and examined its physical and chemical characteristics. However, few scholars have researched the influence of magnetic fluid as hydraulic medium on the performance of hydraulic components. The present research, therefore, attempts to provide a cushioning structure of hydraulic cylinder based on magnetic fluid hydraulic medium, and its cushioning characteristics are studied by simulation.

2 Magneto-Viscous Characteristics of Magnetic Fluid Hydraulic Medium

2.1 Preparation of Magnetic Fluid Hydraulic Medium

2.1.1 Materials

Nanoscale $Ni_{0.5}Zn_{0.5}Fe_2O_4$ particles (Grain diameter: 20–30 nm); Oleic acid (OA); Kunlun L-HM68 anti-wear hydraulic oil.

2.1.2 Instruments and Equipment

Electronic analysis balance, JH-100 digital constant temperature control electric stirrer, KQ ultrasonic dispersion instrument.

2.1.3 Preparation

The magnetic fluid hydraulic transmission medium was prepared by using Kunlun L-HM68 anti-wear hydraulic oil as the base fluid, Nanoscale $Ni_{0.5}Zn_{0.5}Fe_2O_4$ particles as magnetic particles and oleic acid as the dispersing agent. The preparation process is described as follows: The first step was to heat the base liquid to 60 °C, followed by the addition of oleic acid. After stirring the mixed liquid for 30 min, added 2wt% $Ni_{0.5}Zn_{0.5}Fe_2O_4$ particles to the container and stirred it for another 30 min. Lastly, dispersed the fluid for 10 min by using Ultrasound.

The magnetic hydraulic medium with particle dosage of 2wt% were prepared, as shown in Fig. 1. Scanning electron microscopy (SEM) was used to analyze the coating conditions, as illustrated in Fig. 2. As can be seen from the figure, the prepared fluid has uniform particle size, no agglomeration and good coating performance. In addition, the magnetic fluid hydraulic medium has no settlement after standing for 14 days and no obvious settlement after standing for 21 days, indicating its good stability.

Fig. 1. Magnetic fluid hydraulic medium sample

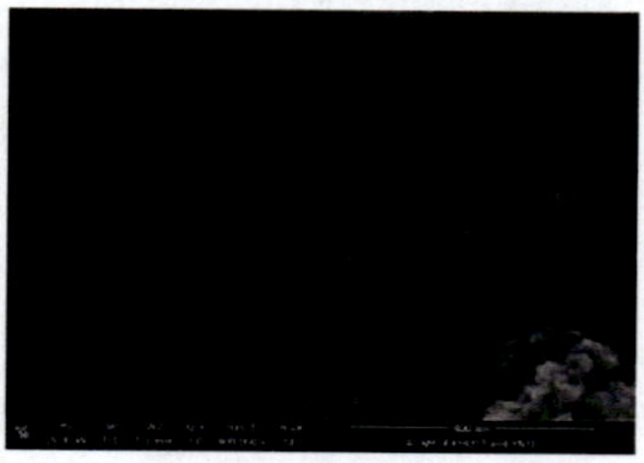

Fig. 2. SEM image

2.2 Magneto-Viscous Properties Test

2.2.1 Test Method

Figure 3(a) shows the testing method of magnetic induction intensity. The magnetic induction intensity is controlled by changing the current input of the electromagnet,

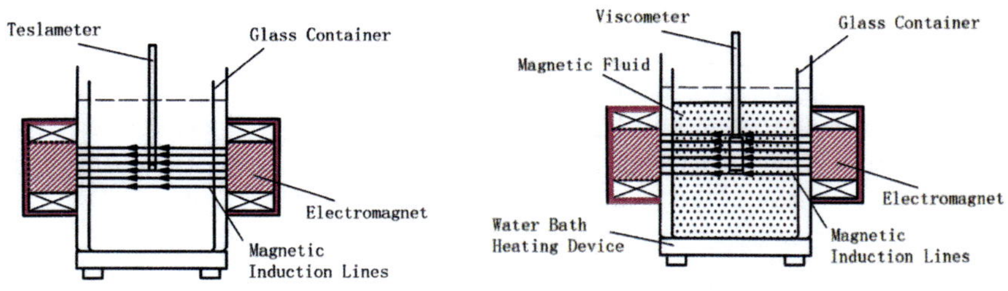

(a) (b)

Fig. 3. Test method for magneto-viscous properties

and the magnetic induction intensity is measured by Tesla meter. Figure 3(b) shows the magneto-viscous characteristics test method. The magnetic induction intensity is controlled by changing the current input of the electromagnet. The viscosity of the magnetic fluid hydraulic medium is tested by using a rotating viscometer. The probe of the tesla meter must be located in the area where the magnetic field is most concentrated as shown.

2.2.2 Results

Based on the above experimental method, the magneto-viscous properties of pure hydraulic oil and 2wt% magnetic fluid hydraulic medium at 40 °C were tested. The result is shown in Fig. 4.

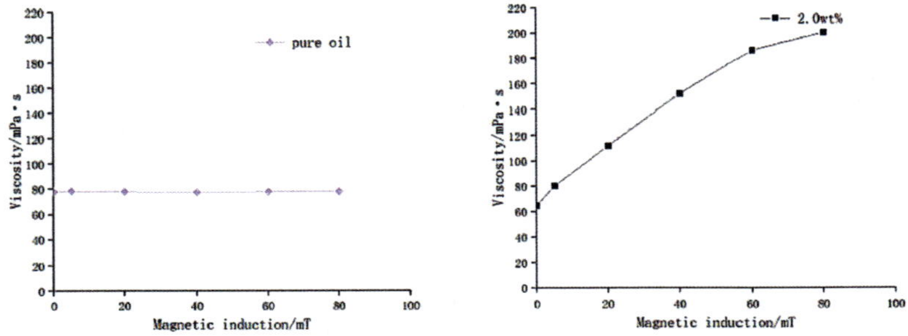

(a) (b)

Fig. 4. Magneto-viscosity curve

As can be seen from Fig. 4(a), the viscosity of pure hydraulic oil oil does not vary with the magnetic induction intensity, which is in line with the reality. And the dynamic viscosity is 80 mPa·s, which is consistent with the viscosity of 68# hydraulic oil at 40 °C.

In Fig. 4(b), when the magnetic induction intensity is 0, the viscosity of the magnetic fluid hydraulic medium is 65 mPa·s, slightly less than that of the pure hydraulic oil, which is caused by the addition of oleic acid with low viscosity. With the increase of magnetic induction intensity of applied magnetic field, the viscosity of magnetic fluid hydraulic medium increases, reaching 200.5 mPa·s at 80 mT.

It can be shown that the prepared magnetic fluid hydraulic medium has good viscosity control performance, and its viscosity can reach 3 times of the initial viscosity at 80 mT. It can meet the viscosity of the hydraulic cylinder requirements.

3 Cushioning Structure and Principle

A hydraulic cylinder cushioning device with electromagnet is proposed, as shown in Fig. 5.

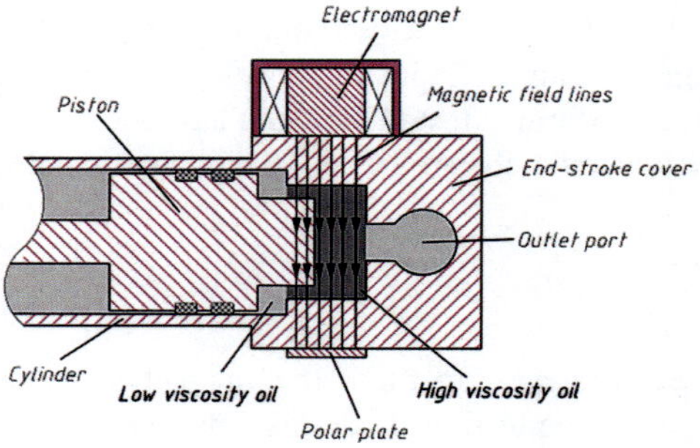

Fig. 5. Cushioning structure of the magnetic fluid hydraulic cylinder

Its working principle is: the magnetic fluid hydraulic medium enters the inlet of the hydraulic cylinder and pushes the piston towards the terminal of the hydraulic cylinder. When the piston is far away from the cushioning chamber, the electromagnet is not energized. At this time, the viscosity of the magnetic fluid hydraulic medium is lower than that of the pure hydraulic oil, which reduces the viscous resistance of the piston and plays an energy-saving effect. When the piston is close to the end-stroke of the cylinder (Fig. 5), the sudden decrease of the cross-sectional area of the hydraulic cylinder, resulting in throttling cushioning effect. At the same time, the electromagnet generates a constant magnetic field that passes vertically through the cushioning chamber, the viscosity of the magnetic fluid hydraulic medium increases under the action of the magnetic field. A high viscosity oil region is formed in the cushioning chamber, as shown in Fig. 5. The flow continuity equation in this process is as follows [11]:

$$Q = \frac{\pi d \delta^3 \Delta p}{12\mu(l - x)} \tag{1}$$

Where, μ—dynamic viscosity of hydraulic medium
l—stroke length of hydraulic cylinder
x—piston displacement
δ—gap between cushion spear and cushion chamber

d—diameter of cushioning chamber (m)

It can be seen from the Eq. (1) that the flow rate Q is related to the dynamic viscosity μ of the hydraulic medium. As μ increases, the flow rate Q will decrease. The the velocity of the piston Vp is related to the flow rate Q as follows:

$$V_p = \frac{Q}{A_0} \tag{2}$$

Where, A_0—piston sectional area

As can be seen from Eq. (2), a decrease in Q will lead to a decrease in the velocity of the piston Vp. From the above relationship, it can be concluded that the viscosity change of the magnetic fluid hydraulic medium can hinder the velocity of the piston at end-stroke of cushioning. Therefore, by increasing the viscosity of the magnetic fluid hydraulic medium, the velocity of the piston at the cushion terminal can be reduced, thus reducing the impact energy.

4 Simulation

As the applied magnetic field changes, the viscosity of the magnetic fluid changes. And the density of magnetic fluid hydraulic medium will be greater than pure hydraulic oil with the addition of magnetic particles. This paper utilizes Amesim software to build a simulation model of the hydraulic cylinder, resulting in the cushioning characteristic curves of the hydraulic cylinder through setting the viscosity and density parameters of the hydraulic medium.

4.1 Simulation Model and Parameters Setting

According to the hydraulic cylinder studied in this paper, a simulation model is established, as shown in Fig. 6.

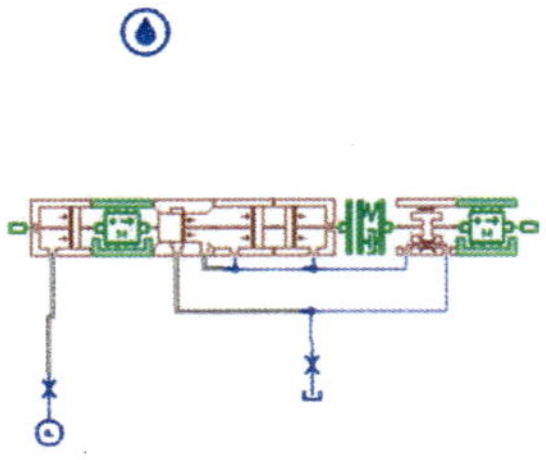

Fig. 6. Simulation model

Parameters of the model are set in Amesim according to the designed size of the hydraulic cylinder. The main parameters are shown in Table 1.

Table 1. Dimension parameters

The parameter names	Values
Pure hydraulic oil density/kg·m^{-3}	875
MF hydraulic medium density/kg·m^{-3}	892
Piston diameter/mm	40
Cushioning spear diameter/mm	30
Length of cushion spear	10
Annular clearance of cushioning chamber/mm	0.3
Piston rod diameter/mm	20
Piston and piston rod quality/kg	2.1
Intake pressure with rod cavity/MPa	2.0
Hydraulic cylinder stroke/mm	230
Diameter of oil inlet/mm	10
Diameter of oil outlet/mm	10

4.2 Results and Discussions

In order to accurately calculate the influence of the prepared magnetic fluid hydraulic medium on the cushion performance of the hydraulic cylinder, parameters must be set according to the measured experimental data in Fig. 4. When the magnetic induction intensity generated by the electromagnet at 80 mT, the viscosity of the magnetic fluid hydraulic medium reaches the maximum value of 200.5 mPa·s, while the viscosity of the pure hydraulic oil remains at 80 mPa·s. Therefore, the viscosity of the hydraulic medium in the simulation model was set to 80 mPa·s and 200.5 mPa·s, and the cushioning characteristics curves were obtained by solving.

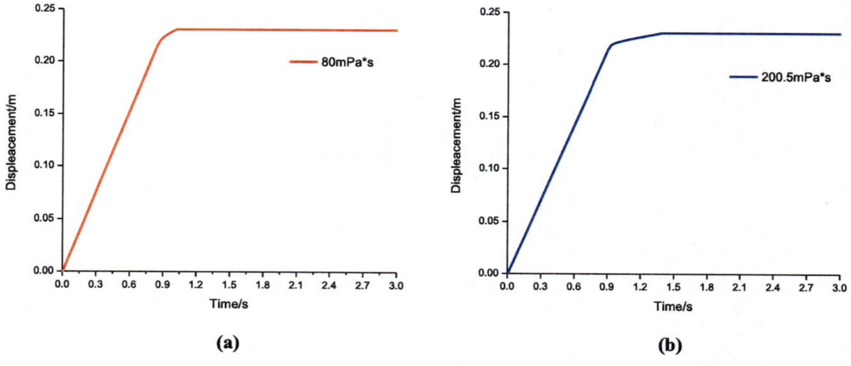

Fig. 7. Time-displacement curves

Figure 7 are time-displacement curves. It can be seen that the displacement has a linear relationship with time within the range of 0–220 mm. This is because before the cushion spear moved into the cushion chamber, the piston is subjected to the steady pressure of the hydraulic medium and does uniform motion. The slope decreases within a displacement of 220–230 mm. This is because when the stroke of the piston reaches 220 mm, the cushion spear enters the cushion chamber and is slowed down by the anti-force. In addition, the slope of the curve in Fig. 7a is greater than that in Fig. 7b within the displacement range of 220–230 mm. This means that Fig. 7(b) has a smoother cushioning process. This shows that the application of magnetic fluid hydraulic medium can effectively improve the cushioning performance. The purpose of cushioning is to reduce the velocity of the piston when it runs to the end-stroke of hydraulic cylinder, so as to reduce the impact kinetic energy. Therefore, the time-velocity curves can better reflect the cushioning effect.

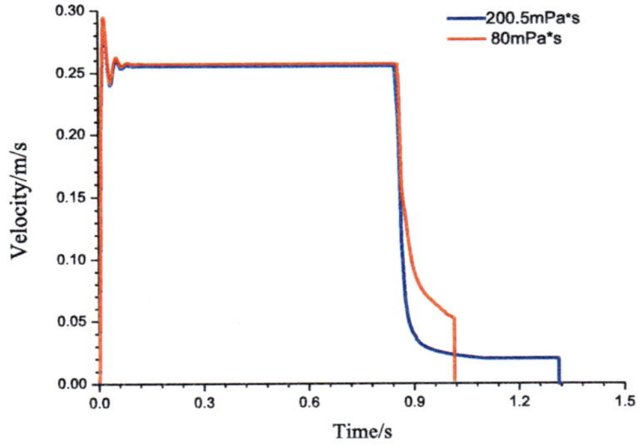

Fig. 8. Time-velocity curves

Figure 8 shows the time-velocity curves. As can be seen from Fig. 8, the downward trend of the blue curve is more stable than that of the red curve. At the end of each curve, the velocity drops precipitously to 0 because the piston reaches the end at this time. It can be seen from Fig. 8. that the terminal cushioning velocity of the blue curve is 0.021 m/s, while the terminal cushioning velocity of the red curve is 0.051 m/s. It shows that increasing the viscosity of the hydraulic medium can improve the cushioning performance of the hydraulic cylinder under the same geometry. And the high viscosity magnetic fluid hydraulic medium under the action of magnetic field can significantly reduce the terminal velocity of hydraulic cylinder,resulting in the reduction of impact kinetic energy.

Through the above analysis, we can conclude that compared with the traditional hydraulic oil, the application of magnetic fluid hydraulic medium can can lead to higher productivity of the hydraulic cylinder cushioning. This paper provides a further reference for the application of magnetic fluid as hydraulic medium within the hydraulic fileds. Next, the research group will prepare more components of the magnetic fluid hydraulic medium and study its magneto-viscous control characteristics, focusing on the establishment of the magnetic fluid hydraulic cylinder cushioning model under multi-field coupling, and verify the accuracy of the model through experiments.

5 Conclusions

Based on the aforementioned analysis, 4 conclusions could be drawn:

(1) An $Ni_{0.5}Zn_{0.5}Fe_2O_4$ magnetic fluid hydraulic medium was prepared. Its viscosity increases with the increase of magnetic induction intensity. When the magnetic induction intensity is 80 mT, the viscosity reaches 200.5 mPa·s.
(2) This paper proposed a hydraulic cylinder cushioning structure with an electromagnet and briefly indicates its working principles. The mechanism of improving the cushion of hydraulic cylinder by magnetic fluid hydraulic medium is also introduced in this paper.
(3) The cushioning process of hydraulic cylinder is simulated and analyzed by AMESim software. The simulation results prove that the control characteristics of the viscosity of magnetic fluid hydraulic medium can improve the cushioning performance of hydraulic cylinder. When the viscosity of the magnetic fluid hydraulic medium is limited at 200 mPas, the impact velocity is reduced by 58.8%, which can effectively reduce the impact energy of the hydraulic cylinder running to the terminal.

The above conclusions and implications may lead to following suggestions: (1) Experiments can further verify the improvement of the cushioning performance of the hydraulic cylinder by the magnetic fluid hydraulic medium; (2) Under the action of the magnetic field, the viscosity of magnetic fluid is anisotropic. In the future, the research group will also establish the viscosity characteristics of magnetic fluid hydraulic medium under multi-field coupling, to describe the viscosity change of magnetic fluid more accurately; and (3) In order to obtain the kinematic characteristics of the magnetic fluid hydraulic cylinder more accurately, it is necessary to study the lubrication performance of the magnetic fluid hydraulic medium.

Acknowledgment. This research was financially supported by the National Natural Science Foundation of China (No. 51165012) and Scientific Research Foundation of Education Department of Yunnan Province (No. 2015Y080).

References

1. Ding, R., Wang, S., et al.: Analysis and optimization of inside-cushion structure in high-speed hydraulic cylinders. J. Beijing Inst. Technol. **28**(4), 799–810 (2019)
2. Feng, J., Gao, Q., et al.: Modeling of telescopic hydraulic cylinder and research on inter-stage buffer. Acta Armamentarii **37**(12), 2268–2276 (2016)
3. Wang, C., Wang, Y., et al.: Comparative study on dynamics performance for cushion process of hydraulic cylinder. Chin. Hydraulics Pneumatics (10), 89–95 (2017)

4. Algar, A., Codina, E., et al.: Experimental study of 3D movement in cushioning of hydraulic cylinder. Energies **10**(6), 746 (2017)
5. Li, Q., Ding, F.: Novel displacement eddy current sensor with temperature compensation for electrohydraulic valves. Sens. Actuators Phys. **122**, 83–87 (2005)
6. Borghi, M., Milani, M.: Mechanical cushion design influence on cylinder dynamics. SAE technical paper 2005-01-3631. SAE International, Warrendale (2005)
7. Pei, L., Xuan, S., et al.: Influence of interparticle friction on the magneto-rheological effect for magnetic fluid: a simulation investigation. Smart Mater. Struct. **29**(11), 115002 (2020)
8. Hu, Z., Wu, Z., et al.: Experimental investigations on viscosity properties of $NiFe_2O_4$ magnetic fluid. J. Eng. Thermophys. **39**(6), 1205–1212 (2018)
9. Mohammadfam, Y., Heris, S.Z., et al.: Experimental investigation of Fe_3O_4/hydraulic oil magnetic nanofluids rheological properties and performance in the presence of magnetic field. Tribol. Int. **142**, 105995 (2020)
10. Sahin, H., Wang, X., et al.: Magneto-rheological fluid flow through complex valve geometries. Int. J. Veh. Des. **63**(2–3), 241–255 (2013)
11. Fan, C.: Study on the buffering process of a new hydraulic cylinder cushioning device. Lanzhou University of Technology (2015)

DEM-CFD Simulation for Recovering Metal Particles with Aero-Electrostatic Separator

Xiangnan Bao[1,2], Zi-qiang Zhou[3](✉), and Guo-hong Dai[4]

[1] Jiangsu Key Laboratory of Recycling and Reuse Technology for Mechanical and Electronic Products, Changshu, China
[2] School of Mechanical Engineering, Jiangsu University of Technology, Changzhou, China
[3] School of Mechanical Engineering, Changshu Institute of Technology, Changshu, China
[4] School of Mechanical Engineering, Changzhou University, Changzhou, China

Abstract. Electro-static separator is most common used in the recycling industry. The particles feed into the Electro-static separator needed to be shredded small enough in order to improve the sorting effect. Not only dust pollution will be caused, but also the energy consumption is evident. The approach of air Electro-static separating help to overcome this problem. The parameters of this kind separator should be selected reasonably. In this paper, the discrete element method and finite element method are used to simulate the separating process. And then, the parameter of air flow should be optimized for certain charge-mass ratio of particle. This is help to determine the control parameters of equipment.

Keywords: Waste printed circuit board · DEM · FEM · Simulation · Electro-static

1 Introduction

The disposal and recycling of waste circuit boards is an important issue in circular economy. There are many ways to deal with waste circuit boards, among which the more efficient and environmentally friendly method is high-voltage electrostatic separation. The sorting process of particles in a high-voltage electric field depends on the balancing process of the forces acting on each particle [1]. Xu used high-voltage electrostatics theory and numerical analysis method to analyze the trajectories of conductive particles under the electric field generated by the coupling of corona electrode, high-voltage static electrode and grounding electrode [2, 3]. Li [4] has studied the new technology of mechanical crushing, screening, drying and corona discharge electrostatic separation.

Although electrostatic separation has been applied to the separation of waste circuit board particles, it has very high requirements for particle size. The particle size needs to be micron in order to have a better sorting effect. In addition to causing greater energy consumption, it will also cause a certain amount of dust pollution [5].

In this study, on the basis of electrostatic separation and wind device, larger particle size can be sorted, which can reduce the decline rate of larger particles, and improve the looseness of particles, so that they can participate in the sorting to the maximum extent.

Y. Wang et al. (Eds.): IWAMA 2021, LNEE 880, pp. 158–165, 2022.
https://doi.org/10.1007/978-981-19-0572-8_21

2 Principle of Air Flow Electrostatic Separation

The air flow electrostatic separator is mainly composed of three parts: charging device, sorting area and collection area. As shown in Fig. 1, the electrostatic electrode of the sorting area is connected with 30 kV DC. The composition of waste printed circuit board particles is complex, mainly divided into metal particles and non-metal particles. The charging device is based on the mechanism of continuous collision charging, and realizes charge transfer through the friction between particles and tube wall [6–8]. Due to different material characteristics, when particles pass through the charging device, metal particles and non-metal particles are fully charged, and metal is positively charged, while non-metal is negatively charged. According to the motion of charged metal particles and non-metal particles in uniform electric field, the charge of metal particles after charging is 3.2 × 10-7C, and that of non-metal particles after charging is - 1.6 × 10-7C [9].

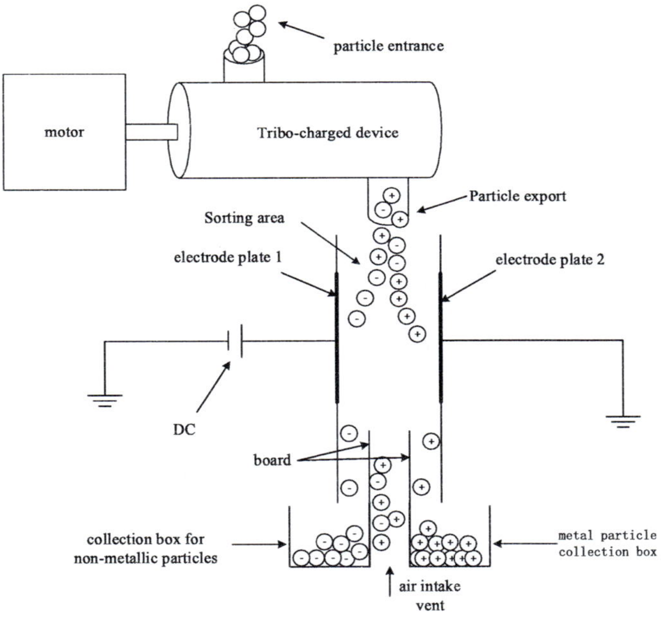

Fig. 1. Airflow-electrostatic separation system diagram

3 Theoretical Model

3.1 Mathematical Model of Particle Phase Trajectory

In this study, the change time of particle velocity is much longer than Stokes relaxation time, so Bassettforce can be ignored. The movement of particles in the separation chamber can be divided into two stages. The first stage is the stress of particles without passing

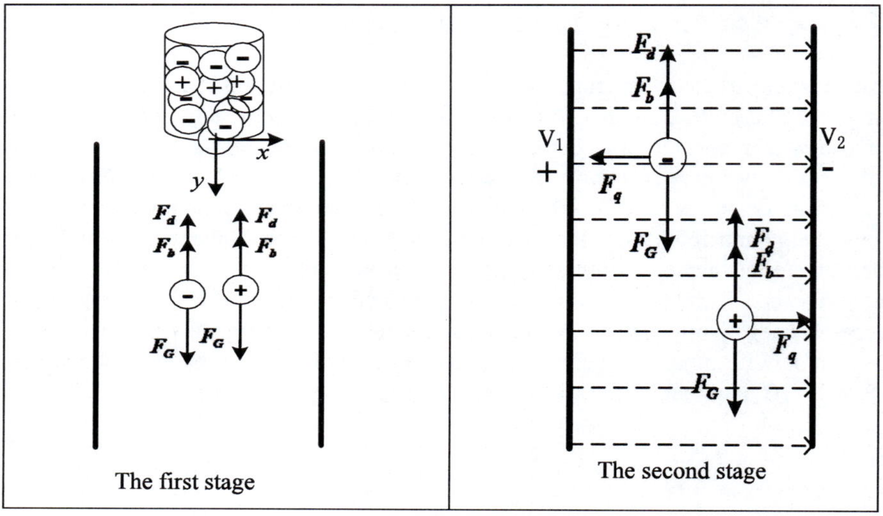

Fig. 2. Force diagram of particles at different stages

through the electrostatic field, and the second stage is the stress of stress of particles in these two stages is shown in the Fig. 2.

In the first stage, the forces loaded on the particles under the influence of air flow are mainly gravity, buoyancy and drag, and their motion trajectory is considered to move downward along the direction of gravity with *a* constant acceleration, which will not be explained here.

In the second stage, the forces loaded on the particles under the influence of air flow are mainly gravity, electric field, buoyancy and drag [10].

$$m\frac{dU_p}{dt} = F_G + F_E + F_d + F_b \tag{1}$$

Where m is the mass of particles; $\frac{d}{dt}$ is the material derivative along the particle flow direction; F_G is the gravity of particles; F_d is the drag force of fluid on particles; F_b is buoyancy; F_E is the electric field force of the particle in the electrostatic field.

The forces in the above formula are as follows:

$$F_E = qE \tag{2}$$

$$F_G = mg \tag{3}$$

$$F_d = C_D A\frac{\rho_f}{2}\left|U_f - U_p\right|(U_f + U_p) \tag{4}$$

$$F_b = \rho_f g v_f \tag{5}$$

Where g is the acceleration of gravity; q is the amount of charge carried by the particles after passing through the charged region; E is the electric field strength in the electrostatic field; C_D is the drag coefficient; A is the projection area of particles; ρ_f is the density of the fluid; U_f is the flow velocity of the fluid at the time increment; U_p is the velocity of particles in the time increment; v_f is the volume of fluid that displaces particles.

Each trajectory Δt corresponds to a time increment. In each time increment, the particle motion is considered to be linear and uniformly accelerated [11]. The acceleration is a, and the velocity increases at $a \cdot \Delta t$. The particles pass through the first stage and enter the second stage. At this time, the particle position changes with each time increment Δt as follows:

$$\begin{cases} x_{j+1} = x_j + v_{jx}\Delta t + \dfrac{1}{2}a_{jx}\Delta t^2 \\ y_{j+1} = y_j + v_{jy}\Delta t + \dfrac{1}{2}a_{jy}\Delta t^2 \end{cases} \tag{6}$$

$$\begin{cases} v_{(j+1)x} = v_{jx} + a_{jx}\Delta t \\ v_{(j+1)y} = v_{jy} + a_{jy}\Delta t \end{cases} \tag{7}$$

$$\begin{cases} a_{(j+1)x} = \frac{1}{m}qE \\ a_{(j+1)y} = \frac{1}{m}(-mg + C_D A \frac{\rho_f}{2}|U_f - U_P|(U_f - U_P) + \rho_f g v_f) \end{cases} \tag{8}$$

3.2 Simulation Model

The Euler model is adopted for the coupling calculation of particle flow and fluid, and the coupling process is shown in the figure below [12] (Fig. 3):

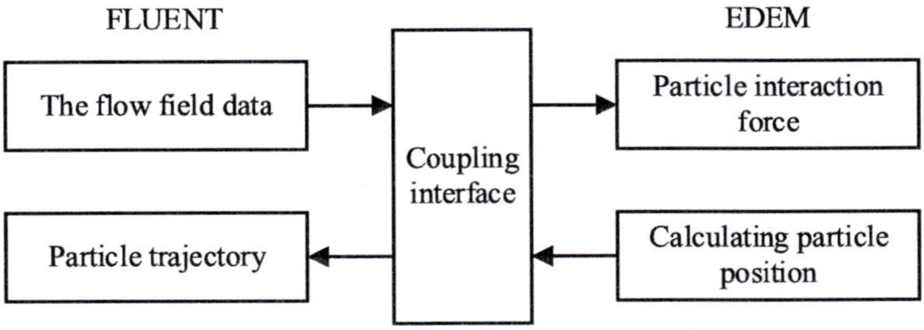

Fig. 3. Coupling flow chart of EDEM and FLUENT

3.3 Flow Field Analysis of Separation Chamber

Three dimensional software is used to model the separation chamber, and ICEM CFD is used to mesh the separation chamber. The fluid medium and device material in the separation chamber are shown in Table 1.

Boundary setting: at the gas inlet, the inlet velocity is 1.5 m/s, 2 m/s, 2.5 m/s and 3 m/s respectively, the turbulence intensity and water drop diameter are set as 5% and

Table 1. Sorting cavity material and medium properties

	Poisson's ratio	Density (kg/m^3)	Viscosity (kg/m-s)	Shear modulus (Pa)
Air		1.293	1.79e-5	
Plexiglass	0.25	1180		1.96e^9
Copper	0.32	8900		3.92e^{10}
Plastic	0.2	2000		3e^9

40 mm respectively (aperture of gas inlet), the gas outlet is set as pressure outlet, the turbulence intensity and water drop diameter are set as 5% and 100 mm respectively. After analyzing the flow field inside the sorting chamber, the velocity distribution cloud diagram of the internal medium is obtained, as shown in Fig. 4.

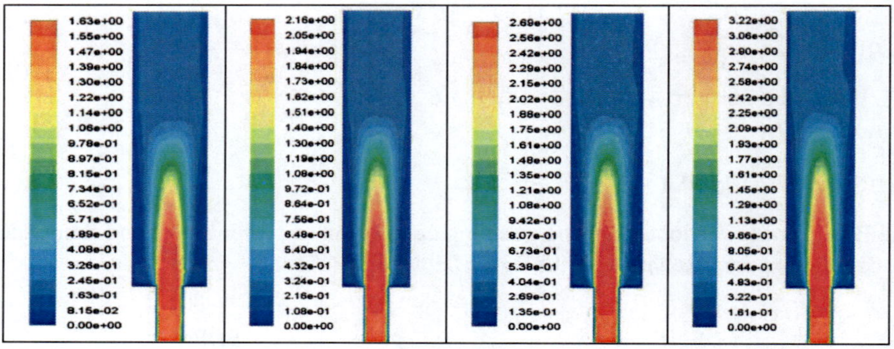

Fig. 4. Velocity distribution cloud image at wind speed of 1.5 m/s, 2 m/s, 2.5 m/s and 3 m/s

3.4 Finite Element Analysis of Particle Phase Dispersion

The metal particles and non-metal particles selected in this study are copper particles and plastic particles with the mass of 250 g and the particle size of 2 mm. The sorting warehouse is made of plexiglass. Through consulting the data, the specific parameter configuration is shown in Table 2.

The simulation process is shown in Fig. 5.

Table 2. Sorting cavity material and medium properties

	Coefficient of restitution	Static friction coefficient	Rolling friction coefficient
Copper-copper	0.15	1	0.3
Copper-plexiglass	0.66	0.5	0.4
Plastic-plastic	0.45	0.5	0.01
Plastic-plexiglass	0.45	0.5	0.01
Plastic-copper	0.4	0.45	0.05

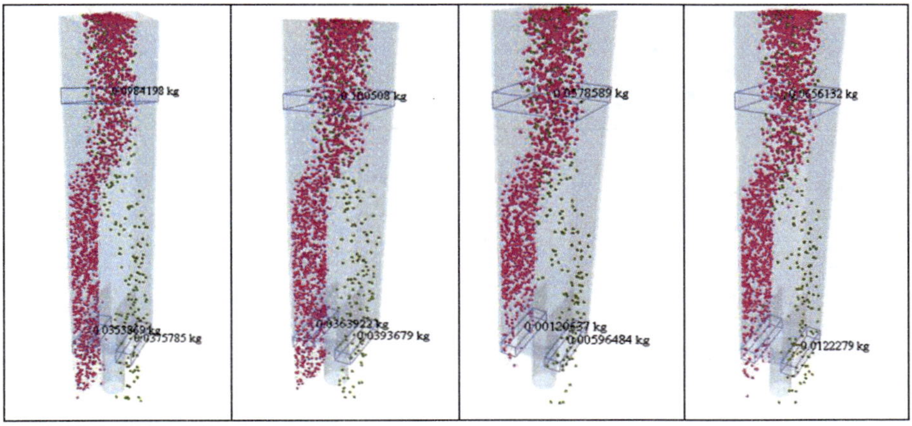

Fig. 5. Diagram of simulation process

4 Analysis of Simulation Results

After the simulation, the distribution data of particles under different wind speeds are exported to the data analysis software origin for analysis and comparison, as shown in Fig. 6.

Figure 6 shows the total mass of the generated metal particles and the sorting metal mass of the metal at different wind speeds with the growth of time between 0 and 1.6 s. When the total metal mass was fixed (0.25 kg), the mass of sorting metal was compared under the four wind speeds. When the total metal mass was 1.3–1.6 s, the metal sorting curve was gradually stable and the sorting of metal particles was completed. When the wind speed is 1.5 m/s, the mass of the metal particles is stable at 0.188488857 kg.When the wind speed is 2 m/s, the mass of the selected metal particles is stable at 0.198927322 kg. It can be seen from the above data that the sorting effect of metal particles is relatively good at the wind speed of 2 m/s under the four wind speeds, and the sorting effect of metal particles is gradually decreasing with the increase of wind speed.

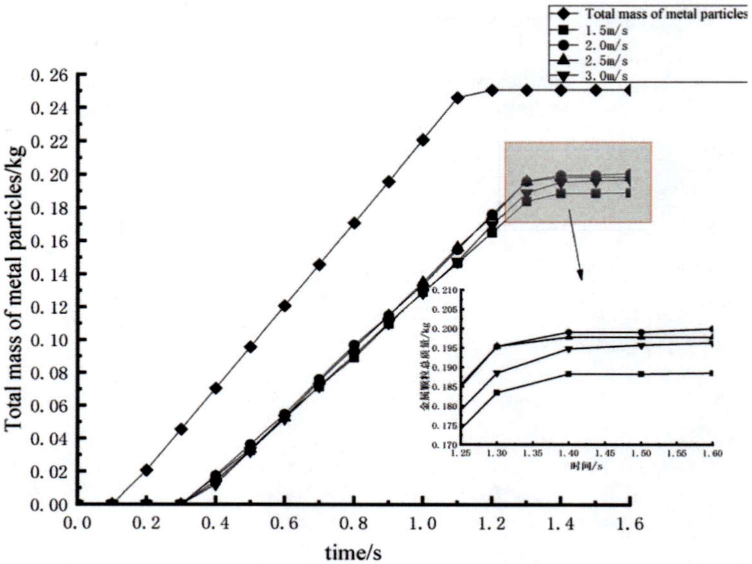

Fig. 6. Separation of metal particles under different wind speeds

5 Conclusion

In this paper, FLUENT software and EDEM software were used to study the separation rate of particles in the separation chamber, and the influence of different wind speeds on the separation rate of metal particles was analyzed. The results show that adding wind in electrostatic separation can improve the feasibility of metal separation rate, and adding appropriate wind speed can greatly improve the metal separation rate. This is will help to Help to determine control parameters of equipment.

Acknowledgment. This paper is supported by Jiangsu Key Laboratory of Green Process Equipment. No. GPE201706.

References

1. Bal, B., Ghosh, S., Das, A.P.: Microbial recovery and recycling of manganese waste and their future application: a review. Geomicrobiol. J. **36**(1), 85–96 (2019)
2. Lulu, Y., Zhenming, X.: Research progress of corona electrostatic separation technology in waste printed circuit boards recycling. Mater. Rev. (2011)
3. Zhang, G., He, Y., Yi, F., Zhang, T., Wang, H.: Recovery of residual metals from fine non-metallic fractions of waste printed circuit boards using a vibrated gas-solid fluidized bed. Sep. Purif. Technol. **207**, 321–328 (2018)
4. Jia, L., Xu, Z., Zhou, Y.: Application of corona discharge and electrostatic force to separate metals and nonmetals from crushed particles of waste printed circuit boards. J. Electrostat. **65**(4), 233–238 (2007)

5. Ito, K., Tamura, R., Zukeran, A., Kawada, Y., Taoka, T.: Simulation and measurement of charged particle trajectory with ionic flow in a wire-to-plate type electrostatic precipitator. Journal of Electrostat. **107**, 103488 (2020)
6. Bilici, M., Dascalescu, L., Dragan, C., Fati, O., Iuga, A., Samuila, A.: Tribocharging and electrostatic separation of mixed granular solids in fluidized bed devices. IEEE Trans. Dielectr. Electr. Insul. **18**(5), 1476–1483 (2011)
7. Benabderrahmane, A., Richard, G., Medles, K., Zeghloul, T., Dascalescu, L.: Dielectric barrier discharge treatment of granular plastic mixtures in view of their triboelectrostatic separation. IEEE Trans. Ind. Appl. **56**(1), 693–703 (2019)
8. Cui, J., Forssberg, E.: Mechanical recycling of waste electric and electronic equipment: a review. J. Hazard. Mater. **99**(3), 243–263 (2003)
9. Richard, G., Touhami, S., Zeghloul, T., Dascalescu, L.: Optimization of metals and plastics recovery from electric cable wastes using a plate-type electrostatic separator. Waste Manag. **60**, 112–122 (2016)
10. Jia, L., Xu, Z., Zhou, Y.: Theoretic model and computer simulation of separating mixture metal particles from waste printed circuit board by electrostatic separator. J. Hazard. Mater. **153**(3), 1308–1313 (2008)
11. Tilmatine, A., Medles, K., Bendimerad, S.E., Boukholda, F., Dascalescu, L.: Electrostatic separators of particles: application to plastic/metal, metal/metal and plastic/plastic mixtures. Waste Manag. **29**(1), 228–232 (2009)
12. Zhou, L.I., Luo, W., Hao, W.U., Liu, B., Zhao, H.: Coupled simulation of sulfur deposition in sour gas reservoir formation based on edem-fluent. Fault-Block Oil & Gas Field (2016)

Check for updates

A Bibliometric Analysis of the Logistical Challenges and Methods for Vaccine Distribution Under the Pandemic

Eugenia Ama Andoh and Hao Yu(✉)

Department of Industrial Engineering, UiT The Arctic University of Norway, Lodve Langesgate 2, 8514 Narvik, Norway
ean099@post.uit.no, hao.yu@uit.no

Abstract. With the emergence and the widespread of the COVID-19 pandemic across the globe, there is an urgent need to effectively control the disease spread through mass vaccination. Several COVID-19 vaccines, e.g., Pfizer/BioNtech and Moderna, etc., have been proven highly effective and have been distributed and administrated in many countries. These vaccines need to be produced in large quantities and transported through dedicated cold chain logistic networks to maintain the quality. Currently, the major logistical challenges are associated with the effective distribution of COVID-19 vaccines to hospitals and healthcare centers in different countries. To better understand and tackle these challenges, we conduct a bibliometric analysis on vaccine supply chains and cold chain logistics for vaccine distribution. The current research landscape is investigated through four main classification analyses including journal co-citation analysis, keyword co-occurrence analysis, country collaborations analysis, and document co-citation analysis. These analyses allow us to identify the publication trends, the most popular journals in this field, the collaborations between countries and to identify the key areas where most attention is given. Finally, the methods are summarized, and the future research opportunities for effective COVID-19 vaccine distribution are identified.

Keywords: Bibliometric analysis · Vaccine supply chain · Cold chain logistics · COVID-19 · Literature review

1 Introduction

From the end of 2019, the COVID-19 pandemic has swept the globe and caused a large number of infections and deaths, which has also dramatically affected the global health care systems, economies, and many industries. As of May 9th, 2021, over 150 million confirmed cases of COVID-19 have been recorded with more than 3.2 million associated deaths, according to the World Health Organization (WHO) [1]. Mass vaccination is considered the most effective way to control infectious disease transmission and to restore people's normal lives. Even though several COVID-19 vaccines, e.g., Pfizer/BioNtech and Moderna, etc., have been developed and used in many countries, there are still

Y. Wang et al. (Eds.): IWAMA 2021, LNEE 880, pp. 166–175, 2022.
https://doi.org/10.1007/978-981-19-0572-8_22

significant supply chain and logistical challenges to satisfy the unprecedented worldwide demands during the pandemic [2].

The vaccine supply chain is a complex system that includes the locations, storage equipment, vehicles, transport routes, and personnel for handling the vaccines from production to administration [3]. Due to the strict temperature requirements for vaccine storage and transportation, cold chain logistics plays the most important role in a vaccine supply chain to ensure the timely and effective connection between the vaccine manufacturers and the points of administration. This has put challenges in logistics infrastructures and operations, especially in developing countries [4]. In this paper, we perform a bibliometric analysis of the research related to vaccine supply chain and cold chain logistics from 2010 to 2020, which aims at understanding the research landscape and the methods developed in this field. Based on the results, four suggestions are given for future research to better tackle the logistical challenges for the effective distribution of COVID-19 vaccines.

2 Research Method and Data Collection

Bibliometric analysis is a comprehensive and rigorous statistical approach to analyze a large amount of research data and to uncover deep insights into a research area [5], and it has been used and has become increasingly popular in the logistics field [6]. Donthu, Kumar, Mukherjee, Pandey and Lim [5] identified the four major steps for performing a bibliometric analysis for a given research area, which include (1) defining the aims and scope; (2) Selecting the techniques; (3) Collecting the relevant data; and (4) analyzing and reporting the results, respectively.

Based on the aims and scope of this paper, we conducted a comprehensive literature search on November 20th, 2020 using the Scopus database. All the literature search was performed on this single day to avoid dynamic updates of the database. Two separate searches were conducted on vaccine supply chain and cold chain logistics, respectively. The results were filtered and combined after several search refinement and eligibility tests, as shown in Fig. 1. Two logical keyword chains were developed for the search, and the initial results yielded a total of 2362 records, where 1470 are related to vaccine supply chains and 892 are related to cold chain logistics.

These records were first limited to the research articles from journals to ensure the quality of the literature. Then, these documents were further refined to show recent papers from the last decade (2010–2020) published in English. A total of 1067 documents remained after this initial refinement. These documents were then subjected to an inclusion and exclusion criterion filter in the subject area. Since the research focus of this paper is within vaccine logistics, supply chain, and operations management, the relevant papers in engineering, computer science, mathematics, decision sciences, energy, material sciences were thus included. On the other hand, the publications in veterinary, animal sciences, food sciences, agriculture sciences, dentistry, immunology, social sciences, neuroscience were excluded.

After this stage, 176 records were left and were screened by titles and by reading the abstracts. Herein, the keywords were defined to include "cold chain", "cold chain logistics" or "vaccine supply chain" in order to select the most relevant papers. After a careful reading, the articles that had none of these keywords in their titles and abstracts were excluded. Also, the topics focusing on vaccine development, vaccination and immunization, risk assessment, fresh foods cold chain, aquaculture cold chain were excluded. In this stage, 57 articles were eliminated, and the other 119 articles were thus selected for the analysis. There were no duplicate articles found when these 119 articles were exported into the Endnote reference manager for duplication checks. Therefore, a total number of 119 articles were used for the bibliometric analysis in this paper.

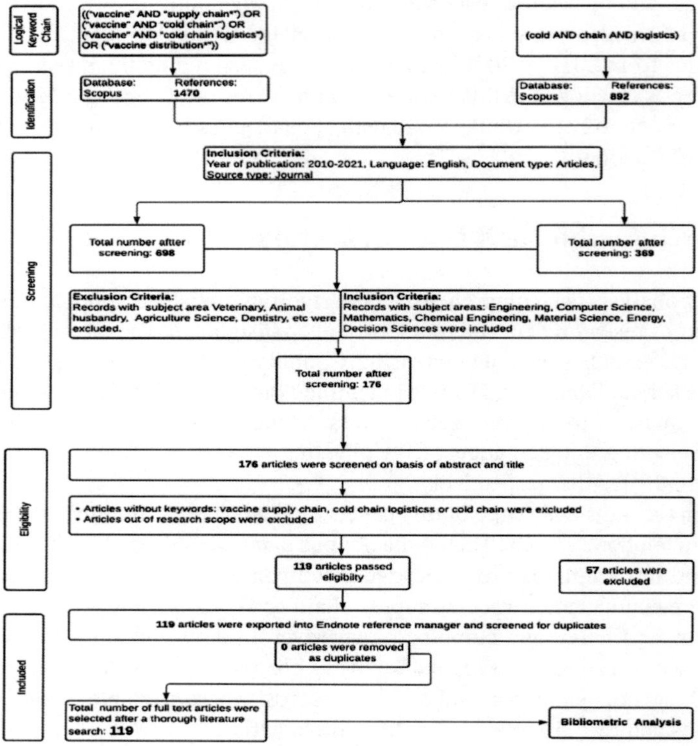

Fig. 1. Data collection procedures.

3 Results of the Bibliometric Analysis

The bibliometric analysis was conducted using a professional software called VOSviewer, and four primary classification analysis were performed, including: (1) journal co-citation analysis; (2) keyword co-occurrence analysis; (3) analysis of the collaborations between countries and regions; and (4) document co-citation analysis. These analyses help to identify the publication trends, the most popular journals in this field, and the key areas that receive the most attention.

The annual and cumulative publications in vaccine supply chain and cold chain logistics are given in Fig. 2. As shown, there has not been a steady growth in the number of articles published each year. In 2014, a peak of 11 articles was published, which is followed by a decrease for the next three years. From 2018 to 2020, an increasing trend is observed with a total of 66 articles published, representing 55% of the total publications in the last decade. This result has shown that increasing focuses have been given to the vaccine supply chain and cold chain logistics, especially during the last three years.

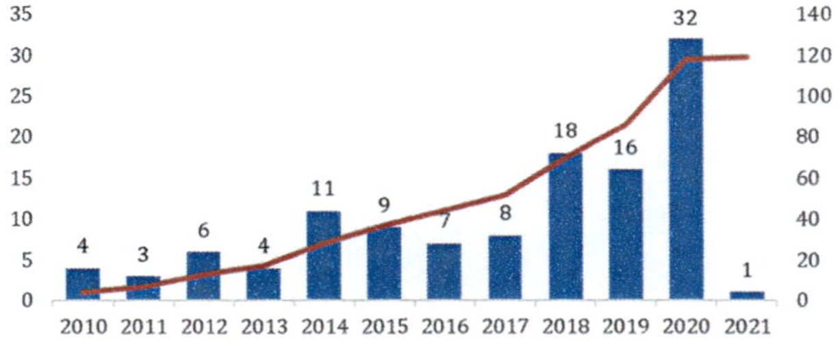

Fig. 2. Publication trends in vaccine supply chain and cold chain logistics.

3.1 Journal Allocation and Co-citation Analysis

The 119 papers selected in this research were published in 91 different journals. Table 1 shows the top 15 journals with more than two papers. These 15 journals have published 40 papers, which account for 34% of the total. As can be seen, *Computers & Industrial Engineering* has the highest number of publications in this field.

A journal co-citation analysis was then performed to identify the most influential and co-cited journals in the vaccine supply chain and cold chain logistics literature. A network of 12 elements with 66 links was created from journals with more than 20 citations. The network shows two clusters of interlinked journals (Fig. 3). Cluster 1 consists of 5 journals, mainly in the field of operations research (OR), where *European Journal of Operational Research* has the most citations and the highest total link strength. Cluster 2 consists of 7 journals representing several research areas of OR and Management Science (MS), where *Management Science* is the most influential journal with a total link strength of 1449. Although OR and MS are often used interchangeably, the journals in Cluster 1 put more focuses on the development of mathematical models and computational algorithms. The journals in Cluster 2, on the other hand, focus on both methodological development and its applications and managerial implications. In addition, it is also observed that *European Journal of Operational Research* and *Management Science* are the most frequently co-cited journals in this field.

3.2 Collaboration Analysis Between Countries and Regions

To identify the countries and regions actively involved in the study of vaccine supply chain and cold chain logistics, a co-authorship network of collaborating countries is

Table 1. Summary of the top 15 journals with the highest number of publications

Journal	Documents
Computers and Industrial Engineering	6
European Journal of Operational Research	4
Annals of Operations Research	3
Omega (United Kingdom)	3
Biologicals	3
Mathematical Problems in Engineering	3
Manufacturing and Service Operations Management	2
International Journal of Multimedia and Ubiquitous Engineering	2
International Journal of Information Management	2
Building and Environment	2
Journal of Biotechnology	2
Neural Computing and Applications	2
International Journal of Refrigeration	2
International Transactions in Operational Research	2
Open Cybernetics and Systemics Journal	2

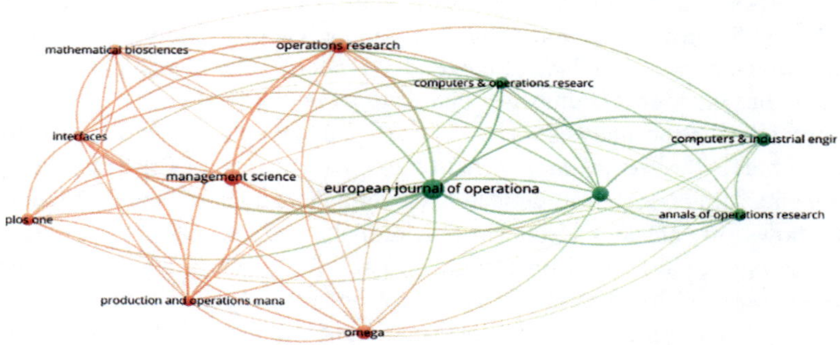

Fig. 3. Journal co-citation analysis.

created in Fig. 4. The minimum number of documents was set to 2, with each country having a minimum citation of 2. This yielded a total of 14 interconnected countries actively collaborating in this research area. As shown, China and the United States are not only the most important countries in this field but also have the most extensive collaborations. Table 2 presents the list of the countries and their contributions to vaccine supply chain and cold chain logistics. The average citation was calculated by dividing the number of citations of each country by the number of documents.

From the list, vaccine supply chain and cold chain logistics are popular research topics in the regions of Europe, North America, and Asia, where China has the highest number of published documents, followed by the United States. Although China has more publications than the United States, the United States has twice as many citations as China, which implies their publications may have a more substantial influence at the time of this study. In the table, the average year column represents the average year in which the documents are published. It means that when comparing China and the United States, the latter has more documents published from 2015 as opposed to 2017 in the case of China. This may explain the reason why the publications from the United States have received more citations than the ones from China, because earlier publications have more influence in a research area and usually receive more cumulative citations.

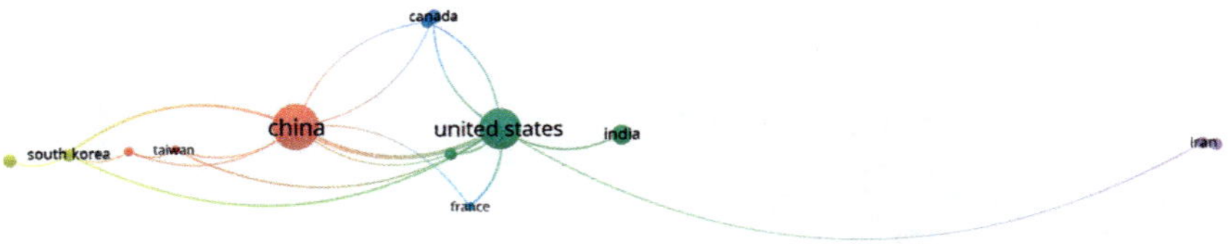

Fig. 4. Mapping of collaborating countries.

Table 2. The countries and regions with the highest number of publications

Country or Region	Documents	Citations	Citations per document	Average year
China	47	182	3.87	2017
United States	36	385	10.69	2015
India	9	36	4.00	2018
Canada	5	17	3.40	2017
South Korea	4	41	10.25	2016
United Kingdom	4	48	12.00	2018
Australia	3	29	9.67	2016
Hong Kong (China)	3	15	5.00	2018
Spain	3	15	5.00	2014
Switzerland	3	26	8.67	2017
France	2	14	7.00	2013
Italy	2	76	38.00	2013
Saudi Arabia	2	19	9.50	2019
South Africa	2	15	7.50	2018

Furthermore, it is noteworthy that several countries, e.g., South Korea, the United Kingdom, Australia, Switzerland, and Italy have a few publications, but the average citations per document are extremely high compared to that of the papers from the other countries. This indicates their strong influences in this research field.

3.3 Co-occurrence Analysis of the Frequent Keywords

The keyword co-occurrence analysis is an assessment to show the main theme and focuses of a particular field. In this analysis, the minimum number of the keyword co-occurrences was set to 3. This criterion leads to 67 qualified ones out of the total 1407 keywords in all publications, which form 7 clusters with their co-occurrence relations. The top 10 keywords in vaccine supply chain and cold chain logistics with the highest frequency of occurrence and inter-connections are vaccines (Freq = 50), cold chain logistics (Freq = 32), cold chain (Freq = 17), optimization (Freq = 15), supply chains (Freq = 15), vehicle routing problems (Freq = 12), vehicle routing (Freq = 11), developing countries (Freq = 10), supply chain management (Freq = 10), and chains (Freq = 10).

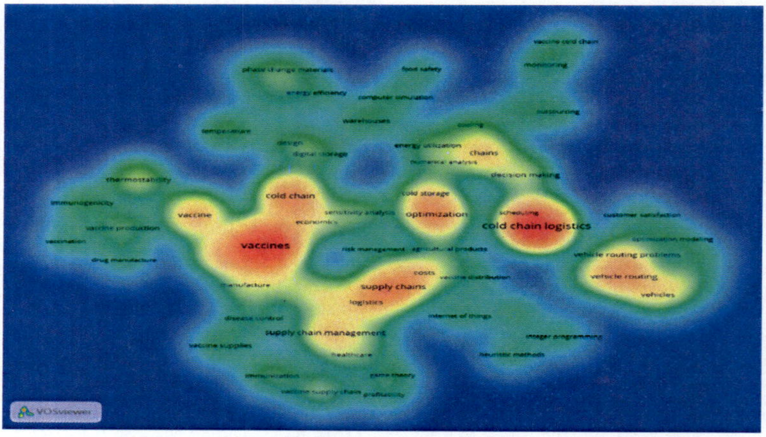

Fig. 5. Density map visualization of keywords.

To further analyze and understand the recurrence of the keywords, a density map was generated in Fig. 5. The density map helps to visualize the most critical areas on the keyword map. The points on the map are represented by a color that indicates the density of the item. The default color coding used in VOSviewer ranges from red to blue, where red indicates the highest density and blue shows the lowest density. Vaccines and cold chain logistics fall in the red zone with the highest research focuses, whereas supply chains, vehicle routing, and optimization fall in an orange zone with a higher density than the keywords in the green zones. The keywords located in the red and orange zones highlight the main research focuses, problems, and methodologies applied, which form the core of the studies related to vaccine supply chain and cold chain logistics. While, on the other hand, the keywords in the green zone are also important relevant fields and may suggest new areas that deserve future research.

3.4 Co-citation Analysis of Documents

The document co-citation analysis is performed to identify the most influential documents in vaccine supply chain and cold chain logistics. The result from the co-citation

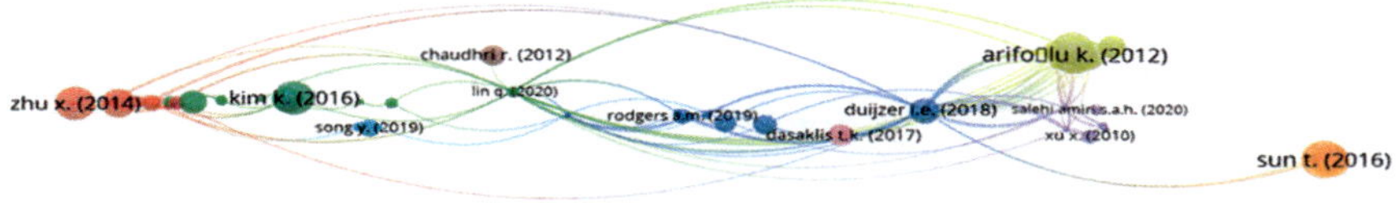

Fig. 6. Mapping of co-cited documents.

analysis of the 119 selected papers is given in Fig. 6. It can be seen that Arifoğlu, Deo and Iravani [7], which studied the influenza vaccine supply chain with a focus on demand and supply, is the most influential document that has the highest number of citations and the strongest interplays in the co-citation map.

4 Discussions and Future Research Opportunities

The results of the bibliometric analysis reveal that the research topics on vaccine supply chain and cold chain logistics have received increasing attention during the last decade across different fields. Extensive studies have been done to support the decision making and to address the planning challenges at strategical, tactical, and operational levels over the four stages of a vaccine supply chain including product, production, allocation, and distribution [8].

4.1 Methods for Vaccine Supply Chain and Cold Chain Logistics

To solve the supply chain and logistical challenges, the methods from OR and MS have been extensively focused and widely adopted, and the most popular methods include optimization, simulation, and combined optimization-simulation.

Optimization is a prescriptive analytics that helps to find the optimal solution through using a model and/or an algorithm under several pre-determined rules and conditions. Optimization models and methods can be used to determine the system configuration of a vaccine supply chain and to make operational decisions related to the vaccine storage, allocation, and transportation of a cold chain distribution network. As logistics costs are directly proportional to customer responsiveness, the network design for vaccine distribution should be modeled to achieve low operating costs, while simultaneously, be capable of rapidly distributing vaccines to end-users, especially during the COVID-19 pandemic. In this regard, several studies have been conducted to determine the strategic location and allocation decisions, and the tactical relocation decisions. Mixed integer linear program (MILP) is the most widely used modeling technique in vaccine supply chain network design, which minimizes the total costs associated with the facility location and demand allocation through optimal decision making. Besides, multi-objective optimization has also been widely used to incorporate with other objectives of vaccine supply chain network design, e.g., carbon emission, customer satisfaction. At the operational level, the research focuses are given to the development of improved heuristics and meta-heuristics for the vehicle routing problems (VRP) and the inventory allocation problems in vaccine distribution systems.

Simulation is a descriptive analytics that helps to test several what-if scenarios and to evaluate the system performance in a dynamic environment. Unlike optimization, simulation focuses primarily on performance measurement but not on decision making. At the country level, the vaccine supply chain consists of four main stages including sourcing, storage, distribution, and administration [9]. Implementing simulation methods, e.g., Monte Carlo simulation, discrete event simulation, and system dynamics, in professional software, e.g., AnyLogic and AnyLogistix, can help to evaluate different decisions in these four stages of vaccine supply chains and cold chain logistics systems. For instance, Shittu, Harnly, Whitaker and Miller [10] addressed the problems related to vaccines storage facilities in Nigeria by applying a simulation model to analyze the effect of the fluctuations from the vaccine supplies and demands on the storage capacity requirements.

Simulation-optimization can use the strengths of both methods and has thus been increasingly focused [11]. On the one hand, simulation can provide inputs and the prediction of parameters under a highly uncertain environment of an epidemic outbreak [12]. On the other hand, the optimization results can be evaluated in a more realistic simulation environment [13]. In this regard, Dillon and Colton [14] used a simulation-optimization method to determine the most economical design of vaccine warehouses in developing countries, and their study showed that this method could yield a more accurate result.

4.2 Future Research Opportunities

Even though increasing efforts have been spent during the last decade, there are still several research opportunities to tackle the challenges of vaccine supply chains and cold chain logistics during the COVID-19 pandemic. The following suggestions are made to guide future research:

1. The combination of advanced predictive analytics with optimization models is a promising direction for future study. The recent development of machine learning, deep learning as well as other artificial intelligence (AI) methods may improve the inputs and the effectiveness of the optimization models. Thus, more studies are needed in this field.
2. The combination of optimization and simulation needs to be enhanced to ensure the viability of a global vaccine supply chain under the pandemic.
3. The implementation of new technologies, e.g., drones, may improve the distribution and last-mile delivery of vaccines, so new models are needed to plan the cold chain logistics system considering the use of new technologies.
4. The mass vaccination will lead to a significant increase in the generation of infectious medical waste, so the reverse logistics issues associated with the mass vaccination should be focused on.

5 Conclusions

This paper presents a bibliometric analysis of 119 documents contributed to vaccine supply chain and cold chain logistics within the past decade. Using a network analysis tool, four classification analyses, namely, journal co-citation analysis, co-occurrence analysis of keywords, country collaboration analysis, document co-citation analysis, were made to understand the research landscape and the most popular methods. The results indicate that there has been a tremendous increase in the number of publications from 2018 with *Computers & Industrial Engineering* having the most publications. Besides, optimization and simulation have been widely used to support different decisions. However, to tackle the unprecedented logistical challenges for the vaccine supply chain during the COVID-19 pandemic, future research needs to focus on methodological integration, incorporation with new technologies, and reverse logistics.

Acknowledgment. This research is partially funded by the Idea Factory project financed by UiT—The Arctic University of Norway.

References

1. WHO. https://covid19.who.int/. Accessed 05 May 2021
2. Kim, J.H., Marks, F., Clemens, J.D.: Looking beyond COVID-19 vaccine phase 3 trials. Nat. Med. **27**, 205–211 (2021)
3. Lee, B.Y., et al.: Re-designing the mozambique vaccine supply chain to improve access to vaccines. Vaccine **34**, 4998–5004 (2016)
4. Yang, Y., Bidkhori, H., Rajgopal, J.: Optimizing vaccine distribution networks in low and middle-income countries. Omega **99**, 102197 (2021)
5. Donthu, N., Kumar, S., Mukherjee, D., Pandey, N., Lim, W.M.: How to conduct a bibliometric analysis: an overview and guidelines. J. Bus. Res. **133**, 285–296 (2021)
6. Bensalem, A., Kin, V.: A bibliometric analysis of reverse logistics from 1992 to 2017. Supply Chain Forum: Int. J. **20**, 15–28 (2019)
7. Arifoğlu, K., Deo, S., Iravani, S.M.R.: Consumption externality and yield uncertainty in the influenza vaccine supply chain: interventions in demand and supply sides. Manag. Sci. **58**, 1072–1091 (2012)
8. Duijzer, L.E., Van Jaarsveld, W., Dekker, R.: Literature review: the vaccine supply chain. Eur. J. Oper. Res. **268**, 174–192 (2018)
9. De Boeck, K., Decouttere, C., Vandaele, N.: Vaccine distribution supply chains in developing countries: a literature review. FEB Research Report KBI_1815 (2018)
10. Shittu, E., Harnly, M., Whitaker, S., Miller, R.: Reorganizing nigeria's vaccine supply chain reduces need for additional storage facilities, but more storage is required. Health Aff. **35**, 293–300 (2016)
11. Yu, H., Sun, X., Solvang, W.D., Laporte, G.: Solving a real-world urban postal service system redesign problem. Sci. Program. **2021**, ID 3058472 (2021)
12. Yu, H., Sun, X., Solvang, W.D., Zhao, X.: Reverse logistics network design for effective management of medical waste in epidemic outbreaks: insights from the coronavirus disease 2019 (COVID-19) outbreak in Wuhan (China). Int. J. Environ. Res. Public Health **17**, 1770 (2020)
13. Sun, X., Yu, H., Solvang, W.D.: Solving the location problem of printers in a university campus using p-median location model and anylogic simulation. In: Wang, Yi., Martinsen, K., Yu, T., Wang, K. (eds.) IWAMA 2019. LNEE, vol. 634, pp. 577–584. Springer, Singapore (2020). https://doi.org/10.1007/978-981-15-2341-0_72
14. Dillon, K., Colton, J.: A design methodology for the economic design of vaccine warehouses in the developing world. Build. Environ. **82**, 160–170 (2014)

Numerical Stress Analysis and Fatigue Life Prediction of the Classical External Geneva Mechanism

Getachew A. Ambaye[1](✉) and Hirpa G. Lemu[2](✉)

[1] Bahir Dar Institute of Technology, Bahir Dar, Ethiopia
Getachew.Admassie@bdu.edu.et
[2] Faculty of Science and Technology, University of Stavanger, Stavanger, Norway
Hirpa.g.lemu@uis.no

Abstract. This study examined the stress distribution and predicting the fatigue life of the classical external Geneva mechanism using analytical and finite element methods for three loading cases. The life cycles were analyzed using FE SAFE software by using the stress output database of ABAQUS software, which has been used for stress analysis of the Geneva wheel mechanism. The maximum stress is found at 45° of the angular position of the driving crank located at the bottom part of the end of the wheel slot and pin. Doubling the angular speed of the driving crank and doubling the torque that applied on the wheel yields almost equal contact pressure and von Mises stress. The wheel has a lower fatigue life cycle compared to the pin.

Keywords: Geneva wheel · Drive crankpin · Active phase · Fatigue life prediction

1 Introduction

Geneva mechanisms are commonly used as an indexing mechanism where intermittent motion is required by transforming rotary motion into intermittent motion. The rotating drive crankpin is usually equipped with a pin that is engaged with the slot of the wheel. The driving crank pin will be engaged with the slot of the wheel only once at a full cycle and this is called the active phase. The Geneva mechanism is extensively used in low-and high-speed automatic machinery. It is deployed for counting mechanisms, indexing, sequencing, motion-picture mechanisms, and others with the limitation of smooth operation due to discontinuities in the acceleration at the beginning and end of the contact between the crankpin and wheel slot. Zhang et al. [1] developed an optimized Geneva mechanism for internal combustion engine applications. Ali and Majid [2] analyzed the stress and deflection of an external four-slot Geneva mechanism using experimental and numerical methods. Lee and Huang [3] made geometry analysis and optimal design of Geneva mechanisms with curved slots. Many other studies were done to improve the dynamic performance to avoid the jerk motion at the beginning and end of the engagement of the wheel slot and drive crankpin [3–6].

Y. Wang et al. (Eds.): IWAMA 2021, LNEE 880, pp. 176–186, 2022.
https://doi.org/10.1007/978-981-19-0572-8_23

Haraga et al. [7] analyzed and modeled the Geneva mechanisms using finite element method. Shekhar et al. [8] analyzed the external Geneva mechanism a conveyor for stress and deflection in addition to the jerk motion analysis. They found that the maximum stress occurred at the bottom part of the driven wheel and the maximum deflection occurred at the tip of the slot. Tuama and Faidh [2] made stress analysis of an external Geneva mechanism made from polycarbonate material experimentally (using a photo-elastic instrument) and numerical at different crank angle positions. The maximum stress is found at 45° of the crank angle position and is located near the end of the slot.

FE SAFE software, which is is a powerful program to analyze the structural fatigue life, was used for the prediction of the fatigue life analysis of the wheel and the driving pin. This software has been used to determine the fatigue life of many mechanical components such as to determine the fatigue life of wind turbine [9], fatigue life analysis of Aluminium joints with multi fasteners interface fit [10] and to determine the fatigue damage of the railway fastener e-clip [11].

In this study, a quarter of the external Geneva wheel mechanism was modeled using Autodesk Inventor software and exported to the ABAQUS software for the stress analysis for different loading cases.

2 Load Distribution and Contact Width Analysis

The 3D quarter Geneva mechanism has been modeled based on the design parameters given in Table 1, as shown colored view in Fig. 1. There are three load cases in this study used to examine the stresses of the wheel and crankpin (Table 2). For numerical analysis, the center point of RP-1 and RP-2 will be constrained in all degrees of freedom except for rotating about their own rotating axes. The torque and angular speed are applied at the center point of the wheel and crank, respectively.

Table 1. Design parameters of external Geneva mechanism

Parameter	Value (mm)
Crankpin radius, $\boldsymbol{r}_1$	75
Thickness of wheel, $\boldsymbol{t}$	4
Number of slots, n	4
Slot center length	44
Drive pin radius, $\boldsymbol{R}_1$	4
Radius of Geneva Wheel and driving crank Shaft	25 and 20
Stop disk radius	63

The normal load varies with the angular position of the contact point. The torque load at the contact point is decomposed into tangential (W_T), radial (W_R), and normal load (W_N), as shown in Fig. 2(a). The normal contact load is given by Eq. (1), and the load distribution for 7000 Nmm with the angular position is shown in Fig. 2(b). The

Table 2. Load cases parameters

Load cases	Torque [N.mm]	Angular speed [rad/s]
Load case 1	7000	5.34
Load case 2	14,000	5.34
Load case 3	700	10.58

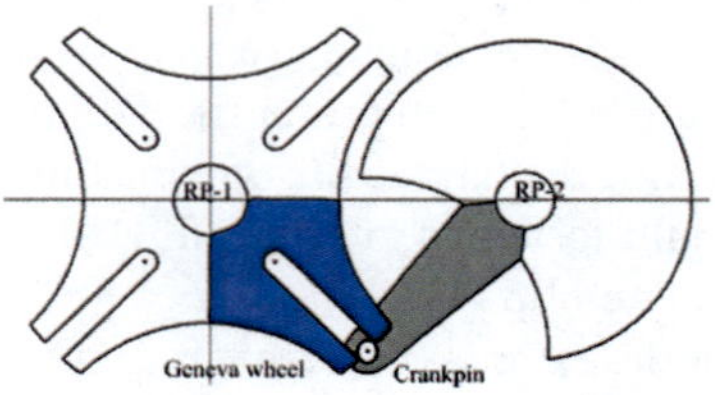

Fig. 1. 2D model of classic Geneva mechanism

maximum load is obtained at half of the angular position of the active phase (which equals to 45°), and the minimum load for load case 1 is approximately 35 N.

$$W_N = \frac{T}{r_1}\cos(\theta_1 + \theta_2) \tag{1}$$

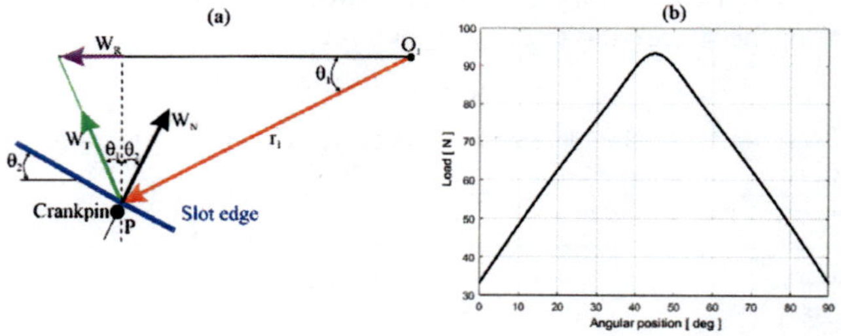

Fig. 2. (a) Schematic diagram of the force decomposition and (b) Normal load distribution for load case 1

According to the Hertzian expression, when the two cylindrical objects come in contact, then the contact width and the maximum contact pressure are given in Eq. 2 (a) and (b) respectively. The contact pressure is the proportion of the normal load of the true contact area. The wheel and crank are made from steel SAE-AISI 1045, whose physical properties are: density, $\rho = 786$ Kg/m^3, modulus of elasticity, E = 210 GPa and poison's ratio, $\nu = 0.3$.

$$b = \sqrt{\frac{4W_N R'}{\pi l E'}}, \quad P_{max} = \frac{W_N}{\pi b l} \tag{2}$$

The above formula is also used for cylindrical and flat surface contact bodies with an infinite radius of curvature on the flat surface (wheel slotted surface). The radius of curvatures for the Geneva mechanism is expressed as $R_{1X} = \infty, R_{1Y} = R_1, R_{2X} = R_{2Y} = \infty$. The reduced radius of curvature of the Geneva mechanism is equal to the radius of the pin and it is given in Eq. 3 (a), while the reduced elastic of modules is given in Eq. 3 (b).

$$\frac{1}{R_1} = \frac{1}{R_X} + \frac{1}{R_Y} = \frac{1}{R_1} + \frac{1}{\infty} + \frac{1}{\infty} + \frac{1}{\infty} = \frac{1}{R_1} \tag{3a}$$

$$\frac{1}{E'} = \frac{1}{2}\left[\frac{1 - \nu_1^2}{E_1} + \frac{1 - \nu_2^2}{E_2}\right] \tag{3b}$$

The contact width and the maximum contact pressure are shown in Fig. 3 (a) and (b), respectively. The maximum contact width and contact pressure are found at 45° of the angular position of the active phase.

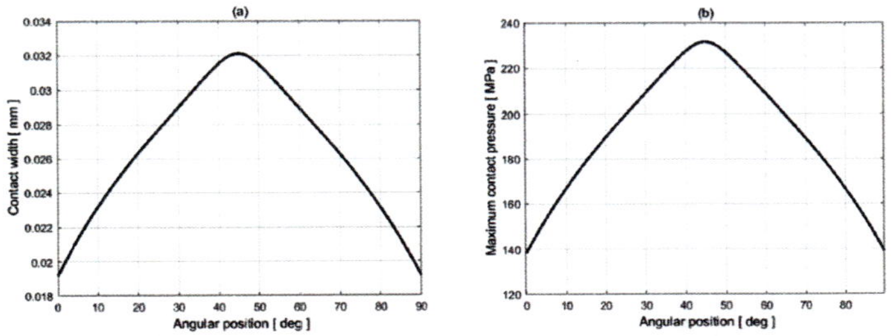

Fig. 3. a) Half contact width, and b) Maximum contact pressure

3 Numerical Stress Distribution Analysis

As described in Sect. 2, both contacting bodies are free to rotate about the rotational direction of the center. There are two steps created for setting the boundary and loading conditions, and in the first step, the wheel is subjected to a torque based on the load case, and the speed will be granted in the second step. The initial temperature or operating temperature was added in the initial step as an initial condition. The analysis contains a C3D8 element type which was used for analysis and the meshed model has 23645 elements and 40427 nodes. A reasonable fine mesh is deployed at the contacting surfaces with a 0.5 mm mesh size, which will enhance the results and the running cost.

The maximum contact pressure distribution occurred between the wheel slot and crankpin against the angular position was analyzed numerically using ABAQUS and analytical equations (Fig. 4). The maximum contact pressure (for load case 1, with 0.15 coefficient of friction) is found at angular position of 45°. The contact pressure of the crankpin varies rapidly with the angular speed, while the Geneva wheel shows a smooth variation of contact pressure with the angular position. In addition, the analytical contact

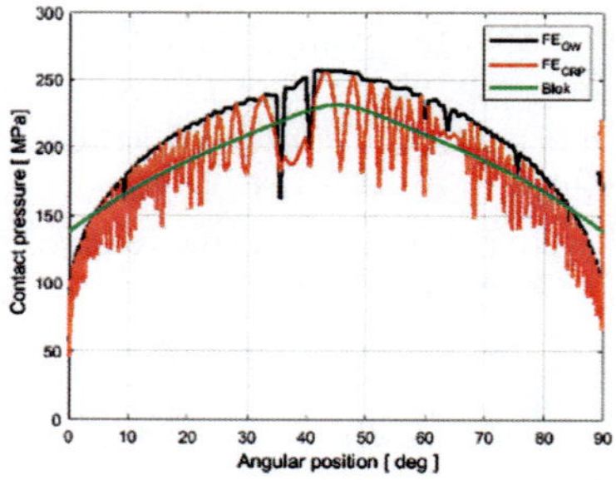

Fig. 4. Maximum contact pressure distribution

pressure looks at the mean contact pressure of the pin that is found using the FE method specifically within 10° to 75°of angular position.

The contour plots are presented in Figs. 5 and 6. The maximum von Mises stress for the wheel occurred at the bottom edge close to the junction of curved and linear slotted surfaces for wheel and bottom edge of the pin as shown in Fig. 5 (a) and (b), respectively. These maximum von Mises stresses occurred when the crank rotated by 45°. The contact pressure of the wheel and the pin is given in Fig. 6 (a) and (b), and like the von Mises stress, the maximum contact pressure is found at the bottom part of the linear slot surfaces and at the bottom part of the pin. The contact pressures of the wheel and pin are nearly the same.

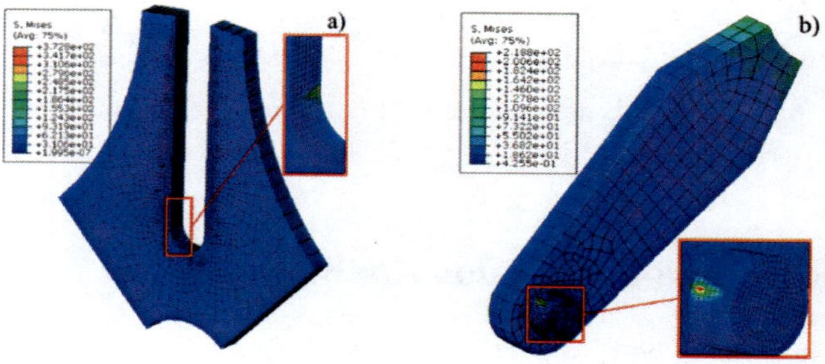

Fig. 5. Von Mises stress: a) Geneva wheel, and b) Crankpin

Figures 7, 8 and 9 (a–b) show the stress distributions for the three different loading cases with 0.15 and 0.3 coefficient of friction. The maximum stress is found in the wheel for all the loading cases. The von Mises stress of the pin is non-smooth as compared with the wheel. The results indicate that the coefficient friction significantly affects the stress of the wheel and pin. The von Mises stress distribution for load case 2 (Fig. 8 (a–b)) and load case 3 (Fig. 9 (a–b)) are nearly identical. But, the load variation is not smooth for load case 2 as compared with load case 3. This clearly indicates that the von Mises stress produced by doubling the torque and doubling angular speed is equal, except for the non-smooth characteristics.

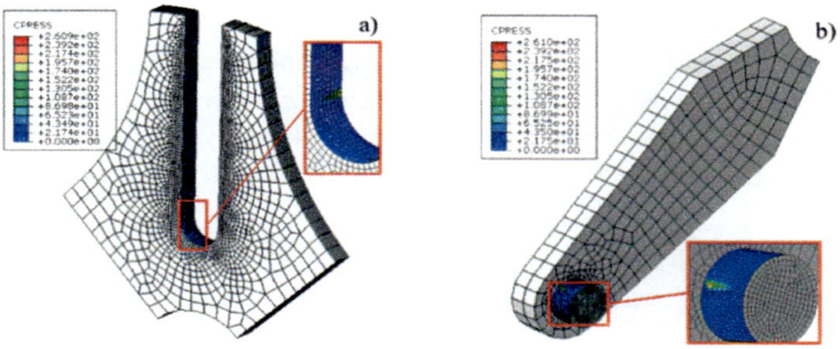

Fig. 6. Contact stress: a) Geneva wheel, and b) Crankpin

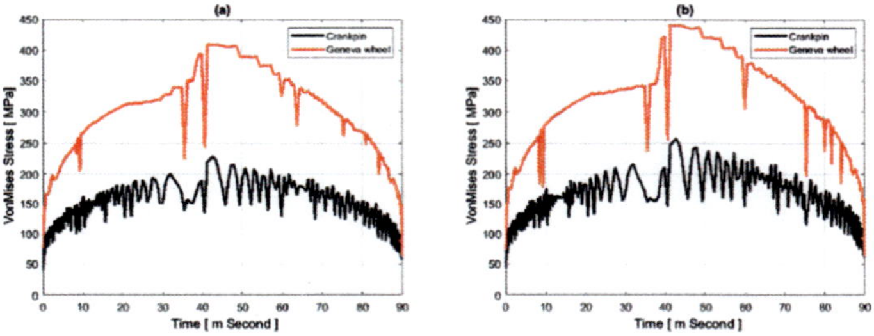

Fig. 7. von Mises stress with load case 1: a) = 0.15, and b) $\mu = 0.3$

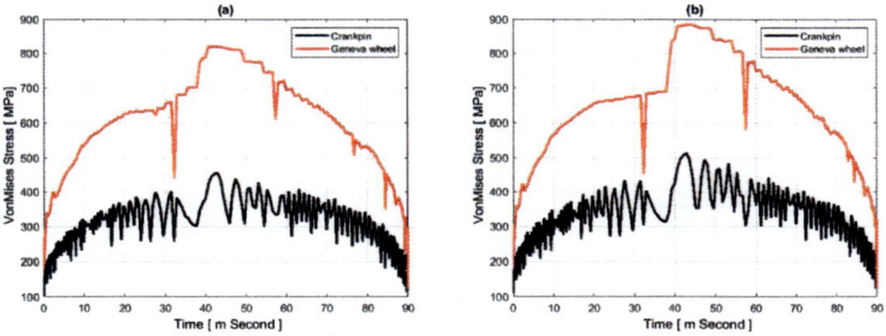

Fig. 8. von Mises stress with load case 2: a) $\mu = 0.15$, and b) $\mu = 0.3$

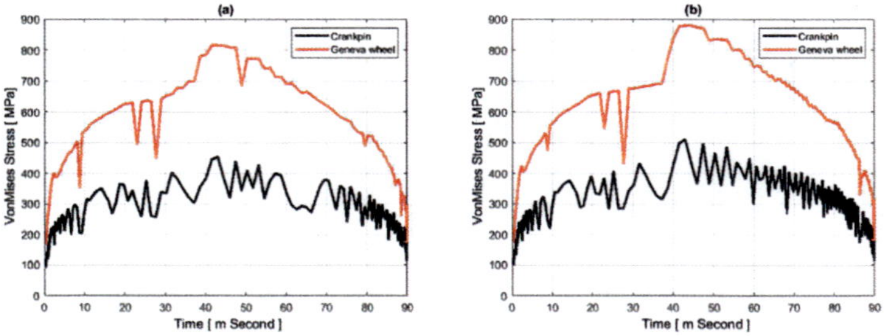

Fig. 9. von Mises stress with load case 3: a) $\mu = 0.15$, and b) $\mu = 0.3$

The maximum contact pressure of the wheel and pin was analyzed, as depicted in Figs. 10, 11 and 12 (a–b), for three loading cases. The maximum contact pressure is found at 45° of angular position and this contact pressure increases as the load increases. The contact pressure of the crankpin is non smooth like the von Mises stress, as shown previously. The friction coefficient of the wheel and crank material also affects the contact pressure.

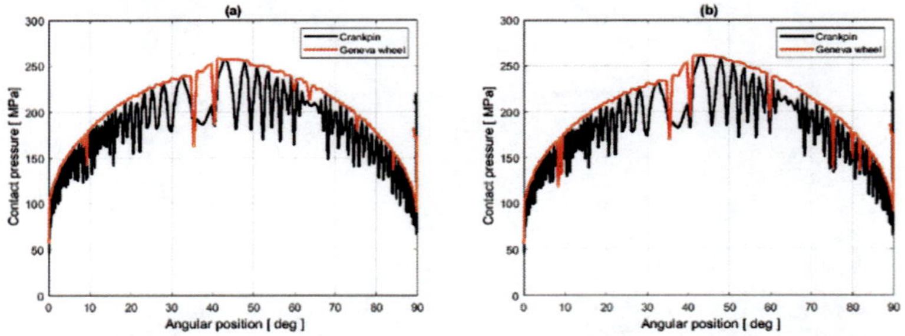

Fig. 10. Contact stress distribution with load case 1: a) $\mu = 0.15$, and b) $\mu = 0.3$

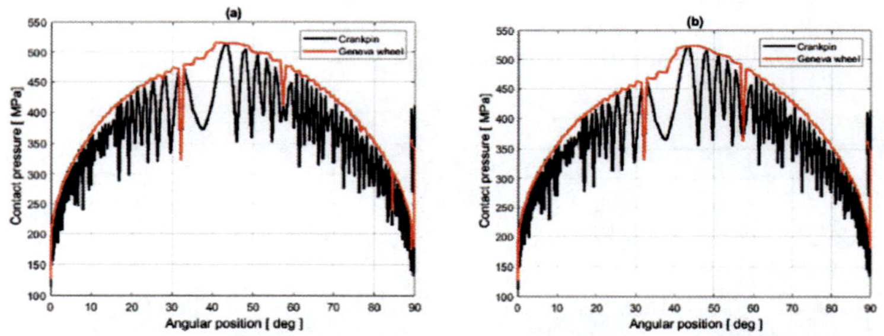

Fig. 11. Contact stress distribution with load case 2: a) $\mu = 0.15$, and b) $\mu = 0.3$

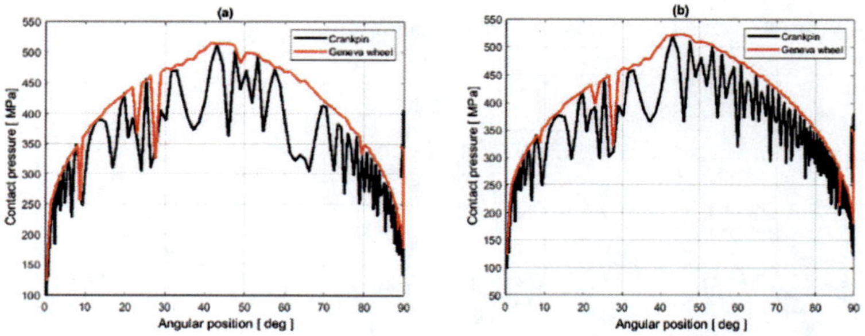

Fig. 12. Contact stress distribution with load case 3: a) $\mu = 0.15$, and b) $\mu = 0.3$

4 Fatigue Life Prediction Using FE-SAFE

The Geneva mechanism transmits the motion and loads repeatedly. The repetitive engagement of the slot and the pin is determined by the number of slots and pin. This study analyzed the mechanism that contacts a four-slot wheel with a single pin. This assures that according to the speed and load, the Geneva mechanism may run more than 1000 cycles (by considering the high-stress fatigue). Stress based fatigue analysis is appropriate for high cycle fatigue life analysis. For the given material, the strength life cycle (S-N) is shown in Fig. 13. From Marin [12] equation, the corrected endurance limit is given by Eq. 4 as:

$$S_e = k_a k_b k_c k_d k_e k_f S_e^{'} \tag{4}$$

Where k_n, $(n = a - f)$ are the surface, size, load, temperature, reliability, and miscellaneous effects modification factors respectively. In this study, this factor is simplified and the endurance limit is given: $S_e = 0.5 S_{ut}$. The fracture strength is given by

$$S_f = aN^b \tag{5}$$

Where $a = \frac{(fS_{ut})^2}{S_e}$ and $b = -\frac{1}{3}\log\left(\frac{fS_{ut}}{S_e}\right)$.

The value of the fatigue strength fraction, $f = 0.89$ is a function of the ultimate tensile strength [12].

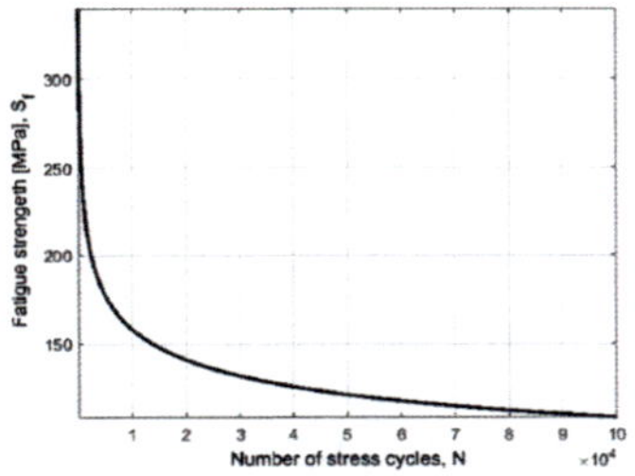

Fig. 13. S-N material response

The fatigue life prediction has been done using the FE-SAFE software, by giving the output database (odb) file of the ABAQUS result. The analysis has been done by scaling the load or stress from 1 to 10 times, and Figs. 14 and 15 (a–b) show the LOG Life by scaling up by 1, 5, and 10 times. There are many data sets for the loading conditions, but in this study, the maximum loading is selected and used as a data set for the prediction of life, and this stress is varies from 69.4 to −306 MPa.

The LOG Life of the wheel for the normal data set is 5.82, which is approximately 0.67 million cycles, as shown in Fig. 14 (a), while for the crank there is no damage, it has more than 10 million cycles. When the load scale increases to 5 times, the LOG Life decreases for the wheel and pin. Figure 15 (a) and (b) show the log life of the wheel and pin, which are 2.201 and 3.742, respectively. In all cases, the minimum life cycle occurred at the bottom part of the linear slot surface and pin for the wheel and pin, respectively. Generally, the life of a pin is larger than the life of the wheel, and as the load or data set scale increases, the life becomes shorter.

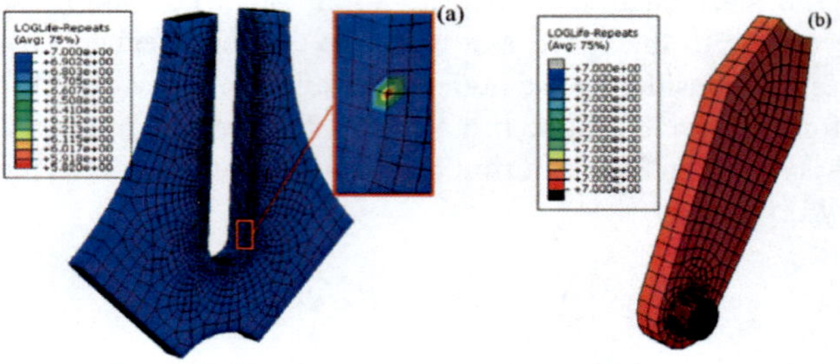

Fig. 14. Fatigue life prediction with scale 1: a) Geneva wheel, and b) Driving crank

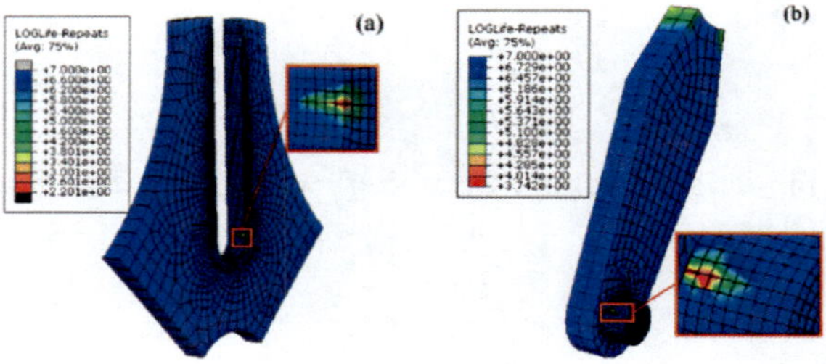

Fig. 15. Fatigue life prediction with scale 5: a) Geneva wheel, and b) Driving crank

The fatigue life of the wheel and pinion for a dataset scale from 1 to 10 times was analyzed. Figure 16 (a) and (b) show the fatigue life of the wheel and crank, respectively. As shown in the figure, the number of stress cycles increases as the fatigue strength decreases. The crank has a maximum fatigue life cycle compared to the wheel.

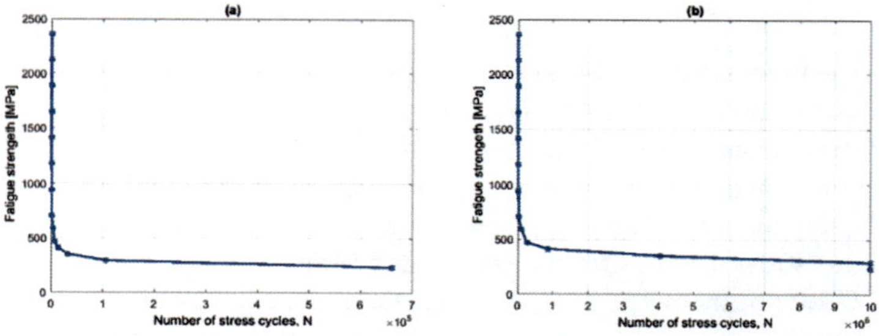

Fig. 16. Fatigue life: a) Geneva wheel, and b) Driving crank

5 Conclusion

In this study, the stress and fatigue life prediction of the classical external Geneva mechanism was studied in three loading cases. The load distribution of the Geneva mechanism was found by resolving the given torque load, and the maximum load was obtained at 45° of the angular position. The analytical calculations were also made for the prediction of the contact width and maximum contact pressure between the wheel slot and the pin. The analytical contact pressure analysis and FE analysis were compared. In addition, the following conclusions were drawn from the numerical results:

- The maximum stress for the wheel occurred at the bottom edge near to the junction of curved and linear slotted surfaces and bottom edge of the pin for the wheel and driving crank respectively. The stress obtained from the pin is non-smooth compared to the wheel set.
- Almost equal contact pressure and von Mises stress are obtained by doubling the torque and doubling the angular speed. In addition, the coefficient of friction significantly affected the stress.
- The fatigue life of the pin is greater than the wheel slot. The minimum fatigue life is obtained at the bottom part of the wheel set near the end of the slot and pin. Therefore, a special attention should be given to the wheel during design and manufacturing. This can possible achieved by modifying the profile of the wheel slot and selecting a better material and making surface treatments.

References

1. Zhang, D., Reed, M., Li, B., et al.: Design optimization of a Geneva mechanism for internal combustion engine application. In: ICIA, pp. 649–654 (2009)
2. Tuama, A.A.H., Faidh-Allah, M.H.: Experimental and numerical investigation of external geneva wheel mechanism with four slots manufacturing from polycarbonate material. IJMME-IJENS **17**(3), 72–81 (2017)
3. Lee, J.J., Huang, K.F.: Geometry analysis and optimal design of Geneva mechanisms with curved slots. Proc. Inst. Mech. Eng. Part C: J. Mech. Eng. Sci. **218**(4), 449–459 (2004)
4. Cheng, C.Y., Lin, Y.: Improving dynamic performance of the Geneva mechanism using non-linear spring elements. Mech. Mach. Theory **30**(1), 119–129 (1995)
5. Heidari, M., Atai, A.A., Shariat Panahi, M.: An improved Geneva mechanism for optimal kinematic performance. Proc. Inst. Mech. Eng. Part C: J. Mech. Eng. Sci. **226**(6), 1517–1525 (2012)
6. Lee, H.P.: Design of a Geneva mechanism with curved slots using parametric polynomials. Mech. Mach. Theory **33**(3), 321–329 (1998)
7. Haraga, G., Ionita, E., Avramescu, A.M.: Analysis and modeling of Geneva mechanism. JIDEG **10**, 23–26 (2015)
8. Shekhar, A.C., Shaik, H.S., Shahab, S.: Analysis of Geneva mechanism using dwell symmetrical coupler curve mechanism. Mater. Today Proc. **39**(4), 1402–1406 (2021)
9. He, P., Hong, R., Wang, H., et al.: Fatigue life analysis of slewing bearings in wind turbines. Int. J. Fatigue **111**, 233–242 (2018)

10. Zhao, Q., Wang, Y., Huang, H., Cheng, S., Liu, F.: Fatigue life simulation and experiment of 2024 aluminum joints with multi-fasteners interference-fit. In: Niepokolczycki, A., Komorowski, J. (eds.) ICAF 2019. LNME, pp. 432–443. Springer, Cham (2020). https://doi.org/10.1007/978-3-030-21503-3_34
11. Hong, X., Xiao, G., Haoyu, W., et al.: Fatigue damage analysis and life prediction of e-clip in railway fasteners based on ABAQUS and FE-SAFE. Adv. Mech. Eng. **10**(3), 1–12 (2018)
12. Budynas, R.G., Nisbett, J.K.: Shigley's Mechanical Engineering Design, 8th edn. McGraw-Hill, New York (2008)

Tensile and Flexural Properties of Acacia Tortilis/Glass Fiber Reinforced Hybrid Composites

Jonathan B. Dawit[1], Hirpa G. Lemu[2(✉)], and Samrawit A. Tewelde[1]

[1] Dire Dawa Institute of Technology, Dire Dawa University, Dire Dawa, Ethiopia
[2] Faculty of Science and Technology, University of Stavanger, Stavanger, Norway
Hirpa.g.lemu@uis.no

Abstract. Natural fiber-reinforced polymer composites become more attractive due to their advantages such as light weight, cheap in cost and renewable but they are also susceptible for degradation, have hydrophilic natures and etc. The main objective of this study is the fabrication and characterization of Acacia tortilis/glass fiber (AT/GF) reinforced hybrid composite. After successful fabrication of Acacia tortili/glass fiber (AT/GF) reinforced hybrid composite using hand lay-up methods, tensile and flexural tests were performed and the promising result were obtained. The comparison of results with previous report and published result indicates that Acacia tortilis/glass fiber reinforced hybrid composite can be considered as a potential composite for different structural and non-structural applications. Also, the authors recommended further studies on different properties of Acacia tortilis/glass fibers reinforced hybrid composited.

Keywords: Acacia tortilis fiber · Glass fiber · Hybrid composite · Tensile property · Flexural property

1 Introduction

Composite material is a material that consists of two or more materials that are combined at a macroscopic level with different physical and chemical properties but the individual components remain separate, not soluble in each other and distinct in final product. The material, which is embedded in, is called reinforcement and the other material in which it is embedded is called the matrix. The reinforcement phase commonly existed in the form of fibers or flakes but sometimes it existed as particles. Composites have a reinforcement phase that is stiffer and stronger than the continuous matrix phase, serve as the principal load carrying members. The nature and property of the reinforcement and matrix used as criteria for the classification of composite materials [1–4].

Natural fiber reinforced composites have many advantageous characteristics over synthetic fiber reinforced composites such as light weight, cheaper, consume less energy during production, renewable, recyclable, and less abrasion on the machine during production and have less health risk [5]. Also, the desired fiber aspect ratio, high specific

Y. Wang et al. (Eds.): IWAMA 2021, LNEE 880, pp. 187–197, 2022.
https://doi.org/10.1007/978-981-19-0572-8_24

strength and stiffness are another features of green composites [6]. But the degree of deterioration over a period of time, suspicious to moisture, lower processing temperature are some of the major limitation of natural fiber reinforced composites [6, 7].

Acacia tortilis plant, locally known as Tedecha (Afaan oromo) and Girar (Amharic) in Ethiopia, is a major vegetation found in the arid and semi-arid parts of Africa and Middle East. The plant is well known as drought resistance species that grows in areas with average annual rainfall as low as 40 mm and as much as 1200 mm, and withstand higher temperature variation of 0 °C to 50 °C. In the arid and semi-arid areas of Ethiopian and other countries, Acacia tortilis plant is used for fuelwood, fodder, shade and also have the potential of improving soil fertility that lead to increased crop productivity [8–10]. Acacia tortilis plant is used as nitrogen fixing and reforestation agent [11–13] pharmaceutical and biological application [14–16] and medicinal application [17].

Alkali treatments, primary on natural fibers, use Sodium hydroxide to remove more lignin and separate fiber bundles into smaller fibers [18, 19]. Previous studies performed on alkali treatment of natural fiber reported similar results. For instance, Abu Bakar et al. [20] and Kian et al. [21] used Alkali treatment to enhance the physical and mechanical strength of Jute and Napier grass fiber respectively. The treatment of sisal fibers by 5–10% concentration of NaOH solution improved its flexural properties [22]. When compared to crude fiber, 5% NaOH treated Kenaf fiber showed better density and tensile property [23]. Alkali treated Luffa sponge fiber showed an improvement on its chemical composition and visible reduction in moisture content of the fiber [24].

Hybridization in composite materials are currently popular techniques for cellulosic/synthetic fiber reinforced polymer composites to get better mechanical properties and to overcome weaknesses of water absorption between hydrophilic fiber and hydrophobic matrix of the composites. The reinforcements can include two types of individual fibers, or a combination of fibers and multiscale particles [25, 26]. Hybrid composite materials have improved properties of flexibility, longer life cycle and cost [27]. A study performed on sisal and glass fibers hybrid composite showed that the composite material can be used as potential choice for the retrofitting of reinforced concreted structures [28]. Study done on sisal/carbon and sisal/glass hybrid fiber reinforced polypropylene (PP) composites to evaluated the tribological and mechanical responses showed improved properties [29]. Hybridization of ramie with other natural fiber and synthetic fiber showed improved mechanical properties as tensile strength, flexural strength, flexural modulus and impact strength significantly when the content of ramie fibers is increased [30, 31]. Those studies implied that hybridization of natural fiber with synthetic fibers yielded improved properties. This study focused on the fabrication and characterization of Acacia tortilis/glass fiber (AT/GF) reinforced hybrid composite. For hand lay-up manufactured composites, the tensile and flexural properties were studied.

2 Materials and Methods

2.1 Materials

The Acacia tortilis fiber used in this study was collected from Modjo, Ethiopia. The fibers were Alkali treatmented by 10 wt. % NaOH solutions to improve the fiber's surface roughness and remove unwanted materials. From previous study, Acacia tortilis

(AT) fiber contained 61.89% of cellulose, 21.26% of Lignin and 17.43% of wax with the moisture content of 6.47% [32]. Unsaturated polyester resin and randomly oriented glass fibers (GF) were purchased from World fiber glass and water proofing engineering plc, Ethiopia. The properties of the fibers and resin are given in Table 1.

Table 1. Properties of fibers and matrix

Materials	Density, g/cm^3	Tensile strength, MPa	Tensile modulus, MPa	Elongation, %
Unsaturated polyester resin [32]	1.2	50	3000	2.5
Acacia tortilis fiber (AT) [33]	0.906	71.63	4.209	1.33
Glass fiber (GF) [current study]	2.45	3450	72.4	4.8

2.2 Composite Fabrication

The Acacia tortilis/glass fiber (AT/GF) reinforced hybrid composite specimens are prepared at the ratio of 20/20 wt.% fibers and 60 wt.% of polyester resin (20:20:60). The composite was fabricated using conventional hand lay-up methods. The treated and chopped Acacia tortillas and randomly oriented glass fibers are placed in mold before unsaturated polyester resin was poured. Then the composite was cured for 24 h under hydraulic press to ensure fiber-matrix adhesions. The dried hybrid composites are then cut into sized specimens according to ASTM standard by motorized hacksaw. The fabrication process is illustrated in Fig. 1.

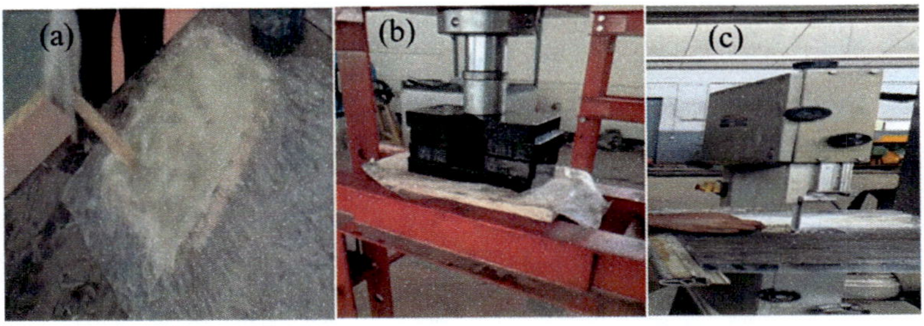

Fig. 1. (a) Hand layup manufacturing, (b) curing process (c) specimen cutting

2.3 Tensile and Flexural Test

The Acacia tortilis/glass fiber reinforced composite specimens for tensile and flexural tests have the dimensions of 250 mm * 25 mm * 5 mm and 135 mm * 20 mm * 5 mm, and followed ASTM D3039 and ASTM D7264-07 standards respectively. Axial tension and three-point bending test procedure was followed to conduct the tensile and flexural tests respectively. Both tests are performed on universal testing machine at the feed speed of 2 mm/s at room temperature as illustrated in Fig. 2.

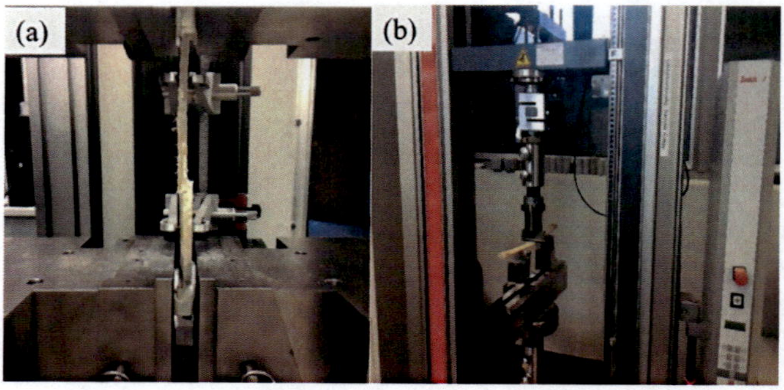

Fig. 2. (a) specimen under tensile test, (b) specimen under flexural test

3 Result and Discussion

3.1 Tensile Properties

The tensile test is used to estimate the tensile strength, Young's modulus and extension of Acacia tortilis/glass fiber reinforced composites. As showed in Table 2, the hybrid composite of Acacia tortilis/glass fiber scored 61.12 MPa of tensile strength, 4.06 GPa of Young's Modulus and 2.53 mm extension.

As shown in Table 2, previous works of the authors conducted on Acacia tortilis fiber reinforced polyester composite reported the tensile strength of 20.14 MPa and 4.82 MPa for 15wt. % and 30wt. % fiber contents respectively [34]. They also reported 3.04 GPa and 3.22 GPa tensile modulus for Acacia tortilis fiber reinforced composites for 15 wt. % and 30 wt. % fiber contents respectively. This result showed that Acacia tortilis/glass hybrid composite have better tensile performance.

Table 2. Mechanical properties of AT20GF20 hybrid composite

Sample designation	Tensile stress	Young's modulus	Extension	Flexural strength	Reference
	MPa	GPa	mm	MPa	
AT1015	20.14	3.04	0.85	25.61	[34]
AT1030	4.82	3.22	1.79	77.5	[34]
AT20GF20	61.12	4.06	2.53	249	Current study

Regarding the common natural fibers reinforced composite as given in Table 3, Acacia tortilis/glass reinforced hybrid composite showed weak tensile performance compared with 40% fiber contained Jute fiber reinforced composite [35], 57 wt. % fiber contained bamboo fiber reinforced epoxy composite [36] and 40 wt. % fiber contained banana fiber reinforced composite [37], but comparable tensile strength with 40 wt. % fiber contained sisal fiber reinforced polyester composite [38]. Compared with natural fiber reinforced hybrid composite, Acacia tortilis/glass hybrid composite exhibited comparable tensile strength with Areka husk/coir hybrid composite [39], but lower tensile strength compared with Kenaf/A and Vera/Jute hybrid composite [40].

Table 3. Tensile strength comparison of different composites

Designation	Composite	Tensile strength, MPa	Reference
AT1015	Acacia tortilis reinforced composite	20.14	[34]
AT1030	Acacia tortilis reinforced composite	4.82	[34]
AT20GF20	Acacia/glass hybrid composite	61.12	Current work
T01	Jute fiber reinforced composite	432	[35]
T02	Bamboo fiber reinforced composite	392	[36]
T03	Banana fiber reinforced composite	102	[37]
T04	Sisal fiber reinforced composite	65.5	[38]
T05	Areca husk/coir hybrid composite	64.91	[39]
T06	Kenaf/Aleo-vera/jute hybrid composite	22.32	[40]
T07	Coir/glass hybrid composite	40.7	[41]
T08	Kenaf/glass hybrid composite	18.4	[39]
T09	Abaca/glass hybrid composite	107.2	[42]
T10	Jute/glass hybrid composite	80	[43]
T11	Jute/pineapple/glass hybrid composite	71.66	[44]

When compared with other glass fiber reinforced hybrid composites, Acacai tortilis/glass fiber reinforced hybrid composite showed better performance than coir/glass fiber reinforced hybrid composite [41] and Kenaf/glass fiber reinforced hybrid composite [39] but it showed less tensile strength regardless of Abaca/glass fiber reinforced hybrid composite [42], Jute/glass fiber reinforced hybrid composite [43] and Jute/Pineapple/glass fiber reinforced hybrid composite (47) as shown in Fig. 3. Regarding tensile modulus, Acacia tortillas/glass fiber reinforced composite showed better tensile modulus respect to Abaca/glass fiber reinforced hybrid composite [42] and Jute/Pineapple/glass fiber reinforced hybrid composite [44].

3.2 Flexural Property

The three-point bending test was used to estimate flexural strength. As tabulated in Table 4, the hybrid composite of Acacia tortilis/glass fiber reinforced composite scored a flexural strength of 249 MPa. Previous studies reported that Acacia tortilis fiber reinforced composites at 15 wt.% and 30 wt.% fiber content showed the flexural strength of

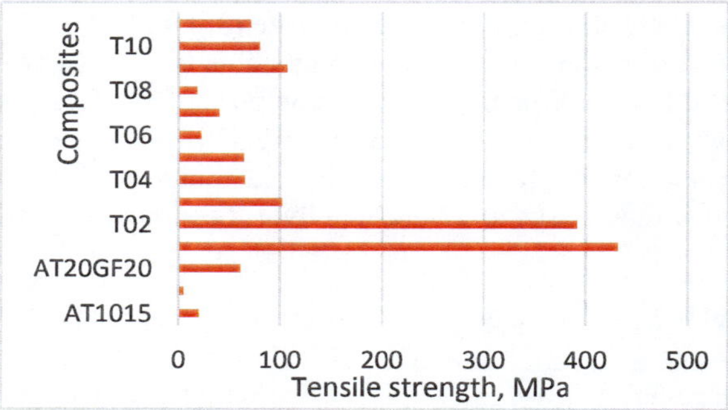

Fig. 3. Tensile strength comparison of different composites.

25.61 MPa and 75.5 MPa [34] that are lower compared with the result that are reported in this study.

When compared with natural fiber reinforced hybrid composite such as sisal/hemp fiber reinforced hybrid composite scored [45], which is lower than acacia/glass hybrid composite. Acacia/glass hybrid composite has higher flexural strength compared to Kenaf/Aloe-Vera/jute fiber reinforced hybrid composite [39], Flax/jute hybrid composite [46], Banana/jute hybrid composite [47] and sisal/banana hybrid composite [48] as shown in Fig. 4.

Table 4. Flexural strength comparison of different composites

Designation	Composite	Flexural strength, MPa	Reference
AT1015	Acacia tortilis reinforced composite	25.61	[34]
AT1030	Acacia tortilis reinforced composite	77.5	[34]
AT20GF20	Acacia/glass hybrid composite	249	Current work
F01	Sisal/hemp hybrid composite	81	[45]
F02	Kenaf/Aleo-vera/jute hybrid composite	94.81	[39]
F03	Flax/Jute hybrid composite	76.55	[46]
F04	Banana/jute hybrid composite	52.3	[47]
F05	Sisal/bamboo hybrid composite	57.47	[48]
F06	Flax/glass hybrid composite	375.35	[49]
F07	false banana/glass hybrid composite	200	[50]

(continued)

Table 4. (*continued*)

Designation	Composite	Flexural strength, MPa	Reference
F08	Rice husk/glass hybrid composite	183.29	[51]
F09	luffa/glass hybrid composite	139.96	[43]
F10	Jute/glass hybrid composite	74	[43]

Regarding the synthetic hybrid composites, Acacia/glass fiber reinforced composite has less flexural strength compared with flax/glass fiber hybrid composite [49]. But this composite has superior flexural strength over false banana/glass hybrid composite [50], rice husk/glass hybrid composite [51], luffa/glass hybrid composite [43] and jute/glass hybrid composite [43]. This showed that Acacia tortilis/glass fiber reinforced hybrid composite has better flexural strength compared with most natural fibers when they hybrid with glass fiber.

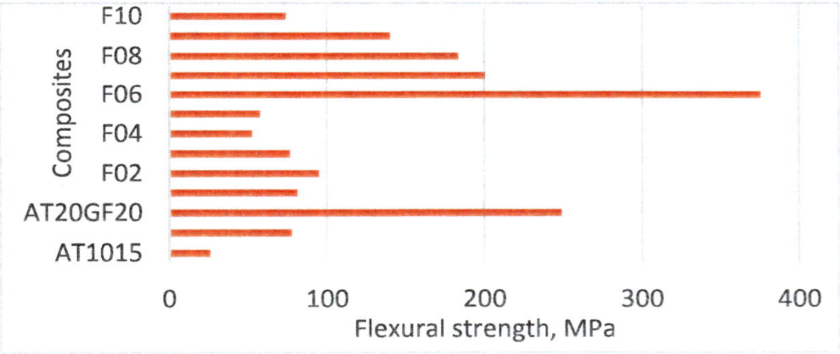

Fig. 4. Flexural strength comparison of different composites

4 Conclusion

This study investigated the tensile and flexural properties of Acacia tortilis/glass fibers hybrid reinforced composites according to ASTM standards that was manufactured successfully using conventional hand Lay-Up methods. Acacia tortilis natural fiber was collected and extracted from its bark manually and then underwent Alkali treatment to improve its surface properties. The result showed that the hybrid composite scored 61.12 MPa of tensile strength, 4.06 GPa of tensile modulus and 249 MPa of flexural strength. This revealed that hybridization of Acacia tortilis natural fiber with glass fiber improved its tensile and flexural properties compared with Acacia torilis fiber reinforced composite. Also, the hybrid composite showed better performance with natural/natural fiber hybrid composite but comparable mechanical properties with some natural fiber/glass fiber reinforced composites. Future works in this direction will be investigating additional properties of Acacia tortilis/glass fibers hybrid reinforced composite such as compression property, wettability and thermal resistance.

Acknowledgment. The authors would like to thank the University of Stavanger (Norway) and Defense University (Ethiopia) for providing their laboratory facilities for this research.

References

1. Javadian, A., Wielopolski, M., Smith, I.F.C., Hebel, D.E.: Bond-behavior study of newly developed bamboo-composite reinforcement in concrete. Constr. Build. Mater. **122**, 110–117 (2016). https://doi.org/10.1016/j.conbuildmat.2016.06.084
2. Elanchezhian, C., Ramnath, B.V., Ramakrishnan, G., Rajendrakumar, M., Naveenkumar, V., Saravanakumar, M.K.: Review on mechanical properties of natural fiber composites. Mater. Today Proc. **5**(1), 1785–1790 (2018). https://doi.org/10.1016/j.matpr.2017.11.276
3. Wei, J.: Degradation behavior and kinetics of sisal fiber in pore solutions of sustainable cementitious composite containing metakaolin. Polym. Degrad. Stab. **150**(February), 1–12 (2018). https://doi.org/10.1016/j.polymdegradstab.2018.01.027
4. Bansal, S., Ramachandran, M., Raichurkar, P.: Comparative analysis of bamboo using jute and coir fiber reinforced polymeric composites. Mater. Today Proc. **4**(2), 3182–3187 (2017). https://doi.org/10.1016/j.matpr.2017.02.203
5. Sreenivasan, S., Sulaiman, S., Ariffin, M.K.A.M., Baharudin, B.T.H.T., Abdan, K.: Physical properties of novel kenaf short fiber reinforced bulk molding compounds (BMC) for compression moulding. Mater. Today Proc. **5**(1), 1226–1232 (2018). https://doi.org/10.1016/j.matpr.2017.11.205
6. Pickering, K.L., Efendy, M.G.A., Le, T.M.: A review of recent developments in natural fibre composites and their mechanical performance. Compos. Part A Appl. Sci. Manuf. **83**, 98–112 (2016)
7. Shekar, H.S.S., Ramachandra, M.: Green composites: a review. Mater. Today Proc. **5**(1), 2518–2526 (2018). https://doi.org/10.1016/j.matpr.2017.11.034
8. AbdElRahman, H.F., Krzywinski, K.: Environmental effects on morphology of Acacia tortilis group in the red sea hills, north-eastern sudan and south-eastern egypt. Forest Ecol. Manag. **255**(1), 254–263 (2008)
9. dos Santos, F.M., de Souza, T.F., Barquete, D.M., Amado, F.D.: Comparative analysis of the sisal and piassava fibers as reinforcements in lightweight cementitious composites with EVA waste. Constr. Build. Mater. **128**, 315–323 (2016)
10. Tomar, S.S., Zafar, S., Talha, M., Gao, W., Hui, D.: State of the art of composite structures in non-deterministic framework: a review. Thin-Walled Struct. **132**, 700–716 (2018). https://doi.org/10.1016/j.tws.2018.09.016
11. Deans, J.D., Diagne, O., Nizinski, J., Lindley, D.K.: Comparative growth, biomass production, nutrient use and soil amelioration by nitrogen-fixing tree species in semi-arid Senegal. Forest Ecol. Manag. **176**(1–3), 253–264 (2003)
12. Gebrekirstos, A., Teketay, D., Fetene, M., Mitlöhner, R.: Adaptation of five co-occurring tree and shrub species to water stress and its implication in restoration of degraded lands. For. Ecol. Manag. **229**(1–3), 259–267 (2006)

13. Abdallah, F., Noumi, Z., Ouled-Belgacem, A., Michalet, R., Touzard, B., Chaieb, M.: The influence of Acacia tortilis (Forssk.) ssp. raddiana (Savi) Brenan presence, grazing, and water availability along the growing season, on the understory herbaceous vegetation in southern Tunisia. J. Arid Environ. **76**(1), 105–114 (2012). https://doi.org/10.1016/j.jaridenv.2011.06.002
14. Alam, P., Alajmi, M.F., Arbab, A.H., Parvez, M.K., Siddiqui, N.A., Alqasoumi, S.I., et al.: Comparative study of antioxidant activity and validated RP-HPTLC analysis of rutin in the leaves of different Acacia species grown in Saudi Arabia. Saudi Pharm. J. **25**(5), 715–723 (2017). https://doi.org/10.1016/j.jsps.2016.10.010
15. Gabr, S., Nikles, S., Pferschy Wenzig, E.M., Ardjomand-Woelkart, K., Hathout, R.M., El-Ahmady, S., et al.: Characterization and optimization of phenolics extracts from Acacia species in relevance to their anti-inflammatory activity. Biochem. Syst. Ecol. **78**, 21–30 (2018). https://doi.org/10.1016/j.bse.2018.03.001
16. Lakhera, A.K., Kumar, V.: Monosaccharide composition of acidic gum exudates from Indian Acacia tortilis ssp. raddiana (Savi) Brenan. Int. J. Biol. Macromol. **94**, 45–50 (2017). https://doi.org/10.1016/j.ijbiomac.2016.09.097
17. Kigondu, E.V.M., Rukunga, G.M., Keriko, J.M., Tonui, W.K., Gathirwa, J.W., Kirira, P.G., et al.: Anti-parasitic activity and cytotoxicity of selected medicinal plants from Kenya. J. Ethnopharmacol. **123**(3), 504–509 (2009)
18. Dayo, A.Q., Zegaoui, A., Nizamani, A.A., Kiran, S., Wang, J., Derradji, M., et al.: The influence of different chemical treatments on the hemp fiber/polybenzoxazine based green composites: Mechanical, thermal and water absorption properties. Mater. Chem. Phys. **217**(May), 270–277 (2018)
19. Zakikhani, P., Zahari, R., Sultan, M.T.H., Majid, D.L.: Extraction and preparation of bamboo fibre-reinforced composites. Mater. Des. **63**, 820–828 (2014). https://doi.org/10.1016/j.matdes.2014.06.058
20. Bakar, N.A., Sultan, M.T.H., Azni, M.E., Hazwan, M.H., Ariffin, A.H.: Extraction and surface characterization of novel bast fibers extracted from the Pennisetum Purpureum plant for composite application. Mater. Today Proc. **5**(10), 21926–21935 (2018). https://doi.org/10.1016/j.matpr.2018.07.052
21. Kian, L.K., Saba, N., Jawaid, M., Sultan, M.T.H.: A review on processing techniques of bast fibers nanocellulose and its polylactic acid (PLA) nanocomposites. Int. J. Biol. Macromol. **121**, 1314–1328 (2019). https://doi.org/10.1016/j.ijbiomac.2018.09.040
22. Rao, K.M.M., Rao, K.M.: Extraction and tensile properties of natural fibers: vakka, date and bamboo. Compos. Struct. **277**(3), 288–295 (2007)
23. Amel, B.A., Paridah, M.T., Sudin, R., Anwar, U.M.K., Hussein, A.S.: Effect of fiber extraction methods on some properties of kenaf bast fiber. Ind. Crops Prod. **46**, 117–123 (2013). https://doi.org/10.1016/j.indcrop.2012.12.015
24. Chen, Y., Su, N., Zhang, K., Zhu, S., Zhu, Z., Qin, W., et al.: Effect of fiber surface treatment on structure, moisture absorption and mechanical properties of luffa sponge fiber bundles. Ind. Crops Prod. **123**(January), 341–352 (2018)
25. Da Silva, L.J., Panzera, T.H., Velloso, V.R., Christoforo, A.L., Scarpa, F.: Hybrid polymeric composites reinforced with sisal fibres and silica microparticles. Compos. Part B Eng. **43**(8), 3436–3444 (2012)
26. Rider, A.N., An, Q., Brack, N., Thostenson, E.T.: Polymer nanocomposite - fiber model interphases: influence of processing and interface chemistry on mechanical performance. Chem. Eng. J. **269**, 121–134 (2015)
27. Sanjay, M., Yogesha, B.: Studies on natural/glass fiber reinforced polymer hybrid composites: an evolution. Mater. Today Proc. **4**(2), 2739–2747 (2017). https://doi.org/10.1016/j.matpr.2017.02.151

28. Padanattil, A., Karingamanna, J., Mini, K.M.: Novel hybrid composites based on glass and sisal fiber for retrofitting of reinforced concrete structures. Constr. Build. Mater. **133**, 146–153 (2017). https://doi.org/10.1016/j.conbuildmat.2016.12.045
29. Aslan, M., Tufan, M., Küçükömeroğlu, T.: Tribological and mechanical performance of sisal-filled waste carbon and glass fibre hybrid composites. Compos Part B Eng. **140**, 241–249 (2018)
30. Jawaid, M., Abdul Khalil, H.P.S.: Cellulosic/synthetic fibre reinforced polymer hybrid composites: a review. Carbohydr. Polym. **86**(1), 1–18 (2011). https://doi.org/10.1016/j.carbpol.2011.04.043
31. Hendra.: A study on cotton-ramie fabric reinforced composites. Int. J. Mater. Sci. **12**(1), 117–125 (2017)
32. Hancox, N.L.: Engineering mechanics of composite materials, vol. 17, 114 p. Materials & Design (1996)
33. Dawit, J.B., Regassa, Y., Lemu, H.G.: Property characterization of acacia tortilis for natural fiber reinforced polymer composite. Results Mater. **5**, 100054 (2020)
34. Dawit, J.B., Lemu, H.G., Regassa, Y., Akessa, A.D.: Investigation of the mechanical properties of Acacia tortilis fiber reinforced natural composite. Mater. Today Proc. (2020). https://doi.org/10.1016/j.matpr.2020.09.308
35. Cruz, J., Fangueiro, R.: Surface modification of natural fibers: a review. Procedia Eng. **155**, 285–288 (2016). https://doi.org/10.1016/j.proeng.2016.08.030
36. Biswas, S., Shahinur, S., Hasan, M., Ahsan, Q.: Physical, mechanical and thermal properties of jute and bamboo fiber reinforced unidirectional epoxy composites. Procedia Eng. **105**, 933–939 (2015). https://doi.org/10.1016/j.proeng.2015.05.118
37. Preet Singh, J.I., Dhawan, V., Singh, S., Jangid, K.: Study of effect of surface treatment on mechanical properties of natural fiber reinforced composites. Mater. Today Proc. **4**(2), 2793–2799 (2017). https://doi.org/10.1016/j.matpr.2017.02.158
38. Ratna Prasad, A.V., Mohana, R.K.: Mechanical properties of natural fibre reinforced polyester composites: jowar, sisal and bamboo. Mater. Des. **32**(8–9), 4658–4663 (2011). https://doi.org/10.1016/j.matdes.2011.03.015
39. Sailesh, A., Arunkumar, R., Saravanan, S.: Mechanical properties and wear properties of kenaf - aloe vera - jute fiber reinforced natural fiber composites. Mater. Today Proc. **5**(2), 7184–7190 (2017). https://doi.org/10.1016/j.matpr.2017.11.384
40. Padmaraj, N.H., Keni, L.G., Chetan, K.N., Shetty, M.: Mechanical characterization of areca husk-coir fiber reinforced hybrid composites. Mater. Today Proc. **5**(1), 1292–1297 (2018). https://doi.org/10.1016/j.matpr.2017.11.214
41. Shrivastava, R., Telang, A., Rana, R.S., Purohit, R.: Mechanical properties of coir/G lass fiber epoxy resin hybrid composite. Mater. Today Proc. **4**(2), 3477–3483 (2017). https://doi.org/10.1016/j.matpr.2017.02.237
42. Punyamurthy, R., Sampathkumar, D., Bennehalli, B., Ranganagowda, R.P., Badyankal, P.V., Venkateshappa, S.C.: Research article abaca fiber reinforced hybrid composites. Int. J. Appl. Eng. Res. **9**(23), 20273–20286 (2014)
43. Sudhabindu, B., Udayakiran, A.C.: The mechanical properties of jute/E-glass fiber reinforced polymer composites influenced by hygrothermal effects. Mater. Today Proc. **5**(2), 6799–6804 (2018). https://doi.org/10.1016/j.matpr.2017.11.339
44. Reddy, M.I., Kumar, M.A., Raju, C.R.B.: Tensile and flexural properties of jute, pineapple leaf and glass fiber reinforced polymer matrix hybrid composites. Mater. Today: Proc. **5**(1), 458–462 (2018)
45. Gupta, N.V., Rao, K.S.: An experimental study on sisal/hemp fiber reinforced hybrid composites. Mater. Today: Proc. **5**(2), 7383–7387 (2018). https://doi.org/10.1016/j.matpr.2017.11.408

46. Karthi, N., Kumaresan, K., Sathish, S., Prabhu, L., Gokulkumar, S., Balaji, D., et al.: Effect of weight fraction on the mechanical properties of flax and jute fibers reinforced epoxy hybrid composites. Mater. Today Proc. (2021). https://doi.org/10.1016/j.matpr.2020.12.1060
47. Ravindran, S., Sozhamannan, G.G., Saravanan, L., Venkatachalapathy, V.S.K.: Study on mechanical behaviour of natural fiber reinforced vinylester hybrid composites. Mater. Today Proc. (2021). https://doi.org/10.1016/j.matpr.2020.12.1077
48. Siva, R., Kesavaram, B., Jones Martin, J., Mathiselvan, G., Navas, K.B., Sangeetha, M.: Mechanical behavior of sisal and banana fiber reinforced hybrid epoxy composites. Mater. Today Proc. (2020). https://doi.org/10.1016/j.matpr.2020.10.805
49. Umapathi, D., Devaraju, A., Harikumar, R.: Exploration of mechanical properties of flax fibre and GFRP hybrid composites. Mater. Today Proc. **39**, 875–878 (2020). https://doi.org/10.1016/j.matpr.2020.10.938
50. Batu, T., Lemu, H.G.: Investigation of mechanical properties of false banana/glass fiber reinforced hybrid composite materials. Results Mater. **8**, 100152 (2020). https://doi.org/10.1016/j.rinma.2020.100152
51. Ismail, M., Rejab, M.R.M., Siregar, J.P., Mohamad, Z., Quanjin, M., Mohammed, A.A.: Mechanical properties of hybrid glass fiber/rice husk reinforced polymer composite. Mater. Today Proc. **27**, 1749–1755 (2020). https://doi.org/10.1016/j.matpr.2020.03.660

Multi Robot Welding Path Planning Based on Improved Ant Colony Algorithm

ZiFeng Xu[1(✉)], Lilan Liu[1], Wei Zou[2], Tao Xu[1], and Shibo Yuan[1]

[1] Shanghai Key Laboratory of Intelligent Manufacturing and Robotics, Shanghai University, Shanghai, China
xuzifeng@shu.edu.cn
[2] Aerospace System Engineering Shanghai, Shanghai, China

Abstract. In view of the unreasonable path planning of multiple robots in the existing BIW Welding process, the welding process of multiple robots is analyzed, the mathematical model is established, and the improved ant colony algorithm is used to calculate, so as to obtain the optimal solder joint task allocation and the optimal welding path. The program is written in Python language to solve the algorithm. The calculation results show that the improved ant colony algorithm can effectively allocate the solder joints of multiple welding robots, and plan a reasonable welding path according to the assigned solder joints.

Keywords: Welding robot · Task assignment · Ant colony algorithm · Path planning

1 Introduction

With the rapid development of automobile industry, intelligent manufacturing has gradually become the production mode of main engine factory. In the whole vehicle production process, industrial robots are widely used in all aspects of production, especially in the BIW Welding process [1]. The welding of vehicle body in white is not completed by a single welding robot on the production line. In the area with dense welding spots, it is generally completed by multiple welding robots. Therefore, in the case of multi robot cooperative work, it is necessary to consider the overall task allocation and path planning of welding robot to achieve the optimal.

Aiming at the research of multi robot cooperative welding process, the ant colony algorithm is introduced and improved to establish the mathematical model of solder joint task allocation and path planning. Finally, the software is used to solve the calculation to obtain the task points and welding path of multi robot, so that the efficiency of multi robot welding process can be optimized.

2 Problem Description

The key station in the welding line of BIW usually uses multiple robots to weld at the same time, which can reduce the resource occupation [2, 3]. When multiple robots work

Y. Wang et al. (Eds.): IWAMA 2021, LNEE 880, pp. 198–206, 2022.
https://doi.org/10.1007/978-981-19-0572-8_25

at the same time, the problem of robot cooperation should be considered. When each welding robot completes the assigned task, each robot does not interfere and operates independently. Therefore, the welding process of multiple robots can be summarized as follows:

(1) The welding points of the station are allocated reasonably [4], and each robot is assigned to the reasonable welding task;
(2) Each robot does not interfere and collide in their own process, and the welding process is the best, that is, the path planning problem of a single welding robot.

3 Mathematical Model

3.1 Mathematical Model of Multi Robot Welding Task Assignment

Solder joint assignment is to assign solder joint to robot reasonably on the premise of considering process factors. Solder joint assignment can be transformed into knapsack problem. Mathematical description of robot welding task assignment problem: given the n solder joints set $C = \{c_1, c_2 ... c_n\}$ and m robots set $R = \{R_1, R_2 ... R_m\}$.The solder joints are assigned to the robot reasonably, and the optimal welding path of the system is found under the constraint conditions. The relationship matrix $X_{m \times n}$ can be used to represent the distribution relationship of solder joints [5], X_{ij} means that i solder joints are assigned to robot j, 1 is assignment, 0 is not allocated, and abstract into mathematical model as follows:

$$\min F = \sum_{j=1}^{m}\sum_{i=1}^{n} X_{ij} V_{ij}$$

$$s.t. \begin{cases} \sum_{i=1}^{n} X_{ij} = W_J \quad \forall j \in \{1, 2, ..., m\} \\ \sum_{j=1}^{m} X_{ij} = 1 \quad \forall i \in \{1, 2, ..., n\} \\ D_{ij} \in Y_j \qquad Y_j \in |r_{\min}, r_{\max}| \\ \qquad X_{ij} \in \{0, 1\} \end{cases} \tag{1}$$

Where V_{ij} is the distance between solder joint c_i and robot R_j, and W_j is the upper limit of the number of points allocated by robot R_i. Constraint 2 means that each solder joint is assigned to only one robot, constraint 3 means the reachable range of the welding robot, Y_j means the working radius of the j-th robot, $r_{\min}, r_{\max}$ means the minimum and maximum working radius. Solder joint task allocation problem is a combinatorial optimization problem, there are many intelligent algorithms to solve, comprehensive algorithm advantages, ant colony algorithm has high efficiency and fast convergence speed [6], so this paper will introduce ant colony algorithm to solve solder.

3.2 Multi Robot Welding Task Assignment Based on Basic Ant Colony Algorithm

Ant colony algorithm is also a kind of simulated evolutionary algorithm. Its basic principle is: in the process of searching for real objects, ants will release pheromones on the path they pass, and other ants can perceive the pheromones when they pass the path, and release their own pheromones at the same time. Therefore, the increase of pheromone concentration on this path will attract more ants to choose this path and form the optimal path for ants to find food [7]. Ant colony algorithm is for the whole, a single ant can not find the best path. The principle of applying ant colony algorithm to solder joint assignment is as follows

(1) Population initialization

Population initialization includes ant colony initialization, pheromone initialization and tabu list initialization.

a. Ant colony initialization

The solder joint is regarded as ant, the number of ant is n, and the ant is placed on the solder joint [8].

b. Pheromone initialization

Solder joint c_i is assigned to robot R_j, which can generate $n \times m$ paths. Therefore, the pheromone released by ants is placed on the path. Assuming that the pheromone concentration on the path connected by solder joint i and robot j at t time is $\tau_{ij}(t)$, initialization is that the pheromone value on each path is the same, i.e. $\tau_{ij}(0) = \tau_0$.

c. Taboo table initialization

Taboo table is the set of elements assigned by solder joints, and the set of solder joints assigned by each robot is recorded, and the initialization taboo table is empty.

(2) The selection strategy of solder joint assignment

After initialization, ant k will select the next path according to the pheromone $\tau_{ij}(t)$ and heuristic information $\eta_{ij}(t)$ on the path, and select solder joints according to the following functions:

$$P_{ij}^k(t) = \begin{cases} \frac{[\tau_{ij}(t)]^\alpha [\eta_{ij}(t)]^\beta}{\sum\limits_{x \in allowed_k} [\tau_{ij}(t)]^\alpha [\eta_{ij}(t)]^\beta} & x \in allowed_k \\ 0 & \text{other} \end{cases} \tag{2}$$

Where $\tau_{ij}(t)$ is the pheromone concentration on the path and $\eta_{ij}(t)$ is the heuristic function, representing the distance intensity factor between the solder joint and the robot. α stands for pheromone heuristic factor, β is pheromone effect degree; It represents the importance of heuristic function, that is, the possibility of selecting a robot with a short distance; $P_{ij}^k(t)$ is the probability that ant k assigns solder joint i to robot j at time t.

(3) Pheromone update

$$\begin{cases} \tau_{ij}(t+1) = (1-\rho)\tau_{ij}(t) + \Delta\tau_{ij} \; 0 < \rho < 1 \\ \tau_{ij} = \sum\limits_{k=1}^{n} \Delta\tau_{ij}^k \end{cases} \tag{3}$$

ρ is the speed of pheromone disappearance on the path, $\Delta\tau_{ij}^{k}$ is the pheromone concentration of the kth ant on the path (i, j), and $\Delta\tau_{ij}$ is the sum of pheromone concentrations released by all ants on the path (i, j). The pheromone is updated by using ant colony cycle strategy model.

$$\Delta\tau_{ij}^{k} = \begin{cases} Q/L_k \text{ The path of ants}(i,j) \\ 0 \end{cases} \tag{4}$$

Where Q is constant and L_k is the length of the whole welding path from the k-th ant. After the solder joint assignment, the path length of each ant is calculated to update the global pheromone.

3.3 Multi Robot Welding Task Assignment Based On Improved Ant Colony Algorithm

Ant colony algorithm has strong adaptability and cooperation in the search performance, mainly in the accumulation of pheromones, constantly optimize the structure to achieve the goal of optimization, without the effect of external factors, the whole ant colony system can develop towards the direction of optimal solution. For multi welding robot cooperative task allocation problem is a combinatorial optimization problem, the basic ant colony algorithm can find the optimal solution, but there are some defects: global search leads to the increase of calculation, slow convergence speed and easy to fall into the optimal solution, so this paper will improve the ant colony algorithm, improve the heuristic function to improve the convergence speed.

(1) Improved heuristic function

$\eta_{ij}(t)$ is a heuristic function, which represents the distance intensity factor between solder joint and robot, and its value is inversely proportional to the distance between nodes. As a self-organized structure, ant colony algorithm has the disadvantage of long search time. Therefore, the heuristic function $\eta_{ij}(t)$ is changed to:

$$\eta_{ij}(t) = 1/d_{ij}^{2} \tag{5}$$

d_{ij} represents the distance between the solder joint and the robot, and the heuristic function is inversely proportional to the square of the distance, which greatly improves the effectiveness of the algorithm, reduces the convergence time, and ensures that the search direction of the shortest path of ant search is towards the shortest target.

(2) Improved pheromone update strategy

Because the basic ant colony algorithm is easy to fall into the local optimum in the pheromone update, more and more ants will choose the path with more pheromones and ignore other paths. Therefore, this paper improves the pheromone update strategy, and applies the longest path and the shortest path of each iteration to the pheromone update, This method not only considers the local search, but also considers the global search, which ensures that the search will not fall into the local optimum, and can ensure the optimal solution and reduce the search time. The improved pheromone updating strategy is as follows:

$$\tau_{ij}(t+1) = (1-\rho)\tau_{ij}(t) + \Delta\tau_{ij} + \sqrt{L_{short}/L_{long}} \tag{6}$$

L_{short} is the shortest path. L_{long} is the longest path. The improved updating strategy makes the pheromone update, increases the amount of pheromone on the path of elite ant, and develops towards the direction of optimal solution.same, the misjudgment rate is a function of k, so the optimal value of k in the bit array set is misjudged.

3.4 Algorithm Flow

The flow chart of solder joint allocation based on improved ant colony algorithm is shown in Fig. 1:

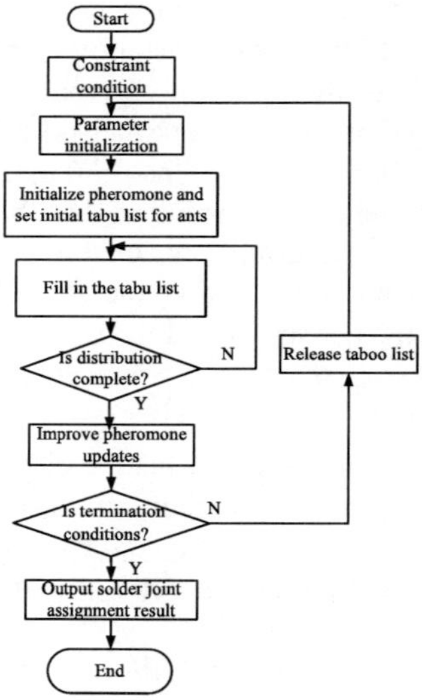

Fig. 1. Solder joint assignment flow chart based on ant colony algorithm

4 Multi Robot Welding Task Assignment and Path Planning

The problem of welding path planning for multiple welding robots is divided into two problems: solder joint assignment and single robot path planning. Genetic algorithm is used to find the optimal welding path. The welding task assignment and path planning of multiple welding robots can be divided into three steps

(1) The objective function and constraint conditions of multi robot welding task assignment problem are designed;

(2) The improved ant colony algorithm is used to solve the knapsack problem of multi robot cooperative solder joint assignment;
(3) Genetic algorithm is used to plan the path of each robot;

Taking two robots for repair welding of front floor in I repair welding 2 of body in white as the research object, the simulation of solder joint distribution and path planning is carried out by using Python language.

In order to verify the effectiveness of the improved ant colony algorithm in the assignment of solder joint tasks and the welding path planning of genetic algorithm, two robot solder joints will be planned. The results show that the shortest path of machine 1 is 33, 30, 29, 27, 22, 8, 11, 6, 20, 19, 5, 9, 25, 15, 14, 1. The digital arrangement represents the welding sequence, with the total path length of 515.047 mm. The shortest path of robot 2 is: 32, 31, 28, 26, 23, 10, 7, 18, 17, 4, 21, 12, 2, 16, 24, 3, and the total path length is 639.66 mm.

From the results of welding point assignment of two robots, the number of points allocated is balanced, and the welding distance is close. The iteration times start to converge at 80 times, which meets the distribution requirements. The results of the two welding robots are shown in Figs. 2 and 3.

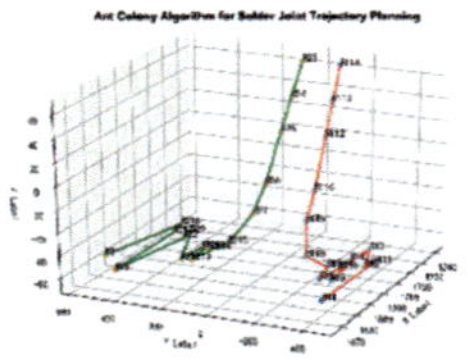

Fig. 2. Welding path assignment results of two robots

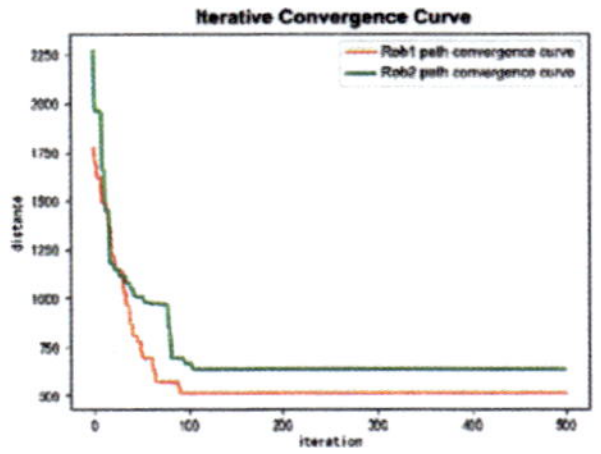

Fig. 3. Convergence curve of solder joint planning

According to the formula, the position and distance of poles of two welding guns are calculated, and the distribution curve is obtained. Set the safety distance threshold of welding torch, and the distance curve value is always greater than the safety threshold, as shown in Fig. 4.

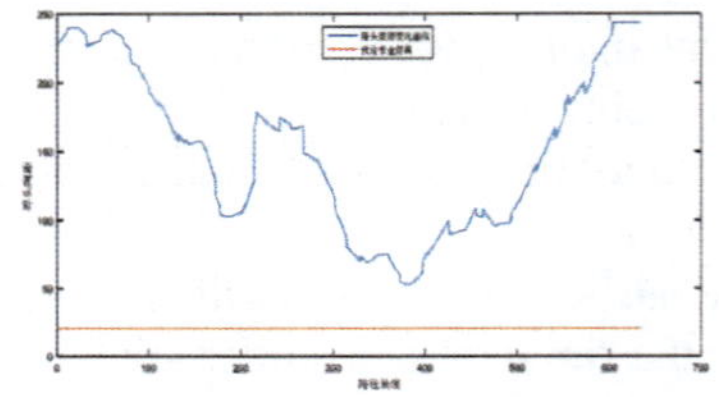

Fig. 4. Pole distance curve of welding torch

Because the two robots' solder joint assignment and path planning can not prove the generality of the algorithm in solder joint assignment and path planning, four robots with two positions of front and rear floor overlapping repair welding in I repair welding 2 of body in white were selected for path planning simulation. After iteration, the distribution results and planning results are shown in Table 1.

Table 1. Solder joint distribution and solder joint path

Robot	Solder joint path	Num	Path (mm)
Rob1	2, 3, 5, 10, 7, 11, 17, 21, 23, 26, 14, 15	12	639.224
Rob2	22, 25, 24, 20, 18, 12, 8, 1, 4, 6, 9, 13, 19, 16	14	663.869
Rob3	46, 44, 36, 41, 37, 33, 32, 29, 30, 27	11	561.807
Rob4	38, 42, 48, 47, 43, 39, 35, 34, 31, 28	11	558.952

According to the results of solder joint assignment of four robots, the number of assigned points is balanced, and the welding distance is relatively close, and according to the distance curve of the poles of two welding guns, they are all within the safety threshold, as shown in Fig. 5, which meets the assignment and safety requirements. The effectiveness and rationality of ant colony algorithm in solder joint distribution and path planning are verified.

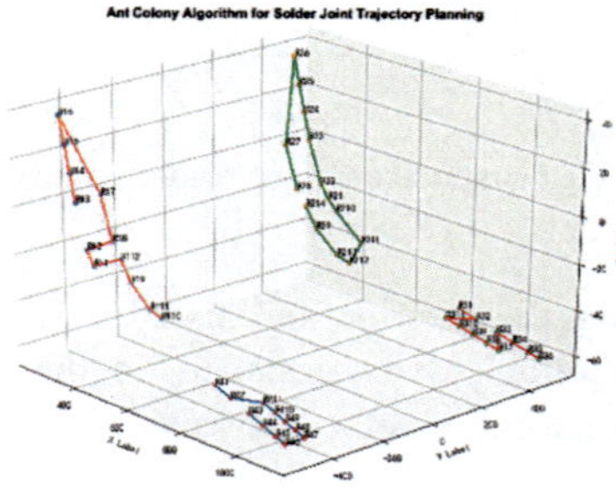

Fig. 5. Welding path planning effect of four robots

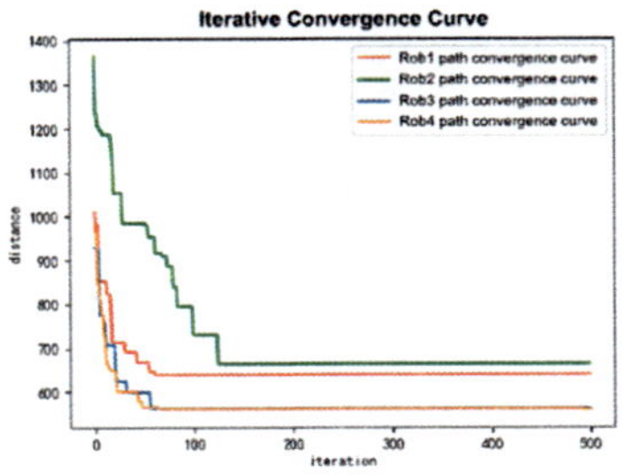

Fig. 6. Iterative curve of welding path planning

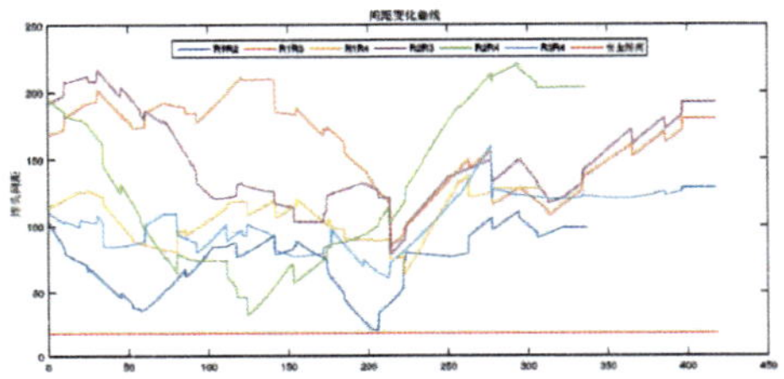

Fig. 7. Pole distance curve of welding torch

Through the planning of the welding path of the above two and four robots, the generality of ant colony algorithm and genetic algorithm in multi robot path planning is preliminarily proved. Therefore, the algorithm is extended to the geo spot location of the BIW lower I assembly welding production line, with 8 robots, the upper limit value of welding point is 5, and the total length of the welding path is 4064.633 mm.

5 Conclusions

In order to optimize the complexity of the industrial model, the current model optimization algorithm is studied, and a triangle folding optimization algorithm is proposed to perform lightweight processing on the production line model. The resulting simplification effect is obvious for the multi-source data collection process of the production line. For the problem of redundant data, an improved Bloom Filter algorithm is proposed to perform redundant data filtering on multi-source data to improve data accuracy and achieve the data requirements in the digital twin virtual-real synchronization system.

Acknowledgements. The authors would like to express appreciation to mentors in Shanghai University for their valuable comments and other helps. Thanks for the program supported by State Administration of Science. Technology and Industry for National Defense (JKCY2020413C002).

References

1. Cao, X., Wang, Z., Feng, J., et al.: Global path planning of robot based on improved ant colony algorithm. Comput. Eng. Sci. **42**(303(03)), 187–193 (2020)

2. Wu, M., Huang, H., Wang, X.: Robot welding path planning based on improved ant colony algorithm. Hanjie Xuebao/Trans. China Weld. Inst. **39**(10), 113–118 (2018)
3. Yi, G., Feng, Z., Mei, T., Li, P., Jin, W., Chen, S.: Multi-AGVs path planning based on improved ant colony algorithm. J. Supercomput. **75**(9), 5898–5913 (2019). https://doi.org/10.1007/s11227-019-02884-9
4. Wang, M., Zhu, C., Wang, F., et al.: Multi-factor of path planning based on an ant colony optimization algorithm. Ann. GIS **18**, 1–12 (2020)
5. Farooq, B., Bao, J., Raza, H., et al.: J. Manuf. Syst. **59**(1), 98–116 (2021)
6. Jinjiaqi, Liu, C., Xuzhenwei: Path planning of welding robot based on improved ant colony algorithm. Heavy Mach. **26**(1), 75–77 (2017)
7. Jiang, M., Wang, F., Ge, Y., et al.: Research on path planning of mobile robot based on improved ant colony algorithm. J. Instrum. **040**(002), 113–121 (2019)
8. Li, R., Teng, Z.: AGV path planning based on improved ant colony algorithm. Mech. Eng. Autom. **27**(1), 70–72 (2020)

Data Acquisition and Processing of Space Docking Physics Experiment Platform

Tao Xu[1(✉)], Lilan Liu[1], Wei Zhou[2(✉)], and Shibo Yuan[1]

[1] Shanghai Key Laboratory of Intelligent Manufacturing and Robotics, Shanghai University, Shanghai, China
tyance@shu.edu.cn
[2] Aerospace System Engineering Shanghai, Shanghai, China

Abstract. In order to ensure the smooth progress of space docking, it is necessary to complete the testing, inspection and acceptance of the ground subsystem of the docking mechanism through physical or semi-physical simulation. In this paper, data collection is carried out for the six-degree-of-freedom space docking physical test bed. The data may be affected by the environment during the collection process, so it is necessary to process the collected data. The denoising autoencoder algorithm is used to process the data, so as to enhance the stability and accuracy of the data set. Provide data assurance for security and feasibility.

Keywords: Space docking mechanism · Data collection · Data filtering · The denoising autoencoder

1 Introduction

Due to the high safety requirements of space spacecraft, the reliability and stability of spacecraft docking structure should be considered first, and various factors should be The kinematics equation is solved and the kinematics characteristics of the full docking process are evaluated safely [1]. The successful docking of spacecraft can greatly extend its service life, reduce the cost of experiments, and expand its range of capabilities through modular replacement. It is particularly important that the space docking mechanism with high success rate needs ground simulation equipment to test a large number of space docking theories and virtual processes, so as to ensure its high reliability [2, 3].

The docking process is often affected by the complex actual environment. Only physical or semi physical simulation can complete the testing, inspection and acceptance of the docking mechanism ground subsystem. This method is a common practice in the world. Moreover, its semi physical simulation research has the advantages of quick adjustment of initial conditions, track parameters and multi environment rapid simulation [4]. It can consider various factors comprehensively and simulate the operation of each subsystem to ensure the safety and reliability of the docking process.

Y. Wang et al. (Eds.): IWAMA 2021, LNEE 880, pp. 207–213, 2022.
https://doi.org/10.1007/978-981-19-0572-8_26

2 Composition of Space Docking Mechanism

The comprehensive test platform of docking mechanism includes: base, support, support frame, various sensing facilities, active docking mechanism, passive docking mechanism, temperature control system and hydraulic servo control system. As shown in Fig. 1.

Fig. 1. Composition of experimental platform

As shown in Fig. 1, Docking test is mainly divided into six steps, which are extension, capture, buffer correction, tension, locking and separation. Firstly, according to the real space docking trajectory, the motion trajectory is generated in the motion simulator. By solving the trajectory, the command is sent to the hydraulic servo mechanism. According to the specified trajectory, the passive docking mechanism is extended, and when it reaches the specified position, it is captured, buffered and corrected, and the docking lock is tightened and locked. When separating, unlock and release the lock in turn after receiving the command. Complete the whole docking process. The warm field system is to configure the docking environment before the whole docking process, and the warm field cover covers all the docking mechanisms.

3 Data Acquisition Module of Experimental Platform

Monitoring the docking test process requires real-time monitoring of the docking test process and key parameters. Real-time monitoring of docking process mainly uses multi-source heterogeneous data during docking test platform test. Various devices of source sensor for multi-source heterogeneous data include gravity balance device, temperature field control system, active docking mechanism, passive docking mechanism and hydraulic servo mechanism. As shown in Fig. 2.

Because the passive docking mechanism of the test bench simulates the vacuum weightlessness condition, the passive docking mechanism is weightless during the docking process due to the tension provided by the hanging rope on the mechanism, so a

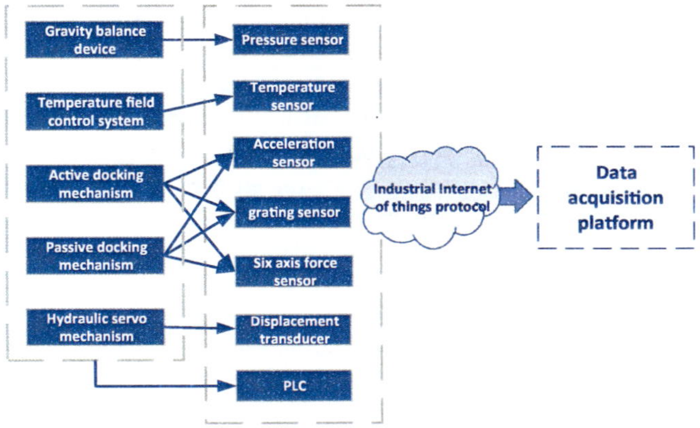

Fig. 2. Sensing facility requirements

pressure sensor is set on the disk contact surface of the mechanism to collect the pressure values during the docking contact process. In the docking mechanism test bench, there is a module to control the movement of the active docking mechanism, that is, the hydraulic servo drive mechanism. The six-degree-of-freedom hydraulic servo mechanism receives the motion track of the simulation software to achieve the specified track movement. In order to detect the stroke and position of the hydraulic pole, a displacement sensor and a six-dimensional sensor are set up. The displacement sensor is used to detect the stroke of the hydraulic rod, and the six-dimensional sensor is used to measure the position and position information at the end of the platform. A six-dimensional probe is mounted on a moving platform and tracked by a laser tracker fixed on the ground to collect six-dimensional position information of the moving platform.

During the docking process, because the rotation angle of the hanger and the displacement of the moving base cannot be measured by the pulses emitted by the motor, a raster sensor is needed to measure the moving signals in three axes which have already been rotated in the Z-axis direction and are fed back to the control device. Feedback control is performed by determining whether a preset array has been reached based on the returned signal information.

The most important parameters in the docking process are the important parameters for the weak collision process of the active docking mechanism and the passive docking mechanism, such as angle, torque, impact force and vibration information. The collision force during the docking process needs to be measured using a force sensor. A force/moment limit sensor allows docking to occur within a controlled range. The process of docking begins when the guide valve of the active and passive docking mechanism starts to touch, so a contact sensor is also required on the guide valve. Real-time monitoring is required from the beginning to the end of the docking lock, so a six-dimensional force sensor is installed on the docking ring to measure and feed back to the control system, and dozens of adjustments are made to the position and posture of the docking ring. After the docking ring is docked, the electromagnet is energized to complete the flexible capture.

The temperature field simulation system of the test bench needs to simulate various high and low temperatures in space. Temperature sensors are needed to monitor the temperature of the temperature field. Temperature information of the temperature field is measured by using thin-film temperature sensor, space point temperature sensor and air tube temperature sensor. Considering the temperature difference between the environment and the inside of the temperature field, the air duct temperature sensor must be well moisturized. Therefore, the installation location of the air duct temperature sensor should also be considered. The space point temperature sensor is connected to the outside using a vacuum plug-in, the inside is fixed with an orifice plate, and is hoisted at the desired temperature point. Dew point temperature sensor is required to meet the use requirements of the measured gas media temperature between −80–80 °C. Therefore, the measured gas needs to be pre-processed first to make the temperature of the gas entering the dew point temperature sensor within the required range.

In addition to some key parameter information obtained by the sensor, some control signals can be obtained directly from the PLC. Specific content is obtained according to the agreement. Other information includes environmental information, design drawings, and so on. Environmental information includes temperature, humidity, atmospheric pressure, and so on. The design drawings are digital descriptions of the physical test bench, including the dimensions of the three-dimensional geometric model of the test bench.

In addition to some key parameter information obtained by the sensor, some control signals can be obtained directly from the PLC. Specific content is obtained according to the agreement. Other information includes environmental information, design drawings, and so on. Environmental information includes temperature, humidity, atmospheric pressure, and so on. The design drawings are digital descriptions of the physical test bench, including the dimensions of the three-dimensional geometric model of the test bench.

When the sensor of the test platform is deployed, the data collection framework of the test platform is shown in Fig. 3, taking into account the internal parameters of the hardware mechanism system.

Based on the above collection framework, the specific process of data collection as shown in the figure can be obtained. For the data needed in digital twin modeling, select the desired signal points in the PLC, write a specific script to get the point signal data in real time through the OPC protocol, and some docking state parameters that need to be detected in the docking process, use a variety of sensors to collect, because the acquisition process involves a variety of protocols. Writing scripts for data acquisition based on different protocols for each sensor is tedious. By using a multi-device Information Association collector based on OPC server, it encapsulates nearly 150 communication protocols internally, which enables real-time multi-node acquisition. Because the OPC protocol can be compatible with most industrial Internet sensor protocols and PLC protocols, using the OPC protocol can greatly reduce the amount of programming. Threads can be reduced and system efficiency improved during transmission.

The actual access flow of the system communication is shown in Fig. 4 below.

The upper layer can read/write to each other through the OPC interface, I/O interface of the PLC and various sensor facilities, in which the OPC interface obtains device status data by using the connected channel, IP address and label address of each data.

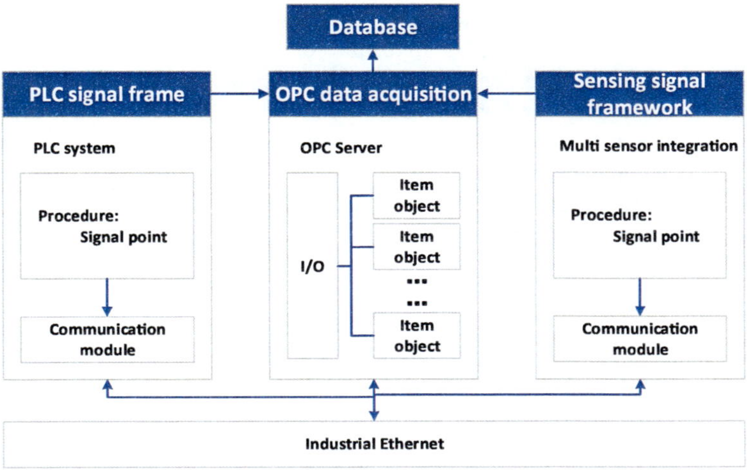

Fig. 3. Collection flowchart

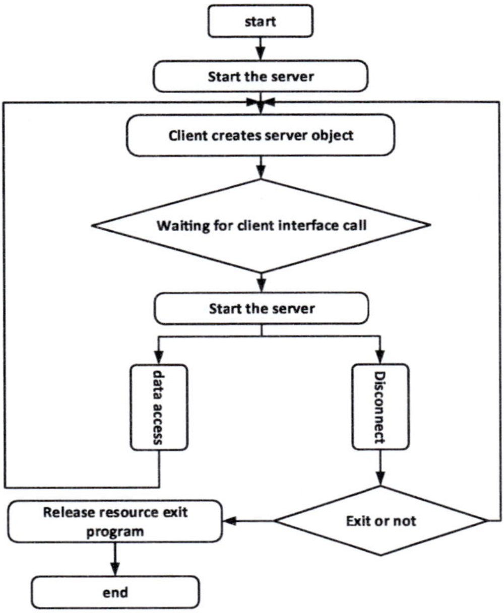

Fig. 4. The actual access process of system data communication

4 Data Processing of Docking Test Based on DAE

The data generated during the experiment is very large, and the values of each characteristic parameter may be different. In the later stage of this paper, we need to search for the similarity of the docking state of the experiment, and need to measure the similarity. In order to reduce the impact of the data scaling on the similarity measurement, we propose to use the noise reduction self-coding algorithm for data denoising.

Self-encoder is an unsupervised learning technology belonging to feed-forward neural network. It uses a back-propagation algorithm to make the output value equal to the input value of the neural network. The self-encoder first compresses the input layer into a low-dimensional eigenvector, and then reconstructs the eigenvector to the output. The basic structure of the self-encoder consists of an input layer, a hidden layer and an output layer. The network structure of Autoencoder is shown in Fig. 5.

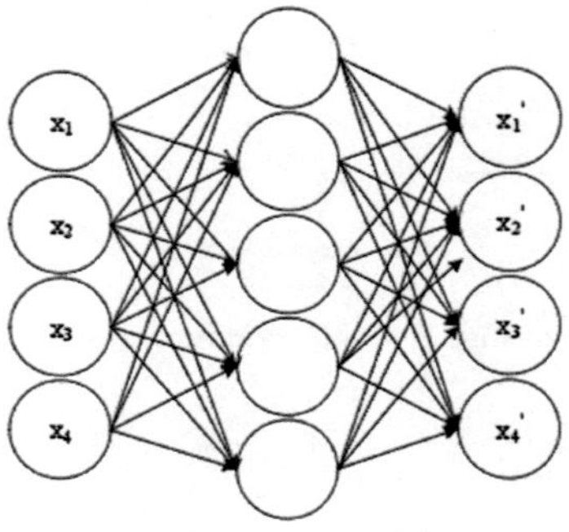

Fig. 5. Coding structural diagram

As you can see from Fig. 5, a self-encoder can be divided into two parts: an encoder (Encoder) and a decoder (Decoder). Encoders encode the input as potential features in a low dimension, and decoders refactor the potential features as output. Formulas 1 and 2 can be used to represent: is the potential feature representation of

$$encoder : z = f(x) = \sigma(W_1 x + b_1) \tag{1}$$

$$decoder : \hat{x} = g(z) = \sigma(W_2 x + b_2) \tag{2}$$

Where $x \in R^{1\times D}$ is the self encoded input, $\hat{x}$ is the output of the encoder. $Z \in R^{1\times K}$ is the potential feature representation of input x. K is the dimension of the latent eigenvector and K << D. W_1 and W_2 is the weight matrix of the network. b_1 and b_2 is the bias vector of the network. $\sigma(\ldots)$ is the activation function.

Although self-encoder structure can be used for feature extraction or data dimensionality reduction, it may not be able to effectively learn features from input data due to insufficient training set or complex model during training. Therefore, in order to solve the above problems and enhance the generalization and robustness of the basic self-encoder, the denoising autoencoder (DAE) is proposed in the paper [63]. The main idea of DAE is to add noise to the input layer, or to restore the input layer as far as possible before adding noise. Assuming that the original input layer is X, the input layer after adding noise is X, and Y is the reconstructed output, the error calculation formula for noise reduction auto coder is shown in formula 3:

$$minL(X, Y) = ||X - Y||^2 \tag{3}$$

The network structure of the noise reduction self encoder is shown in Fig. 6. There are many ways for DAE to add noise to the original input. There are two most commonly used ways. One is to mask the noise. First, a threshold of masking noise is determined $p(0 < p < 1)$, One random floating-point number $r(0 < r < 1)$ is generated at a time. If $r < p$, or let $\overline{X}_i$ is 0, or else let $\overline{X}_i = X_i$. The other is Gaussian noise.

In this paper, a noise reduction self-coding algorithm is used. Data acquisition in temperature field may contain noise, so the input layer of the encoder uses the acquisition values of each sensor device under the high and low temperature experiment. The output of the model is selected as a sensor at normal room temperature for model training.

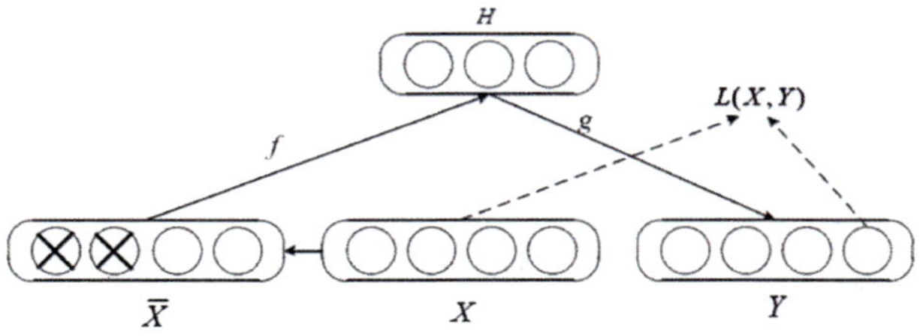

Fig. 6. DAE structure diagram

5 Conclusions

In order to reduce the influence of other factors on the data in physical experiments, a noise reduction self coding algorithm is proposed to reduce the noise of the data in physical experiments. For the multi-source data acquisition process of physical experiment, in order to reduce the impact of data scaling on the similarity measurement, the data is processed to improve the data accuracy and meet the data requirements of digital twin virtual real synchronization system.

Acknowledgements. The authors would like to express appreciation to mentors in Shanghai University for their valuable comments and other helps. Thanks for the program supported by State Administration of Science. Technology and Industry for National Defense (JKCY2020413C002).

References

1. Su, X., Xu, X., Wei, S.: Research on key applications and methods of digital twin technology. China Instrum. (07), 47–53 (2019)
2. Wiedemann, S.M., Kirk, C.L.: Eigenvalue analysis and optimized numerical simulation of the space shuttle remote manipulator system. J. Sound Vib. **307**(1–2), 265–279 (2007)
3. Foust, J.: How clusters of small satellites could replace large spacecraft. Space Rev. 8–20 (2007)
4. Chen, Z.: On orbit operation technology and development analysis abroad. J. Equip. Coll. **25**(6), 63–68 (2014)
5. Vincent, P., Larochelle, H., Bengio, Y., et al.: Extracting and composing robust features with denoising autoencoders. In: Proceedings of the 25th International Conference on Machine Learning, pp. 1096–1103 (2008)

Research on SDN-Based New In-vehicle Network Packet Scheduling Technology

Yong Lv, Yahong Zhai(✉), Peng Li, Junwei Cui, and Wanxu Zhou

School of Electrical and Information Engineering, Hubei University of Automotive Technology, Shiyan 442002, Hubei, China
{201911020,20070026,201911018,202011075,202011064}@huat.edu.cn

Abstract. In view of the problems of low level of centralized network management, low interactivity, and low flexibility in the static deployment of traditional in-vehicle network (IVN), a new type of IVN architecture was designed. At the same time, in order to effectively improve the quality of data transmission under the new IVN architecture and avoid network congestion, a quality of service (QoS) guarantee mechanism under the new architecture has been established. Firstly, the software-defined network (SDN) technology is introduced and combined with the IVN itself to design the new IVN architecture including the forwarding plane, control plane, and application plane. Secondly, the application of four packet scheduling algorithms which are Stochastic Fairness Queueing (SFQ), Priority Queuing (PQ), Class-based Weighted Fair Queuing (CBWFQ) and Low Latency Queue (LLQ) under the new architecture are introduced and analyzed. Finally, the performance of different packet scheduling algorithms are compared and analyzed, and the experimental result shows that: Compared with the other three scheduling algorithms, LLQ can provide better QoS guarantee and network performance for IVN data transmission.

Keywords: Software-defined network · In-vehicle network architecture · Packet scheduling · QoS guarantee

1 Introduction

With the deepening of automotive lightweighting, electrification, intelligence, networking, and sharing, the addition of new services such as active safety, intelligent driving, and intelligent networking also promotes the evolution of automobiles from a single vehicle to a complex distributed Computing system [1]. The application of advanced technology makes the car more intelligent, but also increases the burden of the network in car. The electronic equipment of the car becomes more complex [2]. The demand for communication between different computing units and different domains in the car is increasing, and the demand for high-quality data flow is increasing. The existing in-vehicle network (IVN) architecture is showing its technical limitations [3].

The current IVN architecture design is complex and lacks flexibility. The traditional IVN technology is limited in network bandwidth and network configuration. The update

Y. Wang et al. (Eds.): IWAMA 2021, LNEE 880, pp. 214–222, 2022.
https://doi.org/10.1007/978-981-19-0572-8_27

and addition of corresponding software functions need more flexible architecture [4]. This paper designs a new IVN architecture based on Software-defined Network (SDN) technology, which can integrate multiple heterogeneous data sources into a unified SDN environment, providing network programmability and improving network flexibility.

In the SDN-based new IVN architecture, sensor fusion and artificial intelligence processing require a large amount of data interaction, with different delays and Quality of Service (QoS) constraints. Packet delay or loss will have a greater impact on driving stability and safety. In order to meet the high specification requirements of vehicle using, IVN needs to provide fast and effective data transmission and limited delay for real-time operation [5, 6]. In order to effectively improve the quality of data transmission and avoid network congestion, it is necessary to define a well-defined QoS guarantee mechanism in IVN [7, 8]. Therefore, this paper studies and designs the packet scheduling technology under the new IVN architecture, which has far-reaching practical significance and application value.

The rest of this article is organized as follows: Sect. 2 introduces the design of SDN-based in-vehicle network architecture. Section 3 introduces the application of the four packet scheduling algorithms based on the new architecture in detail. Section 4 conducts an experimental comparative analysis on the performance of the four algorithms, and the discussion and conclusions are summarized in the last part of this article.

2 Design of SDN-Based In-vehicle Network Architecture

With the application of advanced driver assistance systems (ADAS), vehicle-to-everything (V2X) and other technologies, vehicles are becoming more intelligent. In order to meet the development needs, the new open vehicle overall architecture is adopted, as shown in Fig. 1. The new overall architecture of the automobile uses Ethernet as the backbone network. The SDN central controller computing platform can realize the fusion processing of various information, realize the vehicle coordination function, enhance the interaction of IVN, realize the online upgrade of software, the portability of the operating system and other functions.

SDN aims to standardize the control and configuration of network infrastructure. It consolidates network control by migrating the control plane of the network to a logically centralized controller and downgrading the switch to a simple forwarding device. This provides huge advantages for the future development of IVN, including access control or reconfiguration based on a centralized view of the current network state, and provides a reliable method for interoperability between electronic control units. The SDN-based new IVN architecture as shown in Fig. 2 aims to establish a true central processing integration with the vehicle domain controller as the integrated unit, which can overcome the shortcomings of the current in-vehicle communication architecture such as lack of interactivity and flexibility, and realize fully centralized intelligent control of the entire vehicle domain, which is more convenient for extension of services and software upgrades.

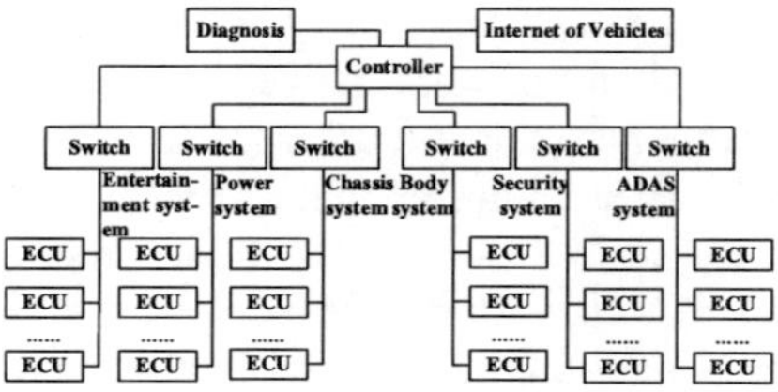

Fig. 1. The new overall architecture of the car

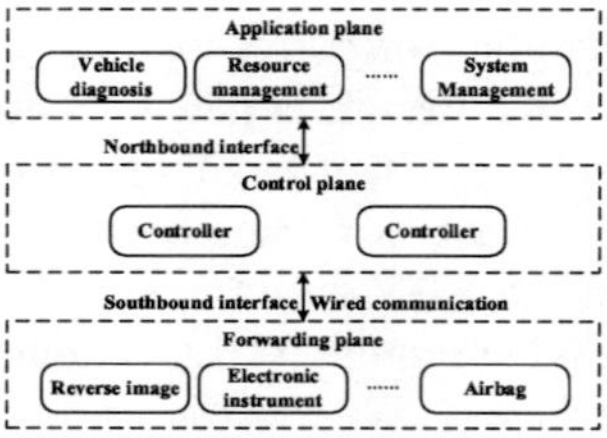

Fig. 2. The SDN-based new IVN architecture

3 Packet Scheduling Technology Based on New Architecture

QoS is a basic network technology used to provide better services for network communication, and it is a network security mechanism used to solve problems such as network delay and congestion [9]. Traffic scheduling in SDN can be divided into traffic scheduling on the data plane and traffic scheduling on the control plane. The former mainly uses the centralized control function of SDN to improve the traffic scheduling method in the traditional network, and the latter mainly focuses on how to solve the traffic scheduling bottleneck problem that appears in the controller.

The data forwarding plane provides an abstract model of the network topology for the SDN controller that hosts the application. Therefore, the controller can use the complete topology of the network to optimize flow management and support user service requirements for scalability and flexibility. The switch queue scheduling model is shown in Fig. 3.

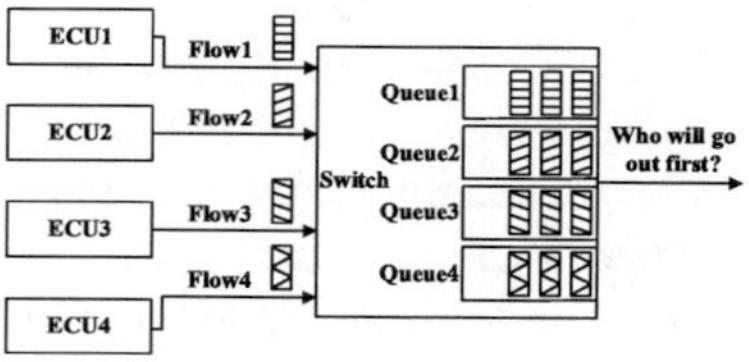

Fig. 3. Switch queue scheduling model

Because traditional switching equipment schedules data flow queues by setting scheduling policies on interfaces, while the SDN matches the output port of the switching equipment through the controller, the data will directly enter the queue if the match is successful [10]. Sometimes the queue is in a state of high load or congestion. In this case, the data flow that will enter the queue can only be queued for waiting, and it is likely that packets will be lost due to the long time being unable to enter the queue. As a result, there will be many queues resulting in packet loss because they cannot get service data, and other queues will be congested because of too many services. Port resources are not fully utilized, resulting in waste of resources, which further intensifies the challenge of ensuring the QoS of various new data services.

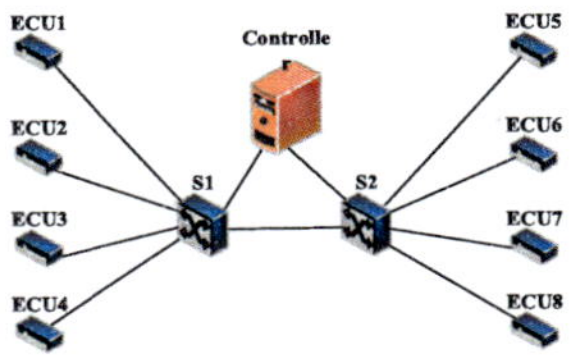

Fig. 4. Experimental topology

Table 1. Data flow parameters

Type	Transmission relationship	Send start time
Flow 1	ECU1-ECU5	0 s
Flow 2	ECU2-ECU6	10 s
Flow 3	ECU3- ECU7	20 s
Flow 4	ECU4-ECU8	30 s

In order to meet the QoS and security performance requirements of various data flows in IVN, we introduced Stochastic Fairness Queueing (SFQ), Priority Queuing (PQ), and Class-based Weighted Fair Queuing (CBWFQ) and Low Latency Queueing (LLQ) four packet scheduling algorithms, and compared their performance in subsequent experiments.

4 Experiment Analysis

In this experimental platform, we built a Mininet network simulation environment on Ubuntu16.04, and verified and compared the performance of the packet scheduling algorithm under the new IVN architecture based on SDN. The network topology of the experiment is shown in Fig. 4. We use hosts to replace different ECUs for simulation experiments. Table 1 shows the specific information of the data flows. These four streams will be allocated to the created queues.

Our model uses Ethernet based on the 100Base-TX standard. Gigabit Ethernet is not a near-term choice because it does not have enough electromagnetic compatibility, the hardware is expensive, and most existing ECUs cannot process data at such high rates. We have implemented four packet scheduling algorithms in OVS, and the queue parameters are shown in Table 2. We configure the switch to have 4 queues (q_1, q_2, q_3, q_4) for each interface. The performance of the packet scheduling algorithm is compared and analyzed from the bandwidth, delay jitter and packet loss rate respectively.

Figures 5, 6, 7 and 8 shows the queue bandwidth usage when different packet scheduling algorithms are used. In the SFQ test, the sending rate of the four streams were all 60 Mbit/s. It can be seen from Fig. 5 that when the sending rate is greater than the limit rate and multiple packets arrive, SFQ forwards packets in turn. When the load reaches the total bandwidth limit, SFQ forwards packets on average. SFQ will evenly allocate bandwidth for the queue, and the sum of all bandwidth is the total bandwidth of the link.

In the PQ test, the sending rate of the four streams are all set to 60 Mbit/s. As shown in Fig. 6, PQ preferentially forwards packets with higher priorities in the queue. Congestion occurs in 10 s. Whenever a high-priority flow enters the interface, the bandwidth of the high-priority data flow is met first and the data flows with the same priority are allocated the same bandwidth.

In the CBWFQ test, the sending rate of the four streams are all set to 60 Mbit/s. As shown in Fig. 7, in CBWFQ, the principle of bandwidth allocation is based on the ratio

Table 2. Queue parameters

Scheduling algorithm	Parameter	q_1	q_2	q_3	q_4	Scheduling algorithm	Parameter	q_1	q_2	q_3	q_4
SFQ	Minimum rate (Mbit/s)	1	1	1	1	CBWFQ	Minimum rate (Mbit/s)	40	30	20	10
	Maximum rate (Mbit/s)	100	100	100	100		Maximum rate (Mbit/s)	100	100	100	100
PQ	Minimum rate (Mbit/s)	1	1	1	1	LLQ	Minimum rate (Mbit/s)	3	2	1	100
	Maximum rate (Mbit/s)	100	100	100	100		Maximum rate (Mbit/s)	100	100	100	100
	Priority	3	2	2	1		Priority	2	2	2	1

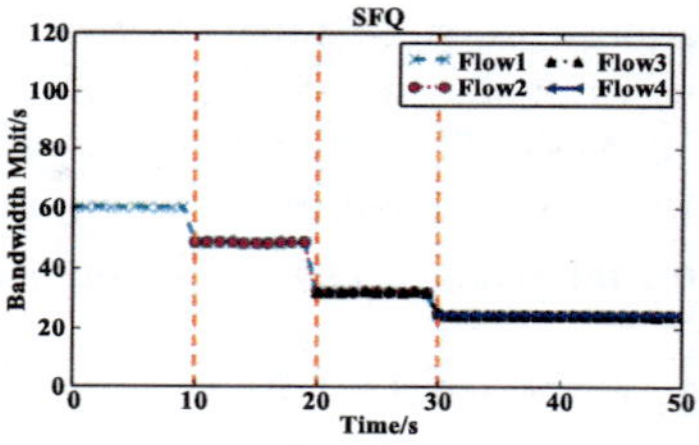

Fig. 5. Bandwidth performance of SFQ

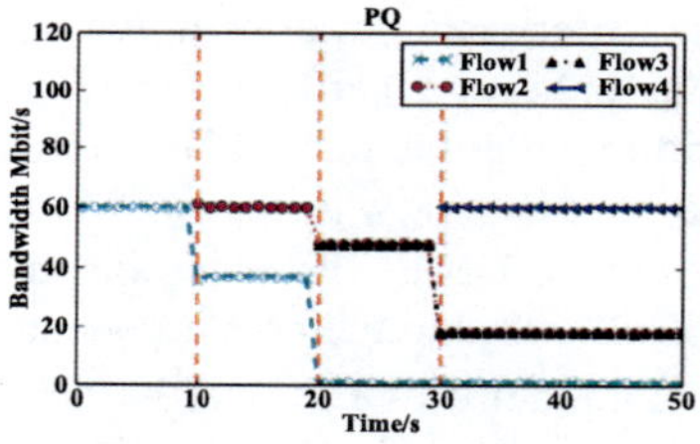

Fig. 6. Bandwidth performance of PQ

of the minimum sending rate of packets in the queue, which is about 4:3:2:1 in Table 1. Therefore, when flow4 enters the interface at 30 s, the bandwidth of the four flows change to 40 Mbit/s, 30 Mbit/s, 20 Mbit/s, and 10 Mbit/s respectively.

In the LLQ test, the sending rate of the four streams are all set to 60 Mbit/s. As shown in Fig. 8, when flow4 with the highest priority arrives at 30 s, due to its higher priority, the bandwidth allocated to flow1, flow2 and flow3 is reduced. The bandwidth occupied ratio is about 3:2:1, and the bandwidth occupied by flow4 reaches 60 Mbit/s.

Jitter usually refers to the degree of packet delay variation. Long queues in data processing usually cause jitter. If the interface is congested, queuing delay will affect the end-to-end delay. It is an important parameter for real-time data transmission in IVN. Figures 9, 10, 11 and 12 shows the delay jitter of the queue when using different packet scheduling algorithms.

The delay jitter of SFQ is shown in Fig. 9. In the Fig, we can see that the jitter change of each queue of SFQ remains stable, and the state of the transmission link is stable. Whenever a new stream is injected, the jitter will increase slightly. With the time getting

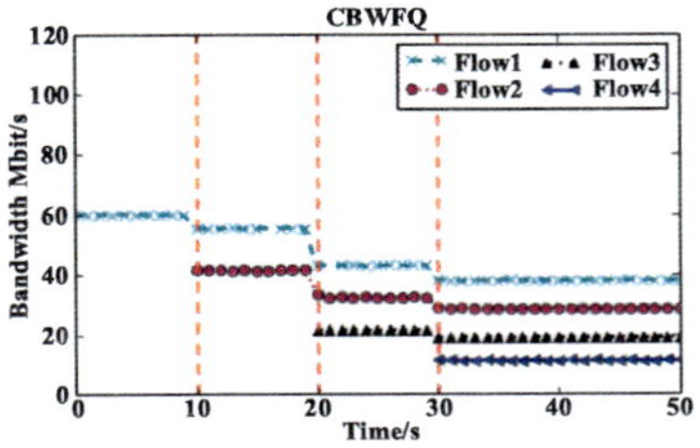

Fig. 7. Bandwidth performance of CBWFQ

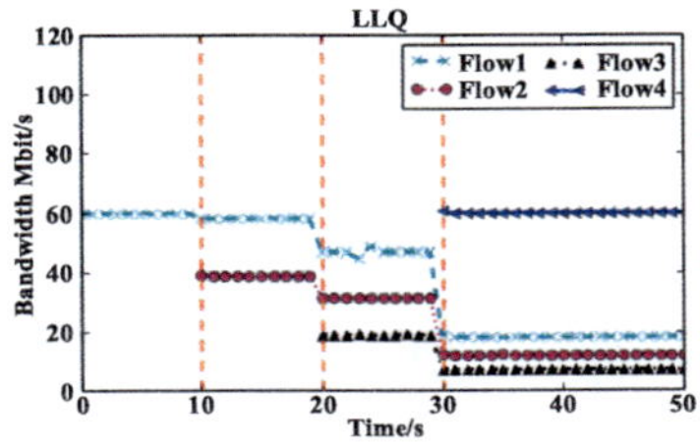

Fig. 8. Bandwidth performance of LLQ

longer, the jitter will be gradually callback. In the end, the four queues show similar jitter changes and tend to be stable.

The delay jitter of PQ is shown in Fig. 10. When the interface is congested, flow1 has the lowest priority, occupies the least bandwidth resources, has a low transmission rate, and has a large jitter. Since flow2 and flow3 have the same priority, they exhibit similar jitter changes. Since flow4 has the highest priority and is in a high-speed forwarding state, the jitter changes are not obvious. PQ gives priority to the transmission of high-priority streams to meet the requirements of real-time transmission in IVN.

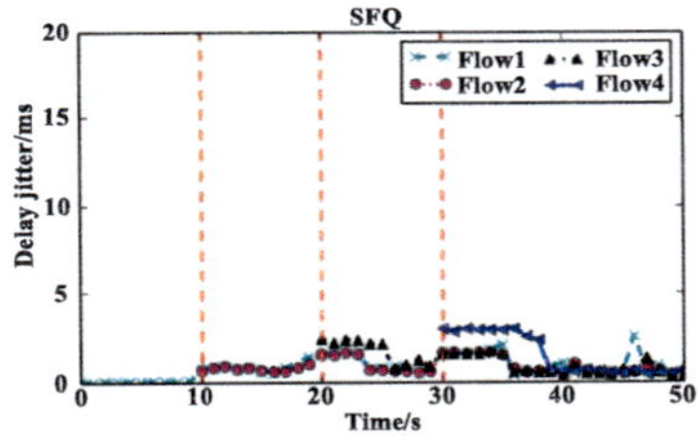

Fig. 9. Jitter performance of SFQ

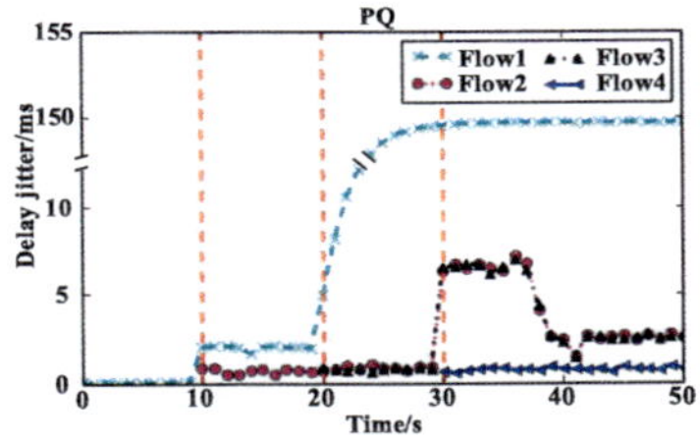

Fig. 10. Jitter performance of PQ

The delay jitter of CBWFQ is shown in Fig. 11. When the interface is congested, the higher the bandwidth occupation, the smaller the delay jitter. In low-speed links, the jitter amplitude changes significantly. There will be a large delay jitter at the initial moment whenever a new flow enters. After a period of time, the jitter decreases and eventually becomes stable. Therefore, it is more suitable for applications that are not sensitive to data stream delay and jitter, such as in-vehicle entertainment systems.

The delay jitter of LLQ is shown in Fig. 12. In the LLQ, flow1, flow2 and flow3 have the same priority. Due to the different occupied bandwidths, the jitter of flow1, flow2 and flow3 increase in turn. The less bandwidth they occupy, the more obvious the jitter. The delay jitter in the first 30s shows a similar trend to CBWFQ. Flow4 has the highest priority, and the jitter change is almost 0, which can further meet the real-time and QoS guarantee of IVN.

The packet loss rate is the ratio of lost packets to the total number of packets. In IVN, packet loss will have a greater impact on driving stability and safety. Figures 13, 14, 15 and 16 shows the packet loss rate of the queue when using different packet scheduling algorithms.

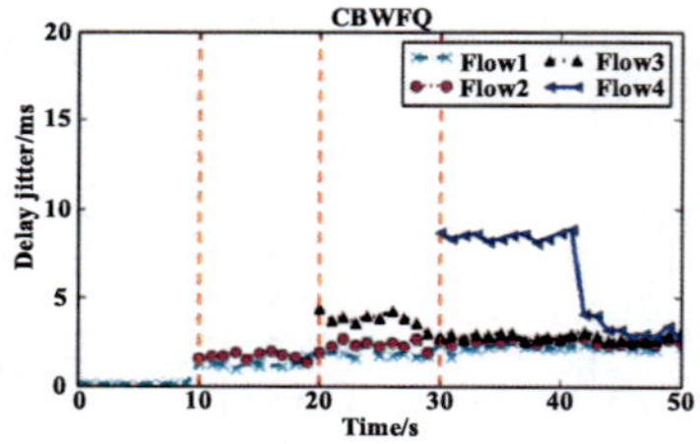

Fig. 11. Jitter performance of CBWFQ

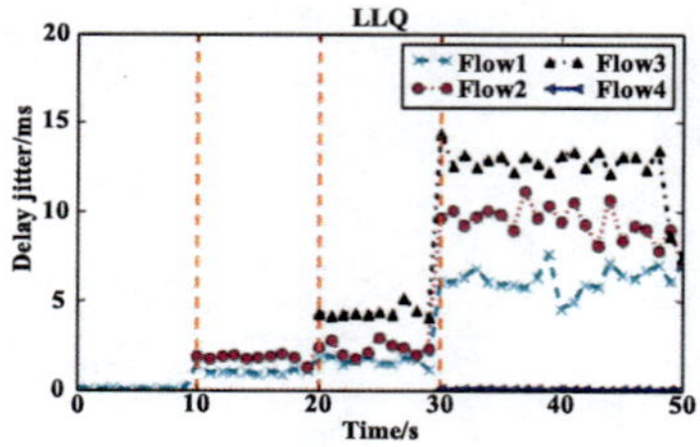

Fig. 12. Jitter performance of LLQ

The packet loss rate of SFQ is shown in Fig. 13. When the interface is congested at 10 s, since SFQ sends each data stream in order, the packet loss rates of the four streams show the same changing trend. When no new flow enters, the packet loss rate is in a stable state. The packet loss rate of the four flows is about 60%. SFQ cannot guarantee that all packets are forwarded.

The packet loss rate of PQ is shown in Fig. 14. When the interface is congested at 10 s, the packet loss rate of flow1 is basically unchanged. When flow3 enters at 20 s, the packet loss rate of flow1 has a substantial increase and basically remains at about 43%. Because flow2 and flow3 have the same priority, the packet loss rate shows a similar trend. When flow4 enters, the packet loss rate of flow2 and flow3 are as high as about 73%. Therefore, in PQ, when the data stream with a higher priority need to be forwarded, it will cause the lower priority data stream to produce higher packet loss, while the highest priority data stream are almost lossless.

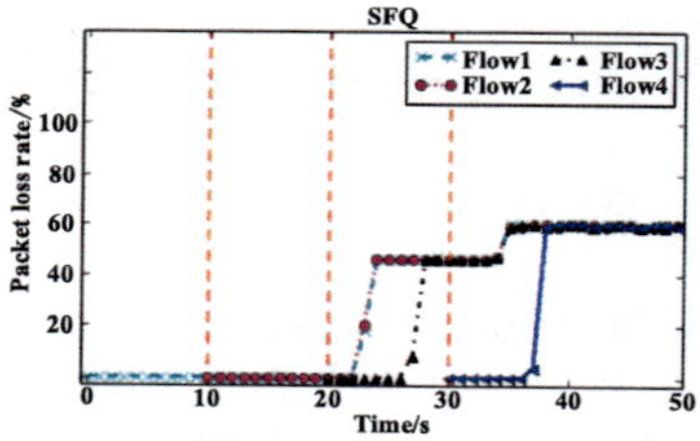

Fig. 13. Packet loss rate of SFQ

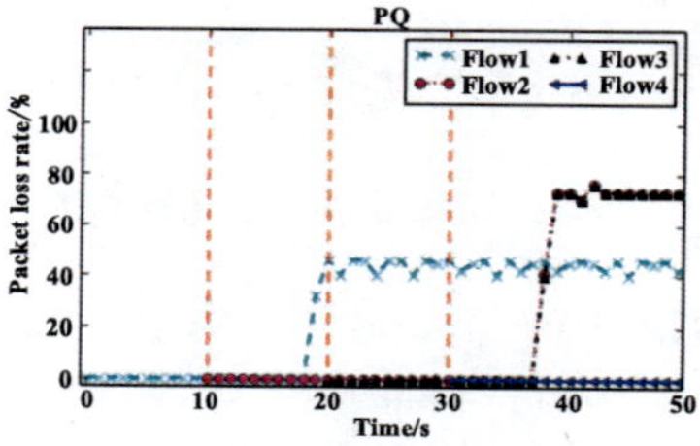

Fig. 14. Packet loss rate of PQ

The packet loss rate of CBWFQ is shown in Fig. 15. When the interface is congested at 10 s, flow1 and flow2 have almost no packet loss before 20 s. In the Fig, we can see that the packet loss rate of flow1, flow2, flow3 and flow4 increase successively, and flow4 has the highest packet loss rate, which is more than 80%. Therefore, in CBWFQ, The data stream have the same priority. The less bandwidth allocated, the higher packet loss rate generated by the data stream.

The packet loss rate of LLQ is shown in Fig. 16. When the interface is congested at 10 s, flow1 has almost no packet loss before 30 s, and flow2 and flow3 have almost no packet loss before the next new flow enters. The change trend of the packet loss rate of flow1, flow2, and flow3 is basically the same as that of CBWFQ. The less bandwidth allocated, the higher the packet loss rate. When a higher-priority flow4 enters at 30 s, the high-priority packet loss rate is almost 0, and the low-priority data flow has a higher packet loss rate. Therefore, in LLQ, streams with higher priority will be forwarded with the lowest delay and have a very low packet loss rate.

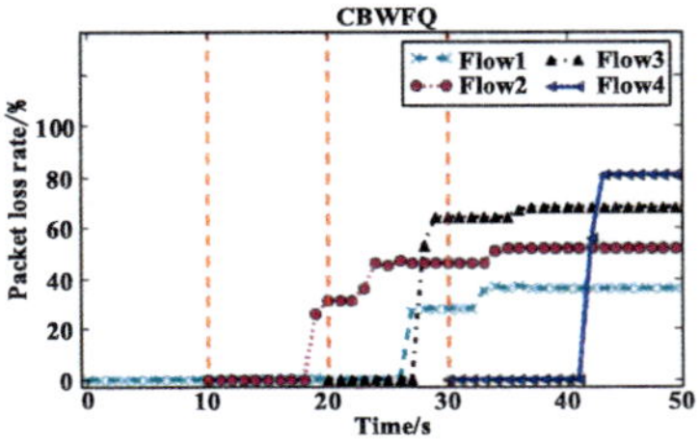

Fig. 15. Packet loss rate of CBWFQ

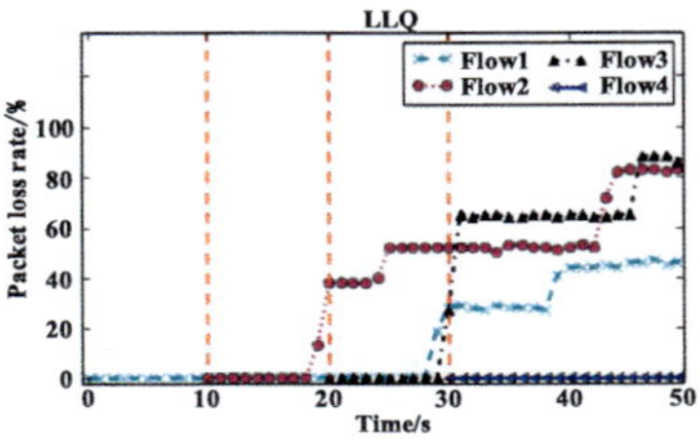

Fig. 16. Packet loss rate of LLQ

5 Conclusion

In the new SDN-based IVN, the four scheduling algorithms have their own advantages and disadvantages under certain network conditions. SFQ treats all data streams equally and is fair in bandwidth and packet loss rate, but does not guarantee bandwidth and real-time performance. PQ provides real-time guarantees for different IVN applications, ensuring the real-time performance and QoS of high-priority data stream transmission, but cannot guarantee the QoS of low-priority queues. CBWFQ can allocate bandwidth for each queue, but the packet loss rate is high, which cannot ensure the stability and security of IVN. LLQ can guarantee the transmission of delay-sensitive traffic and provide low delay guarantee for the highest priority traffic, while the lower priority traffic will tolerate a certain degree of delay.

Compared with the other three scheduling algorithms, LLQ can ensure the bandwidth of the highest priority queue with almost no jitter in the queue and almost no packet loss. Therefore, the LLQ packet scheduling algorithm is more suitable for the new type of IVN.

The current work of this paper mainly studies the application of forwarding plane data flow scheduling technology under the new IVN architecture, and compares and analyzes the performance of four different scheduling algorithms, and draws corresponding conclusions. However, the research on the data traffic scheduling mechanism and strategy of the control plane in the new IVN have not been considered.In the next step of the research work, we will conduct in-depth research and analysis on the dynamic traffic scheduling and resource allocation of the control plane, and further consider the scheduling strategies of different data service requirements to better optimize the network performance of IVN.

Acknowledgment. This work was supported by science and Technology Project 19Y135 of Shiyan city, Hubei Province.

References

1. Zhang, Z.: Research on DoIP technology of vehicle ethernet. Auto Time (01), 14–15 (2021)
2. Walrand, J.C., Turner, M., Myers, R.: An architecture for in-vehicle networks. IEEE Trans. Veh. Technol. (2021)
3. Xiao, W.: In the era of smart cars, research on ECU upgrade technology of in-vehicle ethernet. Automobile Parts (13), 66–69 (2021)
4. Yoon, S., Cho, J.H., Kim, D.S., et al.: Poster: address shuffling based moving target defense for in-vehicle software-defined networks. In: The 25th Annual International Conference on Mobile Computing and Networking, pp. 1–3 (2019)
5. Lo Bello, L., Patti, G., Vasta, G.: Assessments of real-time communications over TSN automotive networks. Electronics **10**(5), 556 (2021)
6. Yan, X., Zhou, K., Zhu, Z., et al.: The summary of automotive real-time ethernet technology development. Softw. Guide **17**(01), 4–7 (2018)
7. Zhou, Z., Lee, J., Berger, M.S., et al.: Simulating TSN traffic scheduling and shaping for future automotive ethernet. J. Commun. Netw. **23**(1), 53–62 (2021)
8. Liu, H., Ma, T.: Research and analysis on scheduler mechanism impact of QoS. Mod. Electron. Tech. **32**(14), 63–65 (2009)
9. Xu, F.: An experimental study of multi-application QoS in a simulated environment. Inf. Technol. Inf. (07), 215–217 (2015)
10. Jia, Z., Chen, X.: SDN based design of network transmission QoS mechanism. Netw. New Media Technol. **10**(03), 58–62 (2021)

Design and Transmission Performance Analysis of a Novel Reducer with Abnormal Cycloidal Gear

Jingyu Mo[1,2](✉), Shanming Luo[1,2], and Kun Liu[3]

[1] School of Mechanical and Energy Engineering, Jimei University, Xiamen 361021, Fujian, China
Mojyxmu@163.com

[2] Department of Mechanical and Electrial Engineering, Xiamen University, Xiamen 361102, China

[3] School of Mechanical and Automotive Engineering, Xiamen University of Technonlogy, Xiamen 361024, Fujian, China

Abstract. Aiming at the problems of the RV reducer in the over-positioning structure, high machining, and assembly accuracy, a novel abnormal cycloidal gearreducer without pin tooth structure is proposed. the transmission principle and structure of the reducer are discussed. And the transmission ratio formula is derived. The transmission efficiency, force of the gear, and transmission error are obtained by the virtual prototype technology. The simulated value of the transmission ratio with the theoretical value is compared, which verifies the feasibility of the transmission principle of the new type of reducer. The results show that the novel reducer has a simpler structure and reduced over-positioning structure, which greatly reduces the requirements for processing and assembly accuracy. The transmission efficiency is similar to that of the RV reducer, and the transmission error is slightly higher than that of the RV reducer.

Keywords: Abnormal cycloidal gear · New type reducer · Transmission performance

1 Introduction

The RV reducer as a new type of two-stage reducer based on the traditional pin-swing planetary transmission mechanism is the most commonly used in industrial robots, machine tools, medical detection equipment and other fields because it has many real advantages [1–3]. The main advantages are as follows: High transmission efficiency, stable operation, high precision, and compact structure. In China, the development speed in the field of the precision reducer is getting faster and faster, from shipment to product quality, are in a big step way to catch up with the Japanese brand. However, the mainstream reducer models represented by RV and harmonic are all from Japan. Although they have been gradually localized through a series of technical means such as reverse engineering, the biggest problem of reverse engineering is 'knowing what it is and not

Y. Wang et al. (Eds.): IWAMA 2021, LNEE 880, pp. 223–230, 2022.
https://doi.org/10.1007/978-981-19-0572-8_28

knowing what it is'. Therefore, it is impossible to quickly absorb the essence of Japanese reducer in the design process, which further leads to the long-term role of catch-up RV and harmonic manufacturers in the market competition in China.

The development of high-quality RV reducers has long been a hot topic in the domestic reducer manufacturing industry. Many scientific research institutes, universities and enterprises have made unremitting efforts to explore and develop new precision reducer products. However, since: **(a)** The structure of the RV reducer is complex, the parts are closely matched, and the three crankshafts act on the cycloid at the same time, which is an over-positioning structure. Moreover, half of the cycloidal pin teeth are mesh at the same time in theory, this requires RV reducer parts have higher mechanical properties, processing accuracy, greatly increased the difficulty of RV reducer parts processing and assembly. **(b)** Transmission bearing is its weak link, it is easy to break through the bearing stress limit and lead to abnormal wear or rupture of the bearing [4–6]. These theoretical and technical problems have not been effectively solved, leading to the domestic high quality cycloid reducer is always blank. The domestic use of high-grade cycloidal pin gear reducer is mostly dependent on imports, and the key equipment needed for the development of emerging industries is not yet self-sufficient, which not only increases the cost of related industries, but also seriously restricts the development of many important industries.

Hence, the domestic reducer to achieve comprehensive catch-up, can only the structure, transmission principle and other aspects of comprehensive innovation. Therefore, this paper based on existing research reducer precision, a novel abnormal cycloidal gear (i.e. ACG) reducer is proposed [7]. Without using less tooth difference, the large transmission ratio can be achieved by reasonable tooth number matching. It has the characteristics of no needle teeth, simple structure, small size and low assembly difficulty.

2 Transmission Principle of ACG Reducer

2.1 Basic Structure and Transmission Principle

ACG reducer is mainly composed of input shaft H, case 1, first abnormal cycloid gear 2, adaptor 3, second abnormal cycloid gear 4, and output gear 5, the structure of ACG reducer is shown in Fig. 1. The transmission principle is as follows:

Firstly, the driving force of the motor is input from the input shaft H, and the 1-st abnormal cycloid gear 2 and 2-st abnormal cycloid gear 4 are driven simultaneously by the input shaft H. Then, 1-st abnormal cycloid gear 2 rotates at the same time with eccentric motion around the case 1 on the shell, and the 2-st abnormal cycloid gear 4 rotates at the same time with eccentric motion around the output gear 5.The mass of 1-st abnormal cycloid gear 2 and 2-st abnormal cycloid gear 4 is basically consistent, and the phase difference is 180°. Finally, due to the difference of tooth number between 1-st abnormal cycloid gear 2 and 2-st abnormal cycloid gear 4, the rotation speed is different.the pin shaft on the adaptor 3 is used to drive the1-st abnormal cycloid gear 2 and 2-st abnormal cycloid gear 4 at the same time, forcing the rotation speed of the gear to be consistent.And then the output gear 5 is set as an active part, and the speed difference between gear 2 and gear 4 is transformed into the output speed of the output gear 5, and the output speed of the reducer is the speed of the output gear 5.

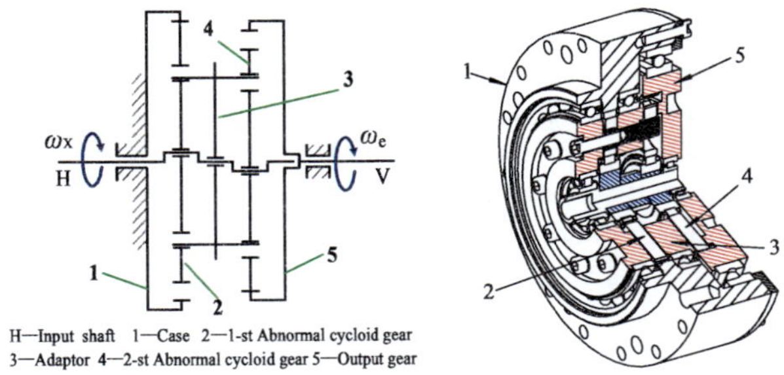

Fig. 1. The transmission principle of ACG reducer

2.2 Transmission Ratio

ACG reducer is a 2K-H type combined planetary cycle gear system. The transmission ratio of ACG reducer is calculated by the conversion mechanism method. Assuming that the whole reducer rotates counterclockwise with angular velocity $-\omega_H$, the relative transmission ratio of each component is:

$$i_{21}^{H} = \frac{\omega_2 - \omega_H}{\omega_1 - \omega_H} = \frac{z_1}{z_2}\; i_{45}^{H} = \frac{\omega_4 - \omega_H}{\omega_5 - \omega_H} = \frac{z_5}{z_4}\; i_{15}^{H} = \frac{\omega_1 - \omega_H}{\omega_5 - \omega_H} = \frac{z_2 z_5}{z_1 z_4} \tag{1}$$

When thecase 1 is fixed, H is input and the gear 5is output, the transmission ratio of the ACG reducer is

$$i_{H5}^{1} = \frac{1}{1 - (1/i_{15}^{H})} = \frac{z_2 z_5}{z_2 z_5 - z_1 z_4} \tag{2}$$

The number of teeth above satisfies: $z_1 > z_5$, $z_2 > z_4$. By Eq. (2) can be seen: When $0 < i_{15}^{H} < 1$, then $-\infty < i_{H5}^{1} < 0$, the input and output components of the reducer turn opposite.When $1 < i_{15}^{H} < \infty$, then $1 < i_{H5}^{1} < \infty$, The input and output components of the reducer turn the same. There are two values for the tooth number difference of two internal gear pairs. One is that the tooth number difference of two internal gear pairs is the same, and the other is that the tooth number difference of two gear pairs is different. The problem discussed in this paper belongs to the latter.

Assuming that the tooth difference between gear 1 and gear 2 is G1, the tooth difference between gear 4 and gear 5 is G2, and the error tooth difference between gear 2 and gear 4 is J, then

$$G_1 = z_2 - z_1 \; G_2 = z_5 - z_4 \; G_2 = G_1 \pm 1 \; \mathrm{J} = z_4 - z_1 \tag{3}$$

According to Eq. (3), Eq. (2) is simplified as a quadratic function of an element about z_4, namely

$$z_4 = \frac{1}{2}(-(J + G_1 - G_1 i_{H5}^{1} + G_2 i_{H5}^{1}) \pm\sqrt{(J + G_1 - G_1 i_{H5}^{1} + G_2 i_{H5}^{1})^2 - 4G_1 J((1 - i_{H5}^{1}))}) \tag{4}$$

When the transmission ratio is positive, take '+'. When the transmission ratio is negative, take '–'. z_i ($i = 1, 2, 4, 5$) represents the number of teeth of gear i.

As the z_4 is 90 and 40, G_1 is 3, 4, 5 and 6, the change of transmission ratio of cycloid gear reducer is shown in Fig. 2. It can be seen from the diagram that the transmission ratio of ACG reducer is very wide, and the large transmission ratio can be achieved by reasonably selecting the number of teeth.

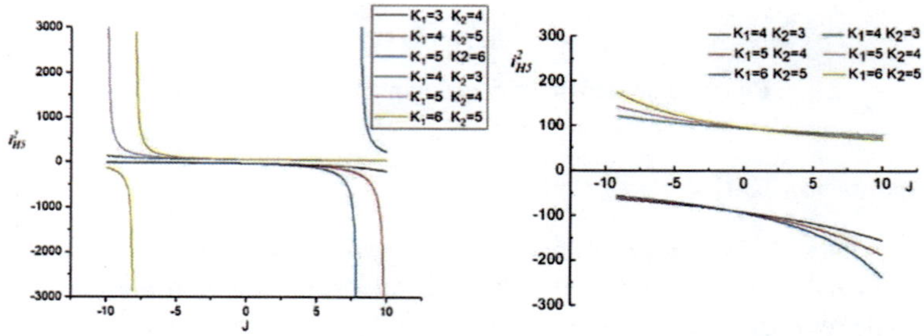

Fig. 2. Influence of different tooth number difference on transmission ratio

As mentioned above, comparedwith the traditional cycloid pin gear reducer, the ACG reducer has the following characteristics:

1) Simple structure and lower requirements for manufacturing and assembly accuracy. The 2K-H combined gear train transmission structure with single input shaft input and inner gear ring output, and eliminates the needle tooth, which makes the structure simpler and reduces the difficulty of processing and assembly.
2) Large stiffness and the large range of transmission ratio. The inner gear ring output directly without additional output structure increases the stiffness of the reducer. Meanwhile, without increasing the radial and axial dimensions of the reducer, a large transmission ratio can still be obtained by reasonably selecting the number of teeth.

3 Tooth Design and Geometric Modeling of ACG Reducer

The tooth profile shape directly affects the transmission performance of cycloid gear reducer. The new tooth profile is based on the ordinary cycloidal gear tooth profile. The original cycloidal tooth profile of the cycloidal gear tooth profile at the base circle is replaced by the inner equidistant line of the outer cycloid and the outer equidistant line of the inner cycloid. This tooth profile is called abnormal cycloidal gear. The tooth profile curve equation is [7]

$$\begin{cases} x = (R + r)\cos\theta - r\lambda\cos(k\theta + \theta) - C\frac{\cos\theta - \lambda\cos(k\theta+\theta)}{\sqrt{1+\lambda^2-2\lambda\cos k\theta}} \\ y = (R + r)\sin\theta - r\lambda\sin(k\theta + \theta) - C\frac{\sin\theta - \lambda\sin(k\theta+\theta)}{\sqrt{1+\lambda^2-2\lambda\cos k\theta}} \\ R_1\varphi_1 = R_2\varphi_2,\quad R_1\rho_1 = R_2\rho_2 \end{cases} \tag{5}$$

Where C is the isometric value, r is the generating circle, R is the fixed base circle, θ is the angle of pure rolling of r along R, $k = R/r$.

Based on the above the structure and tooth shape of the ACG reducer, Taking the number of teeth combination $z_1 = 58$, $z_2 = 54$, $z_4 = 41$, $z_5 = 44$ as an example, the three-dimensional model of the new reducer is designed. The rated output torque of the reducer is $T_c = 100$ N·m, the rated input speed is $n_1 = 1450$ rpm, and the transmission ratio is $i = -1188$. The geometric structure of the reducer as shown in Fig. 3.

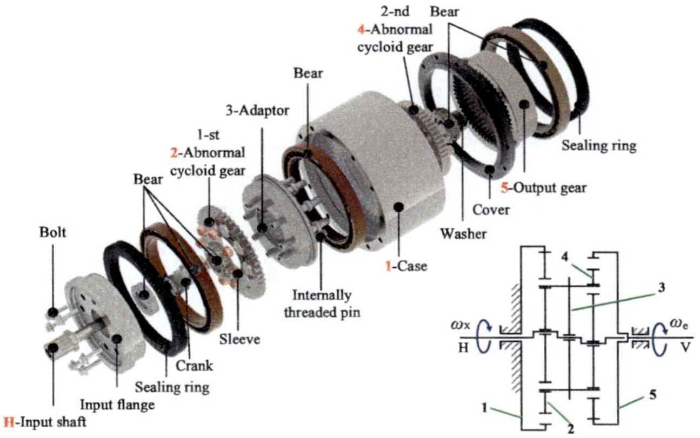

Fig. 3. Explosion diagram of the ACG reducer

4 Transmission Performance Analysis of ACG Reducer

4.1 Assembly and Simulation of Virtual Prototype of Reducer

The 3D model of the ACG reducer is imported into ADAMS for motion simulation. The following constraints are added: (1) The rotation pair is between the input shaft and the earth. (2) Eccentric shaft and input shaft are fixed pairs. (3) The contact pair is between the pin and the gear. (4) Case 1 is a fixed pair with the earth. (5) Gear 5 and earth are rotational pairs. The input shaft speed is 11880 d/s, and the output load is 1.0×10^5 N·mm.

4.2 Analysis of Simulation Results

(1) Transmission Ratio. When the constant speed of the reducer input shaft is set to 11880 d/s, the angular velocity of the adaptor 3 and the output gear 5 are shown in Fig. 4 and Fig. 5, respectively. According to Fig. 4, it can be seen that the speed of the adaptor fluctuates between −895 d/s and −865 d/s, and the fluctuation amplitude is small. The average rate is −880 d/s, which is consistent with the theoretical value. Fig. 5 is the local amplification diagram of the output speed. It can be seen that the output speed fluctuates between −9 d/s and −11 d/s, with slight fluctuation, and the average speed is −10 d/s. The transmission ratio obtained from the input speed and output speed are −1188, which is consistent with the theoretical design.

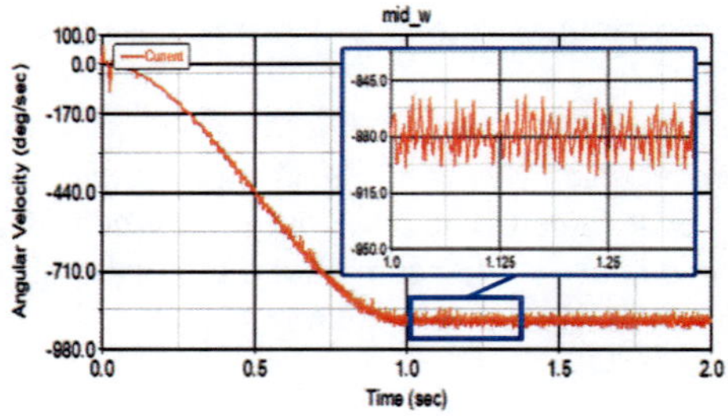

Fig. 4. Angular velocity of the adaptor

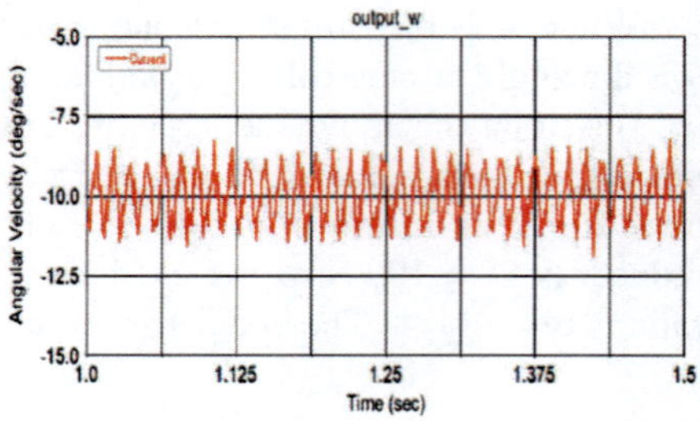

Fig. 5. Angular velocity of the output gear 5

(2) Transmission Efficiency. The simulation results of transmission efficiency for ACG reducer are shown in Fig. 6 and Fig. 7. It can be seen that the input torque fluctuates between 90 N·mm and 115 N·mm, and the average torque is 100 N·mm. According to the simulation results, when the input torque is 100 N·mm, the input speed is 11880 d, the output torque is 1×10^5 N·mm. Thus, the output speed is −10 d/s, the transmission efficiency of the reducer can be calculated to be about 85%, which is close to the efficiency of the existing RV reducer.

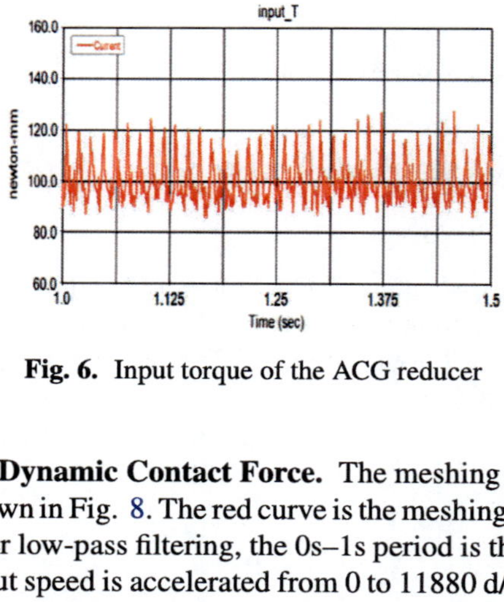

Fig. 6. Input torque of the ACG reducer

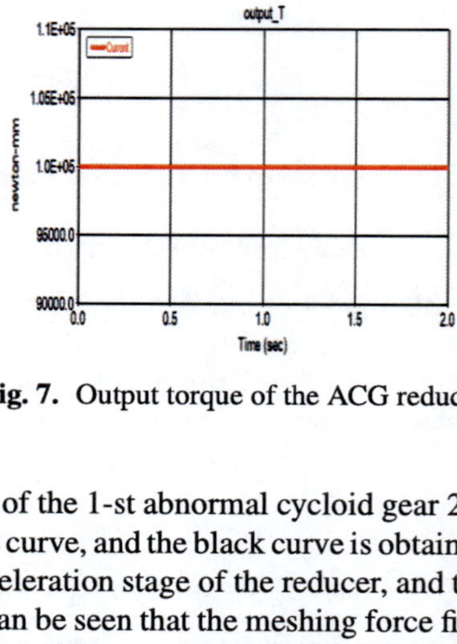

Fig. 7. Output torque of the ACG reducer

(3) Dynamic Contact Force. The meshing force of the 1-st abnormal cycloid gear 2 is shown in Fig. 8. The red curve is the meshing force curve, and the black curve is obtained after low-pass filtering, the 0s–1s period is the acceleration stage of the reducer, and the input speed is accelerated from 0 to 11880 d/s. It can be seen that the meshing force first increases suddenly, then tends to be stable, and then increases slowly. The main reason is that due to the influence of system stiffness and gear material, and other factors, the input power of the reducer cannot be evenly distributed in each power branch, lead to different contact forces at each meshing point of the reducer. After 1 s, the input speed was maintained at a constant rate of 11880 d/s, and the black curve fluctuated between 1750 N and 2100 N. The theoretical normal force of the 1-st abnormal cycloid gear 2 was between 1705 N and 2011 N, and the simulation value was very close to the theoretical calculation value.

The meshing force of the 2-st abnormal cycloid gear4 is shown in Fig. 9. The black curve is the curve after low-pass filtering. The 0 s–1 s period is the acceleration stage of the reducer, and the input speed is accelerated from 0 to 11880 d/s. It can be seen that the meshing force increases suddenly. They tend to be stable and slow. 1 s later, the input speed is maintained at a constant rate of 11880 d/s, and the black curve fluctuates between 1750 N and 2100 N, which is very close to the theoretical calculation of the gear 2's normal force.

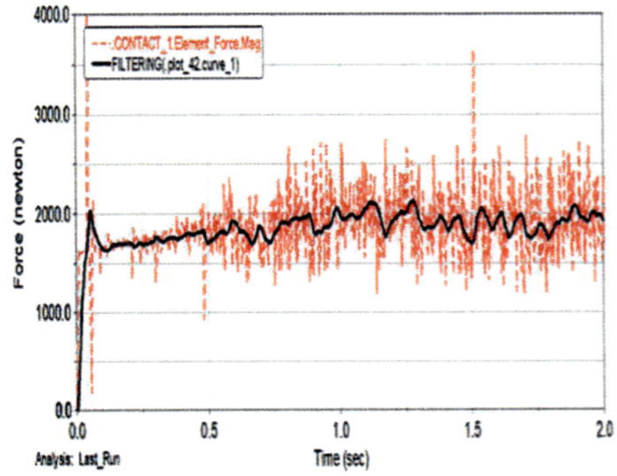

Fig. 8. Meshing force of gear 2

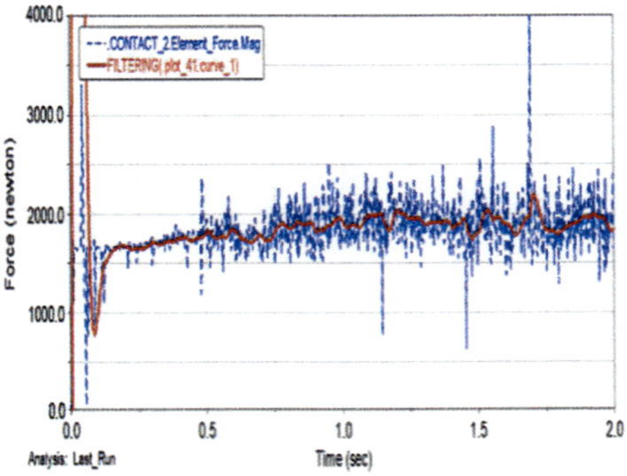

Fig. 9. Meshing force of gear 4

(4) Transmission Error. More than 20,000 sets of data are exported from ADAMS software to Excel for data processing. A collection of data is extracted every 50, and 437 sets of data with the output angle between 60° and 420° are obtained. The output end rotates 360° and experiences a considerable period. The transmission error curve as shown in Fig. 10. It can be seen that the transmission error of the reducer is 63.199″, while the transmission error of the existing high precision RV reducer is less than 60″, and the transmission error of the cycloid gear reducer is slightly higher than that of the RV reducer.

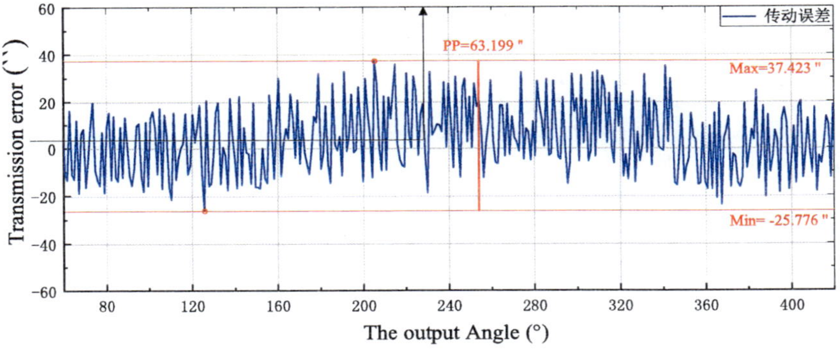

Fig. 10. Transmission error of the ACG reducer

5 Conclusions

1) This paper designs a new type of abnormal cycloidal gear reducer, mainly analyzes its transmission principle and derives the calculation formula of transmission ratio. This study has identified that the new type reducer has the characteristics of simple structure, a wide range of transmission ratios, and no pin-tooth structure.
2) The simulation results of the transmission ratio of the abnormal cycloidal gear reducer are consistent with the theoretical design, indicating that the reducer can realize the predetermined transmission ratio design, and verifying the correctness of the reducer model.
3) In ADAMS dynamic simulation, it can be found that the contact force of cycloid-like gear fluctuates, but the whole tends to be stable. The transmission efficiency of the abnormal cycloidal gear reducer is similar to that of the RV reducer. but the transmission error of the new reducer is slightly higher than that of the RV reducer.

Acknowledgment. The work was supported by the National Natural Science Foundation of China under No. 51975499, the Program for Innovative Research Team in Science and Technology in Fujian Province University. The financial and technique supports are gratefully acknowledged.

References

1. Huang, X., He, W.J., Fu, Y.X.: Summary of precision speed reducer of industrial robots. Mach. Tool Hydraulics **43**(13), 1–6 (2015)
2. Zhou, J.: Intelligent manufacturing-main direction of "made in China 2025". Chin. J. Mech. Eng. **17**, 2273–2284 (2015)
3. Chen, B.K., Tan, L., Zhong, H., et al.: Finite element contact analysis of cycloid enveloping planetary transmission and the development of analysis program. J. Mech. Strength **35**(3), 366–371 (2013)
4. Yang, Y.H., Chen, C., Wang, S.Y.: Response sensitivity to design parameters of RV Reducer. Chin. J. Mech. Eng. **31**, 49 (2018)
5. Li, S.T.: Design and strength analysis methods of the trochoidal gear reducers. Mech. Mach. Theory **81**, 140–154 (2014)
6. He, W.D., Li, X., Li, L.X.: Study on new pin-cycloid drive with high load-capacity and high transmission efficiency. Chin. J. Mech. Eng. **7**, 565–569 (2005)
7. Luo, S.M., Mo, J.Y., Xu, J.M., et al.: A new type of abnormal cycloidal gear reducer. Patent: CN110630720A (2019)
8. Fan, J.M.: Design and performance analysis of reducer with the abnormal cycloidal gear and large transmission ratio. Xiamen University of Technology (2019)

Noncontact Clearance Measurement Research Based on Machine Vision

Kai Che[1], Dongli Lu[2], Jun Guo[2], Yufeng Chen[1](✉), Guosheng Peng[1], and Lianbing Xu[1]

[1] School of Electrical and Information Engineering, Hubei University of Automotive Technology, Shiyan 442002, Hubei, China
chenyf_dy@huat.edu.cn
[2] Zhejiang Hong Cheng Computer Systems Co., Ltd., Hangzhou 311100, China
{ludl,guojun}@zjhcsoft.com

Abstract. Parts assembly clearance measurement is facing a trend towards high-precision and noncontact. This work aims to measure clearance by image processing based on machine vision. The machine vision system is to highlight the assembly clearance region. Hence, clearance regions are segmenting, and rotating to vertical, then get the geometric center of the region and the inclination relative to the horizontal direction, the two points intersecting the boundary of the region can be obtained through the linear relationship, the clearance width is the pixel distance between two points mapped to the actual width in the world coordinate system. Results of the measurement results show that the system works effectively and meets the requirements, which makes it suitable for industrial applications.

Keywords: Clearance measurement · Image process · Machine vision system

1 Introduction

Size measurement is used to ensure that the parts or the products conformity to quality requirements. These requirements may vary depending on the application, and the precision industry is the most demanding ones. Noncontact methods is of particular research interest in industrial intelligent manufacturing [1, 2].

Nowadays, clearance measurement research mainly faces the following problems: with different appearance of products, different measurement schemes need to be designed for the products, such as forging measurement based on laser scanning, size measurement based on electrical characteristics. Also, the accuracy makes that the same measurement system is difficult to be widely used in industrial production. The measurement is greatly influenced by external factors, such as temperature and humidity in laser measurement, and subjective factors in manual measurement.

In order to deal with the above-mentioned measurement problems, this paper proposes a general clearance measurement method focuses on the image measurement of the assembly clearance. Test results of this method on the prototype system shows that the measure accuracy is 0.0067 mm@200 ms, which is relative to the 0.02 mm@2000 ms of manual vernier calipers.

Y. Wang et al. (Eds.): IWAMA 2021, LNEE 880, pp. 231–238, 2022.
https://doi.org/10.1007/978-981-19-0572-8_29

2 Related Works

Industrial application is one of the most challenging problems in clearance measurement. The conventional methods of measurements use indirect sensor such as probe meter, capacitance meter, grating interferometer, and filler gauge [3]. These contact measurement methods are complicated to operate and do not apply to large-scale measurement data acquisition and analysis. For example, probe measurement method has high requirements on the conductivity of the measured object, and the measurement result is easily affected by the applied voltage fluctuation [4, 5]; capacitance measurement is greatly affected by parasitic capacitance, and the output characteristics are nonlinear [6, 7], Grating interference method has high requirements for the measured environment and poor universality, filler gauge must be operated manually, the measurement of the sameness is poor [8].

Currently, image processing is becoming an extremely useful tool for Industrial measurement [8–12]. Most of them are used for the measurement and inspection of large-scale structures on the surface, such as laser three-dimensional imaging to model the surface of large-scale workpieces, blue light scanning measurement for large-scale construction. But these systems can only achieve offline measurement and high equipment costs.

3 Clearance Measurement Algorithm

All the assembly parts images undergo the same processing steps. First, an image enhancement algorithm is performed, eliminate noise and highlight the clearances. Second, extracting the region of clearance and fitting the edge of sub-pixel precision. Third, measuring width based on geometrical characteristics of the region. Finally, compare with the standard size. its main steps will be detailed hereafter.

3.1 Measurement Task Description

There is a variety of clearances for assembling industrial parts. Furthermore, without a university accepted standard for clearances. Figure 1 shows an example of such clearances which can be attributed to different forms.

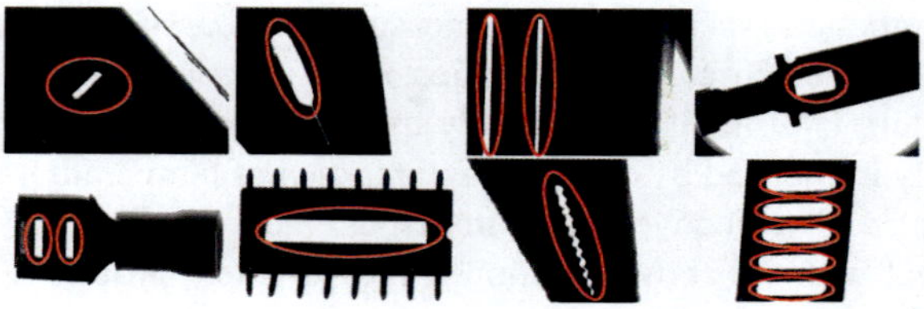

Fig. 1. Different kinds of clearances

As a general rule, clearances are very narrow, presents different geometries and with a homogeneous gray value. Their visual appearance depends strongly on the illumination system. A good lighting solution can greatly highlight the clearance from the background and help to segment the clearance regions in the image processing.

An additional complication for machine vision measurement systems is that the processing time must strictly match the running speed of the assembly line. In general, clearance of assembly part must be measure within 400 ms, and depending on its sizes. Our proposal focuses on clearance measurement.

Finally, the width of the clearance to be measured must be at least 0.02 mm, since smaller clearances are not easily measured by contact methods. Also, we can use the large-array camera to improve measurement accuracy. Processing speed should be within 500 ms as this is matching the speed of the assembly line.

3.2 Preprocessing

The goal of preprocessing is to reduce the design complexity of the measurement algorithm. Because of the backlighting technique, the primary colors of image are black and white, edges and corners are the high frequency areas of the image to be segmented. To enhance clearance, a mask with m $\times$ n is created, while m is half width of the image and n is half height of the image. The enhanced grey value $G_{(r,c)}$ depends on the grey value $g_{(r,c)}$ in input image at the same position and the size of mask. The clearance to be measured in enhanced image appears sharper. $G_{(r,c)}$ is implemented by:

$$G_{(r,c)} = \left(g_{(r,c)} - \frac{1}{m \times n}\sum\nolimits_{i=m,j=n}^{m,n} g(i,j)\right) \times \alpha + g_{(r,c)} \tag{1}$$

Where α denotes adjustment factor.

3.3 Extract Region

The region of clearance is of high gray-scale, and background is of low gray-scale. By this characteristic of clearance, we could extract edges region for morphological processing. Before threshold operation, smooth image $G_{(r,c)}$ by a mask averaging. The operation is implemented by:

$$S_{(r,c)} = f_n\left(G_{(r,c)}\right) \tag{2}$$

Where f [] denotes the operation of smooth, n denotes the size of smooth mask. Then, extracting edge regions by local thresholds based on $G_{(r,c)}$ and $S_{(r,c)}$. In $G_{(r,c)}$, if the gray value of position (r, c) is less than the same position of $S_{(r,c)}$ it is reserve, otherwise, delete it. The operation is implemented by:

$$T_{(r,c)} \leq S_{(r,c)} + \beta \tag{3}$$

Where β denotes compensation factor.

By dynamic threshold, we extracted edges region from background, but also exhibit too much noise region correspond to the edge attachments like metal scraps and should be eliminated. By the characteristic of noise region, setting the region area as a threshold could filter noise region. To avoid arbitrary parameter of threshold, an adaptive algorithm that determines the threshold has been developed. This algorithm has the following steps:

Step 1. Region Area Histogram Statistics
Statistical the number of region areas and calculate region area of all the connected parts in the binary image. Store the results in the form of a region area histogram with $H = h_1, \ldots h_n$, which in turn increases from h_1 to h_n.

Step 2. Threshold Selection
The five largest regions $(h_{n-4}, h_{n-3}, h_{n-2}, h_{n-1}, h_n)$ are selected. Comparing the area difference of adjacent region from h_n to h_{n-4}, reserve two regions if the difference does not exceed 0.1, otherwise, reserve the bigger one and end the procedure. The operation is implemented by:

$$R = \begin{cases} h_n, & \left(\frac{h_n - h_{n-1}}{h_n} \leq 0.1\right) \\ h_n, h_{n-1}, & \left(\frac{h_n - h_{n-1}}{h_n} > 0.1\right) \end{cases} n \in (1, 2, 3, 4, 5) \tag{4}$$

Step 3. Calculate the Circularity of Regions
According to the geometric characteristics of the region R, the region of clearance in the shape of narrow bars and other irrelevant regions are similar to a circle or other shapes. To get the clearance region, calculate the circularity of region. If the circularity is less than 0.2 it is reserve, otherwise, delete it. The circularity C is obtained from:

$$C = \frac{S}{\pi l^2} \tag{5}$$

Where S denotes the area of region, l denotes the half length of minimum enclosing rectangle.

Once the estimation of the threshold and circularity has been computed, region of clearance remain in the binary image. Note that there is usually one clearance region.

We segment the assembly clearance region in the above steps. To measure the width of the clearance, we need to get the center of the region, the inclination relative to the horizontal direction θ and the sub-pixel edge contour. According to the inclination angle θ, a line equation l_k orthogonal to the direction of the region could be obtained, and the intersect distance between the sub-pixel contour and l_k is the pixel-level width of the clearance. The measure algorithm procedure has the following steps:

Step 1. Center of the Region and Inclination
Since the density of the clearance region can be considered to be uniform in the binary image, the center of the region (C_x, C_y). According to the characteristics of the clearance region, fitting the minimum circumscribed ellipse of region, taking the long axis of the ellipse as l_r by connecting, calculate the angle θ between the line l_r and the horizontal direction. As illustrated in Fig. 2.

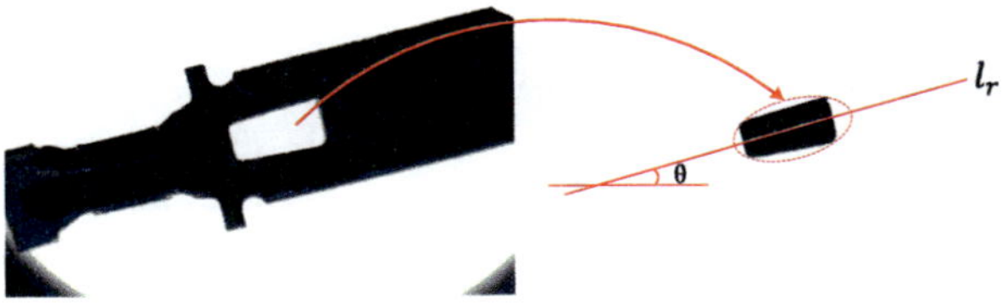

Fig. 2. Angle of region and direction

Step 2. Affine Transformation and Sub-pixel Contour of the Region

To avoid the starting point of measurements is not only, an affine transformation is needed to the clearance region. A 3 × 3 of identity matrix H is created. The rotation angle matrix $M_{rotation\_angle}$ is created byθin step 1. Create rotation factor RF by H and $M_{rotation\_angle}$.

The rotation matrix RM is created by the center point (C_x, C_y) and rotation factor RF. Since the rotation matrix RM and clearance region R are known, the affine transformation region R_{affine} can be obtained.

Extracting the edge of the region R_{affine} is the premise of measuring the clearance width. Edge fitting algorithms such as sobel, canny and roberts could only provide pixel-level precision, which means that the edge information between integer pixels is lost and cannot meet the high-precision measurement. To improve the measurement accuracy, we extracted the sub-pixel outline of the region by reference, as illustrated in Fig. 3.

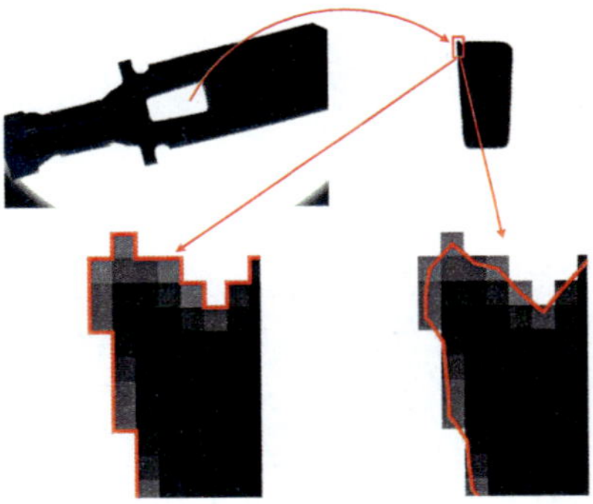

Fig. 3. Comparison between pixel point and sub-pixel

Step 3. Width Measurement

Extracting the edge of the region is the premise of measuring the clearance width. From the center point of the region R_{affine} $(C_{affine\_x}, C_{affine\_y})$, a straight line equation l_b crossing the center of R_{affine} could be given by:

$$l_b : x = C_x \tag{6}$$

Points P_{bottom} and P_{top} are two intersections of l_b with the contour, P_{bottom} is the bottom of the contour, and P_{top} is the top of the contour. The straight line equation l_n through A and perpendicular to l_b intersects the sub-pixel contour at one point, and the width of clearance is 0, indicating that the measurement starts from the bottom of the clearance region R_{affine}. The ordinate value of P_{bottom} changes abscissa in steps of one pixel and a set of straight lines equation $\{l_n\}$ perpendicular to l_b are generated until to the abscissa of P_{top}.

There are n line equations in the set $\{l_n\}$, $\{P_m\}$ represents the set of point pairs where the line equations intersect the contour, as shown in Eq. (7).

$$P_m = \{((r_{11}, c_{11}), (r_{12}, c_{12})), \ldots, ((r_{m1}, c_{m1}), (r_{m2}, c_{m2}))\} \quad (7)$$

Each brace in the set represents an intersecting point pair. The distance between each point pair is the set of pixel level distance $\{D_m\}$.

The distance obtained in the above steps is the pixel level distance $\{D_k\}$. The actual width of the clearance is given by Eq. (8).

$$D_{rel} = \tau\{D_k\} \quad (8)$$

Where τ denotes the corresponding coefficient between the actual distance of the system calibration and the pixel.

4 Experimental and Analysis

Some parts are to test the clearance method. Under the same conditions, 80 width data measured for every part. The measured pixel-level widths are shown in Table 1.

Table 1. Measured pixel-level width

Part	Measured width (pixel-level)					
Part1	221.261	221.368	222.566	222.912	224.517	226.707
	229.000	232.695	233.170	234.367	235.645	236.672
	237.743	238.436	239.411	240.236	241.287	243.030
	245.548	246.139	246.125	247.726	248.506	248.227
Part2	28.2782	55.471	74.331	86.3234	95.0390	102.577
	118.802	121.413	124.559	126.317	127.528	128.274
	129.602	130.566	131.088	128.616	130.015	129.339
	129.518	129.071	129.269	129.482	129.896	130.239
Part3	613.200	612.506	612.602	610.647	611.104	611.751
	613.279	613.721	614.647	614.215	613.209	612.280
	614.933	613.896	614.060	613.446	613.755	612.847
	613.353	611.731	612.746	613.633	613.847	613.645

The width conversion coefficient $\tau = 0.0067912$. Taking the irregularity of the part into account and the fact of the corner position cannot be measured, the middle 32 data points in the pixel-level distance of each parts are selected corresponding to 32 manual measurements with vernier caliper accuracy of 0.01 mm, the manual measured widths are shown in Table 2.

Table 2. Manual measured width

Part	Manual measurement (mm)					
Part1	1.68	1.69	1.72	1.69	1.65	1.75
	1.8	1.82	1.83	1.87	1.85	1.82
	1.82	1.81	1.80	1.79	1.85	1.86
	1.90	1.91	1.91	1.88	1.90	1.85
Part2	0.79	0.76	0.72	0.78	0.75	0.80
	0.80	0.79	0.77	0.85	0.86	0.86
	0.84	0.78	0.78	0.78	0.78	0.79
	0.8	0.79	0.76	0.75	0.76	0.78
Part3	4.11	4.15	4.12	4.12	4.13	4.15
	4.13	4.15	4.15	4.16	4.15	4.15
	4.12	4.13	4.15	4.15	4.15	4.16
	4.14	4.12	4.14	4.15	4.15	4.16

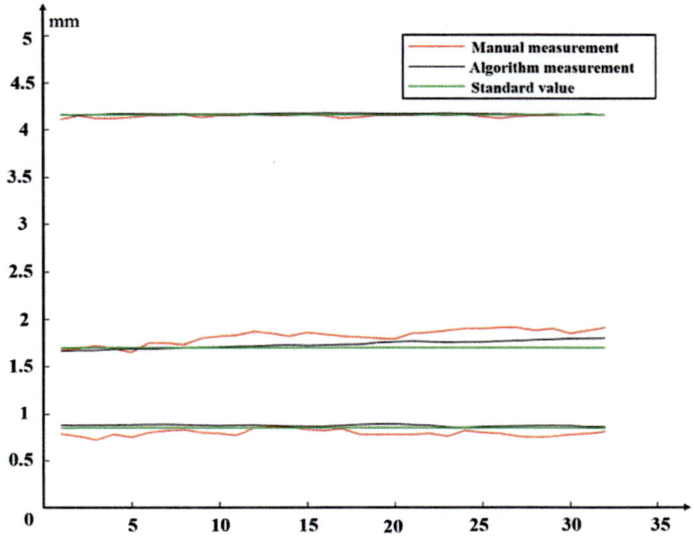

Fig. 4. Comparison of measurement results

Figure 4 shows that the deviation of the algorithm measurement results is smaller and more stable than the manual measurement. The lack of the proposed algorithm lies in the irregular assembly edges and clearance that cannot be backlit. During the image measurement process, if the center point of the segmented region is not on the vertical line between its center of gravity and the image coordinate axis, it will easily lead to deviations in the measurement.

5 Conclusion

This paper has been concerned with the assembly clearance width measurement. Clearance width is the pixel distance between two points mapped to the actual width in the world coordinate system. The algorithm is assessed in terms of accuracy and speed, and also compared with the manual measurement. According to the obtained results, the algorithm meets the requirements which makes it suitable for industrial applications.

Acknowledgment. This work was supported by The Guiding project of of Hubei Provincial Department of Education (No. B2020080).

References

1. Fan, Q.C., Tian, J., Zhu, F.: The statistical control methods of the simplified measurement process. Acta Metrol. Sin. **33**(3), 284–288 (2012)
2. Fu, X.B., Liu, B., Zhang, Y.C.: An optical non-contact measurement method for hot-state size of cylindrical shell forging. Measurement **45**(6), 1343–1349 (2012)
3. Groover, M.P.: Fundamentals of Modern Manufacturing: Materials, Processes and Systems, Forth. Wiley (2010)
4. Hopwood, J.: Langmuir probe measurements of a radio frequency induction plasma. J. Vac. Technol. **11**(1), 152–156 (1993)
5. Yang, P., Takamura, T., Takahashi, S.: Development of high-precision micro-coordinate measuring machine: multi-probe measurement system for measuring yaw and straightness motion error of XY linear stage. Precis. Eng. **35**(3), 424–430 (2011)
6. Zurek, S., Meydan, T.: A novel capacitive flux density sensor. Sens. Actuators A Phys. **129**(1–2), 121–125 (2006)
7. Kojima, T., Kroug, M., Uzawa, Y.: Contribution of quantum susceptance in SIS junction capacitance measurement. IEEE Trans. Appl. Supercond. **29**, 1 (2019)
8. Markushov, Y., Evtihiev, N., Grezev, N.: Multipass narrow gap of heavy gauge steel with filler wire. Phys. Procedia **71**, 267–271 (2015)
9. Mohamed, S., Kaseko, S.: A neural network-based methodology for pavement crack detection and classification. Transp. Res. Part C Emerg. Technol. **1**, 275–291 (1993)
10. Jahanshahi, M.R., Kelly, J.S., Masri, S.F.: A survey and evaluation of promising approaches for automatic image-based defect detection of bridge structures. Struct. Infrastruct. Eng. **5**(6), 455–486 (2009)
11. Zhang, Y.P., He, T., Wen, C.J.: Application and research of machine vision in industrial measurement. Opt. Precis. Eng. **9**(4), 324–329 (2001)
12. Fan, X., Lu, D.X., Wang, J.: Instrument technology: research and application of machine vision in industrial field. China Acad. J. Abs. **13**(22), 5 (2007)

Effect of Building Orientation in Mechanical Properties of Ti6Al4V Produced with Laser Powder Bed Fusion

Endre V. Nes[1], Even Wilberg Hovig[1], Leandro Feitosa[2], and Knut Sørby[1](✉)

[1] Department of Mechanical and Industrial Engineering, Norwegian University of Science and Technology, Trondheim, Norway
knut.sorby@ntnu.no
[2] Sandvik Machining Solutions AB, Additive Manufacturing, Sandviken, Sweden

Abstract. The paper presents tensile tests and Charpy tests of Ti6Al4V samples made with different build orientations in a laser powder bed fusion process. Seven different build orientations from 0° to 90° were tested. The test specimens were heat treated at 850 °C. Fracture surfaces were investigated with SEM. The materials showed a high degree of isotropy when measuring yield strength, tensile strength and elastic modulus. Elongation measurements indicated some degree of anisotropy, but not statistically significant. A similar trend was found in Charpy impact energy. Pores, microstructure and melt pool boundaries were investigated to explain the obtained results. The impact of pores was found to be insignificant. Melt pool boundaries and columnar prior-β grains are not a significant factor in the anisotropic trends observed. The anisotropy in mechanical properties was found to be caused by microstructure and/or crystallographic texture.

Keywords: Additive manufacturing · Ti6Al4V · Anisotropy · Tensile tests · Charpy tests · Powder bed fusion

1 Introduction

Additive manufacturing (AM) is a process of joining materials to make parts from 3D model data. It increases design freedom, manufacturing flexibility, product customization, enables shorter time to market, fast prototyping, direct repair of metallic parts, and decreases the traditional economy-of-scale constraints. Laser-based powder bed fusion of metals (PBF-LB/M) is one of the AM technologies that has received a lot of attention due to its ability to produce geometrical complex metallic structures. One of the challenges in PBF-L/M is the reduced ability to generate repeatable mechanical properties with different build orientations [3].

PBF-LB/M of Ti6Al4V without post heat treatments is inherently anisotropic. This is due to the formation of martensite (α'), which has preferred slip systems dependent on the build orientation [15]. Heat treatment above 800 °C have shown to completely decompose the martensite, removing the anisotropy introduced by the martensite [3, 14, 18].

Y. Wang et al. (Eds.): IWAMA 2021, LNEE 880, pp. 239–245, 2022.
https://doi.org/10.1007/978-981-19-0572-8_30

With respect to build orientation mechanical properties of PBF-LB/M Ti6Al4V are influenced by columnar β-grains orientated along the build direction, which occurs because of epitaxial growth [18]. Kumar et al. [5] found that with the right scanning parameters, including the 67° rotational scanning strategy, the effect of columnar β-grains on mechanical properties could be mitigated, without the use of heat treatment above the β-transus temperature of 995 °C.

The influence of pores on the mechanical properties are dependent on their morphology and size [2, 12]. The influence of spherical pores on the mechanical properties are independent of the build orientation as their morphology makes it indifferent in which direction they are loaded [4]. Elongated pores are much more detrimental to the mechanical properties than spherical pores, due to sharp rims and crack tips that results in areas of high stress concentration [2, 4]. Elongated pores due to lack of fusion are oriented perpendicular to the build direction [2, 4]. This will give a different mechanical response in the horizontal (0°) direction compared to the vertical (90°) direction [11, 13].

Most published literature on the effect of build orientations of the Ti6Al4V built with AM consider few orientations. Often only the horizontal (0°) and vertical (90°) orientations are considered, with some publications also including 45°. In this paper the effect of build orientations on mechanical properties is investigated in seven different build orientation. Both tensile and Charpy tests are performed in the 0°, 15°, 30°, 45°, 60°, 75° and 90° build orientation. The purpose of this paper is to broaden the understanding of the effect of build orientations in PBF-LB/M of Ti6Al4V. The fracture surfaces are examined with scanning electron microscope (SEM) and electron backscatter diffraction (EBSD).

2 Material and Method

Initially a total of two sets of specimens for tensile tests (14 specimens, two for every angle) and two sets of Charpy test specimens were built. One of the tensile tests in the 30° orientation was discarded due to a failure in the test equipment. All other tests were performed correctly.

All tensile specimens were machined to a surface roughness *Ra* smaller than 0.5 μm. Only one of the sets of Charpy specimens were machined while the other set was left with an as-built surface. All samples received a post heat treatment at 850 °C for two hours under argon atmosphere, and then air cooled.

The material was supplied by Sandvik Osprey. The chemical composition is provided in Table 1. All specimens were manufactured in an EOS M290 machine. The process parameters are provided in Table 2.

The dimensions of tensile specimens are based on ASTM E8/E8M standard, but adapted for AM. To minimize powder usage, the tensile specimens are made as small as possible, without the thickness affecting the results. The dimensions of the tensile test samples are shown in Fig. 1. Charpy specimens were produced according to ASTM E23/ISO148 (type A).

Tensile tests were carried out at room temperature in an MTS 809 Axial Test System with a 100 kN load cell. The machine was set to move at a constant speed of 1 mm per minute. Digital Image Correlation (DIC) from Vic3D was used to capture the strain.

Table 1. Nominal chemical composition

Element	C	N	O	Al	Ti	V	Fe
Wt-%	0.009	0.0165	0.109	6.2	89.5	4	0.2

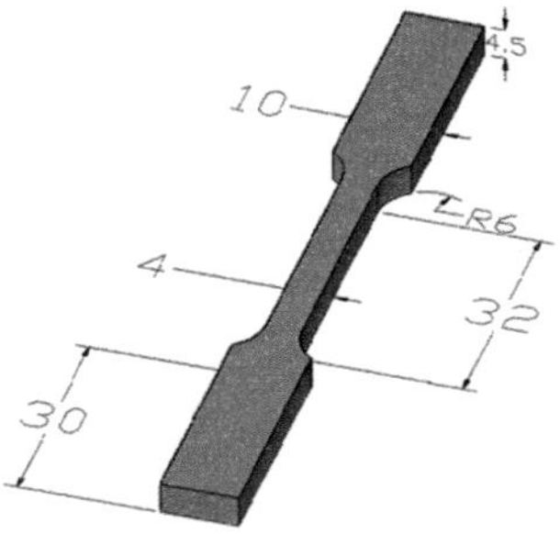

Fig. 1. Test sample dimensions

Table 2. Process parameters

Laser power	280 W
Layer thickness	30 μm
Scan velocity	1200 mm/s
Hatch spacing	140 μm
Substrate temperature	80 °C
Atmosphere	Argon

Output data from the tensile tests were post-processed with an in-house developed Matlab script. The yield strength was determined at 0.2% strain.

Charpy impact tests were carried out according to ASTM E23/ISO148 with self-centering tongs in an Instron MXP450 machine. The initial potential energy of the machine is 450 J. The machine has an resolution of 0.023 J at an impact energy absorption of 15 J.

Fracture surfaces of the tensile and Charpy specimens were investigated with a scanning electron microscope (SEM), on a Quanta FEG 650 system.

3 Results

3.1 Tensile Properties

Of the tensile specimens the fracture surfaces of 0°, 45° and 90° build orientations were investigated with SEM. The whole fracture surfaces were carefully examined.

Some spherical pores were found. The sum of all pores found on the tensile specimens were 5 pores. The total area of the pores were approximately 0.004 mm^2 on a total area of 270 mm^2. No elongated pores was found.

The fracture surfaces had dimples due to nucleation of voids and shear lips, indicating ductile fracture.

3.2 Charpy Properties

Machined v-notch radii were measured to 250 ± 30 μm. It was difficult to determine the v-notch radius on as-built specimens due to fused powder particles on the surface obstructing the sight of the v-notch radius. As-built v-notch radii were measured to 130 ± 60 μm.

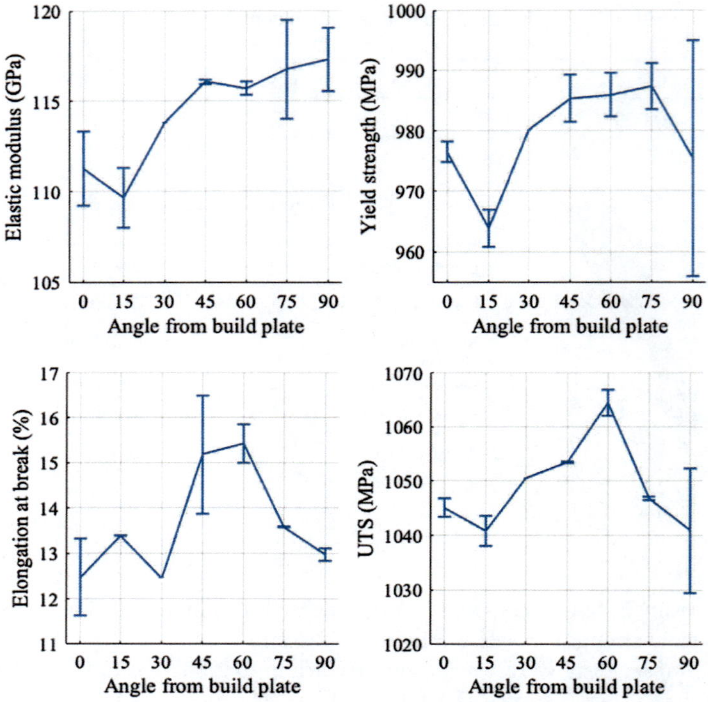

Fig. 2. Tensile results.

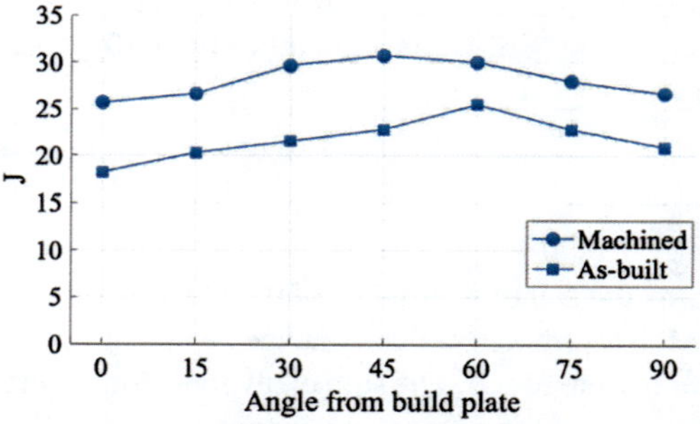

Fig. 3. Charpy impact results, absorbed energy [J].

The fracture surfaces from the Charpy tests for 0°, 45°, and 90° build orientations, both machined and as-built specimens, were investigated with SEM. These fracture surfaces also had dimples due to nucleation of voids. The area of the fracture surfaces of the Charpy specimens are considerably larger than the surfaces area from the tensile tests, and they were therefore only briefly examined in the SEM. There were few small spherical pores, matching findings from the tensile tests, with the exception of two large elongated pores found on the 0° as-built specimen. The pores had a length of 70–100 μm and a width of approximately 45 μm.

4 Discussion

There are small differences between mechanical tensile results with respect to build orientation. An analysis of variance was performed to investigate if there is any statistically significant difference between build orientations. The results showed that the *p*-values were above 0.05 for all the mechanical tensile properties, meaning that there is no significant difference between build orientations. Still, it can be argued that a trend can be seen in the elongation property. A similar trend of the same form can also be seen in Charpy impact energy with respect to build orientation.

Cracks preferentially propagate along alignment of pores [9]. As there were few pores overall on the fracture surfaces of the specimens it indicates that pores played a small role in the fracture mechanism and influence on mechanical properties. Since all pores, except for two of the 0° Charpy as-built specimen, were spherical, they should not influence the mechanical properties with respect to build orientation as their spherical morphology make it indifferent which direction they are pulled upon [4].

Melt pool boundaries (MPBs) have been found to greatly affect mechanical properties [8]. In Ti6Al4V, MPBs consist of precipitated Ti_3Al [10]. The Ti_3Al solvus temperature for Ti6Al4V is 550–600 °C, depending on the exact aluminum and oxygen content [7]. This means that the heat treatment would have dissolved the melt pool boundaries. Lack of flat, brittle, surfaces on the fracture surface also indicates that MPBs were not a part of the facture process.

Columnar prior-β grains in PBF-LB/M occur due to epitaxial growth. These grains often lead to anisotropy in elongation between 0° and 90° tensile specimens, as they yield higher elongation when load is applied parallel to the grains (90°). Use of the right scanning parameters and a 67° scanning strategy has shown to make prior-β grains equiaxed in the horizontal plane and jagged and discontinued in the vertical plane [5]. This is because of the slight mismatch between the melt pools from layer to layer due to the 67° rotational scanning. EBSD results coincides with Kumar et al. [5], which used the exact same scanning parameters, mitigated the anisotropic behavior in elongation between 0° and 90° tensile specimens. Elongation and EBSD results showed no anisotropy between 0° and 90° specimens and jagged and discontinued prior-β grains for 0°, 45° and 90° planes, indicating that columnar prior-β grains did not affect the mechanical properties with respect to build orientation.

The anisotropic nature of the HCP α-phase can generate different mechanical properties regarding the crystallographic orientation which can lead to mechanical anisotropy of PBF-LB/M Ti6Al4V. Pole figures shows a weak texture which may be too small to produce any significant effect on mechanical properties with respect to build orientation. Still, crystallographic orientations have shown to influence both elongation and Charpy impact energy [1, 15], and therefore may be the cause of the anisotropy that is observed for the elongation.

Microstructural texture of the α-phase may also affect mechanical properties through the size, width, organization and orientation. The α colony size of the lamella structure determines the effective slip length and the fracture toughness, increasing with lamella size [7]. The colony boundaries also work as strong obstacles for microcrack propagation [7]. Yang et al. [16] found that more than 95% of the major axes of the α' (martensite) had an angle of $-30°$ to $-60°$ or $30°$ to $60°$ from the building direction (vertical axis), with the highest peaks found at $-45°$ and $45°$, and correlating with the trends seen in elongation. This could also be a viable explanation for the anisotropic trend in elongation, as the pulling along the major axis of the α colonies gives a longer slip surface. For lamellar Ti6Al4 V, the crack propagation in Charpy specimens propagates either at the prior-β grain boundary or between two α platelets [1]. It can then be speculated that the crack propagation takes a longer route along the α platelets for the specimens in accordance with the trends observed for the impact energy, which are not yet investigated.

The importance of the microstructure and crystallographic texture in PBF-LB/M has not yet been investigated for Ti6Al4V at a detailed level without the influence of other defects such as pores and columnar grains. It is therefore difficult to determine to what extent the microstructure and crystallographic texture contribute to the observed results.

The difference between the Charpy test results for the machined and as-built surfaces may be explained by the difference between the notch-radius sizes between the two sample sets, as a smaller notch-radius has been shown to lead to a lower absorbed energy at room temperature [6]. However, the general difference in surface roughness between the machined and as-built samples will not affect the results in the Charpy tests [17].

5 Conclusion

The analysis of variance showed that there were no statistically significant difference for the tensile properties between build orientations. Still a trend in elongation can be seen graphically. A similar trend can be seen in Charpy impact energy.

Pores played a small role in the fracture process and did not influence the mechanical properties with respect to build orientation.

Effects of Columnar prior-β grains on the mechanical properties with respect to build orientation was found to be mitigated due to the use of the right scanning parameters and a 67° rotational scanning strategy.

The direction of the major axis of the lamellar structure and the crystallographic texture was found to most likely influence the mechanical properties with respect to build orientation, but to what degree could not be determined from literature.

Lastly, the energy difference between the Charpy trends for the machined and as-built surfaces can be explained by the difference between the notch-radius sizes between the two batches.

Acknowledgment. The authors thank the *Interreg Sweden-Norway* programme for supporting this work through the *TROJAM* project.

References

1. Buirette, C., Huez, J., Gey, N., Vassel, A., Andrieu, E.: Study of crack propagation mechanisms during Charpy impact toughness tests on both equiaxed and lamellar microstructures of Ti–6Al–4V titanium alloy. Mater. Sci. Eng. A. **618**, 546–557 (2014)
2. Gong, H., Rafi, K., Gu, H., Ram, G.J., Starr, T., Stucker, B.: Influence of defects on mechanical properties of Ti–6Al–4 V components produced by selective laser melting and electron beam melting. Mater. Des. **86**, 545–554 (2015)
3. Hartunian, P., Eshraghi, M.: Effect of build orientation on the microstructure and mechanical properties of selective laser-melted Ti-6Al-4V alloy. J. Manuf. Mater. Proces. **2**(4), 69 (2018)
4. Kasperovich, G., Haubrich, J., Gussone, J., Requena, G.: Correlation between porosity and processing parameters in TiAl6V4 produced by selective laser melting. Mater. Des. **105**, 160–170 (2016)
5. Kumar, P., Prakash, O., Ramamurty, U.: Micro-and meso-structures and their influence on mechanical properties of selectively laser melted Ti-6Al-4V. Acta Mater. **154**, 246–260 (2018)
6. Kurishita, H., Kayano, H., Narui, M., Yamazaki, M., Kano, Y., Shibahara, I.: Effects of V-notch dimensions on Charpy impact test results for differently sized miniature specimens of ferritic steel. Mater. Trans. JIM **34**(11), 1042–1052 (1993)
7. Lütjering, G., Williams, J.C.: Titanium. Springer, Heidelberg (2007)
8. Shifeng, W., Shuai, L., Qingsong, W., Yan, C., Sheng, Z., Yusheng, S.: Effect of molten pool boundaries on the mechanical properties of selective laser melting parts. J. Mater. Process. Technol. **214**(11), 2660–2667 (2014)
9. Stef, J., Poulon-Quintin, A., Redjaimia, A., Ghanbaja, J., Ferry, O., De Sousa, M., et al.: Mechanism of porosity formation and influence on mechanical properties in selective laser melting of Ti-6Al-4V parts. Mater. Des. **156**, 480–493 (2018)
10. Thijs, L., Verhaeghe, F., Craeghs, T., Van Humbeeck, J., Kruth, J.-P.: A study of the microstructural evolution during selective laser melting of Ti–6Al–4V. Acta Mater. **58**(9), 3303–3312 (2010)
11. Vilaro, T., Colin, C., Bartout, J.-D.: As-fabricated and heat-treated microstructures of the Ti-6Al-4V alloy processed by selective laser melting. Metall. Mater. Trans. A. **42**(10), 3190–3199 (2011)
12. Voisin, T., Calta, N.P., Khairallah, S.A., Forien, J.-B., Balogh, L., Cunningham, R.W., et al.: Defects-dictated tensile properties of selective laser melted Ti-6Al-4V. Mater. Des. **158**, 113–126 (2018)
13. Wu, M.-W., Lai, P.-H., Chen, J.-K.: Anisotropy in the impact toughness of selective laser melted Ti-6Al-4V alloy. Mater. Sci. Eng. A. **650**, 295-299 (2016)
14. Xu, W., Lui, E.W., Pateras, A., Qian, M., Brandt, M.: In situ tailoring microstructure in additively manufactured Ti-6Al-4V for superior mechanical performance. Acta Mater. **125**, 390–400 (2017)
15. Yang, J., Yu, H., Wang, Z., Zeng, X.: Effect of crystallographic orientation on mechanical anisotropy of selective laser melted Ti-6Al-4V alloy. Mater. Charact. **127**, 137–145 (2017)
16. Yang, J., Yu, H., Yin, J., Gao, M., Wang, Z., Zeng, X.: Formation and control of martensite in Ti-6Al-4V alloy produced by selective laser melting. Mater. Des. **108**, 308–318 (2016)
17. Yasa, E., Deckers, J., Kruth, J.-P., Rombouts, M., Luyten, J.: Charpy impact testing of metallic selective laser melting parts. Virtual Phys. Prototyping **5**(2), 89–98 (2010)
18. Zhang, X.-Y., Fang, G., Leeflang, S., Böttger, A.J., Zadpoor, A.A., Zhou, J.: Effect of subtransus heat treatment on the microstructure and mechanical properties of additively manufactured Ti-6Al-4V alloy. J. Alloy. Compd. **735**, 1562–1575 (2018)

Optimal Design of Truss Based on LA-GSA

Xiao Zhang and Mingjian Liu(✉)

College of Information Engineering, Dalian Ocean University, Dalian, China
liumingjian@dlou.edu.cn

Abstract. Gravity search algorithm is a competitive swarm intelligence optimization algorithm proposed in recent years. However, the standard gravity search algorithm has the disadvantage of slow convergence. In order to effectively use the optimization algorithm to solve the problem of truss structure optimization, this paper proposes a gravity search algorithm (LA-GSA) based on Learning Automata (LA) and applies it to truss optimization design. The algorithm uses LA to automatically adjust the gravitational constant G(t) in GSA to improve the optimization accuracy of GSA. In the experiment, this paper first tests LA-GSA algorithm and other methods on six reference functions. The results show that GSA-LA algorithm is more effective in finding the optimal solution and has higher searching accuracy. Then, this algorithm is applied to optimize the size of the 72 bar space truss. Compared with other algorithms, the results show that the iterative times of LA-GSA to obtain the optimal value are significantly reduced, and the optimal solution obtained by LA-GSA is obviously better than the general algorithm.

Keywords: Gravitational search algorithm · Learning automata · Gravitational constant · Reference function · Search accuracy · Truss optimization

1 Introduction

For truss structure, under the condition of given structure form, material and shape, the size of each bar is optimized so that the structure can meet the constraints and achieve the lightest total structure mass, the best design rationality and the lightest engineering cost mass, which is the structural optimization [1]. There are many variables in structural optimization design, and the traditional computing methods to solve these problems are faced with high computational complexity and long computational time. Therefore, the research on heuristic computational intelligence methods is more and more active. In the past decade, a series of intelligent algorithms have emerged, such as genetic algorithms, particle swarm optimization, ant colony optimization, simulated annealing, harmony search and Firefly Algorithm (FA) [2–4]. These methods do not require traditional mathematical assumptions and therefore increase the globally optimal location probability compared with traditional methods. The random nature of meta-heuristic optimization methods allows a larger portion of the search space to be explored than traditional optimization methods.

Scholars at home and abroad have also applied intelligent algorithms to truss structure optimization. For example, enhanced chaotic colliding body algorithms [5] and improved

Y. Wang et al. (Eds.): IWAMA 2021, LNEE 880, pp. 246–253, 2022.
https://doi.org/10.1007/978-981-19-0572-8_31

binary bat flexible sampling Algorithm [6], simulated plant growth algorithm, etc., all carry out structural optimization of trusses [7], effectively promoting the development of truss structure optimization. However, there is still room for research in the convergence speed and stability of the algorithm.

GSA was proposed inspired by the law of gravity and has been widely applied in various engineering applications [8]. In GSA, parameter fine-tuning plays a decisive role in the efficiency of the algorithm, and when the parameters are not properly adjusted, GSA's local search performance will be poor [9]. In order to obtain better results when solving practical engineering problems, improve the search performance of GSA algorithm and avoid its falling into local optimal solution, this paper proposes a gravity search algorithm based on learning automata (LA-GSA) and applies it to the size optimization of truss structures.

2 Related Theoretical Basis

2.1 Gravitational Search Algorithm

Gravitational Search Algorithm (GSA) is a new group intelligence optimization algorithm proposed by Iranian scholar Esmat in 2009 [10]. Standard GSA to gravitation and Newton's second law as the foundation, said: gravity between particles, the greater the quality and the greater the nearer the distance between the particles of gravity, the particles are produced by other particles on the direction of the force of gravity acceleration, the direction of the acceleration is always point to the quality of the larger particles, making different particle moving in the direction of quality of the larger particles. Ultimately, the position of the most massive particle becomes the optimal solution.

2.2 Learning Automaton

Learning Automata (LA) was first proposed by Narendra and Thathachar in 1974. It is a method to use automata to find the best action among potential limited actions and has been widely applied in different types of projects [11]. The basic steps of the method are as follows: first, a possible action is randomly selected from a limited set of potential actions; Then, the selected actions are used to evaluate the process random environment, and the feedback signal is calculated. Finally, the feedback signal is used to update the probability of each action. In LA, it is the signal update in the iterative process that enables it to obtain the best results through learning. Figure 1 shows the learning process of learning automata.

Where, the learning automata is defined by a quad $\{\alpha, \beta, \rho, T\}$, where $\alpha = \{\alpha_1, \alpha_2, ..., \alpha_r\}$ represents the set of optional actions, $\beta = \{\beta_1, \beta_2, ..., \beta_r\}$ represents the feedback value of the environment, $\rho = \{\rho_1, \rho_2, ..., \rho_r\}$ represents the probability vector of each action being selected, and T represents the update strategy of the learning automata.

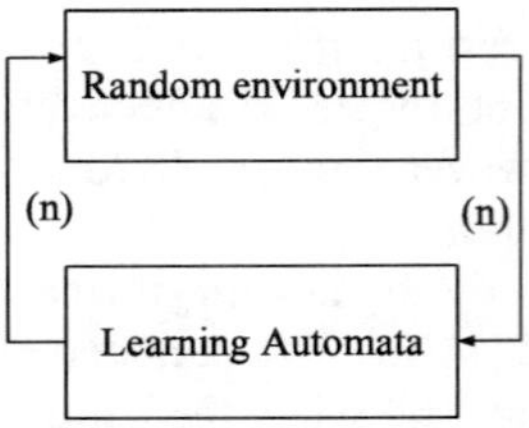

Fig. 1. The relationship between learning automata and random environment

3 Gravity Search Algorithm Based on Learning Automata

The current research shows that better optimization performance and faster convergence rate can be achieved by fine-tuning the parameters of the evolutionary algorithm. In this paper, a gravitation search algorithm based on learning automata (LA-GSA) is proposed to automatically adjust the gravitation constant G(t) in GSA based on learning automata. The specific implementation process is similar to the original GSA process, except that G(t) is no longer calculated according to Eq. (5) at the end of each iteration of GSA.

In LA-GSA, G(t) is updated with LA with N actions, where each action corresponds to a discrete value that divides the range of G(t) into N equal parts In each iteration of GSA, the new value of G(t) corresponds to the most likely action in LA's set of actions. The implementation process of LA-GSA is shown in Fig. 2.

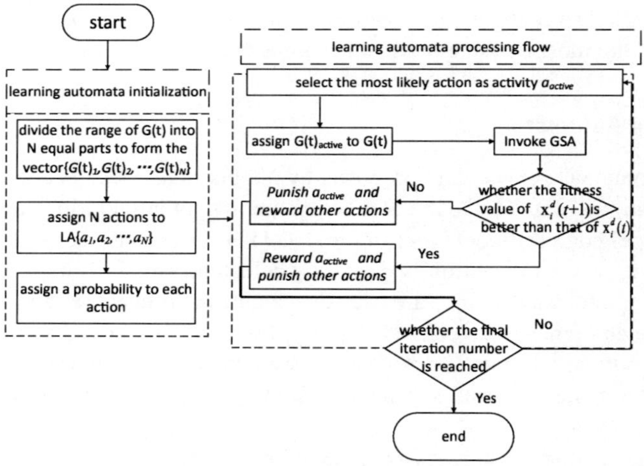

Fig. 2. Flowchart of LA-GSA algorithm

After G(t) is obtained in each iteration of GSA, the new position $X_i^d(t+1)$ of individual particle i can be obtained according to the position calculation formula of GSA. At this time, the newly obtained position is substituted into the fitness value calculation function (reference function). If $X_i^d(t+1)$ has a better fitness value than

$X_i^d(t)$, it indicates that G(t) obtained at this time is effective. In this case, learning automatically gets rewarded; Otherwise, the learning automata will be punished.

4 Comparative Analysis of Experiments

4.1 Optimization Comparison of Reference Functions

4.1.1 Benchmark Functions

In order to evaluate the optimization performance of LA-GSA algorithm, this paper compares its optimization with GSA, GA [12] and LA-PSO (particle swarm optimization algorithm based on automatic learning machine) on the six reference functions in Table 1. Each of the test functions in Table 1 has unique characteristics that make them suitable for verifying the performance of the new algorithm. Among them, functions Ackley, Griewank, Schwefel's Problem 2.26 and Rastrigin are multi-mode functions, and the number of local optimal values of these functions increases exponentially with the increase of the Problem dimension. The functions Sphere and Rotated hyper-ellipsoid are single-mode functions [13, 14].

Table 1. Benchmark functions

Benchmark functions	Value range
F_1 (Ackley)	[−32, 32]
F_2 (Griewank)	[−600, 600]
F_3 (Rastrigin)	[−6.12, 6.12]
F_4 (Sphere)	[−100, 100]
F_5 (Rotated hyper-ellipsoid)	[−100, 100]
F_6 (Schwefel's problem 2.26)	[−500, 500]

4.1.2 Comparison of GA, LA-PSO, GSA and LA-GSA

In order to make a comprehensive comparison of the performance of LA-GSA, GA, GSA and LA-PSO algorithms, the running times of each algorithm were set to 30, the dimensions were set to 15 and 30, and the maximum iteration times were set to 500 in the experiment. The average error and minimum error of the running results of each algorithm were calculated respectively, and the results were shown in Table 2–3.

As can be seen from Table 2–3, when the dimension is 15, THE performance of GA algorithm is the worst on each benchmark function, while the performance of LA-GSA algorithm is better than THAT of GSA on all test functions, and the results of four of the six test functions are better than that of LA-PSO. When the dimension is 30, LA-GSA has obvious superiority, which is significantly better than GA and GSA, while LA-PSO only performs better in function F3. The above results are because the number of local optimal values increases with the increase of dimension. However, in LA-PSO and LA-GSA, LA is used to make these two algorithms can be well separated from the local optimal. However, GA and GSA algorithms can not search for the global optimal position because they fall into the local optimal position.

Table 2. The average error and minimum error of the four algorithms on each reference function when the dimension space is 15

	GA		GSA		LA-PSO		LA-GSA	
	Average error	Minimum error	Average error	Minimum error	Average error	Minimum error	Average error	Minimum error
F_1	3.6905	1.2498	3.19E−12	5.39E−26	5.04E−10	7.42E−18	1.91E−19	8.48E−28
F_2	8.13E−03	7.12E−04	3.84E−03	6.21E−06	8.16E−08	5.23E−22	2.59E−07	2.52E−28
F_3	1.1396	7.82E−03	9.27E−15	5.18E−20	7.12E−07	8.18E−25	6.12E−14	4.32E−21
F_4	0.8791	2.97E−03	3.77E−12	3.56E−15	6.13E−05	4.72E−19	6.23E−11	7.13E−27
F_5	0.4613	2.53E−06	8.4852	7.23E−02	4.29E−03	7.29E−12	3.16E−09	7.95E−17
F_6	2.5412	0.4149	6.78E−16	6.81E−25	3.58E−10	7.32E−18	7.14E−14	5.29E−26

Table 3. The average error and minimum error of the four algorithms on each reference function when the dimension space is 30

	GA		GSA		LA-PSO		LA-GSA	
	Average error	Minimum error	Average error	Minimum error	Average error	Minimum error	Average error	Minimum error
F_1	10.1202	3.7272	8.81E−05	9.21E−08	8.15E−05	7.27E−12	9.58E−11	2.59E−14
F_2	17.3547	3.991	0.8492	3.62E−03	3.25E−05	8.29E−06	6.09E−14	6.25E−16
F_3	9.1947	0.3639	14.4721	3.6174	4.11E−05	3.72E−08	2.76E−05	9.13E−07
F_4	3.9847	8.19E−02	9.11E−10	4.59E−12	6.21E−05	4.29E−14	5.26E−09	2.76E−18
F_5	27.8499	20.9124	38.1483	21.7943	0.5921	5.39E−03	7.68E−04	9.22E−04
F_6	9.1414	2.3152	3.68E−05	6.81E−06	7.99E−07	8.27E−08	6.33E−12	7.42E−14

4.1.3 Compared with Other New Optimization Algorithms

In Sect. 4.1.2, LA-GSA is compared with the original GSA, GA and La-PSO of LA based particle swarm optimization algorithm, and the results clearly show the superiority of GSA-LA over GSA. In addition, as a powerful basic optimization algorithm, LA-GSA is also significantly better than GA, and GSA is significantly more suitable for LA than PSO and LA.

In addition, this paper also compares LA-GSA with two new optimization algorithms, Gray Wolf Optimization (GWO) [15] and Teaching and Learning Optimization (TLBO) [16]. The experimental results are shown in Table 4. As can be seen from Table 4, LA-GSA is superior to GWO in all test functions, while TLBO has smaller errors in F3 and F5, which fully demonstrates the validity of LA-GSA proposed in this paper.

Table 4. The average error error and minimum error of the three algorithms

	GWO		TLBO		LA-GSA	
	Average error	Minimum error	Average error	Minimum error	Average error	Minimum error
F_1	4.29E−06	3.11E−10	6.12E−11	9.29E−12	9.58E−11	2.59E−14
F_2	9.37E−09	7.31E−12	8.14E−08	7.13E−11	6.09E−14	6.25E−16
F_3	2.02E−05	2.06E−09	2.66E−03	3.46E−10	2.76E−05	9.13E−07
F_4	6.17E−08	7.81E−11	4.19E−11	5.13E−16	5.26E−09	2.76E−18
F_5	7.91E−03	5.51E−05	9.28E−03	4.72E−06	7.68E−04	9.22E−04
F_6	3.52E−11	9.73E−09	5.15E−05	6.24E−09	6.33E−12	7.42E−14

4.2 Truss Analysis

A typical truss optimization example is selected to verify the effectiveness and reliability of LA-GSA in solving the shape and size optimization problems of constrained trusses. The optimization results of LA-GSA were compared with those of GSA and other optimization techniques. The proposed IGSA is implemented by Python and the operating system is Microsoft Window 10 professional edition.

72-bar space truss structure, the bars are divided into 16 groups, the maximum displacement limit of nodes 1–6 along x and y is 6.35 mm, the maximum allowable stress is [−172.375, 172.375] MPa, the density is 2678 kg/m^3, the elastic modulus E = 68950 MPa, the length and width are 3.048 m, the maximum allowable stress is [−172.375, 172.375] MPa, the density is 2678 kg/m^3, the elastic modulus E = 68950 MPa, the length and width are 3.048 m. The height of each section is 1.524m, and the load condition and action position are the same as in literature [17]. The control parameters of the algorithm are set as follows: the maximum number of iterations is 500, the search space dimension is 16, and the total population is 75. Under the same constraints, the results are compared with those in other literatures, and the optimization results are shown in Table 5.

As can be seen from Table 5, the optimized weight of the 72-bar space truss obtained by IGSA is 168.933 kg. Compared with the results obtained by GSA and reference [18–20], the weight is reduced to different degrees. Meanwhile, it is found in the experiment that LA-GSA algorithm converges to the optimal solution at the 240th iteration, while GSA cable algorithm converges to the corresponding optimal solution at the 440th iteration. It can be seen that the improved gravity search algorithm is better than the standard gravity search algorithm in searching for the optimal solution, and it converges to the optimal solution faster.

Table 5. Comparison of optimization results of 72 bar space truss

Group number	Bar	Review [18]	Review [19]	Review [20]	Review [19]	GSA	LA-GSA
1	1–4	98.098	99.001	96.743	100.614	98.633	97.711
2	5–12	340.291	379.71	341.904	356.097	339.724	347
3	13–16	261.001	217.001	306.42	239.998	268.663	260.033
4	17–18	364.033	388.742	332.226	323.194	358.373	339.775
5	19–22	323.646	167.259	293.517	380.936	322.793	327.968
6	23–30	332.226	350.291	338.678	339.968	336.863	61.259
7	31–34	61.259	61.259	74.162	61.259	61.259	61.259
8	35–36	61.259	94.098	103.194	61.259	61.259	61.259
9	37–40	822.612	710.742	741.903	800.612	807.903	816.806
10	41–48	328.871	370.484	374.162	334.807	322.783	329.065
11	49–52	61.259	61.259	61.259	61.259	61.259	61.259
12	53–54	61.259	61.259	61.259	61.259	61.259	61.259
13	55–58	1220.805	1337.6441	1128.9991	1169.643	1232.033	1210.095
14	59–66	329.517	321.517	322.549	335.259	334.223	328.484
15	67–70	61.259	61.259	61.259	61.259	61.259	61.259
16	71–72	61.259	61.259	61.259	61.259	61.259	61.259
Weight (kg)		168.953	173.023	171.723	169.653	169.223	168.933

5 Discussion and Conclusion

In this paper, a new GSA algorithm, LA-GSA algorithm (LA-GSA), is proposed for solving continuous unconstrained global optimization problems. The algorithm uses LA to automatically adjust the gravitational constant G(t) in the GSA running process, so as to avoid falling into local optimum during the GSA running process and improve the optimization accuracy. In the experiment, LA-GSA is tested on six benchmark functions, and the simulation results show that LA-GSA is more effective in finding the optimal solution than the original GSA and other swarm intelligence based algorithms, and its performance is better than other algorithms. Through the structural optimization of the truss, it is shown that the algorithm is suitable for the optimization design of the section size of the space truss structure.

References

1. Shakya, A., Nanakorn, P., Petprakob, W.: A ground-structure-based representation with an element-removal algorithm for truss topology optimization. Struct. Multidiscip. Optim. **58**(2), 657–675 (2018)

2. Nguyen, H., Moayedi, H., Foong, L.K., et al.: Optimizing ANN models with PSO for predicting short building seismic response. Eng. Comput. Int. J. Simul. Based Eng. **35**, 3 (2020)
3. Kumar, S., Solanki, V.K., Choudhary, S.K., et al.: Comparative study on ant colony optimization (ACO) and K-means clustering approaches for jobs scheduling and energy optimization model in Internet of Things (IoT). Int. J. Interact. Multimedia Artif. Intell. **6**(1) (2020)
4. Delahaye, D., Chaimatanan, S., Mongeau, M.: Simulated annealing: from basics to applications. In: Gendreau, M., Potvin, J.-Y. (eds.) Handbook of metaheuristics. ISORMS, vol. 272, pp. 1–35. Springer, Cham (2019). https://doi.org/10.1007/978-3-319-91086-4_1
5. Kaveh, A., Dadras, A., Montazeran, A.H.: Chaotic enhanced colliding bodies algorithms for size optimization of truss structures. Acta Mech. **229**(7), 2883–2907 (2018)
6. Hamzehkolaei, N.S., Miri, M., Rashki, M.: An improved binary bat flexible sampling algorithm for reliability-based design optimization of truss structures with discrete-continuous variables. Eng. Comput. **32**, 641–671 (2018)
7. Degertekin, S.O., Lamberti, L., Ugur, I.B.: Sizing, layout and topology design optimization of truss structures using the Jaya algorithm. Appl. Soft Comput. **70**, 903–928 (2018)
8. Tung, N.S., Chakravorty, S., Bhullar, H.S.: Gravity local search inspired particle swarm algorithm for economic power dispatch planning problem in small scale system. Int. J. Grid Distrib. Comput. **9**(5), 111–124 (2016)
9. Wang, X., Zhang, G., Wang, X., et al.: Output-only structural parameter identification with evolutionary algorithms and correlation functions. Smart Mater. Struct. **29**(3) (2020)
10. Chen, L., Sun, H., Zhao, W., et al.: AI based gravity compensation algorithm and simulation of load end of robotic arm wrist force. Math. Probl. Eng. **2021**(8), 1–11 (2021)
11. Goodwin, M., Yazidi, A., Jonassen, T.M.: Distributed learning automata-based S-learning scheme for classification. Pattern Anal. Appl. **23**(9) (2020)
12. Katoch, S., Chauhan, S.S., Kumar, V.: A review on genetic algorithm: past, present, and future. Multimedia Tools Appl. **80**(5), 8091–8126 (2021)
13. Beigvand, S.D., Abdi, H., Scala, M.L.: Optimal operation of multicarrier energy systems using time varying acceleration coefficient gravitational search algorithm. Energy **114**(1), 253–265 (2016)
14. Priya, R.D., Sivaraj, R., Anitha, N., et al.: Forward feature extraction from imbalanced microarray datasets using wrapper based incremental genetic algorithm. Int. J. Bio-Inspired Comput. **16**(3), 171 (2020)
15. Shakarami, M.R., Davoudkhani, I.F.: Wide-area power system stabilizer design based on Grey Wolf Optimization algorithm considering the time delay. Electr. Power Syst. Res. **133**(4), 149–159 (2016)
16. Elroby, M.M.H., Mekhamer, S.F., Talaat, H.E.A., et al.: Population based optimization algorithms improvement using the predictive particles. Int. J. Electr. Comput. Eng. **10**(3), 3261 (2020)
17. Zhou, M., Rozvany, G.I.N.: DCOC: an optimality criteria method for large systems Part II: algorithm. Struct. Optim. **6**(4), 250–262 (1993)
18. Wang, Y., Gao, S., Yu, Y., et al.: A gravitational search algorithm with chaotic neural oscillators. IEEE Access **8**, 25938–25948 (2020)
19. Pervez, I., Sarwar, A., Tayyab, M., et al.: Gravitational search algorithm (GSA) based maximum power point tracking in a solar PV based generation system. In: 2019 Innovations in Power and Advanced Computing Technologies (i-PACT), vol. 1, pp. 1–6 IEEE (2019)
20. Zhao, F., Xue, F., Zhang, Y., et al.: A hybrid algorithm based on self-adaptive gravitational search algorithm and differential evolution. Expert Syst. Appl. **113**, 515–530 (2018)

Cutting Forces in Machining of Low-Lead and Lead-Free Brass Alloys

Magdalena S. Müller(✉) and Knut Sørby

Department of Mechanical and Industrial Engineering, Norwegian University of Science and Technology, 7491 Trondheim, Norway
magdalena.s.muller@ntnu.no

Abstract. Knowledge of cutting forces and the influence of the tool geometry on lead-free brass is necessary due to the increasing restrictions of the use of lead. Today, research on the cutting of lead-free brass has mainly focused on comparing different cutting conditions and different alloys. This study aims to explain the differences in cutting forces using the orthogonal cutting model and Merchant equation. A turning method aiming for cutting conditions as close as possible to the orthogonal cutting model was used in the tests. In summary, the findings in this study suggest that the Merchant equation is not applicable in the cutting of brass, even though the cutting force decreased when increasing the rake angle. Measurements of the chip thickness gives a good indication about the size of the shear plane angle.

Keywords: Brass · Machinability · Orthogonal cutting model · Tool geometry

1 Introduction

Brass is an alloy commonly used in drinking water supply systems. Lead is often used as an alloying element, which improves machinability. Lead is not solvable in solid brass. Therefore, it will form globular precipitation at the grain boundaries and improve chip breakage. Additionally, lead has a low melting point compared to brass. During cutting, the lead melts and acts as a lubricant at the tool point (Nobel et al. 2014). The lead content correlates with the cutting forces and the chip form (Vilarinho et al. 2005). These favorable effects lead to low cutting forces and low tool wear (Nobel et al. 2014).

Since lead is a heavy metal, which is harmful to human health and the environment, authorities such as the EU restrict its use. Likely, these restrictions will be increased in the future and thereby affect the whole production chain (Estelle 2016).

In recent years, several new low-lead and lead-free brass alloys were developed. However, compared to the traditional leaded brass alloys, these alloys typically show higher cutting forces, increased tool wear, and lower chip breakability. These lead to increased costs for machining processes in lead-free brasses (Schultheiss et al. 2016). Different approaches are described in the literature to overcome the drawbacks of lead-free and low-lead brass alloys. Toulfatzis et al. (2018) used heat treatment before cutting to change the microstructure of the alloys. Cutting forces were only marginal affected by

Y. Wang et al. (Eds.): IWAMA 2021, LNEE 880, pp. 254–261, 2022.
https://doi.org/10.1007/978-981-19-0572-8_32

this, the surface roughness was not affected at all, but the chip morphology improved for all alloys investigated. Klocke et al. (2016) compared the tangential/normal-force ratios for cutting different brass alloys with various tools. The leaded brass alloy CW614N showed the lowest force ratios, nearly independent of the tool and cutting speed. The silicon alloyed brass CW724R showed a higher force ratio than CW614N independent of the tool and cutting speed but a lower force ratio than CW510L and CW511L. The force ratios in the lead-free brasses CW510L and CW511L were significantly higher but decreased at higher cutting speeds. Uncoated carbide tools gave the highest force ratios, while a TiB_2 coating gave the lowest force ratio. Nobel et al. (2015) investigated the influence on chip breaking of different tool geometries. The depth of the chip breaking geometry influenced chip formation in CW724R, but not on CW511L. The rake angle had a lower influence on chip breaking.

This article aims to compare the cutting forces in different alloys and investigate the influence of the rake angle on the cutting force, the shear plane angle, and the friction angle using the orthogonal cutting model and the Merchant equation. The lead-free brass alloy CW511L, the lead-free silicon-alloyed brass CW724R, and the low-lead brass CW625N are investigated. Section 2 of this paper describes the alloys used and the method to analyze the cutting forces. Section 3 will give the results and discussion, and Sect. 4 provides the conclusion.

2 Materials and Methods

2.1 Investigated Alloys

The lead-free brass alloys CW511L and CW724R were compared to the lead-containing alloy CW625N. Nordic Brass Gusum supplied CW511L and CW625N in October 2020, Diehl supplied CW724R. CW511L was used in a variant called AquaNordic, which means that ceramic Al_2O_3 nanoparticles are added to the microstructure. CW724R is a silicon alloyed brass. It should be commented that the CW511L alloy has been slightly altered after these tests were conducted. Table 1 gives an overview of the chemical composition of the alloys. Table 2 lists the mechanical properties.

2.2 Experimental Setup

A Colchester V53250 lathe with a 7.5 kW spindle motor was used to machine the alloys. This machine is a manual lathe with constant cutting speed control. During cutting, the forces were measured using a Kistler 9257B dynamometer. A LabVIEW application calculated the average forces measured during one cut. A CoroCut RF123H25-2525BM tool holder from Sandvik Coromant was used. Two different uncoated cemented carbide tool inserts by Sandvik Coromant were used, as presented in Table 3. All tests were conducted in dry cutting conditions.

The lathe was operated at a cutting speed of 200 m/min for all tests. The tests were conducted at a depth of cut of 2.0 mm. Feeds between 0.10 mm and 0.25 mm were used. Three repetitions of each combination of feed, depth of cut, and tool insert were conducted. The initial cutting tests were conducted as a face turning operation with a radial feed. Average values from the three repetitions were calculated.

Table 1. Chemical compositions of the alloys according to the supplier's data sheets (*Brass Alloys for Drinking Water Applications* (2017); Aqua Nordic Lead free brass 2019; Rod CW625N 2019).

	CW625N	CW511L AquaNordic	CW724R
Cu	63.0–64.0%	62.5–63.5%	75.0–77.0%
Zn	Remainder	Remainder	Remainder
Pb	1.2–1.5%	<0.1%	≤0.1%
Sn	≤0.3%	≤0.1%	≤0,3
Fe	≤0.1%	≤0.1%	≤0.3%
Al	0.5–0.6%	≤0.05%	≤0.05%
Ni	≤0.2%	≤0.1%	≤0.2%
Mn	≤0.1%	<0.1%	≤0.05%
As	0.02–0.05%	0.02–0.05%	–
Si	–	–	2.7–3.5%
P	–	–	0.02–0.01%
Others	≤0.2%	<0.2%	<0.02%

Table 2. Mechanical properties of the alloys according to the data sheets by the suppliers (*Brass Alloys for Drinking Water Applications* (2017); Rod CW625N 2019; Aqua Nordic Lead free brass 2019).

	CW625N	CW511L AquaNordic	CW724R
Tensile strength R_m	>400 MPa	>330 MPa	500 MPa
Yield strength Rp02	~300 MPa	~240 MPa	200 MPa
Fracture elongation A5	>30%	>40%	22%
Brinell hardness	~110 HB	~95 HB	135 HB

Table 3. Tool inserts.

#	Designation	Material	Width	Chip breaking geometry	Rake angle
1	N123H2-0520-0002-BG H13A	Uncoated cemented carbide	5 mm	No	0°
2	N123H2-0400-0004-TMH13A	Uncoated cemented carbide	4 mm	Yes	8°

Additional tests with cutting conditions closer to the orthogonal cutting model were conducted as follows. Instead of radial cutting from the whole rod, grooves were cut to get thin disks. Each disk was 2.0 mm thick. The disks were cut with the same parameters and tools as used in the other tests. A schematic drawing of face turning and disk turning is shown in Fig. 1. The chips were collected after each cut, and the chip thickness and width were measured. Figure 2 shows the orthogonal cutting model. Based on this model, the results of the tests are evaluated. All forces shown in Fig. 2 can be calculated from the measured cutting forces and the rake angle α. The shear plane angle is named φ, and the friction angle is named β in the drawing.

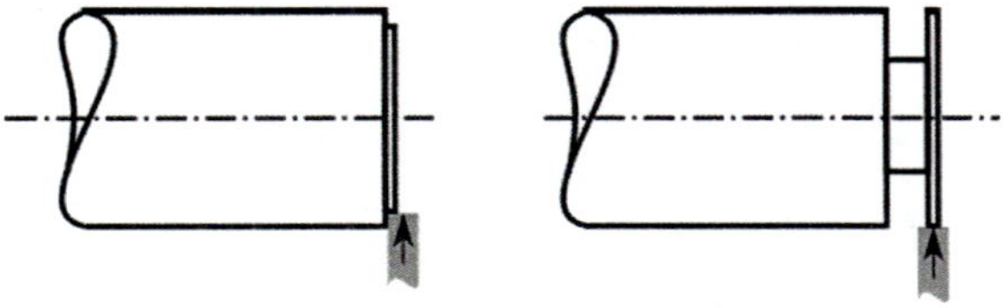

Fig. 1. Schematic drawing of the two cutting modes: a) face turning, b) disk turning. The arrow indicates the direction of the feed.

3 Results and Discussion

All materials were cut with the same cutting parameters to investigate the differences between the alloys and the tools. First, average values of the forces measured were calculated from the three repetitions per configuration. The main cutting force F_c (in the direction of the cutting speed) was plotted against the feed for each of the materials and tools, as shown in Fig. 3 for face turning and in Fig. 4 for disk turning.

Lead-alloyed CW625N showed for both tools the lowest cutting force. In face turning, the cutting force for CW724R was on average 15% higher than in CW625N, while it was only 1% higher in disk tuning. The differences varied for the different tools. The highest cutting force overall was measured for cutting CW511L. On average, the main cutting force was 42% higher than for CW625N in face turning and 43% higher in disk turning.

For face turning of CW625N, changing from tool #1 to tool #2 decreased the cutting forces on average by 8%. For disk turning in CW625N, the cutting force decreased on average by 11% when changing from tool #1 to tool #2. For face turning of CW724R, on average, only a slight decrease of 3% in cutting force was noticed when changing to tool #2. On the other hand, for disk turning in CW724R, changing to tool #2, decreased the cutting forces by 13%. For face turning of CW511L, the cutting forces were remarkably decreased by 37% when using tool #2 instead of tool #1. For disk turning only a decrease of 9% was noticed for CW511L. Since tool #2 has an increased rake angle and chip-breaking geometry, a combination of both likely leads to the decreased force. Overall, disk turning showed lower cutting forces. Disk turning brings the cutting conditions closer to the orthogonal cutting model since the measured passive force is almost 0 N.

Nevertheless, different factors will influence the cutting forces measured. For example, using only the middle part of the cutting tool in disk turning instead of the edge in face turning will affect the rigidity. Due to the different tool geometries, different vibrations may occur.

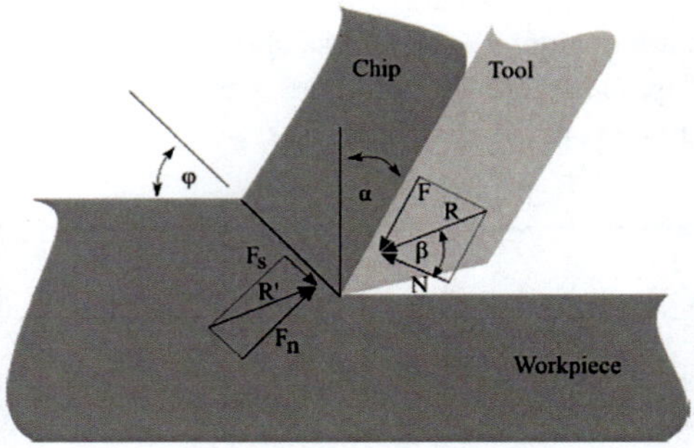

Fig. 2. Orthogonal cutting model of the forces acting on the chip.

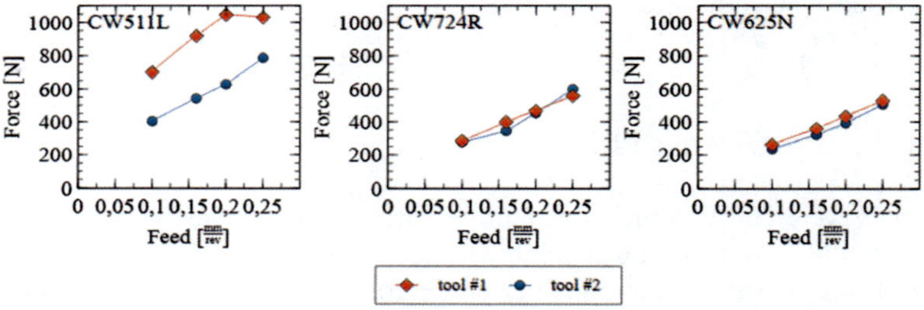

Fig. 3. Main cutting force F_c in face turning at different feeds using different tools.

Only the disk turning tests are used for further evaluation with the Merchant equation since the cutting conditions in these tests were closer to the orthogonal cutting model. These results are compared to the results gained from measuring the chip geometry. The shear plane angle and the friction angle for each of the measurements were calculated using the orthogonal cutting model and the Merchant equation. For comparison, the shear plane angle was calculated from the chip measurements as well. This shear plane angle was used to calculate the friction angle from the Merchant equation. The shear plane angels for the different tools, materials, and methods of determination are plotted against the feed in Fig. 5. Figure 6 shows plots of the friction angle against the feed. From the Merchant equation, it was expected to get a higher shear plane angle when increasing the rake angle. Hence, tool #2 should give a higher shear plane angle compared to tool #1. This effect is visible for all tested alloys in the analysis with Merchant's equation. For the chip analysis, this effect was not visible for CW625N. A possible explanation for this is errors in the measurement of the chip thickness since the chips were very thin and discontinuous. Additionally, the shear plane angle for a feed of 0.25 mm/rev in CW724R calculated from the chip measurements seems to be very high, which might be due to an error in the measurement of the chip thickness as well. The chip width was measured as well. Comparing the measured chip width to the depth of cut, it was visible that side flow was present during machining, especially for CW511L and CW625. This reduces the accuracy of the Merchant equation since this model analyzes only in two dimensions and does not take a side flow into account. As a result of this, the shear plane angles calculated from the chip measurements are probably more accurate.

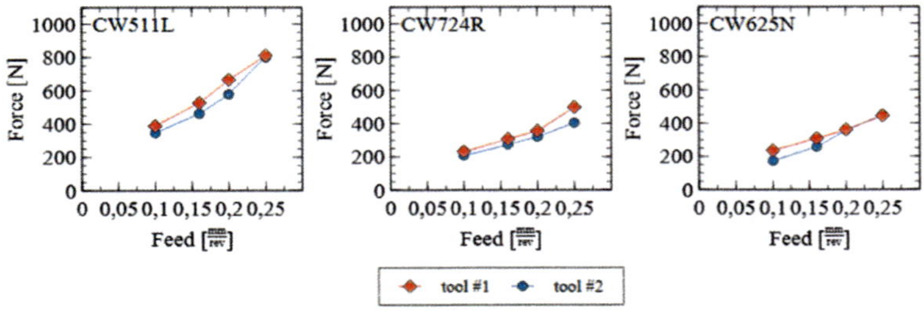

Fig. 4. Main cutting force F_c at different feeds for thin disk turning using different tools.

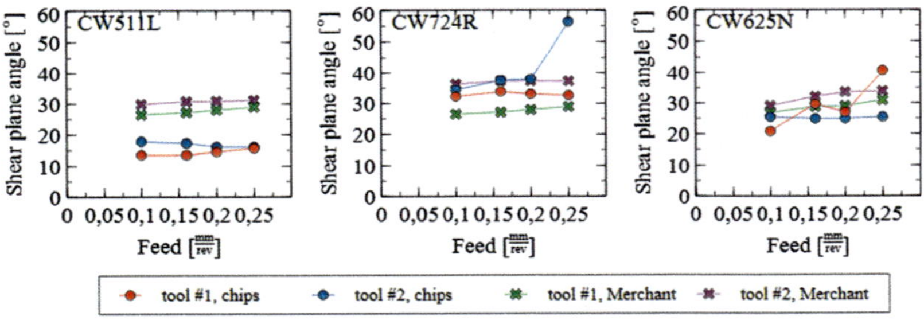

Fig. 5. Shear plane angles for cutting of thin disks with tool #1 and tool #2, calculated from chip measurements and from Merchant equation.

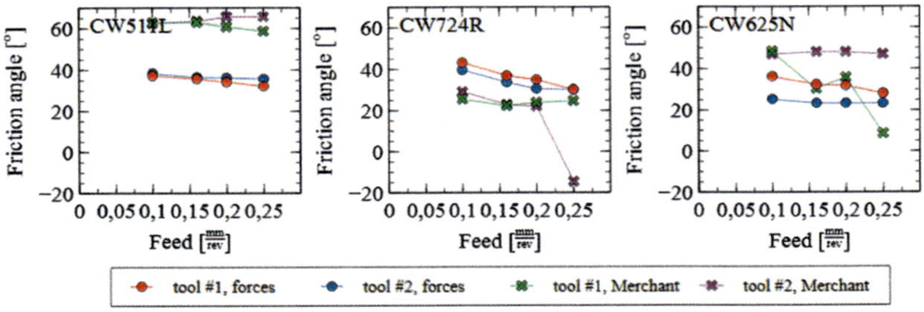

Fig. 6. Friction angles calculated according to orthogonal cutting model and Merchant equation for disk turning.

The friction angle was calculated from the measured forces and for comparison with the shear plane angle from the chip thickness measurements using the Merchant equation. The latter might be less accurate since the Merchant equation is based on a two-dimensional model, where the side flow is not recognized. Due to the error of measurement in the chip thickness and the model, which does not consider the side flow, only the force-based results for the friction angle are discussed further. The friction angle was on average 20% higher in CW724R and 21% higher in CW511L compared to CW625N. The friction angle varied as well when the tool was changed. When changing from tool #1 to tool #2, the friction angle decreased by 26% for CW625N and by 7% for CW724R, but it increased by 5% for CW511L. Tool #2 has a chip-breaking geometry which probably helped to remove the discontinuous chips in CW625N and CW724R faster from the cutting zone and thereby reduced the friction. The increase in CW511L could possibly be explained by the long continuous chips, which were pushing on the rake face and causing more friction. On average, CW625N showed the lowest friction angle, which is probably due to the lubricating effect of the lead in the microstructure described in the literature (Nobel et al. 2014). CW724R contains abrasive κ-phase, which increases the abrasiveness of the material (Schultheiss et al. 2016) and might thereby increase the friction between the tool and workpiece and cause the increased friction angle compared to CW625N. Additionally, the absence of lead will play a major role in this. The friction in cutting lead-free brass needs to be investigated further.

4 Conclusion

Overall, the results for the force measurements are in good accordance with the literature. The lead-free alloy CW511L showed the highest cutting force and lead alloyed CW625N had the lowest cutting force. In summary, it was shown that the tool geometry influences the cutting forces. Following the Merchant equation, the cutting forces decreased, and the shear plane angle increased for a tool with a higher rake angle. The decrease of the forces might be due to a combined effect of the higher rake angle and the chip breaking geometry in tool #2. The reduction of the force was especially noticeable in cutting CW511L. Additional investigation on the effect of the tool geometry is necessary. A limitation of this study is that brass is a ductile material, which leads to side flow during cutting, even though the ratio between the depth of cut and feed was high and the cutting conditions in disk turning were very close to the orthogonal cutting model. Since the Merchant equation does not consider the side flow, the results analyzed with the Merchant equation are not that accurate. Future studies should find a proper method to investigate the friction between the tool and the workpiece. Another aim should be to examine the influence of the tool geometry on cutting forces in brass in more detail.

Acknowledgment. The authors thank the Research Council of Norway for supporting this work through the research project *LOBUS – Low Lead Brass for Sustainable Community Development.*

References

Aqua Nordic lead free brass: Data Sheet (2019). https://www.nordicbrass.se/en/products/rod-aquanordic-lead-free-brass. Accessed 26 May 2021

Brass Alloys for Drinking Water Applications: Data Sheet. Röthenbach a.d. Pegnitz (2017). https://www.diehl.com/cms/files/Diehl_Metall_Messing_Brass_Alloys_for_Drinking_Water_Applications.pdf. Accessed 26 May 2021

Estelle, A.A.: Drinking water lead regulations: impact on the brass value chain. Mater. Sci. Technol. (United Kingdom), 1763–1770. https://doi.org/10.1080/02670836.2016.1220906

Klocke, F., Nobel, C., Veselovac, D.: Influence of tool coating, tool material, and cutting speed on the machinability of low-leaded brass alloys in turning. Mater. Manuf. Process. **31**(14), 1895–1903 (2016). https://doi.org/10.1080/10426914.2015.1127944

Nobel, C., et al.: Machinability enhancement of lead-free brass alloys. In: Procedia CIRP, pp. 95–100. Elsevier (2014). https://doi.org/10.1016/j.procir.2014.03.018

Nobel, C., et al.: Application of a new, severe-condition friction test method to understand the machining characteristics of Cu-Zn alloys using coated cutting tools. Wear **344–345**, 58–68 (2015). https://doi.org/10.1016/j.wear.2015.10.016

Rod CW625N: Data Sheet (2019). https://www.nordicbrass.se/en/products/rod-cw625n. Accessed 26 May 2021

Schultheiss, F., et al.: Machinability of CuZn21Si3P brass. Mat. Sci. Technol. (United Kingdom) **32**(17), 1744–1750 (2016). https://doi.org/10.1080/02670836.2016.1189199

Toulfatzis, A., et al.: Machinability of eco-friendly lead-free brass alloys: cutting-force and surface-roughness optimization. Metals **8**(4), 250 (2018). https://doi.org/10.3390/met8040250

Vilarinho, C., et al.: Influence of the chemical composition on the machinability of brasses. J. Mater. Process. Technol. **170**(1–2), 441–447 (2005). https://doi.org/10.1016/j.jmatprotec.2005.05.035

Numerical Simulation and Experimental Research on Novel Hydrocyclone Desanding System for Offshore Platform

Hongbo Fang[1,2](✉)

[1] Petroleum Engineering Co., Ltd., SINOPEC, Dongying 257026, China
[2] Shandong Provincial Key Laboratory of Oilfield Produced Water Treatment and Environmental Pollution Control, Dongying 257026, China

Abstract. Sand in produced water is a crucial factor in reducing oil production and increasing unnecessary energy loss. In this paper, due to the high sand concentration in the produced water, a new type of offshore high concentration solid phase sand separation system was proposed. First and foremost, taking the standard hydrocyclone as the research object, an external characteristic experiment with similar parameter criteria was carried out, by measuring the mass flow rate and pressure of the outlet and inlet, the separation efficiency and pressure drop are obtained. Afterward, using the method of numerical simulation, the formation mechanism of the air core of the standard hydrocyclone, the change of the internal flow field and the separation efficiency are discussed. The results evidence that the inlet flow range of the standard hydrocyclone is 30 L/min–50 L/min, the energy loss of the new type of desanding system is small, and the separation efficiency reaches more than 90%.

Keywords: Desanding system · Hydrocyclone · Internal flow field · Inlet flow rates · Separation efficiency

1 Introduction

With the continuous exploitation of offshore oil resources, the probability of discovery of onshore integrated oil fields is gradually decreasing, and offshore oil resources will be an essential source of crude oil production [1]. Petroleum mining is often accompanied by the production of solid phase sand particles, and the concentration of different mining stages is not the same. There are two main factors in the production of solid sand: (1) Human factors such as production pressure difference, flow rate, improper use of the auxiliary active agent. (2) Natural factors such as fluid properties and wellbore plastic deformation [2]. Although the corresponding measures were taken in production, the high viscosity of crude oil will carry sand particles into the production system, which will lead to pipeline blockage, wear of flow passage components, the decline of production capacity of heating device and separation device, which seriously affects the production of the oilfield [3]. According to the field investigation, the Bohai oilfield,

Y. Wang et al. (Eds.): IWAMA 2021, LNEE 880, pp. 262–269, 2022.
https://doi.org/10.1007/978-981-19-0572-8_33

Penglai 19-3 oilfield and the South China Sea oilfield are all facing sand production problems [4]. A certain oilfield in the South China Sea has installed a desanding system at 10 wellheads and made it the world's largest sand treatment system. The processing capacity is 800–1200 m^3/d and can be used for 24-h sand production with separation size less than 25 μm and about 1–2 tons of solid sand [5]. Some wells in the Penglai 19-3 oilfield failed to control sand. This is due to the high sand concentration in the produced fluid of the oilfield and no corresponding measures have been taken in advance. Therefore, mechanical filtration and manual filtration are mainly used to prevent damage to subsequent equipment. The described method is challenging to meet the criteria for discharging sand [6].

In this paper, a 50 mm standard hydrocyclone is selected as the desanding and separating device, the separation performance of a standard hydrocyclone under a wide range of inlet flow rate was simulated by using the Reynolds stress model (RSM) and the mixture model. The relationship between the inlet flow rate and the air core, pressure, axial velocity and separation efficiency is pointed out. The mass separation efficiency between the underflow outlet mass flow rate and the inlet mass flow rate in the external characteristics is discussed. Through the study of the experimental data of inlet flow rate and external characteristics, the relationship between the two was obtained, and the expected separation effect was achieved by changing the inlet flow rate, which provided technical guidance for solid-liquid separation of solid-liquid sand separation system under the condition of high concentration (more than 1 vol.%).

2 Design and Treatment of Solid Sand Separation System

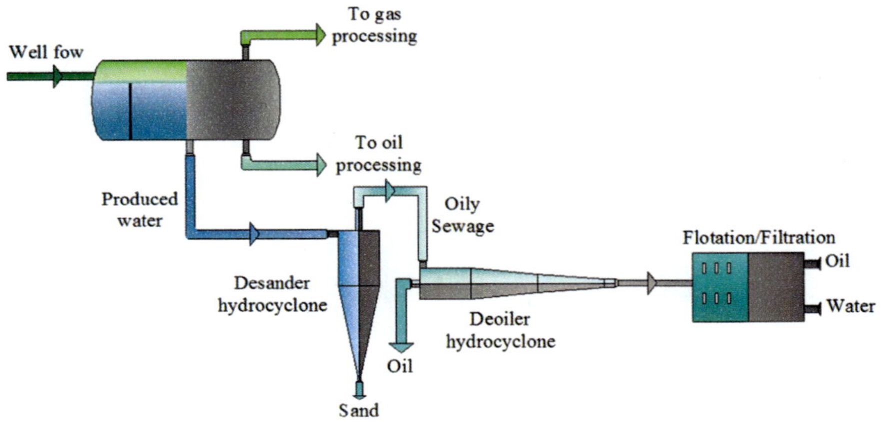

Fig. 1. Process-flow schematic of desanding

Based on the standard hydrocyclone, the separation system of high concentration solid sand at sea is designed. The specific flow schematic is shown in Fig. 1. After passing through the pipeline, the oil well fluid first passes through the first stage three-phase separator, where the coarse sand settles at a low speed and is removed by manual removal or jet equipment. However, some fine sand particles adhere to the surface of the oil phase, float to the oil accumulation layer or oil-water interface, and finally flow into the oil treatment system. The other part flows into the produced water system. At the same time, the gas phase and most of the oil phase are filtered out by the separation device, and a small part of the oil phase will adhere to the sand surface and enter the next stage. Sand and oil are further removed by sand blasting and injected into hydrocyclone, and finally into the flotation filtration device. Because hydrocyclone is a component without power device, it is necessary to install injection device in the separator to make the produced water to carry out cyclone separation at a certain pressure [7].

3 Numerical Simulation and Experiment

The geometric proportion of the standard hydrocyclone meets Rietema's standard, and the diameter of the body is related to the separation particle size [8]. Therefore, the body diameter of 50 mm is selected in this study. There is a positive correlation between the number of inlets and the stability of the internal flow field, but too many inlets will have higher requirements on the application site, so single inlet is selected. The reason why the rectangular inlet is chosen is that the larger angular momentum is produced when the material enters and is conducive to the stability of the flow field. The cone angle is the standard type. The model structure is shown in Fig. 2, and the specific structure size is shown in Table 1. The 3D computational domain is meshed by ICEM software. At the same time, in order to improve the calculation accuracy, the hexahedral mesh is utilized to make the mesh boundary and the fluid flow direction as vertical or coincident as possible. In addition, in order to better capture the boundary features and improve the accuracy of the mesh generation, the local encryption method is used to mesh the wall face and the vortex finder.

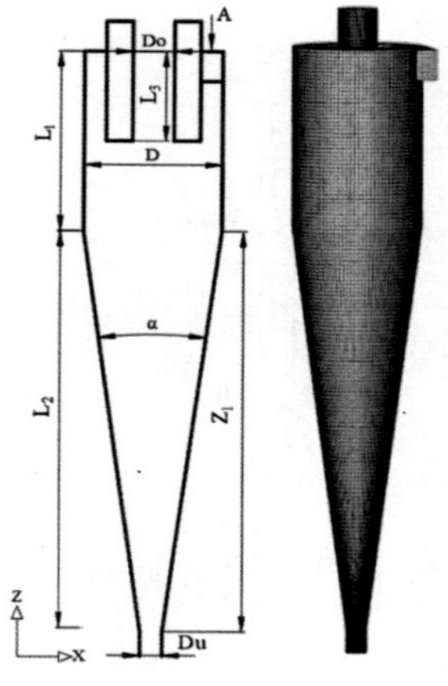

Fig. 2. Physical model and meshing

Table 1. Structural parameters of open Hydrocyclone

Parameter	Value
Diameter of body (D)/mm	50
Diameter of inlet (A)/mm^2	12 (L) × 8 (W)
Diameter of overflow outlet (D_o)/mm	15
Diameter of underflow outlet (D_u)/mm	8
Length of cylindrical part (L_1)/mm	72
Length of cone part (L_2)/mm	159.5
Length of vortex finder (L_3)/mm	36
Cone angle (α)/(°)	15°
Feed solid concentration/(vol. %)	3

In the numerical simulation, the boundary condition at the inlet is the velocity inlet, and the outlet is the pressure outlet. The gauge pressure of which was 0 MPa. A "no-slip" is considered as a wall boundary condition, which ensures that a zero velocity is imposed on the wall. In the present study the standard wall function is used for the simulation at the wall. The standard wall function is a logarithmic function of y^+. The value of y^+ is between 32 and 180. "SIMPLE" is regarded as the pressure-velocity coupling. "QUICK" is set at the discretization scheme of the volume fraction and momentum. "PRESTO!" is used at the discretization scheme of pressure. Water is treated as continuous phase and solid particles as discrete phase. Samples were taken from the treatment site of produced liquid from an offshore platform, the viscosity of the mixture measured by the super rotary rheometer was 0.97 mPa·s, the density of the mixture is calculated by measuring its volume and mass, which is 1047.75 kg/m^3, the BT-9300S laser particle size analyzer was used to measure the diameter of 90% particles between 1~30 μm. The particle size distribution was shown in Table 2.

Table 2. Particle size distribution of sand particle

Particle size/(μm)	1	3	5	10	15	20	25	30
Volume fraction/(%)	0.14	0.19	0.35	0.61	0.59	0.55	0.41	0.16
Size distribution/(%)	4.67	6.33	11.67	20.33	19.67	18.33	13.67	5.33

4 Results and Discussion

4.1 Air Core Distribution

The formation process of the air core is understood from the macroscopic view by using numerical simulation method. The simulation results are presented in Fig. 3. When t = 0 s, the fluid begins to enter the hydrocyclone along the wall under the action of pressure. Before t = 0.75 s, due to the air in the hydrocyclone, the original air began to be squeezed out as the fluid entered, at this time, the internal flow field in the hydrocyclone has not been completely formed and the liquid phase will be discharged from the underflow outlet, so the liquid phase occupies most of the effective area of the underflow outlet, resulting in the air mainly discharged from the overflow outlet. After t = 0.75 s, most of the air in the cyclone have been expelled, and the three-dimensional motion of the fluid gradually accelerates with the decrease of the rotating radius, so that the static pressure of the fluid changes into the dynamic pressure. This process continues until the remaining static pressure does not have enough energy to compensate for the energy consumption, and the flow rate in this process continues to increase. When the radius is reduced to a certain threshold, the hydrostatic pressure is zero, which will form a negative pressure area with this radius. Because the underflow outlet and overflow outlet are equal to atmospheric pressure, the external air will return to the overflow and underflow outlets. When t = 1.25 s, a relatively stable axial through air core is formed, which indicates that the time required for the formation of an air core in the hydrocyclone is 1.25 s and the working condition is stable.

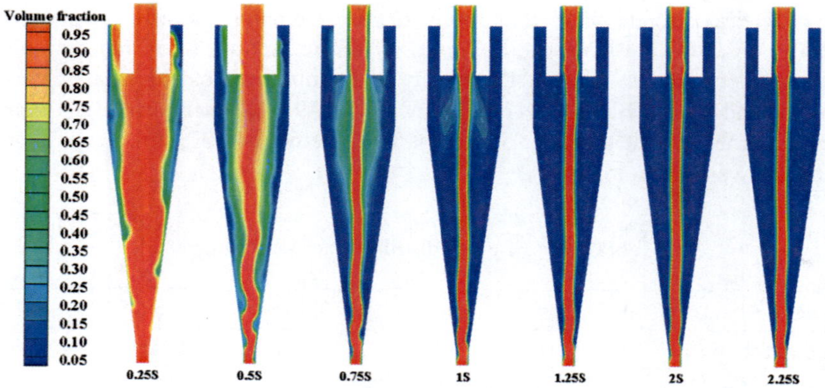

Fig. 3. Air core formation process

4.2 Pressure Distribution

The pressure drop of standard hydrocyclone is the result of vortex energy consumption, solid sand resistance and fluid wall friction. The separation performance is directly related to the vortex energy consumption, and the macroscopic performance is the pressure drop of inlet, underflow and overflow. Figure 4 and Fig. 5 show the comparison of

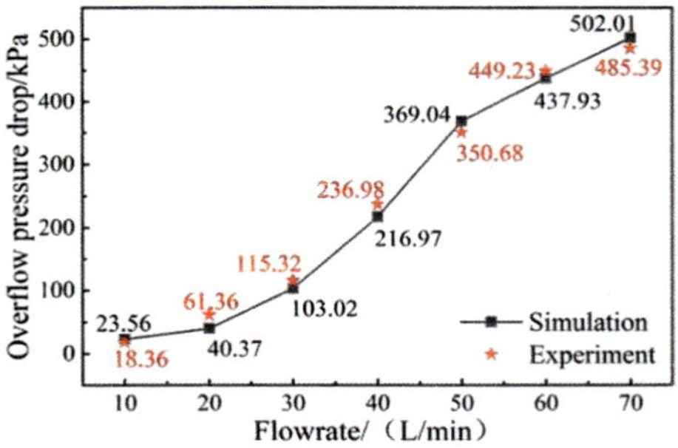

Fig. 4. Overflow pressure drop

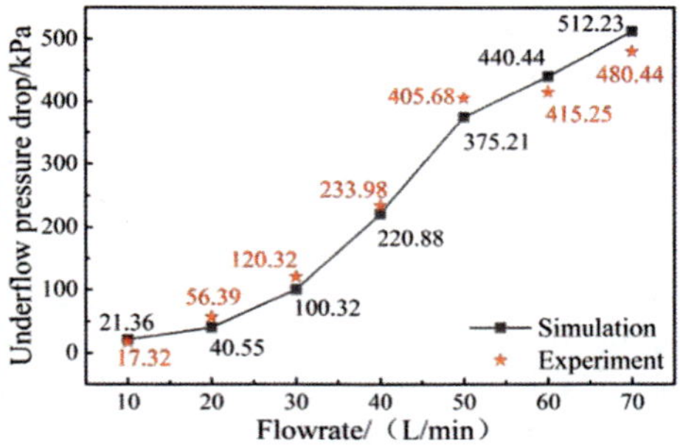

Fig. 5. Underflow pressure drop

overflow pressure drop and underflow pressure drop at different inlet flow rates. From the overall analysis, with the increase of the inlet flow rate, overflow pressure drop and underflow pressure drop gradually increase. When the inlet flow rate is less than 30 L/min, the increase amplitude of overflow pressure drop and underflow pressure drop is small, which is due to the smaller inlet flow rate, resulting in the lower pressure of the whole flow field. When the inlet flow rate increases from 30 L/min to 50 L/min, the increase amplitude of overflow pressure drop and underflow pressure drop is 266.02 kPa and 274.89 kPa, which are similar to the wall pressure distribution, and the pressure difference is about the wall pressure value. When the inlet flow rate is greater than 50 L/min, the local loss near the outlet increases with the increase of the inlet flow rate.

4.3 Separation Efficiency

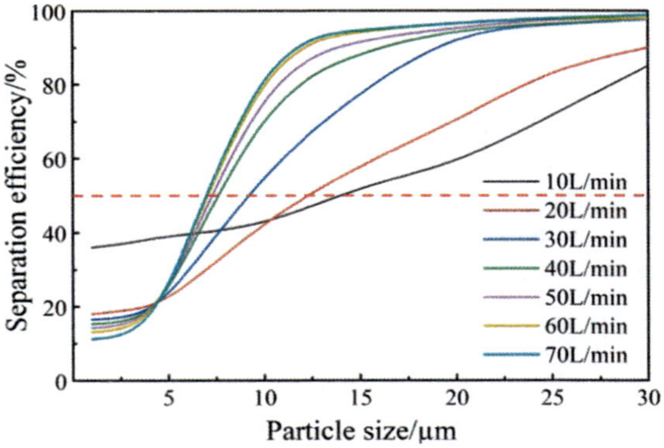

Fig. 6. Relationship between separation efficiency and particle sizes

The variation relationship of different particle sizes at different inlet flow rates is shown in Fig. 6. From the overall analysis, under the condition of constant inlet flow rate, the separation efficiency increases with the increase of solid particle size, and the separation efficiency increases faster when the inlet flow rate is greater than 30%. When the inlet flow rate is 10%, the separation effect of standard hydrocyclone is not ideal and the growth is slow due to the high turbulence intensity [9]. When the inlet flow rate is more than 50%, the increase of separation efficiency becomes smaller, which is due to the increase of flow rate and the increase of fluid velocity, which makes the solid sand collide with the wall of hydrocyclone, and then rebounds into the inner swirl and is discharged from the overflow outlet. On the other hand, when the flow rate increases, the centrifugal force will increase, which will aggravate the collision between solid sand particles and solid sand particles, sand and water, and sand particles and the wall of hydrocyclone, so that the large particle size of solid sand particles can be broken into small particle size sand particles, thereby reducing the separation efficiency.

The cutting particle size decreases with the increase of inlet flow rate, and the minimum is 7.5 μm. When the particle size is less than 5 μm, the separation efficiency decreases with the increase of inlet flow rate, which means that the hydrocyclone no longer has the classification function. The reason for this change is that when the particle size is small, it is mainly affected by the dispersion force and is proportional to the split ratio of water phase [10]. When the solid particle size is in the range of 5–20 μm, the separation efficiency increases with the increase of the inlet flow rate, which is mainly due to the change of the dispersion force into centrifugal force. When the solid particle size is more than 20 μm and the inlet flow rate is greater than 30 L/min, the separation efficiency tends to be stable and the maximum difference is 0.98%.

5 Conclusion

Numerical simulation and laboratory experiments were carried out by the standard hydrocyclone in a new offshore high concentration of solid sand separation system. The influence of inlet flow rate on air core, pressure, axial velocity and separation efficiency was analyzed. The analysis leads to the following conclusions:

(1) For the air core of standard hydrocyclone, the intake air from underflow and overflow rendezvous near the bottom of the vertex finder, resulting in the maximum radius of the air core here. When the inlet flow rate is more than 30 L/min, the difference of stable distribution of air core is small.
(2) The inlet flow rate is positively correlated with the pressure loss of standard hydrocyclone. With the increase of the inlet flow rate, the increase of wall pressure of standard hydrocyclone increases linearly, but when the inlet flow rate is greater than 50%, the increase amplitude is slightly reduced due to the energy consumption needed to form an air core. The experimental and numerical results demonstrate that the increase amplitude of underflow pressure drop and overflow pressure drop reaches the maximum at the inlet flow rate of 30 L/min–50 L/min.
(3) When the inlet flow rate increases, the separation efficiency changes significantly for the particle size in the range of 5–20 μm. The experimental results show that the separation efficiency is stable when the inlet flow rate is more than 30%, and the value is about 93.90%. Based on the analysis of internal flow field energy consumption and separation efficiency of standard hydrocyclone, the optimal inlet flow rate range of standard hydrocyclone in solid phase sand separation system is 30 L/min–50 L/min under the condition of high concentration (more than 1 vol.%).

Acknowledgment. This work has been supported by open fund of Shandong Provincial Key Laboratory of Oilfield Produced Water Treatment and Environmental Pollution Cotrol (grant numbers: 10205363-20-ZC0607-0001).

References

1. Cai, L.X., Chen, J., Liu, M., Ji, Y., An, S.: Numerical studies on dynamic characteristics of oil-water separation in loop flotation column using a population balance model. Sep. Purif. Technol. **176**, 134–144 (2016)
2. Sang, Y., Chen, J., Zhang, L., Wang, Y.: Discussion on sand removal technologies for oil-gas field systems. Petrochemical Saf. Environ. Prot. Technol. **04**, 5–10 (2007)
3. He, Y.W.: Study on optimal operation of emulsification to reduce viscosity and transportation of heavy oil. Chem. Eng. **34**(02), 40–45 (2020)
4. Feng, B.: Sand production causes and treatment measures in oil wells. Chem. Eng. Des. Commun. **46**(03), 270–271 (2020)
5. Rawlins, C.H.: Sand Management Methodologies for Sustained Facilities Operations. In: The North Africa Technical Conference and Exhibition. Cairo, Egypt (2013)
6. Li, C.: Discussion on sand removal from sandy crude and purification of oily sand in offshore viscous oilfield. China Offshore Oil Gas **04**, 263–265 (2007)
7. Wang, H., et al.: Effect of internal flow field and separation in hydrocyclone of flow-back fluids under different degrees of gel breaking. J. Shandong Univ. Sci. Technol. (Nat. Sci.) **38**(03), 91–99 (2019)
8. Cui, B., et al.: Study on interaction effects between the hydrocyclone feed flow rate and the feed size distribution. Powder Technol. **366**, 617–628 (2020)
9. Vakamalla, T.R., Mangadoddy, N.: Numerical simulation of industrial hydrocyclones performance: role of turbulence modelling. Sep. Purif. Technol. **176**, 23–39 (2017)
10. Nageswararao, K.: Modelling of hydrocyclone classifiers: a critique of 'bypass' and corrected efficiency. Powder Technol. **297**, 106–114 (2016)

Numerical Simulation of Free Ascension and Coaxial Coalescence of Bubbles in Gas-Liquid System

Bing Liu[1], Huanxin Zhao[1], Heng Guan[1], Qun Gao[1], Zhen Wu[1], Jianliang Xue[2], and Peishan Huang[3](✉)

[1] College of Mechanical and Electronic Engineering, Shandong University of Science and Technology, Qingdao 266590, China
[2] College of Safety and Environmental Engineering, Shandong University of Science and Technology, Qingdao 266590, China
[3] Qingdao Deepblue Subsea Engineering Co., Ltd., Qingdao 266500, China
Subseausa@163.com

Abstract. In this paper, VOF (Volume of Fluid) method is used to simulate the free ascension and coaxial coalescence of bubbles in gas-liquid system. First, the correctness of the method was verified by experiment. Then, the process of coaxial and free rising bubble coalescence in aqueous solutions was simulated. And the influence of liquid density changes on the bubble shape, rising velocity and coalescence time was studied. The results show that with the increase of liquid density, the deformation rate of bubbles increases and the bubble shape becomes flatter. With the increase of liquid density, the initial time of bubble coalescence is obviously advanced, the height and velocity of coalescence decrease gradually, the drainage time and fusion time of the two bubbles extend, and the coalescence time increases. In addition, when the liquid density exceeds 1400 $kg{\cdot}m^{-3}$, the interface oscillation of the coalescated bubble is obvious, the movement trajectory is not straight, and it will gradually deform until broken when rising.

Keywords: Bubble coalescence · VOF method · Numerical simulation · Liquid density

1 Introduction

Bubble coalescence is an important research content of gas-liquid two-phase flow movement, and it is closely related to engineering problems such as mineral flotation, oil removal by air flotation and dangerous chemical reaction simulation. Bubble coalescence generally refers to the process that two bubbles contact each other to form a liquid film, then the liquid film gradually thins until it breaks, and finally the two bubbles merge into one [1]. Bubble coalescence not only changes the shape of two bubbles, but also has a great influence on interphase mass transfer and energy transfer in gas-liquid system [2]. In fact, due to the difference of liquid solute, such as saline wastewater, the physical parameters of liquid, especially the liquid density, have a large gap, which will have a

Y. Wang et al. (Eds.): IWAMA 2021, LNEE 880, pp. 270–277, 2022.
https://doi.org/10.1007/978-981-19-0572-8_34

significant effect on the bubble shape and coalescence behavior, and ultimately affect the production efficiency.

The bubble coalescence process is stochastic and multi-scale. The momentum and energy exchange process are very complex, which brings great difficulties to the experiment and numerical simulation [3]. Scholars at home and abroad have carried out a lot of research on it. Zhang Jinya et al. [4] used VOF (Volume of Fluid) method to conduct numerical simulation study on the floating and coalesting process of bubbles, and the results showed that the rising velocity of bubbles depended on the viscosity and density of the liquid. Zhang Huahai et al. [5] showed that the surfactant had obvious inhibitory effect on bubble coalescence through experiments. W. Abbassi et al. [6] conducted a numerical simulation study on the coalescence process of coaxial bubbles with the VOF method, and the results showed that the coalescence time and height of two continuous bubbles generally increased with the increase of liquid surface tension and the decrease of viscosity. Sandra Orvalho et al. [7] found through experiments that electrolyte concentration had an inhibitory effect on bubble coalescence, which was similar to that of liquid viscosity.

Current studies mainly focus on the behavior of bubbles, and the influence of liquid physical parameters on bubble coalescence is not perfect. Most studies only consider the change of liquid viscosity. In fact, taking the common gas-liquid two-phase flow as an example, the viscosity of aqueous solution does not change much, while the density gap is large. However, few studies have focused on the influence of density change on bubble coalescing. Therefore, VOF method was used to simulate the free rising coalescence of two coaxial bubbles, and the influence of liquid density on the shape,velocity and coalescence time of bubbles was studied.

2 Model and Simulation

2.1 Physical Model

The VOF model is used to simulate the free rising and coalescence process of two coaxial bubbles in Fluent software. To simplify the calculation process, the simulated gas-liquid two-phase flow is set as an isothermal system, the fluid is homogeneous and incompressible, and the initial state of the liquid is static [8].

Figure 1 shows the physical model and meshing. In order to save the computational resources, a two-dimensional model was used in this simulation, and the calculation domain was a rectangular region of $15d \times 32d$. The initial state of the bubble is spherical, and the diameter d is 6 mm. At the initial moment, the center of the lower bubble is 20 mm away from the bottom boundary, the bubble spacing $h = 2d$. The influence of the wall can be ignored [9]. The mesh is structured with 0.2 mm spacing, and the total number of the mesh is about 400 thousand.

2.2 Simulation Settings

Select the transient mode, turn on the gravity. The upper boundary is the pressure outlet, and the pressure is atmospheric pressure. The wall surface has no slip. The liquid phase is the main phase and air is the secondary phase. And the surface tension coefficient of the

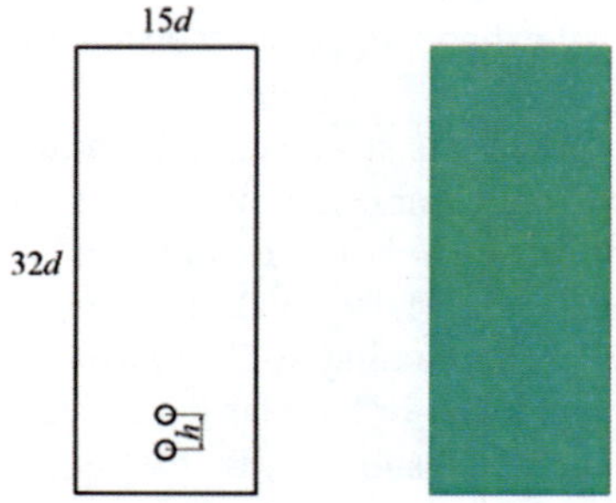

Fig. 1. Physical model and meshing

gas-liquid interface σ is 0.0705 N/m. Select SIMPLEC algorithm. PRESTO algorithm and the first order upwind format are selected for discrete format. The time step is 0.001 s. The viscosity of the liquid remains unchanged, and the initial temperature is 25 °C (298.15 K).

The density of liquid water is set to 1000 kg·m^{-3}, and the liquid density changes with the material density, which is set as 800, 1000, 1200, 1400 and 1600, respectively. The influence rules of the liquid density changes on the bubble shape, rising velocity and coalescence time were analyzed.

3 Analysis of Simulation Results

3.1 Bubble Shape

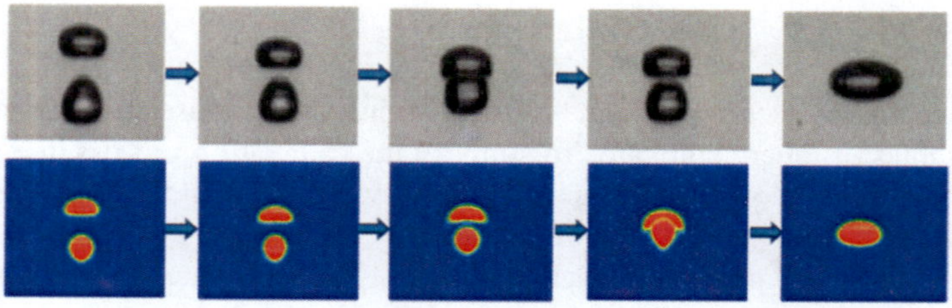

Fig. 2. Experimental results and simulation results

Figure 2 shows the comparison between the experimental results and the simulation results of bubble coalescence in glycerol. The change of bubble shape is consistent, which indicates that the simulation method is correct. The rising and coalescence of bubbles can be divided into three stages: "*approach*", "*drain*" and "*coalescence*". In the process, the bubble gradually deforms, the upper bubble shape flattens gradually, and the lower stretches along the vertical direction. The rising of the bubbles drives the movement of the surrounding liquid, generating upward vortexes on both sides (Fig. 3). The wake flows form at the bottom of the bubbles, and a pressure gradient is generated between the bubbles. Under the action of vortex and pressure gradient, the lower bubble will accelerate upward, the bubbles begin to approach. Then, the bottom of the upper bubble gradually depresses and wraps the lower bubble, forming a connected bubble, and a thin liquid film is formed between the bubbles. With the narrowing of the distance, the liquid in the film is gradually discharged, the thickness of the film decreases until it is broken, and the bubbles contact with each other, and rise in a straight line at a steady speed.

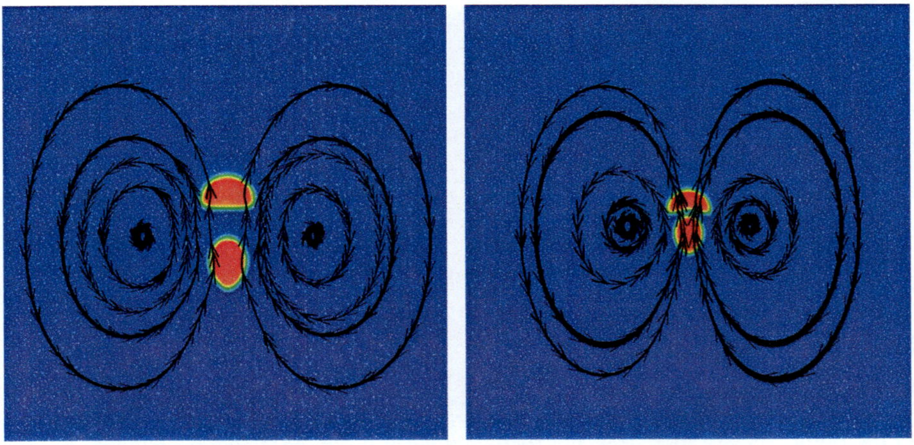

Fig. 3. The swirl on either side of bubbles as they rise

The deformation rate of bubbles coalesce can be described by the bubble length-diameter ratio E ($E = L/D$), which is the ratio of the longitudinal length L of two bubbles to the projected diameter D [10]. To facilitate description and analysis, a few moments need to be defined. The time when the vertex of the lower bubble and the bottom of the upper bubble are at the same height is taken as the beginning time t_1. The time when the two bubbles begin to fuse is taken as the beginning time of fusion t_2. The time when the two bubbles coalesce and finish rising steadily is taken as the coalescing completion time t_3.

Figure 4 is a line graph of E changing with liquid density at t_2. It can be seen that with the increase of liquid density, the bubble deformation rate increases significantly, and the bubble aspect ratio decreases from 0.88 to 0.575. The change of bubble shape is the result of the interaction of various forces [11]. With the increase of liquid density, the Weber number and Bond number increase obviously, the action of inertial force increases obviously, and the action of surface tension weakens relatively. This makes it easier for the bubbles to deform as they move. Therefore, when the liquid density increases, the bubble shape is difficult to maintain stability, and the bubble deformation rate increases. In addition, when the liquid density exceeds 1400 $kg{\cdot}m^{-3}$, the shape of large bubbles formed by coalesce cannot continue to maintain an ellipse. When it rises, the interface oscillates obviously, and its motion trajectory is no longer a straight line, but approximately a zigzag. In the process of rising, the bubbles will gradually deform until broken.

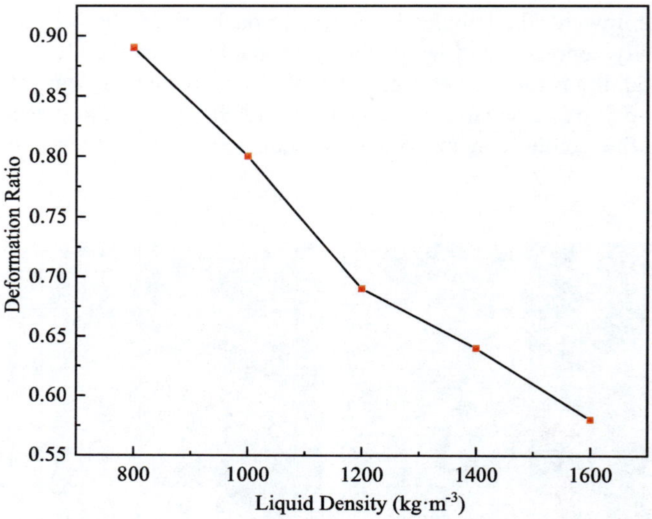

Fig. 4. The broken line graph of E changing with liquid density

3.2 Rising Velocity

As the liquid density increases, the rising velocity of the two bubbles changes. As can be seen from Fig. 5, with the increase of liquid density, the rising velocity of the lower bubble increases, while the rising velocity of the upper bubble decreases.

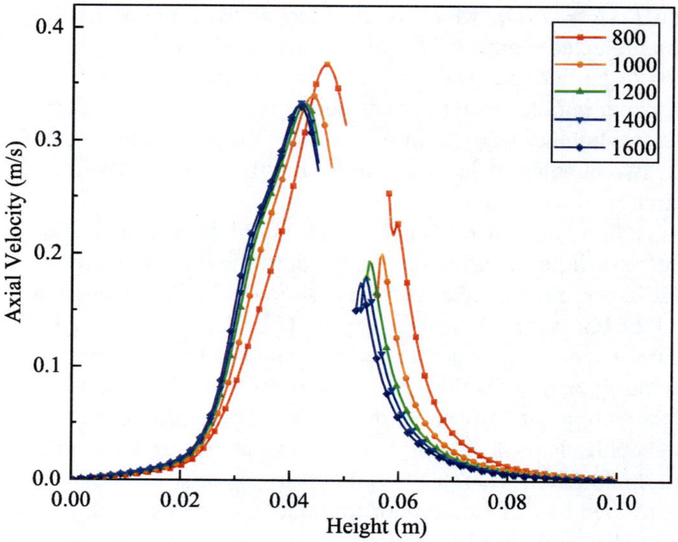

Fig. 5. Comparison of axial velocity of instantaneous flow field around the bubbles

This is because when the liquid density increases, the forces of the two bubbles change differently [12]. Although the increase of liquid density increases the buoyancy of the bubble, due to the larger deformation of the upper bubble, the drag force and resistance increases, and the rising acceleration decreases, so the rising velocity gradually decreases. However, the deformation of the lower bubble is relatively small, and the resistance changes little. Under the acceleration of the eddy currents on both sides and the pressure gradient between the bubbles, the rising acceleration increases, so the rising velocity increases. In addition, with the increase of liquid density, the two peaks of the curve decrease, and the axial height corresponding to the peak decreases gradually, indicating that the velocity and height of the two bubbles when they were coalesced also decrease gradually.

3.3 Coalescence Time

With the increase of liquid density, the velocity and height of bubble coalescence decrease gradually, so the coalescence time changes. Bubble coalescence time can be divided into two parts: drainage time T_1 and fusion time T_2. Drainage time $T_1 = t_2 - t_1$, that is, the time from the beginning of coalescence to the liquid film breaking. Fusion time $T_2 = t_3 - t_2$, that is, the time from the liquid film breaking to the complete fusion of two bubbles. The coalescence time T can be expressed as $T_1 + T_2$.

As can be seen from the two figures below, with the increase of liquid density, the initial time t_1 decreases from 0.165 s to 0.145 s significantly. And the drainage time T_1 increases slightly from 0.007 s to 0.009 s. The fusion time T_2 increases from 0.022 s to 0.028 s. The coalescence time T increases from 0.029 s to 0.037 s (Figs. 6 and 7).

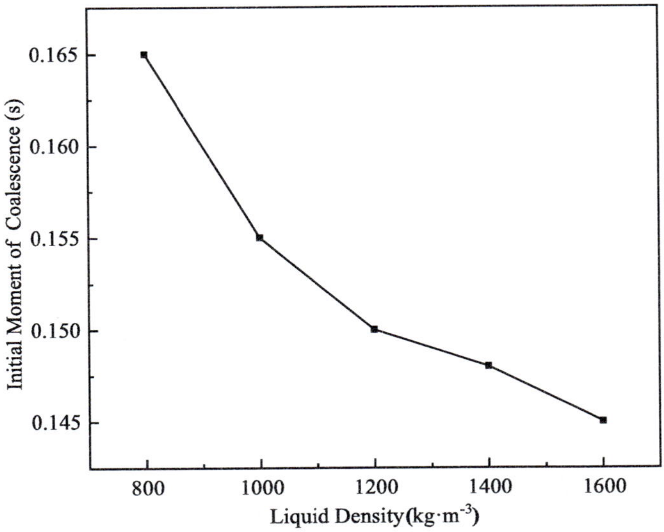

Fig. 6. The initial moment of bubble coalesce

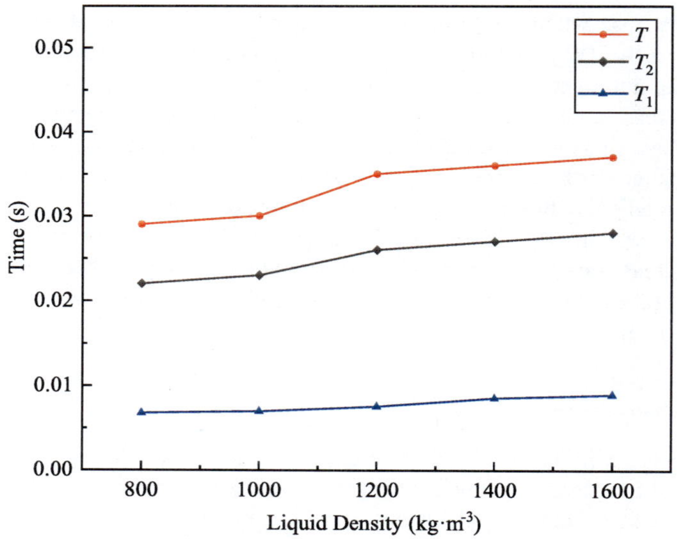

Fig. 7. The time of each stage of bubble coalescence

These changes indicate that the beginning time of bubble coalescence is earlier, the drainage time and fusion time of two bubbles are prolonged, and the time of bubble coalescence is increased. This is because when the liquid density increases, the acceleration of the upper bubbles decreases, while the acceleration of the lower bubbles increases, and the time required to approach decreases. Therefore, the initial moment of bubble coalesce decreases significantly, and the coalescing time is advanced. On the other hand, with the increase of liquid density, the deformation of the bubbles intensifies, the contact area between the two bubbles increases, the length of the liquid film increases, so that the drainage time and fusion time increase. In addition, as the velocity and height of the coalesce decrease, the kinetic energy of the bubbles decreases. The drainage velocity and fusion velocity decrease, so that the drainage time and fusion time are prolonged, and the coalesce time increases accordingly.

4 Conclusion

In this paper, VOF method was used to simulate the coalescence process of coaxial free rising bubbles in gas-liquid system, and the influence of liquid density on the coalescence shape, rising velocity and coalescence time of bubbles was studied. The main conclusions are as follows:

(1) With the increase of liquid density, the bubble shape is difficult to maintain stability, the deformation rate increases, the aspect ratio E decreases. When the liquid density exceeds 1400 kg·m^{-3}, the interface oscillation of the large bubble formed by coalesce is obvious, and it will gradually deform until broken.

(2) When the liquid density increases, the acceleration of the upper bubble decreases and its velocity decreases. The acceleration of the lower bubble increases and its velocity increases relatively. Therefore, the height and velocity of bubble coalescence decrease with the increase of liquid density.
(3) With the increase of liquid density, the approach process of two bubbles is accelerated, and the initial moment of bubble coalescence is obviously advanced. In addition, the length of liquid film between bubbles increases, the drainage velocity and fusion velocity decrease, which will prolong the drainage time and fusion time, and increase the coalescence time.

Acknowledgment. The work described in this article has been conducted as part of the Natural Science Foundation of Shandong Province (ZR2020ME224).

References

1. Zhan, J., et al.: Simulation study of helium bubble coalescence in tungsten at various temperatures relevant to fusion conditions.Comput. Mater. Sci. **187**, 1–6 (2021)
2. Jain, D., Kuipers, J., Deen, N.G.: Numerical study of coalescence and breakup in a bubble column using a hybrid volume of fluid and discrete bubble model approach. Chem. Eng. Sci. **119**(119), 134–146 (2014)
3. Bo, W., Hao, Z., Chen, T., Wu, D.: Numerical simulation of underwater bubble motion. China Sci. Technol. Pres. Online **5**(08), 647–650 (2010)
4. Jinya, Z., et al.: Numerical simulation of the interaction between oil droplet and bubble in air flotation process. China Sci. Technol. Paper **8**(06), 525–529 (2013)
5. Huahai, Z., et al.: Experimental study on the effect of alcohol surfactant on bubble coalescence in full concentration range. CIESC J. **71**(09), 4161–4167 (2020)
6. Abbassi, W., et al.: Numerical simulation of free ascension and coaxial coalescence of air bubbles using the volume of fluid method (VOF). Comput. Fluids **161**, 47–59 (2017)
7. Orvalho, S., Stanovsky, P., Ruzicka, M.C.: Bubble coalescence in electrolytes: effect of bubble approach velocity. Chem. Eng. J. **406**, 1–18 (2021)
8. Xiaogang, J., Zhijiang, Y., Yuting, Z., Zhilin, Z.: Simulation of bubble coalescence behavior based on COMSOL. Comput. Simul. **36**(07), 191–194 (2019)
9. Qi, L., Luo, Z.H.: CFD-VOF-DPM simulations of bubble rising and coalescence in low hold-up particle-liquid suspension systems. Powder Technol. **339**, 459–469 (2018)
10. Swart, B., Zhao, Y., Khaku, M., et al.: In situ characterisation of size distribution and rise velocity of microbubbles by high-speed photography. Chem. Eng. Sci. **225**, 1–15 (2020)
11. Chakraborty, I., Biswas, G., Ghoshdastidar, P.S.: A coupled level-set and volume-of-fluid method for the buoyant rise of gas bubbles in liquids. Int. J. Heat Mass Transf. **58**(1–2), 240–259 (2013)
12. RyskinG, L.: Numerical solution of free-boundary problems in fluid mechanics. Part 1. The finite-difference technique. J. Fluid Mech. **148**, 37–43 (1984)

Study on the Influence of Oil Droplet Physical Parameters on the Separation Performance of Hydrocyclone

Bing Liu[1], Qun Gao[1], Zhen Wu[1], Hongbo Fang[2,3](✉), Xiaolong Xiao[2,3], and Mingxiu Yao[2,3]

[1] College of Mechanical and Electronic Engineering, Shandong University of Science and Technology, Qingdao 266590, China
[2] Petroleum Engineering Co., Ltd., SINOPEC, Dongying 257026, China
[3] Shandong Provincial Key Laboratory of Oilfield Produced Water Treatment and Environmental Pollution Control, Dongying 257026, China

Abstract. In order to study the influence of oil density and particle size on the separation performance of hydrocyclone, the tangential velocity and axial velocity distributions were studied with the help of fluid dynamics software Fluent, and the influence of oil density and particle size on the separation efficiency of hydrocyclone was analyzed. The results show that within a certain range, the separation efficiency of the hydrocyclone decreases with the increase of oil phase density. When the oil phase density is greater than 880 kg/m^3, the separation efficiency is less than 85%. When the size range of oil droplets is less than 10 um, the separation efficiency of the hydrocyclone is less than 70%. By increasing the size of oil droplets, the migration trajectory of oil droplets can be changed so that more oil droplets are captured by the overflow port, thus improving the separation efficiency of the hydrocyclone.

Keywords: Oil-water separation cyclone · Oil phase density · Oil droplet size · Separation efficiency

1 Introduction

At the beginning of the 21st century, the integrated water content of oil fields in China reached 80.2%, and the oil fields entered the high water cut period. With the sharp increase in the water content of the produced fluid in the oilfield, the traditional oil and gas gathering and transportation processing equipment has the disadvantages of low separation efficiency and greatly increased operating energy consumption [1]. As for the application of liquid-liquid hydrocyclone for oil-water separation, its internal flow field characteristics and oil-water separation process are very complex, leading to many parameters affecting the separation efficiency of liquid-liquid hydrocyclone. Therefore, the study of relevant parameters is of great significance to improve the efficiency of oil-water separation and the sustainable development of China's petroleum exploitation industry. H. H. Al-Kayiem [2] improves the performance of the axial flow liquid-liquid

Y. Wang et al. (Eds.): IWAMA 2021, LNEE 880, pp. 278–285, 2022.
https://doi.org/10.1007/978-981-19-0572-8_35

cyclone separator by optimizing the cyclone blade Angle; Zeng [3] proposed a new type of axial oil-water separation separator based on droplet trajectory analysis. Experiments showed that the separation efficiency was related to the flow rate and oil intake fraction. Liu [4] performed numerical simulation on the internal flow field of oil-water cyclones with different structures, and it was found that the flow field performance of the double-cone structure was higher than that of the tubular structure. In addition to the structure and operation parameters that affect the separation performance of the hydrocyclone, the physical properties of oil droplets are also important factors affecting the separation performance of the hydrocyclone. Xing Lei et al. [5] carried out numerical simulation of the migration trajectory of oil droplets in the discrete phase in the swirling flow field with different incidence positions, velocities and particle sizes, and summarized the influence law of the particle size of oil droplets on their migration trajectory. Jin Song et al. [6] studied the influence of particle size of oil phase in an oil-water separation hydrocyclone on the separation efficiency, and obtained the variation rule of different characteristic particle size with the flow rate. Wang Sheng [7] observed the change of oil-water distribution in the hydrocyclone by changing the oil phase density, and found that the oil-water separation hydrocyclone could only be effective within a certain range of oil phase density. Based on the above analysis, it is found that the density and particle size of the oil phase have an influence on the change of the oil droplet migration trajectory, and then affect the separation performance of the oil-water separation cyclone. However, there are relatively few studies on this.

In this paper, Reynolds Stress Model (RSM) and Discrete Phase Model (DPM) are used to comprehensively analyze the effects of different oil density and particle size on the separation efficiency [8], and determine the influence law of the physical parameters of oil droplets on the separation performance of hydrocyclone. The results further enrich the study on the influence parameters of hydrocyclone separation performance and provide guidance for the application of oil-water separation hydrocyclone. The results further enrich the study on the influence parameters of hydrocyclone separation performance and provide guidance for the application of oil-water separation hydrocyclone.

2 Numerical Simulation

2.1 Geometric Model

The transport trajectories and separation performance of oil droplets with different oil phase densities and particle sizes in the cyclone field were numerically analyzed using a conventional bicone hydrocyclone. The main fluid domain model of the cyclone is shown in Fig. 1 in millimeters. In addition, the sections selected in this section shall be selected in accordance with Fig. 1 if no special instructions are given. In order to make up for the lack of structured grid that can't solve the mesh subdivision of arbitrary shape and arbitrarily connected region, and hexahedral grid has the characteristics of high quality and relatively small number of grids generated, this paper adopts hexahedral unstructured grid, with 746840 grid units and 768558 nodes. The grid division is shown in Fig. 2.

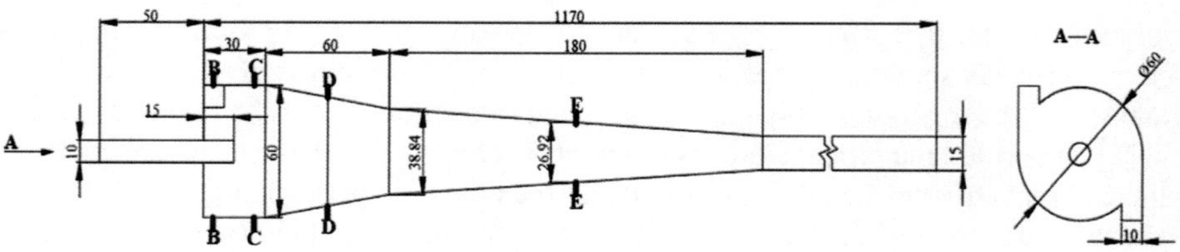

Fig. 1. Biconical hydrocyclone fluid domain model

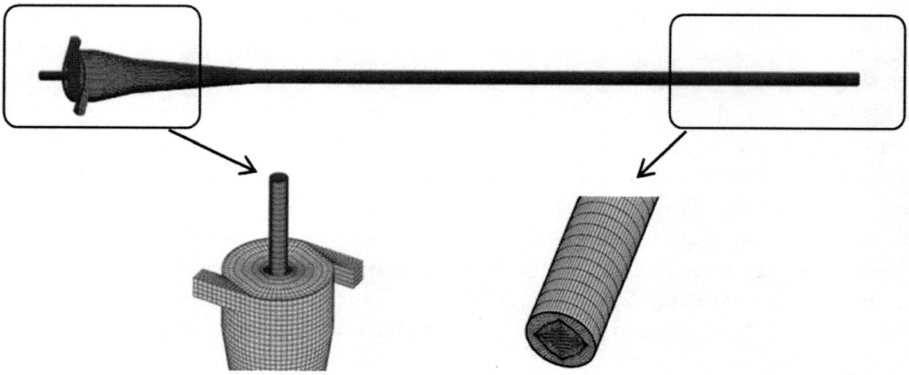

Fig. 2. Grid structure

2.2 Boundary Conditions and Numerical Calculation Algorithms

(1) Inlet boundary conditions: speed inlet, hydraulic diameter 10 mm, turbulence intensity 4.34%; The discrete phase is set to Escape at both entrances, and the jet source is the point source.
(2) Outflow boundary condition: free exit is used for both overflow and underflow, and the trajectory of discrete phase is calculated when it moves to the overflow boundary [5]. The discrete phase is set as trap at the overflow port and escape at the underflow.
(3) Parameter setting: the inlet velocity of continuous phase water is 10 m/s, and the initial velocity of discrete phase oil drops is the same as that of continuous phase water.
(4) Wall boundary conditions: no leakage, no sliding wall boundary.
(5) Numerical calculation method: the internal flow field of the hydrocyclone is anisotropic incompressible flow, the solver adopts the pressure-based solver, the algorithm adopts the SIMPLE algorithm, and the gradient adopts the element-based least square method, the Presto pressure difference and the Quick discrete format, and then initializes the calculation domain to start the iterative solution calculation.

3 Simulation Results

3.1 Velocity Field Analysis

Tangential velocity determines the magnitude of centrifugal force, and has a certain influence on radial and axial velocities [9]. Due to the density difference between the heavy continuous phase and the light discrete phase, the centrifugal force is different, which causes the droplet of the discrete phase to move to the axis. Therefore, the study of the tangential velocity inside the hydrocyclone of the target structure is a basic work to analyze the migration trajectory of oil droplets in the discrete phase. Figure 3 shows the tangential velocity distribution curves of the cyclone at different sections. It can be seen from the structural characteristics and theory of the cyclone that the tangential velocity distribution is circumferentially symmetric.

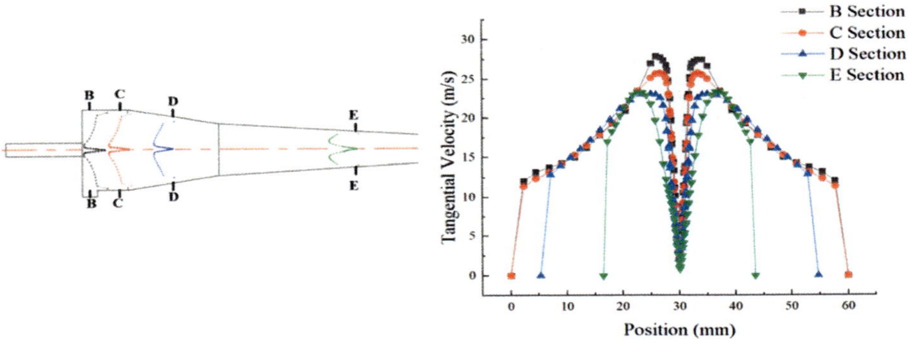

Fig. 3. The tangential velocity distribution curve of the hydrocyclone at different sections

Combined with the above figure and the tangential velocity distribution cloud map in Fig. 4, it can be concluded that the tangential velocity gradually increases from the wall of the device, and reaches the maximum value at the axis circle with a radius of 5 mm and then drops rapidly. Thus, the tangential velocity of the outer swirl is greater than that of the inner swirl.

3.2 Influence of Oil Phase Density on Separation Efficiency of Hydrocyclone

The density difference between oil phase and water phase is one of the main factors affecting the separation performance of hydrocyclone. Therefore, in order to study the influence of oil phase density on the separation performance of hydrocyclone, The oil phase densities of 820 kg/m^3, 840 kg/m^3, 860 kg/m^3, 880 kg/m^3, 900 kg/m^3, 920 kg/m^3 were analyzed respectively. In the numerical simulation, the DPM model was used to inject oil droplets with particle size of 15 um by point-injection method to observe the movement trajectories of oil droplets with different oil phase densities and the corresponding separation efficiency.

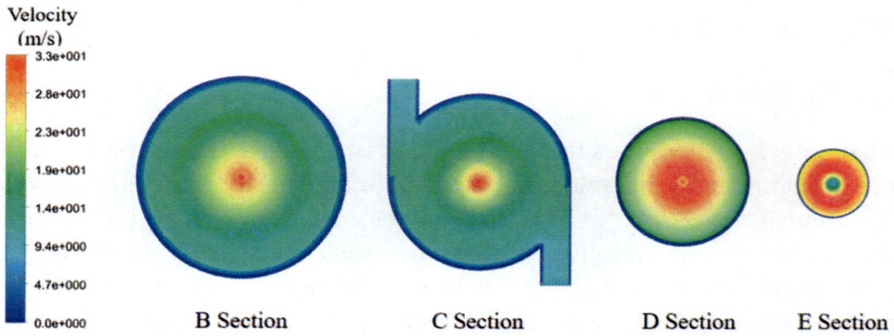

Fig. 4. Cloud images of tangential velocities at different sections

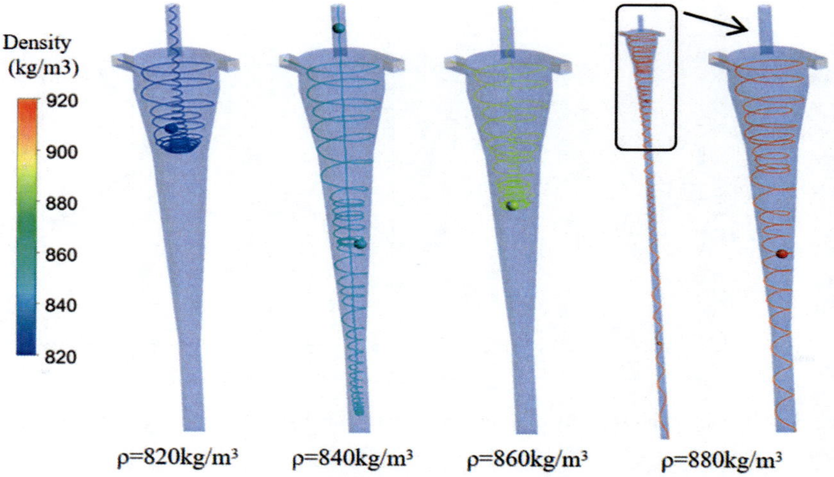

Fig. 5. Migration trajectories of oil droplets with different oil phase densities

As can be seen from Fig. 5, the migration trajectories of oil droplets with different oil phase densities show several different migration trajectories of oil droplets due to different oil phase densities. The oil droplets enter the swirl cavity at a certain speed through the inlet. At first, the centrifugal force is stronger than the turbulence and the oil droplets enter the swirl. Due to the different oil phase densities and the random characteristics of the flow field, the oil droplets with the oil phase density of 820 kg/m^3 first pass through the zero-axis velocity envelope at the large cone end and enter the inner swirl. For oil droplets with oil phase densities of 840 kg/m^3 and 860 kg/m^3, the density between the oil droplets and the continuous phase water is reduced, so that the oil droplets continue to move downward with the axial velocity under the action of the outer swirling flow [10]. When the oil droplets migrate to the small cone end, the turbulence effect is greater than the centrifugal force effect [11], so that the oil droplets generate radial velocity and slowly rotate to the center, and finally are carried into the overflow outlet by the inner swirl under the action of turbulence. The oil droplets with the oil phase density of 880 kg/m^3 are subjected to sufficient centrifugal force after entering the swirling flow field and under the action of wall attached, so that the oil droplets maintain good tangential and axial following and are discharged along the inner wall of the cyclone to the underflow. As can be seen from Fig. 6, as the oil phase density gradually increases, more oil droplets escape from the underflow. The separation efficiency of the hydrocyclone decreases gradually with the increase of oil phase density, especially when the oil phase density is greater than 880 kg/m^3, the separation efficiency decreases significantly to below 85%.

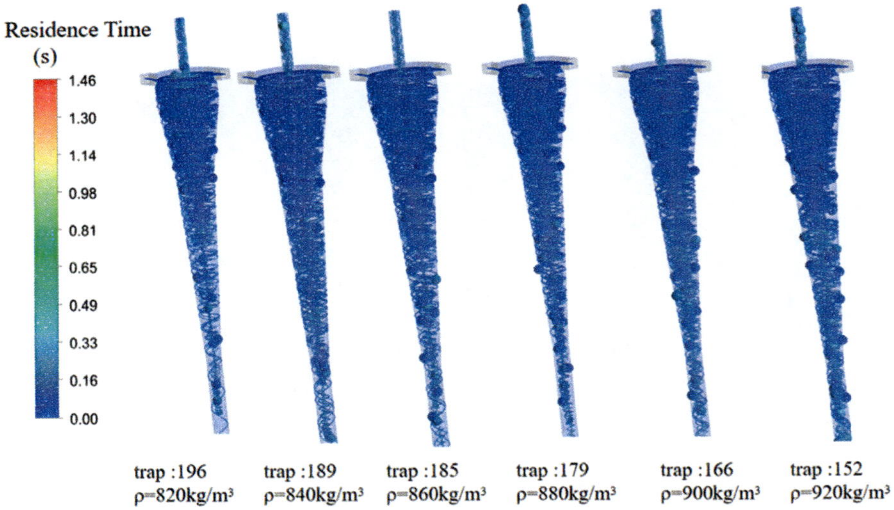

Fig. 6. Particles migration trajectories of different oil phase densities

3.3 Influence of Droplet Size on Separation Efficiency of Hydrocyclone

In the process of hydrocyclone oil-water separation, the produced liquid will have different particle sizes, so it is necessary to study the separation performance of hydrocyclone under different particle sizes of oil droplets, in order to determine the suitable particle size range for hydrocyclone treatment. In this paper, on the premise that the oil droplets in the discrete phase will not be broken in the flow field of the target structure, 50 oil droplets are injected into each of the two entrances and the effects of particle sizes of 5 um, 8 um, 15 um and 20 um on the separation performance of the hydrocyclone are simulated and calculated respectively.

According to the separation efficiency of oil droplets with different particle sizes in Fig. 7, within the order of magnitude of particle size less than 10 um, more oil droplets are discharged from the underflow under the action of outer swirling flow, and the separation efficiency is below 70%. However, with the increase of particle size, the separation efficiency still shows an upward trend. When the particle size increases to 15 um, the trajectories of oil droplets with this particle size change significantly in the swirling flow field. After entering the cyclone cavity, more oil droplets first rotate along the wall under the action of the outer cyclone and produce axial displacement and migrate toward the bottom flow direction. After entering the large cone section, they break through the envelope surface of zero axial velocity and enter the inner cyclone. Under the action of the inner cyclone, they flow out of the overflow outlet with the separation efficiency of 76%. When the particle size of oil drops continues to increase to 21 um, the track of oil drops at the axis of the lower end of the cyclone becomes less and less, that is, the oil droplets discharged from the underflow port gradually decrease, and the separation efficiency is 94%.

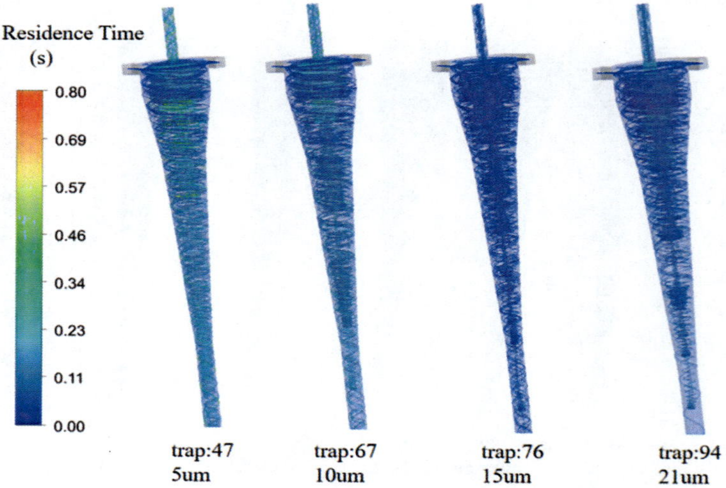

Fig. 7. Separation efficiency of oil droplets with different particle sizes

4 Conclusion

In this paper, fluid dynamics software Fluent is used to study the migration trajectory and separation performance of discrete phase oil droplets in an oil-water separation cyclone, and the following conclusions are drawn within the research scope:

(1) Oil phase density will affect the migration trajectory of oil droplets. With the gradual increase of oil phase density, more oil droplets will flow out of the bottom flow port. When the oil phase density is greater than 880 kg/m^3, the separation efficiency drops to less than 85%.
(2) The hydrocyclone did not separate the oil droplets with particle size less than 10 um well, and a large number of oil droplets escaped from the bottom flow under the action of the outer swirl, and the separation efficiency was less than 70%. Therefore, the average particle size of the produced liquid should be greater than 10 um.
(3) Increasing the particle size of the oil droplets can improve the separation efficiency of the cyclone. When the particle size of the oil droplets increases to 15 um, the

migration trajectory of the oil droplets changes significantly, and more oil droplets are captured by the overflow outlet under the action of the inner cyclone, and the separation efficiency is 76%. When the particle size of oil droplets was increased to 21 um, very few oil droplets escaped from the underflow port, and the separation efficiency was as high as 94%.

Acknowledgment. This study was supported by open fund of Shandong Provincial Key Laboratory of Oilfield Produced Water Treatment and Environmental Pollution Cotrol.

References

1. Jiaqing, C., et al.: Development status and prospect of pre-water separation technology for produced fluid from high water cut oil Wells. Acta Petrolei Sinica **41**(11), 1434–1444 (2020)
2. Al-Kayiem, H.H., Hamza, J.E., Lemmu, T.A.: Performance enhancement of axial concurrent liquid–liquid hydrocyclone separator through optimization of the swirler vane angle. J. Petroleum Explor. Prod. Technol. **10**(7), 2957–2967 (2020). https://doi.org/10.1007/s13202-020-00903-7
3. Zeng, X., Zhao, L., Zhao, W., et al.: Experimental study on a novel axial separator for OilWater separation. Indus. Eng. Chem. Res. **59**(48), 21177–21186 (2020)
4. Bo, L., Xu, J., Yue, W.: The numerical simulation of the inside flow field of oil–water hydrocyclone with different structures. In: 2016 2nd Workshop on Advanced Research and Technology in Industry Applications (WARTIA-16) (2016)
5. Lei, X.: Research on oil droplet migration trajectory of discrete phase in swirling flow field. Northeast Petroleum University (2016)
6. Jinsong, Z., Shuchu, F., Yuxing, L., Miaoer, L.: Research on flow mechanism and application of hydrocyclone for oil-water separation. Filtr. Sep. **03**, 15–18 (2001)
7. Sheng, W., Wei, W., Shiying, S., Hui, T., Yingxiang, W.: Research on the influence of the separation performance of axial flow swirl oil and water separator. China Petroleum Mach. **46**(05), 99–104 (2018)
8. Qiaoduo, Y.: Numerical Simulation of Oil Water Separation and Droplet Breakage in Hydro cyclone. Huazhong University of Science and Technology, Wuhan (2014)
9. Lixin, Z., Minghu, J., Feng, L., Jie, H., Jinling, L., Haijing, S.: Force analysis of dispersed phase droplet in hydrocyclone – study on velocity field of liquid-liquid hydrocyclone, part 5. Petroleum Mach. **05**, 24–27 (1999)
10. Zhonghe, Z.: Study on the influence of the rheological properties of produced fluid with polymerization on the separation performance of hydrocyclone. Northeast Petroleum University (2018)
11. Rietema, K.: Performance and design of hydrocyclones parts I-IV. Chem. Eng. Sci. **15**(3), 298–302 (1961)

Study on the Mechanism of Bubble Coalescence in an Air-Lift Loop Reactor

Bing Liu[1], Heng Guan[1], Huanxin Zhao[1], Zhen Wu[1], Qun Gao[1], Jianliang Xue[1(✉)], and Peishan Huang[2]

[1] College of Mechanical and Electronic Engineering, Shandong University of Science and Technology, Qingdao 266590, China
wenyu@sdust.edu.cn
[2] Qingdao Deep Blue Subsea Engineering Technology Co., Ltd., Qingdao 266500, China

Abstract. Based on the premise that two bubbles can merge online, this study uses a combination of experiment and numerical simulation to study the influence of the initial bubble size and initial spacing on the difference mechanism of bubble rise speed. It is found that when the bubble diameter is 4–6 mm, the initial bubble size has little effect on the bubble rising velocity difference, and the ultimate velocity of the two bubbles before coalescence is basically the same. The initial bubble spacing has a great effect on the bubble rising velocity difference, and the increase of the initial bubble spacing will increase the ultimate velocity before coalescence, larger ultimate velocity will produce velocity fluctuations when the bubbles collide and contact, which reduces the velocity stability. This study systematically analyzed the influence of initial bubble size and initial bubble spacing on bubble coalescence behavior, and provided a theoretical reference for the design of the orifice plate structure of the airlift loop reactor.

Keywords: Airlift loop reactor · Bubble · Merge · Different ascent speed · Numerical simulation

1 Introduction

Airlift loop reactor is widely used in hydrometallurgy and air flotation for oil removal. The core components affecting gas distribution in the reactor are porous resistance plate and gas distributor [1]. At present, the design of porous resistance plate or gas distributor is mainly based on empirical principle, and the coalescence and interaction between bubbles cannot be considered [2]. In addition, because the coalescence behavior of bubbles is very complex, the obtained model is mainly a semi theoretical and semi empirical model. It is difficult to ensure the efficient operation of gas-liquid equipment only based on semi theoretical and semi empirical models [3]. Therefore, further research on coalescence and interaction in the process of bubble rising has important theoretical significance for the design of gas-liquid equipment. In airlift loop reactor, the relative position arrangement and the number of openings of porous resistance plate and gas distributor directly affect the gas-liquid two-phase interaction. Based on the porous resistance plate and gas

Y. Wang et al. (Eds.): IWAMA 2021, LNEE 880, pp. 286–294, 2022.
https://doi.org/10.1007/978-981-19-0572-8_36

distributor shown in Fig. 1, the coalescence and interaction of two bubbles in the rising process were studied by combining experimental and numerical simulation methods, and the effects of initial bubble size and initial spacing on coalescence behavior were investigated. Through the analysis of the bubble coalescence phenomenon in the airlift loop reactor, a more in-depth and effective study of the reactor is carried out.

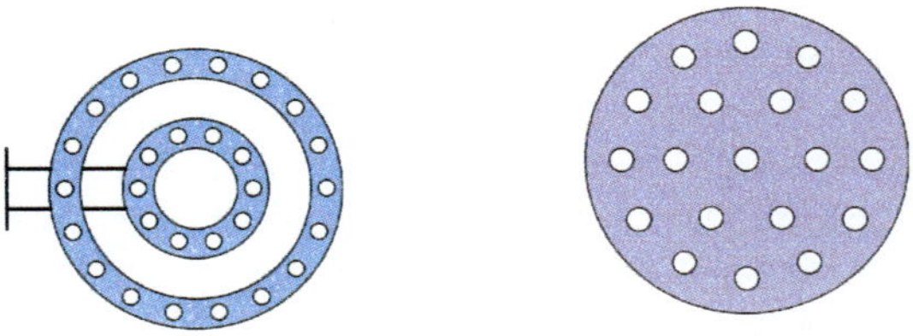

(a) Gas distributor (b) Porous resistance plate

Fig. 1. Gas distributor and porous resistance plate

2 Numerical Simulation Method and Experimental Verification

2.1 Governing Equations

Because the bubble diameter is small, the effect of surface tension is not significant, the momentum equation neglecting the surface tension can be simplified as:

$$\rho\left[\frac{\partial u}{\partial t} + (u \cdot \nabla)u\right] = -\nabla p + \nabla \cdot (2\mu D) + \kappa \delta_s n + \rho g \tag{1}$$

Where: p is the pressure, μ is the dynamic viscosity, κ is the curvature of the interface, δ_s is the Dirac distribution related to the interface, and n is the unit normal vector of the interface. D is the stress tensor

$$D_{ij} = \frac{1}{2}\left[\frac{\partial u_j}{\partial x_i} + \frac{\partial u_i}{\partial x_j}\right] \tag{2}$$

Continuity equation of incompressible fluid

$$\nabla \cdot u = 0 \tag{3}$$

VOF method is used to simulate the volume function C of bubble interface

$$\frac{\partial C}{\partial t} + \nabla \cdot (uC) = 0 \tag{4}$$

For two-phase flow, both μ and ρ in Eq. (1) are determined by volume function C

$$\rho = \rho_1 C + \rho_2 (1 - C) \tag{5}$$

$$u = u_1 C + u_2 (1 - C) \tag{6}$$

Where: ρ_1, ρ_2, μ_1, μ_2 is the density and dynamic viscosity of the two fluids.

2.2 Physical Model and Solution

According to the conclusion of reference [4], when the distance between the center of the bubble and the wall is greater than 5 times the radius, the influence of the wall can be ignored. Consider the above factors, establish the two-dimensional model and area size as shown in Fig. 2, the size of the two-dimensional computational domain is 80 mm × 190 mm, the upper part is 80 mm × 10 mm separation area. Gambit is used for grid division, and the number of grids divided is 441148. After initialization, the computational domain is filled with static liquid. The outlet is set as the pressure outlet, the pressure outlet is atmospheric pressure, bubbles with different sizes and spacing are placed at the bottom of the calculation domain, several other stationary walls are assumed to be slip free. PISO method is used to solve the pressure velocity coupling equation; presto method and first-order upwind scheme are used to discretize the pressure and momentum terms. When the time step is 1.0 times 1.0×10^{-5} s and the residual error of velocity and momentum is less than 0.001, it is considered that the calculation converges. TECPLOT 360 software and MATLAB program are used for post-processing of calculation results.

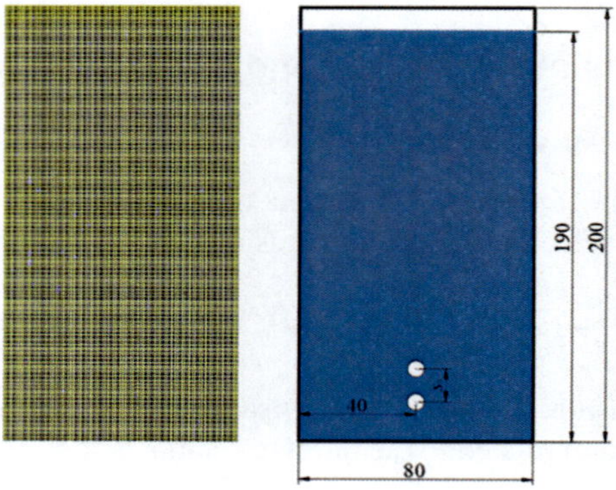

Fig. 2. 2D model and area size

2.3 Experimental Equipment

Based on the airlift external loop reactor, a simple experimental model as shown in Fig. 3 is established. The high-speed camera used in the experiment is phantom v1211, and the shooting area is 512 × 512 pixels, 9000 frames per second, exposure time of 1 μs. The height of the test bench is 800 mm, and the size of the bottom surface is 100 mm × 100 mm. A separation area is left at the top to connect with the atmosphere, and a bubble generating device is connected at the bottom. Because it is difficult to control the number of pores in the orifice plate, and it is difficult to track the bubbles based on the specified bubble, the air jet method is used to generate bubbles in the experiment. The size and initial spacing of bubbles can be changed by controlling the time of gas injection and the time interval of bubble generation by PC. Screw pump is used for external circulation of liquid. The flow of external circulation liquid is controlled by flowmeter and regulating valve, so that the liquid flow is always in low-speed laminar flow state.

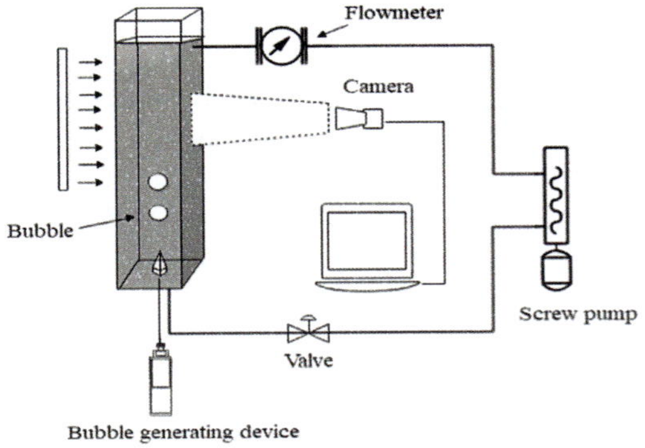

Fig. 3. Schematic diagram of experimental device

2.4 Comparison Between Experiment and Simulation

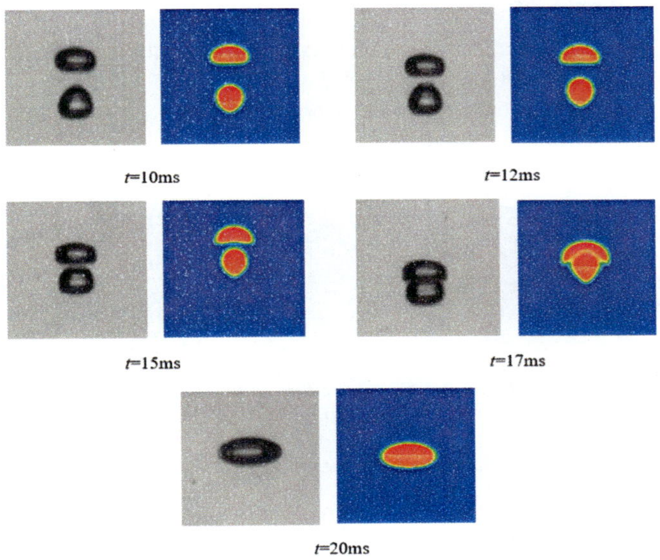

Fig. 4. Comparison of 2D simulation and experiment

Experiments and simulations were carried out at room temperature (298.15 K) and atmospheric pressure (0.1013 MPa), with an initial diameter of 6 mm and an initial spacing of 5 mm. In order to facilitate image comparison, both experiments and simulations used glycerin as the fluid medium. According to the experimental part of references [5, 6], a high-speed camera was used to capture the upward coalescence of two coaxially linearly arranged bubbles in glycerin. In Fig. 4, the experiment simulates the rising process of two bubbles online at different times. The two bubbles coalesced after the following, deformation and liquid film draining. Due to the wake and shear thinning effect of the leading bubble, the latter bubble accelerates and gradually catches up with the leading bubble. The shape of the two bubbles is deformed under the action of the flow field. The shape of the former bubble becomes slightly flat, while the latter bubble stretches in the vertical direction. In general, the simulation results are in good agreement with the experimental results, it shows that when the number of grids is 441148, the rising behavior of bubbles can be accurately simulated.

3 Results and Discussion

The rising speed difference between bubbles has an important impact on gas residence time and bubble distribution, and is one of the important parameters of reactor design. In this study, in the water-air two-phase medium, the change of the axial velocity difference between the two bubbles caused by the change of bubble size and bubble spacing was discussed.

3.1 Effect of Initial Spacing on Velocity Difference

In this study, the bubble diameter is 6 mm, and the trailing bubble coordinates are (0.04, 0.02). By changing the distances between the leading bubble and the trailing bubble, the axial velocity changes of two bubbles from free rising to coalescing contact under different initial spacing S_0 are studied.

S_0=0.01m S_0=0.015m S_0=0.02m

Fig. 5. Simulation of two bubbles with different initial spacing

Figure 5 shows the axial position changes of the two bubbles when the trailing bubble rises to the maximum velocity (Ultimate velocity) under three different initial distances when the bubble diameter $d = 6$ mm. It is found that with the increase of the initial spacing, the position of the bubble reaching the ultimate velocity increases gradually, the shape of the trailing bubble changes from spherical cap to flat ellipsoid, the convexity of the upper edge and the concavity of the lower edge decrease with the increase of the initial spacing, the shape of the leading bubble is ellipsoidal, and the change of its shape is less affected by the change of the initial spacing. The lower edge of the bubble gradually protrudes with the increase of the initial spacing.

Figure 6 shows the axial velocity curve of two bubbles from free rising to coalescing contact under three different initial distances when the bubble diameter $d = 6$ mm.

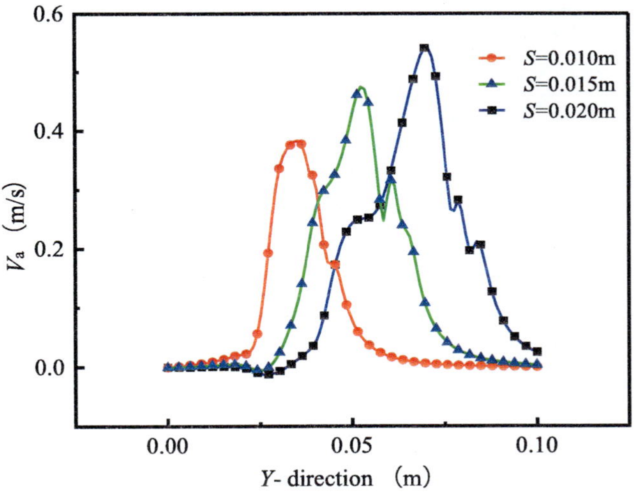

Fig. 6. Axial velocity variation

The above velocity curves have experienced four processes: bubble acceleration, bubble discharge deceleration, bubble collision and bubble fusion. When the bubble reaches the highest point, it is the ultimate velocity of the bubble. After the bubble reaches the ultimate velocity, due to the influence of the drainage resistance, the velocity begins to decline. After the end of the drainage, the bubbles will collide with each other. Due to the velocity difference between the two bubbles, the velocity fluctuation occurs during the collision and contact process. After a short period of velocity fluctuation, the bubbles begin to fuse slowly, the bubble velocity will continue to decrease until the two bubbles coalesce. By comparing the axial velocity curves under the three different initial distances in Fig. 6, with the increase of the initial space between bubbles, the free rise and acceleration phases of the bubbles become longer, the axial position of the bubbles reaching the ultimate velocity before collision and contact is delayed, and the ultimate velocity of bubbles gradually increases. In the deceleration stage of bubble discharge, the increase of the initial spacing makes the velocity decrease increase, when the bubbles collide with each other, the increase of the initial distance makes the frequency and degree of velocity oscillation increase, and the stability of velocity decrease. In the process of two bubbles fusion, the change of initial distance has little effect on the change rate of fusion velocity, and the change trend of velocity is the same.

3.2 Effect of Bubble Size on Velocity Difference

In this section, based on the given initial spacing, the release position coordinates of the leading bubble and the trailing bubble are (0.04, 0.02), (0.04, 0.03), respectively, to explore the axial velocity changes of different bubble sizes from free rising to coalescing contact.

Figure 7 shows the position changes of two bubbles with three different sizes when they are released at rest and accelerated to the maximum speed (ultimate speed). It is

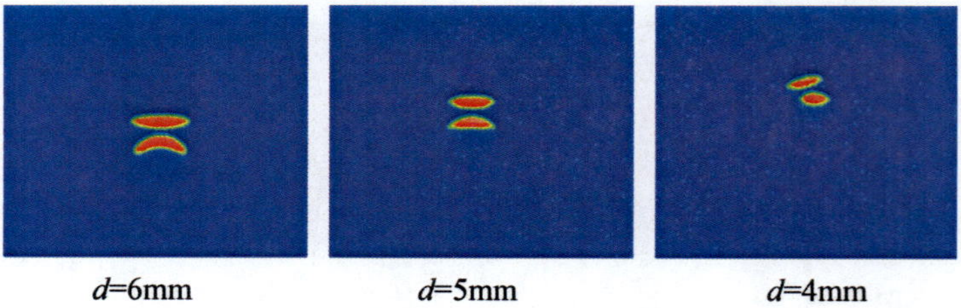

Fig. 7. Simulation of two bubbles with different sizes

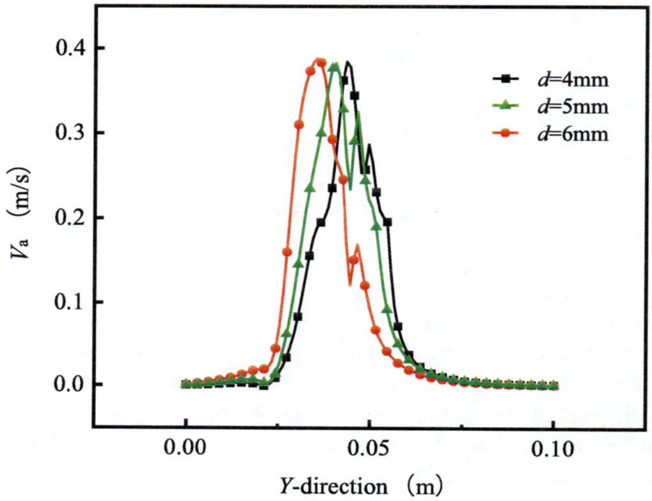

Fig. 8. Axial velocity variation

found that when the initial spacing is constant, the smaller the bubble, the higher the position to reach the ultimate velocity. When the bubble diameter is 4 mm, compared with the other two kinds of bubbles, the position to reach the maximum velocity is not on the same axis, and the first bubble appears to tilt to the left. For 4–6 mm diameter bubbles, with the increase of bubble volume, the deformation increases when the bubble reaches the ultimate velocity.

Figure 8 shows the axial velocity curves of three different bubble sizes (4 mm, 5 mm, 6 mm) from free rising to coalescing contact based on the same initial spacing. The results show that when the initial spacing is fixed, the ultimate velocity of bubbles is basically the same when the bubble size is changed, but the axial position of bubbles reaching the ultimate velocity is different. The larger the bubble, the lower the axial position to reach the ultimate velocity. During the deceleration of the bubble discharge, as the bubble diameter increases, the speed drop during the discharge is greater. When the bubble collides with each other, the difference of bubble velocity amplitude is small when the bubble size is 4 mm and 6 mm, while the bubble velocity amplitude is larger when the bubble diameter is 5 mm. It can be concluded that in the range of 4–6 mm in diameter, the velocity amplitude will first increase and then decrease when the bubble collides with each other. The velocity amplitude will reach the maximum when it is close to 5 mm, and it will be in the stage of velocity amplitude transition when the diameter is 5 mm. In the process of two bubbles fusion, the change of bubble size has a great influence on the change rate of fusion velocity. With the increase of bubble diameter, the change rate of fusion velocity decreases.

4 Discussion and Conclusion

In this paper, using 2D VOF method, by changing the initial size and initial spacing of the bubbles, analyzed the mechanism of the difference in bubble rising speed, the conclusions reached are as follows:

(1) When the bubble diameter is 4–6 mm, the initial size of the bubble has little effect on the bubble rising speed difference, and the ultimate speed reached by the trailing bubble rising is basically the same. During the bubble fusion process, the change of bubble size has a greater impact on the rate of change of the fusion speed, as the bubble diameter increases, the rate of velocity change decreases.
(2) When the bubble diameter is 6 mm, the increase of the initial bubble spacing will increase the ultimate speed of bubble rise, which will affect the velocity stability of the bubble during collision and contact. During the bubble fusion process, the change of the initial spacing will change the fusion speed. The change of the initial spacing has little effect on the rate of change of the fusion speed; the speed change trend is the same.
(3) When the bubble size is 4 mm and 6 mm, the velocity amplitude difference is small when the bubbles collide and contact. When the bubble diameter is about 5 mm, the velocity amplitude generated by the collision of the two bubbles will have the maximum peak. When the diameter is 5 mm, the velocity amplitude is in the transitional stage.

Acknowledgment. The work described in this article has been conducted as part of the National Natural Science Foundation of China (Grant No. 52070123).

References

1. Wang, T.: Experimental research and numerical simulation of the hydrodynamic behavior of gas-liquid (slurry) reactor. Ph.D. dissertation, Tsinghua University (2004)
2. Simcik, M., Mota, A., Ruzicka, M.C., et al.: CFD simulation and experimental measurement of gas holdup and liquid interstitial velocity in internal loop airlift reactor. Chem. Eng. Sci. **66**(14), 3268–3279 (2011)
3. Kitagawa, A., Denissenko, P., Murai, Y.: Behavior of bubbles moving along horizontal flat plates with different surface wettability. Exp. Thermal Fluid Sci. **104**, 141–152 (2019)
4. Chen, R.H., Tian, W.X., Su, G.H., et al.: Numerical investigation on coalescence of bubble pairs rising in a stagnant liquid. Chem. Eng. Sci. **66**(21), 5055–5063 (2011)

5. Liao, J., Ziegenhein, T., Rzehak, R., et al.: Bubbly flow in an airlift column: a CFD study. J. Chem. Technol. Biotechnol. **91**, 2904–2915 (2016)
6. Besagni, G., Inzoli, F., et al.: Two-phase bubble columns: a comprehensive review Chem. Eng. **2**(2), 3 (2018)

A Review of Vibration-Based Piezoelectric Energy Harvester

Yunchao Wang(✉) and Wenying Yang

School of Marine Equipment and Mechanical Engineering, Jimei University, Xiamen 361021, China

ychaowang@jmu.edu.cn

Abstract. The state of the art in piezoelectric energy harvesting (PEH) at home and abroad is reviewed from the aspects of piezoelectric theory, material properties, structure of device and practical applications. PEH has great development prospects in the field of Micro-Electro-Mechanical System (MEMS) due to the simple structure and ease of integration. However, the large power output can be achieved only if the resonant frequency is reached which restricts the development of PEH. In order to break through the core technical issues, researchers are devoted to the development of new piezoelectric materials and optimization of structures to enhance the efficiency of PEH. The trend and perspective of PEH are discussed and references are provided for further research on PEH in this paper.

Keywords: Piezoelectric energy harvesting · Vibration · Power output

1 Introduction

Global energy crisis and environmental issues have increasingly become the focus that the world is paying more attention to over the several decades. Recent studies have shown that the combustion products of conventional fossil fuels are harmful to the environment and human health [1]. For instance, carbon dioxide emission will aggravate the global greenhouse effect. Effective methods to alleviate the energy crisis have been considered from two different aspects. On one hand, fossil fuels are gradually replaced by renewable energy sources. On the other hand, researchers are also beginning to turn their attention to the recovery of dissipated energy. There are various unutilized energy sources from the surrounding environment, such as solar energy [2], thermal energy [3], mechanical vibration, wind energy [4], sound energy [5], wave energy [6], among which the power density of solar energy is the highest. The power density of mechanical vibration is lower than that of solar energy, but it is more sustainable. Mechanical vibration exists in all kinds of machines. The recovery of mechanical vibration energy enables the dissipated energy to be recycled, thus raising the efficiency of machine. On a conclusion, the energy recovery technology of mechanical vibration has an extensive application space and high utilization value.

Vibration frequently occurs in the environment around us, and it is feasible to capture energy from mechanical vibration. Researchers utilize different energy conversion

Y. Wang et al. (Eds.): IWAMA 2021, LNEE 880, pp. 295–302, 2022.
https://doi.org/10.1007/978-981-19-0572-8_37

mechanisms to recycle and store the wasted energy [8]. At present, the three main forms to recover vibration energy are electrostatic [12], piezoelectric and electromagnetic [13]. Wide attention has been attracted by the PEH as the simple structure, high power density [14], easy to integrate and suitable for application in MEMS. Wireless sensors benefit from the rapid development of MEMS technology. Nowadays, wireless sensors still adopt battery-powered mode. When the battery of sensors are exhausted, they need to be replaced in sequence. It is inefficient to replace batteries under the condition of applying numerous sensors [15]. As a result, it is feasible and potential to harvest vibration energy for powering wireless sensors [16].

This paper introduces the mechanism of PEH, and then systematically presents the current development status from the aspects of the material and structure and the practical application of PEH. The piezoelectric energy regenerative suspension is presented detailed, in that its importance and development potential are worthy of our attention.

2 Mechanism of Piezoelectric Energy Harvesting

2.1 Piezoelectric Effect

Piezoelectric effect, the interconversion of mechanical and electrical energy, was first discovered by the Curie brothers, including the direct piezoelectric effect and inverse piezoelectric effect. When an external force is applied to piezoelectric material, the deformation occurs and the phase of the crystal inside material is shifted. The electric potential balance is broken, and the internal polarization occurs. In the view of macroscopic, it appears electric potential difference between two relatives surfaces, which is called as the direct piezoelectric effect. On the contrary, when an electric field is applied in the polarization direction of the material, the inside of the piezoelectric material is polarized and deformed, which is defined as the inverse piezoelectric effect. The piezoelectric effects are governed by the piezoelectric constitutive equations as follows:

$$\begin{bmatrix} \delta \\ D \end{bmatrix} = \begin{bmatrix} s^E & d^t \\ d & \varepsilon^T \end{bmatrix} \begin{bmatrix} \sigma \\ E \end{bmatrix} \tag{1}$$

Where δ and σ represent the strain and stress components; D and E refer to the electric displacement and electric field components; s, ε and d are the elastic compliance, the dielectric constant, and the piezoelectric coefficient, respectively; the superscripts E and T denote that the respective constants are evaluated at the constant electric field and constant stress, respectively; and the superscript t stands for the transpose.

2.2 Mode of Piezoelectric Coupling

Figure 1 illustrates two piezoelectric coupling modes [28], and the piezoelectric material was polarized along the "3" direction. The "33" mode [18] means compress stress and strain are applied in the polarization direction, and the "31" mode [19] means stress and strain are applied perpendicular to the polarization direction of the material. Both modes generate electric charges on the electrode surface, and the electro-mechanical coupling coefficient of the "33" mode is higher than that of the "31" mode. For the "33" mode, the piezoelectric stack structure can be configured to improve the generating power. This structure is not easy to reach the resonant frequency and has high stiffness, which is appropriated for recovering higher energy. The "31" mode of power generation is to integrate piezoelectric materials into a structure that is susceptible to tension and compression causing piezoelectric materials deformed to generate power. This mode is applied to collecting tiny energy. The cantilever beam structure is widely applied. For a given force, this structure can produce a larger strain and has a lower resonant frequency.

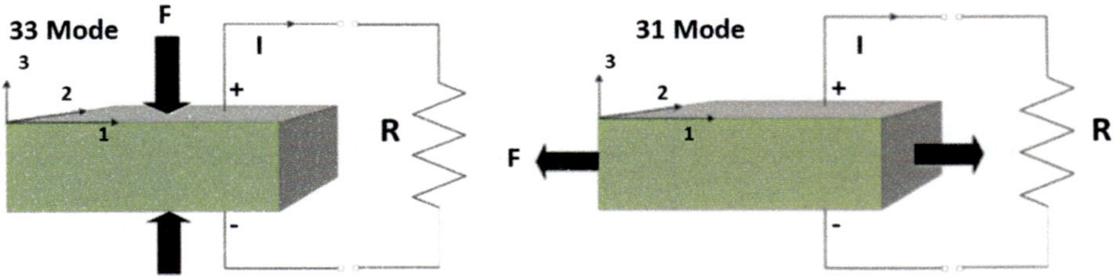

Fig. 1. Illustration of 33 mode and 31 mode operation for piezoelectric material

2.3 Effect of Piezoelectric Stacking

Studies have shown that by mechanically connecting multiple piezoelectric layers in series, and making piezoelectric stacks in series or in parallel on the circuit, the power generation can be increased obviously. The manufacturing technique of piezoelectric stacks adopted is to bond piezoelectric plates through binder. The piezoelectric stack can withstand a greater mechanical force, and the output power it captures is higher than that of the piezoelectric cantilever beam.

In the case of series connection, the total charge output Q' is equal to the charge of single layer Q, and the output voltage U' is equal to n multiply the voltage of single layer U.

$$Q' = Q, U' = nU \tag{2}$$

In the case of parallel connection, the total charge output Q' is equal to n multiply the charge of single layer Q, and the output voltage U' is equal to the voltage of single layer U.

$$Q' = nQ, U' = U \tag{3}$$

3 Literature Survey

3.1 Property of Piezoelectric Materials

The performance of piezoelectric materials has a crucial impact on the efficiency of power generation. According to the structural characteristics, the piezoelectric can be divided into four categories, namely ceramics, polymers, composite materials and monocrystals. In case of ceramics and polymers, the direction of the polarization axis "3" depends on the polarization direction. But crystalline materials, such as aluminum nitride or zinc oxide, the direction of the polarization axis is along the crystal orientation. Currently, the most widely used piezoelectric materials are piezoelectric ceramics, such as PZT and barium titanate. Due to the properties of ceramic, the material is prone to be cracked when subjected to large stresses and cannot produce large deformation, lead to the application limited.

3.2 Design of Piezoelectric Energy Harvester

On account of the simple structure, it is easy to obtain large strain under a given force input, so the piezoelectric cantilever beam structure has been widely used [21]. The resonant frequency of the cantilever beam is much higher than that of the vibration of surrounding environment, thus the resonant frequency is reduced by adding a proof mass, so that the recovered power is greatly increased [22].

The rectangular piezoelectric cantilever beam is the most common structure in piezoelectric energy harvester. Scholars have also conducted a lot of research on energy harvester of other structures and shapes. Baker [23] et al. customized rectangular and trapezoidal piezoelectric cantilever beam. The analysis results show that adjusting the shape of the cantilever beam to make the strain distribution more homogeneous can increase the average strain. The experimental results show that the actual power output of the power generation device of the trapezoidal beam is 30% higher than that of the rectangular beam.

Pradeesh et al. [24] presented the effect of placement of piezoelectric materials over the length of cantilever beam, and the influence of mass material, volume, shape, size and position on power generation efficiency. The analysis found that the material and volume of the proof mass have no significant impact on the power output and resonant frequency, but the placement has a 10.9% influence on the resonance frequency of the beam. Figure 2 shows the rectangular cantilever beam with PZT-5A at fixed at a distance from free end.

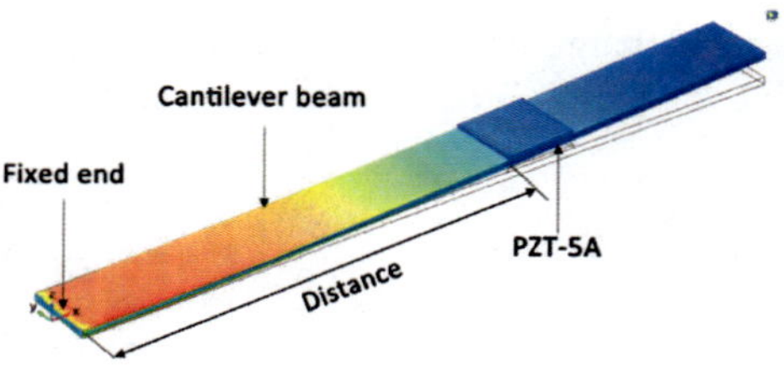

Fig. 2. Rectangular cantilever beam.

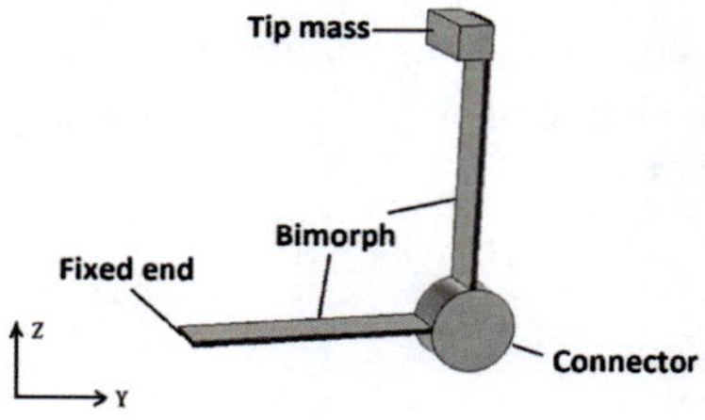

Fig. 3. Schematic of the V-shaped energy harvester

Yue Zhao [25] et al. proposed a new type of V-shaped piezoelectric energy harvester based on the traditional piezoelectric bimorph cantilever structure, as shown in Fig. 3.

The energy harvesting efficiency has been successfully improved, and the resonant frequency of the two piezoelectric bimorph beams can be easily adjusted by changing the angle between the two piezoelectric bimorph beams. Simulations and experiments have concluded that the structure can not only improve the output performance but also adjust the working frequency band conveniently.

Platt [26] et al. compared a homogeneous cylindrical piezoelectric block with a piezoelectric stack of the same volume. The results show that the same-shaped stack and homogeneous piezoelectric elements produce the same power in a matched load circuit. The stack element should be preferred to adopt, as a result it is easier to reduce the output voltage to the expected level.

Kim [27] et al. studied the power generation capacity of a piezoelectric transducer with a "cymbal" structure. The "cymbal" transducer consists of a piezoelectric disc stuck in the middle of two concave metal caps. The metal caps enhance the piezoelectric stack's ability to bear the load and amplify the stress. The structure increases the piezoelectric coefficient and improves the conversion efficiency. However, due to the complicated manufacturing process, the cost is out of control and it is difficult to put into production and application.

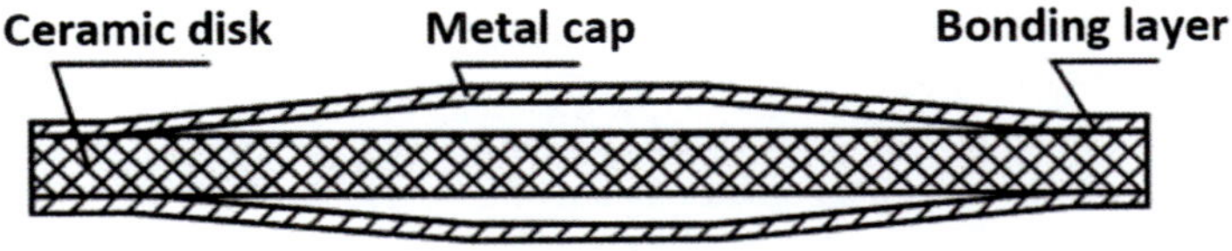

Fig. 4. Structure of "cymbal" transducer

Shahruz [20] designed a broadband piezoelectric energy harvester. This structure can broaden the working frequency by connecting single-degree-of-freedom cantilever beams of different sizes in series on the structure. The results show that, compared with a single cantilever beam, the structure effectively broadens the working frequency band and improves the adaptability to the environment.

3.3 Application of Piezoelectric Energy Harvesting Technology

Han Xiao [17] et al. proposed a dimensionless analysis method to predict the output voltage and power of a quarter vehicle model, and evaluated the performance of the regenerative suspension. The analysis of time domain and frequency domain proved the feasibility of converting vibration energy into electrical energy through piezoelectric materials and applying it to suspension systems. Figure 4 shown a two degree of freedom piezoelectric vibration energy harvesting system model.

Wiwiek [11] et al. studied the design, modeling and analysis of multilayer piezoelectric vibration energy harvester in the suspension. The principle of this structure is to amplify the force applied to piezoelectric bimorph through a specific mechanical structure. The results show that the mechanism can produce 2.75 times the output voltage and 7.17 times the power than directly installed on the suspension (Fig. 5).

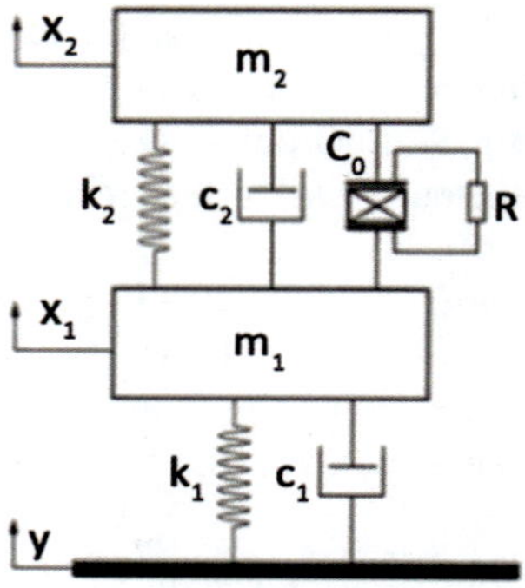

Fig. 5. A two degree of freedom piezoelectric vibration energy harvesting system model.

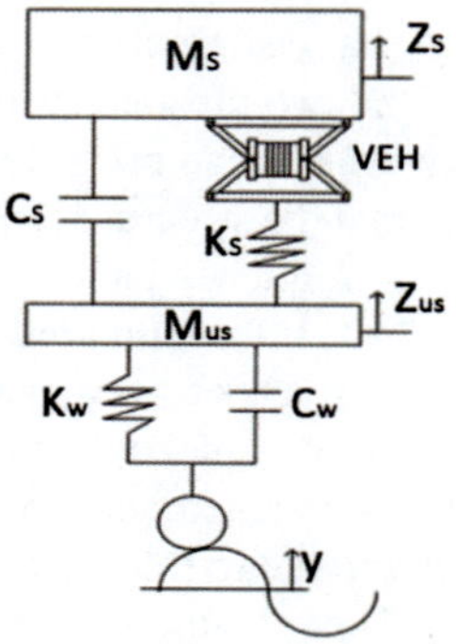

Fig. 6. Schematic of multilayer piezoelectric energy harvesting in the suspension

Furthermore, the influence of piezoelectric generator on suspension system performance needs to be focused emphatically in the following research (Fig. 6).

John [10] et al. studied two types of piezoelectric power-generating shoes [7]. PZT and PVDF are installed on the heel and sole of the foot respectively. Compared with other forms of energy harvesting devices, the electromagnetic power generation device has been tested to be two orders of magnitude higher than the piezoelectric power generation system, but compared to the piezoelectric energy harvester, it is difficult to be integrated into the shoes without affecting the appearance.

Jiang [9] et al. developed a road-based energy recovery device. By inserting piezoelectric materials into the ground, energy is recovered from the vibration caused by traffic. The proposed road harvester uses a set of piezoelectric recovery devices to convert mechanical energy into electrical energy. Each recovery device contains three piezoelectric stacks, and theoretical analysis and experiments prove that the output power depends not only on the device itself, but also on the external resistive load. There is an optimal resistance to make the output power reach the maximum.

Taylor [33] et al. invented a piezoelectric energy harvester that recovers liquid energy, named it the energy harvesting eel, which converts the energy of the liquid flowing in sea or river into electrical energy. The principle is similar to the flag behind the flagpole on a windy day. The piezoelectric material used is PVDF.

4 Discussion and Conclusion

The development and research achievements of piezoelectric generator have been reviewed in this paper. The noteworthy advantage of PEH is the ease of application attributed to the simple structure. According to the literature reviewed so far, the power output provided by piezoelectric generators still can not be used as power supply for mobile electronic devices. Low output power is an important factor restricting the development of PEH, but the improvement prospects are positive. The future work is to design and manufacture a new type of vibration energy recovery device, which can be widely applied in the field of micro-electromechanics, medical monitoring and transportation. Researchers have proposed a series of solutions to the issues discussed above. These solutions have not thoroughly overcome the shortcomings of PEH, but they have provided researchers with more creative ideas and methods. Facts have proved that piezoelectric energy harvesting is a promising development direction in the future.

Acknowledgment. The authors would like to thank support from the National Natural Science Foundation of China, China (No. 51575233) and the National Natural Science Foundation of China, China (No. 51975122).

References

1. Heard, B.P.: Burden of proof: a comprehensive review of the feasibility of 100% renewable-electricity systems. Renew. Sustain. Energy Rev. **76**, 1122–1133 (2017)
2. Bastien, P.: Practical PV energy harvesting under real indoor lighting conditions. Sol. Energy **224**, 3–9 (2021)
3. Yinxiang, B.: Hierarchically structured lead-free barium strontium titanate for low-grade thermal energy harvesting. Ceram. Int. **47**(13), 18761–18772 (2021)
4. Wu, N., Wang, Q., Xie, X.: Wind energy harvesting with a piezoelectric harvester. Smart Mater. Struct. **22**(9), 095023 (2013)
5. Salem, S., Fraňa, K., Nová, I.: Design of acoustic energy harvesting unit using piezo-electric diaphragm. Mater. Sci. Forum **5954**, 109–115 (2020)
6. Nan, W., Wang, Q., Xie, X.: Ocean wave energy harvesting with a piezoelectric coupled buoy structure. Appl. Ocean Res. **50**, 110–118 (2015)
7. Choi, Y.M., Lee, M.G., Jeon, Y.: Wearable biomechanical energy harvesting technologies. Energies **10**, 1483 (2017)
8. Cassidy, I.L., Scruggs, J.T., Behrens, S.: Design of electromagnetic energy harvesters for large-scale structural vibration applications. In: Proceedings of SPIE - The International Society for Optical Engineering, vol. 7977 (2011)
9. Zheng, P., Wang, R., Gao, J.: A comprehensive review on regenerative shock absorber system. J. Vib. Eng. Technol. **8**, 225–246 (2019)
10. Kymissis, J., Kendall, C., Paradiso, J.A.: Parasitic power harvesting in shoes. In: Second International Symposium on Wearable Computers (2002)
11. Hendrowati, W., Guntur, H.L., Nyoman Sutantra, I.: Design, modeling and analysis of implementing a multilayer piezoelectric vibration energy harvesting mechanism in the vehicle suspension. Engineering **4**(11), 728–738 (2012)
12. Dragunov, V.P., Sinitskiy, R.E., Dragunova, E.V.: Single-capacitor electrostatic vibration energy converter based on the bennet doubler. Russ. Microlectron. **50**(3), 178–188 (2021)
13. Muhammad, F.F., Ket, T.C., Daniil, Y.: Structural optimisation through material selections for multi-cantilevered vibration electromagnetic energy harvesters. Mech. Syst. Sig. Process. **162** (2022)
14. de Oliveira, F.A.C., de Lima Monteiro, D.W., Martini, C.D.: Design, modeling, characterization and analysis of a low frequency micro-fabricated piezoelectric cantilever for vibration sensing and energy harvesting applications. Sens. Actuators A Phys. **326** (2021)
15. Yi Jun Min: Energy-aware data compression and transmission range control scheme for energy-harvesting wireless sensor networks. IEMEK J. Embed. Syst. Appl. **11**(4), 243–249 (2016)

16. Ondrej, R.: Development and experimental validation of self-powered wireless vibration sensor node using vibration energy harvester. Mech. Syst. Sig. Process. **160** (2021)
17. Han Xiao, X., Wang, S.J.: A dimensionless analysis of a 2DOF piezoelectric vibration energy harvester. Mech. Syst. Sig. Process. **58–59**, 355–375 (2015)
18. Caliò, R., Rongala, U., Camboni, D.: Piezoelectric energy harvesting solutions. Sensors **14**(3), 4755–4790 (2014)
19. Lu, F., Lee, H.P., Lim, S.P.: Modeling and analysis of micro piezoelectric power Generators for micro-electromechanical-systems applications. Smart Mater. Struct. **13**(1), 57 (2004)
20. Shahruz, S.M.: Design of mechanical band-pass filters for energy scavenging. J. Sound Vib. **292**(3), 987–998 (2005)
21. Saadon, S., Sidek, O.: A review of vibration-based MEMS piezoelectric energy harvesters. Energy Convers. Manag. **52**(2011), 500–504 (2011)
22. Lumentut, M.F., Howard, I.M.: Parametric design-based modal damped vibrational piezoelectric energy harvesters with arbitrary proof mass offset: numerical and analytical validations. Mech. Syst. Sig. Process **68–69**, 562–586 (2016)
23. Baker, J., Roundy, S., Wrigh, T.P.: Alternative geometries for increasing power density in vibration energy scavenging for wireless sensor networks. In: Proceedings 3rd International Energy Conversion Engineering Conference (2005)
24. Pradeesh, E.L., Udhayakumar, S.: Effect of placement of piezoelectric material and proof mass on the performance of piezoelectric energy harvester. Mech. Syst. Signal Process. **130**, 664–676 (2019)
25. Zhao, Y.: Modeling and experiment of a V-shaped piezoelectric energy harvester. Shock Vib. (2018)
26. Platt, S.R., Farritor, S., Haider, H.: On low-frequency electric power generation with PZT ceramics. IEEE ASME Trans. Mechatron. **10**, 240–252 (2005)
27. Hyeoung, W.K.: Energy harvesting using a piezoelectric "Cymbal" transducer in dynamic environment. Jpn. J. Appl. Phys. **43**(9A), 6178–6178(2004)
28. Wei, C., Jing, X.: A comprehensive review on vibration energy harvesting: modelling and realization. Renew. Sustain. Energy Rev. **74**, 1–18 (2017)

An Adaptive Sliding Mode Observation for Vehicle Lateral Force

Shurong Zhou[1], Yunchao Wang[1](✉), Xin Liu[2], and Jinxi Liu[1]

[1] School of Marine Equipment and Mechanical Engineering, Jimei University, Xiamen 361021, Fujian, China
ychaowang@jmu.edu.cn

[2] Mechanical and Automotive Engineering, Xiamen University of Technology, Xiamen 361024, China

Abstract. The tire lateral force is crucial to vehicle lateral stability control which aims at enhancing vehicle handing safety. Accurate observation of lateral force can effectively prevent vehicle side slip and out of control when the vehicle is steering at high speed. However due to high costs of sensors, it is difficult for vehicles to measure the lateral force directly. Therefore, the accurate and robust observation algorithm of lateral force is essential. A co-simulation model of vehicle is built by using the software of CarSim and Simulink, and some simulations are carried out under single-line shifting and double-line shifting conditions. Results show that: compared with the sliding mode observation (SMO) algorithm,the error of adaptive sliding mode observation (ASMO) algorithm is reduced 30.01%. That indicates that ASMO can effectively improve the robustness and accuracy of the system.

Keywords: Tire lateral force · Adaptive sliding mode observer · Compensation algorithm

1 Introduction

The lateral force of vehicle tires is an important index that characterizes the lateral stability of the vehicle, and it is also an important variable that controls the lateral stability of the vehicle [1]. In real life, due to the cost or accuracy of on-board sensors, and the complexity of the algorithm in practical applications, the lateral force measurement of vehicle tires has certain incompleteness and inaccuracy [2, 3], Thus, it is necessary to observe the tire lateral force accurately. A lot of research work has been carried out for the key technologies and difficulties of vehicle lateral stability control when observing tire lateral force. In order to better control the lateral stability of the vehicle, how to improve the accuracy and robustness of the observation algorithm becomes important.

Zong Changfu et al. proposed a vehicle state estimation information fusion algorithm based on EKF. The information fusion system performs hierarchical filtering on the input signal to reduce the impact on road noise and enables better observation of the yaw rate [4]. Xu Jinli et al. introduced adaptive control to the unscented Kalman filter algorithm in observing the side slip angle and yaw rate, so that the lateral velocity and lateral force of the algorithm are in good match with the CarSim output [5]. Lin Fen et al.

Y. Wang et al. (Eds.): IWAMA 2021, LNEE 880, pp. 303–308, 2022.
https://doi.org/10.1007/978-981-19-0572-8_38

designed a tire force observer structure based on UKF. By establishing the mapping relationship between the typical road adhesion coefficient and S1ip-slope, the tire force estimation has a higher accuracy [6, 7]. Wang Qidong et al. proposed a non-dominated sorting genetic algorithm based on an elite strategy for optimization, which improved the stability of car driving [8]. Nie Xiaobo and others used fuzzy PID control algorithm to observe and evaluate the lateral displacement, yaw rate, and lateral acceleration of the entire vehicle. The results show that the real-time and self-adaptability is better, and it is effectively improved the observation accuracy [9].

In the above method, although the filter of the EKF-based information fusion algorithm reflects the smoothing effect on the signal, the deviation of the fusion result of the centroid side deviation angle and the experimental measurement value is slightly larger. When the UKF designs the sample weight distribution, the calculation is more complicated, and it is necessary to assume that the road input noise is white noise. Fuzzy logic is more complicated to calculate the expected deviation membership function of the side slip angle of the center of mass, and it is more cumbersome to adjust the gain value of PID and has little applicability. Therefore, in order to improve the accuracy of the tire lateral force observation algorithm, this paper proposes an adaptive sliding mode observation algorithm and selects two typical working conditions of single-line shifting and double-line shifting for observation verification. The test shows that the method is low in cost and accurate. To meet application requirements, provide theoretical support for the study of tire lateral force observation models under different working conditions.

2 Vehicle Dynamics Model

In order to simultaneously estimate the tire lateral force, the two-degree-of-freedom (2DOF) single-track vehicle model should be established, as shown in Fig. 1.

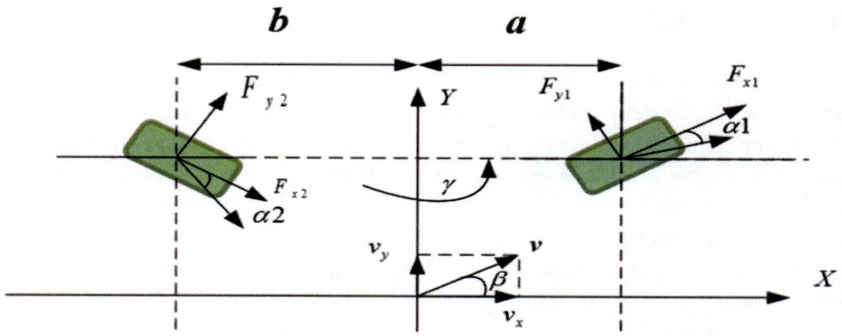

Fig. 1. The single-track vehicle model

$$\begin{cases} m(\dot{u} - v_y\gamma) = F_{x2} - F_{y1}\sin\delta + F_{x1}\cos\delta \\ mu(\dot{\beta} + \gamma) = F_{y2} + F_{y1}\cos\delta + F_{x1}\sin\delta \\ I_z\dot{\gamma} = F_{y1}a\cos\delta - F_{y2}b + F_{x1}a\sin\delta \end{cases} \quad (1)$$

Where: m is the mass of the car; F_{xi} and F_{yi} (i = 1, 2) are the longitudinal force and lateral force of the tire's front and rear axles, a, b are the distance from vehicle gravity center to front and rear axle, respectively.

3 Design of Adaptive Sliding Mode Observer (ASMO) for Lateral Force

3.1 Design of Sliding Mode Observer (SMO)

Sliding mode control has good robustness to system parameter uncertainty and external disturbance [10], so it has a wide range of applications.

Actually, The sliding mode system can be considered as follows:

$$\begin{cases} \dot{x} = Au + Bd + \Gamma \\ y = x \end{cases} \tag{2}$$

Where, x is the state of the system, u is the input of the system, y is the measurement output, and Γ is other system interference item, A, B which are real constants. d is an unknown and bounded input variable. Moreover, it is also the variable which need to be observed, respectively.

This paper selects the error $\sigma = x - \hat{x}$ as the sliding mode surface. Then, we can design the Lyapunov function as follows:

$$V = \frac{\sigma^2}{2} \tag{3}$$

According to the state observer theory, and based on the (2) and (3). The sliding mode observer can be considerd as follows:

$$\begin{cases} \dot{\hat{x}} = Au + B\hat{d} + \Gamma + K(x - \hat{x}) \\ \hat{d} = B^{-1}\varepsilon sign(\sigma) \end{cases} \tag{4}$$

Where, $sign(\sigma)$ is the switch function as follows:

$$sign(\sigma) = \begin{cases} 1, & \sigma > 0 \\ 0, & \sigma = 0 \\ -1, & \sigma < 0 \end{cases} \tag{5}$$

According to the stability of the Lyapunov system, the final observation of the unknown input can be expressed as:

$$\hat{d} = B^{-1}Lsign(\sigma) + B^{-1}K(x - \hat{x}) \tag{6}$$

Where L represents sliding mode gain, K represents feedback gain.

Due to the time delay and system intertia, the SMO may occur the trmble, and big tremble may influences the observation accuracy therefore. In this paper, using the hyperbolic tangent function to replace these discontinuous switching functions can effectively reduce trmble in sliding mode control. Then, the hyperbolic tangent function can be as follows:

$$\tanh(\lambda) = \frac{e^{\lambda} - e^{-\lambda}}{e^{\lambda} + e^{-\lambda}} \tag{7}$$

3.2 Design of Lateral Force ASMO

Based on (1), (4), (6), and (7) The front wheel lateral force sliding mode observer designed in this paper is as follows:

$$\hat{F}_{y1} = \frac{I_Z}{(a+b)\cos\delta} L_{y1} \tanh(\gamma - \hat{\gamma}) + \frac{I_Z}{(a+b)\cos\delta} K_1(\gamma - \hat{\gamma}) \tag{8}$$

Where L_{y1} represents the front wheel lateral force sliding mode gain, K_1 represents the front wheel lateral force feedback gain.

Similarly, the observed value of the rear wheel lateral force can be expressed as:

$$\hat{F}_{y2} = -\frac{I_Z}{(a+b)} L_{y2} \tanh(\gamma - \hat{\gamma}) - \frac{I_Z}{(a+b)} K_2(\gamma - \hat{\gamma}) \tag{9}$$

Moreover, in order to improve the accuracy of observed F_{y1} and e F_{y2}, a adaptive feedback law is designed. Thus, the observed value of lateral force can be expressed as;

$$\hat{F}_{y1} = \frac{I_Z}{(a+b)\cos\delta} L_{y1} \tanh(\gamma - \hat{\gamma}) + \frac{I_Z}{(a+b)\cos\delta} K_1(\gamma - \hat{\gamma}) + \Phi_1 \tag{10}$$

$$\hat{F}_{y2} = -\frac{I_Z}{(a+b)} L_{y2} \tanh(\gamma - \hat{\gamma}) - \frac{I_Z}{(a+b)} K_2(\gamma - \hat{\gamma}) + \Phi_2 \tag{11}$$

$$\Phi_i = \eta_i \left(m a_y - \hat{F}_{y_1} - \hat{F}_{y_2} \right) \quad i = 1, 2 \tag{12}$$

Where η_1 and η_2 are the adaptive observer feedback gain.

4 Simulation and Comparative Analysis

In order to verify the effectiveness of the algorithm, based on the co-simulation of CarSim and Simulink for vehicle dynamics is established. The basic parameters of the vehicle are shown in Table 1:

Table 1. The basic parameters of the vehicle

Parameters	Values
Distance from vehicle gravity center to front *a/m*	1.20
Distance from vehicle gravity center to rear *b/m*	1.59
Vehicle mass *m/kg*	2780
Inertia moment about the vehicle vertical axis *IZ/(kg·m2)*	2280

4.1 Single-Line Shifting Condition

As a typical operating condition for vehicle driving, the single-line shifting condition can effectively stimulate the lateral characteristics of the tire. The road friction coefficient is 0.55 and longitudinal velocity of the vehicle is shown in Fig. 2.

The simulation effect diagram of vehicle lateral force is shown in Fig. 3 and 4.

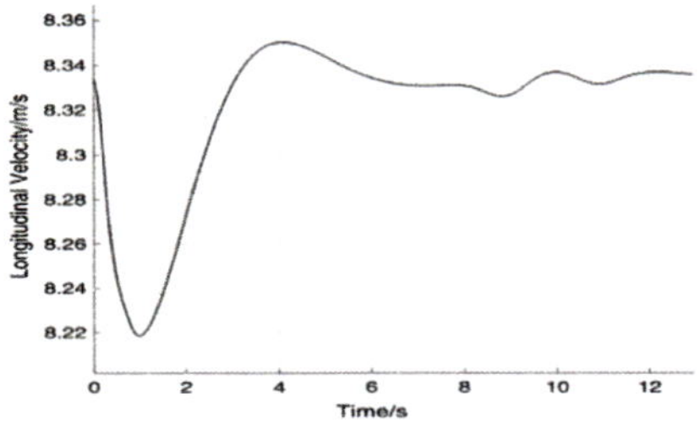

Fig. 2. The longitudinal velocity

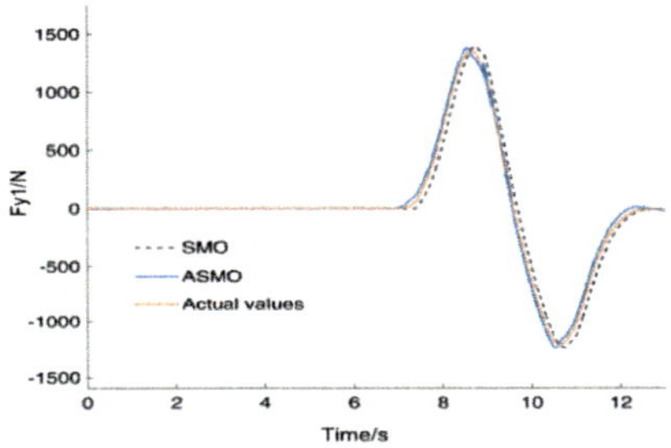

Fig. 3. Front wheel lateral force

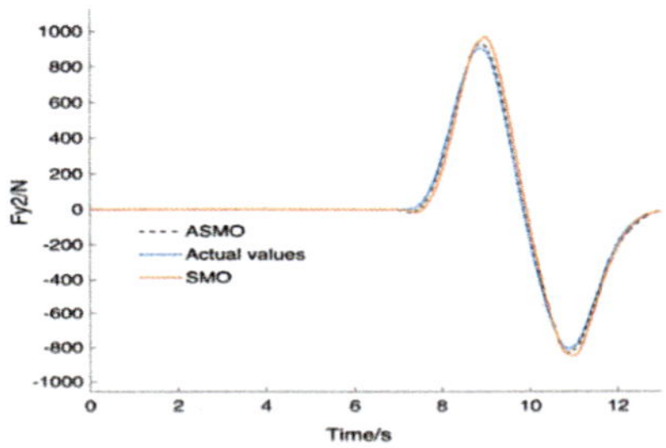

Fig. 4. Rear wheel lateral force

4.2 Double-Line Shifting Condition

In order to further verify the effectiveness of the algorithm, the vehicle double-line-shifting simulation condition was carried out and road friction coefficient is 0.5. The simulation effect is shown in Fig. 5 and Fig. 6.

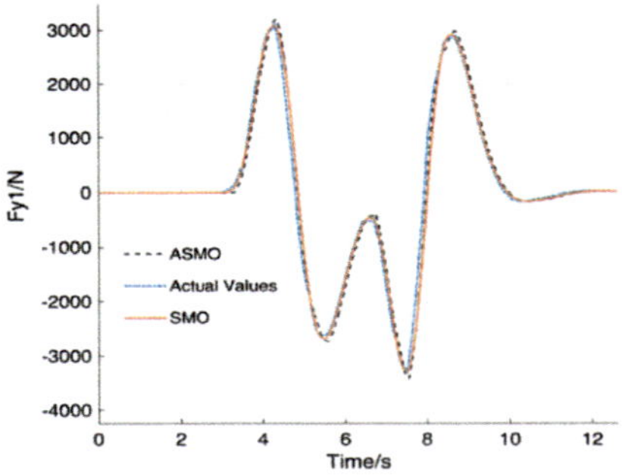

Fig. 5. Front wheel lateral force

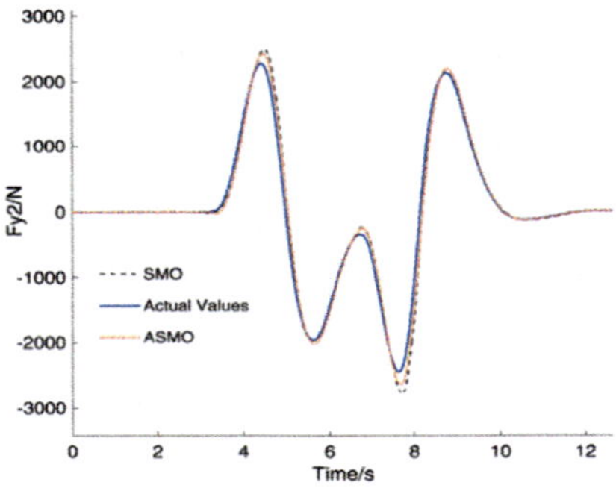

Fig. 6. Rear wheel lateral force

From Fig. 3 to Fig. 6 show that when the vehicle enters a turning state, the sliding mode observation error is significantly higher, and the time lag is also more serious. The errors of the ASMO are 8.68%, 7.09%, 8.44%, 9.27%, respectively, while errors of SMO is larger than that of ASMO, which is 10.77%, 10.13%, 10.24%, 11.55%, respectively.

compared with the SMO, ASMO increases the robustness of the system, it effectively reduce the error of the system, and improve the observation accuracy, respectively.

5 Conclusion

An adaptive sliding mode observation algorithm for vehicle lateral force is proposed, and a vehicle co-simulation is established based on CarSim and Simulink.

By increasing the adaptive compensation, the robustness of the algorithm is enhanced, and the observation accuracy of the algorithm is improved, which can better meet the real-time requirements.

Acknowledgement. The authors would like to thank support from the National Natural Science Foundation of China, China (No. 51575233) and the National Natural Science Foundation of China, China (No. 51975122).

References

1. Zhang, Y.Q., et al.: Vehicle dynamic system simulation and optimization using multibody dynamics. J. Dyn. Control (01), 68–74 (2007)
2. Chen, W.W., Liu, X.Y., Huang, H., et al.: Research on side slip angle dynamic boundary control for vehicle stability control considering the impact of road surface. J. Mech. Eng. **48**(14), 112–118 (2012)
3. Gao, Z.H., Zheng, N.N., Cheng, H.: Soft sensor of vehicle state based on vehicle dynamics and Kalman filter. J. Syst. Simul. **16**(1), 24–26 (2004)
4. Zong, C.F., Pan, Z., Hu, D., et al.: Information fusion algorithm for vehicle state estimation based on extended Kalman filtering. J. Mech. Eng. **45**(10), 272–277 (2009)
5. Xu, J.L., Jiang, Y.J., Wei, Y.Y., et al.: Research on estimation of key parameters of automotive ESP system. Mech. Sci. Technol. Aerosp. Eng. **40**(1), 125–131 (2021)
6. Lin, F., Huang, C.: Unscented Kalman filter for road friction coefficient estimation. J. Harbin Inst. Technol. **45**(7), 115–120 (2013)
7. Zhou, W.Q., Qi, X., Chen, L., et al.: Vehicle state estimation based on the combination of unscented Kalman filtering and genetic algorithm. Autom. Eng. **41**(2), 198–205 (2019)
8. Wang, Q.D., Wang, J.B., Chen, W.W., et al.: EPS and ESP coordinated control strategy based on vehicle driving safety boundary. J. Mech. Eng. **52**(6), 99–107 (2016)
9. Nie, X.B., Xiong, Y., Pan, Y.J.: Multi-condition co-simulations of vehicle stability control via fuzzy PID algorithm. J. Dyn. Control (2021)
10. Ma, Y.J., et al.: Cascade tire-force estimation method based on sliding-mode observer. Inf. Control **45**(02), 177–184 (2016)

Characteristic Analysis of Hydro-Pneumatic Suspension Roll Motion Based on a Novel Real Gas State Equation

Yunchao Wang(✉) and Zixu Li

School of Marine Equipment and Mechanical Engineering, Jimei University, Xiamen 361021, China
ychaowang@jmu.edu.cn

Abstract. Experimental researches show that the gas polytropic process does not follow the ideal gas state equation. And the accuracies of the commonly applied real gas models are limited because the impact of gas hysteresis is neglected. In this paper, a novel real gas state equation considering gas hysteresis is derived. The main factor is the compression ratio of gas and the gas hysteresis time is introduced into this model. Based on that, the dynamic model of roll motion property of suspension system is derived combining with the geometric model of 6 cylinders hydro-pneumatic suspension. Sinusoidal excitation tests under different working conditions are carried out and the simulation and experimental results are compared. The maximum discrepancy of roll moment is 5.3%. The results indicate that the model can accurately reflect the roll property of hydro-pneumatic system.

Keywords: Hydro-pneumatic suspension · Gas hysteresis · Real gas state equation · Dynamic property

1 Introduction

Hydro-pneumatic suspension system have nonlinear characteristics and high energy density so that the application is becoming more extensive. It is generally composed of an accumulator and an oil cylinder. Oil in the cylinder is used as a force transmission medium and inert gas in the accumulator is used as an elastic medium and to absorb excessive force [1].

The multivariate characteristic of the gas in the accumulator is one of the key factors affecting the performance of the overall suspension system. Most of the previous research results are based on the multivariate characteristics of ideal gas. The multivariate index was selected as a constant. And the gas pressure was predicted by multivariate gas equation. However, the real gas behavior in an actual hydro-pneumatic suspension system does not exactly follow the regularity of ideal gas in terms of compressibility and heat-transfer performance, which will cause errors between the performance of prediction models and real suspensions. Many pieces of researches were taken in order to explore

Y. Wang et al. (Eds.): IWAMA 2021, LNEE 880, pp. 309–315, 2022.
https://doi.org/10.1007/978-981-19-0572-8_39

the polytropic process of real gas and to build models that reflect the real gas behavior. Many researches applied different real gas stateequation to fix this problem. Deprez [2] applied Van der Waals equation to establish suspension model. However, the number of coefficients is not enough so that the model cannot accurately predict pressure in a wide range of temperature variations. Yan [3] improved the Van der Waals gas equation, established the gas state equation, and analyzed its advantages and disadvantages in describing real gas behavior. Zhan [4] expanded the application fields of Van der Waals equation by using the molecular motion and molecular mean field theories. Another model that is widely used to reflect real gas characteristics is the Benedict-Webb-Rubin (BWR) equation which was proposed in 1951 by Benedict [5]. This equation contains 8 parameters which make it have the ability to predict gas pressure in a wide range. Els [6] introduced a thermal time constant model into BWR equation and analyzed the effects of time and temperature on gas pressure. And experimental results showed that the thermal time constant would vary with the changes of working conditions. Westhuizen [7] introduced the BWR equation into the hydro-pneumatic suspension model. And the performance of ideal gas models and BWR real gas models were compared in this study. The results showed that the accuracy of BWR model significantly better than the ideal gas isothermal model. Yang [8] also analyzed the influence of the temperature change on the hydro-pneumatic suspension system based on the BWR equation. However, the accuracy of BWR model in high frequency range will decrease and the influence of gas hysteresis on the gas pressure cannot be reflected.

In addition, most of the previous studies on accumulator properties were aimed at piston accumulators. And the working conditions in these researches were different from those on suspension systems in actual application. More researches need to be taken in medium and high vibration frequency range and high pressure.

In this article, the gas pressure in the bladder accumulator is predicted through a novel real gas model considering gas hysteresis. Expressions of gas pressure and output force during compression and expansion of the accumulator with bladder were derived. Combining with the geometric model of 6 cylinders hydro-pneumatic suspension test platform, the dynamic model of vehicle suspension system is deduced. Sinusoidal excitation tests under different frequencies and amplitudes were carried out and the results of roll motion properties were compared with the model simulation results.

2 Modelling of Roll Motion Property

2.1 Modelling of Gas Pressure

The gas pressure in the bladder accumulator is predicted through the gas polytropic process state equation and the multivariate index value is determined by the method in the former study of the research group.

$$P = P_0(V_0/V)^n \tag{1}$$

$$n = 1 + k\eta_t \tag{2}$$

P is the actual gas pressure in accumulator, P_0 is the initial charging gas pressure, V is the actual volume of gas, V_0 is the initial volume of gas. n is the multivariate index.

Generally, the value was selected as a constant between 1–1.4 because the working condition is between isothermal and adiabatic conditions. However, the research group found that the value varies with the change of volume compression ratio η_t. k_1 are the coefficient.

$$\eta_t = -V_t/V_0 \tag{3}$$

$$k = a\eta_t + b \tag{4}$$

V_t is the volume change considering the gas hysteresis time. The time is the lag time of pressure change compared to volume change and it can be obtained by test. a and b are the constants of volume compression ratio under different excitation frequencies, and they can be identified by the experimental data.

2.2 Modelling of Suspension Test Platform

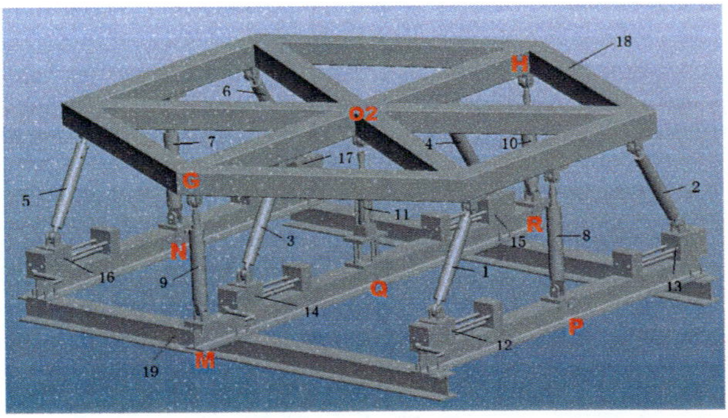

1-6: suspension cylinder 7-10: excitation cylinder

11: guide cylinder 12-17: sliding rail 18: frame body 19: pedestal

Fig. 1. 3D schematic of 6-cylinder vehicle hydro-pneumatic suspension platform

The 6-cylinder vehicle hydro-pneumatic suspension platform is used to simulate different driving conditions. By controlling 4 excitation cylinders in 4 direction, roll, pitch and bounce motion of the suspension under different frequencies and amplitudes. And sensors are applied to detect the displacements of excitation cylinders and the gas pressures in suspension cylinders.

In this article, the vector method is applied to model the platform. This method can simplify the expression of derivation formula. Every node of the platform is given a coordinate and the geometric characteristics of the platform, for example the displacements of nodes and the lengths of force arms, can be evaluated through vector algorithms.

As shown in Fig. 2, the coordinate system was established with Q as the origin. The line where MR located is the x-axis, and the direction from Q to R is positive. The line

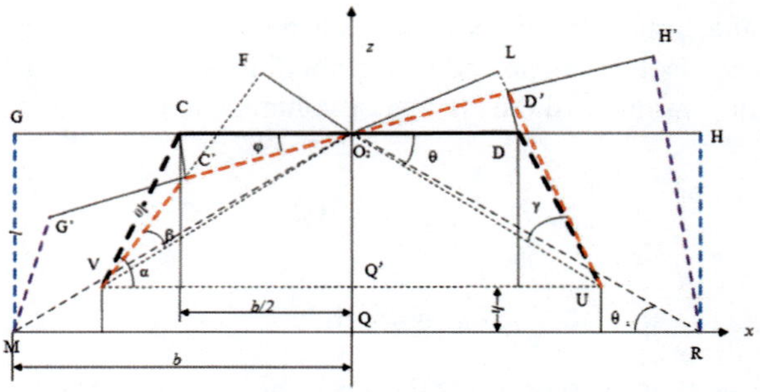

Fig. 2. Platform roll motion diagram (Plane GHRM in Fig. 1)

where PN is located is the y axis, and the direction from Q to N is positive. The line where QO_2 located is the z-axis, the direction from Q to O_2 is positive.

Applying the vector calculation rules, the length variation of suspension cylinder VC x_{i1} was deduced as:

$$x_1 = \left|\overrightarrow{VC}\right| - \left|\overrightarrow{VC'}\right| \tag{5}$$

Assuming d_1 is the arm length of the output force of suspension cylinder VC, which is the length of O2F in Fig. 2. Platform roll motion diagramb, then the formula of it can be deduced as:

$$d_1 = \frac{\left|\overrightarrow{O_2V} \times \overrightarrow{VC'}\right|}{\left|\overrightarrow{VC'}\right|} \tag{6}$$

Applying this method, the length variation of suspension cylinder UD x_{i2} and the arm length d_2 can also be obtained (by Eq. (7) and Eq. (8).):

$$x_2 = \left|\overrightarrow{UD}\right| - \left|\overrightarrow{UD'}\right| \tag{7}$$

$$d_2 = \frac{\left|\overrightarrow{O_2U} \times \overrightarrow{UD'}\right|}{\left|\overrightarrow{UD'}\right|} \tag{8}$$

2.3 The Output Force of Suspension Cylinder and Roll Motion of Suspension System

The output force of single suspension cylinder in the suspension system (shown in Fig. 3) is evaluated by Eq. (9):

$$F_i = (P_{1i} - P_{10i})A_1 - (P_{2i} - P_{20i} + P_{or})A_2 \tag{9}$$

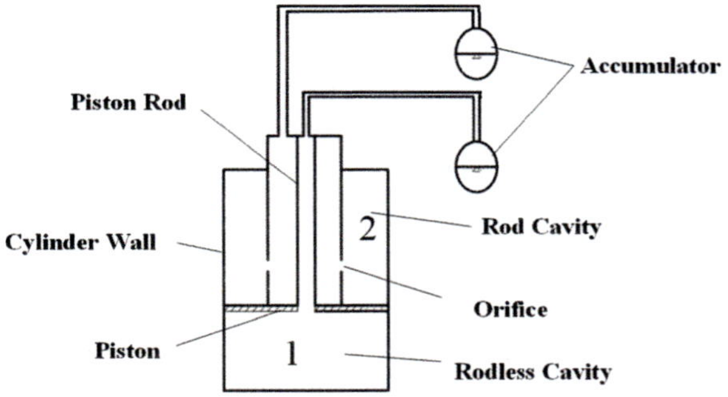

Fig. 3. Dual-chamber cylinder structure

i subscripts the left or right cylinder in Fig. 2. P_{1i} and P_{2i} are respectively the actual pressure of rodless and rod cavity, which can be calculated by Eq. (1). P_{1i0} and P_{2i0} are the initial pressures of corresponding cavities. A_1 and A_2 are the equivalent area of piston. P_{or} is the pressure local pressure loss when hydraulic oil flow through orifice.

The roll motion M_R can be evaluated by Eq. (10):

$$M_R = \sum F_i d_i \tag{10}$$

And the stiffness K_R and damping coefficient C_R can be put as:

$$K_R = \frac{\partial M_R}{\partial \varphi} \tag{11}$$

$$C_R = \frac{\partial M_R}{\partial \dot{\varphi}} \tag{12}$$

φ and $\dot{\varphi}$ are respectively the roll angle and the roll angular velocity.

3 Result Comparison and Discussion

Sinusoidal excitation tests under different working conditions were carried out and the comparison of roll moment between simulation and experimental results are shown below.

From the results shown in Fig. 4, the average discrepancy of moment-angle is 3.26% and the maximum discrepancy is 5.3% in Fig. 4(d). It can be seen that the novel vehicle hydro-pneumatic suspension model can accurately reflect the roll motion properties of the suspension. The hysteresis loop in the predicted moment curve well reflects the influence of the gas hysteresis phenomenon in bladder accumulator on the full-vehicle suspension performance, and the prediction can be verified by test data.

The results indicate that the gas hysteresis has a significant impact on the gas pressure change in the accumulators. Thus, the application of ideal gas state equation can cause error on the properties of hydro-pneumatic suspension system.

Furthering, from the establishment of model it can also be noticed that the hydro-pneumatic suspension system has highly nonlinear characteristics which make it adapt the road conditions well.

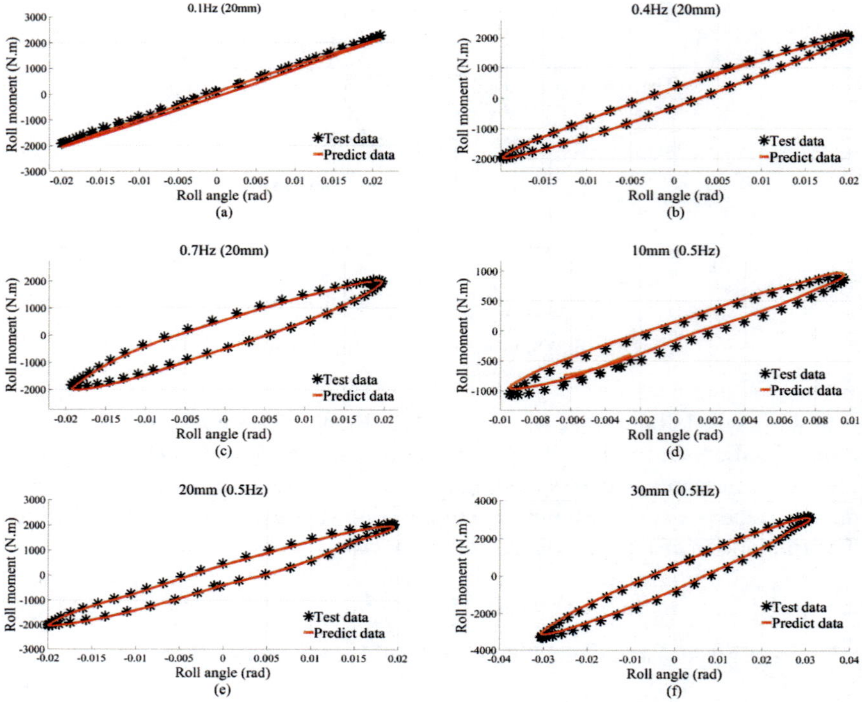

Fig. 4. Comparison of roll moments under different working conditions

4 Conclusion

In this study, a novel model of hydro-pneumatic suspension system considering the influence of gas hysteresis was established and the roll property was predicted and verified by experiments. The results indicate that the actual gas behavior does not follow the rule of ideal gas. The compression ratio of gas in bladder accumulators and the gas hysteresis time are two main factors of the gas pressure. The results show that the novel model can reflect roll property with high accuracy. Furthermore, the model can be well applied to improve the design of multi-axle suspension and the accuracy of suspension control.

Acknowledgement. The authors would like to thank support from the National Natural Science Foundation of China, China (No. 51575233) and the National Natural Science Foundation of China, China (No. 51975122).

References

1. Jiao, N., Guo, J., Liu, S.: Hydro-pneumatic suspension system hybrid reliability modeling considering the temperature influence. IEEE Access **5**, 19144–19153 (2017)

2. Deprez, K., Moshou, D., Ramon, H.: Comfort improvement by passive and semi-active hydropneumatic suspension using global optimization technique. In: Proceedings of the American Control Conference, pp. 8–10 (2002)
3. Yan, J.: A new generalized equation of state for real gases. J. Harbin Inst. Technol. **38**, 802–803 (1980)
4. Zhan, S.: Derivation of Van der Waals equation by molecular mean field theory. Coll. Phys. **28**(02), 3–5 (2009)
5. Benedict, M., Webb, G.B., Rubin, L.C.: An empirical equation for thermodynamic properties of light hydrocarbons and their mixtures: constants for twelve hydrocarbons. Chem. Eng. Prog. **47**, 419–430 (1951)
6. Els, P.S., Grobbelaar, B.: Investigation of the time and temperature dependency of hydro-pneumatic suspension systems, SAE Tech. Paper Ser. **3**, 55–62 (1993)
7. Van der Westhuizen, S.F., Schalk Els, P.: Comparison of different gas models to calculate the spring force of a hydropneumatic suspension. J. Terramech. **57**, 41–59 (2015)
8. Yang, Y., Zhu, H.: Analysis based on benedict-Webb-Rubin equation of state for temperature change characteristics of hydro-pneumatic suspension. Chin. Hydraul. Pneum. **8**, 44–53 (2018)

Research on Disorderly Grasping System Based on Binocular Vision

Hao Feng, Ning Chen(✉), Qinfeng Wang, and Haodong Liu

School of Marine Equipment and Mechanical Engineering, Jimei University, Xiamen 361021, Fujian, China
cn1972@jmu.edu.cn

Abstract. In order to make up for the shortcomings of traditional sorting methods, a disordered sorting system was built based on binocular vision. The system is based on machine vision to realize that when disordered workpieces are sent into the field of view of the binocular camera by the conveyor belt, the image information of the workpiece is first collected by the binocular camera, the 3D point cloud of the workpiece is obtained, the target workpiece is identified and positioned, and the end of the manipulator is obtained. The relative pose relationship with the workpiece can guide the robotic arm to sort and grasp the workpiece in real time, realize the sorting task, can form an automated workpiece sorting detection line, and improve the efficiency of workpiece sorting.

Keywords: Disorderly sorting · Binocular vision · Three-dimensional point cloud · Relative pose

1 Introduction

Machine vision technology refers to the use of cameras to simulate the visual function of the human eye to measure and judge target objects. Vision technology has been more and more widely used in industry, and it plays a vital role in improving production efficiency and achieving the purpose of intelligent production [1]. In recent years, machine vision technology has continued to deepen in the field of automation, which has significantly improved the degree of automation in the world's automobile factories and greatly improved the quality of products put on the world market. Foreign research on industrial robots has been conducted for many years, and the combination of vision technology and robots has been widely used in electronic appliances, automobiles and other fields. Nagata applied genetic algorithm to robot visual servo technology [2] to realize the control of robot operations; University of Western Australia developed Australia's Telerobot robot [3], which is a six-degree-of-freedom robot with a vision system.my country started late in the field of vision, but with the country's strong support for the robotics industry and the concept of "Made in China 2025", as well as domestic colleges, research institutes, and enterprises all participating in research in the field of vision, they gradually made up for it. The technical gap in the field of vision is closed.

Y. Wang et al. (Eds.): IWAMA 2021, LNEE 880, pp. 316–324, 2022.
https://doi.org/10.1007/978-981-19-0572-8_40

As a major automobile manufacturing country [4], China urgently needs to change the traditional manual mode to liberate and develop productivity. The application of machine vision technology to the auto parts sorting system has the advantages of high quality, high efficiency and high intelligence compared with manual sorting. Therefore, the development and research of a sorting system based on machine vision is of great significance to the development of my country's industry.

2 Workflow of the Sorting System

Based on EPSON C4-A601S robot and BASLER industrial camera, this paper designs and builds a set of industrial robot sorting system platform based on binocular vision. The flow chart of the system work is shown in Fig. 1. The first step of the sorting system is binocular camera calibration and hand-eye calibration. When the workpiece is conveyed to the camera's field of view through the conveyor belt, the camera takes pictures of the target workpiece to obtain the workpiece. The image data is transferred to the PC, and the disparity map is obtained through the stereo matching algorithm after image preprocessing. At the same time, the disparity of the corresponding point is used to calculate the stereo information of the object, so as to obtain the three-dimensional point cloud data of the target workpiece and provide the workpiece The position and posture information of the robot is transferred from the PC to the robot controller to guide the robot arm to position the workpiece, and the electric gripper at the end of the robot arm is controlled to track and clamp the target workpiece within its working range [5], Put different kinds of workpieces into the designated position, and the sorting system will repeat the above steps until the sorting task of the workpieces is completed.

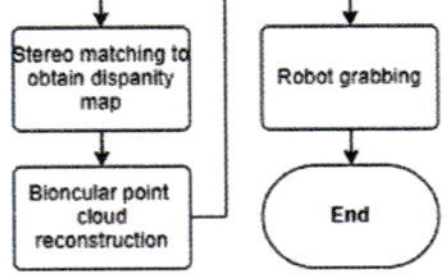

Fig. 1. Unordered fetching system workflow

3 Sorting System Hardware Composition

In this paper, based on Epson robots and binocular cameras, an experimental platform for the sorting system as shown in Fig. 2 is created. The robot sorting experiment platform is mainly composed of a robot system, a vision system, a network switch and a transmission device.

The robot system is mainly composed of an EPSON robot and a control cabinet. At the same time, the robot base is equipped with a rack and pinion sliding table driven by a Leisai motor. In order to realize the grasping of the workpiece, the end effector of the robot is selected as the electric gripper, which is used to grasp the workpiece on the conveyor belt and complete the sorting action.

The vision system consists of two Basler acA2500-14gm industrial cameras, a camera mount and an LED light source, which are fixed above the conveyor belt. Image acquisition is performed on the workpiece that enters the camera's field of view through the conveyor belt to obtain the image information of the workpiece.

The transportation system mainly consists of a belt conveyor and a placement platform. Its main function is for the transportation of sorting objects.

The role of the network switch In the debugging process of the sorting system, in order to solve the problem of a single machine and a single network, the switch is used to complete the information interaction between the PC, the robot and the vision system in the local area network.

The conveyor uses a belt conveyor to transport the sorting objects, and the sorting objects are auto parts.

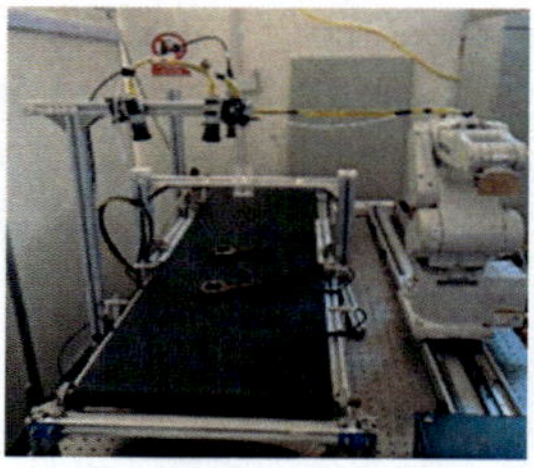

Fig. 2. Experimental platform

4 Binocular Camera Calibration

4.1 Binocular Camera Construction

Ordinary monocular cameras have multiple points corresponding to one point in the projection, and any point on the projection line corresponds to the same image point, and it is impossible to obtain the coordinates of the spatial point in the camera coordinate system, because the monocular camera lacks the depth information of the spatial point. Only the straight line where the space point lies in the camera coordinates can be obtained. The binocular camera can eliminate the many-to-one relationship, and calculate the depth information of the space point according to the coordinates of the same space point in the two camera images, so as to determine the value of the image point on the three-dimensional coordinate Z axis. There are two main ways to build a binocular camera, namely the parallel camera axis and the non-parallel camera axis. The camera axis is not parallel to the construction method, that is, only the two camera axes need to intersect. The construction method is relatively simple but the calculation amount of the three-dimensional coordinates of the solution space is large, so this paper adopts the camera axis parallel type. The parallel construction of the camera axis can get a public field of view, the geometric distortion error is small, and it is relatively easy to calculate the three-dimensional coordinates of the space, and the parallel construction is also easier to install, so that the angle between the two camera axes is 0°. The model of the camera axis parallel construction scheme is shown in Fig. 3 below.

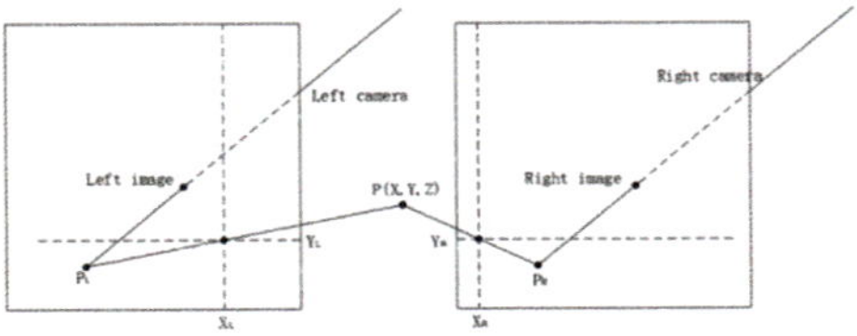

Fig. 3. Parallel axis design scheme model

4.2 Calibration of Binocular Camera

The purpose of camera calibration is to determine the relationship between the three-dimensional geometric position of a point in the space object table and the corresponding point in the image. A geometric model must be established. The geometric model is the camera parameter. The calibration process is to find the camera's internal parameters, external parameters, and distortion. The process of parameters. Distortion parameters and internal parameters are inherent characteristics of the camera itself, and can be used all the time after one calibration.

The process of binocular camera calibration is as follows: first load the calibration file of the calibration board used, and then use the monocular camera calibration operator to perform the internal and external camera parameter calibration of the binocular system. After calibrating the two cameras, the internal parameters of the left and right cameras are obtained. The calibration parameters are shown in Table 1 below. Then use the two installed cameras to shoot 20 pairs of images, select 12 pairs of images with better image quality, and then set the initial parameter values of the two cameras, such as focal length, pixel size, etc. Load all the images of the calibration board that meet the standards into the dual target calibration model, call the segmentation calibration board area operator and extract the two-dimensional mark point operator to obtain the pose relationship and circular mark points of the calibration board in the left and right cameras The coordinates and other information are shown in Table 2, and the obtained parameters are saved in the dual target fixed model [6].

The two maps generated by the calibration of the monocular camera generate a new conversion map, which describes the image of a binocular camera pair to a common corrected image. Then select a group of image pairs to perform correction experiments

Table 1. Calibration results of internal parameters of left and right cameras

Camera	Camera internal parameters					
	f	k	Sx	Sy	Cx	Cy
Left camera	0.008120	−1068.42	2.2e−006	2.2e−006	1358.47	1003.59
Right camera	0.008046	−1046.36	2.2e−006	2.2e−006	1361.30	1009.28

Table 2. Right camera position relative to left camera position

Position	Rx (°)	Ry (°)	Rz (°)	Tx (m)	Ty (m)	Tz (m)
	0.319591	0.728335	0.181898	0.038982	0.000764	0.00720

to obtain the relative pose parameters between the two cameras, as shown in Table 2. The results of calibration before and after correction [7] obtained at the same time are shown in Fig. 4 and Fig. 5. The purpose of the correction is to make the projection of the two pictures on the imaging plane only have the parallax in the left and right directions, so as to prepare for the spatial pose extraction.

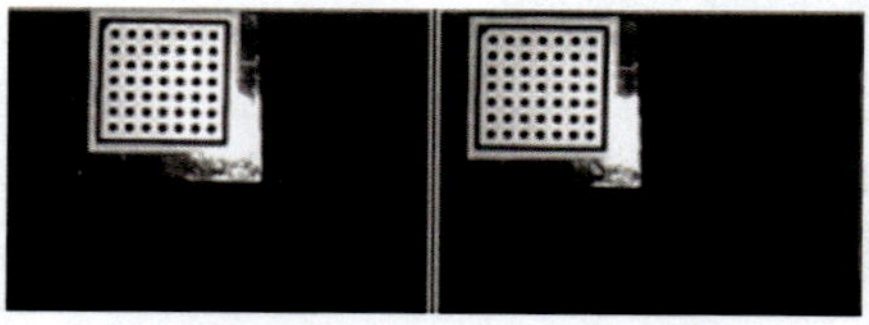

Fig. 4. Calibration plate before correction of the image

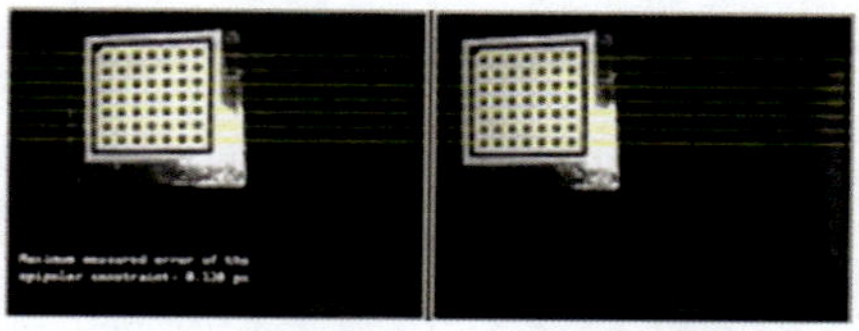

Fig. 5. Modified image of calibration plate

4.3 Hand-Eye Calibration

Hand-eye calibration is an important prerequisite for realizing robot hand-eye coordination [8]. The so-called hand-eye calibration is to unify the coordinate system of the robot and the coordinate system of the vision system, so that the posture of the object determined by the vision system can be transformed into the robot coordinate system, and the robot drives the end effector to complete the work on the object. Common hand-eye systems include Eye-to-Hand and Eye-in-Hand. In this paper, Eye-to-Hand calibration system is adopted. The position of the camera and the robot base is relatively fixed. The binocular camera is placed on the conveyor belt through a fixed frame and does not move with the robot arm, which is convenient for positioning the workpiece target in the camera's field of view.

In the Eye-to-Hand system, it is mainly to solve the robot and make the conversion relationship between the coordinate system and the camera coordinate system [9]. The specific calibration process is as follows: First, establish the coordinate system, and use the base of the manipulator as the coordinate origin. The direction is the Z axis, and the transverse direction is the X axis. According to the right-hand screw rule, the Y axis is perpendicular to the X and Z axes to establish a base coordinate system (the base of the robotic arm) [10]. By analogy, a tool coordinate system is established on the manipulator tool flange, the Z axis is vertically downward, and a hand coordinate system is established on the gripper, with the Z axis vertically upward. Establish a (camera) cam coordinate system with the CCD target surface of the camera as the reference surface. The Z axis is perpendicular to the CCD target surface, the X axis is the image left, and the Y axis is perpendicular to the XZ axis. During the calibration process, guide the gripper of the manipulator to grab the calibration board and move it to obtain the coordinates of the calibration board in different poses relative to the base. The center of the calibration board is the origin, the Z axis is vertical downwards, and XY is the plane of the calibration board, according to the calibration board. Corner information is established (calibration board) cal coordinate system. The two-target calibration can obtain the coordinates of the calibration plate relative to the cam, so as to obtain the conversion relationship between the cam and the base coordinate system. By extracting the calibration board information and the manipulator coordinates one-to-one correspondence, a matrix transformation is completed, and the hand-eye calibration is completed. Figure 6 is a schematic diagram of the coordinate system of Eye-to-Hand [11].

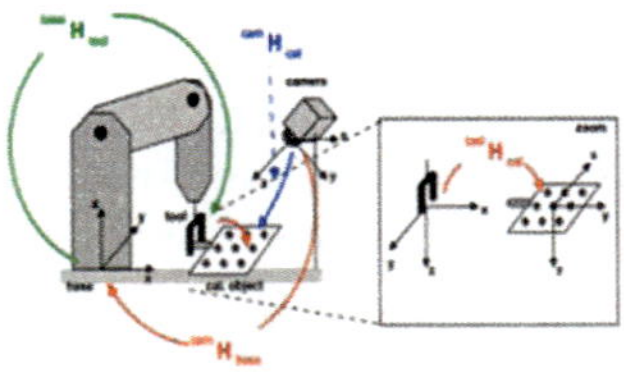

Fig. 6. Eye-to-Hand coordinate system [12]

5 Binocular Point Cloud Reconstruction

5.1 Acquisition of 3D Point Cloud

The three-dimensional point cloud acquisition used in this article uses the binocular stereo parallax method, which uses a binocular camera with parallel axes to shoot the workpiece. Since the binocular camera and the workpiece conform to the principle of triangulation, two images taken at different positions are acquired. When a proper number of workpiece image pairs are collected, the 3D information of the target scene is measured according to the principle of parallax and triangulation [13], so as to obtain the 3D point cloud data of the workpiece.

5.2 Three-Dimensional Display

Enter each set of camera calibration parameters to generate a binocular corrected map, and use a multi-grid algorithm to calculate the parallax of the corrected stereo pair for a pair of images. Finally, four different modes of three-dimensional display are carried out. As shown in Fig. 7 below, a is displayed as a closed surface and is colored by texture mapping. The texture is passed as the second channel (for grayscale value texture) or the second to fourth channel (for color texture) of the image to be displayed; b is displayed as a closed surface, and the current LUT (LUT Up) is used. Table coloring; c is displayed as a hidden line graph, and is colored using the current LUT. d is the height line is extracted and displayed at their actual height, these lines are colored using the current LUT.

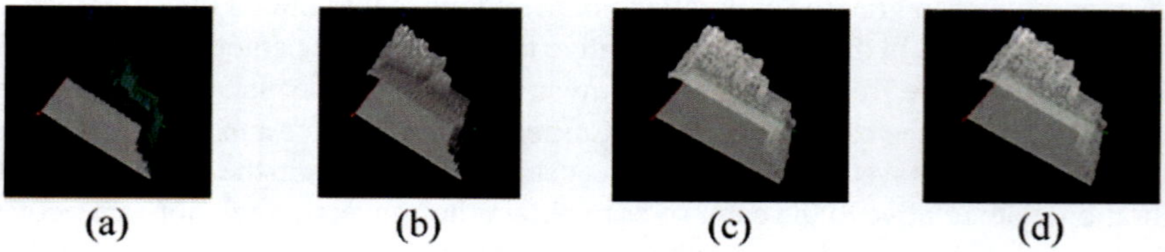

Fig. 7. Four different modes of 3D display

6 Positioning and Crawling

The robot arm is connected with the host computer through TCP/IP communication, and TCP/IP communication is realized by setting the IP address of the host computer and the IP address of the controller in the same network segment. First, obtain the three-dimensional position and posture information of the workpiece through the vision system collection map and transfer it to the PC end of the upper computer. The QSocket module of the upper computer QT software sends commands and accepts the return data to the robotic arm. After the robotic arm controller receives the data, After processing the data, the robot arm is controlled to complete the related movement. At the same time, write a program on the Epson manipulator to accept the command from the host computer and execute the movement under this command. After the manipulator completes the motion task, the state data of the manipulator is returned to the host computer.

The end-of-manipulator gripper-electric gripper communicates with the host computer through RS23 [14]. When the manipulator reaches the target position, QT receives the return data, judges whether it has reached the target position, and then guides the manipulator to move to the target workpiece position. Grab the workpiece. The following Fig. 8 shows the three-dimensional positioning map of auto parts.

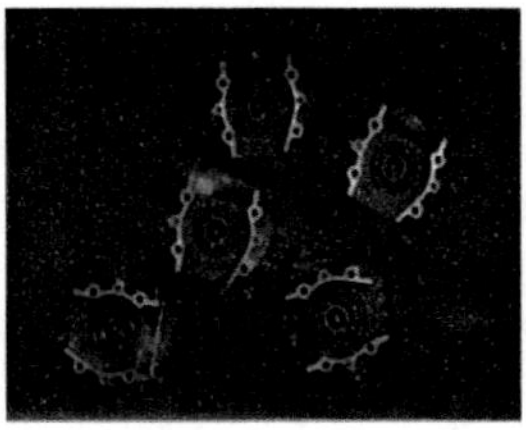

Fig. 8. 3D positioning diagram of auto parts

7 Conclusion

In summary, this article mainly analyzes the application of binocular vision in the sorting system, designs a parallel axis camera installation plan, and performs internal and external parameter calibration of the binocular camera and adopts Eye-to-Hand Hand-eye calibration method. The influence of different parameters in the stereo matching algorithm on the matching result is analyzed, and the image segmentation method and the image processing method of the workpiece pose detection are studied, and the pose detection of the messy workpiece is realized. Compile the software part of the host computer and build the hardware experimental system, and conduct experiments on the robotic arm grabbing and sorting. The experimental results show that the system can achieve a higher level of automatic detection and sorting of parts, which is of great significance for reducing enterprise costs, improving product quality and reducing staff labor intensity.

References

1. Zhang, J.: Research and design of real-time detection scheme of zipper tooth number distribution based on machine vision. Doctoral dissertation, Shantou University (2006)
2. Murakami, S., Takemoto, F., Fujimura, H., Ide, E.: Weld-line tracking control of arc welding robot using fuzzy logic controller. Fuzzy Sets Syst. **32**(2), 221–237 (1989)
3. Nagata, T., Konshi, K., Zha, H.B.: Cooperative manipulations based on genetic algorithms using contact information. In: International Conference on Intelligent Robots & Systems. IEEE (1995)
4. Ji, Z.: Manufacturing – the main direction of "made in China 2025". China Mech. Eng. **26**(17), 2273–2284 (2015)
5. Shiyu, W., Hu, L., Yilan, S., Pin, W.: Design and implementation of robot sorting system based on machine vision. Comb. Mach. Tool Autom. Process. Technol. **3**, 125–129 (2017)
6. Xie, N.: Research on key technologies of sorting system based on binocular vision. Doctoral dissertation, Harbin Institute of Technology (2006)
7. Lin, Y.: Research on target localization and hand eye calibration technology for 3D disordered grasping. Doctoral dissertation, Yantai University (2020)
8. Angeles, J., Soucy, G., Ferrie, F.P.: The online solution of the hand-eye problem. IEEE Trans. Rob. **16**(6), 720–731 (2000)
9. Cheng, Y.: Robot hand eye calibration and object positioning for industrial applications. Doctoral dissertation, Zhejiang University (2016)
10. Xu, G.: Research on robot guidance and positioning technology based on point cloud processing. Doctoral dissertation, Jiangnan University (2020)

11. Desmurget, M., Pélisson, D., Rossetti, Y., Prablanc, C.: From eye to hand: planning goal-directed movements. Neurosci. Biobehav. Rev. **22**(6), 761–788 (1998)
12. Liu, P.: Research on robot 3D vision system calibration and target recognition technology. Doctoral dissertation, Huazhong University of Science and Technology (2006)
13. Jia, Y.: Research on target recognition technology based on point cloud. Doctoral dissertation, North China University (2006)
14. Liu, Z., Li, Z., Zhao, X., Zou, F.: Research on industrial robot sorting technology based on machine vision. Manuf. Autom. **26**(17), 25–30 (2013)

A Deep Learning Based Object-Level Semantic Loop Closure Detection Algorithm

Qilin Li[1], Qi Shang[2], Ning Chen[1](✉), and Qinfeng Wang[1]

[1] School of Marine Equipment and Mechanical Engineering, Jimei University, Xiamen 361021, Fujian, China
cn1972@jmu.edu.cn

[2] Kehua Data Co., Ltd., Huli District, Xiamen, Fujian, China

Abstract. SLAM (Simultaneous Localization and Mapping) is a key technology for mobile robots to have spatial perception capabilities. At the same time, the rapid development of deep learning has led to the maturation of technologies such as target detection and semantic segmentation, which provide technical support for robots to acquire semantic perception capabilities. In this paper, we design a closed-loop detection algorithm based on object semantic information, using YOLO v3 to extract the semantic information and location information of objects in historical key frames and current frames, filtering out interfering objects with a static semantic library, and designing a similarity calculation function to complete the calculation of similarity. The experiments show that the method can complete the closed-loop detection while excluding the interference of dynamic objects.

Keywords: Deep learning · Visual SLAM · Loop closure detection · Semantic information

1 Introduction

The purpose of SLAM is to eliminate the accumulated errors in the visual odometry and back-end calculation parts of the robot after a long period of motion and to detect loops by obtaining the similarity of the images. The back-end optimisation algorithm then optimises the loop closure detection information to achieve more accurate positioning and consistent global mapping.

In the early days feature points would be used for closed-loop detection, but such loop closure detection methods are too computationally expensive to be used in real-time SLAM systems [1–3]. Currently, a widely used algorithm in the loop closure detection problem is based on BOW (Bag of Word) [4–7]. However, BOW requires a large amount of memory to store visual words. Therefore this approach only works very well when performed under conditions where the environment is known. And these artificially designed feature points are often very sensitive to changes in light intensity [8].

In recent years, deep learning techniques are also evolving and deep learning is making a splash in the field of image recognition. The loop closure detection technique is essentially a feature matching problem between images, which is an image processing

Y. Wang et al. (Eds.): IWAMA 2021, LNEE 880, pp. 325–331, 2022.
https://doi.org/10.1007/978-981-19-0572-8_41

problem, so this paper uses deep learning techniques to improve the loop closure detection algorithm. This paper uses deep learning techniques to improve the loop closure detection algorithm so that it can overcome the above disadvantages.

2 YOLO v3 Network Structure

YOLO v3 [9] is an end-to-end target detection algorithm implemented on the darknet network architecture, as shown in Fig. 1 for the YOLO v3 network structure.

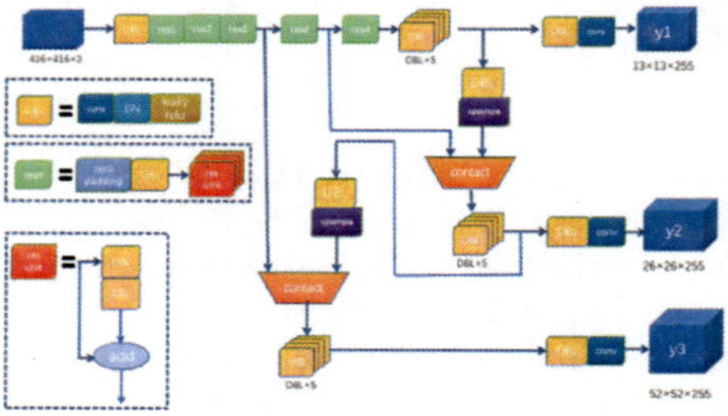

Fig. 1. YOLO v3 structure diagram

The detection process of YOLO v3 consists of the following steps. First, the input image is scaled to a specified scale in order to meet the requirements of the network architecture. The image is then divided into different cells, where each lattice is responsible for the object whose centre point falls on that lattice. Finally, to prevent multiple grids from responding to the same object, YOLO v3 rejects unwanted results by non-maximal value suppression. The non-maximum suppression first obtains the target frame with the highest confidence level, then calculates the IOU between the other target frames and that target frame, and rejects the corresponding target frame when the IOU is greater than a certain threshold, resulting in a target frame with no overlap and the highest confidence level.

3 Deep Learning Based Loop Closure Detection Algorithm Design

3.1 Effective Semantic Information Extraction

After the image information collected by the camera is processed by the network, the semantic information of the current frame is obtained, including the object category contained in the current frame, the confidence level of a single object, and the position of the object in the picture. In this paper, object information with a confidence level greater than 70% is selected to characterize the current picture and improve the fault tolerance rate of the YOLO v3 target detection network. The filtered semantic information in the picture is formed into a string, and the current picture is judged whether the current picture is the same as the picture in the historical key frame by comparing the string, so as to judge whether the types and quantities of objects contained in the two pictures are the same. The valid information database object types are shown in the Table 1.

Table 1. Types of objects in the active information base

Name of object type				
Computer	Desk	Table	Fridge	Chair
Lamp	Trashcan	Side table	Door	Bookshelf
Keyboard	Window	Computer monitor	Sofa	Printer
Bench	Clock	Tv	Telephone	Bed

3.2 Closed Loop Detection

After completing the semantic information extraction and similarity matching, the system will make a decision on whether the current image is closed-loop to complete the closed-loop detection: if the similarity between the current frame and the historical keyframe is greater than 0.7, it will be judged as closed-loop and the semantic information will be discarded, otherwise it will be regarded as historical keyframe and its semantic information will be saved (Fig. 2).

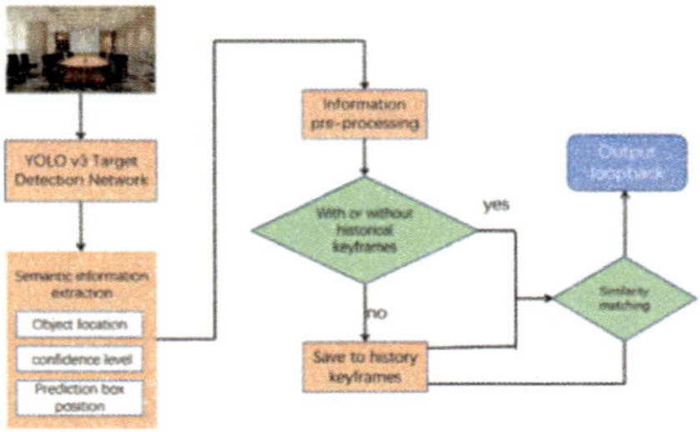

Fig. 2. Loop closure detection flow

The flow chart of the loop closure detection algorithm for YOLO v3 is shown in Fig. 9.

4 Experimental Results of Closed-Loop Detection Algorithms Based on Semantic Information About Objects

Three control groups were set up and tested in similar positions, different lighting and dynamic scenes, and were recorded as control group 1, control group 2 and control group 3. The results were recorded as follows.

4.1 Proximity Groups

The average IOU is 0.48, which is below the set threshold of 0.7 and is not considered to be a loop closure detection. The results show that the network can exclude the false recognition of loop closure detection when the two shooting points are close to each other and the same scene is shot (Tables 2 and 3 and Figs. 3 and 4).

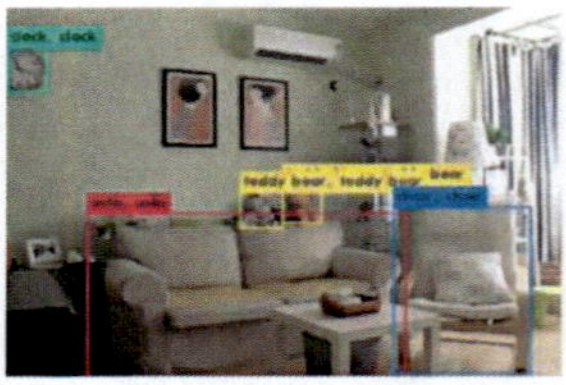

Fig. 3. Historical frame semantic information

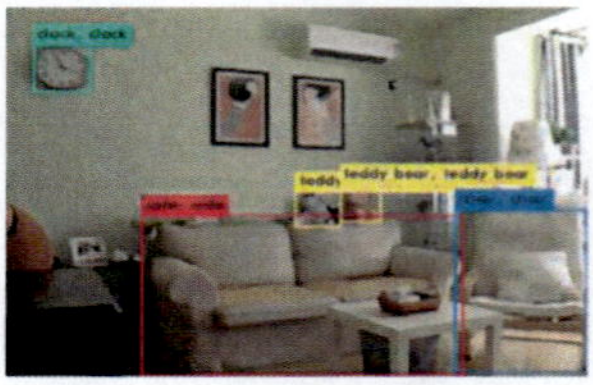

Fig. 4. Current frame semantic information

Table 2. Semantic information for historical frames and current frames

Category name	Historical position coordinates	Current position coordinates
Clock	(0.04 0.17 0.07 0.12)	(0.10 0.17 0.10 0.13)
Sofa	(0.43 0.78 0.58 0.46)	(0.51 0.79 0.55 0.45)
Chair	(0.82 0.76 0.25 0.45)	(0.89 0.77 0.23 0.44)

Table 3. Average IOU and IOU values for various objects

Category name	Calculation of IOU	The average IOU
Clock	0.16	
Sofa	0.73	
Chair	0.56	0.48

4.2 Dynamic Object Groups

The image below shows the image taken when the camera position is constant and the environment changes, the left image is the history frame and the right image is the current frame (Tables 4 and 5 and Figs. 5 and 6).

Fig. 5. Historical frame semantic information

Fig. 6. Current frame semantic information

Table 4. Semantic information for historical frames and current frames

Category name	Historical position coordinates	Current position coordinates
Bed	(0.53 0.57 1.02 0.90)	(0.54 0.57 1.00 0.96)
Chair	(0.22 0.40 0.14 0.38)	(0.22 0.40 0.15 0.40)
Handbag	(0.48 0.24 0.11 0.27)	–

Table 5. Average IOUs and IOUs for each type of object

Category name	Calculation of IOU	The average IOU
Bed	0.92	
Chair	0.89	0.905

The table above shows that the average IOU of the above control group is 0.905, which is higher than the set threshold 0.8, which is higher than the set threshold, and is judged to be a loop closure detection. The results show that the network can reject dynamic objects in order to exclude the interference of dynamic objects in the scene to loop closure detection.

4.3 Light Variation Group

The image below shows the camera at the same location at the same shooting point under the change of ambient light, the left image as the history frame, the right image as the current frame (Tables 6 and 7 and Figs. 7 and 8).

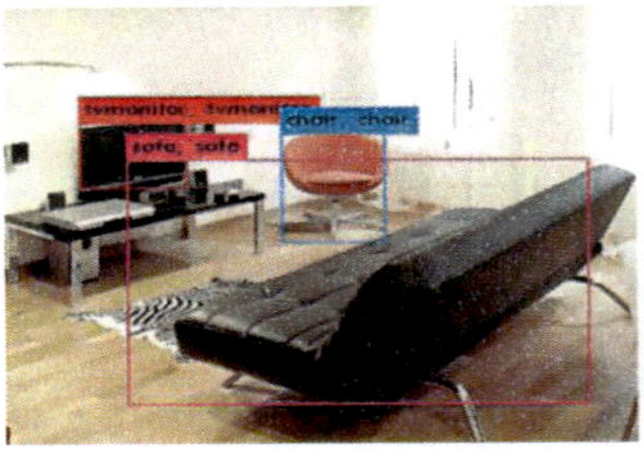

Fig. 7. Semantic information of historical frame 3

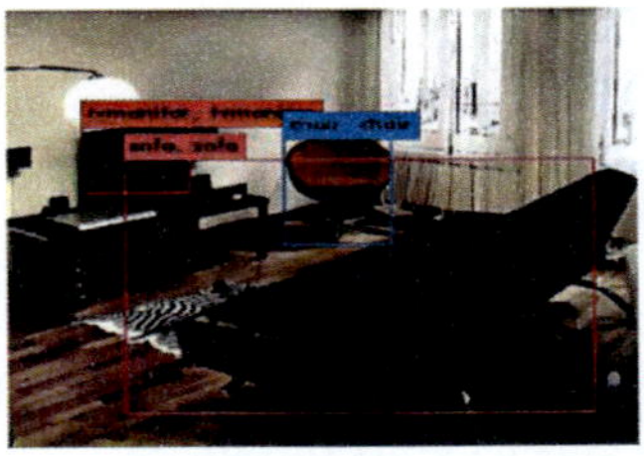

Fig. 8. Semantic information of the current frame 3

From the above table, it can be seen that the average IOU of the above control group can be calculated to be 0.95, which is higher than the set threshold and is judged to be a loop closure detection. The results show that the network can avoid the effect of light intensity changes on loop closure detection.

The IOU values for each control group are summarised below.

Table 6. Semantic information for historical frames and current frames

Category name	Historical position coordinates	Current position coordinates
Tv monitor	(0.20 0.34 0.17 0.15)	(0.20 0.34 0.17 0.15)
Sofa	(0.56 0.63 0.73 0.56)	(0.56 0.63 0.74 0.58)
Chair	(0.52 0.42 0.16 0.24)	(0.52 0.42 0.17 0.25)

Table 7. Average IOU and IOU for each type of object

Category name	Calculation of IOU	The average IOU
Tv monitor	1	
Sofa	0.95	0.95
Chair	0.90	

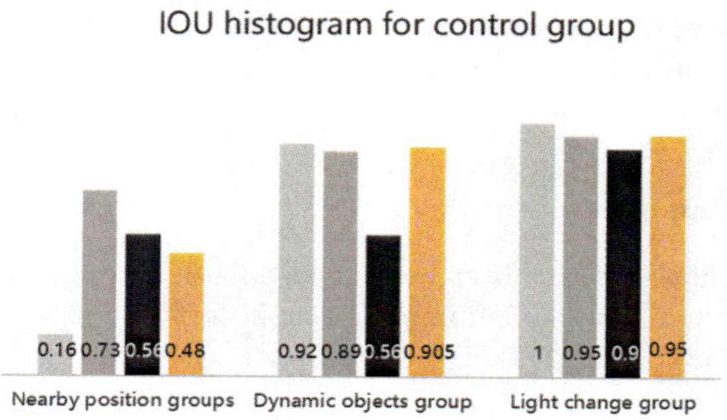

Fig. 9. IOU histogram for each control group

5 Conclusion

In this paper, an object-level closed-loop detection method is designed. Experimental results show that this method can effectively detect the loop closure detection and, by introducing a library of valid semantics, can exclude the interference of dynamic objects in the scene. At the same time, the object detection method based on deep learning, free from the dependence on stable light intensity, can also have better determination results when the light changes, and is more stable and accurate than the traditional BOW method.

References

1. Lowe David, G.: Distinctive image features from scale-invariant keypoints. Int. J. Comput. Vis. (2004). https://doi.org/10.1023/B:VISI.0000029664.99615.94

2. Bay, H., Ess, A., Tuytelaars, T., Gool, L.: Speeded-Up Robust Features (SURF). Comput. Vis. Image Underst. **110**, 346–359 (2008)
3. Li, S., Zhang, T., Gao, X., Wang, D., Xian, Y.: Semi-direct monocular visual and visual-inertial SLAM with loop closure detection. Robot. Auton. Syst. **112**, 201–210 (2019)
4. Stumm, E., Mei, C., Lacroix, S., Nieto, J., Hutter, M., Siegwart, R.: Robust visual place recognition with graph kernels. In: 2016 IEEE Conference on Computer Vision and Pattern Recognition (CVPR), pp. 4535–4544 (2016)
5. Cummins, M., Newman, P.: FAB-MAP: probabilistic localization and mapping in the space of appearance. Int. J. Robot. Res. **27**, 647–665 (2008)
6. Gálvez-López, D., Tardós Juan, D.: Bags of binary words for fast place recognition in image sequences. IEEE Trans. Robot. **28**, 1188–1197 (2012)
7. Garcia-Fidalgo, E., Ortiz, A.: iBoW-LCD: an appearance-based loop-closure detection approach using incremental bags of binary words. IEEE Robot. Autom. Lett. **3**, 3051–3057 (2018)
8. Hou, Y., Zhang, H., Zhou, S.: Convolutional neural network-based image representation for visual loop closure detection. In: 2015 IEEE International Conference on Information and Automation, pp. 2238–2245 (2015)
9. Redmon, J., Farhadi, A.: YOLOv3: an incremental improvement. ArXiv arXiv:1804.02767 (2018)

Densely Connected Image Classification Algorithm Combining with Self-attention

Yupeng Chen, Ning Chen(✉), and Qinfeng Wang

School of Marine Equipment and Mechanical Engineering, Jimei University, Xiamen 361021, Fujian, China
cn1972@jmu.edu.cn

Abstract. In each dense block of DenseNet, the outputs of all previous layers are stitched in the channel dimension as the input of the next layer, but the correlation of features between different channels is not fully considered, and a large number of redundant features are also generated, leading to overfitting of the network. Therefore, an image classification algorithm combining dense connection and attention mechanism is proposed. The algorithm improves the expression ability of features in the process of downsampling through Steam Block; at the same time, it uses branches with different perceptual fields to extract features at different scales, which improves the discriminative ability of the model for multi-scale objects; finally, it introduces a self-attention mechanism to learn the interdependence between different feature channels and adaptively reduces the interference of non-essential features as well as noise. Using the improved DenseNet algorithm to test on the Cifar-10 dataset, the experimental results show that the classification accuracy can reach 94.31%, which is 2.21% higher than that of classical DenseNet, speeding up the model fitting speed, enhancing the generalization ability of the model and improving the classification effect.

Keywords: Image classification · Self-attention · DenseNet

1 Introduction

Image classification refers to determining the category to which an image belongs by some classification algorithm, and is a popular direction of research in the field of computer vision [1]. In the past decade, deep learning has driven the rapid development of artificial intelligence technology, and image classification algorithms based on deep convolutional neural networks can quickly obtain the underlying and deep features of targets and obtain a level of image classification beyond human cognition, providing new ideas for target detection, semantic segmentation, pose estimation, and other fields, which have been widely used [2–4].

In traditional deep neural networks, as the depth of the network deepens, the gradient explosion, gradient disappearance and network degradation problems tend to occur in the backpropagation process, leading to network difficulty in convergence [5]. Instead of using ResNet [6] to deepen the number of network layers and Inception [7] to widen

Y. Wang et al. (Eds.): IWAMA 2021, LNEE 880, pp. 332–340, 2022.
https://doi.org/10.1007/978-981-19-0572-8_42

the network structure to improve the network performance, the DenseNet proposed by GAO enhanced the information flow of features between networks by setting up feature multiplexing and bypass connections to effectively alleviate the degree disappearance problem, which enriches the scale of features by learning to acquire features of different sizes through branches with different perceptual fields [8].

In recent years, some researchers have directly added attention modules to the network for direct training [9] or used existing networks as a basis for fusing attention [10], adaptively adjusting the weights of each channel by learning methods to verify the performance and accuracy of the network.

2 Improved DenseNet Network

As shown in Fig. 1, DenseNet is mainly composed of two modules, Dense Block and Translation Layer. Each Dense Block consists of l Dense Layers. The Dense Layer reduces the number of input features to 4k by a 1×1 convolution as a bottleneck structure before each 3×3 convolution, and splices the outputs of all previous layers in the channel dimension as the input of the next layer, which enhances the back propagation capability of the gradient. By adding the Translation Layer between the two adjacent blocks to serve as a transition, the feature map size and channel dimension are compressed by half through convolution and pooling operations, which reduces the network parameters while deepening the number of layers, and the training effect of the model is more generalized.

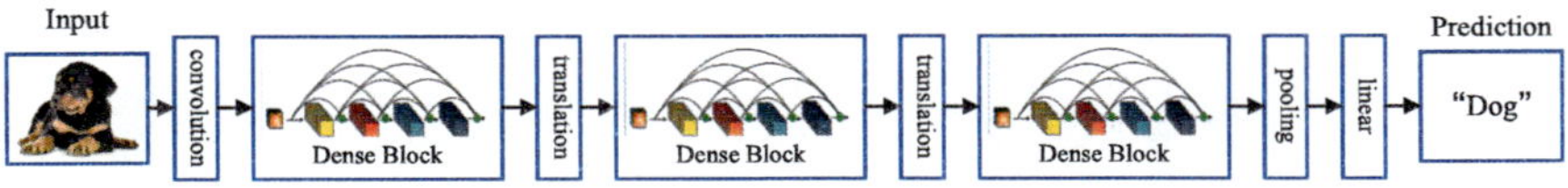

Fig. 1. Architecture of DenseNet

2.1 Steam Block

As in Fig. 2, the original feature map is firstly copied in 4 copies, then each feature map is sampled separately in interline sampling, and the 12 slices obtained are stitched in the channel dimension, and finally the 3×3 convolution is used for feature extraction and adjusting the number of output channels to retain the complete image feature information while increasing less computation.

The initial layer of the original DenseNet network extracts the shallow information of the image by 7×7 convolution with a step size of 2, which has limited feature representation capability. As in Fig. 3, we reduce the feature map resolution to 1/2 of the original image by Focus, then use two different downsampling methods to reduce the features in parallel, and finally adjust the output channel dimension by 1×1 convolution after stitching the results of the two branches by channel. After Steam Block feature map resolution is compressed to 1/4 of the original image, the feature representation is effectively improved with a small increase in computational cost.

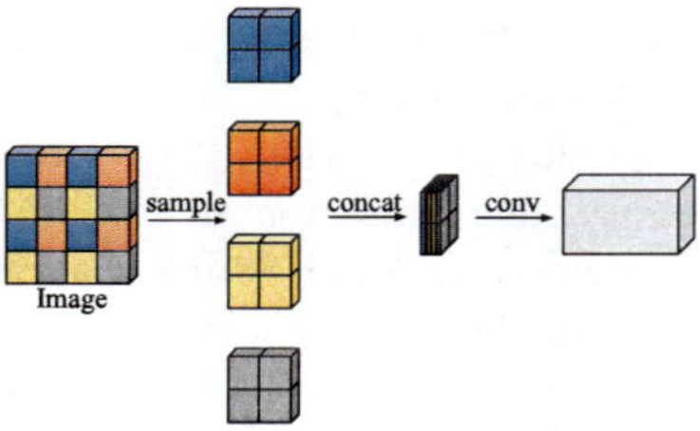

Fig. 2. Focus sampling principle

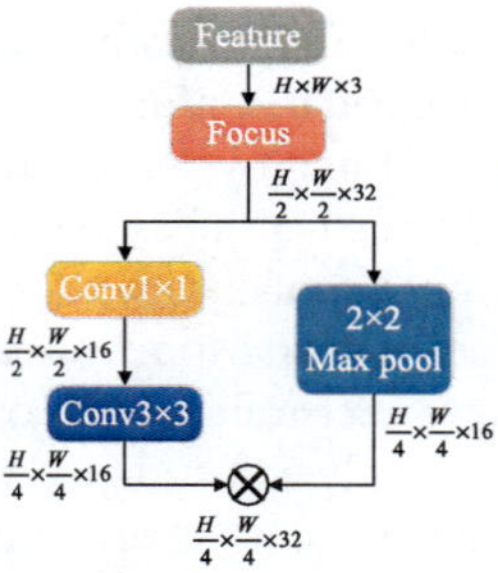

Fig. 3. Structure of steam block

2.2 Two Way Dense Layer

The bottleneck layer of the original Dense Layer adopts the structure of Convolution-BatchNormalization-ReLU. As in Fig. 4, we compare several commonly used loss functions and finally choose to use the Mish activation function with no boundary and higher training stability and accuracy, and change the bottleneck mechanism to the structure of Convolution-BatchNormalization-Mish.

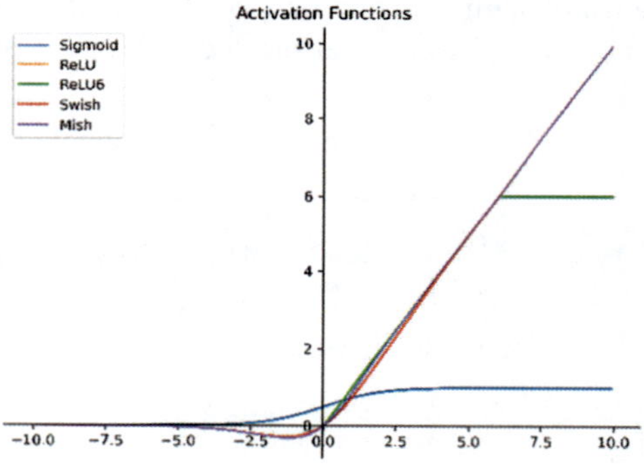

Fig. 4. Comparison diagram of common activation functions

In addition, as shown in Fig. 5 (a), the original dense layer first compresses the channels to 4k by 1 × 1 convolution, and then extracts the features by 3 × 3 convolution, however, the number of channels in the first few bottleneck layers of DenseNet is much larger than the number of input channels, which increases the computational overhead of the middle layer. As shown in Fig. 5 (b), the number of channels in the middle layer is dynamically adjusted according to the number of input channels c and b, so as to reduce the number of model parameters. In addition, a new branch consisting of two 3 × 3 convolutions is added to the original dense layer to extract multiscale features from two different perceptual fields, while taking into account the targets at different scales. Finally, a "dense connection" is achieved by stitching the front and back layers on the channels to ensure the same size and number of channels of each feature map inside the block.

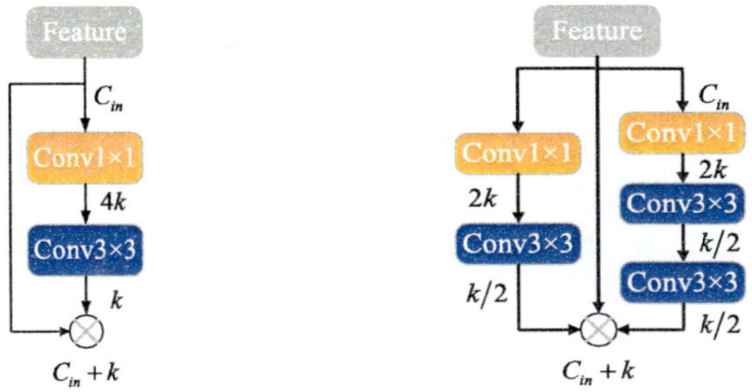

(a) original dense layer (b) two way dense layer

Fig. 5. Dense layer improvement comparison

2.3 Self-attention Translation Layer

The nature of deep neural networks to perceive local context through convolutional kernels makes it difficult to capture the global contextual information of images, and this global perception capability is necessary for the network to understand the high-level semantic information of images. For this reason, a self-attention mechanism is introduced at the end of each transition layer, which adaptively learns the weights assigned to each feature channel based on the captured pixel features, enhancing the important channels and suppressing the unimportant ones, bringing a huge improvement to the deep neural network.

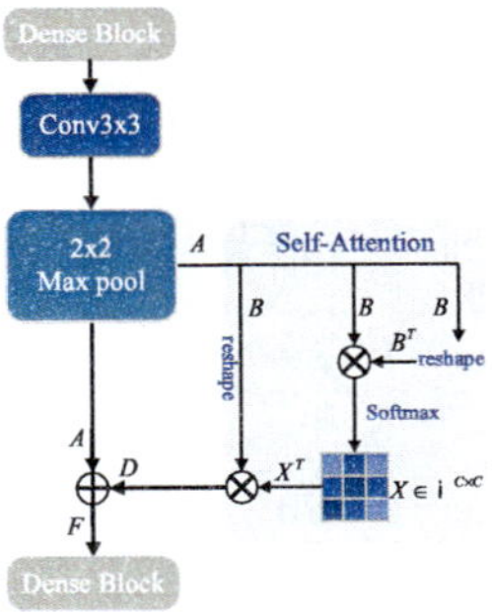

Fig. 6. Architecture of improved self-attention translation layer

As in Fig. 6, for a given input feature $A \in \mathbb{R}^{C\times H\times W}$, the feature maps $B \in \mathbb{R}^{C\times N}$, $B^T \in \mathbb{R}^{N\times C}$ are obtained after reconstruction, and then B and B^T are multiplied by the softmax layer to obtain the feature map $X \in \mathbb{R}^{C\times C}$. The influence factor of the i-th channel on the j-th channel is:

$$x_{ji} = \frac{\exp(A_i \cdot A_j)}{\sum_{i=1}^{N} \exp(A_i \cdot A_j)} \tag{1}$$

In addition, a matrix multiplication of X^T and B is done and the result is reconstructed into $D \in \mathbb{R}^{C \times H \times W}$, and finally the feature map D is summed element by element with the feature map A.

$$F_j = \sum_{i=1}^{C} (x_{ji}A_i) + A_i \tag{2}$$

The final feature map with feature weighting for all channels is obtained $F \in \mathbb{R}^{C \times H \times W}$.

3 Experiments

The experiments were conducted on the Cifar-10 dataset, whose training and validation sets contain 50,000 and 10,000 32 × 32 RGB color images with 10 categories, respectively. The images are first resized to 128 × 128, then the images are randomly flipped, edge-filled and then randomly cropped to generate 128 × 128 pixel images.

3.1 Experimental Setup

The growth rate k in DenseNet network is set to 32, and the number of Dense Layers for each Dense Block is (6, 12, 24). The experiments were performed using computer hardware and software configuration as in Table 1, training with cross-entropy loss function, gradient update using Ranger as the optimizer, initial learning rate of 0.1, momentum of 0.9, batch size of 128, and iteration rounds epoch of 150, and the best test model for the validation set was saved during training.

Table 1. Experimental hardware and software settings

Experiment platform	Parameter setting
CPU	Intel(R) Core (TM) i7-9750H @2.60 GHz
RAM	32 GB
GPU	8 GB RTX 2080-Max
Operation system	Windows 10 64 bit
Deep learning framework	PyTorch1.8.1

3.2 Evaluation Index

In order to accurately and comprehensively judge the generalization performance of this model, this paper uses Params, FLOPs, Accuracy, Precision and Recall as evaluation metrics to evaluate the network performance respectively.

For a convolutional layer, suppose the size of the input feature map is $H \times W \times C_{in}$ and the size of the output feature map is $H' \times W' \times C_{out}$, then:

$$Params \triangleq C_{out} \times (H \times W \times C_{in} + 1) \tag{3}$$

$$FLOPs \triangleq H' \times W' \times C_{out} \times (H \times W \times C_{in} + 1) \tag{4}$$

Suppose TP is the number of positive samples predicted to be positive, TN is the number of negative samples predicted to be negative, FP is the number of negative samples predicted to be positive, and FN is the number of positive samples predicted to be negative, expressed as follows.

$$Accuracy \triangleq \frac{TP + TN}{TP + TN + FP + FN} \tag{5}$$

$$Precision \triangleq \frac{TP}{TP + FP} \tag{6}$$

$$Recall \triangleq \frac{TP}{TP + FN} \tag{7}$$

3.3 Test Results and Performance Analysis

In order to measure the performance of the improved DenseNet network, the improvement effect of different improvements on the model was verified by ablation experiments on the Cifar-10 dataset according to the model parameter settings and experimental parameter settings. Three sets of experiments are set up in Table 2. Among them, DenseNet-A replaces the initial 7 × 7 convolution of DenseNet with Steam Block; DenseNet-B replaces the original dense layer with two way dense layer on the basis of DenseNet-A. DenseNet-C adds the attention mechanism based on DenseNet-B.

Table 2. Test indicators of different DenseNet models

Models	Feature dim	Params	FLOPs	Accuracy
DenseNet	128 × 128	4.24M	1.72G	92.10%
DenseNet-A	128 × 128	4.23M	1.66G	93.12%
DenseNet-B	128 × 128	12.94M	2.78G	93.22%
DenseNet-C	128 × 128	14.42M	2.93G	94.31%

Compared with DenseNet, the evaluation metrics of the three improved methods improved by about 1.02%–2.21%. It can be seen from DenseNet-A that the addition of Steam Block obtained higher accuracy and enriched the expressiveness of the features, while reducing the number of parameters and operations. Compared to DenseNet-C, the

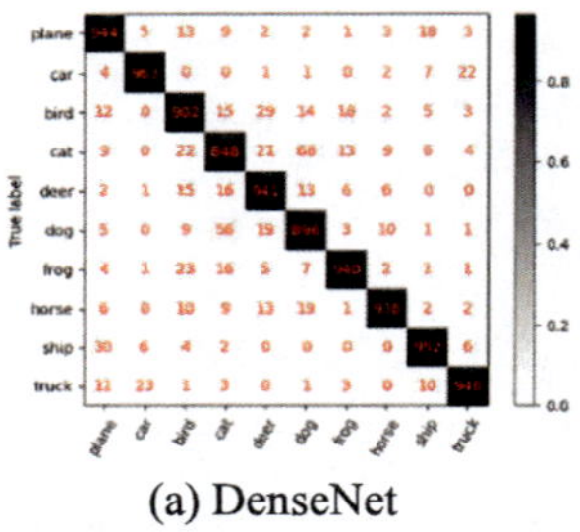

(a) DenseNet

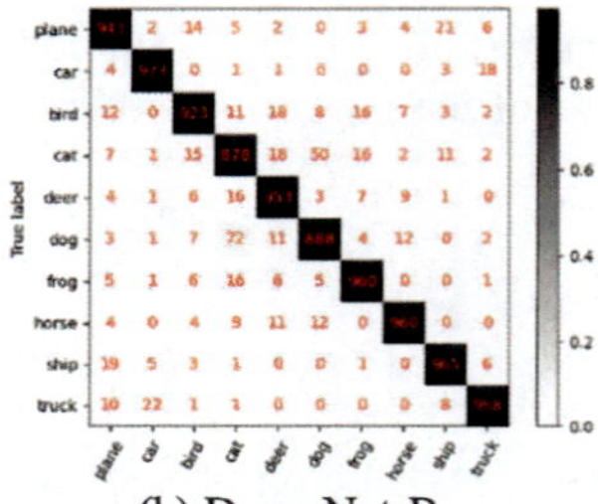

(b) DenseNet-B

Fig. 7. Confusion matrix of cifar-10 (validation)

introduction of self-attention improves the classification performance with negligible loss compared to the performance improvement.

The confusion matrix obtained from testing on the cifar-10 validation set is given in Fig. 7. The number of elements on the main diagonal of this matrix provides a visual representation of the classification performance of the network, as well as an analysis of the presence or absence of misclassification in each category and the classification effectiveness of objects in different categories, DenseNet-B DenseNet-B has higher classification accuracy for larger categories.

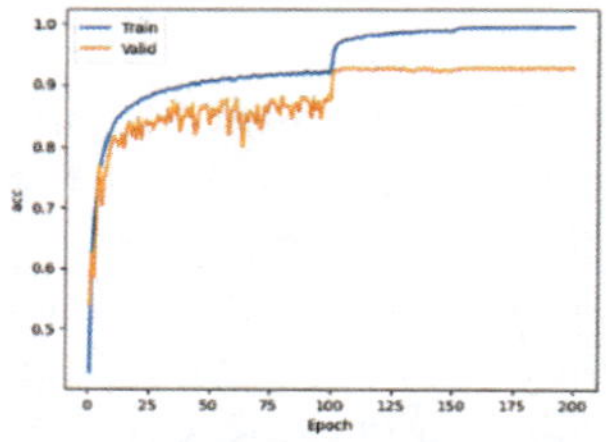

(a) Training and verification accuracy curve after improvement

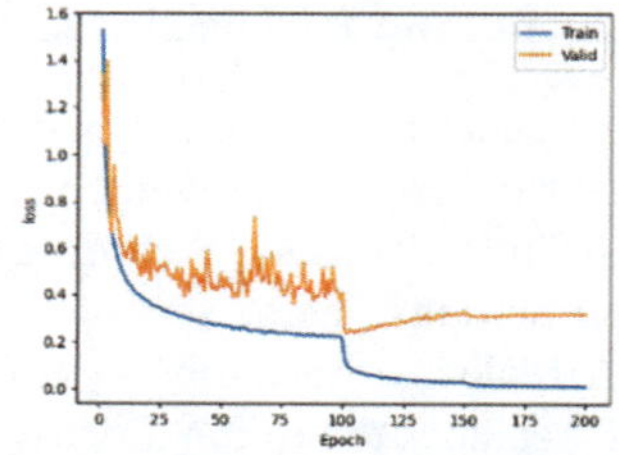

(b) Training and verification of loss function curve before improvement

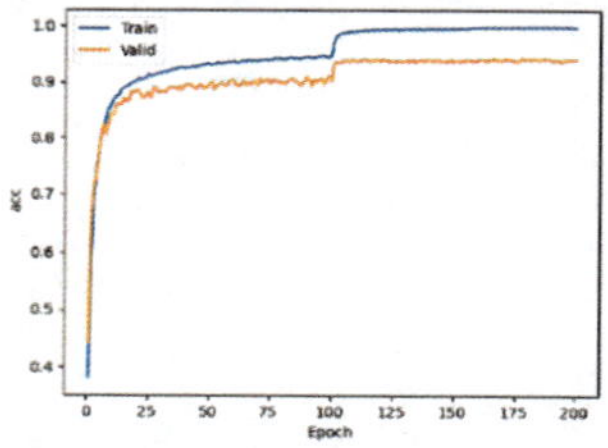

(c) Training and verification accuracy curve after improvement

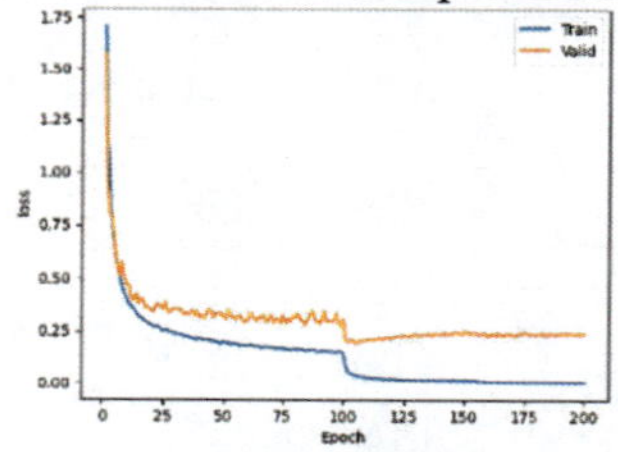

(d) Training and verification of loss function curve after improvement

Fig. 8. Training accuracy and loss function curve before and after improvement

The loss function and accuracy curves of the DenseNet model and DenseNet-C model for training and testing are shown in Fig. 8 (a) and (b), we can see that the training error of the original DenseNet model on the data set gradually decreases as the training proceeds, but after the training reaches 100 epochs, the error of the model on the validation set However, after the training reaches 100 epochs, the error of the model on the validation set increases with the increase of model complexity, and the accuracy rate remains basically the same, and an obvious overfitting phenomenon occurs. From Fig. 8 (c) and (d), we can see that the curve fluctuations level off during training and testing, and the overfitting phenomenon is significantly suppressed. This is due to the fact that the self-attention mechanism we introduced in can adaptively extract useful features within the channel and weaken the role of non-essential features, which makes the classification performance of the improved DenseNet network more superior, under the condition that the number of iterations is increasing, the convergence speed and recognition rate are significantly improved, the structure is optimized, and the loss function value of the network model is decreasing, with better generalization ability.

4 Conclusion

In this paper, we combine the self-attentive mechanism to improve the structure of DenseNet network. Firstly, the initial 7 × 7 convolution of the original network is replaced by Stem Block, which improves the feature expression ability while reducing the computation; secondly, a large perceptual field branch composed of two 3 × 3 convolutions is added on the basis of Dense Layer to enrich the level of extracted features; finally, an attention mechanism is added in the Finally, an attention mechanism is added to the transition layer to adaptively adjust the weights of different channel features to suppress the interference of non-essential features as well as noise. The experimental results show that the improved DenseNet network has a significant effect on the accuracy of image classification and can effectively reduce the phenomenon of overfitting image classification.

Reference

1. Tokarev, K.E., Zotov, V.M., Khavronina, V.N., Rodionova, O.V.: Convolutional neural network of deep learning in computer vision and image classification problems. In: IOP Conference Series: Earth and Environmental Science, vol. 786, no. 1, p. 012040. IOP Publishing, June 2021
2. Wang, B., Gao, J., Si, S.: Image classification and application based on convolutional neural network. Electron. Packag. **21**(5), 50503 (2021)
3. Liu, B., Zhao, Y., Yang, B., Zhao, S., Gu, R., Gahegan, M.: A gastric cancer recognition algorithm on gastric pathological sections based on multistage attention-DenseNet. Concurr. Comput. Pract. Exp. **33**(10), e6188 (2021)
4. Cao, H., Chen, G., Li, Z., Lin, J., Knoll, A.: Lightweight convolutional neural network with Gaussian-based grasping representation for robotic grasping detection. arXiv preprint arXiv: 2101.10226 (2021)
5. Betere, J.I., Kinjo, H., Nakazono, K., Oshiro, N.: Investigation of training performance of convolutional neural networks evolved by genetic algorithms using an activity function. Artif. Life Robot. **25**(1), 1–7 (2019). https://doi.org/10.1007/s10015-019-00561-x
6. He, K., Zhang, X., Ren, S., Sun, J.: Deep residual learning for image recognition. In: Proceedings of the IEEE Conference on Computer Vision and Pattern Recognition, pp. 770–778 (2016)

7. Huang, G., Liu, Z., Van Der Maaten, L., Weinberger, K.Q.: Densely connected convolutional networks. In: Proceedings of the IEEE Conference on Computer Vision and Pattern Recognition, pp. 4700–4708 (2017)
8. Wang, R.J., Li, X., Ling, C.X.: Pelee: a real-time object detection system on mobile devices. arXiv preprint arXiv:1804.06882 (2018)
9. Li, X., Wang, W., Hu, X., Yang, J.: Selective kernel networks. In: Proceedings of the IEEE/CVF Conference on Computer Vision and Pattern Recognition, pp. 510–519 (2019)
10. Zhang, H., et al.: ResNeSt: split-attention networks. arXiv preprint arXiv:2004.08955 (2020)

Research on Binocular Vision Pose Tracking and Detection Algorithm Based on Deep Learning

Shaopeng Wu, Ning Chen(✉), Xian Xu, and Qinfeng Wang

School of Marine Equipment and Mechanical Engineering, Jimei University, Xiamen 361021, Fujian, China
cn1972@jmu.edu.cn

Abstract. In this paper, the target is segmented by instances based on deep learning. The real-time tracking and detection method of inter-vessel poses is investigated, combining binocular vision techniques to achieve normal estimation of the target, and finally obtaining the target poses by positional decomposition. The algorithm used in this paper does not need to add markers or known target 3D models, so it has good scene applicability and robustness. The main research includes a single-image deep learning pose detection method to address the problem of adding specific markers for traditional visual pose detection. The detection results are tracked using the IOU Tracker tracking algorithm, and the results show that the accuracy of binocular vision pose detection based on deep learning is significantly improved, and the real-time performance meets the requirements.

Keywords: Deep learning · Visual pose detection · Binocular vision · Wave compensation

1 Introduction

Under the complex wave environment, the ship will produce six degrees of freedom motion of transverse swing, longitudinal swing, lift and sink, transverse rocking, longitudinal rocking and bow rocking under the action of wind and wave surge, resulting in violent ship rocking, which seriously affects the safety and efficiency of offshore operations [1–3]. By accurately detecting the relative motion posture between ships in real time, the wave compensation system can be controlled to compensate the motion of the offshore replenishment equipment, so that the replenishment material and the replenished ship can generate a follow motion, weaken or eliminate the possible collision during the replenishment process and avoid the dangerous situation [4]. At present, the commonly used methods for ship position detection include inertial sensor detection [5], global positioning system (GPS) detection [6], geomagnetic field method [7], laser detection method, visual detection method, etc., among which the more mature application is the contact measurement method based on inertial sensors, However, the contact measurement method is not suitable for detecting the position of unmodified supply

Y. Wang et al. (Eds.): IWAMA 2021, LNEE 880, pp. 341–347, 2022.
https://doi.org/10.1007/978-981-19-0572-8_43

vessels. Therefore, this paper investigates the visual inspection method of inter-ship attitude.

Current research shows that traditional visual pose detection algorithms need to add specific markers on the target and are not suitable for multi-target arbitrary pose detection; deep learning-based pose detection algorithms can output their corresponding poses by training end-to-end deep neural networks [8–10], which take images as input.

2 Posture Tracking Detection Algorithm Design

First, binocular stereo matching is performed to obtain an image parallax map; then a deep learning network is used to segment a single image instance. The coordinate points are selected on each plane instance and the 3D point coordinates are calculated using the image parallax map; a plane fitting algorithm is used to fit the plane parameters to a series of 3D points and perform plane normal estimation. Combining the planar instance segmentation and normal information for planar pose solution, the real-time pose of the planar instance is obtained. Finally, the planar poses are tracked using the IOU Tracker tracking algorithm. The coordinate points can be selected using farthest point sampling and averaging sampling, and k points are selected on each plane instance. The 3D coordinates of the k points are obtained by combining the binocular depth information, and the plane parameters are obtained by fitting the Ransac plane to the k points to obtain the plane normals, and then the plane instances are solved to obtain the attitude detection results.

2.1 Deep Learning Single Image Pose Detection

The deep learning-based single-image bit-pose detection algorithm uses an encoder-decoder network structure. C1 to C5 use a bottom-up pathway, i.e., the ordinary convolutional features are condensed from the C1 to C5 use a bottom-up pathway, i.e., the process of expressing features in a condensed manner from the bottom up layer by layer. P1 to P5 use a top-down pathway, where each layer of information is processed with reference to the higher dimensional information of the previous layer as input. The decoder network uses 1 × 1 convolution, which can effectively reduce the number of channels in the intermediate levels, so that the output has the same number of channels for each feature map in different dimensions.

2.2 Binocular Vision Normal Estimation

As shown in Fig. 1(a), Farthest point sampling is a common sampling algorithm, and the sampling points obtained by Farthest point sampling are uniformly distributed and generally distributed near the boundary first.

The average sampling is to divide each plane image area in x, y and z directions with equal spacing, and select the intersection points belonging to the plane instance area as sampling points. As shown in Fig. 1(b), the green points are the sampling points, and the spacing of the sampling points is adjusted by setting the division spacing. The average

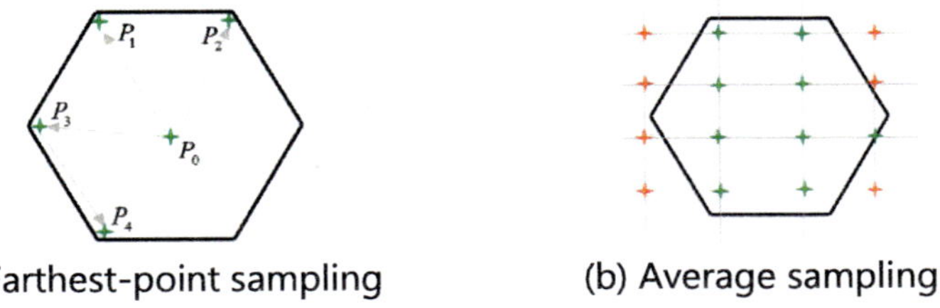

Fig. 1. Planar coordinate sampling method

sampling method is very efficient and can adjust the sampling points and spacing, but it cannot precisely control the number of sampling points.

The 3D coordinates of n points on the plane are known, and the plane normals can be solved by fitting the plane parameters. The plane parameter fitting methods can be divided into least squares, Hoff transform and Ransac methods. The least squares method is mainly used in the case of more interior points and fewer noise points. The Hoff transform uses a voting mechanism to effectively reduce the interference of noise, but it is generally only used in two-dimensional or three-dimensional processing; the Ransac method is simple and general, and works well in practice, but requires setting hyperparameters. Considering the practical application effect, this paper uses the Ransac method for planar parameter fitting. The plane normal can be obtained by fitting n points to the plane using Ransac.

2.3 Displacement Solution

As can be seen from Sect. 2.2, the deep learning-based method has already found the planar instance segmentation and its corresponding normals, and the planar pose solution is performed next to obtain the real-time pose of the plane. Similar to PoseCNN, we estimate the 3D translation of the object by locating the center of the plane instance in the image and combining the normal to get the distance from the camera, and obtain the pose information of the plane by defining the vertical plane up as the Z-axis of the plane coordinate system and solving it with the plane normal information.

2.4 Multi-target Tracking

IOU Tracker [11] is a simple and effective tracking model in multi-target tracking algorithm, which is fast and does not require image information.The basic idea of IOU Tracker is that when the detection frame rate is high enough and the detection effect is good enough, the IOU (Intersection Over Union) between each target border of the two frames before and after can be used as a strong basis for association. IOU Tracker does not consider appearance information, does not predict motion trajectories, and does not use replicated matching algorithms, but directly matches all the borders of two frames using a greedy strategy.

3 Experiment and Analysis

3.1 Deep Learning Environment Configuration

This chapter uses the deep learning framework PyTorch to implement the network model. The entire model is trained using the ScanNet dataset generated by PlaneNet, where an extended version of ResNet-101-FPN is used for the encoder, and ResNet-101 implemented on ImageNet is used as pre-training weights.

3.2 Example Segmentation Results Analysis

Both PlaneRCNN and PlanarReconstruction are trained using the ScanNet dataset, and the detection of NYU dataset and user images using the training. PlaneRCNN has better detection results for small planes, but will misidentify the redundant small planes. In contrast, PlanarReconstruction has poorer detection results for small planes, but is able to detect more complete planes. Therefore, this chapter performs planar pose estimation based on PlanarReconstruction.

The planar instance segmentation results on the ScanNet dataset compared with the real planar instance segmentation basically yielded planar instance segmentation results with consistent number and shape, but there were cases of inaccurate edge segmentation and errors in identifying planes of similar depth as the same plane. Similar to PlaneRCNN and PlanarReconstruction, we use the planar and pixel recall rates as evaluation metrics. Planar recall is the percentage of correctly predicted standard planes, and pixel recall is the percentage of correctly predicted pixels in a plane. A standard plane is considered correctly predicted if i) one of the predicted planes has an intersection-to-merge ratio (IOU) score greater than 0.5, and ii) the average depth difference of the overlapping regions is less than a threshold, which varies from 0.05 to 0.6 m in 0.05 m increments.

3.3 Binocular Vision Normal Estimation and Analysis

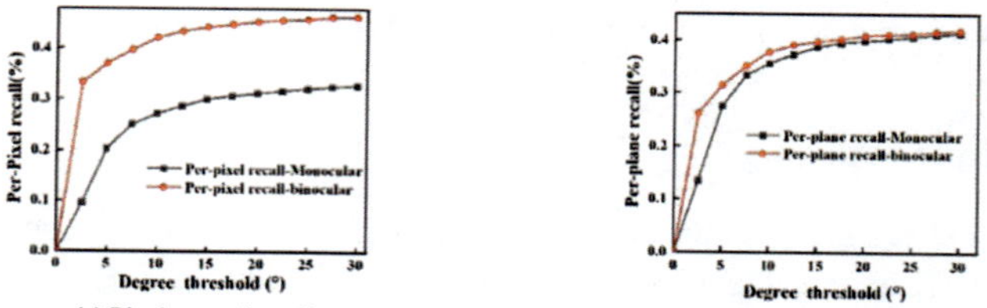

(a) Pixel normal recall

(b) Planar normal recall

Fig. 2. Normal recall solved using real depth information on the ScanNet test dataset

Normal estimation using binocular depth information is mainly performed using parallax maps for 3D point computation and then further planar parameter fitting and normal estimation. As shown in Fig. 2(a) for the pixel normal recall solved using real depth information on the ScanNet test dataset (red) and the pixel normal recall predicted using the single-image deep learning network (black), it can be seen that the mean pixel normal recall using image depth information for fitting planar normals is 0.465, and the pixel normal recall for single images is 0.328. as shown in Fig. 2(b) The average recall of the plane normals using the real depth information solved on the ScanNet test dataset (red) and the recall of the pixel normals predicted using the single-image deep learning network (black) can be seen to be 0.419 for the plane normals fitted using the depth information and 0.414 for the single-image pixel normals. normal estimation predicted by the deep learning network, the accuracy of the planar normal prediction based on stereo depth information solving is significantly improved.

Fig. 3. Results of planar pose estimation on ScanNet dataset using real depth information

The results of normal estimation on ScanNet dataset using real depth information for planar instances are shown in column 2 of Fig. 3, and it can be seen that the normal results obtained by using real depth information for normal estimation are basically correct, and the planar pose estimation is basically consistent with the real results in simple scenarios, and the larger the planar area and the flatter the boundary, the better the normal estimation effect. As can be seen from the figure, since the normal estimation using depth information is done by using the Ransac iterative fitting method, the effect of normal estimation will be different each time, and the difference can be reduced by setting a higher threshold value. As shown in Fig. 3, column 3, the results of normal estimation using real depth information (red) and normal estimation using single-image depth neural network (green) are compared, and it is found that the accuracy of normal estimation using real depth information is higher and the reference is better.

3.4 Planar Posture Tracking Detection Results

Figure 4, column 1, shows the first frame of binocular camera pose tracking detection, image 2 shows the result of tracking detection using IOU Tracker, and image 3 shows the result of detection without the tracking algorithm. As can be seen from the figure,

Fig. 4. Planar pose tracking detection results on a binocular camera

the IOU Tracker tracking algorithm is able to achieve tracking of most of the planar instances compared to the case without the tracking algorithm, as shown by matching and correlating the same planar instances in different frames.

The results of the real-time evaluation of the binocular vision pose tracking detection algorithm are presented. When using the ZED binocular camera for pose tracking detection, the average time to predict the pose of each frame is 0.106 s with CPU only, and the real-time performance is 9.4 fps; using an RTX3080 GPU, the average time to predict the pose of each frame is 0.0368 s, and the real-time performance is 27.2 fps, which can meet the real-time performance requirements.

4 Conclusion

A binocular vision pose tracking detection algorithm based on deep learning is studied. The experimental results show that the detection accuracy is greatly improved by using stereo depth information for normal estimation and pose detection, which verifies the effectiveness of the algorithm; the real-time performance of the binocular vision pose tracking detection algorithm running on GPU can reach 27 fps, which meets the real-time requirements. Focus on the introduction of binocular vision to improve the accuracy of pose detection, the introduction of deep learning algorithms to solve the limitations of pose detection requires specific markers and requires a known three-dimensional model of the target, to achieve the purpose of real-time tracking and detection of the target pose.

References

1. Wang, Z.J., et al.: Current status and development tendency of waves compensation system. Ship Science & Technology (2014)
2. Bai, Y. Yong-Pan, H.U.: Discussion on the development trends of wave compensation technology of offshore alongside replenishment. Naval Architecture and Ocean Engineering (2016)
3. Cui, G.: Development of underway replenishment technology. Ship Engineering (2018)
4. Lindal, H.J.: Development of a Vision-Based Measurement System for Relative Motion Compensation. University of Agder, Grimstad (2014)

5. Chengyi, Y.: Research on attitude measurement system based on the combination of geomagnetic and micro-inertial devices. Harbin: Harbin Inst. Technol. (2013)
6. Ji, D.: MTi micro-inertial heading system_GPS combination technology, Harbin: Harbin Engineering University (2009)
7. Ji, C., Yang, X.: Automatic measurement of ship attitude in earth magnetic field. Ship Engineering (1999)
8. Tordal, S.S., Pawlus, W., Hovland, G.: Real-time 6-DOF vessel-to-vessel motion compensation using laser tracker. In: OCEANS 2017 – Aberdeen (2017)
9. Kehl, W., Manhardt, F., Tombari, F., et al.: SSD-6D: making RGB-based 3D detection and 6D pose estimation great again. In: 2017 IEEE International Conference on Computer Vision (ICCV) (2017)
10. Yu, X., et al.: PoseCNN: a convolutional neural network for 6D object pose estimation in cluttered scenes. Robotics: Science and Systems (2018)
11. Bochinski, E., Eiselein, V., Sikora, T.: High-speed tracking-by-detection without using image information. In: International Workshop on Traffic and Street Surveillance for Safety and Security at IEEE AVSS 2017(2017)

Influence of the Grinding Passes on Microstructure and Its Uniformity of Iron QT400

Xiao-feng Zhao, Ju-dong Liu(✉), and Kai Xu

College of Marine Equipment and Mechanical Engineering, Jimei University, Xiamen 361021, Fujian, China

Abstract. Grind-hardening experiment on iron QT400 was done with a conventional surface grinder. The influence of the grinding passes on the depth of hardened layer and its uniformity were investigated. The results show that, when using a smaller depth of cut a_p and a larger table speed v_w, under the condition of increasing grinding passes, the hardening phase in the surface layer continues to increase and the softening phase gradually decreases until the completely hardened of the workpieces in length direction; The hardness zone is higher and widen as well as the uniformity of the grinding hardening depth can be improved. When the material machining allowance is constant, reducing the grinding passes appropriately can increase the depth of hardened layer and ensure the accuracy and production efficiency.

Keywords: Grinding pass · Grind-hardening · Nodular cast iron QT400 · The uniformity of the hardened layer

1 Introduction

Grind-Hardening is a new technology that integrates the surface heat treatment process of machining parts in the processing production line, it has the advantages of saving resources, improving production efficiency and reducing environmental pollution(such as reducing the emission of exhaust gas, waste water, and waste discharge), therefore, creating a very wide application prospect [1].

Since Brinksmeier E and Brockhoff et al. [2] proposed the technology of grinding-hardening in the 1890s, a large amount of theoretical and experimental researches have been attracted to the grinding-hardening of 45 steel, 40Cr Steel, 42CrMo steel, 65Mn steel, GCr15 steel and other medium-high carbon steels in surface grinding, internal grinding, and external grinding by many scholars at home and abroad [13–17]. The influence of grinding conditions, grinding wheel character and pre-stress on hardening mechanism, properties of hardened layer, microhardness, residual stress and burrs have been systematically studied. So far, the materials in the research are concentrated on carbon steel or alloy steel with better hardenability, but brittle materials such as cast iron have not been involved. As a widely used hardenable material, cast iron has good

Y. Wang et al. (Eds.): IWAMA 2021, LNEE 880, pp. 348–355, 2022.
https://doi.org/10.1007/978-981-19-0572-8_44

wear resistance, toughness, and heat resistance, but due to differences in chemical composition, original structure and mechanical properties from steel, it inevitably lead to differences in grinding mechanism, hardened layer structure and hardened layer depth. Based on surface grinding experiments, this paper systematically study the influence of the grinding passes on the microstructure, micro-hardness, depth of hardened layer and its uniformity of the ductile iron QT400.

2 Experimental Material and Method

2.1 Experimental Material

The experimental material selected is nodular cast iron QT400 with the size of 80 mm × 6 mm × 30 mm (length × width × height), whose chemical composition (in mass fraction) is shown in Table 1. The original structure of the specimen matrix is ferrite and spherical graphite, and the matrix structure and distribution are shown in Fig. 1. The substrate hardness is 9HRC–11HRC (190 $HV_{0.2}$–230 $HV_{0.2}$).

Fig. 1. Microstructure of nodular cast iron QT400

Table 1. Chemical composition of nodular cast iron QT400 (mass fraction, %)

C	Si	Mn	P	S	Mg	Re
3.4	3.2	0.2	0.06	0.01	0.04	0.02

2.2 Method

The equipment for the experiment is a modified M7130 with a horizontal axis and a rectangular table; the grinding conditions are shown in Table 2.

After grind-hardened, the workpiece was cut along the length of the workpiece to make test samples with size of 5 mm × 6 mm × 5 mm, which were inlaid and polished to make metallographic specimens. After etching with 4% nitric acid alcohol solution for 20S, the macroscopic and microstructure of depth of hardened layer was observed with VK-X1000 laser confocal microscope and PHENOM-XL scanning electron microscope. The Falcon511 automatic vickers microhardness tester is used to measure the microhardness along the depth direction with 1.96N load and 10S load time.

Table 2. Grinding process conditions

Grinding wheel	Wheel speed V_s (m/s)	Depth of cut a_p (mm)	Table speed V_w (m/min)	Grind mode	Coolant
A60L6V	25.6	0.1	0.15, 0.2, 0.3, 0.4	(Two-passes) up grinding + down grinding (Three-passes) up grinding + down grinding + up grinding (Four-passes) up grinding + down grinding + up grinding + down grinding (Five-passes) up grinding + down grinding + up grinding + down grinding + up grinding	Dry grinding
		0.1, 0.3, 0.4, 0.5	0.4		

3 Experimental Result and Analysis

3.1 Grinding Hardened Layer Structure

3.1.1 Typical Morphology

Figure 2 shows the typical morphology of different grinding passes when the grinding parameters are $a_p = 0.1$ mm and $v_w = 0.40$ m/min. It shows that the two-passes surface layer is basically unchanged. The three-passes and four-passes surface layer is composed of a small amount of black and white mixture, and the five-passes surface layer is composed of dark materials. Analysis, with the grinding passes increasing, abrasive particles of the grinding wheel become dull, grinding heats up the surface layer and raises its temperature, the higher the temperature cause the better grinding results.

3.1.2 Microstructure

Figure 3 shows the microstructure of the surface layer under different grinding passes. It shows that the microstructure of the two-passes, three-passes and four-passes surface layers consist of ferrite + acicular martensite + retained austenite. In the process of grinding, the temperature of the surface layer is between A_{c1}–A_{c3}. The microstructure of the five-passes surface layer is composed of acicular martensite + retained austenite + spheroidal graphite. In the process of grinding, the temperature of the surface layer is

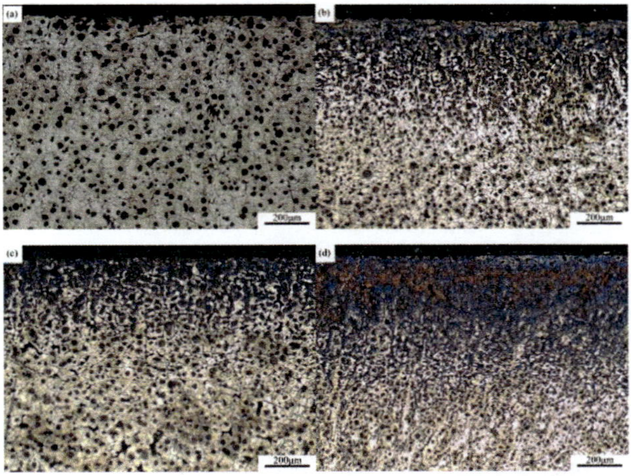

Fig. 2. Typical morphology (a) Two-passes (b) Three-passes (c) Four-passes (d) Five-passes

above A_{c3}. Besides, within the same field of view, with the grinding passes increasing, the ferrite content of the surface layer gradually decreases, and the martensite content gradually increases until the microstructure of the surface layer is completely composed of acicular martensite + retained austenite + spherical graphite.

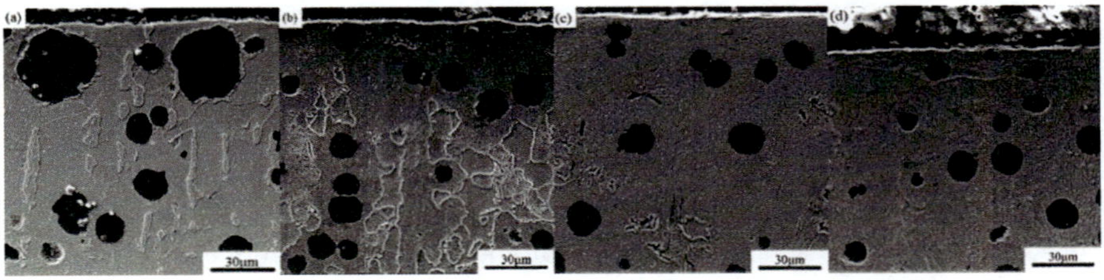

Fig. 3. The microstructure of the surface layer under different grinding passes. (a) two-passes (b) three-passes (c) four-passes (d) five-passes

The analysis shows that with the increasing of the grinding passes, the heats of last grinding make the microstructure of the next grinding surface layer constantly change. For example, when the grinding parameters are $a_p = 0.1$ mm and $v_w = 0.40$ m/min, in the first grinding, the microstructure of the surface layer is mainly ferrite; in the second grinding, the microstructure of the surface layer is mainly a large amount of ferrite + a small amount of martensite; in the third grinding, the microstructure of the surface layer is mainly ferrite + a small amount of martensite, and the microstructure is shown in Fig. 3(a); in the fourth grinding, the structure of the surface layer is mainly ferrite + martensite, and the microstructure is shown in Fig. 3(b); in the fifth passes, the microstructure of the surface layer is mainly a small amount of ferrite + a lot of martensite, and the microstructure is shown in Fig. 3(c). It can be seen that during the next grinding process, the hardning phases in the surface layer continue to increase, the softening phases gradually decrease. With the grinding passes increases, the grinding heat raises the temperature of the surface layer, keeps it above Ac3 and realizes austenization. Therefore, the ferrite content continues to decline, and the martensite content gradually increases, until the microstructure of the surface layer is completely acicular martensite + retained austenite + spherical graphite.

3.2 Microhardness

Since there are no marked changes microhardness distribution of depth of hardened layer though the different grinding pass conditions are given. Thus only the microhardness distribution curve at the condition of $a_p = 0.5$ mm and $v_w = 0.40$ m/min were given as shown in Fig. 4.

It shows that the average microhardness of the high-hardness zone is about 800 $HV_{0.2}$, which is about 2.7 times higher than that of the parent metal. In the hardness decrease area, the microhardness gradually decreases, and the microhardness drop is about 580 $HV_{0.2}$.

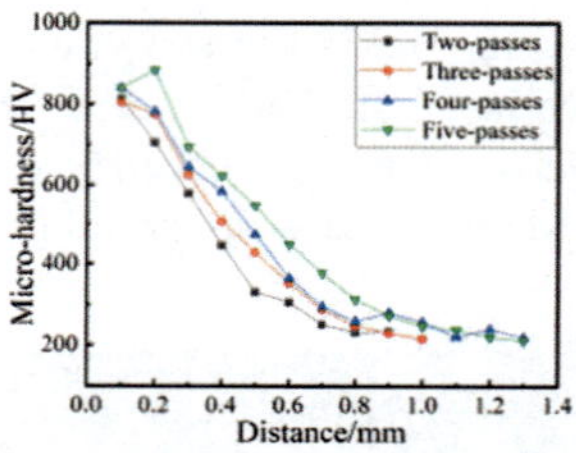

Fig. 4. Microhardness distribution curve ($a_p = 0.5$ mm, $V_w = 0.40$ m/min)

It can also be seen from the figure that increasing the grinding passes can increase the height and width of the high-hardness zone and increase the width of the decreasing-hardness zone. By increasing the grinding passes, the heat and the heat time are increased to make the carbon molecules can fully diffuse and dissolve into the austenite. The higher the carbon mass fraction in the austenite, the carbon content of the martensite after cooling. The higher the amount of the carbon molecules led to the higher the microhardness. Owing to the higher heat and the longer heat time cause the wider range of the heat-affected zone, width of the high-hardness zone and the decreasing-hardness zone are increased.

3.3 Depth of Hardened Layer and Its Uniformity

In engineering practice, in order to improve the wear resistance of ductile iron, the hardness after quenching needs to be higher than 55 HRC, which is equivalent to 600 $HV_{0.2}$. In this paper, the area with microhardness greater than 600 $HV_{0.2}$ is defined as the hardened layer.

3.3.1 The Influence of the Grinding Passes on the Depth and Uniformity of the Hardened Layer Under the Same Grinding Parameters

Figure 5 shows the influence of different grinding passes on the depth and range of the hardened layer. It shows that increasing the grinding passes, the depth of hardened layer gradually increases. From two-passes to five-passes, the depth of hardened layer increases by about 0.15 mm on average, with an average increase of about 52%. It can also shows that as the grinding passes increases, the range of the depth of hardened layer becomes smaller. From two-passes to five-passes, the range decreases by 0.00947, accounting for 47.35% of the two-passes. Therefore, increasing the grinding passes can not only increase the depth of hardened layer, but also significantly improve the uniformity of the hardened layer. Because only the grinding passes had increased. On the one hand, the workpiece is repeatedly heated and the abrasive grains of the grinding wheel are passivated to create more heat and longer heating times, higher heat longer heating times leads to the depth of hardened layer getting thicker and its uniformity can be observably enhanced [6].

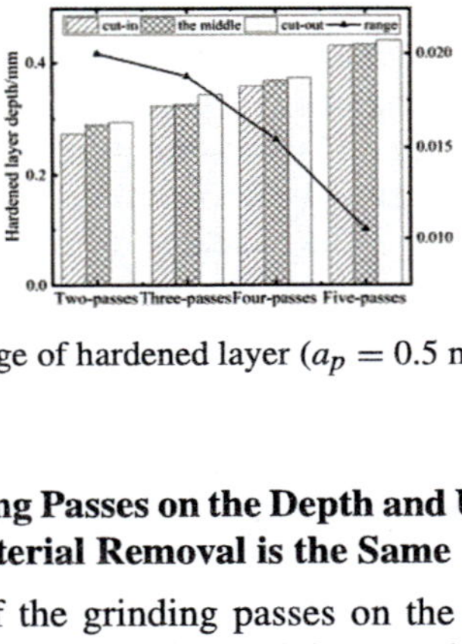

Fig. 5. Depth and range of hardened layer ($a_p = 0.5$ mm, $v_w = 0.40$ m/min)

3.3.2 The Effect of Grinding Passes on the Depth and Uniformity of the Hardened Layer When the Material Removal is the Same

Figure 6 shows the effect of the grinding passes on the depth and uniformity of the hardened layer when the table speed and material removal are the same. It shows that as the grinding passes increases, the grinding depth of each pass decreases. After grinding, the depth of hardened layer decreases and its uniformity can be improved.

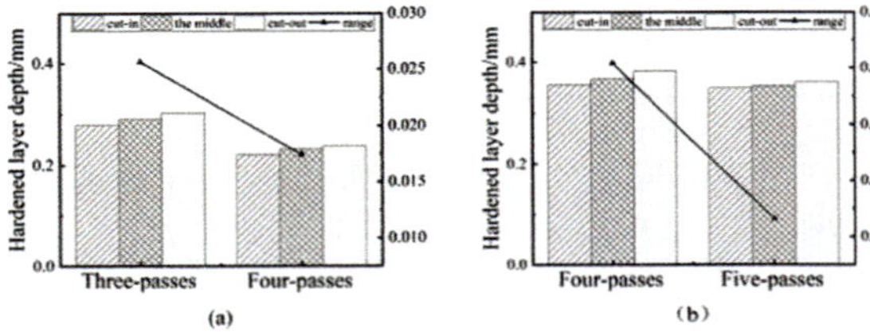

Fig. 6. Depth and range of hardened layer (a) Material removal amount 1.2 mm (b) Material removal amount 2 mm

It is analyzed that when the material removal amount is the same as the workpiece table speed v_w, with the grinding passes increasing, the depth of hardened layer after grinding decreases, and the uniformity of the depth of hardened layer increases. This is the amount of material removed is the same. As the grinding passes increase, the grinding depth of each pass decreases, the grinding force decreases, and the heat generated by the grinding decreases. Therefore, the increase of the grinding passes can reduce the depth of hardened layer. The range decrease in the depth of hardened layer is due to the increase in the grinding passes, and the uniformity of the hardened layer after grinding is improved. Therefore, in the actual production process, when the material machining allowance is constant, appropriately reducing the grinding passes can increase the depth of hardened layer and improve the production efficiency.

4 Discussion and Conclusion

(1) When using a smaller grinding depth and a larger table speed, the completely hardened of the workpieces in length direction by increasing the grinding passes, and the uniformity of the grinding hardening depth can be improved. (2) With the increase of the grinding passes, the hardning phase in the surface layer continues to increase, and the softening phase gradually decreases. After cooling, the martensite content in the hardened layer gradually increases. (3) With the grinding passes increases, the height and width of the high hardness zone in the microhardness curve can be increased, and the width of the hardness drop zone can be increased. (4) In the actual production process, when the material machining allowance is constant, appropriately reducing the grinding passes can increase the depth of hardened layer and improve the production efficiency.

Acknowledgment. The authors would like to thank support from the National Natural Science Foundation of China, China (No. 51676085) and The Natural Science Foundation of Fujian Province, China (No. 2014J01199).

References

1. Brinksmeier, E., Brockhoff, T.: Utilization of grinding heat as a new heat treatment process. CIRP Ann. Manuf. Technol. **45**(1), 283–286 (1996)
2. Brinksmeier, E., Brockhoff, T.: Randschicht-wärmebehandlung durch schleifen. HTM. Härterei-technische Mitteilungen **49**(5), 327–330 (1994)
3. Brockhoff, T., Brinksmeier, E.: Grind-hardening: a comprehensive view. CIRP Ann. Manuf. Technol. **48**(1), 255–260 (1999)
4. Zarudi, I., Zhang, L.C.: Modelling the structure changes in quenchable steel subjected to grinding. J. Mater. Sci. **37**(20), 4333–4341 (2002)
5. Zarudi, I., Zhang, L.C.: Mechanical property improvement of quenchable steel by grinding. J. Mater. Sci. **37**(18), 3935–3943 (2002)
6. Wang, G.C., Liu, J.D., Pei, H.J., et al.: Study on forming mechanism of surface hardening in two-pass grinding 40Cr steel. Key Eng. Mater. **304**, 588–592 (2006)
7. Liu, J.D., Wang, G.C., Wang, Z., et al.: Experimental research on grind-hardening of 65Mn steel. Mater. Sci. Forum **505**, 787–792 (2006)

8. Judong, L., Guicheng, W., Kangmin,C.: Effect of grinding parameters on the grind-hardening layer of steel 40Cr. China Mech. Eng, **17**, 1842–1845 (2006) (in Chinese)
9. Liu, J.D., Zhuang, J.Z., Huang, S.W.: Influence of the Grinding pass on microstructure and its uniformity of the depth of hardened layer. Adv. Mater. Res. **211–212**, 36–39 (2011)
10. Xiu, S., Shi, X.: Transformation mechanism of microstructure and residual stress within hardening layer in PSHG. J. Adv. Mech. Des. Syst. Manuf. **9**(3) (2015)
11. Zhang, Y., Ge, P.Q., Be, W.B.: The study for variable grinding depth to control plane grind-hardening layer depth distribution. Int. J. Adv. Manuf. Technol. **84**(5–8), 1269–1276 (2016)
12. Liu, M., Zhang, K., Xiu, S.: Mechanism investigation of hardening layer hardness uniformity based on grind-hardening process. Int. J. Adv. Manuf. Technol. **88**(9–12), 1–10 (2016)
13. Guo, Y., Xiu, S.C., Liu, M.H., et al.: Uniformity mechanism investigation of hardness penetration depth during grind-hardening process. Int. J. Adv. Manuf. Technol. **89**(5–8), 2001–2010 (2017)
14. Zhang, J., Wang, G.C., Pei, H.J.: Effects of grinding parameters on residual stress of 42CrMo steel surface layer in Grind-hardening. In: International Symposium on Mechanical Engineering and Material Science (ISMEMS 2017) (2018)
15. Kolkwitz, B., Kohls, E., Heinzel, C., et al.: Correlations between thermal loads during grind-hardening and material modifications using the concept of process signatures. J. Manuf. Mater. Process. **2**(1), 20 (2018)
16. Wang, Y., Xiu, S., Zhang, S.: Controlling grain sizes of 42CrMo steel by pre-stress hardening grinding. Materials **12**(19), 3124 (2019)
17. Huang, X., Ren, Y., Wu, W., et al.: Research on grind-hardening layer and residual stresses based on variable grinding forces. Int. J. Adv. Manuf. Technol. **103**, 1045–1055 (2019)

Optimization and Feedback of Assembly Experiment Scheduling Based on Digital Twin

Kai Guo, Lilan Liu(✉), Zenggui Gao, and Muchen Yang

Shanghai Key Laboratory of Intelligent Manufacturing and Robotics, Shanghai University, Shanghai, China
lancy@shu.edu.cn

Abstract. With the increasing demand for personalized customization, the randomness of personalized customization tasks is strong, the state of the production line is unstable when the order arrives, and the scheduling model is difficult to determine. One of the basic requirements of intelligent manufacturing is to respond to individual needs and efficiently and flexibly produce small batches of multiple varieties. This paper proposes to use genetic algorithm to solve the resource scheduling in the face of multi-variety and small-batch tasks, and upload the scheduling results to the digital twin platform, and use the synchronous mapping of the digital twin to verify with the high-fidelity model, which reduces production accidents. Finally, the research results were applied to a miniature customized production line in a laboratory, which proved the feasibility of the method.

Keywords: Personalized customization · Digital twin · Multi-variety and small batch

1 Introduction

With the in-depth application of a new generation of information technology such as cloud computing, big data, and the Internet of Things to the manufacturing industry, a large number of related industrial applications have been implemented. At present, the world's major manufacturing countries have proposed their own advanced manufacturing strategies, such as Made in China 2025, German Industry 4.0 and the Industrial Internet in the United States. Although the advanced manufacturing strategies introduced by various countries are not the same, their common goal is the same, which is to achieve intelligent manufacturing [1].

However, the current production line production process of manufacturing enterprises is very complicated, and the production method is more and more inclined to personalized customized production. The traditional workshop scheduling method is no longer suitable for this new production method. On the one hand, due to the numerous processing and production links of modern manufacturing enterprises, the interrelationships are complex, the production continuity is strong, and the changes are rapid, relevant information cannot be interconnected and shared between production resources, and the information in the production process cannot be automatically and timely sensed. And

Y. Wang et al. (Eds.): IWAMA 2021, LNEE 880, pp. 356–364, 2022.
https://doi.org/10.1007/978-981-19-0572-8_45

deal with it, causing the changes caused by a certain partial response delay to affect the operation of the entire production system; on the other hand, the production and processing process is usually dynamic, and real-time information such as product orders, raw material inventory, equipment operating status, and processing process is also constantly changing. Adjustment, the dispatch center cannot integrate and update the changed information in a timely manner, and make timely and dynamic decision-making responses to adjust the dispatch plan. Therefore, in the actual workshop production and manufacturing process, the scheduling plan often has deviations, and the real-time response is slow, causing serious waste of production resources, resulting in low production efficiency [2].

Digital Twin (DT) is a technology that dynamically maps the state of the physical space through a digital model. By constructing a virtual model of the physical workshop, the real production state of the workshop is mapped to the virtual workshop. Using the virtual workshop, before production, virtual production can be carried out through the digital twin model, the production process can be simulated, and the parameters can be adjusted to realize early warning of possible problems and reduce unexpected conditions in the production process [3, 4].

The overall framework of this article is shown in Fig. 1. It describes the main structure of the twin personalized production line scheduling mechanism, and also clarifies the specific flow of the personalized production line scheduling operation under this mechanism. The main operation process is as follows: The actual physical production line production execution system can perceive and obtain information related to scheduling in personalized production line production in real time, such as the pose signal of the Dobot-CR5 robot arm, the position signal of the AGV car, and the assembly parts. The position and attitude information of the Dobot robot arm is obtained by calling the API of the Dobot control cabinet. Haikang's AGV trolley obtains the position information of the AGV trolley through the supporting RCS control system. The assembly parts are passed through the Unity3D Radiographic detection technology to obtain information. Establish a three-dimensional model model of the personalized production line, including the three-dimensional model of the robot arm, AGV, product and data display board, and the digital twin will perform remote monitoring and real-time simulation. After the personalized assembly line of the digital twin is established, the virtual model Perform virtual debugging at the layer level. Through the genetic algorithm used in this paper, the intelligent scheduling method of multi-variety and small batch tasks is studied, and the virtual model layer is fed back to the physical workshop to execute the scheduling plan after virtual simulation verification, forming a continuous iterative optimization of the virtual and real production line scheduling process.

The rest of this paper is organized as follows: Sect. 2 explains genetic algorithm and mathematical model of multi-variety and small batch workshop scheduling. Section 3 details experimental verification. Conclusions are summarized in the last section of this paper.

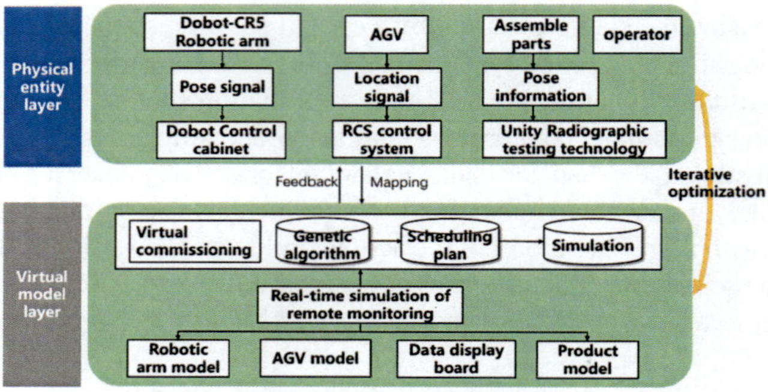

Fig. 1. Overall architecture diagram

2 Methodology

2.1 Overview of Genetic Algorithm

Genetic Algorithm (GA) [5–7] is a highly parallel, random and adaptive search algorithm that draws on the natural selection and natural genetic mechanisms of the biological world. It was first proposed by Professor J. Holland of the United States in the book "Adaptability in Natural and Artificial Systems" in 1975. It is a multi-parameter, multi-group simultaneous optimization method that imitates the principle of "natural selection, survival of the fittest" in the evolution of natural organisms. Its main feature is the group search strategy and the information exchange between individuals in the group. The search does not depend on gradient information. It is especially suitable for dealing with complex and nonlinear problems that are difficult to solve by traditional search methods. Because of its characteristics, its theoretical and application research has become a very hot topic since the 1980s, especially its application research has been extremely active, not only has its application fields expanded, but also uses genetic algorithms for optimization and rule learning. The ability of the computer has also been significantly improved. At present, it has been widely used in combinatorial optimization, machine learning, adaptive control, planning and design, and artificial life. It is considered to be one of the key technologies in intelligent computing in the 21st century.

The main processing steps of genetic algorithm are:

1) Coding: First, encode the solution of the optimization problem. The purpose of coding is to make the manifestation of the solution of the optimization problem suitable for the genetic operation in the genetic algorithm;
2) Construct fitness function: The construction and application of fitness function are basically based on the objective function of the optimization problem. When the fitness function is determined, the law of natural selection is to determine which chromosomes are suitable for the probability distribution determined by the value of the fitness function Survival, which are eliminated, the surviving chromosomes form a population, forming a population that can reproduce the next generation;

3) Crossover of chromosomes: The genetic combination of the father's chromosomes reaches the next generation of individuals through the cross between the chromosomes of the father. The generation of offspring is a reproductive process, which produces a new solution;
4) Chromosome variation: gene variation may occur in the process of generating new solutions. The variation changes the coding of some solutions and makes the solution more ergodic.

Figure 2 describes the process and key steps of a typical genetic algorithm, including population initialization, selection operations, fitness functions, gene coding, crossover operations, mutation operations, and parameters related to the termination conditions of the algorithm.

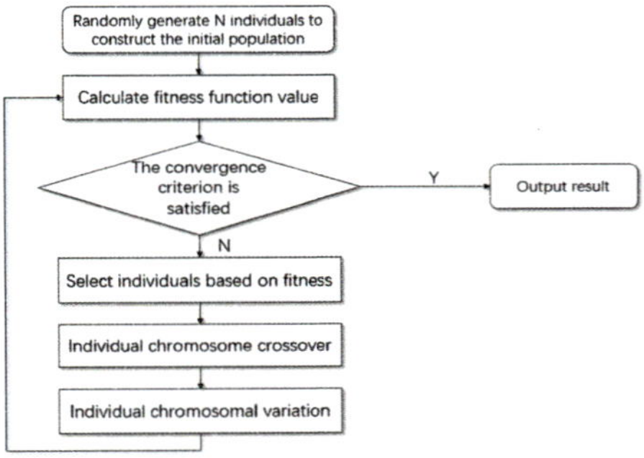

Fig. 2. Genetic algorithm flow chart

2.2 The Mathematical Model of Multi-variety and Small-Batch Workshop Scheduling

Multi-variety and small-batch manufacturing is one of the production modes of discrete manufacturing, usually multiple workpieces are processed together, and the process paths are different [8, 9]. Since the multi-variety and small-batch production process has the characteristics of discrete manufacturing, after simplification and abstraction, the mathematical model of production scheduling needs at least the following basic constraints:

1. If the process has priority, the workpiece is processed according to the priority of the process;
2. Each processing equipment can only process one workpiece at any time;
3. Scheduling is based on the same batch of orders, and the objects of scheduling are all tasks of the same batch of orders, which are indivisible;
4. The preparation time of the workpiece on the equipment (including grasping operation, workpiece inspection, etc.) and AGV transportation time are all added to the assembly time of this assembly process;

5. The workpiece will not be interrupted during the assembly process, that is, once the workpiece enters the assembly, it will not be unexpectedly terminated before the end of the assembly;

Suppose there are a total of n workpieces $j_i(i = 1, 2, \ldots, n)$ in the production line, and each workpiece has n_i processes $O_{ij}(j = 1, 2, \ldots, n_i)$, and there are a total of m units in the production line Equipment $e_i(j = 1, 2, \ldots, m)$, each equipment can process w_i processes, and each process is denoted as $p_{ij}(j = 1, 2, \ldots, w_i)$.

Objective function

$$min(max(c_i)), i \in (1, n) \tag{1}$$

In formula 1, c_i represents the completion time of the workpiece j_i. The optimization goal of this article is the latest completion time.

Constraint 1:

$$c_{ij} - q_{ij} \geq c_{i(j-1)} \tag{2}$$

In formula 2, c_{ij}—the completion time of process O_{ij}; p_{ij}—the assembly time of process O_{ij}. Formula 2 indicates that the start time of the subsequent process cannot be earlier than the completion time of the previous process.

Constraint 2:

$$e_i(t) \neq e_j(t), i \neq j \tag{3}$$

In formula 3, $e_i(t)$——the equipment where the workpiece j_i is located at time t. Equation 3 shows that different workpieces are assembled on different equipment at the same time, and one equipment can only assemble one workpiece at any time.

Constraint 3:

$$\left[st_{ij}^k, ed_{ij}^k\right] \cap \left[st_{ij}^k, ed_{ij}^k\right] = \emptyset \tag{4}$$

$$i, i' \in [1, n], j, j' \in [1, n], k \in [1, m], i \neq i, j \neq j$$

In formula 4, S_{ij}^k—the starting time of the j process of the i workpiece on the k equipment; ed_{ij}^k—the completion time of the j process of the i workpiece on the k device; formula 4 indicates that the two processes processed on one device cannot overlap in time.

3 Experiment Analysis

This section uses the miniature customized car assembly line in our laboratory as the hardware foundation to construct the corresponding system to verify the method mentioned in this article. The experiment in this section simulates the task of assembling a car with a single batch and multiple orders. Since the type and quantity of candies required for each order are completely specified by the user, the task of the experiment is a multi-variety and small-batch manufacturing.

In order to simulate the characteristics of multi-variety and small-batch production, the experimental platform provides 7 types of car assembly. At the same time, the 8 robot arms are divided into 4 groups. The robot arm 1 appearing later is the robot arm combination 1. Each trolley is assembled by a group of robot arms, and each group of robot arms is assigned 3 types of assembleable cars. The Table 1 reflects the functions of each group of robotic arms.

Table 1. Types of cars that can be assembled with each robotic arm

	Car1	Car2	Car3	Car4	Car5	Car6	Car7
Robot arm 1	✓			✓			✓
Robot arm 2	✓	✓			✓		
Robot arm 3			✓	✓			✓
Robot arm 4		✓			✓	✓	

This intelligent manufacturing production line provides users with personalized customized car services, and provides users with a high degree of freedom for the production and packaging of car types and number of cars. The experimental production line is divided into 7 types of assembled models. If there are users who place orders (5 users) as shown in the Table 2, each user specifies the type of car and the corresponding quantity. At this time, the intelligent production line needs to complete all the processing tasks of the order as soon as possible under the premise of ensuring quality. In order to make better use of resources, the order is best to be produced in parallel. And each car has the same priority order.

Table 2. Model and number required for multiple orders

	Car1	Car2	Car3	Car4	Car5	Car6	Car7
User 1	2		5			1	
User 2		2		2	2		4
User 3	1		1	1	2		
User 4		1				4	3
User 5	3		1	2	3	2	2

The genetic algorithm parameters used in this experiment are set as follows: population size 300, maximum genetic algebra 50, crossover rate 0.8, mutation rate 0.6. The Fig. 3 describes the transformation process of the optimal solution in the optimization process of the genetic algorithm. It can be seen that the optimal solution is almost reached in about 15 generations.

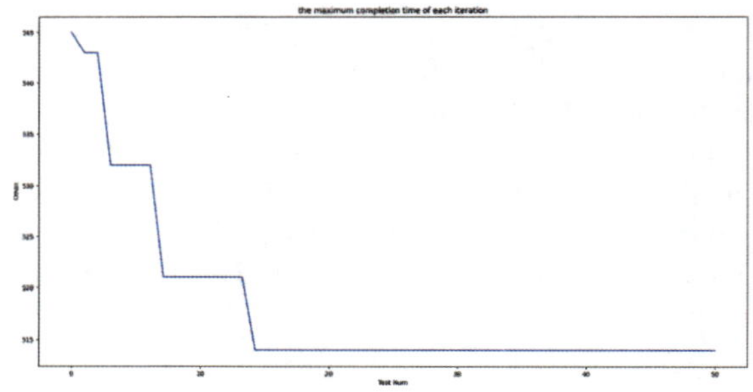

Fig. 3. Transformation of optimal solution in genetic algorithm

Figure 4 describes the Gantt chart for the task of multi-variety and small-batch orders. It can be seen from the figure that the optimal solution is 514 s. The colors in the Fig. 7 represent 7 different models, and the quantity of each color block represents the order needs. The number of trolleys.

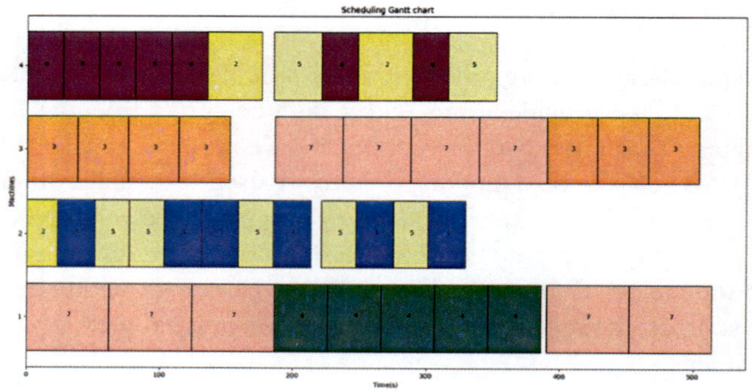

Fig. 4. Resource scheduling Gantt chart based on genetic algorithm

Finally, after obtaining the optimal solution through the task list issued by the user, we verify it in the digital twin system to avoid accidents during actual operation. Figure 5 shows the digital twin system of the smart production line. Perform algorithm testing and virtual simulation of personalized task order scheduling in the virtual production line. The simulation results of personalized task orders are analyzed. After the end, the operation status of the production line is judged, and the next production step of the physical production line is determined. The results can also be displayed through the UI display panel in Unity visualization technology. Through the algorithm model to optimize the visualization of the analysis results, the effect of real-time monitoring of the running status of the assembly line is realized. The picture shows the digital twin scene of the intelligent production line.

Fig. 5. Digital twin system

4 Discussion and Conclusion

This article takes the personalized custom car assembly line as the object, faces the task of personalized customization, combines the verification and analysis of the digital twin system, and proposes an assembly line architecture oriented to personalized customization. After analyzing the particularity of personalized customization tasks, based on genetic algorithms, the resource scheduling analysis is performed on the multi-variety and small-batch orders issued by users, and finally the results obtained are verified by the digital twin platform, which greatly reduces the actual Accidents in the production line.

Acknowledgment. The authors would like to express appreciation to Shanghai Key Laboratory of Intelligent Manufacturing & Robotics and all members of the CIMS laboratory for their support.Thanks for the funding from Shanghai Science and Technology Committee of China (No. 19511105200).

References

1. Lee, J., Bagheri, B., Kao, H.-A.: A cyber-physical systems architecture for industry 4.0-based manufacturing systems. Manuf. Lett. **3**, 18–23 (2015)
2. Rusk, J.: Connecting the digital twin: from idea through production, to customer and back. NASA Tech. Briefs **42**(6) (2018)
3. Hao, Z., Qiang, L., Xin, C., et al.: A digital twin-based approach for designing and multi-objecive optimization of hollow glass production line. IEEE Access **5**, 26901–26911 (2017)
4. Sun, X., Bao, J., Li, J., Zhang, Y., Liu, S., Zhou, B.: A digital twin-driven approach for the assembly-commissioning of high precision products. Robot. Comput. Integr. Manuf. **61** (2020)
5. Park, B.J., Choi, H.R., Kim, H.S.: A hybrid genetic algorithm for the job shop scheduling problem. Comput. Ind. Eng. **167**(4), 77–95 (2003)
6. Hartmann, S.: A competitive genetic algorithm for resource-constrained project scheduling. Nav. Res. Logist. **45**(7), 733–750 (2015)
7. Kobayashi, S.: An efficient genetic algorithm for job shop scheduling problems. In: ICGA **24**(4) (2017)

8. Hai, L., Li, L.Q., Li, X.D: Research on collaboration-oriented production management in multi-variety and Small-batch environment. In: 2012 9th International Conference on Fuzzy Systems and Knowledge Discovery, IEEE (2012)
9. Jiang, X., Zhang, Y., Zhao, K., et al.: Research on multi-type & small batch oriented process quality control system under network environment. In: IEEE International Conference on Automation and Logistics, pp. 1050–1056 (2008)

Research on Semi-automatic Coronary Artery Centerline Extraction Based on Deep Learning

Chuanhong Zhou, Yang Xu(✉), and Qiuyi Ye

School of Mechatronic Engineering and Automation, Shanghai Key Laboratory of Intelligent Manufacturing and Robotics, Shanghai University, Shanghai, China
shuxuyang@shu.edu.cn

Abstract. In recent years, Cardio vascular disease (CVD) has become a major cause of premature morbidity and mortality in our population, and the prevention and treatment of CVD have become urgent. Currently, the mainstream method to diagnose coronary artery disease is Coronary Computed Tomography Angiography (CCTA). The processing and analysis of CCTA images can help doctors understand the connection and trend of a coronary artery. Aiming at the problem that the radius error of the centerline obtained by the traditional semi-automatic coronary centerline extraction network is large, and it is easy to extend to the non-coronary tissue at the coronary ostium and end, this paper proposes a new three-dimensional convolutional neural network: MST-Net (Multi-Scale Tracking Net), through the multi-scale feature extraction layer, enables the network to extract more local and global features, to obtain more accurate predictions in the coronary centerline extraction task.

Keywords: CCTA · CNN · Coronary artery centerline extraction

1 Introduction

Accurate information about the geometry and topology of patients' blood vessels is essential for many medical applications. In patients with suspected coronary artery disease, coronary computed tomography angiography (CCTA) can be used to obtain non-invasive information of cardiac vessels [1]. CCTA uses the principle that X-ray cannot penetrate the contrast agent to display the image of the coronary artery by intravenous injection of contrast agent.

Centerline extraction plays an irreplaceable role in three-dimensional reconstruction of coronary artery and acquisition of vascular cross-section. So far, many centerline extraction algorithms have been designed. According to the degree of automation, these algorithms can be divided into three categories, from initial manual depiction to semi-automatic coronary centerline extraction, and automatic coronary centerline extraction.

The automatic 3D thinning algorithm developed by Ma et al. [2] and Holland et al. [3] is an effective algorithm. The algorithm can maintain the topological structure of the target well and make the skeleton line have good accuracy, but it depends on the quality

Y. Wang et al. (Eds.): IWAMA 2021, LNEE 880, pp. 365–373, 2022.
https://doi.org/10.1007/978-981-19-0572-8_46

of the target image and is not effective for complex targets. In addition, most refinement methods are a gradual iterative process, which takes a long time for larger targets.

The algorithm based on a minimum path is to calculate the minimum cost path between the start point and the endpoint defined manually or automatically. Benmansou et al. [4] proposed a minimum path algorithm using key points, which can extract the centerline only by determining a starting point in advance. The advantage of the minimum path method is that the generated path has a high coincidence degree with the reference centerline, and the calculation amount is low and the calculation speed is fast. However, its disadvantage is that there may be a shortcut at the bend of the centerline. Therefore, it is essential to design a cost function that is low at the centerline and high at other locations.

The tracking-based algorithm extracts the blood vessel centerline by the iterative judgment of the position, direction, and radius of the blood vessel. The semi-automatic coronary artery centerline extraction method proposed by Wolterink et al. [5] can be trained with limited training data. Once the model training is completed, the complete coronary artery centerline can be quickly, automatically or interactively extracted from CCTA images. However, the network structure adopted by this method is relatively simple, and it is difficult to extract sufficient feature information, resulting in a large error in the radius of the extracted centerline, and it is easy to extend to non-coronary tissues at the mouth and end of the coronary artery.

2 Coronary Artery Centerline Extraction Network Based on Multi-scale Feature Extraction Layer

The traditional coronary centerline extraction network Origin-Net is composed of a simple stacked expanded convolution layer. There are seven layers in the network, the first two and the last three layers do not use the extended convolution, that is, the ordinary convolution layer. $|D|+1$ indicates that the number of output channels in the last layer is the number of reference directions plus 1, that is, the number of channels corresponding to a radius. In this paper, $|D| = 500$ is adopted.

2.1 MST-Net Based on Multi-scale Feature Extraction Layer

The image information near the cutting block is also important for the prediction of the reference direction. In addition, the radius of proximal and distal coronary artery is different, and the size of the block is fixed. If only a single scale is used for feature extraction, the neural network will be difficult to accurately determine the radius of the coronary artery at different locations.

This paper proposes a multi-scale 3D CNN model MST-Net based on 3D CCTA images, which connects the large receptive field convolutional network (CNN1) and the small receptive field convolutional network (CNN2) in parallel so that the network can extract more local and global features. At the same time, the network takes isotropic 3D image segmentation as input, which can extract more spatial structure information than 2D section view or 2.5D three orthogonal section view, to improve the prediction accuracy of direction category and radius. Among them, the size of Patch is $19 \times 19 \times 19$, and the voxel spacing is resampled to 0.5 mm. To integrate monitoring data and failure information, we used the recorded timestamps in each database as connections. Since both monitoring data and failure information are necessary for fault identification, we included only the records which have been recorded in both datasets as observations in the analysis. The network structure of MST-Net is shown in Fig. 1.

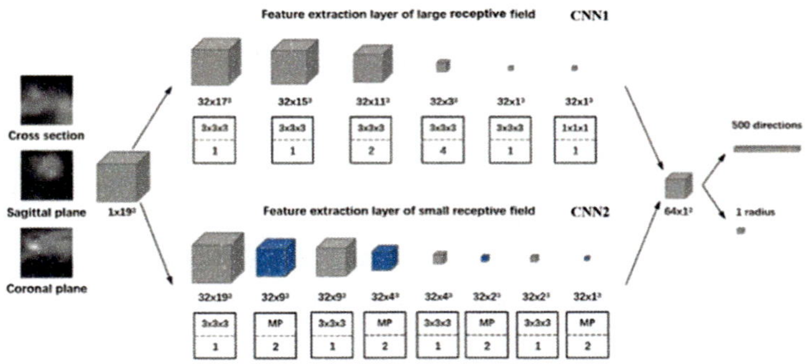

Fig. 1. MST-Net network structure

CNN1 is composed of stacked $3 \times 3 \times 3$ extended convolution layers whose convolution kernel has different degrees of expansion. Among them, the expansion degree of the first two convolution kernels and the last two convolution kernels is 1, and the expansion degree of the third and fourth convolution kernels is 2 and 4 respectively. The step length increased between convolution kernel elements enables these layers to extract image features at a large scale.

CNN2 consists of a $3 \times 3 \times 3$ convolution layer and a Max Pooling (MP) layer with an expansion degree of 1. The step length of all MPs is 2. CNN2 can obtain smaller feature maps through MP, thus extracting image features at a smaller scale.

In the final layer of MST-Net, the feature maps extracted by CNN1 and CNN2 are spliced and processed by $1 \times 1 \times 1$ convolution layer to finally obtain the probability distribution and radius value of the direction category.

2.2 Network Training Strategy

The training updates the network parameters θ through the Adam optimizer to minimize the loss function, as shown in Eq. (1).

$$l(\theta) = l_c(\theta) + \lambda_r l_r(\theta) + \lambda_\omega \|\theta\|^2 \tag{1}$$

Here, l_c is the classification cross-entropy between the probability distribution on the reference direction set D and the posterior probability distribution. l_r is the square mean error between the reference radius and the predicted radius, which is weighted by the parameters λ_ω, and $\lambda_\omega \|\theta\|^2$ is the L2 regularization term of the network parameters. To balance the impact of each loss item on the overall loss function, there use $\lambda_r = 10$ and $\lambda_\omega = 0.01$.

This paper uses the exponential decay strategy of learning rate: the initial learning rate $\eta = 0.001$, the decay rate $d = 0.1$, and the total decay step $n = 3200$. Each batch e carries out the decay of the learning rate. The calculation of learning rate $\eta_{(e+1)}$ in the $e + 1$st batch is shown in Eq. (2).

$$\eta_{e+1} = \eta_e d^{e/n} \tag{2}$$

2.3 Centerline Iterative Extraction Strategy

Since the centerline extraction network can only predict the direction of the coronary artery at a certain point, it must be iteratively predicted to obtain a complete coronary centerline. The centerline iterative extraction strategy is divided into five steps:

1. The centerline extraction algorithm needs to start from the seed point x_0, intercept isotropic 3D cutting block P_0 with x_0 as the center, and input it into the centerline extraction neural network.
2. To determine the two initial directions d_0 and $d_0^{'}$ from the seed point to the proximal and distal end of the coronary artery, multiple local maxima are identified in the posterior probability distribution $p(D|P_0)$. The corresponding reference direction can be found through D, and the reference direction with an angle greater than or equal to 90° is selected, that is, d_0 and $d_0^{'}$. Randomly select two seed points to obtain the posterior probability distribution and visualize it, as shown in Fig. 2. The gray scattered points represent 500 standard directions. The closer the color of the scattered points is to black, the greater the probability. The two green scattered points represent the extracted two centerline reference directions, namely d_0 and $d_0^{'}$.

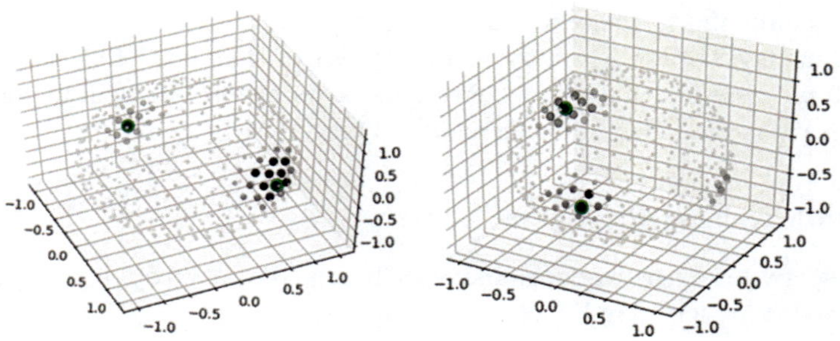

Fig. 2. Predictive probability distribution of reference direction of centerline

3. The centerline extraction network will first extract along the d_0 direction. To extract the centerline along the d_0 direction, the point x_0 needs to be offset by a certain distance to the d_0 direction to reach point x_1. The offset distance depends on the

predicted radius r_0 of the coronary artery. The position $L(x_1)$ of x_1 is calculated as Eq. (3):

$$L(x_1) = L(x_0) + d_0 * r_0 \tag{3}$$

4. A new slice P_1 is extracted at x_1, and processed by the centerline extraction network again to provide $p(D|P_1)$ and r_1. Choose the direction with the highest probability in $p(D|P_1)$ as the new centerline reference direction d_1. To prevent the centerline extractor from moving backward, the direction with an angle of more than 60° from the previous centerline reference direction d_1 is excluded first. Repeat these four steps until the termination conditions are met.
5. The centerline extraction network is along the d_0 direction, starting from x_0 point again, and repeating the first to fourth steps until the termination condition is met.

Among them, the normalized entropy $H(p(D|P)) \in [0, 1]$ of posterior probability distribution calculated on each P is used as the termination condition, as shown in Eq. (4).

$$H(p(D|P)) = \frac{\sum_{d \in D} -p(d|P)\log_2 p(d|P)}{\log_2 |D|} \tag{4}$$

If the normalized entropy of the posterior probability distribution of the current block P exceeds the threshold $Q_H = 0.7$, the centerline extraction is terminated. To cross these areas for centerline extraction and prevent the extraction process from being terminated early, this paper calculates the moving average of the entropy values obtained from the past three centerline extractions as the current entropy value, similar to the probability centerline extraction scheme proposed by Wang et al. [6]. When the extracted centerline is too close to the extracted center line, the centerline extraction process will be terminated.

3 Experimental Results of Coronary Centerline Extraction

This paper uses two data sets, one from the open coronary artery centerline extraction evaluation framework, namely CAT08, which is part of the Rotterdam coronary artery evaluation framework. A total of 32 CCTA images were collected on 64-slice CT scanner or dual-source CT scanner.

The other data set comes from Shanghai Xingmai Technology. It includes a total of 32 CCTA images, which are randomly divided into 12 training sets, 4 verification sets, and 16 test sets. Unlike the CAT08 dataset, the radius reference annotation of the centerline is not marked. Here uses the semi-automatic coronary artery centerline extraction network trained on the CAT08 data set to predict the centerline of the apricot vein dataset, and obtain the radius corresponding to all the center points as a reference.

3.1 Evaluation Criterion

All evaluation methods are based on the fact that points on the centerline are marked as True Positive (TP), False Negative (FN) or False Positive (FP).

If the distance from a point on the extracted centerline to at least one point on the reference centerline is less than the annotation radius, the point is marked as True Positive Track (TPT), otherwise as False Positive Track (FPT). If the distance from at least one point on the reference centerline to the marked point is less than the annotation radius at the reference point, the point is marked as True Positive Reference (TPR), otherwise it is marked as False Negative Reference (FNR). The most distal point with a diameter greater than 1.5 mm on the extraction centerline and reference centerline is shown in Fig. 3 The reference centerline with annotation radius is represented by a red dotted line, and the extracted centerline is represented by a green dotted line. The English names above the green dotted line and below the red dotted line are the mark partitions of the extracted centerline and the reference centerline.

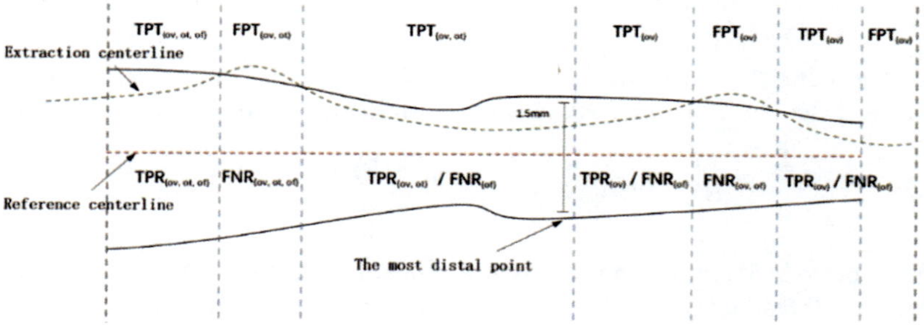

Fig. 3. Evaluation results of Tanh function

According to the marker division of the points on the centerline, four evaluation criteria for the extraction of semi-automatic coronary artery centerline can be obtained, as shown below.

(1) Overlap (OV), which represents the ability to extract the centerline of a complete coronary artery, is defined as the ratio of the number of correctly predicted points on the extracted centerline and reference centerline to the total number of points on the extracted centerline and reference centerline, as in Eq. (5):

$$OV = \frac{\|TRT_{ov}\| + \|TPR_{ov}\|}{\|TRT_{ov}\| + \|TPR_{ov}\| + \|FRT_{ov}\| + \|FNR_{ov}\|} \tag{5}$$

(2) Overlap until first error (OF), which characterizes how many center points have been correctly extracted when the first error occurs in the centerline extraction. The measured value is defined as the ratio of the number of correctly predicted points on the reference centerline to the total number of points on the reference centerline before the first error, as in Eq. (6):

$$OF = \frac{\|TPR_{of}\|}{\|TPR_{of}\| + \|FNR_{of}\|} \tag{6}$$

(3) Overlap with the clinically relevant part of the vessel (OT), which provides the index that this method can extract the blood vessels assumed to be clinically relevant.

The calculation equation of OT is (7):

$$OT = \frac{\|TPT_{ot}\| + \|TPR_{ot}\|}{\|TPT_{ot}\| + |TPR_{ot}| + \|FPT_{ot}\| + \|FNR_{ot}\|} \tag{7}$$

(4) Average inside (AI) is the average Euclidean distance of all connection points between the reference centerline and the extraction centerline, representing the accuracy of the extraction of the centerline.

3.2 Experimental Results and Analysis

To quantify the semi-automatic centerline extraction effect of Origin-Net and MST-Net, this paper lists the subjective image quality, OV, OF, OT, AI and the average value of all data. Table 1 and Table 2 are respectively the semi-automatic centerline extraction result quantization table of Origin-Net and MST-Net on the CAT08 data set.

Table 1. Centerline extraction results of Origin-Net on the CAT08 dataset

Dataset	Image quality	OV	OF	OT	AI
0	Moderate	0.845861	0.695699	0.911544	0.292105
1	Moderate	0.904549	0.832893	0.904549	0.307739
2	Good	0.960592	0.905508	0.960445	0.245638
3	Poor	0.729346	0.544256	0.732200	0.344307
4	Moderate	0.978051	0.962574	0.983478	0.220648
5	Poor	0.991538	0.980645	0.991538	0.239278
6	Good	0.910526	0.840496	0.917518	0.32893
7	Good	0.846039	0.458109	0.868569	0.288229
Average		0.902949	0.7775225	0.908730	0.283359

Table 1 and Table 2 can be seen the average values of OV, OF, OT and AI are increased by 3.3%, 8.1%, 3.1% and 1.8% respectively. Table 3 is quantitative tables of semi-automatic centerline extraction results of origin net and MST net on Xingmai dataset respectively.

The results in Table 3 show that the centerline extraction network MST-Net performs well in OV and OT on the Xingmai dataset, reaching 0.856 and 0.900.However, the OF and AI indexes of the two centerline extraction networks are poor. This is because the radius annotation data on the Xingmai dataset is not expert annotation, but is predicted by the centerline extraction network trained on the CAT08 dataset.

Table 2. Centerline extraction results of MST-Net on CAT08 dataset

Dataset	Image quality	OV	OF	OT	AI
0	Moderate	0.90078125	0.719256	0.900781	0.28924
1	Moderate	0.9450655	0.905	0.9450655	0.277964
2	Good	0.993287	0.988726	0.993287	0.249331
3	Poor	0.83801	0.644116	0.838965	0.309166
4	Moderate	0.934102	0.763937	0.93857	0.253262
5	Poor	0.966457	0.940753	0.9664565	0.299242
6	Good	0.955354	0.877686	0.962863	0.269559
7	Good	0.928929	0.889917	0.952782	0.277770
Average		0.932748	0.841174	0.937346	0.278192

Table 3. Centerline extraction results of Origin-Net and MST-Net on Xingmai dataset

Dataset	OV	OF	OT	AI
AVG (Origin)	0.839435	0.378801	0.886131	0.558457
AVG (MST)	0.855525	0.382509	0.900348	0.550471

4 Summary

Aiming at the problem that the radius error of the centerline obtained by the traditional semi-automatic coronary centerline extraction network is large, and the centerline is easy to extend to non-coronary tissues at the mouth and end of the coronary artery, this paper proposes a new multi-scale coronary centerline extraction network: MST-Net. The network combines the stacked expansion convolution layer with the down-sampling feature extraction layer. Through multi-scale feature extraction, the network can extract more local features and capture more global features, thus improving the network's ability to predict the direction and radius of coronary artery extension and alleviating the above problems.

The experimental results of semi-automatic coronary artery centerline extraction show that the new semi-automatic coronary artery centerline extraction network MST-Net extracts the centerline on CAT08 and Xingmai data sets compared with the traditional network Origin-Net, and its OV, OF, OT and AI indexes are greatly improved, which effectively reduces the error of radius prediction and improves the accuracy of centerline extraction.

Acknowledgment. We thank Shanghai Key Laboratory of Intelligent Manufacturing and Robotics for assistance with our work.

References

1. Abbara, S., Arbab-Zadeh, A., Callister, T., et al.: SCCT guidelines for the interpretation and reporting of coronary CT angiography: a report of the Society of Cardiovascular Computed Tomography Guidelines Committee. J. Cardiovasc. Comput. Tomogr. **8**(3), 342–358 (2014)
2. Ma, C., Sonka, M.: A fully parallel 3D thinning algorithm and its applications. Comput. Vis. Image Underst. **64**(3), 420–433 (1996)
3. Holland, L., Olivier, L.: Implementation of a 3D thinning algorithm. J. Sci. World **2014**(2), 599–605 (2014)
4. Benmansour, F., Cohen, L.: Fast object segmentation by growing minimal paths from a single point on 2D or 3D images. J. Math. Imag. Vision **33**(2), 209–221 (2009)
5. Jelmer, M., Wolterink, et al.: Coronary artery centerline extraction in cardiac CT angiography using a CNN-based orientation classifier. Med. Image Anal. **51,** 46–60 (2018)
6. Wang, X., Heimann, T., Lo, P., et al.: Statistical tracking of tree-like tubular structures with efficient branching detection in 3D medical image data. Phys. Med. Biol. **57**(16), 5325 (2012)

Research on 3D Model Search Technology Based on Sketches

Chuanhong Zhou, Youquan Tan(✉), and Lei Ding

School of Mechatronic Engineering and Automation, Shanghai Key Laboratory of Intelligent Manufacturing and Robotics, Shanghai University, Shanghai, China
quanh@shu.edu.cn

Abstract. In the field of intelligent manufacturing, how to construct the parts of 3D model has become a major problem in the process of modeling. At present, the major industrial parts manufacturers have developed an available 3D model part library, so an effective search method is needed to help users find the required 3D model from the large 3D model database. The most commonly used 3d model search method is by name, although this method is very accurate, some of the models are only a general name, unable to achieve accurate positioning, so I need another way which has the function of automatic classification search, together with the name of the search method, according to actual needs to choose a suitable method to use. Due to the rapid development of deep learning technology, this paper proposes a method to realize 3D model search by hand sketching. The application of deep learning to the classification and search of 3D models can significantly improve the searching ability of 3D models. In this way, it can realize the search of 3D models more intelligently.

Keywords: CNN · Discriminant feature extraction · Mutual learning strategy

1 Introduction

Calculate and analyze the similarity between the sketch and the extracted two-dimensional image. The detailed process is shown in Fig. 1:

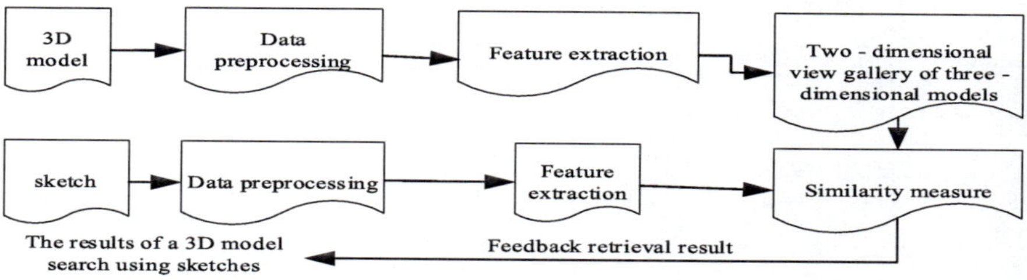

Fig. 1. Sketch-based 3D model search flowchart

Y. Wang et al. (Eds.): IWAMA 2021, LNEE 880, pp. 374–381, 2022.
https://doi.org/10.1007/978-981-19-0572-8_47

2 Two-Dimensional Display Method of Three-Dimensional Model

Inspired by the MVCNN [1] model, to improve the reduction of the 3D model when using the projection method, this paper proposes to circumscribe a sphere outside the 3D model, and then divide the outer surface of the sphere into 5 layers at equal distances in the vertical direction. As the acquisition perspective layer, and place an acquisition camera at an interval of 18° on the semicircular area of the sphere, and ensure that the lens of the camera is perpendicular to the line between the center of mass of the model and the camera. After the end of one shooting, the semi-circle is rotated by 18°, and the shooting is continued until the semi-circle returns to the initial position of 0°. In addition, images are extracted vertically upwards and vertically downwards respectively. Figure 2 is a schematic diagram of the projection points when using the projection method, and $20 \times 5 + 2$ two-dimensional images will be extracted, as shown in Fig. 3:

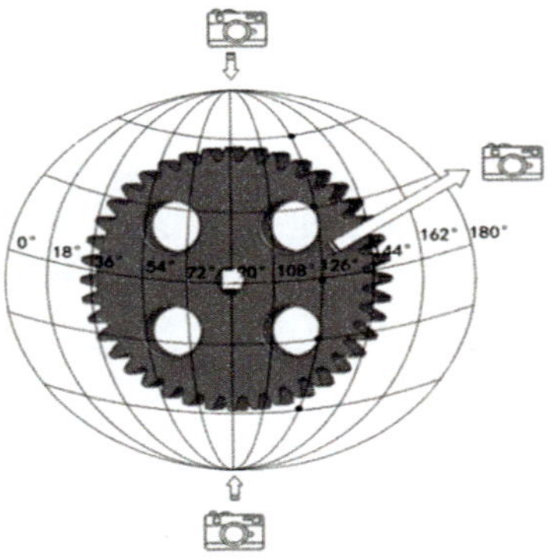

Fig. 2. Schematic diagram of three-dimensional model projection

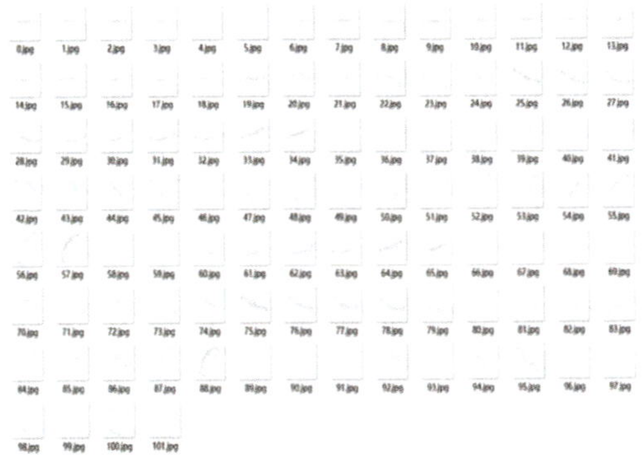

Fig. 3. The three-dimensional model is projected to form a two-dimensional image

3 Discriminative Feature Extraction Based on Convolutional Neural Network

3.1 Discriminant Feature Extraction of Three-Dimensional Model Convolutional Neural Network

The process of discriminant learning of 3D model is shown in Fig. 4:

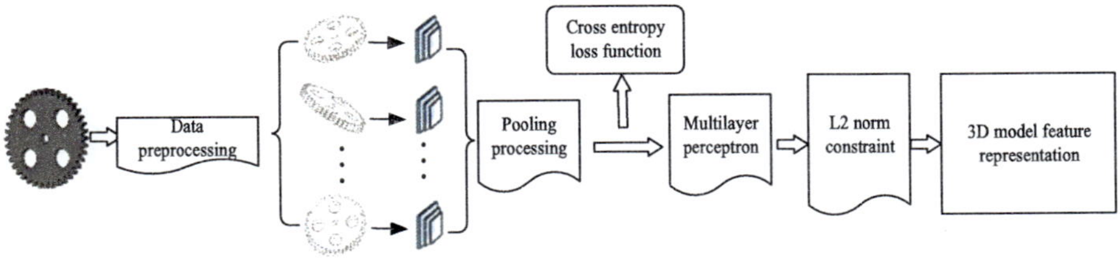

Fig. 4. Discretionary learning flow chart of 3D model

If a 3D model x_m^i is given, suppose that the set of projected views obtained after data preprocessing is:

$$V^t = \left\{v_j; 1 \leq j \leq l, v_j \in R^{H \times W \times 3}\right\} \quad (1)$$

Where l represents the number of projected views obtained, and H and W represent the height and width of projected views respectively. Then, the weighted convolutional neural network is used to extract the features of these views respectively, and the characteristic response graph set obtained is as follows:

$$F^i = \left\{f_j; 1 \leq j \leq l, f_j \in R^{C \times H' \times W'}\right\} \quad (2)$$

Where C, H' and W' represent the number of channels, height and width of the characteristic response graph respectively. Then, these eigenresponse graphs are sent to the pooling part for processing, and the eigenresponse graph set F^i is converted into a eigenvector representation to obtain a more compact feature representation.

The set of characteristic response graph F^i was respectively averaged pooled in the channel direction, and the pooled feature vectors were obtained to represent the set $P^i = \left\{p_j; 1 \leq j \leq l, p_j \in R^{C \times 1 \times 1}\right\}$. Then, a more compact feature vector $F_m^i \in R^{C \times 1 \times 1}$ was obtained for classification through the maximum pooled treatment in the feature direction. The whole process of pooled characteristic response graph is shown in Fig. 5:

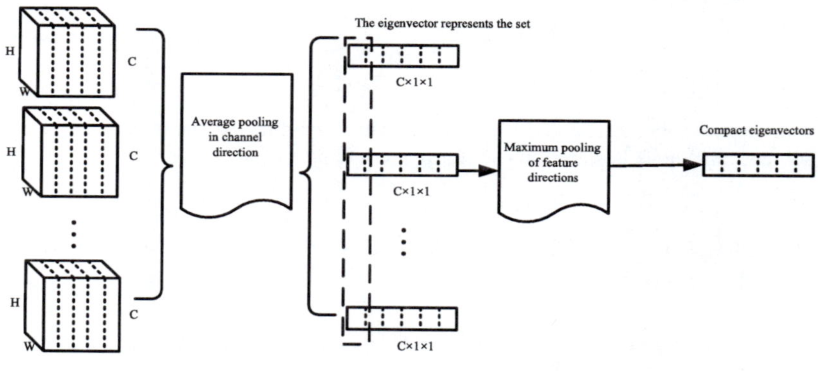

Fig. 5. Pool processing flow chart of characteristic response graph

Since the convolutional neural network used contains the ReLU nonlinear activation function (the output value is greater than or equal to 0), it will cause the feature vector F_m^i to be distributed in a certain quadrant, which limits the expression of the network.In order not to modify the convolutional neural network used and avoid affecting the performance of the network, after obtaining the feature vector F_m^i, use a multi-layer perceptron operation to convert all the features into any quadrant. The specific operation process is shown in Fig. 6. It is represented as the feature vector F_m^i is processed by the full connection layer FC, batch normalization layer BN, nonlinear activation function *LeakyReLU*, and is processed by the full connection layer FC again, and then is processed by the normalization of L2 norm to obtain the final output feature for searching.

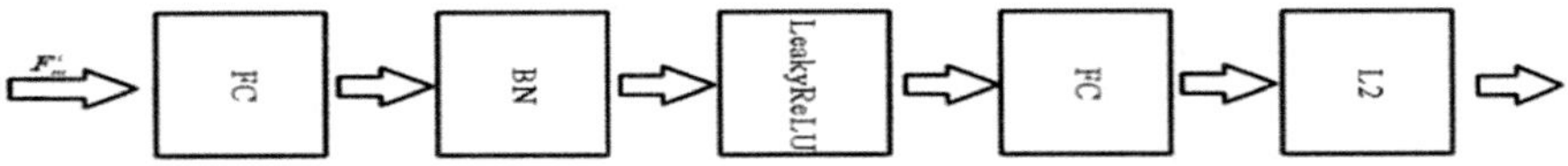

Fig. 6. Operation flow chart of multilayer perceptron

For a given 3D model training set $\left\{\left(x_m^i, y_m^i\right)\right\}_{i=1}^{N_m}$, it contains N_m samples, and each sample $x_m^i \in S_{train}$ is associated with the sample label $y_m^i \in \{2, 3, \ldots |C|\}$ to which it belongs, where $|C|$ represents the number of categories in the 3D model training set, and S_{train} represents the training sample set. These samples will be embedded into A higher dimensional space by the Convolutional Neural Network $f_\theta(\cdot)$. However, in high dimensional space, the value obtained by Euclidean distance operation on the eigenvector is often very large, which will lead to very unstable values and easy to produce numerical overflow errors. Therefore, we need to standardize $f_\theta^m\left(x_m^i\right)$ to get $f_m^i \in R^d$ through transformation. The transformation process is as follows:

$$f_m^i = \frac{f_\theta^m\left(x_m^i\right)}{\left\|f_\theta^m\left(x_m^i\right)\right\|_2} \tag{3}$$

Assume that features from the same type of three-dimensional model share a corresponding category feature representation vector. We can get $|C|$ feature vectors, which are $C = \{c_1, c_2, \ldots c_{|c|}\}$ respectively. When, $C_{y_m^i}^2 = 1, C_{y_m^i} \in R^d$ represents the category feature vector associated with category y_m^i.

In the training stage, the feature vector is represented by the updated parameter class in the data set. At this point, it is assumed that there are M batch of training samples, and the discriminant loss function L_d is defined as follows.

$$L_d = \frac{1}{m}\sum\nolimits_{i=1}^{M} max\left(0, D\left(f_m^i, c_{y_m^i}\right) + m - \min\nolimits_{j \neq y_m^i} D\left(f_m^i, c_j\right) + D\left(f_m^i, c_{y_m^i}\right)\right) \tag{4}$$

Here, m represents the interval threshold, and $D(,)$ represents the cosine distance. Since the eigenvectors f_m^i and $c_{y_m^i}$ need to be processed by Formula (3), the cosine distance between two vectors can be calculated through the cross product operation:

$$D\left(f_m^i, c_{y_m^i}\right) = 1 - \frac{\left\langle f_m^i, c_{y_m^i}\right\rangle}{\left\|f_m^i\right\| \cdot \left\|c_{y_m^i}\right\|} = 1 - \left\langle f_m^i, c_{y_m^i}\right\rangle \tag{5}$$

The learning goal of objective function L_d is to make the sum of the distance between sample x_m^i and its corresponding category representation vector $c_{y_m^i}$ and the interval threshold m smaller than the distance between sample x_m^i and its closest class representation vector $c_j(j \neq y_m^i)$, and let sample x_m^i and its The corresponding category indicates that the distance between the vectors $c_{y_m^i}$ is as close as possible. If only L_d is used to supervise the convolutional neural network, since the value of L_d may be very small, the 3D model features extracted by the convolutional neural network will all be close to the category representation features of the same class, and there will be no distinction between the sample features of the same class.Using the cross-entropy loss function can make the sample features within the class have greater differences. Therefore, the cross-entropy loss function loss L_s and the objective function L_d can be used for joint supervision, and the convolutional neural network can be trained to learn the discriminative features of the three-dimensional model. Then the objective function L_{3DMN} used in the discriminative learning phase of the entire 3D model is as follows:

$$L_{3DMN} = L_S + \lambda L_d = -\sum_{i=1}^{m} log \frac{e^{w_{y_m^i}^T f_\theta^m(x_m^i) + b_{y_m^i}}}{\sum_{j=1}^{|C|} e^{w_j^T f_\theta^m(x_m^i) + b_j}} + \lambda L_d \tag{6}$$

Where λ is a hyperparameter used to balance the cross-entropy loss function loss L_s and the objective function L_d,W and b represent the parameters of the classifier.

3.2 Discriminant Feature Extraction of Sketch Convolutional Neural Network

Yang et al. [2] proposed a feature algorithm SketchGCN based on convolution and global branch network structure to extract features within and between strokes for feature annotation, which makes the semantic segmentation of hand-drawn sketches better. In this paper, the sketch is segmented according to the edges that make up the sketch, and the features that form the partial sketch after segmentation are extracted separately, and then the segmented partial sketch is complementary to the original sketch through the mutual learning strategy, and the feature label of the sketch is extracted more effectively. The discriminative feature learning process of the sketch is shown in Fig. 7:

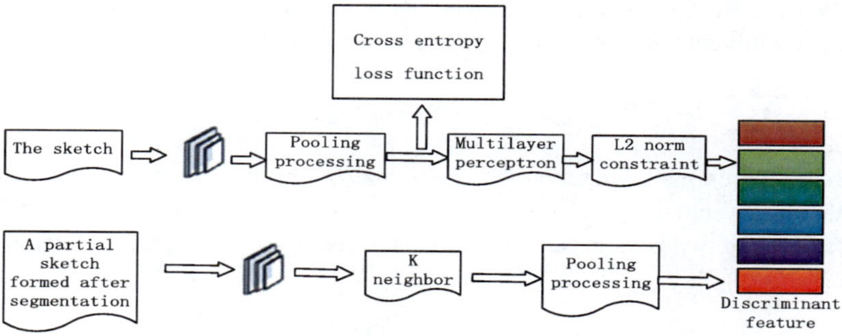

Fig. 7. Discriminant learning flow chart of the sketch

3.3 Mutual Learning Between Convolutional Neural Networks of Sketches

Ying Zhang [3] proposed a deep mutual learning strategy so that only two untrained model networks need to learn from each other to promote the improvement of network performance. Since the convolutional neural network used for sketch recognition is more complex, the mutual learning strategy needs to introduce category consistency loss and visual attention consistency loss [4, 5] to constrain the inaccuracy between hand-drawn sketches and partial sketches, and then use hand-drawn sketches and segmented partial The convolutional neural network of the sketch complements the information to complete the learning and migration of the two networks. The calculation process of the learning loss function for hand-drawn sketches and partial sketches is shown in Fig. 8:

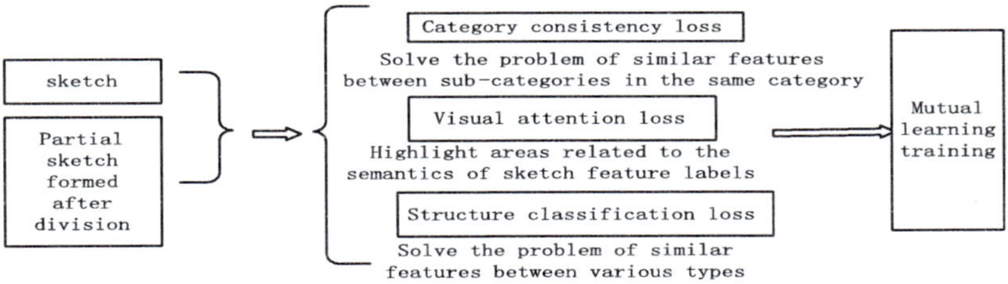

Fig. 8. The calculation flow chart of mutual learning loss between hand-drawn sketches and partial sketches

4 Similarity Measurement Between Sketch and 3D Model

For any given two-dimensional sketch $x_s^i \in S$, set the feature descriptor obtained through the convolutional neural network as $f_s^i = f_\theta^s(x_s^i) \epsilon R^d$, For any three-dimensional model x_m^i, its feature descriptor can be expressed as $f_m^i = f_\theta^m(x_m^i) \epsilon R^d$, then the two-dimensional sketch x_s^i to the three-dimensional model The distance between x_m^i can be expressed as $D(f_s^i, f_m^i)$, and the similarity between the sketch and the 3D model can be calculated according to the cosine distance. The calculation formula is:

$$D\left(f_s^i, f_m^i\right) = 1 - \frac{\left\langle f_s^i, f_m^i\right\rangle}{\left\|f_s^i\right\| \cdot \left\|f_m^i\right\|} \tag{7}$$

Among them, $\| \bullet \|$ represents the L2 norm, and $\langle , \rangle$ represents the vector product operation between two vectors. Since f_s^i and f_m^i can be standardized by the formula, the formula can be equivalent to:

$$D\left(f_s^i, f_m^i\right) = 1 - \left\langle f_s^i, f_m^i\right\rangle \tag{8}$$

When matching the 3D model, first, calculate the hand-drawn sketch and the 2D image generated in the parts library according to the method of similarity measurement, find the projection view with the highest similarity calculation result, and then find the corresponding projection view 3D model.

On the 3D graphics application interface, first draw a sketch of the part by hand, and then calculate it according to the similarity measurement method. According to the degree of similarity, find the 2D projection view that best matches the hand-drawn sketch, and display it in the right area. At this time, select The projection view will display the 3D part model corresponding to this 2D image on the lower left side of the visible area, as shown in Fig. 9:

Fig. 9. Use the visual interface to search for 3D models

5 Discussion and Conclusion

For the judgment of the accuracy of sketch recognition of 3D models, this paper uses the mAP (mean Average Precision) index commonly used in the field of 3D model search to judge the accuracy of recognition. The calculation method is as shown in the formula:

$$mAP = \frac{1}{N_q} \sum\nolimits_{q \in N_q} \sum\nolimits_{r \in R} \frac{R_i / L_i}{N_R} \tag{9}$$

Among them, q is the set of input hand-drawn sketches, $N_q$is the number of images in the set, R_i is the ranking number of the sample in the search output list, L_i is the ranking number of the sample in the image database list to be searched, and N_R is the number of samples.

In order to verify the performance of the proposed dual-branch feature extraction network based on mutual learning strategy, a comparative analysis of 3D model recognition was conducted, and the search precision curve was drawn, as shown in Fig. 10:

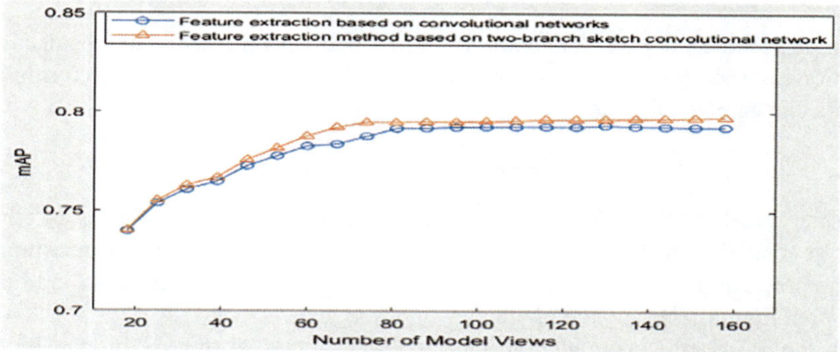

Fig. 10. Comparison of recognition accuracy of two sketch feature recognition algorithms

It can be seen from Fig. 10 that the sketch feature recognition algorithm used in this paper is better than using a convolutional neural network alone in terms of the effect of recognizing 3D models. After the number of model views is greater than 95, mAP is almost unchanged, so this paper selects 102 views to meet the requirements.

Acknowledgment. We thank Shanghai Key Laboratory of Intelligent Manufacturing and Robotics for assistance with our work.

References

1. Wang, C., Cheng, M., Sohel, F., Bennamoun, M., Li, J.: NormalNet: a voxel-based CNN for 3D object classification and retrieval. Neurocomputing **3**(23), 139–147 (2019)
2. Zhu, F., Xie, J., Fang, Y.: Learning cross-domain neural networks for sketch-based 3D shape retrieva. In: Conference of the American Association for Artificial Intelligenc, pp. 3683–3689 (2016)
3. Zhang, Y., Xiang, T., Hospedales, T.M., et al.: Deep mutual learning. In: Proceedings of the IEEE Conference on Computer Vision and Pattern Recognition, pp. 4320–4328 (2018)
4. Guo, H., Zheng, K., Fan, X., et al.: Visual attention consistency under image transforms for multi-label image classification. In: Proceedings of the IEEE Conference on Computer Vision and Pattern Recognition, pp. 729–739 (2019)
5. Selvaraju, R.R., Cogswell, M., Das, A., et al.: Grad-CAM:visual explanations from deep networks via gradient-based localization. In: Proceedings of the IEEE International Conference on Computer Vision, pp. 618–626 (2017)

The Research on Image-Based Chinese Ink Painting NPR Based on Deep Learning

Chuanhong Zhou, Kunpeng Chen(✉), and Shiyu Pan

School of Mechatronic Engineering and Automation, Shanghai Key Laboratory of Intelligent Manufacturing and Robotics, Shanghai University, Shanghai, China
shuchenkunpeng@shu.edu.cn

Abstract. Image stylization technology can render the content photos taken by users in a specified style, reducing the artistic threshold. But the colour of the ink style image is light, the texture is not clear, the style is difficult to grasp, and the direct conversion effect is not good. This paper aims at the style of Chinese ink painting and realizes the ink style rendering based on the quantitative analysis of the style. To grasp the style of ink and wash, quantitative analysis of the relationship between low-level visual features and style, and propose the preprocessing method of style pre-matching. Aiming at freehand ink painting, this paper proposes a method of image ink stylization based on residual Wasserstein Gan. RESNET is used to extract features. Wasserstein Gan network adaptively approximates the feature distribution of ink style images and adjusts the style weight according to the image entropy, which greatly reduces the difficulty of parameter adjustment. Finally, realize the transformation of ink stylization.

Keywords: Ink stylization · Style pre-matching · ResNet · WassersteinGAN

1 Introduction

With the rapid development of artificial intelligence and other cutting-edge technology. The tentacles of deep learning technology have penetrated the art field. Some remarkable achievements have been made in the synthesis of paintings. Gatys et al. [1] first proposed a method of separating and recombining the content and style features of an image through a style transfer algorithm based on CNN to achieve image stylization rendering. This traditional stylization algorithm [1] achieves good rendering effects on oil paintings and watercolours with strong colours and clear textures. Due to the differences between Chinese and Western cultures, the ink painting style is elegant. The direct effect of using traditional stylization algorithms for ink painting style rendering is poor.

To make the image ink painting style rendering effect is good, need to in-depth analysis of ink painting style rendering law, quantitative analysis of ink style, and then the ink style rendering. On the quantification of style, Li et al.'s [2, 3] research shows that style can be simply interpreted as the distribution of features. In this paper, the image one-dimensional entropy, image moment, image brightness and image two-dimensional entropy are used to quantify the low-order visual features of the image to quantify the

Y. Wang et al. (Eds.): IWAMA 2021, LNEE 880, pp. 382–389, 2022.
https://doi.org/10.1007/978-981-19-0572-8_48

ink style. and constructs the style pre-matching module by using the law of ink style presentation. This module can predict the effect of style conversion and realize the pretreatment before ink style.

Secondly, the feature extraction of ink style is carried out by using the depth residual network. After the selection of content-style image is completed by the style pre-matching module, the image ink style method based on Residual-WassersteinGAN [4, 5] is proposed. It is proved that the rendering effect of freehand ink style is good.

2 Style Pre-matching Based on Quantitative Analysis of Ink Style

2.1 Ink Painting Style Analysis

Due to the different historical stages of the Northern Song, Southern Song, and Yuan Dynasties, their painting styles also showed obvious differences. To achieve the desired goal of image ink style conversion, it is necessary to determine which style of ink painting matches the image to be converted. In the absence of high artistic appreciation, it is difficult to determine the style of different paintings and images. To further grasp the law of style presentation, a quantitative analysis of style is carried out. First, it is necessary to clarify the style label. According to the composition characteristics and artistic conception of ink landscape painting, the artistic conception of ink landscape painting is divided into the following five kinds: quiet and distant, fresh and beautiful, peaceful and slowly, Strong and vigorous and magnificent. Select the famous paintings of the Song Dynasty for research. For further quantitative analysis of style, the low-level visual characteristics are quantified by statistics of painting amplitude, image brightness, image entropy, and image moment, and the relationship between these features and style is analyzed. The statistical results are shown in Table 1.

The ratio K of different types of ink painting varies with the change of artistic conception. From the perspective of image entropy, the more style tends to be majestic, the greater the information entropy, that is, the more informative the picture contains. When the style tends to be quieter and more distant, the information entropy is relatively less, but the two-dimensional entropy is greater. Regarding the brightness of the image, the brightness of the quiet painting is higher, and the painting of extraordinary and magnificent style is often lower in brightness.

Therefore, the image style prediction can be carried out preliminarily based on the rendering rules of different style images. It is easier to have a better conversion effect to select the content with a similar style based on the style category.

2.2 Image Style Pre-matching Module

An image style pre-matching module is proposed, which selects content-style images with similar style or high similarity to achieve better style conversion effect and improve style conversion efficiency.

1. Image style matching based on naive Bayes: For the style prediction of new art photos and paintings, the content-style images with consistent style categories will inevitably have a better conversion effect.

Table 1. Style quantitative analysis

Painting name	Volume ratio K	One-dimensional entropy	Two-dimensional entropy	Brightness	Style
Xiaoxiang	0.35	6.02	17.27	141.80	Quiet and distant
Xiashan	0.16	6.08	14.99	149.75	Quiet and distant
Xiajing Mountain Pass Waiting for Ferry	0.15	6.26	4.94	149.75	Quiet and distant
Idyllic Scenery	1.25	7.04	6.49	162.19	Fresh and beautiful
Spring Outing and Evening Return	2.02	6.91	4.12	161.32	Peaceful and slowly
Bleak Temple in Snow Mountain	1.69	6.89	5.423	133.19	Strong and vigorous
Kuanglu	1.74	7.10	4.99	127.99	Magnificent
Xishan Xinglu	2.00	6.93	4.08	136.86	Magnificent
Xishan Lanruo	3.22	7.08	4.71	105.23	Magnificent

2. Image Similarity Matching Based on Image Moment: For the weak artistic life picture, the similarity of content image and style image is calculated. According to the principle of image stylization [1], the Gram matrix quantizes the correlation between features, and the effect of stylization depends on the content image and style. The more similar the image is, the better the style conversion is.

2.2.1 Image Style Matching Based on Naive Bayes

Based on the data in Table 1, to realize the style prediction of the new image, several points need to be considered: The data set is small and the feature dimension is not high. Considering that this is the pre-processing for deep learning style conversion, it is required to spend as little time as possible to complete the matching. The naive Bayesian classification algorithm is still effective in the case of small data volume and has the advantages of simple operation and high classification accuracy, which is suitable for the preliminary analysis and classification of styles.

Naive Bayes is a supervised learning algorithm based on the naive assumption that the features are independent [6, 7]. The Bayesian algorithm is based on the Bayesian theorem, the formula of the Bayesian theorem is as follows:

$$P(A|B) = \frac{P(A|B)P(B)}{P(B)} \tag{1}$$

Among them, P(A) is the prior probability, P(B|A) is the conditional probability, and P(A|B) is the posterior probability. The Bayesian classifier is transformed according to Bayes' theorem. For data set $D = \{(x^{(1)}, y^{(1)}), (x^{(2)}, y^{(2)}), \ldots, (x^{(n)}, y^{(n)})\}$. There is a total of N data, and each data x has n features, and y is the category corresponding to x. Suppose there are a total of k categories. Then, given a new x, according to Bayes' theorem:

$$P(C_k|x) = \frac{P(x|C_k)P(C_k)}{P(x)} \tag{2}$$

The x in this formula is a vector, which has n features, from x [1] to x [n]. The Naive Bayes algorithm has a hypothesis that these n features are independent of each other and do not affect each other.

Expand P(x) in formula (2) according to the full probability formula to obtain the naive Bayes model:

$$P(C_k|x) = \frac{P(C_k)\prod_{i=1}^{n} P(x|C_k)}{\sum_{k=1}^{k}\left[P(x)\prod_{i=1}^{n} P(x_i|C_k)\right]} \tag{3}$$

The implementation process of the image style matching module based on Naive Bayes is shown in Fig. 1.

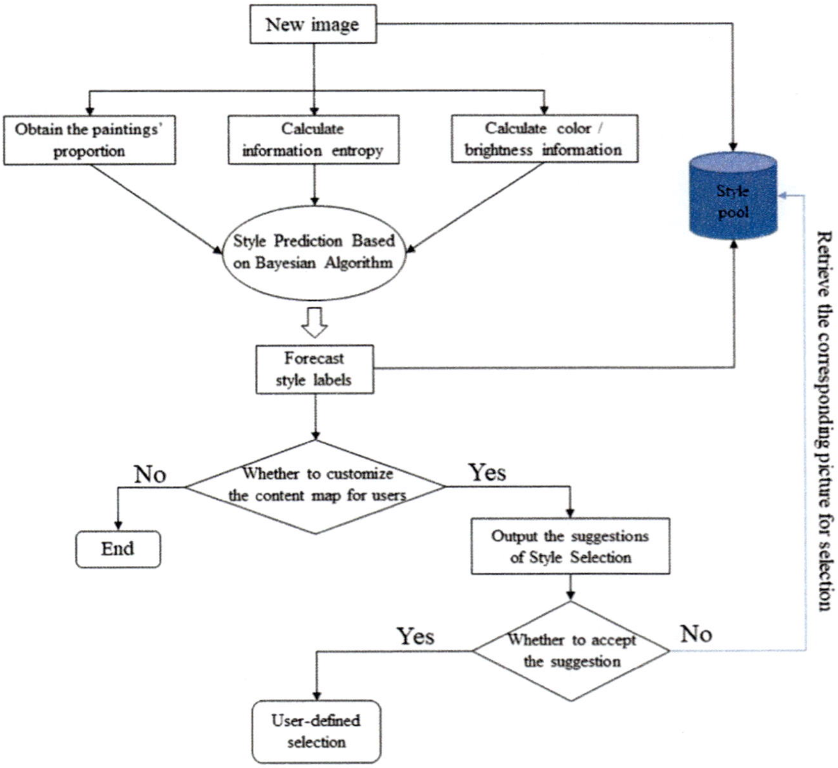

Fig. 1. Image style matching based on naive Bayes

2.2.2 Image Similarity Matching Based on Image Moment

In the image moments, Hu constructed seven invariant moments by using second and third order normalized center moments. The feature quantity composed of Hu moments has a low recognition rate for images, but it has the absolute advantage of fast speed, which is very suitable for the early analysis of content and style images. Therefore, Hu moment degree calculation is used to select similar content style image for stylization conversion.

The image similarity matching process based on image moments is shown in Fig. 2.

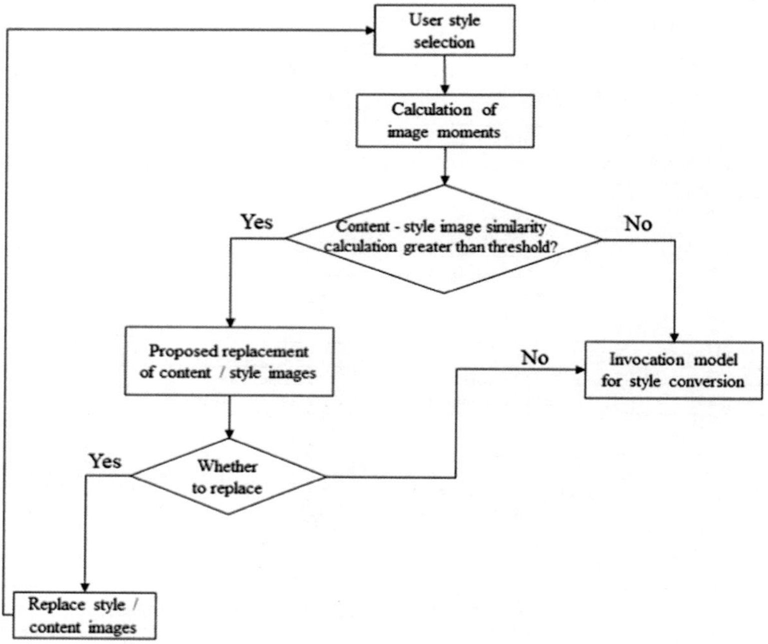

Fig. 2. Image similarity matching based on image moment

3 Freehand Ink Stylization Based on Image Iteration

3.1 Image Stylization Based on Different Styles

3.1.1 Style Pre-matching

Before style conversion, to predict the effect of style conversion, style pre-matching is performed. The calculation results of image similarity based on moment features are shown in Table 2.

According to the similarity calculation results, the replacement effect will be obtained by selecting the style-content combination 1 with less difference.

Table 2. Content-style pre-matching

Style-content combination	Style image	Content image	Similarity
Style-content combination 1			0.32
Style-content combination 2			0.53

3.2 Image Stylization Based on Wasserstein Generative Adversarial Network

Acknowle Different from the strong western painting art, the ink style images use light colors and unclear textures. The artistic characteristics of the combination of virtual and real make the style more difficult to characterize. Direct representation of ink style is difficult, instead of using GAN network to adaptively approximate the feature distribution of ink style image.

Eddie Huang et al. [8] hypothesized that it would be a better choice to redefine the style loss under the popular distribution distance metric-Wasserstein distance. In Wasserstein-GAN, the discriminator D outputs a binary classification result to determine the true and false of the input image and reflects the similarity of the true and false data distribution by fitting the Wasserstein distance.

Use x^*, x_c, x_s to represent generated image, content image, and style image respectively. Let the x^*, x_c, x_s feature maps of the layer l of CNN use $F^l \in R^{N_l \times M_l}$,$S^l \in R^{N_l \times M_l}$. N_l is the number of the feature maps in Layer l, and M_l is the height multiplied by the width of feature map. Now, the column of F^l, P^l, S^l is called a feature. Intuitively, features can be interpreted as 'pixels' of feature mapping.

In neural type transfer, the generated image is optimized according to the following loss:

$$L = \alpha L_{style} + (1 - \alpha) L_{content} \tag{4}$$

Among them, the content loss $L_{content}$ is defined as mean square error between spatially corresponding features:

$$L_{content} = \frac{1}{2} \sum\nolimits_{i=1}^{N_l} \sum\nolimits_{j=1}^{M_l} \left(F_{ij}^l - P_{ij}^l \right)^2 \tag{5}$$

The style loss L_{style} is defined as the weighted sum of the distribution distance between the style feature and the generated image feature.

$$L_{style}^l = D\left[F^l, S^l\right], L_{style} = \sum\nolimits_l w_l L_{style}^l \tag{6}$$

Among them, L^l_{style} is the style loss of l layers in the style layer. L_{style} is the total style loss which calculates the weighted sum of the losses of each style layer. D is a certain distribution distance metric, where Wasserstein distance is used as the distance metric.

Different from Eddie Huang et al. [8], to solve the problem of insufficient feature extraction ability of traditional image stylization methods, Resnet 158 is used as feature extractor, and the features of standardized layer, conv1 layer of a convolutional neural network, layer1, layer2, layer3 and layer4 of the residual network are used to characterize the style. The content is described by the standardized layer, conv1 layer of the convolutional neural network, and layer1, layer2, layer3 of the residual network.

To compute style loss using Wasserstein distance, a discriminator network is attached to the corresponding layer. These discriminators approximate the Wasserstein distance between style features and generated features under Wasserstein-gp. Each discriminator has 3 hidden layers, 256 dimensions, and ReLU activation. Before these features are input into the discriminator, they are input through the unit-level Tanh activation, which is conducive to the regularization and boundedness of the training loss.

3.3 Experimental Results and Analysis

The experiment uses the style-content combination 1 in Table 2. In the experiment, the style weight is adjusted based on the one-dimensional entropy in Table 1, and the sensitivity of the network to the style is strengthened by increasing the style weight for the style image with low one-dimensional entropy. Since the larger the one-dimensional entropy of the style image is, the larger the amount of information contained in the image is, and the style feature is more obvious. Therefore, it is completely reasonable to adjust the style weight based on image one-dimensional entropy, and greatly reduce the difficulty of network parameter adjustment.

Due to freehand ink painting more advocate godlike, it does not strictly demand the form and structure. The grasp of the relationship between virtuality and reality in his artistic creation is an important method to express the content of the composition. Considering the style based on the pixel distribution of the whole image is conducive to grasping the artistic characteristics of the combination of virtuality and reality in Chinese painting. The ink image stylization based on Residual-WassersteinGAN achieves a good ink stylization conversion effect, as shown in Fig. 3.

Fig. 3. Ink image stylization based on Residual-Wasserstein GAN

4 Summary

In this paper, the quantitative analysis of ink style is carried out, and the automatic conversion of ink style is realized based on the image style algorithm of deep learning. The main work and innovations are as follows:

1. For ink style is difficult to grasp the problem, quantify the low-level visual features of the image to analyze the ink style. Based on the idea that style is an artistic bridge between photos and paintings, a style pre-matching method is proposed, the effect of style conversion is estimated, the selection of content images and style images is realized, and the pre-processing of ink stylization is completed.
2. When traditional stylization algorithms are applied to ink style images, the feature extraction ability is insufficient and the style representation is difficult. Therefore, an image ink stylized method based on Residual-Wasserstein GAN is proposed. And put forward the idea of adjusting the style weight according to the image entropy, reducing the difficulty of parameter adjustment, and finally realized the style conversion of ink painting.

Acknowledgment. We thank Shanghai Key Laboratory of Intelligent Manufacturing and Robotics for assistance with our work.

References

1. Gatys, L., Ecker, A., Bethge, M.: A neural algorithm of artistic style. J. Vis. **16**(12), 326 (2016)
2. Li, Y., Wang, N., Liu, J., et al.: Demystifying neural style transfer. arXiv preprint arXiv:1701.01036 (2017)
3. Li, Y., Fang, C., Yang, J., et al.: Universal style transfer via feature transforms. arXiv preprint arXiv:1705.08086 (2017)
4. Arjovsky, M., Chintala, S., Bottou, L.: Wasserstein GAN, pp. 124–129 (2017)
5. Gulrajani, I., Ahmed, F., Arjovsky, M., et al.: Improved training of wasserstein gans. arXiv preprint arXiv:1704.00028 (2017)
6. Kendall, A., Gal, Y.: What uncertainties do we need in bayesian deep learning for computer vision? arXiv preprint arXiv:1703.04977 (2017)
7. Su, Y., Zhang, Y., Hu, P., Tu, X.: Sentiment analysis research based on combination of naive Bayes and latent Dirichlet allocation. J. Comput. Appl. 555–561 (2016)
8. Huang, E., Gupta, S.: Style is a Distribution of Features (2020)

Design of a Comprehensive Experimental Platform for Intelligent Robots Based on Machine Vision

Haishu Ma(✉), Lixia Li, Fengxiang Shao, and Xidong Liu

Department of Mechanical Engineering, Henan University of Engineering, Zhengzhou, China
haishuma@haue.edu.cn

Abstract. With the increasing demand for personalized customization, the randomness of personalized customization tasks is strong, the state of the production line is unstable when the order arrives, and the scheduling model is difficult to determine. One of the basic requirements of intelligent manufacturing is to respond to individual needs and efficiently and flexibly produce small batches of multiple varieties. This paper proposes to use genetic algorithm to solve the resource scheduling in the face of multi-variety and small-batch tasks, and upload the scheduling results to the digital twin platform, and use the synchronous mapping of the digital twin to verify with the high-fidelity model, which reduces production accidents. Finally, the research results were applied to a miniature customized production line in a laboratory, which proved the feasibility of the method.

Keywords: Personalized customization · Digital twin · Multi-variety and small batch

1 Introduction

With the continuous development of artificial intelligence, robots have evolved from traditional online teaching to intelligence. Intelligent robots that have the ability to perceive and can interact with the surrounding environment are the development trend [1]. Machine vision refers to the use of cameras, image processing units, and machine vision algorithms to give robots the ability to see, so as to perform target recognition, detection, measurement and other functions. It is a necessary means to realize intelligent manufacturing [2].

Mechanical engineering is a popular new engineering major that meets the national strategic needs. It has strong novelty, comprehensiveness and practicality. It is one of the new majors supported by the Ministry of Education [3]. In the course of undergraduate teaching, adding practical teaching links that integrate machine vision, embedded programming and robot control will help college students to master solid theoretical knowledge and cultivate their comprehensive practical ability [4].

Y. Wang et al. (Eds.): IWAMA 2021, LNEE 880, pp. 390–396, 2022.
https://doi.org/10.1007/978-981-19-0572-8_49

In order to meet the teaching and training goals of robotics engineering students, a comprehensive experimental platform for intelligent robots based on machine vision has been developed. At the same time, taking into account the needs of class grouping in the future, this platform uses a desktop-level small four-axis robotic arm to realize sorting operations, which not only meets the requirements of teaching practice but also takes into account the cost of open design.

2 Overall Design of the Platform

The experimental platform is composed of Dobot robotic arm, image acquisition system, and Raspberry Pi ARM Cortex-A53. Among them, the Raspberry Pi development board has 4 USB expansion ports, a network port and a Wi-Fi, which are connected to the robotic arm through a USB cable and can be used as the core control board of the robotic arm. The operation process of the system is as follows: First, the camera collects images, locates the target through opencv image processing, and the neural network model performs image recognition to determine the target category. After object recognition and positioning, the robotic arm is controlled to place the target in the designated sorting position. The overall framework of this experimental platform is shown in Fig. 1.

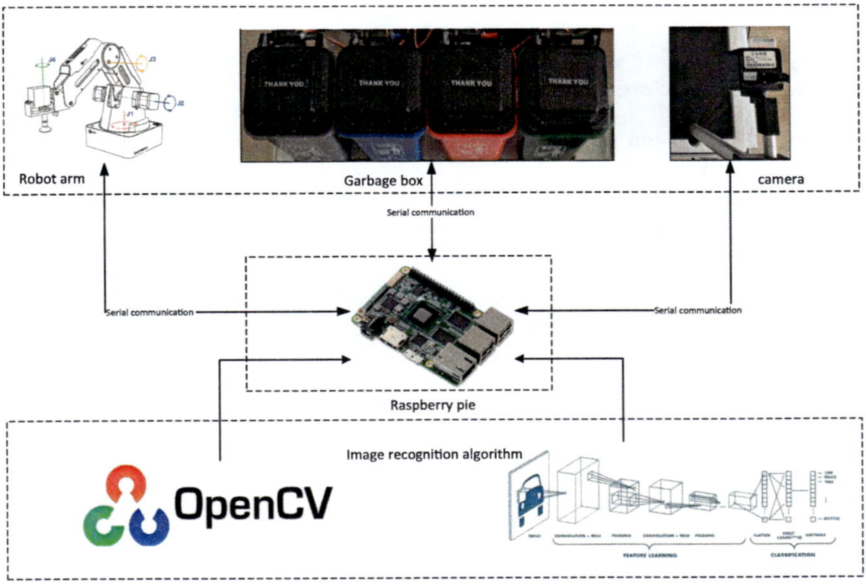

Fig. 1. The overall framework of the experimental platform

2.1 Robot Arm

The DOBOT robotic arm used in this experimental platform is a multi-functional, high-precision, lightweight, intelligent experimental robotic arm (Fig. 2), which is a one-stop STEAM education integrated platform [5]. The 13 expansion interfaces reserved by the robotic arm support secondary development, and more practical scenarios can be developed through software programming combined with hardware expansion, such as 3D printing, laser engraving, writing and drawing, and other functions to meet various experimental requirements.

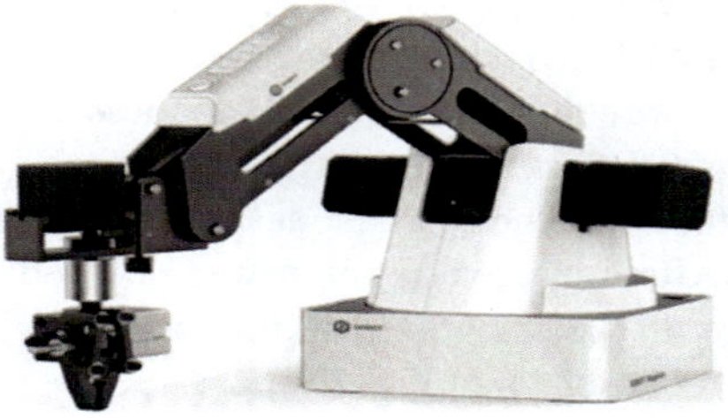

Fig. 2. Dobot robot arm

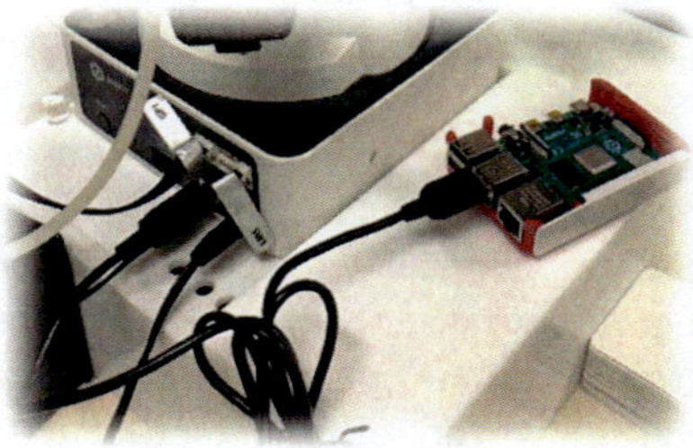

Fig. 3. Connection between raspberry pie and robot

2.2 Pyserial Serial Communication

pySerial encapsulates the serial communication module, has a unified interface on the supported platforms, and accesses the serial settings through python properties. Serial communication experiments can be carried out based on pyserial. The core class serial of the pySerial module realizes the establishment, disconnection, data transmission and other functions of serial communication [6]. The following steps are required to use the serial object for serial port data transmission: import the serial port module, create the serial port object, set the serial port parameters such as port name, transmission rate, open the serial connection, read and write data, and close the serial connection. Connect the robotic arm and the Raspberry Pi through a USB cable, and students can control the robotic arm through the dobot_serial module of the robotic arm serial communication, as shown in Fig. 3.

2.3 Robot Arm Motion Control

The motion modes of the manipulator include jog mode, point mode (PTP), and arc motion mode (ARC) [7]. Based on this experimental platform, students can practice robotic arm motion control. The robotic arm uses the point mode by default, that is, point-to-point motion. The Dobot robotic arm point mode uses the MOVJ motion mode, namely joint motion, moving from point A to point B, and each joint runs from the corresponding joint angle of point A. To the joint angle corresponding to point B [8]. During the joint movement, the running time of each joint axis must be the same and reach the end point at the same time. The end of the Dobot robot arm is installed as a suction cup (Fig. 4). After the target is identified and positioned, the coordinates are returned, and then the robot arm motion command can be executed. Target and place the sorting position.

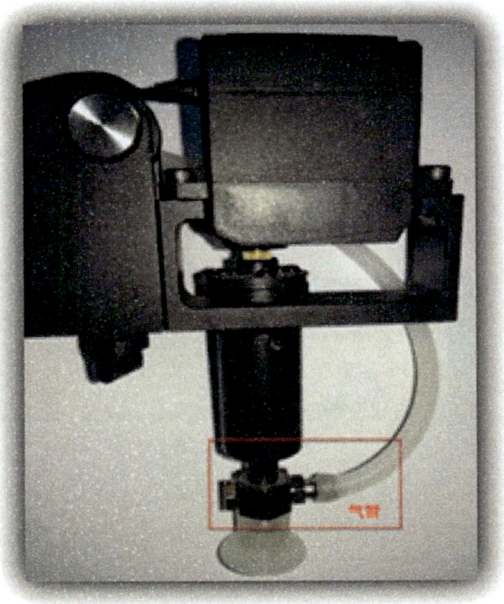

Fig. 4. Robotic arm end mechanism

3 Opencv Visual Inspection

Based on this experimental platform, opencv can be used for target detection experiments. First, open the camera to take a picture and read the image, perform image preprocessing to remove distortion, convert it into a grayscale image, and use Gaussian filter smoothing to facilitate the later target contour extraction [9]. Then perform edge detection, find the contour line corresponding to the target in the picture, and calculate the area of the contour. If the size of the contour area is not large enough, we will discard the area, thinking it is the noise left over from the edge detection process. If the waiting area is large enough, the center of the contour area is calculated and fed back to the robotic arm for suction operation. The Opencv visual inspection process is shown in Fig. 5.

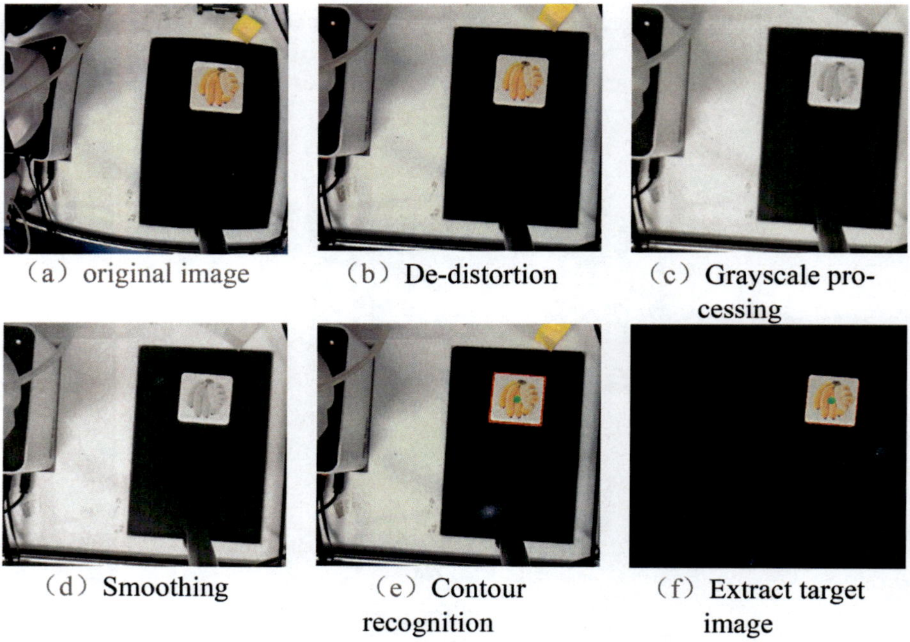

Fig. 5. Target image processing

4 Image Recognition

4.1 Data Set Processing

In order to facilitate students' practice, four common fruits are selected as the target of image recognition. Part of the data set source is obtained through web crawlers, and part of it is obtained through data enhancement to increase the diversity of training samples, improve model robustness, and avoid overfitting. Moreover, data enhancement can reduce the model's dependence on certain features, thereby improving the generalization ability of the model. Specific data enhancement methods include image flipping, rotation, translation, zooming, random cropping, color dithering, etc. Therefore, students can practice data collection and data preprocessing.

4.2 Neural Network Model

The residual network resnet has been widely used as the mainstream neural network model in the field of image recognition. Among them, the lightweight model Resnet18 network has low hardware requirements and can be calculated on the Raspberry Pi, so it is used for teaching on the experimental platform [10]. Through the understanding and learning of mainstream models, students can have a deeper understanding of neural networks.

4.3 Model Training

It is not necessary to randomly initialize the network to train the complete neural network, because the training data set is not large enough, and training the complete network is time-consuming and labor-intensive. The migration learning method is usually adopted to freeze the weights of all networks except the final fully connected layer. The last fully connected layer is replaced with a new layer with random weights, and only this layer is trained. That is, the parameters of the convolutional layer are fixed, and the parameters of the last fully connected layer are adjusted [11]. The Resnet18 migration learning program is shown in Table 1. Due to the pre-training model, high accuracy can be achieved after 10 training cycles (Fig. 6). The final model prediction result is shown in Fig. 7. The transfer learning is achieved by fixing the convolutional layer and only modifying the fully connected layer. The convolutional layer parameters are unchanged during training. Cross-Entropy is uesd as the loss function and momentum SGD is selected as optimizer.

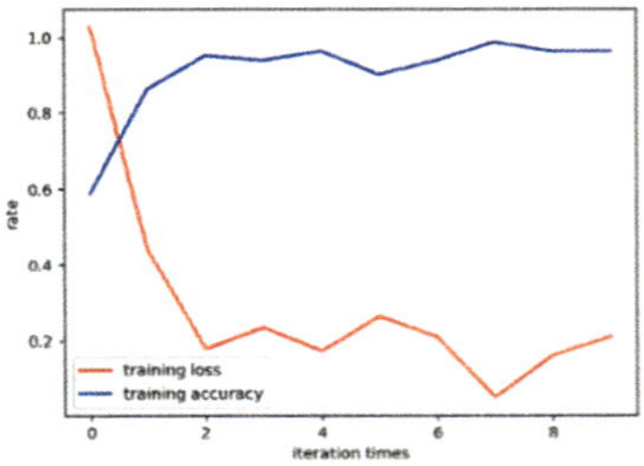

Fig. 6. Model training curve

Fig. 7. Model prediction result

Table 1. Resnet18 transfer learning

```
model_conv = torchvision.models.resnet18(pretrained=True)
for param in model_conv.parameters():
        param.requires_grad = False
model_conv.fc = nn.Linear(num_ftrs, 2)
criterion = nn.CrossEntropyLoss()
optimizer_conv =
optim.SGD(model_conv.fc.parameters(), lr=0.001, momentum=0.9)
model_conv = train_model(model_conv, criterion, optimizer_ft,
num_epochs=10)
```

5 Conclusion

This paper designs a comprehensive experimental platform for intelligent robots based on machine vision, which collects, recognizes, locates, and controls the robot arm to perform sorting operations on the target image.Students of related majors can carry out experiments in communication, control, and programming.The platform uses Raspberry Pi as the core control board, based on a desktop four-axis robotic arm to achieve target sorting, and its design and development cost is moderate, which can meet the requirements of class students to carry out practice.The platform can also be used to carry out scientific research work in artificial intelligence, machine vision, embedded programming, and robotic arm control.

References

1. Williams, H.A., et al.: Robotic kiwifruit harvesting using machine vision, convolutional neural networks, and robotic arms. Biosys. Eng. **181**, 140–156 (2019)
2. Zhihong, C., et al.: A vision-based robotic grasping system using deep learning for garbage sorting. In: 2017 36th Chinese Control Conference (CCC), IEEE (2017)
3. Chaudhury, A., et al.: Machine vision system for 3D plant phenotyping. IEEE/ACM Trans. Comput. Biol. Bioinf. **16**(6), 2009–2022 (2018)
4. Abad, A.C., et al.: Fuzzy logic-controlled 6-DOF robotic arm color-based sorter with machine vision feedback. Int. J. Adv. Comput. Sci. Appl. **9**(5), 21–31 (2018)
5. Hock, O., Šedo, J.: Forward and inverse kinematics using pseudoinverse and transposition method for robotic arm dobot. In: Kinematics. BoD–Books on Demand (2017)
6. Dudak, J., et al.: Serial communication protocol with enhanced properties–securing communication layer for smart sensors applications. IEEE Sens. J. **19**(1), 378–390 (2018)
7. Huang, Y., et al.: Performance evaluation of a foot interface to operate a robot arm. IEEE Robot. Autom. Lett. **4**(4), 3302–3309 (2019)
8. Kim, J., et al.: Dynamic model and motion control of a robotic manipulator. J. Robot. Netw. Artif. Life **4**(2), 138–141 (2017)
9. Chandan, G., Jain, A., Jain, H.: Real time object detection and tracking using Deep Learning and OpenCV. In: 2018 International Conference on Inventive Research in Computing Applications (ICIRCA), IEEE (2018)
10. Zhuang, F., et al.: A comprehensive survey on transfer learning. Proc. IEEE **109**(1), 43–76 (2020)
11. Ou, X., et al.: Moving object detection method via ResNet-18 with encoder–decoder structure in complex scenes. IEEE Access **7**, 108152–108160 (2019)

Influence of Guides on Dynamic Characteristics of Elevator

Xiaomei Jiang[1(✉)], Michael Namokel[2(✉)], Chaobin Hu[1], and Fusheng Zhang[1]

[1] Jiangsu Key Laboratory of Elevator Intelligent Safety and School of Mechanical Engineering, Changshu Institute of Technology, Changshu 215500, Jiangsu, China
[2] Nordhessen University of Applied Sciences, Kassel, Germany
michael.namokel@diploma.de

Abstract. With the development of high and super-high buildings, elevator speed is now increasing faster and faster. The elevator is the necessary vertical transport of high-rise building. The faster speed and more serious horizontal vibrations can reduce the passengers' ride comfort and even the elevator's service life and also increase energy consumption. The analysis of elevator horizontal vibration and the simulation of guides' friction are studied; the mathematical model of elevator horizontal vibration is established based on the rigid body kinematics theory. The motion of elevator car is divided into translation and rotation around the centre of mass. According to Newton's law of motion and Euler's equation, the dynamic model of elevator car horizontal vibration is deduced and analyzed, and the change of guide rail stiffness on elevator horizontal vibration is studied, the opposite and the same direction bending of the guide rail will intensify the car horizontal vibration, and the same direction bending affect much more than the different direction bending through comparison.

Keywords: Guides · Horizontal vibration · Dynamic modeling · Simulation

1 Introduction

With the emergence of high-rise buildings in the city, the demand for elevators is increasing, which requires further improvement of the speed. The increase of elevator speed aggravates the horizontal vibration and guides friction, which affects the ride comfort and service life. Therefore, reducing horizontal vibration can improve speed, reduce friction and energy consumption of guides. It is very significant to study the horizontal vibration and guide friction of elevator.

When the elevator is running, in addition to the up and down vibration, passengers will also feel the horizontal direction of the left and right, back and forth shaking. Although the vibration in the horizontal direction is smaller than that in the vertical direction, due to its low vibration frequency and relatively sensitive human body, it is more likely to cause discomfort of passengers in the car. There are many main factors that cause elevator horizontal vibration. The guide rail and guide shoe system are the main excitation sources for horizontal vibration of elevator system [1].

Y. Wang et al. (Eds.): IWAMA 2021, LNEE 880, pp. 397–405, 2022.
https://doi.org/10.1007/978-981-19-0572-8_50

Li Lijing et al. [2] proved that the straightness of guide rail is the main cause of car vibration. Utsunomiya et al. [3] established the vibration model by taking the car body and the car frame as a rigid body connected by spring and damper. Guo Lifeng [4] regards the car body and car frame as two rigid bodies with six degrees of freedom, and establishes a model with twelve degrees of freedom to explain the influence of guide rail unevenness on horizontal vibration.

2 Horizontal Vibration Dynamics Modeling

At present, most elevators on the market use type "T" guide rail. In the process of production and installation, the guide rail will inevitably produce various geometrical tolerances and dimensional deviations, which will have an important impact on the horizontal vibration of the elevator. As the main influence factor of elevator horizontal vibration, the guide excitation mainly displays in two aspects, the irregularity of guide rail and the variable rigidity of guide rail. The guide rails on both sides of the car support and guide the elevator car. The guide shoes are installed on both sides of the car and closely connected with the guide rails. The car moves up and down in the shaft along the guide rails under the drag of the steel wire rope [5]. The whole guides of the elevator contain: elevator guide shoe, elevator guide rail and elevator guide rail bracket. The main factors that affect the horizontal vibration of the elevator car are the processing roughness of the contact surface of the guide rail, the irregularity of the guide rail and the installation deviation of the guide rail [6].

The elevator guide rail is installed in the shaft, which provides guidance and ensures the car and counterweight move up and down. The elevator guide rail is connected and installed in the shaft section by section. The contact position of the rolling guide shoe is constantly changing. The rigidity of the guide rail is larger when it is close to the support, smaller when it is far away from the support. The contact rigidity of the rolling guide shoe and the guide rail is constantly changing with the change of the position.

In order to facilitate the analysis, it is assumed that the car and the car frame are rigidly connected, all the forces are uniformly applied on the car bottom, and the car is regarded as a symmetrical body [7]. Considering the car motion as translation and rotation around the center of mass, the horizontal vibration model of the car is established according to Newton's law of motion. According to the right-hand Cartesian coordinate system rules, the coordinates are defined. In Fig. 1, Oxyz is the inertial coordinate system, and $O'x'y'z'$ is the continuum coordinate system. Before the elevator does not run, the two coordinate systems coincide. When the elevator starts to run, the two coordinate systems operate in Z direction with the elevator car. Among them, the direction of each coordinate axis of coordinate system Oxyz and the coordinate origin O are fixed in the horizontal plane, the Z axis is parallel to the elevator shaft; the coordinate origin O' of $O'x'y'z'$ is equivalent to the gravity center of the elevator car, the positive direction of x' axis is perpendicular to the car door, the y' axis is parallel to the car door, and the positive direction of z' axis is perpendicular to the car roof; m is the total mass of the elevator; L1, L2 and L3, L4 are the horizontal distance and vertical distance from each guide shoe to the car centroid respectively; L5, L6 … L10 is the size of car respectively; FL, F2 … F12 is the force of each guide shoe on the car. The motion of the car can

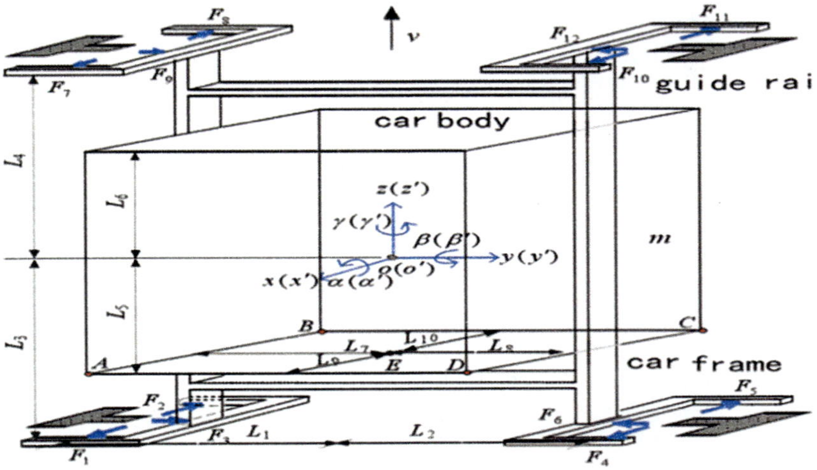

Fig. 1. Horizontal vibration model of elevator

be divided into the translation of the car's center of mass and the rotation around the center of mass [8]. Correspondingly, the car model is represented by the car's position equation and the car's attitude equation. The translation of the car is represented by the displacement vector R of the origin O′ of the coordinate system O′x′y′z′ in the inertial coordinate system Oxyz:

$$R = [x\ y\ z]^T \tag{1}$$

In the inertial coordinate system oxyz, Newton's second law of motion is used to analyze the elevator car, and a translational equation is established, from which the position dynamic equation of the elevator car can be obtained as follows:

$$\begin{cases} (m_c + \Delta m)x = F_x + \Delta F_x \\ (m_c + \Delta m)y = F_y + \Delta F_y \\ (m_c + \Delta m)z = F_z + \Delta F_z \end{cases} \tag{2}$$

In formula (2), Fx, Fy and Fz are respectively the control forces along the x, y and z axes, and Δ Fx, Δ Fy and ΔFz are respectively the disturbance forces along the x, y and z axes.

It is set that the car rotation is represented by the Cardan angle coordinate H, and the attitude of the elevator car is divided into the rotation angles α, β, γ around the x′, y′, z′ axes of the continuum coordinate system, which are recorded as follows:

$$H = [\alpha\ \beta\ \gamma]^T \tag{3}$$

Set the angular velocity of the car relative to the center of mass in the coordinate system o′x′y′z′ as:

$$\omega = \left[\omega_{x'}\ \omega_{y'}\ \omega_{z'}\right] \tag{4}$$

Establish the kinematic equation of elevator car attitude:

$$\omega = g(H)\begin{bmatrix}\alpha\\ \beta\\ \gamma\end{bmatrix},\ g(H) = \begin{bmatrix}\cos\beta\cos\gamma & \sin\gamma & 0\\ -\cos\beta\sin\gamma & \cos\gamma & 0\\ \sin\beta & 0 & 1\end{bmatrix} \tag{5}$$

From Eqs. (4) and (5),

$$\begin{bmatrix}\alpha\\ \beta\\ \gamma\end{bmatrix} = g(H)^{-1}\omega = \begin{bmatrix}\cos\gamma\sec\beta & -\sin\gamma\sec\beta & 0\\ \sin\gamma & \cos\gamma & 0\\ \cos\gamma\tan\beta & \sin\gamma\tan\beta & 1\end{bmatrix}\begin{bmatrix}\omega_{x'}\\ \omega_{y'}\\ \omega_{z'}\end{bmatrix} \tag{6}$$

Because α, β and γ are small angles, there are:

$$\sin\theta \approx \theta,\ \tan\theta \approx \theta\ \ \cos\theta \approx 1\ \ (\theta = \alpha, \beta, \gamma) \tag{7}$$

$$\begin{bmatrix}\alpha\\ \beta\\ \gamma\end{bmatrix} = \begin{bmatrix}\omega_{x'} - \gamma\omega_{y'}\\ \gamma\omega_{x'} + \omega_{y'}\\ \beta\omega_{x'} + \omega_{z'}\end{bmatrix} \tag{8}$$

Plug Eq. (7) into Eq. (6) and remove the secondary term, we can get:

$$\begin{bmatrix}\alpha\\ \beta\\ \gamma\end{bmatrix} = \begin{bmatrix}\omega_{x'} - \gamma\omega_{y'}\\ \gamma\omega_{x'} + \omega_{y'}\\ \beta\omega_{x'} + \omega_{z'}\end{bmatrix} \tag{8}$$

The rotation equation of the car can be expressed by the Euler equation based on the rigid body dynamics and the motion of the rigid body around the mass center:

$$\begin{cases} I_{x'}\omega_{x'} - (I_{y'} - I_{z'})\,\omega_{y'}\omega_{z'} = M_{x'} + \Delta M_{x'}\\ I_{y'}\omega_{y'} - (I_{z'} - I_{x'})\,\omega_{z'}\omega_{x'} = M_{y'} + \Delta M_{y'}\\ I_{z'}\omega_{z'} - (I_{x'} - I_{y'})\,\omega_{x'}\omega_{y'} = M_{z'} + \Delta M_{z'} \end{cases} \tag{9}$$

In formula (9), Mx′, My′, Mz′ respectively represent the control torque of axis x′, y′, z′, and ΔMx′, ΔMy′, ΔMz′ respectively represent the interference torque of axis x′, y′, z′; Ix′ = Jx′ + ΔJx′, Iy′ = Jy′ + ΔJy′, Iz′ = Jz′ + ΔJz′, where Ix′, Iy′, Iz′ are the total moment of inertia of the load, Jx′, Jy′, Jz′ are the moment of inertia of the car, ΔJx′, ΔJy′, ΔJz′ are the moment of inertia of the load.

The position, velocity and acceleration of any point of rigid body are determined by the position and attitude of rigid body. Set point w as any point on the elevator car, and its position vector, speed vector and acceleration vector are respectively expressed as:

$$r_w = [r_{wx}\ r_{wy}\ r_{wz}]^T\ \ v_w = [v_{wx}\ v_{wy}\ v_{wz}]^T\ \ a_w = [a_{wx}\ a_{wy}\ a_{wz}]^T \tag{10}$$

The radius vector from point O′ to point W is expressed in the continuum coordinate system O′x′y′z′:

$$\begin{cases} \rho_W = \left[\rho_{wx'}\ \rho_{wy'}\ \rho_{wz'}\right]^T \\ r_W = R + \rho_W \\ v_W = R + \omega\rho_W \\ a_W = R + \omega\rho_W + \omega^2\rho_W \end{cases} \tag{11}$$

For vector ω, the skew symmetric matrix of ω is definined:

$$[\omega_X] = \begin{bmatrix} 0 & -\omega_{z'} & \omega_{y'} \\ \omega_{z'} & 0 & -\omega_{x'} \\ -\omega_{y'} & \omega_{x'} & 0 \end{bmatrix} \tag{12}$$

The direction cosine matrix of inertial coordinate system for continuum coordinate system is as follows:

$$\begin{bmatrix} \cos\beta\cos\gamma & -\cos\beta\sin\gamma & \sin\beta \\ \sin\alpha\sin\beta\cos\gamma + \cos\alpha\sin\gamma & -\sin\alpha\sin\beta\sin\gamma + \cos\alpha\cos\gamma & -\sin\alpha\cos\beta \\ -\cos\alpha\sin\beta\cos\gamma + \sin\alpha\sin\gamma & \cos\alpha\sin\beta\sin\gamma + \sin\alpha\cos\gamma & \cos\alpha\cos\beta \end{bmatrix} \approx \begin{bmatrix} 1 & \gamma & \beta \\ -\gamma & 1 & \alpha \\ \beta & \alpha & 1 \end{bmatrix} \tag{13}$$

The coordinate expression of formula (11) is as follows:

$$\begin{bmatrix} r_{wx} \\ r_{wy} \\ r_{wz} \end{bmatrix} = \begin{bmatrix} x \\ y \\ z \end{bmatrix} + \begin{bmatrix} 1 & \gamma & -\beta \\ -\gamma & 1 & \alpha \\ \beta & -\alpha & 1 \end{bmatrix} \begin{bmatrix} \rho_{wx'} \\ \rho_{wy'} \\ \rho_{wz'} \end{bmatrix} \tag{14a}$$

$$\begin{bmatrix} v_{wx} \\ v_{wy} \\ v_{wz} \end{bmatrix} = \begin{bmatrix} x \\ y \\ z \end{bmatrix} + \begin{bmatrix} 1 & \gamma & -\beta \\ -\gamma & 1 & \alpha \\ \beta & -\alpha & 1 \end{bmatrix} \begin{bmatrix} 0 & \omega_{z'} & -\omega_{y'} \\ \omega_{z'} & 0 & -\omega_{x'} \\ -\omega_{y'} & \omega_{x'} & 0 \end{bmatrix} \begin{bmatrix} \rho_{wx'} \\ \rho_{wy'} \\ \rho_{wz'} \end{bmatrix} \tag{14b}$$

$$\begin{aligned} \begin{bmatrix} a_{wx} \\ a_{wy} \\ a_{wz} \end{bmatrix} = \begin{bmatrix} x \\ y \\ z \end{bmatrix} + \begin{bmatrix} 1 & \gamma & -\beta \\ -\gamma & 1 & \alpha \\ \beta & -\alpha & 1 \end{bmatrix} \begin{bmatrix} 0 & -\omega_{z'} & -\omega_{y'} \\ \omega_{z'} & 0 & -\omega_{x'} \\ -\omega_{y'} & \omega_{x'} & 0 \end{bmatrix} \begin{bmatrix} \rho_{wx'} \\ \rho_{wy'} \\ \rho_{wz'} \end{bmatrix} \\ + \begin{bmatrix} 0 & -\omega_{z'} & \omega_{y'} \\ \omega_{z'} & 0 & -\omega_{x'} \\ -\omega_{y'} & \omega_{x'} & 0 \end{bmatrix} \begin{bmatrix} 0 & -\omega_{z'} & \omega_{y'} \\ \omega_{z'} & 0 & -\omega_{x'} \\ -\omega_{y'} & \omega_{x'} & 0 \end{bmatrix} \begin{bmatrix} \rho_{wx'} \\ \rho_{wy'} \\ \rho_{wz'} \end{bmatrix} \end{aligned} \tag{14c}$$

Let ax and ay be the horizontal vibration accelerations in X and Y directions at point E, the radius vector of point E in the coordinate system O′x′y′z′ is:

$P_E = \left[P_{Ex'}\ P_{Ey'}\ P_{Ez'}\right]^T$, from Eq. (14c), we can get:

$$\begin{bmatrix} a_x \\ a_y \end{bmatrix} = \begin{bmatrix} x + \left(-\omega_{z'}^2 - \omega_{y'}^2\right)P_{Ex'} + \left(-\omega_{z'} + \omega_{x'}\omega_{y'}\right)P_{Ey'} + \left(\omega_{y'} + \omega_{x'}\omega_{z'}\right)P_{Ez'} \\ y + \left(\omega_{z'} - \omega_{x'}\omega_{y'}\right)P_{Ex'} + \left(-\omega_{z'}^2 - \omega_{x'}^2\right)P_{Ey'} + \left(-\omega_{x'} + \omega_{y'}\omega_{z'}\right)P_{Ez'} \end{bmatrix} \tag{15}$$

State space expression and force analysis of car model.

(1) State equation of car position

Position state variable: $X_p = [x_1\ x_2\ x_3\ x_4\ x_5\ x_6]^T = \left[x\ x\ y\ y\ z\ z\right]^T$, Control variable: $U_p = \left[F_x\ F_y\ F_z\right]^T$, Disturbance variable: $W_p = \left[\Delta F_x\ \Delta F_y\ \Delta F_z\right]^T$.

From Eq. (2), the car position state equation is obtained:

$$X_p = A_p X_p + \left(B_p + \Delta B_p\right) U_p + C_p W_p \tag{16}$$

(2) State equation of car attitude

Attitude state variable:

$X_a = [x_7\ x_8\ x_9\ x_{10}\ x_{11}\ x_{12}]^T = \left[\alpha\ \beta\ \gamma\ \omega_{x'}\ \omega_{y'}\ \omega_{z'}\right]^T$,

Control variable: $U_a = \left[M_{x'}\ M_{y'}\ M_{z'}\right]^T$,

Disturbance variable: $W_a = \left[\Delta M_{x'}\ \Delta M_{y'}\ \Delta M_{z'}\right]^T$.

From Eq. (9), the attitude state equation of the car is obtained:

$$X_p = f(X_a) + \Delta(X_a) + [G(X_a) + \Delta G(X_a)]U_a \tag{17}$$

(3) Output equation

Take the output variable $Y = \left[\ a_x\ a_y\right]^T$ and get from Eq. (15):

$$Y = h\left(X_p,\ X_a\right) \tag{18}$$

$$h\left(X_p,\ X_a\right) = \begin{bmatrix} x_2 + \left(x_{12}^2 - x_{11}^2\right)P_{Ex'} + (-x_{12} + x_{10}x_{11})P_{Ey'} + (x_{11} + x_{12}x_{10})P_{Ez'} \\ x_4 + (x_{12} + x_{10}x_{11})P_{Ex'} + \left(-x_{12}^2 - x_{10}^2\right)P_{Ey'} + (-x_{10} + x_{11}x_{12})P_{Ez'} \end{bmatrix}$$

(4) Force of guide shoe on car

As the car is an axisymmetric structure, the force on the car is simplified as:

$$F_i = [K_s(X_{ki} - X_{ki}) + C_s(X_{ci} - X_{ci})] \tag{19}$$

3 Horizontal Vibration Dynamics Analysis

Irregularity of guide rail include installation joint bulge between two guide rails, bending caused by insufficient rigidity of guide rail and maladjustment of guide rail. The influence is mainly studied from the specific excitation forms such as sawtooth, step and sine excitation. The guide rail itself is a rigid beam, when the two connected guide rails are not in the same direction, it will affect the irregularity of the guide rail.

Figure 2 shows the horizontal vibration acceleration of the car considering the variable stiffness of the guide rail itself when the speed is 3 m/s, and Fig. 3 shows the horizontal vibration acceleration of the car without considering the variable stiffness of the guide rail itself when the speed is 3 m/s. Through comparison, the trend of horizontal vibration acceleration of the elevator is the same. The maximum peak-peak value of car horizontal vibration acceleration is 0.0544 m/s^2 when considering the variable stiffness, 0.0504 m/s^2 when not considering the variable stiffness of the guide rail itself. The maximum peak-peak value of horizontal vibration acceleration when considering the variable stiffness is slightly greater than the maximum peak-peak value of the horizontal vibration acceleration when not considering the variable stiffness, and the variable stiffness of the guide rail has little effect on the elevator.

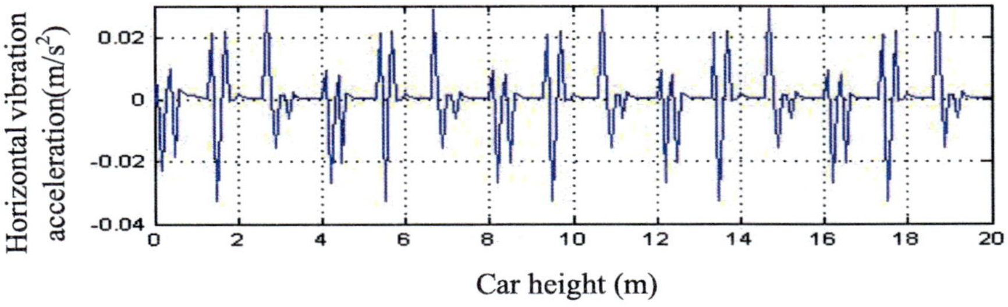

Fig. 2. Horizontal vibration acceleration (3 m/s, consider guide variable stiffness)

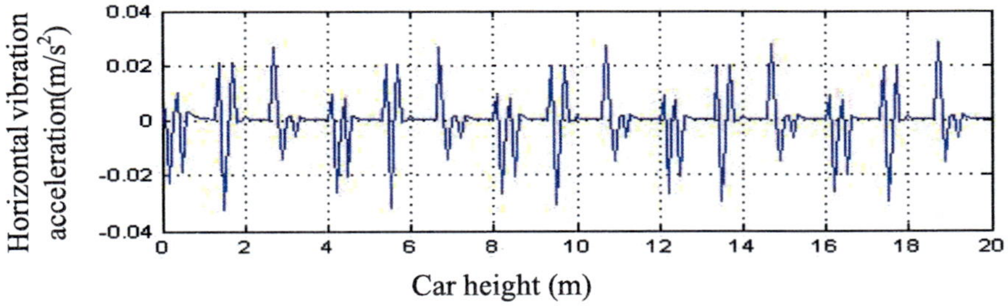

Fig. 3. Horizontal vibration acceleration (3 m/s, ignore guide variable stiffness)

The selected speed range is from 1 m/s to 8 m/s, and the corresponding maximum peak value of horizontal vibration acceleration is obtained, and the maximum peak-peak value of acceleration - elevator running speed curve is generated. As shown in Fig. 4, with the increase of elevator running speed, the maximum peak-peak value of elevator horizontal vibration acceleration changes more and more. When the speed is 3 m/s, the peak-peak value of vibration acceleration has a bulge. The main reason may be that the excitation frequency of guide rail at this time is close to the natural frequency of elevator system itself, which is in resonance state [9]. Therefore, it is necessary to pay attention to the selection of elevator running speed.

When the elevator is running, the rolling guide shoe contacts with the guide rail, and there is mutual force between them. The guide rail will bend and deform under the action of the force, which is the external excitation that causes the car vibration. The guide rail is installed in the elevator shaft section by section, just like the beam, the guide rail will produce certain bending when it is stressed, and the bending direction of different guide

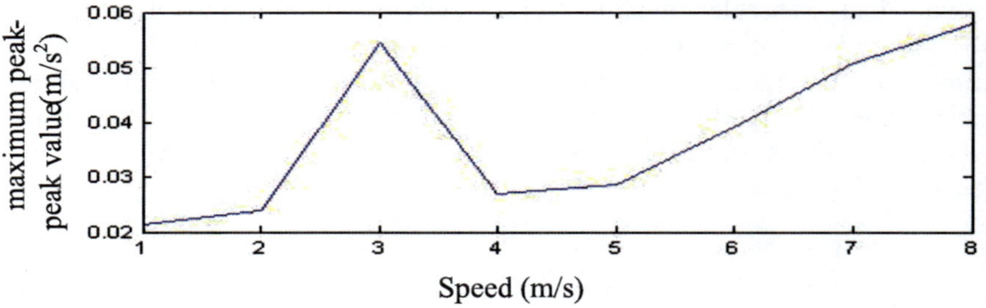

Fig. 4. Horizontal vibration acceleration maximum peak-peak value vs. speed

rails is different. During the analysis, it is assumed that the bending of the guide rail is ideal bending, and the bending direction of the adjacent guide rails may be the same or the opposite, and the same direction bending and different direction bending of the guide rail are respectively compared and simulated here. Figures 5 and 6 show the change of elevator horizontal vibration acceleration with elevator running height.

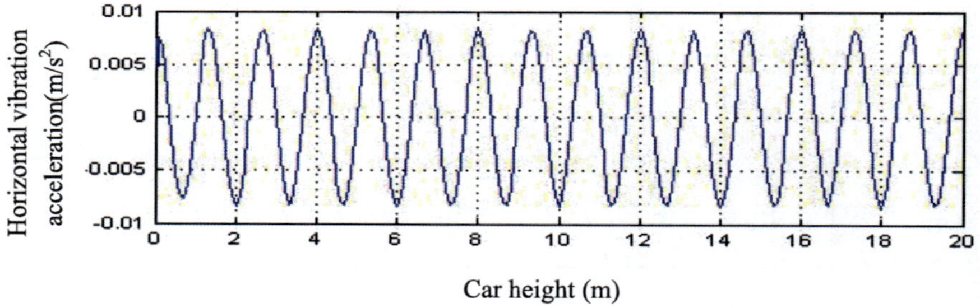

Fig. 5. Horizontal vibration acceleration (3 m/s) under guide bend excitation (opposite direction)

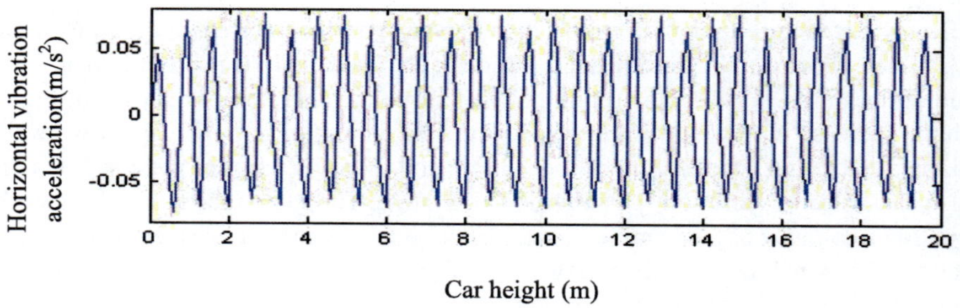

Fig. 6. Horizontal vibration acceleration (3 m/s) under guide bend excitation (same direction)

4 Discussion and Conclusion

Through the comparative analysis of the influence of the same and different direction bending of the guide rail on the horizontal vibration acceleration, it can be concluded that when the running speed and the excitation amplitude are the same, the influence of the different direction bending of the guide rail on the horizontal vibration acceleration is less than the influence of the same direction bending of the guide rail on the horizontal vibration acceleration. The main reason is that when the guide rail is bent in the same direction, it is equivalent to adding a step excitation to the guide rail joint, and the vibration acceleration increases; when the guide rail is bent in different directions, the transition is smooth and the first derivative is continuous, and the peak-peak value of vibration acceleration is small. Therefore, during the installation of guide rail, special attention shall be paid to the smooth transition of guide rail to avoid the same direction bending of guide rail.

Acknowledgment. The work described in this article has been conducted as part of the research project Natural Science Research Major Project of higher education institution of Jiangsu Province (grant numbers: 20KJA460011). Special thanks to all those who helped me during the writing of this paper.

References

1. Yin, J., Rui, Y.-N., Jiang, L.-M., et al.: Research on multi degree of freedom horizontal dynamic characteristics and simulation of high-speed elevator. Mech. Des. **28**(10), 70–73 (2011)
2. Lijing, L., Xingfei, L., Guoxiong, Z., et al.: Horizontal vibration model of elevator car. Lift. Transp. Machin. **5**, 3–5 (2002)
3. Utsunomiya, K., Okamoto, K.-I.: Active roller guide system for high-speed elevators. Elevator World **50**(4), 86–93 (2002)
4. Wujun, F., Changming, Z., Changyou, Z., et al.: Horizontal vibration modeling and dynamic response analysis of high-speed elevator. Mech. Des. Res. **19**(6), 65–67 (2003)
5. Jiang, X.-M.: Research on vibration Control of traction elevator. In: International Industrial Informatics and Computer Engineering Conference, pp. 2144–2147 (2015)
6. Guo, L., Jiang, X.: Research on horizontal vibration of traction elevator. In: Wang, K., Wang, Y., Strandhagen, J.O., Yu, T. (eds.) IWAMA 2018. LNEE, vol. 484, pp. 131–140. Springer, Singapore (2019). https://doi.org/10.1007/978-981-13-2375-1_18
7. Jiang, X.M., Namokel, M.: Stress analysis of hoist auto lift car frame. In: International Workshop of Advanced Manufacturing and Automation, pp. 151–159 (2019)
8. Liu, H.-W.: Material Strength, 5th edn. Higher Education Press, Beijing (2010)
9. International Standard ISO 18738. Lifts (Elevators)-Measurement of Lift Ride Quality (2003)

Design of Safety Warning Device for Escalator Handrail Based on ARM

Xiaomei Jiang[1](✉), Michael Namokel[2](✉), Chaobin Hu[1], and Fusheng Zhang[1]

[1] Jiangsu Key Laboratory of Elevator Intelligent Safety and School of Mechanical Engineering, Changshu Institute of Technology, Changshu 215500, Jiangsu, China
[2] Nordhessen University of Applied Sciences, Kassel, Germany
michael.namokel@diploma.de

Abstract. As a kind of special equipment for escalator to carry personnel, its safety performance is very important. in order to prevent the passengers from leaning on the handrail belt and the wall protection plate of the escalator, the pressure sensor installed under the escalator and the wall protection plate and the ARM microprocessor constitute a voice warning and shutdown protection device, so as to effectively protect the personal safety of passengers. It can be widely used in the safety retrofitting of various escalator facilities.

Keywords: Escalator · ARM · Pressure sensor · Voice warning

1 Introduction

Escalator is a special equipment to transport passengers in shopping malls, hotels, airports, stations and other crowded public places. It has become an indispensable public facility in urban life and its safety performance is very important. At present, escalator safety accidents occur frequently [1]. In the process of passengers taking the escalator, a small number of passengers can often be seen directly leaning on the wall panel, which may lead to instability of the wall panel and cause accidents. Due to this kind of leaning behavior, it leads to falling accident, the case of stuck and then injury occur from time to time. The number of passenger injuries caused by improper use of escalators has increased significantly, with 10–20 cases per month, "among them, 70% of the injuries were caused by passengers leaning against the handrail and being brought down by escalators." Therefore, it is particularly important to design and develop devices to prevent such dangerous accidents. If the detailed diagnosis information can not be provided to eliminate the fault and quickly evacuate the passenger flow, when the elevator stops, it may cause serious congestion safety problems [2].

Escalator handrail safety warning device is a kind of safety device which can provide voice warning and emergency stop. The pressure sensor is used to sense the pressure on the side of the handrail and its duration time, and send an electrical signal to the pressure analysis device [3]. The pressure analysis device analyzes the received electrical signal, and sends the command to the voice warning device, the voice warning device will send a prompt voice, which will prompt the passengers to correct the non-standard behavior. If

Y. Wang et al. (Eds.): IWAMA 2021, LNEE 880, pp. 406–411, 2022.
https://doi.org/10.1007/978-981-19-0572-8_51

the prompt is still invalid, the elevator stops until the passengers correct the non-standard behavior.

2 Design Scheme

The traditional escalator handrail mostly uses proximity switch to detect the opening on the speed measuring roller, and the signal pulse frequency obtained corresponds to the running speed of the handrail [4]. This kind of proximity switch can not effectively sense the passenger's squeeze on the handrail. The design of signal acquisition is to use the pressure sensor to collect the signal of escalator passengers leaning on the handrail. The sensor is a film type contact sensor, which is evenly distributed on the handrail guide rail directly contacted by the handrail. When the sensor receives the pressure from the outside to reach a certain pressure value, a trigger signal will be generated, and the pressure value lasts for 3 s, The trigger signal is sent to the voice warning device after being processed by the pressure analysis device. At this time, the voice warning device will send out the prompt sound of "please do not lean on the handrail", so as to timely remind the relevant passengers by sensing the non-standard behavior of passengers, such as leaning on, which can effectively improve the safety of the escalator. The shape of the voice warning device is flat cylinder, which is evenly embedded on both sides of the wall panel, saving space. Figure 1 shows the hardware layout of escalator handrail safety warning device, and Fig. 2 shows the location of pressure sensor.

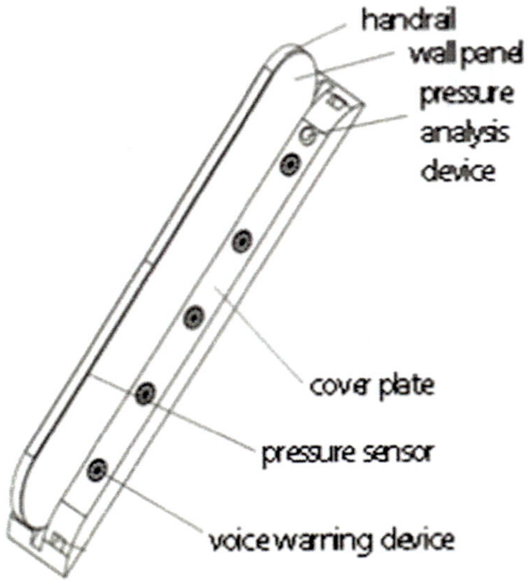

Fig. 1. Hardware layout of escalator handrail safety warning device

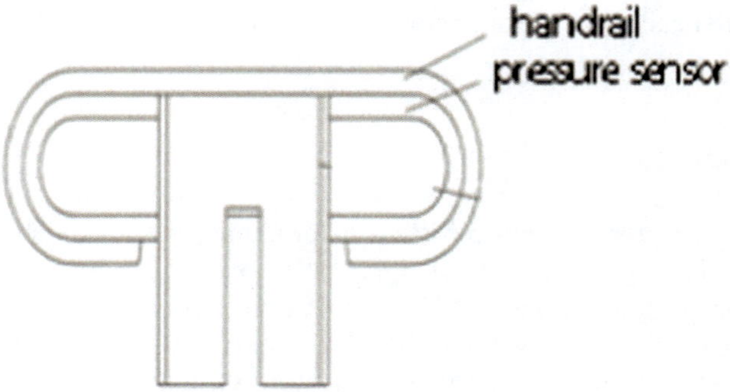

Fig. 2. Installation location of pressure sensor

3 Hardware Component

The pressure detection is completed by bridge strain gauge. Figure 3 is the circuit of detection sensing part. Because the impedance of the strain gauge is relatively large, when the two input terminals of the pressure sensor have input constant current ($I_{IN} = V_R/RR$), the input voltage of the pressure sensor will be relatively high, plus the voltage drop of RR and the change of bridge resistor with temperature. The power supply voltage is 15 V.

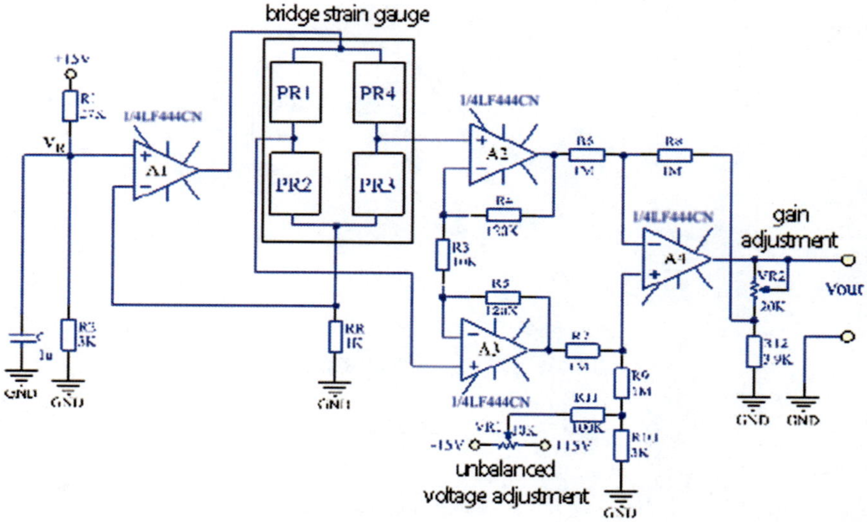

Fig. 3. Signal processing circuit of pressure sensor

The most common bridge measuring circuit is used in this design, which pastes the pressure strain sensor evenly on the inner surface of the escalator handrail. When the pressure of the escalator handrail changes, the sensitive grid of the pressure sensor strain gauge will also deform, and the resistance value will change accordingly. Because the resistance change is difficult to measure directly and accurately, it is also inconvenient to deal with directly. Therefore, a conversion circuit must be used to convert the resistance change of the strain gauge into a voltage or current change, but the voltage or current signal is very small, and a gain amplifier circuit is needed to convert the voltage or current signal into a signal that can be received by the A/D converter. The actual circuit is shown in Fig. 3. The conversion circuit converts the resistance change of the strain gauge into a voltage signal that can be received by the A/D converter.

The parallel differential amplifier is composed of high precision and low drift operational amplifiers A2 and A3, and the pre-processing circuit is composed of A1; In ideal case, the input impedance and common mode rejection ratio of the parallel differential amplifier are infinite. The CMRR has nothing to do with the precision and value of the peripheral resistance.

The latter stage amplifier is composed of A4 instrument amplifier, which converts the two terminal signal into single terminal signal. Because the output impedance of the preamplifier is very low and the common mode driving technology is adopted, the instrument amplifier of the latter stage can achieve high gain and high common mode rejection ratio. In Fig. 3, VR1 is the unbalanced voltage adjustment potentiometer and VR2 is the gain adjustment potentiometer.

The escalator safety handrail warning system based on ARM, mainly using the powerful data calculation and processing function of ARM, the analog signal generated by the pressure strain sensor is converted into digital signal, and the escalator warning and shutdown protection are realized through software programming. The main circuit board adopts LPC2129, which has low power consumption, high performance and rich peripheral resources. It is a 16/32 bit ARM7TDMI-S CPU microprocessor based on real-time simulation and tracking, with 128 K byte high-speed flash memory. The unique acceleration structure and 1228 bit wide memory interface make 32 bits run at maximum clock. If the application scale of code is strictly controlled, the 16 bit thumb mode can be used to reduce the code scale by less than 30%, but the performance loss is very small. LPC2129 adopts 64 pin package, multiple 32-bit timers, 4-way 10 bit ADC, 2-way CAN or 8-way 10 bit ADC, 2-way CAN and 9 external interrupts, making it particularly suitable for elevator controller [5]. The microcontroller itself optimizes the conversion between power consumption and CPU speed. The high-precision analog-to-digital converter has high driving ability and reduces the demand for external components. The unique event system simplifies the signal transmission between peripheral devices. The pressure strain sensor is distributed on one side of the mobile handrail and transmits the pressure information to the transmission unit. After isolation, it is sent to ARM (LPC2129) microprocessor for conversion and analysis [6]. When the pressure reaches the alarm setting value, the processor outputs the alarm signal, and the signal is transmitted to the voice announcer to send out the voice alarm through the horn. When the alarm time or pressure reaches the stop setting value, the microprocessor will send a stop signal to stop the escalator for protection, and send the fault code to the escalator fault display panel through CAN bus. At this time, the escalator can not be started again, and it must be checked and confirmed by the elevator maintenance personnel before it can be started again. Figure 4 is the system architecture diagram.

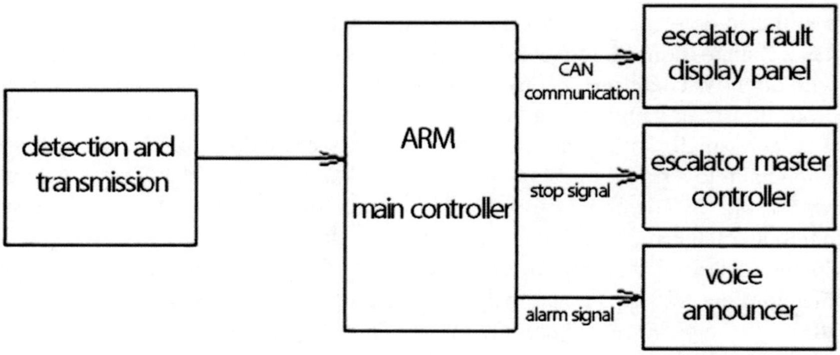

Fig. 4. System architecture diagram

4 Software Settings

The escalator safety handrail early warning system mainly uses the powerful data calculation and processing function of ARM to convert the analog signal generated by the pressure strain sensor into digital signal, and realizes the escalator early warning and shutdown protection through software programming [7]. The sensing pressure information is transmitted to the pressure processing and analysis device. A pressure rating of 200 N is set in the pressure processing and analysis device, which is the pressure value that ordinary passengers rest on. When the pressure is greater than this pressure value, the pressure lasts for more than 3 s. The pressure processing and analysis device transmits the signal to the voice warning device, and then sends out a voice alarm through the horn. When the pressure is greater than 500 N, the pressure lasts for more than 8 s, and the escalator stop signal is started.

5 Discussion and Conclusion

The design of escalator safety handrail warning system based on ARM, the pressure sensors are evenly distributed in the handrail guide rail, when passengers lean on the handrail, they will apply a certain pressure on the pressure sensor on the upper surface of the handrail guide rail through the handrail, so as to monitor the non-standard behavior of passengers, and timely remind the relevant passengers to correct their behavior, to avoid the occurrence of safety accidents caused by passengers leaning on the handrail for a long time. It has strong practical significance for large flow occasions, and can be widely used in the use of various escalators. It is of great significance to develop a practical, convenient, accurate and highly automated safety warning device for escalator handrails to ensure the safe operation of escalators and moving sidewalks and to protect the personal safety of passengers.

Acknowledgment. The work described in this article has been conducted as part of the research project Natural Science Research Major Project of higher education institution of Jiangsu Province (grant numbers: 20KJA460011). Special thanks to all those who helped me during the writing of this paper.

References

1. Jiang, X., Niu, S., Guo, L.: Braking distance monitoring system for escalator. In: Wang, K., Wang, Y., Strandhagen, J.O., Yu, T. (eds.) IWAMA 2017. LNEE, vol. 451, pp. 197–205. Springer, Singapore (2018). https://doi.org/10.1007/978-981-10-5768-7_22
2. Wang, X.Y.: Discussion on escalator accident model and the emergency measures. China Elevator **30**(6), 52–53 (2019)
3. Zhang, X.X., Yang, C.Y., Cheng, Y.: Application of intelligent escalator monitoring system based on open pose. J. Heilongjiang Univ. Technol. **20**(10), 49–55 (2020)
4. Shao, L.: Sensors on elevators and escalators. Sensor World **25**(2), 14–19 (2019)
5. Tong, H.: Design of Elevator Control System based on ARM Microprocessor. China University of Geosciences, Wuhan (2011)
6. Xu, F.: Research and Design of Intelligent Elevator Control System Based on Linux. Shandong University, Jinan (2020)
7. Jiang, X.-M.: Research on intelligent elevator control system. Adv. Mater. Res. **605–607**, 1802–1805 (2013). https://doi.org/10.4028/www.scientific.net/AMR.605-607.1802

Gauge Deviation Measuring Instrument for Elevator

Xiaomei Jiang[1(✉)], Michael Namokel[2(✉)], Chaobin Hu[1], and Fusheng Zhang[1]

[1] Jiangsu Key Laboratory of Elevator Intelligent Safety and School of Mechanical Engineering, Changshu Institute of Technology, Changshu 215500, Jiangsu, China
[2] Nordhessen University of Applied Sciences, Kassel, Germany
michael.namokel@diploma.de

Abstract. As China possesses the largest number of elevators in the world, the accidents of elevators are also increasing. In the process of elevator operation, the car runs along the direction of the guide rail. Whether the distance and deviation of the guide rail meet the design requirements directly determines whether the elevator can run safely. At the same time, the distance and deviation of the guide rail of the elevator also indirectly affect the user's comfort. Manual inspection is not easy to operate, inefficient and wasteful of resources. In order to solve this problem, an advanced measuring instrument of elevator gauge deviation is proposed, which can detect gauge deviation with many obstacles and narrow space at the elevator car top, the difficulty of equipment installation and measurement. This instrument can be widely used in elevator installation, inspection, maintenance, testing and other occasions. It can significantly improve the measurement accuracy and work efficiency, and the product market demand is large, which will produce considerable economic and social benefits. It is small in size, light in weight. It can be free carry-on and measure the gauge deviation of the elevator efficiently, conveniently and accurately.

Keywords: Measuring instrument · Laser displacement sensor · STC15 SCM

1 Introduction

With the penetration of elevators into people's lives, the incidence of elevator safety accidents is gradually increasing. Although people's knowledge on elevator safety has been increasing in recent years, elevator accidents still emerge in an endless stream. Therefore, in the strength of elevator detection and the avoidance of elevator safety accidents, how to achieve scientific prevention, scientific detection and scientific avoidance has become an inevitable important topic [1].

In the process of elevator operation, the guide rail is an important indicator of the safe operation of the car. Because the car moves up and down along the guide rail, the installation quality of the guide rail will directly determine the safe and stable operation of the elevator. 《requirements and methods for elevator supervision and inspection contents》 and 《TSG T7001-2009 elevator supervision and inspection and regular inspection rules - traction and forced drive elevator》 provide the measurement method of guide

Y. Wang et al. (Eds.): IWAMA 2021, LNEE 880, pp. 412–418, 2022.
https://doi.org/10.1007/978-981-19-0572-8_52

gauge: control the elevator to run up and down, and measure the gauge with a steel tape at the guide rail bracket near the shaking position [2, 3]. At present, the detection method used in China is mainly manual measurement. Recently, laser displacement instrument has been used to measure the gauge, that is, the use of high-precision laser beam instead of the tape gauge has greatly improved. However, these two measurement methods need to fix and dismantle the instruments artificially, and human factors are easy to cause measurement error. Due to the lack of devices that can independently run along the elevator guide rail, only one point of gauge value can be measured each time. If the gauge value of the whole section of elevator guide rail is to be measured accurately, it needs to be measured in sections and batches, which has the disadvantages of cumbersome operation, low efficiency and poor accuracy. Therefore, the design of an automatic gauge deviation tester for elevator guide rail is the trend to realize the automation, convenience and intelligence of modern inspection and detection technology.

In order to solve the above problems, a new type of measuring instrument for measuring gauge and deviation is designed on the basis of considering the narrow space on the top of elevator car, the difficulty of equipment installation and the difficulty of measurement. This measuring instrument is small in size, light in weight, accurate in measurement. It can measure the gauge deviation of the elevator efficiently, conveniently and accurately. Through comparative analysis, the inspector can avoid the possible risk in time [4].

2 Working Principle Analysis

As a part of the elevator operation system, each elevator has at least four rows of guide rails. The guide rail not only plays a guiding role in the up and down operation of the elevator car, but also plays a good auxiliary role in the braking process of the safety gear. Therefore, whether the rail connection is intact, whether the rail surface is smooth, whether the rail is horizontal and vertical, and whether the distance between the two rails meets the standard will directly affect the safe operation of the elevator system. The measuring principle of the elevator gauge deviation measuring instrument is shown in Fig. 1 below.

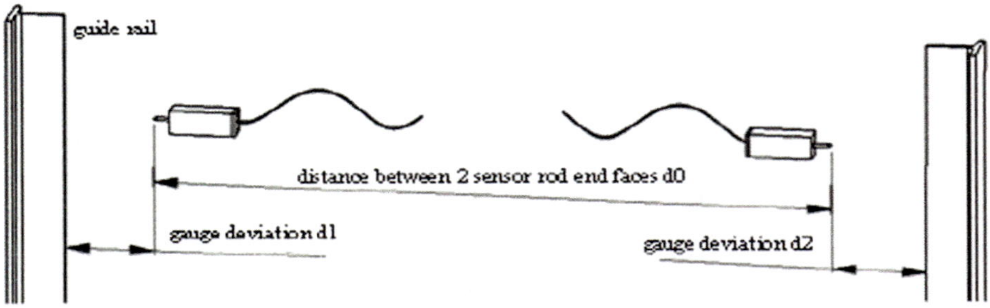

Fig. 1. Schematic measuring principle of gauge deviation

The displacement sensor I and displacement sensor II are respectively fixed on the left and right guide shoes or other structures on the top of the elevator car. In the measurement process, the distance d0 between the displacement sensor I and the displacement sensor II remains unchanged on the horizontal plane from the beginning to the end, d1 and d2 respectively represent the measured gauge deviation. Then, the gauge D of the elevator can be obtained by the following formula:

$$D = d0 + d1 + d2 \quad (1)$$

The relationship between the preset gauge D0 of elevator and the gauge deviation △ D is: △ D = D–D0.

In the process of measuring the elevator gauge, only the relative distance between the two sensors remains unchanged from the beginning to the end, can the accuracy of the measured data be ensured. Before using the elevator gauge deviation measuring instrument for measurement, two probes with displacement sensor shall be fixed on the structural component of the car top which is stable and not easy to shake during the operation of the elevator, the measuring rod end of the sensor shall be facing the front and pressed on the end face of the guide rail appropriately, and the cable connection shall be used between the host and the probe.

The surveyor is located at the top of the car, holding the main machine, or placing the main machine at a suitable position on the top of the car for operation. According to the actual situation, the instrument can be set to manual mode and timing mode (in this mode, the measuring personnel should input the measurement time interval). The surveyor shall input the preset gauge of the elevator (if it is not input, the instrument defaults to 0), and the distance between the rod end faces of the measured by two displacement sensors. The surveyors control the elevator maintenance operation and start the measurement of the elevator gauge and deviation. The measurement personnel can read the measurement data in real time, or check the measurement data stored in the instrument when the elevator is finished running. The specific placement of the instrument is shown in Fig. 2.

Because of the bad environment on the top of the elevator car, the laser displacement sensor is not easy to install, so some parts need to be designed to fix the sensor and the designated position, so that it will not shake and not affecting the measurement results. Here, the clips are used to fix the sensor. In addition, a shell for the sensor is needed to protect the sensor from easy damage and then an L-shaped support plate designed to fix the sensor on it through clips.

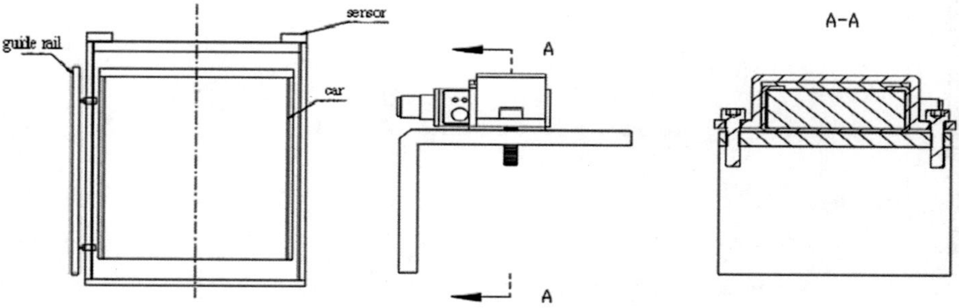

Fig. 2. Schematic diagram of instrument placement

3 System Design Scheme

3.1 Hardware Design

The development of elevator gauge deviation measuring instrument needs to select the appropriate hardware circuit. A stable hardware circuit design can ensure the smooth operation of the instrument. This paper designs the hardware circuit from the following aspects: first, choose 12 V power supply to make the circuit work normally. Second, the selection of circuit components can make the instrument work normally and have enough anti-interference. Third, the circuit can not have too many components, which will lead to excessive energy consumption, and high energy consumption will produce too high temperature, which will adversely affect the stability of the circuit.

The hardware circuit design includes the following aspects; The selection and circuit design of laser ranging sensor; The selection and circuit design of STC15 MCU; OLED screen selection and peripheral communication design; The selection of power supply and the design of Buck module.

The principle of laser displacement sensor is shown in Fig. 3, which is composed of LED, emission control circuit, emission optical system, receiving optical system, timing module, avalanche diode and other components.

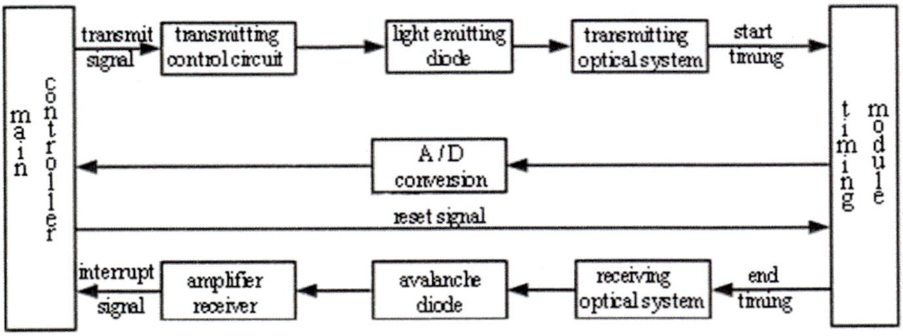

Fig. 3. Sensor internal schematic diagram

The working principle of the laser sensor is as follows: the main controller sends out the signal, and the transmitting control circuit transmits the command to the led to emit the laser to the target after receiving the signal. After the laser emitted by the LED reaches the target, the target laser will be scattered in all directions, and the scattered laser will be transmitted back to the sensor receiver, The image is transmitted to the avalanche diode through the receiving optical system, and the avalanche diode converts it into an electrical signal. In this process, the distance between the laser sensor and the target is divided by two and multiplied by the speed of light [5].

This system uses STC15F2K60S2 single chip microcomputer, µVision5 integrated development environment produced by keil 5 company as the software design platform, including initialization program design, display module design, displacement sensor data acquisition and transmission module program design [6].

STC15F2K60S2 is a dual serial port MCU, which realizes the system function by time-sharing port mapping. Among them, the serial port 1 is P1.0 and P1.1, serial port 2 is p0.0 and P0.1, and serial port 3 is p2.1 and p2.2. In this design, the key communicates with p2.1, p2.2, reset key and p5.4, and realizes the complete communication of the system. The schematic diagram of the system hardware circuit is shown in Fig. 4.

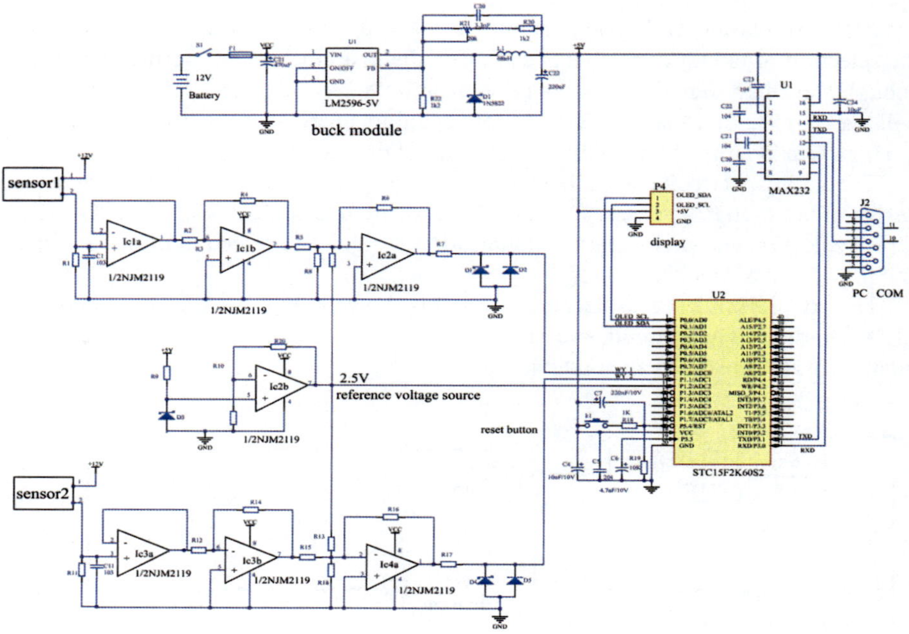

Fig. 4. Schematic diagram of system hardware circuit

The selected display is 0.91 in. OLED screen produced by RISYM company. The display is driven by SSD1306, and its display resolution is 128 * 32, which is simple and convenient, and can well meet the display requirements of the measuring instrument. The display uses IIC communication interface, which only needs four wires to realize the communication connection between OLED display and STC15F2K60 MCU [7]. IIC interface is a widely used serial interface, which uses two data and clock lines to complete data transmission, and has weak anti-interference.

As a necessary part of equipment operation, the stability and reliability of power supply are very important [8]. Because the measuring place of the equipment is located on the top of the elevator car, the common power supply can not meet the requirements of the measuring instrument, so it is necessary to supply power to each component by self-contained power supply. In this design, the voltage range of the selected lithium battery is 9 V–12.6 V. In order to prevent the voltage fluctuation from affecting the normal operation of each module and ensure that it works within the allowable voltage range, it is necessary to use the buck voltage stabilizing circuit to stabilize the stable operation of important modules. Before the voltage is reduced, the voltage is stabilized at about 12 V, which can supply power to the laser ranging sensor. After a step-down, the voltage drops to 5 V to supply power for OLED display, STC12 MCU and keys [9].

3.2 Software Design

As the core controller, SCM not only needs to complete the data acquisition and processing, but also needs to complete the system internal hardware initialization, measurement interval, idle and hardware debugging mode time setting and other functions. The main program flow chart of the system is shown in Fig. 5. Since the hardware is still in the state of unknown parameters when the MCU starts, if the system directly runs the program at the beginning, it will cause the processor to issue wrong instructions and perform wrong operations. So in order to prevent this situation, we design the system initialization program before the program starts to ensure that the initial state of the hardware is in a known state, so that the program can run normally. The software mainly includes ADC initialization, delay function initialization, I/O interface initialization, key initialization and so on.

The measuring instrument is in an unknown state after power on. If the program is run without system initialization, the output result will be undefined (random value), which seriously affects the normal use of the measuring instrument. In order to ensure that the value is fixed after power on, the variable definition is initialized according to the initialization mechanism.

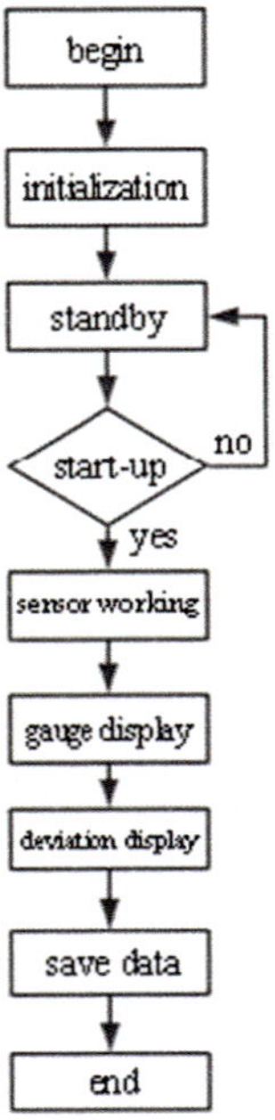

Fig. 5. System main program flow chart

In the system initialization program design, each parameter of the measurement operation is initialized, such as delay function initialization, pin initialization, key initialization, ADC initialization, OLED initialization, laser displcement sensor accept character count set to 0.

4 Conclusion

In this design, after understanding the working principle of the displacement sensor, through the selection of the measurement distance, the laser ranging sensor is finally selected. Laser ranging sensor has high precision and long life, and can realize non-contact measurement. In addition, because SCM is the core part of the system, it is necessary to select a SCM that can meet the requirements of the circuit, so the final SCM model is stc15f2k60s2. Select OLED display screen for data display, because the working environment of the measuring instrument can not supply power to the equipment through the ordinary power supply, so we have to select the appropriate power supply for power supply, and finally choose the 12 V power supply. In order to better design the software part of the system and make the design program more easy to understand, combined with the selected type of MCU, the system software part is designed by using µVision5 integrated development environment by Keil company. It has the advantages of small size, easy to carry, simple operation, high efficiency and accuracy.

Acknowledgment. The work described in this article has been conducted as part of the research project Natural Science Research Major Project of higher education institution of Jiangsu Province (grant numbers: 20KJA460011). Special thanks to all those who helped me during the writing of this paper.

References

1. Jiang, X., Namokel, M., Hu, C., Tian, R.: Research on lift fault prediction and diagnosis based on multi-sensor information fusion. In: Wang, Y., Martinsen, K., Yu, T., Wang, K. (eds.) IWAMA 2019. LNEE, vol. 634, pp. 160–168. Springer, Singapore (2020). https://doi.org/10.1007/978-981-15-2341-0_20
2. Ye, W., Huang, Y., Hezhe, Z.: Technical research on Intelligent elevator measuring instrument. Sci. Technol. Innov. Guide **15**(2), 54–56 (2018)
3. TSG T7001–2009 elevator supervision and inspection and regular inspection rules - traction and forced drive elevator (2009)
4. Zhang, Y.: Development of Elevator Gauge Deviation Measuring Instrument. Suzhou University (2014)
5. Chen, Y.: Research on High Precision Pulsed Laser Rangefinder. Xi'an Engineering University of Technology, Xi'an (2014)
6. Zhang, S., Hu, Y., Zhang, L.: Principle and Application of Embedded SCM STM32. China Machine Press, China (2019)
7. Shi, J., Wang Chen, H., Zeyuan, W.Y.: Design of OLED display driver based on STM32. Electron. World **12**, 127–128 (2018)
8. Xiao, L.: Type selection and performance evaluation of UPS system.Electric. Technol. **21**(1), 117–120 + 125 (2020)
9. Liang, C., Zilong, Z.: Control circuit design of monolithic buck DC-DC converter. Electron. Dev. **43**(1), 205–209 (2020)

Microstructure Analysis of Sintered Metal with Iron Powder and Tin Powder

Ya Gao[1,2](✉), Jingchao Zhang[3], Qingsong Zhao[3], Yubo Meng[2], and Lixia Li[2]

[1] Zhengzhou University, Zhengzhou, China
[2] Henan University of Engineering, Zhengzhou, China
[3] SIPPR Engineering Group Co., Ltd., Zhengzhou, China

Abstract. Fe-based matrix segments were prepared by hot-pressing sintering with Fe powder and Sn powder. The microstructures of the sintered matrixes were analyzed by SEM, EDS and XRD. The effects of the composition ratio of Fe powder and Sn powder and sintering temperature on the microstructure were studied. The results show that the microstructure of FeSn matrix sintered at 620 °C is mainly composed of α-Fe and FeSn phase. There is no metallurgical bonding between the Fe powders, and the FeSn compounds in the matrix are mainly distributed along the interface between α-Fe. The microstructure of FeSn matrix sintered at 760 °C mainly contains α-Fe, (Fe, Sn) solid solution and Fe_3Sn_2 compound. Metallurgical bonding occurs between α-Fe phase. The increase of Sn content boosts metallurgical bonding between (Fe, Sn) solid solution. Fe_3Sn_2 compounds mainly distribute along the pores between Fe powders.

Keywords: Hot-pressing sintering · Metal powder · Microstructure

1 Introduction

Sintered diamond tools are composed of diamond particles and metal matrix. The tools are obtained by uniformly mixing diamond particles into metal powder, sintering into diamond segments. The metal matrix has the function of embedding diamond and matching the properties of processed materials and determines whether diamond can play an excellent performance [1]. Fe powder has excellent formability, compactibility and sintering activity, and Fe-based matrix has suitable mechanical properties [2]. Fe powder has a lower price, so the diamond tools of Fe-based matrix can obtain better economic effect [3, 4]. The diamond tools of Fe-based matrix need high sintering temperature, so the diamond in matrix is easy to be eroded by high temperature. In order to solve this problem, the matrix generally contains low melting point components, which are melted in the sintering process to reduce the sintering temperature and improve density of the matrix. Sn is usually added to iron matrix as a low melting point element. In the process of sintering, Sn powder forms liquid phase. In the sintering system liquid phase has a lubricating effect on the powder, which reduces the friction between the powder particles and promotes the particle rearrangement [5]. The liquid phase of the sintering system can improve reaction rate, while it can fill the pores of matrix so as to improve the matrix

Y. Wang et al. (Eds.): IWAMA 2021, LNEE 880, pp. 419–425, 2022.
https://doi.org/10.1007/978-981-19-0572-8_53

density and mechanical properties [6]. Besides, Sn can also improve the wettability and reduce the interfacial tension between Fe powder and diamond [7–9], so Sn is widely used in Fe-based matrix. In this paper, the microstructure of sintered matrix with Fe powder and Sn powder was studied. The proper composition and sintering temperature of Fe powder and Sn powder are discussed in order to provide guidance for the selection of composition and process parameters of sintered Fe-based diamond tools.

2 Experimental Procedures

High purity of Fe powder which is 300 mesh and Sn powder which is 200 mesh were used in the test. The mixed powder is composed of different proportion of Fe powder and Sn powder. In the mixed powder, weight contents of Sn powder were 3 wt.%, 6 wt.%, 9 wt.%, 12 wt.%, 15 wt.%. The powder mixture was mixed in a three-dimensional mixer for 50min to make the metal powder well-distributed. The powder mixture was weighed to the proper quality, and then loaded into graphite molds, each of which has four patterns. The powder mixture was produced into metal segments by hot-pressing. The sintering pressure was 22 kN, while the sintering temperature is 620 °C, 760 °C and 800 °C respectively. After sintering, the temperature was reduced to 550 °C, and then the pressure is unloaded and the molds were cooled naturally to room temperature.

The cross section of the sintered segment was selected as the sample, and then the sample is mechanically ground and polished. The microstructure of the matrix was observed by scanning electron microscope (SEM), and the phases in the matrix were analyzed by energy dispersive spectrometer (EDS) and X-ray diffraction (XRD).

3 Results and Discussion

3.1 Microstructure of FeSn Matrix Sintered at 620 °C

Figure 1 illustrates the microstructures of FeSn matrixes which have different content of Sn. The EDS results of the phase in the sintered matrix in Fig. 1 are listed in Table 1. The microstructures and EDS analysis of the matrixes show that the sintered segments are mainly composed of two phases and the dark gray phase distributes along the gray black interphase gap. The EDS analysis showed that the composition of the dark regions is mainly Fe (99.89 at.%) and the bright matrix is composed of Fe (50 at.%) and Sn (50 at.%). Combining with the XRD pattern in Fig. 2, it is estimated that the phase of dark regions in the structure is α-Fe phase and the phase of bright matrix is FeSn compounds.

By comparing the microstructure of sintered matrix with different Sn content, it can be concluded that there is no particle bonding between the Fe powders, and there are many pores in the microstructure so that the matrixes are low density. When the content of Sn in the matrix changes, the phase composition in the sintering matrix remains α-Fe phase and FeSn compounds. During hot-pressing sintering, Sn powder melts into liquid phase and fills in the pores of Fe powder particles, so it is easy to react with the surface of Fe powder to form FeSn compound. In result, the FeSn compounds in the matrix are mainly distributed along the interface between α-Fe.

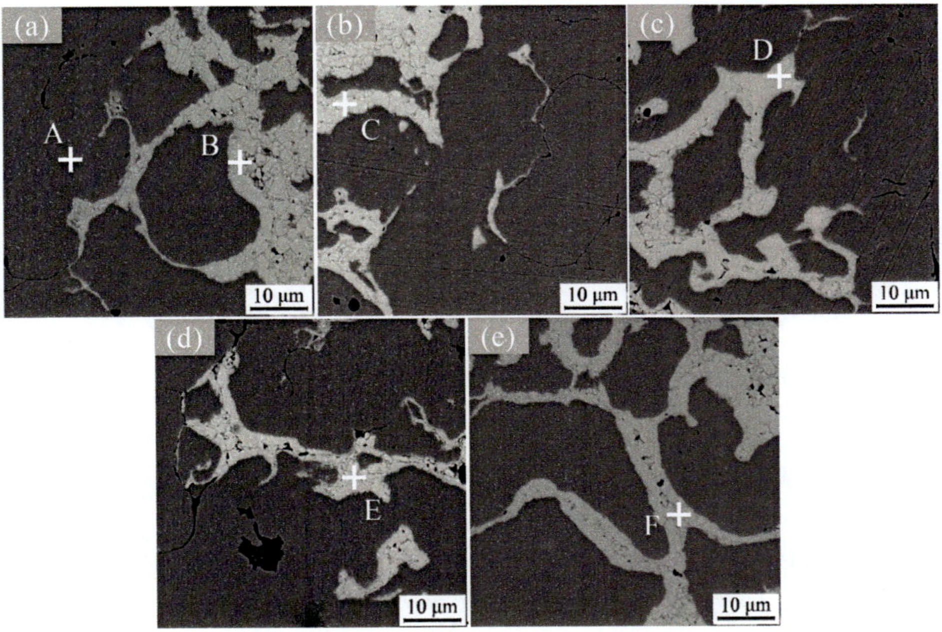

Fig. 1. The microstructures of FeSn matrix sintered at 620 °C. (a) Sn 3wt.% (b) Sn 6wt.% (c) Sn 9wt.% (d) Sn 12wt.% (e) Sn 15wt.%

Table 1. EDS results of the points in Fig. 1

Sample	Point	wt.%		at.%	
		Fe	Sn	Fe	Sn
Sn 3wt.%	A	99.77	0.23	99.89	0.11
	B	36.59	63.41	55.09	44.91
Sn 6wt.%	C	32.41	67.59	50.48	49.52
Sn 9wt.%	D	31.15	68.85	49.02	50.98
Sn 12wt.%	E	31.54	68.46	49.48	50.52
Sn 15wt.%	F	32.30	67.70	50.35	49.65

3.2 Microstructure of FeSn matrix sintered at 760 °C

Figure 3 illustrates the microstructures of FeSn matrixes sintered at 760 °C which have different content of Sn. The EDS results of the phase in the sintered matrix in Fig. 3 are listed in Table 2. The microstructures and EDS analysis of the matrixes show that the sintered segments are mainly composed of two phases and the pale gray phase distributes along the dark gray interphase gap. The EDS results show that the composition of the dark regions of the dark gray phase is mainly Fe (99.65 at.%) and the bright matrix of

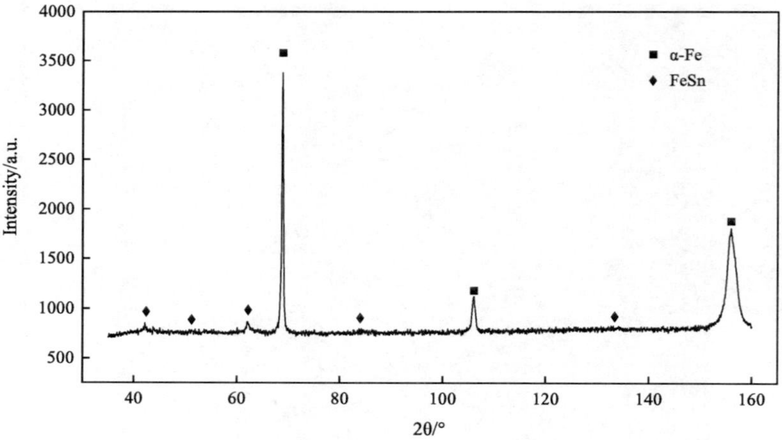

Fig. 2. XRD result of the matrix of Sn 12wt.% sintered at 620 °C

the dark gray phase is composed of Fe (90.73 at.%) and Sn (9.27 at.%). The composition of pale gray phase contains Fe (61.23 at.%) and Sn (38.77 at.%).

Combining with the XRD results in Fig. 4, it is estimated that dark gray phase in the structure is α-Fe phase and pale gray phase is Fe_3Sn_2 compounds. The results indicate that the matrixes only contain α-Fe when the content of Sn is 3 wt.% and 6 wt.%. The matrix is composed of α-Fe and Fe_3Sn_2 compound when the content of Sn is 9 wt.%. The matrix consist of α-Fe, (Fe, Sn) solid solution and Fe_3Sn_2 compound. Sn is dissolved in α-Fe to form (Fe, Sn) solid solution. The crystal structure of (Fe, Sn) solid solution is still the same as that of α-Fe. When the content of Sn is 3 wt%, 6 wt% and 9 wt%, the solid solubility is small, and the crystal structure and lattice constant do not change greatly. For this reason, There are no characteristic peak in XRD results of them. When the content of Sn is 12 wt.%, 15 wt.%, the solid solubility of Sn in solid solution is large that result in the change of lattice constant. The characteristic peak of solid solution appears in the low angle position of characteristic peak of α-Fe.

In summary, the content of Sn changes continuously from α-Fe to (Fe, Sn) solid solution. It means that solid solubility of Sn changes continuously from the edge to the center of Fe powder particles. Due to the higher sintering temperature, the diffusion speed between Fe powder particles is faster, with the result that metallurgical bonding occurs between Fe phases so that the interface between α-Fe phase is intermittent. In addition, the higher the content of Sn is, the more liquid phase is produced, which can promote the diffusion and migration of elements in the system. Thus, the easier metallurgical bonding between (Fe, Sn) solid solution interfaces is, and the less interface area is. When the content of Sn in the matrix is low, Sn elements are all soluble in α-Fe. The matrixes mainly consist of α-Fe phase and (Fe, Sn) solid solution. When the content of Sn is higher, The matrixes mainly consist of α-Fe phase, (Fe, Sn) solid solution and Fe_3Sn_2 compound. The Fe_3Sn_2 compounds in the matrix are mainly distributed along the pores between the α-Fe phase.

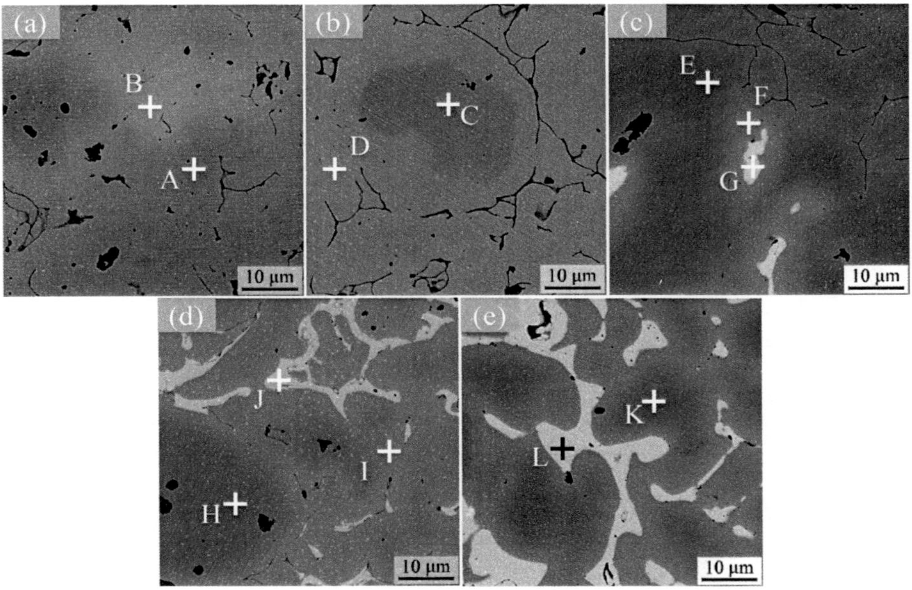

Fig. 3. The microstructures of FeSn matrix sintered at 760 °C. (a) Sn 3wt.% (b) Sn 6wt.% (c) Sn 9wt.% (d) Sn 12wt.% (e) Sn 15wt.%

Table 2. EDS results of the points in Fig. 2

Sample	Point	wt.%		at.%	
		Fe	Sn	Fe	Sn
Sn 3wt%	A	99.26	0.74	99.65	0.35
	B	82.16	17.84	90.73	9.27
Sn 6wt%	C	99.85	0.15	99.93	0.07
	D	95.76	4.24	97.96	2.04
Sn 9wt%	E	99.874	0.26	99.88	0.12
	F	90.48	9.52	95.28	4.72
	G	42.62	57.38	61.23	38.77
Sn 12wt%	H	99.12	0.88	99.59	0.41
	I	83.36	16.64	91.42	8.58
	J	42.96	57.04	61.55	38.45
Sn 15wt%	K	99.66	0.34	99.84	0.16
	L	44.16	55.84	62.70	37.30

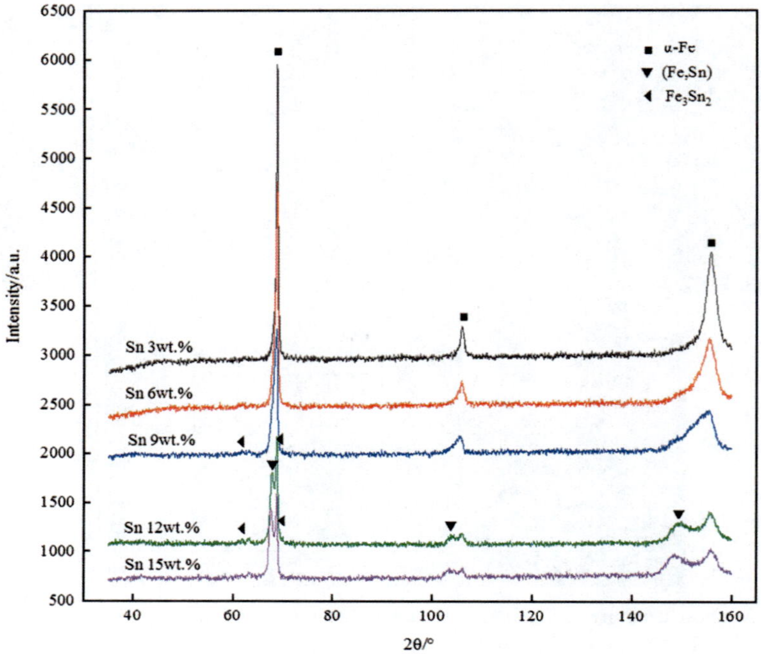

Fig. 4. XRD result of the matrixes sintered at 760 °C

4 Conclusion

The effects of temperature and component on the microstructure of FeSn matrixes were investigated. Based on the experimental work performed here, the following conclusions can be reached.

(1) The sintering temperature has a great influence on the microstructure of the matrix. Microstructure of FeSn matrix sintered at 620 °C is mainly composed of α-Fe and FeSn compounds, and there is no metallurgical bonding between the α-Fe phase. Microstructure of FeSn matrix sintered at 760 °C is mainly composed of α-Fe, (Fe, Sn) solid solution and Fe_3Sn_2 compounds, and there is partly metallurgical bonding between the α-Fe phase.

(2) The content of Sn has a great influence on the microstructure of the matrix. The higher the tin content, the more compounds in the structure. The increase of Sn content boosts metallurgical bonding between (Fe, Sn) solid solution.

Acknowledgment. The authors gratefully acknowledge the financial support received from the Doctoral Foundation for Henan University of Engineering (No. DKJ2019024).

References

1. Tyrala, D., Romanski, A., Konstanty, J.: The effects of powder composition on microstructure and properties of hot-pressed matrix materials for sintered diamond tools. J. Mater. Eng. Perform. **29**(3), 1467–1472 (2019)
2. Mechnik, V.A., Bondarenko, N.A., Kolodnitskyi, V.M., Zakiev, V.I., Zakiev, I.M., Ignatovich, S.R., et al.: Formation of Fe-Cu-Ni-Sn-VN nanocrystalline matrix by vacuum hot pressing for diamond-containing composite. mechanical and tribological properties. J. Superh. Mater. **41**(6), 388–401 (2019)
3. Bulut, B., Tazegul, O., Baydogan, M., Kayali, E.S.: Investigation and application of Fe–Co—Cu based diamond cutting tools with different bronze content used in marble production. In: Silva, Lucas F M. (ed.) Materials Design and Applications. ASM, vol. 65, pp. 307–314. Springer, Cham (2017). https://doi.org/10.1007/978-3-319-50784-2_23
4. Barbosa, A.P., Bobrovnitchii, G.S., Skyry, A.L., Guimars, R.S., Filgueira, M.: Structure, microstructure and mechanical properties of PM Fe-Cu-Co alloys. Mater. Des. **31**(1), 522 (2010)
5. Wang, L., Pouchly, V., Maca, K., Shen, Z., Xiong, Y.: Intensive particle rearrangement in the early stage of spark plasma sintering process. J. Asian Ceramic Soc. **3**(2), 183–187 (2015)
6. Azevedo, J., Serrenho, A.C., Allwood, J.M.: The deformation of metal powder particles: hardness and microstructure. Procedia Eng. **207**, 1200–1205 (2017)
7. Schubert, W.D., Fugger, M., Wittmann, B., Useldinger, R.: Aspects of sintering of cemented carbides with fe-based binders. Int. J. Refract Metal Hard Mater. **49**, 110–123 (2015)
8. Martin, W., David, L., John, Å., et al.: Diffusion modeling in cemented carbides: solubility assessment for Co, Fe and Ni binder systems. Int. J. Refrac. Metals Hard Mater. **68** (2017)
9. Li, B., Zheng, Z., Yu, H., Zeng, D.: Improved permeability of fe based amorphous magnetic powder cores by adding permalloy. J. Magnet. Magnet. Mater. **438**, 138–143 (2017)

Cavitations Behavior of a LZ82 Mg-Li Alloy During Superplastic Deformation

Xuhe Liu(✉), Xidong Liu, Fengxiang Shao, Haoming Zhang, and Hongsong Zhang

Department of Mechanical Engineering, Henan University of Engineering, Zhengzhou, Henan, China
liuxh@haue.edu.cn

Abstract. Specimens of a two-pass extruded LZ82 alloy were tensile tested at the optimum superplastic temperature of 290 °C and an initial strain rate of $1.0 \times 10^{-4}s^{-1}$ to predetermined strain, and the cavitations behavior during the deformation were investigated. The nucleation of the cavities is generally observed to be associated with the grain triple junctions and mainly controlled by diffusion, the nucleation is continuously during deformation. When the true strain is low, the mechanism of cavity growth is controlled by diffusion. When the true strain is high, the mechanism of cavity growth is mainly controlled by plastic deformation, however the diffusion still react. Cavities grow and coalesce during deformation, resulting in the materials fracture. And under tension deforming conditions, T = 290 °C, $\varepsilon 0 = 1.5 \times 10^{-4}s^{-1}$, the cavity tolerance of present alloy is about 10.5%. Cavities grow and coalesce result in the materials fracture.

Keywords: Mg-Li alloy · Superplasticity · Cavitation

1 Introduction

Mg-Li alloy is the lightest metal structural material [1], compared to the normal magnesium alloys, due to the added of Li; Mg-Li alloys have lower density and good ductility at room temperature [1, 2]. It has broad application prospects in the aerospace, automotive, electronics and the defense fields as structural materials [3]. When the Li content is greater than 5.7%, β-phase appears (the solid solution of Mg in Li), which improve the plasticity significantly of the alloy [4, 5], even result in superplastic under certain conditions [6, 7].

The realization of superplastic in Mg alloys is significant; It enables alloys to be used in stamping or die forging processes to manufacture complex parts [8]. There are several studies demonstrating the occurrence of superplasticity in Mg-Li alloy two-phase alloys containing various amounts of Li and these reports show that excellent ductility can be attained at temperature in the range of 200 °C–400 °C [9–11].

It is known that the cavitations can occur in all kinds of materials during the tensile superplastic flow, and the development of interred cavitations during superplastic flow was first reported in a Zn-22% Al eutectoid alloy many years ago. From a commercial point of view, the understanding of cavitations behavior in superplastic flow is very

Y. Wang et al. (Eds.): IWAMA 2021, LNEE 880, pp. 426–432, 2022.
https://doi.org/10.1007/978-981-19-0572-8_54

important because they are likely to influence many post-forming properties of the alloys and limit varies of applications of superplastic formed parts: a high volume of cavitations is not appropriate for safety [12, 13]. Although cavity is a key feature of superplastic alloy failure, So far, there seems to be no information on the role of the degree of cavitation development in the Mg-Li alloy, especially under the optimum deformation condition.

Superlasticitiy has been observed in a Mg-Li alloy LZ82 at temperature of 290 °C with an initial strain rate $1.5 \times 10^{-4}s^{-1}$, and the dominate deformation mechanism is grain-boundary sliding controlled by grain-boundary diffusion [9–11].

In this paper, the two-pass extruded LZ82 alloy was prepared and the cavitations of the alloy at different strain were studied.

2 Experimental Methods

The LZ82 alloys with a chemical composition of Mg-8wt. %Li-2wt. %Zn were produced by the vacuum induction furnace under argon shield. The ingot was heat-treated at 573K for 24 h and then extruded for two pass at 553K with annealing in process at 473K for 1 h. The obtained plates have a width of 30 mm and a thickness of 3 mm, with an extrusion rate of 100:1.

Tensile specimens with a gauge length of 10 mm and a gauge width of 6 mm were processed from the extrusion plate, and the loading axis is parallel to the extrusion direction. Tensile tests were conducted at 290 °C and initial strain rate of $1.5 \times 10^{-4}s^{-1}$, using an WDW3050 Tensile testing machine, the temperature was controlled within ±2 °C during testing. The specimen is deformed to a predetermined strain. After tensile test, selected specimens were prepared for metallographic sample.

Quantitative measurement of the cavity is performed using an optical microscope connected to a computer through an image monitor. All of the detailed cavity information was collected using software (IPP6.0) where the software can count and measure the cavity in the selected area.. The area and length of each cavity was measured and autolined, 15 different domains were selected for each sample for cavity measurement.

3 Results and Discussion

Figure 1 shows the microstructure of the alloy after deformation. In Fig. 1(a), the light gray and dark gray parts are the α phase and β phase of the LZ82 alloy respective, and the black parts are the cavities formed during deformation. Figure 1(b) is the SEM image of the cavities. It can be seen from the image, the structure of the cavities is not the spherical shape, but of complex structure, and there are several grains around a cavity, large cavities are resulted from the linkage of small cavities which nucleate at grain boundaries due to the stress affect.

The evolution of cavities with strain can be inspected in Fig. 2 which shows OM images of the microstructures after deformation to the predetermined strains. The tensile axis is perpendicular in the Fig. 2. When the strain is lower, the cavities are small and less, and the distributions are random, Cavities less than 2 μm in diameter are roughly circular. When the strain is greater, the cavities are large, and cavities elongated along the tensile direction.

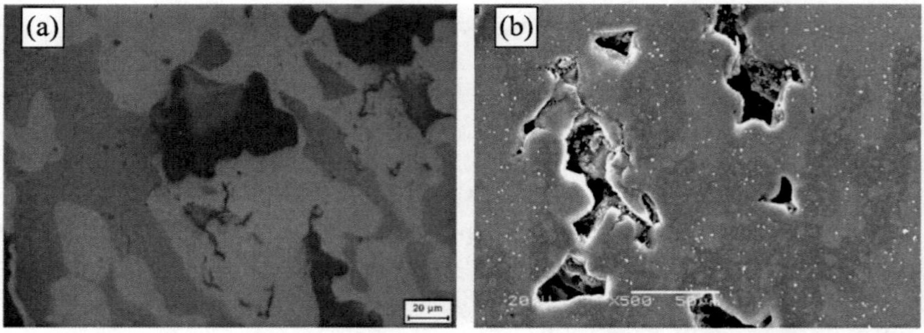

Fig. 1. OM and SEM image of the cavities (ε = 2.05)

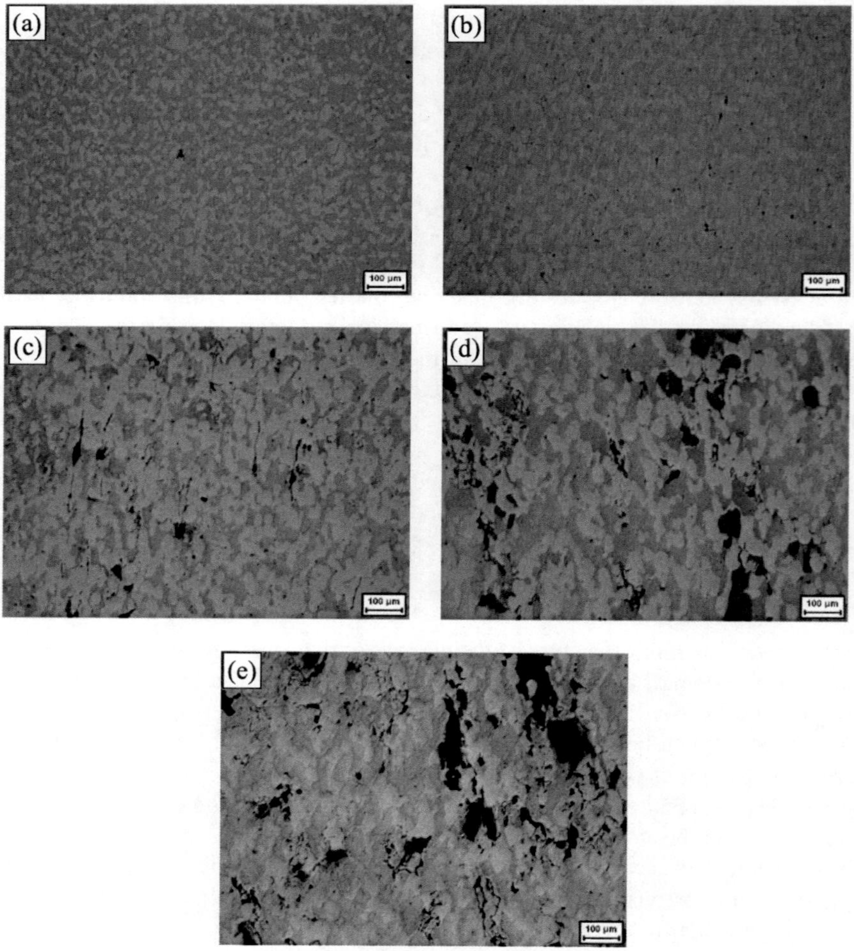

Fig. 2. OM images of cavitations at different strain. (a) 0.42, (b) 1.11, (c) 1.66, (d) 2.05 and (e) 2.15.

With the increase of strain, cavities grow up and more new voids form, leading to an increase in the density and volume fraction of the cavity.

At the strain of 2.05, the growth, merging and connection of cavities are becoming more and more obvious, resulting in the increase of cavity volume fraction.

It is noted that the nucleation and growth of cavity are related to the triple connection of grains and grain boundaries. Furthermore, the distribution of cavities in the whole area is relatively uniform of LZ82 alloy.

The high magnification microstructure of maximum strain specimen is shown in Fig. 3. The morphology inside the cavity is clearly displayed, indicating that the cavity grows along the grain boundary. The observed morphologies strongly suggest that cavity growth is plasticity controlled, but may also be interpreted to indicated diffusion growth at the small cavity size.

Figure 4 show the variation of cavity density with strain for the LZ82 alloy deformation at 290 °C and an initial strain rate of $1.5 \times 10^{-4}s^{-1}$. The number of cavities increases with the increase of strain, this indicates that the new cavity is continuously nucleated. In the middle stage of superplastic deformation, the void density increases rapidly. At this stage, cavity growth and the nucleation of new cavities are helpful to improve the cavity density. However, When the strain reaches a certain value, the cavity density does not increase significantly and remains near a certain value.

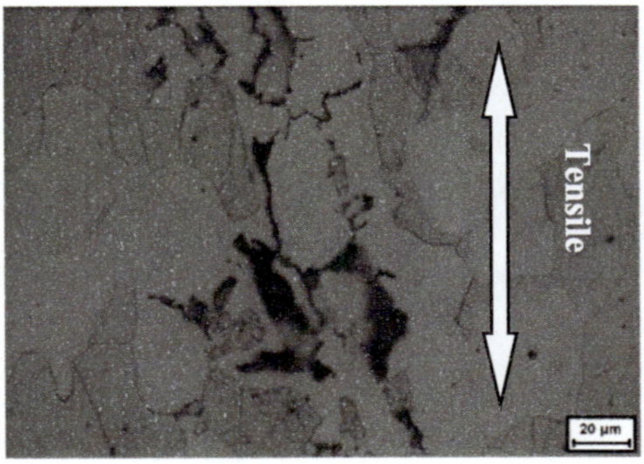

Fig. 3. Microstructure of the alloy at true strain 1.2

Figure 5 shows the variation of cavity density with cavity diameter at two different strains of 1.11 and 1.87 correspond to the small strain and large strain. It can be seen from the image, with the increase of strain, the number of large-size holes increases significantly. This means that, while new cavities were continuously nucleated during superplastic deformation, the increase of strain leads to the significant growth, consolidation and connection of cavities.

Generally speaking, With the increase of strain, the proportion of small-size voids decreases. With the increase of strain level, the hole size tends to move to a higher level, and the effect of strain is relatively large.

According to conclusion of the investigations of superplastic deformation, there are two theories of nucleate and growth of cavities during superplastic deformation, (1)

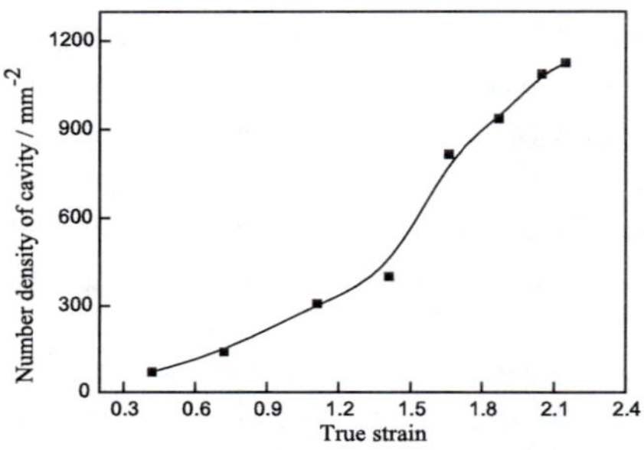

Fig. 4. Variation of cavity density with true strain

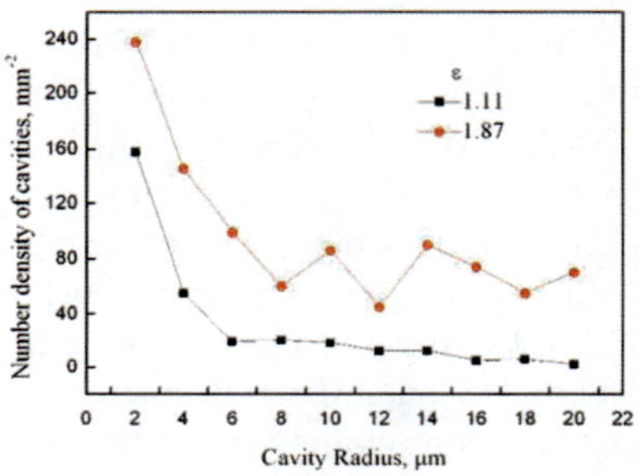

Fig. 5. Statistics of cavity size with different strains

the diffusion mechanism (DIF) and (2) the plasticity deformation mechanism (PLA). The authors will explain the cavitations in the two-pass LZ82 alloy during superplastic deformation using the two mechanisms.

For the cavity growth controlled by DIF and PLA, The variety of ϕ_V with strain ε can be expressed by formula (1) and formula (2) [14–17]:

$$\varphi_v = \varphi_0 + \frac{N4\pi\Omega\delta_{gb}D_{gb}}{5kT}\frac{\sigma}{\dot{\varepsilon}}\varepsilon \tag{1}$$

$$\varphi_v = \varphi_0 \exp(\eta\varepsilon) \tag{2}$$

where Ω is the atom volume, δ_{gb} is the grain boundary thickness, usually taken as twice the Burgers vector, ε is the true strain, σ is the flow stress, T is the absolute temperature, $\dot{\varepsilon}$ is the strain rate, k is the Boltzmann's constant, D_{gb} is the grain-boundary diffusion coefficient, ϕ_0 is the cavity radius at zero strain, η is the cavity growth rate parameter.

In this experiment, the deformation condition is 290 °C and $1.5 \times 10^{-4}\text{s}^{-1}$, under this condition, the superplastic elongation is 758%, the value of m is ~0.55, and a theoretic predicted valueof η is 1.6. The Comparison of experimentally measured cavity volume fraction with theoretical predictions based on diffusion model (DIF) and plastic model (PLA) was shown in Fig. 6.

At the initial stage of deformation, the experimental curve and the predictions for diffusive growth are coincided, indicating that cavity growth is controlled by diffusion mechanism in this stage. When the strain ε is larger than 1.4, the cavity volume fraction increases quickly with increasing strain, the experimental curve deviate from the predictions for diffusive growth, approach to the predictions for plasticity controlled growth gradually, indicating the plasticity mechanism play a major role, and the diffusion mechanism also exist.

In Fig. 6, the slop of the curve turn to larger, compared with the Fig. 4, it can be affirmed that the speed of new cavity nucleation and the original empty growing is larger, contribute to the enormous increasing of the volume fraction. From the measured curve, when the strain reaches 1.8, the cavity volume fraction increased dramatically, this is because when the strain reaches a certain value, cavities coarsening began to appear,

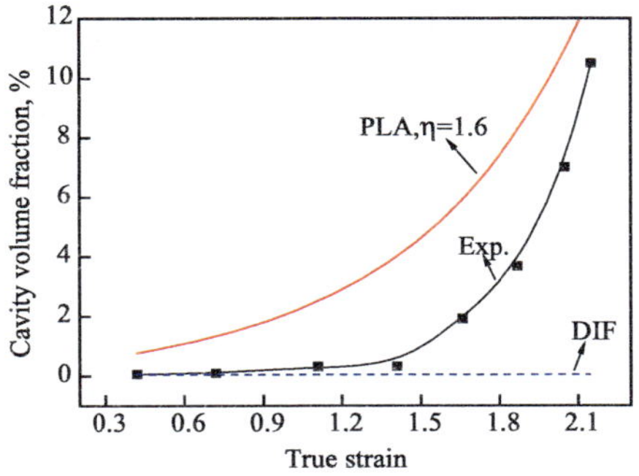

Fig. 6. Comparisons of volume fraction experimentally measured of cavity with theoretical prediction results based on diffusion model (DIF) and plasticity model (PLA)

leading to the cavity volume fraction increases rapidly. This shows that, at large strain, the growth and connected of cavities contribute more than the new cavity nucleation do, in this case, void volume fraction increased primarily by the growth of cavities, coarsening and connected.

When the strain is 2.15, the maximum cavity volume fraction is about 10.5%. However, When the cavity volume fraction reaches 2%, its growth rate increases sharply. In the superplastic alloys, The maximum cavity volume fraction is called the cavity tolerance, so in this experiment, the cavity tolerance is 10.5%. If the cavity concentration limit is assumed to be 2%, the maximum allowable value sometimes used in hot forming of structural parts is only 1.6, which is expected from Fig. 6.

With the increase of strain, cavities begin to join adjacent ones, and finally result in the failure of specimens, as shown in Fig. 7.

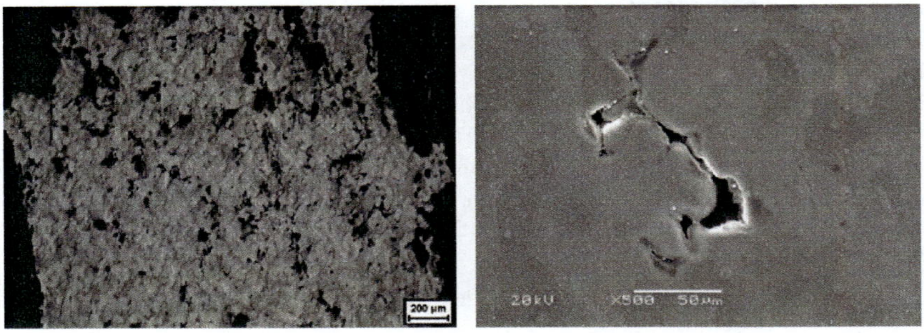

Fig. 7. OM and SEM images of the cavities

4 Conclusion

The nucleation of the cavities was generally observed to be associated with the grain triple junctions. In the process of deformation, new cavity nucleate continuously. when the true strain ε is lower than 1.4, the mechanism of cavity growth is controlled by diffusion, when the true strain ε is higher than 1.4, the mechanism of cavity growth is main controlled by plastic deformation, and the diffusion still react. The cavity tolerance of present alloy is about 10.5%. Cavities grow and coalesce during deformation and result in the materials fracture.

References

1. Rahulan, N., et al.: Mechanical behavior of Mg-Li-Al alloys. Mater. Today Proc. **5**(10), 17935–17943 (2018)
2. Kim, Y.H., et al.: Effects of Li addition on microstructure and mechanical properties of Mg–6Al–2Sn–0.4Mn alloys, Trans. Nonferrous Met. Soc. China **26**(3), 697–703 (2016)
3. Rahulan, N., et al.: Mechanical behavior of Mg-Li-Al alloys. Mat. Today Proc. 5(10), 17935-17943 (2018)
4. Wu, R.Z., et al.: Recent progress in magnesium-lithium alloys. Int. Mater. Rev. **60**(2), 65–100 (2015)
5. Li, C.Q., et al.: Influence of the lithium content on the negative difference effect of Mg-Li alloys. J. Mater. Sci. Technol. **57**(22), 138–145 (2020)
6. Zhao, J., et al.: Influence of Li addition on the microstructures and mechanical properties of Mg–Li alloys. Met. Mater. Int. **27**(6), 1403–1415 (2019). https://doi.org/10.1007/s12540-019-00528-4
7. Sivakesavam, O., et al., Characteristics of superplasticity domain in the processing map for hot working of as–cast Mg–11.5Li–1.5Al alloy. Mater. Sci. Eng. A, 212(1), 178–181 (2018)
8. Dong, S., et al.: Superplasticity in Mg–Li–Zn alloys processed by high ratio extrusion. Adv. Manuf. Process. **23**(4), 336–341 (2008)
9. Guo, X.F., et al.: Superplastic behavior of reciprocating extruded Mg-6Zn-1Y-0.6Ce-0.6Zr from rapidly solidified ribbons. J. Wuhan Univ. Technol.-Mater Sci. Ed, **27**(6), 1033–1037 (2012)
10. Liu, X., et al.: Superplasticity at elevated temperature of an Mg-8%Li-2%Zn alloy. J. Alloy. Compd. **541**, 372–375 (2012)
11. Liu, X., et al.: Deformation and microstructure evolution of a high strain rate superplastic Mg–Li–Zn alloy. J. Alloy. Compd. **509**(39), 9558–9561 (2011)
12. Liu, X., et al.: Superplasticity in a two-phase Mg-8Li-2Zn alloy processed by two-pass extrusion. Mater. Sci. Eng. A **528**(19–20), 6157–6162 (2011)
13. Hosokawa, H., et al.: Effects of Si on deformation behavior and cavitation of coarse-grained Al-4.5Mg alloys exhibiting large elongation. Acta Mater. **47**(6), 1859–1867 (1999)
14. Aguirre, J.V., et al.: Experimental investigation of cavitation behavior in AZ61 magnesium alloy. Mater. Trans. **46**(3), 626–630 (2005)
15. Chokshi, A.H., et al.: A model for diffusional cavity growth in superplasticity. Acta Metall. **35**(5), 1089–1101 (1987)
16. Hancock, J.W.: Creep cavitation without a vacancy flux. Met. Sci. **10**(9), 319–325 (1976)
17. Beere, W., et al.: Creep cavitation by vacancy diffusion in plastically deforming solid. Met. Sci. **12**(4), 172–176 (1978)
18. Chokshit, A.H., et al.: An analysis of cavity nucleation in superplasticity. Acta Metall. **37**(11), 3007–3017 (1989)

Vehicles Detection Based on Improved YOLOv3

Xudong Dong and Liangwen Yan(✉)

School of Mechatronic Engineering and Automation, Shanghai University, Shanghai 200444, China
lw_yan@staff.shu.edu.cn

Abstract. Real-time detection of road vehicles is a hot research topic in the field of computer vision. Aiming at the problems of low detection accuracy and slow speed in road vehicles detection algorithms, a road vehicles target detection method based on improved YOLOv3 is proposed in this paper. Firstly, by embedding the Convolutional Block Attention Module (CBAM) in the feature layer after the fusion of low-level features and high-level features, the ability of the neural network to obtain effective information is improved. Secondly, the adaptive multi-scale anchor frame size is determined based on the K-means clustering algorithm, which is beneficial to improve the detection ability of large, medium and small vehicles targets. Finally, the DIoU loss function is introduced to improve the ability to recognize occluded targets. The experimental results show that the average accuracy of the proposed method on the standard data set KITTI reaches 94.36%, which is 2.2% higher than the traditional YOLOv3, and the detection speed reaches 52.0f/s, in other words, the model can still maintain a better real-time detection speed when detection accuracy is improved.

Keywords: Vehicles detection · YOLOv3 · Attention mechanism · Loss function

1 Introduction

Road vehicles detection technology is an important research content of machine vision, and vehicles detection technology based on deep learning is the main research hotspot at present. Vehicles detection technology can be applied to many fields such as unmanned driving, intelligent transportation and traffic safety. In recent years, with the improvement of computer computing performance and the growth of storage space, Convolutional Neural Network has gradually become the focus of research in the field of computer vision, and has achieved great success with its powerful feature expression ability [1]. Vehicles detection algorithms based on deep learning include two-stage target detection algorithm and one-stage target detection algorithm. The two-stage target detection algorithm usually includes two parts: candidate region generation and classification regression. Representative algorithms include R-CNN [2], Fast R-CNN [3], Faster R-CNN [4], Mask R-CNN [5], Cascade R-CNN [6], etc. The advantage of such methods is high precision, but the disadvantage is that they cannot meet the demand of real-time application due to excessive computation. The one-stage target detection algorithm does not need the candidate region generation part, but treats the detection problem as a regression

Y. Wang et al. (Eds.): IWAMA 2021, LNEE 880, pp. 433–440, 2022.
https://doi.org/10.1007/978-981-19-0572-8_55

problem, directly predicting the target location and category. Representative algorithms include YOLO [7], SSD [8], YOLOv2 [9], YOLOv3 [10], etc. The advantages of such methods are clear structure and good real-time performance, while the disadvantages are lower precision compared with R-CNN series methods.

In this paper, based on the YOLOv3 network, a Convolutional Block Attention Module (CBAM) [11] was embedded into the network to select information more critical to the current task goal and suppress other useless information from a large number of information. At the same time, the K-means clustering algorithm is used to calculate the anchor boxes suitable for the vehicles target, and the DIoU loss function [12] which fully considers the overlap ratio between the candidate boxes and the real boxes, the distance from the center point and the aspect ratio is introduced to replace the original frame position loss function.

The rest of this paper is organized as follows: Sect. 2 explains the basic principle of YOLOv3 algorithm. Section 3 details the related improvement work. Section 4 introduces the experimental environment and experimental results in detail. Discussion and conclusions are summarized in the last section of this paper.

2 The Basic Principle of YOLOv3 Algorithm

The basic principle of YOLOv3 algorithm is to take the fixed-size image as the input of the network and obtain the position of the bounding boxes and its category based on regression, so as to achieve end-to-end target detection. The YOLOv3 model is mainly divided into two parts: Darknet-53 Convolutional Neural Network(CNN) feature extraction and YOLO layer feature map prediction. Yolov3 proposed two methods to deal with the difficulty of identification caused by the change of target size. First, three types of downsampling were used, which were 32 times, 16 times and 8 times downsampling. Three different scale feature maps were output: 13×13, 26×26 and 52×52. Second, based on the idea of Feature Pyramid Networks, YOLOv3 model uses the method of multi-scale fusion to perform feature fusion on three feature map of different scales output from Darknet-53 backbone network. The method of multi-scale fusion can obtain better fine-grained features and more meaningful semantic information. At the same time, the multi-scale training of the network model can be realized by randomly changing the size of the input image during the training process, so as to improve the sensitivity and detection accuracy of the algorithm to small targets.

2.1 Darknet-53

The structure of Darknet-53 feature extraction network is shown in Fig. 1.

Darknet-53, as the basic network of Yolov3, is different from the traditional CNN network model. Darknet 53 abandoned the usual pooling layer and full connection layer and used a lot of residual network to build a deeper network. Each of the smallest convolutional layers consists of convolution operations, Batch Normalization (BN), and the activation function Leaky-Relu. The dimensionality of the tensor is controlled by changing the step size of the filters kernel, so as to gradually achieve the purpose of dimensionality reduction and channel dimensionality increase. Finally, the prediction of the target object is output on three scales.

	Type	Filters	Size	Output
	Convolutional	32	3x3	512x512x32
	Convolutional	64	3x3/2	256x256x64
	Convolutional	32	1x1	
1x	Convolutional	64	3x3	
	Residual			256x256x64
	Convolutional	128	3x3/2	128x128x128
	Convolutional	64	1x1	
2x	Convolutional	128	3x3	
	Residual			128x128x128
	Convolutional	256	3x3/2	64x64x256
	Convolutional	128	1x1	
8x	Convolutional	256	3x3	
	Residual			64x64x256
	Convolutional	512	3x3/2	32x32x512
	Convolutional	256	1x1	
8x	Convolutional	512	3x3	
	Residual			32x32x512
	Convolutional	1024	3x3/2	16x16x1024
	Convolutional	512	1x1	
4x	Convolutional	1024	3x3	
	Residual			16x16x1024

Fig. 1. Darknet-53 network structure diagram

2.2 Feature Map

Each output tensor of Darknet-53 was used to obtain the real feature map for target positioning through the corresponding convolution, Batch Normalization, activation function and upsampling techniques. Yolov3 adopts the idea of multi-scale classification and divides the YOLO layer into three scales. In the same scale YOLO layer, convolution operation is used and the convolution kernel (3×3 and 1×1) is used to complete the interaction between feature map and local features. Target is detected on the feature map, and three bounding boxes are predicted for each cell with the help of three anchor boxes.

Each pixel in the feature map was used as a grid cell to judge the confidence of an object in a given candidate box. Non-maximum Suppression (NMS) algorithm was used to extract the candidate box with the highest confidence in each cell and output the final result.

3 Related Improvement Work

In order to improve the accuracy of road vehicles target detection without affecting the real-time performance, this paper improves it on the basis of YOLOV3. There are three main improvement points. One is to add a CBAM module after the main feature extraction network Darknet-53 and after the feature layer after the fusion of low-level features and high-level features, so as to improve the ability of neural network to obtain effective information with a few additional training parameters. Second, according to the distribution characteristics of the data set, K-means clustering algorithm is carried out on the real boxes size of the target in the data set to determine the adaptive multi-scale anchor boxes size, so as to improve the detection ability of vehicles targets of large, medium and small sizes. Third, the coordinate error loss function DIoU is introduced, which fully considers the overlap ratio between the candidate boxes and the real boxes, the distance from the center point and the length-width ratio.This measure can improve the accuracy of vehicles detection without increasing the model size and inference speed. The improved YOLOv3 network structure is shown in Fig. 2.

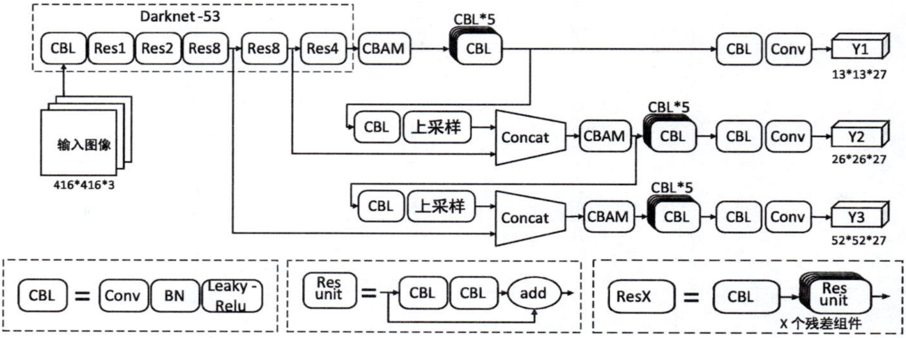

Fig. 2. The improved YOLOv3 network structure

3.1 CBAM

In order to select the information that is more critical to the current task objective from numerous feature information and improve the efficiency and accuracy of image information processing, the attention mechanism of deep learning is introduced in this paper. In essence, the attention mechanism module in deep learning is similar to the selective visual attention mechanism of humans. Both are to obtain the target area that needs to be focused on by quickly scanning the global image, and then devote more attention resources to this area, quickly filter high-value information from a large amount of information, and suppress other useless information. CBAM is a simple and effective attention mechanism module designed for convolutional neural networks. The overview of CBAM is shown in Fig. 3.

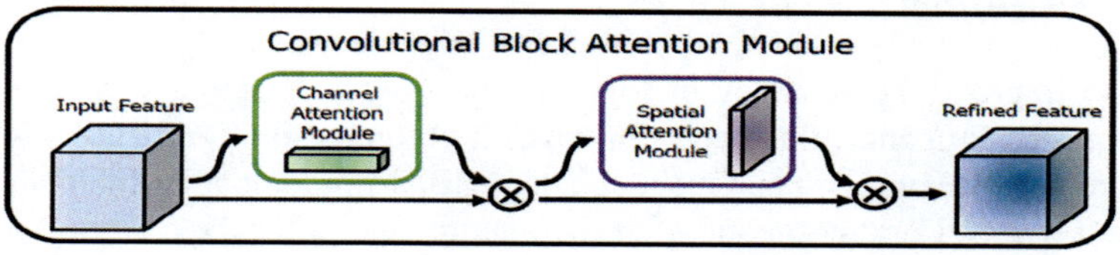

Fig. 3. The overview of CBAM

CBAM is a kind of attention mechanism module that combines spatial and channel. Channel attention module produces a channel attention map by exploiting the interchannel relationship of features. As each channel of a feature map is considered as a feature detector, channel attention focuses on 'what' is meaningful for a given input image. Spatial attention module generates a spatial attention map by utilizing the interspatial relationship of features. Different from the channel attention, the spatial attention focuses on 'where' is an informative part, which is complementary to the channel attention.

CBAM computes the attention map of the feature map generated by CNN from channel and spatial dimensions, and multiplies the attention map with the input feature map for feature adaptive learning.As a lightweight general module, CBAM can be integrated into convolution layer of target detection network for training. In this paper, CBAM module is embedded in the multi feature fusion network, and the information of vehicles target is selected from many information to suppress other useless information.

3.2 Anchors Boxes

For 416 $\times$ 416 image input, the fusion scales of Yolov3 are 13 $\times$ 13, 26 $\times$ 26, and 52 $\times$ 52. Each scale is assigned three anchor boxes corresponding to three bounding boxes, and each bounding box predicted a target. This can not only solve the problem that one grid cell in the original YOLO algorithm can only predict one target, but also make the prediction results of multiple scales more accurate.

Appropriate size and number of anchor boxes are more beneficial to improve the accuracy of model detection. The number of anchors and aspect ratio of the original YOLOv3 algorithm were clustered based on the COCO data set of 80 types of targets including vehicles. However, for the KITTI data set used in this paper, the original size of the anchor boxes is no longer applicable. Therefore, this paper takes KITTI data set as the statistical object and uses K-means clustering algorithm to obtain the new number and aspect ratio of anchor boxes. Nine anchor boxes sizes corresponding to three types of vehicles targets in the three scales feature maps obtained are shown in Table 1.

Table 1. Clustering box dimensions on KITTI

Feature map	Anchor boxes
13 $\times$ 13	(62, 39), (97, 53), (138, 82)
26 $\times$ 26	(19, 15), (31, 20), (49, 24)
52 $\times$ 52	(9, 7), (14, 10), (26, 11)

3.3 Loss Function

This paper introduces the DIoU (Distance-IoU) loss function as the position error loss function, which can better optimize the position information of the border. The definition of DIoU loss is as follows:

$$L_{DIoU} = 1 - \frac{|B \cap B^{gt}|}{|B \cup B^{gt}|} + \frac{\rho^2(b, b^{gt})}{c^2} \tag{1}$$

where ρ denotes the Euclidean distance between the center point b of the candidate box and the center point b^{gt} of the real box; c denotes the diagonal distance of the smallest rectangle that can cover both the candidate box and the real box. Based on the IoU loss function, the DIoU loss function fully considers the overlap ratio between the candidate box and the real box, the distance from the center point and the aspect ratio, so that the regression of the target box becomes more stable. In addition, DIoU can avoid the situation that the two boxes have no interection, that is, the IoU value is 0, so it is more in line with the mechanism of target box regression.

4 Experiment and Result Analysis

4.1 Experimental Environment

The training and testing environment of this paper is: Matpool Cloud GPU cloud computing platform, using NVIDIA GeForce RTX 2080Ti graphics card, video memory size 11 GB, using PyTorch framework.

4.2 Datasets and Evaluation Methodology

The 2D target detection part of KITTI data set is used in this paper, which contains 7481 real image data (including labels) collected from scenes such as urban areas, rural areas and highways. According to the actual needs, a total of 6820 images containing the label categories of Truck, Tram, Van and Car were obtained after screening. The filtered KITTI data set was divided according to the ratio of training set: validation set: test set = 8:1:1. Finally, the number of images used for training is 5456, the number of images in the validation set is 682, and the number of images in the test set is 682. Table 2 shows the actual number of four different types of vehicles tags after the data set is divided.

Table 2. Each set contains the actual number of vehicles type tags

Category	Numbers	Car	Van	Tram	Truck
Traing set	5456	22860	2306	417	902
Validation set	682	2906	301	34	103
Test set	682	2976	307	60	89

Mean Average Precision (mAP) and Frames Per Second (FPS) are used to evaluate the model. The definition of mAP and FPS is as follows:

$$P = \frac{T}{S} \tag{2}$$

$$AP = \frac{1}{N}\sum_{i=1}^{N} P_i \tag{3}$$

$$mAP = \frac{1}{M}\sum_{j=1}^{M} AP \tag{4}$$

$$FPS = \frac{Frame_{count}}{Detection_{time}} \tag{5}$$

where, T denotes the number of correct detection that satisfies IoU threshold in a single image, S denotes the total number of this category on the image, N denotes the number of images used for testing, M denotes the total number of detection categories, $Frame_{count}$ denotes the number of images processed, and $Detection_{time}$ denotes the time of image processed.

4.3 The Experimental Results

In order to verify the performance of the improved YOLOv3 vehicles detection method proposed in this paper, this paper compares it with the original YOLOv3. The KITTI data set filtered in this paper was used for training and testing respectively. The experimental results are shown in Table 3.

Table 3. Comparison between various methods

Method	FPS	AP%				mAP%
		Car	Van	Tram	Truck	
YOLOv3	56.8	91.42%	87.01%	94.43%	95.86%	92.16%
Our method	52.0	92.16	92.46%	96.77%	96.04%	94.36%

The improved YOLOv3 on the KITTI vehicles test set improved by 2.2% compared to the original YOLOv3 mAP, and AP values of each category has increased. In terms of real-time performance, the complexity of the algorithm is slightly improved due to the addition of CBAM module, but the number of image transmission frames per second can still be maintained at about 52, which fully meets the real-time requirements for vehicles target detection. It can be concluded that the improved YOLOv3 vehicle detection algorithm in this paper can improve the accuracy of vehicles detection tasks without affecting the real-time performance, which proves the effectiveness of the improvement.

5 Discussion and Conclusion

Based on the YOLOv3 algorithm, this paper improves the target detection model by embedding the CBAM module in the YOLOv3 network, obtaining the adaptive anchor boxes size based on the K-means clustering algorithm, and introducing the DIoU loss function. By comparing the experimental results on the public data set KITTI with the existing models, it is proved that the proposed model has high detection accuracy and real-time performance.

The limitation of this study lies in that the subject of the data set is foreign vehicles and the perspective of image acquisition has a certain regularity, which leads to the problem of simplification of feature extraction in model training. Therefore, in the follow-up work, we should mainly expand the vehicles data, so that the balance and comprehensiveness of domestic and foreign vehicles data can reach a higher level. In this way, the detection ability of the model to various types and angles of vehicles in the image can be improved.

References

1. Du, J., He, N.: Real-time Road vehicles detection based on improved YOLOv3. Comput. Eng. Appl. **56**(11), 26–32 (2020)
2. Girshick, R., Donahue, J., Darrell, T.: Rich feature hierarchies for accurate object detection and semantic segmentation. In: Proceedings of IEEE Conference on Computer Vision and Pattern Recognition, pp. 580–587 (2014)
3. Girshick, R.: Fast R-CNN. In: Proceedings of IEEE Conference on Computer Vision and Pattern Recognition, pp. 1440–1448 (2015)
4. Ren, S., He, K., Girshick, R., et al.: Faster R-CNN: towards real-time object detection with region proposal networks. IEEE Trans. Pattern Anal. Mach. Intell. **39**(6), 1137–1149 (2017)
5. He, K., et al.: Mask R-CNN. In: Proceedings of IEEE Conference on Computer Vision and Pattern Recognition, pp. 2961–2969 (2017)
6. Cai, Z., Vasconcelos, N.: Cascade R-CNN: delving into high quality object detection. In: Proceedings of IEEE Conference on Computer Vision and Pattern Recognition, pp. 6154–6162 (2018)
7. Redmon, J., Divvala, S., Girshick, R., et al.: You only look once: unified, real-time object detection. In: Proceedings of IEEE Conference on Computer Vision and Pattern Recognition, pp. 779–788 (2016)
8. Redmon, J., Farhadi, A.: YOLO9000: better, faster, stronger. In: Proceedings of IEEE Conference on Computer Vision and Pattern Recognition, pp. 6517–6525 (2017)
9. Liu, W., Anguelov, D., Erhan, D., et al.: SSD: Single shot MultiBox detector. In: Proceedings of European Conference on Computer Vision, vol. 9905, pp. 21–37 (2016)
10. Redmon, J., Farhadi, A.: YOLOv3: an incremental improvement. In: Proceedings of IEEE Conference on Computer Vision and Pattern Recognition (2018)
11. Woo, S., Park, J., Lee, J.-Y., Kweon, I.S.: CBAM: convolutional block attention module. In: Ferrari, V., Hebert, M., Sminchisescu, C., Weiss, Y. (eds.) ECCV 2018. LNCS, vol. 11211, pp. 3–19. Springer, Cham (2018). https://doi.org/10.1007/978-3-030-01234-2_1
12. Zheng, Z., Wang, P., Liu, W., et al.: Distance-IoU loss: faster and better learning for bounding box regression. Proc. AAAI Conf. Artif. Intell. **34**(07), 12993–13000 (2020)

Research on Sensor Fusion Map Building Algorithm in High Similarity Environment

Gui-juan Lin[1](✉), Hou-de Dai[2], Xin Hu[1], and Ke-yu Liu[1]

[1] Xiamen Key Laboratory of Intelligent Manufacturing Equipment, School of Mechanical and Automotive Engineering, Xiamen University of Technology, Xiamen 361024, Fujian, China
linguijuan@xmut.edu.cn

[2] Quanzhou Institute of Equipment Manufacturing, Haixi Institute, Chinese Academy of Sciences, Jinjiang 362215, Fujian, China

Abstract. Simultaneous localization and mapping (SLAM) based on laser radar (LiDAR) is the core technology to realize mobile robot navigation, but a single LiDAR scan-matching map construction method does not meet the application requirements in the case of a single environmental feature. In this paper, based on the graph optimization approach, we use a fusion algorithm based on odometry and IMU data to improve the relative positioning accuracy, and further build the environmental map based on the ROS platform in a promenade environment with high shape similarity and single features. Data acquisition and accuracy analysis experiments verify that the algorithm can achieve great map building results.

Keywords: Mobile robot · SLAM · Sensor fusion · Promenade environment

1 Introduction

In mobile robots, Simultaneous Localization And Mapping (SLAM) based on LiDAR is the core technology to achieve navigation. The commonly used 2D SLAM building algorithms include Gmapping, Hector, and Cartographer. Gmapping is an algorithm based on the FastSLAM algorithm proposed by Michael Montemerlo et al. at Carnegie Mellon University [1, 2], which proposes distribution and selective resampling to reduce the number of particles and prevent particle degradation. However, the SLAM method obtains distribution from the odometer, so it relies on the accuracy of the odometer and cannot get a good closed loop. Hector is a Gaussian Newton method to solve the target equations and construct a map based on matching laser data points with the environmental map, but the SLAM method fails where there are few environmental features [3]. Cartographer is a new graph optimization based laser SLAM algorithm proposed by Google, which established submaps by front-end scanning matching, and optimized global maps by back-end closed-loop detection and sparse pose adjustment [4]. However, the Cartographer algorithm lacked high positioning accuracy in the long path environment. In this paper it is used a Cartographer's SLAM mapping based on graph optimization, which can accurately build maps with great closed-loop and will correct measurement errors in the mapping.

Y. Wang et al. (Eds.): IWAMA 2021, LNEE 880, pp. 441–449, 2022.
https://doi.org/10.1007/978-981-19-0572-8_56

In order to reduce the accumulation error of odometer and make the position estimation of trolley more accurate, it is used the fusion algorithm of odometer and IMU in the algorithm based on graph optimization to improve the relative positioning accuracy, and also construct the map in the promenade environment. It is proved that the method can achieve better map building effect under the condition of less environmental features.

In view of the above situation, a graph optimization-based SLAM mapping algorithm were presented for mobile robots in the paper. Meanwhile those were studied that EKF fusion IMU and odometer relative localization algorithm. The excellence of these algorithms were verified by subsequent experiments of mapping and navigation in real environments.

2 EKF Algorithm for Fusing IMU and Odometer

2.1 Principle of EKF Fusion Algorithm

Autonomous mobile robots explore unknown environments and perceive their surroundings environments. Not only an accurate map is needed, but also a great motion model and control inputs is needed to achieve the estimation of the robot's poses [5]. The relative positioning of the robot based on odometry is achieved by trajectory extrapolation in which the previous moment's poses x_{k-1}, the control input u_k at this moment and the Gaussian noise w_k are known [6]. The trajectory extrapolation model of the autonomous mobile robot is expressed as.

$$x_k = x_{k-1} + \begin{bmatrix} \int_{k-1}^{k} v(t)\cos\theta(t)dt \\ \int_{k-1}^{k} v(t)\sin\theta(t)dt \\ \int_{k-1}^{k} \omega(t)dt \end{bmatrix} + w(k) \tag{1}$$

v is the linear, ω is angular velocities of the robot, and θ means the rotation of the robot relative to the world coordinate system, as shown in Fig. 1.

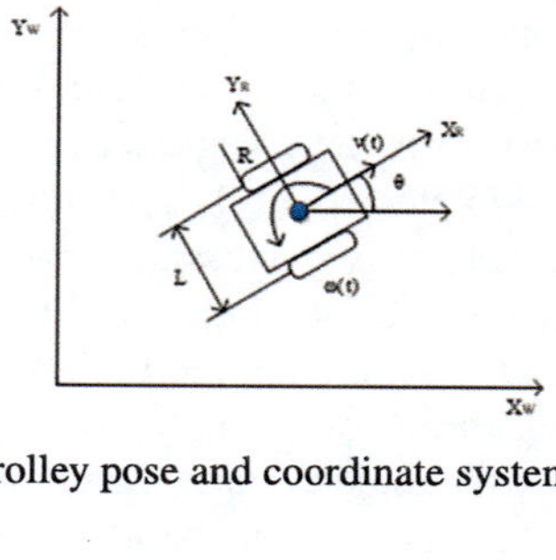

Fig. 1. Trolley pose and coordinate system diagram

The mobile robot uses an optical encoder to obtain the rotational speed of the wheels and deduces the trajectory to determine posture and position. It is assumed that the mobile robot is a pure sliding rigid body, which is not affected by the accuracy error of optical encoder and side slippage of the wheels during walking, etc. In order to reduce the accumulation of odometer errors and make robot's positional estimation more accurate, the extended Kalman filter algorithm is used to fuse the encoder and IMU [7].

During a period of time interval Δt, the encoder reading is sampled and the speed ω_l of the left wheel and the speed ω_r of the right wheel are obtained. The displacement Δs_l and Δs_r for this period of time are obtained as in Eq. (2).

$$\begin{cases} \Delta s_{\mathrm{r}} = R w_{\mathrm{r}} \Delta t \\ \Delta s_{\mathrm{l}} = R w_{\mathrm{l}} \Delta t \end{cases} \tag{2}$$

For a similar period of time, i.e., when Δt is smaller, the robot also turns a smaller angle, when the change of the robot center from the previous moment is approximated linearly, as in Eq. (3).

$$\begin{cases} \Delta s = \frac{\Delta s_{\mathrm{l}} + \Delta s_{\mathrm{r}}}{2} \\ \Delta \theta = \frac{\Delta s_{\mathrm{r}} - \Delta s_l}{L} \end{cases} \tag{3}$$

The measurement error $w(k)$ is known to be Gaussian white noise with zero mean, $w_k \sim N(0, Q_k)$, Q_k refers to the covariance matrix of w_k, the w_k is subdivided into the left wheels error $w_{\mathrm{L}n}$ and right wheels error $w_{\mathrm{R}n}$. The error vector and covariance matrix generated by the odometer motion is expressed as Eqs. (4) and (5).

$$w(k) = \left[w_{L\eta}(k), w_{R\eta}(k) \right]^{\mathrm{T}} \tag{4}$$

$$Q_k = \begin{bmatrix} w_{L_n}^2(k) & 0 \\ 0 & w_{\mathrm{R}_n}^2(k) \end{bmatrix} \tag{5}$$

The variation of the robot's heading angle (yaw), ω_z, is obtained from the IMU gyroscope and accelerometer measurements, and the increment of the angle is obtained by integrating the angular acceleration, as in Eq. (6):

$$\theta_k = \theta_{k-1} + \int_{k-1}^{k} \omega_z(t) \mathrm{d}t \tag{6}$$

Based on the IMU measurement data. The robot's positional update at moment k is:

$$x_k = \begin{bmatrix} x_{k-1} \\ y_{k-1} \\ \theta_{k-1} \end{bmatrix} + \begin{bmatrix} \Delta s_{k-1} \cos\left(\theta_{k-1} + \frac{\omega_z \Delta t}{2} \right) \\ \Delta s_{k-1} \sin\left(\theta_{k-1} + \frac{\omega_z \Delta t}{2} \right) \\ \omega_z \Delta t \end{bmatrix} + v_k \tag{7}$$

v_k is the measurement error of the IMU and conforms to a Gaussian distribution with zero mean, $v_k \sim N\{0, R_k\}$, and R_k refers to the random wandering of the IMUs gyroscope, means the IMU measurement covariance [8]. Prediction phase:

$$\begin{cases} \hat{x}_k = f(\overline{x}_{k-1}, u_{k-1}, 0) \\ \hat{p}_k = F_k \overline{p}_{k-1} F_k^{\mathrm{T}} + W_k Q_k W_k^{\mathrm{T}} \end{cases} \tag{8}$$

$\hat{x}_k$ is the prediction value, which is obtained from the state transfer matrix at moment k−1, $\hat{p}_k$ is the covariance matrix of the estimation error on the state vector prediction value $\hat{x}_k$, the matrix F_k refers to the Jacobi matrix derived from the function f with respect to the bit pose, and the matrix W_k refers to the Jacobi matrix of the function f with respect to the disturbance noise.

Update phase: K_k refers to the Kalman gain obtained from the observation matrix, $\overline{x}_k$ refers to the updated state estimate after the measurement. H_k refers to the Jacobi matrix of the measurement model with respect to x, and R_k is the random wander of the gyroscope.

$$\begin{cases} K_k = \hat{p}_k H_k^{\mathrm{T}} \left(H_k \hat{p}_k H_k^{\mathrm{T}} + R_k \right)^{-1} \\ \overline{x}_k = \hat{x}_k + K_k \left(z_k - h(\hat{x}_k) \right) \\ \overline{p}_k = (I - K_k H_k) \hat{p}_k \end{cases} \tag{9}$$

3 Graph Construction Method Based on Graph Optimization

3.1 Front-End Scan Matching

When the data of one scan is obtained, this scanning frame is inserted into the best position of the subgraph by Correlation Scan Match (CSM) method. The scanning frame is matched more finely with subgraph by the gradient optimization of Gauss Newton. So the multiple scanning frames constitute a subgraph. After the construction of one subgraph is completed, the subsequent scanning frames constitute a new subgraph, as shown in Fig. 2.

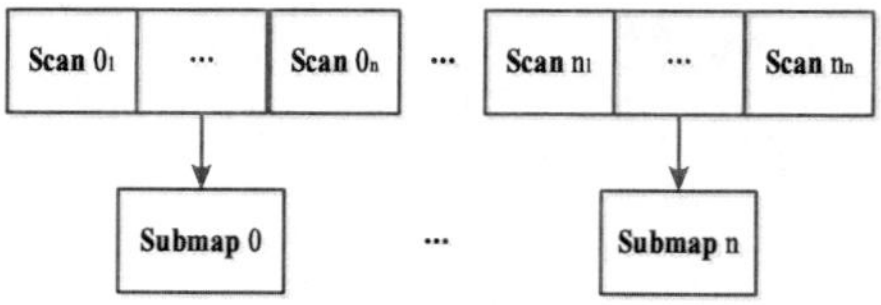

Fig. 2. The process of building subgraphs

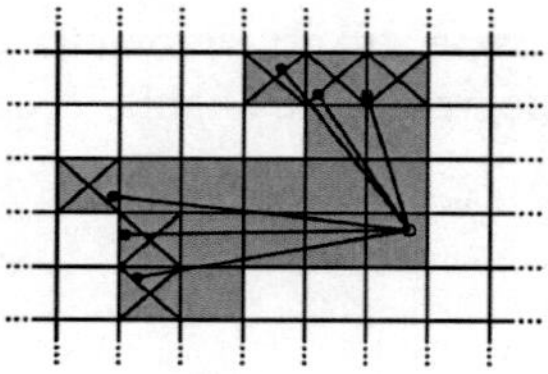

Fig. 3. Occupancy grid generated for each scann

A set of scanning forms a laser spot gathers. If a total of k laser beams are distributed in a cycle, a sector-shaped scanning area is formed from the starting point K = 1 to the end point K = k, which is expressed as following Eq. (10).

$$H = \{h_k\}_{k=1,\ldots,K} \tag{10}$$

H denotes the laser spot gathers, h_k denotes each subset of laser spot gathers, each laser beam. The bit pose of each scanning frame is defined as $\xi = \left(\xi_x, y_y, \xi_\theta\right)$ in Cartographer, ξ_x, ξ_y represents the car translation in the world coordinate system, ξ_θ represents the rotation. When one scan frame of a LIDAR is completed, it needs to be inserted into the subgraph, so the coordinate transformation process is needed to do as follows.

$$T_{\xi}p = \begin{pmatrix} \cos\xi_{\theta} & -\sin\xi_{\theta} \\ \sin\xi_{\theta} & \cos\xi_{\theta} \end{pmatrix} p + \begin{pmatrix} \xi_x \\ \xi_y \end{pmatrix} \tag{11}$$

p denotes the coordinates of the sensor relative to the car, the first matrix denotes the rotation matrix R_{ξ}, and the second denotes the translation matrix T_{ξ}. The subgraph is constructed from consecutive scanning frames, and the occupied values of the grid points are updated for each scanning frame data. The probability value of occupation is represented by the interval $[p_{min}, p_{max}]$, where less than this interval means idle, in this interval means unknown, and greater than this interval means occupied. Thus each frame of the laser generates a set of hits and misses for the grid points, as shown in Fig. 3.

The shaded part indicates the known idle grid points, namely $p_{miss} < p_{min}$. The end with a fork refers to these points that the laser point beam hits an obstacle and returns, namely $p_{hits} > p_{max}$. The optimal placement of each scanning frame into the subgraph requires a final step, which transforms the optimization of $T_{\xi}p$ into a nonlinear least squares problem to be solved by Ceres.

$$\underset{\xi}{\operatorname{argmin}} \sum_{k=1}^{K} \left(1 - M_{\text{smooth}}\left(T_{\xi}h_k\right)\right)^2 \tag{12}$$

M_{smooth} is a smoothing function that outputs the resultant number between (0,1) using a dual cubic interpolation method to transform the laser beam h_k to the optimal position. It can provide better accuracy than the raster resolution.

3.2 Back-End Optimization

To solve the resource-intensive and front-end accumulated errors due to the oversized map in large scene maps, the back-end finds the bit-pose deviation between the current frame and the historical frame by closed-loop detection. solves the whole map is solved globally by an optimization function to reduce the error of submaps and scanned frames in the map [9].

The entire process of the closed-loop optimization problem is solved by Sparse Pose Adjustment (SPA) to achieve nonlinear least squares problems, as follows.

$$\underset{\Xi_m \Xi_s}{\arg min} \frac{1}{2} \sum_{ij} \rho\left(E^2\left(\xi_i^m, \xi_j^s, \sum ij, \xi_{ij}\right)\right) \tag{13}$$

Ξm, Ξs is respectively the poses of the subgraph and each scanned frame in the world coordinate system, denoted by $\xi_i^m$and ξ_j^s respectively. ξ_{ij} denotes the pose of the current scanned frame in the corresponding subgraph, and $\sum ij$ is the estimated covariance matrix. E refers to the residuals. E is solved by the following equation:

$$\begin{cases} E^2\left(\xi_i^m, \xi_j^s, \Sigma ij, \xi_{ij}\right) = e\left(\xi_i^m, \xi_j^s, \xi_{ij}\right)^T \Sigma_{ij}^{-1} e\left(\xi_i^m, \xi_j^s, \xi_{ij}\right) \\ e\left(\xi_i^m, \xi_j^s, \xi_{ij}\right) = \xi_{ij} - \begin{pmatrix} R_{\xi_i^{\mathrm{m}}}^{-1}\left(t_{\xi_i^{\mathrm{m}}} - t_{\xi_j^{\mathrm{s}}}\right) \\ \xi_{i;\theta}^{\mathrm{m}} - \xi_{j;\theta}^{\mathrm{s}} \end{pmatrix} \end{cases} \tag{14}$$

In the formula, the $e\left(\xi_i^m, \xi_j^s, \xi_{ij}\right)$ term represents loss function. The function is used to reduce the effect of deviation points, which are generated during the SPA caused by adding incorrect constraints during scan matching. The above is the closed-loop optimization problem. For closed-loop detection, the features of the laser scan can be extracted to detect the closed-loop by Histogram-based matching; or the detection of laser point cloud features can be done by deep learning [10]. The general matching method is used in the paper.

4 Corridor Map Building Experiment and Accuracy Analysis

4.1 Long Distance Diagram Building Experiment

The environment created in this building experiment includes several offices with a long corridor. In the long corridor environment, structural similarity is high and no distinct features. Therefore, in order to ensure that the length of the promenade is accurate, it is not possible to rely only on LiDAR data, but also on relative positioning data, i.e. data based on the EKF fusion odometer and IMU mentioned in Sect. 1. The simultaneous positioning and map building is achieved by manipulating the trolley's, and the process is shown in Fig. 4 and Fig. 5.

Fig. 4. Corner building map

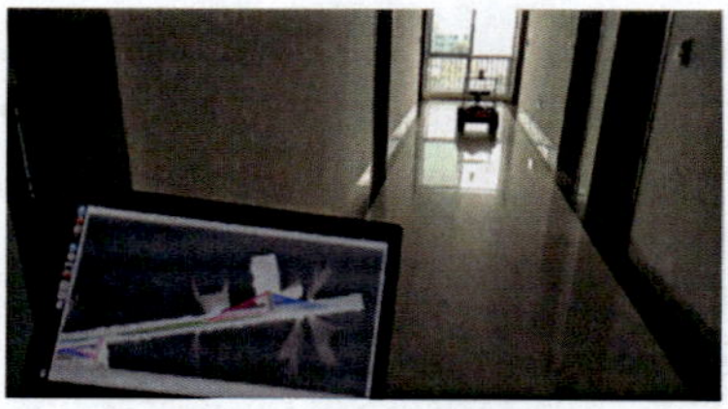

Fig. 5. The end of the corridor building map

The closed loop is detected by offline, and the relative positional relationship between the current scan frame and the key frame before closed loop is obtained using the detected closed loop, and the global map is optimized by the constraint function, and the optimized map is shown in Fig. 6.

4.2 Accuracy Analysis

To verify the accuracy of the map, the maximum length of the two labs, the width of the corridor were measured respectively, while the coordinates of the corresponding pixel points on the map were read through the library provided by Opencv, as shown in the red box in Fig. 7. The absolute error of the distance between the two coordinates is obtained by the coordinate Euclidean distance formula $dis(a, b) = \sqrt{(x_b - x_a)^2 + (y_b - y_a)^2)}$, and the coordinate length scale is transformed by 0.05 m/pix. The error is obtained between the real environment and the map.

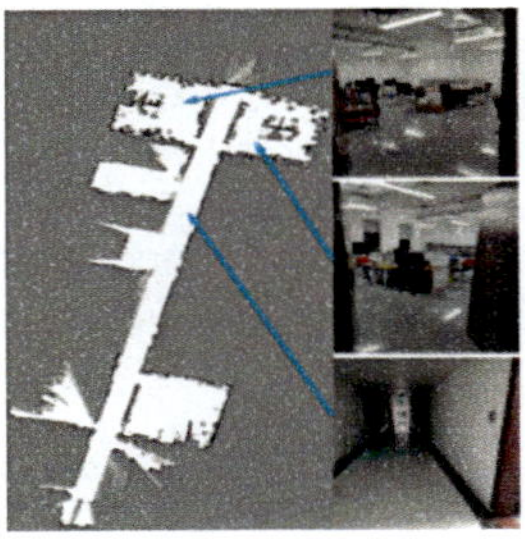

Fig. 6. Optimized map

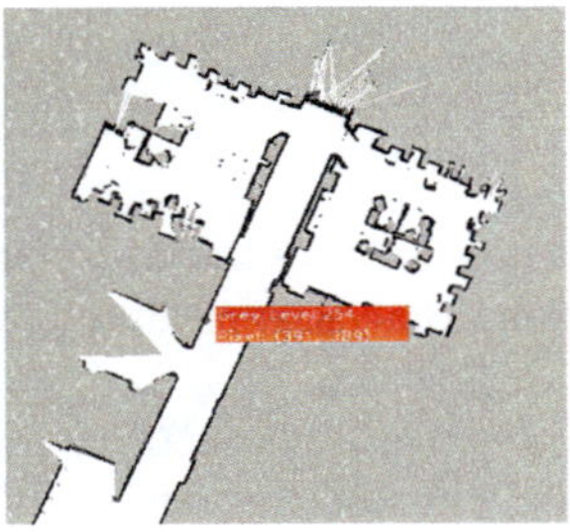

Fig. 7. Read map pixel coordinates

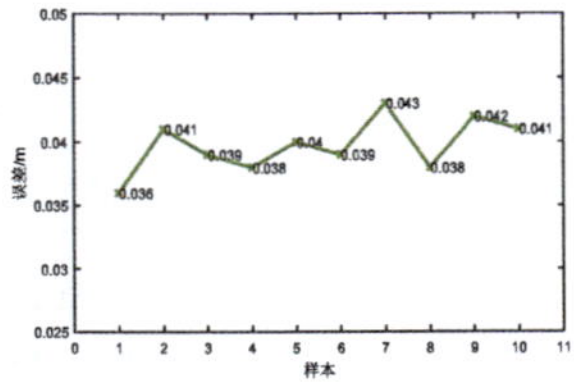

Fig. 8. Error of real environment and map

The width of the corridor was measured as 2.15 m in the actual environment. The length and width of the two laboratories were 8.30 m and 7.60 m, respectively. In the constructed map, the Euclidean distance between the two pixel coordinates of the corridor width were measured to be 42.28. The laboratory length and width were 165.16 and 151.22 respectively, which is the distance between the two corresponding pixels coordinate. The error from the actual distance values after conversion was 0.036 m, 0.042 m and 0.039 m respectively. The error with the actual distance values were 0.036 m, 0.042 m and 0.039 m. Similarly, by comparison of other samples in the map, the error line graph is obtained as shown in Fig. 7 and Fig. 8.

When the size of each pixel in the raster map corresponds to the actual environment is 0.05 m * 0.05 m, there will be an error of 0.036 m–0.43 m between the map and the actual environment. If a smaller resolution raster is used, the accuracy of the map will be further improved, but the computational effort in building the map will be correspondingly increased and the efficiency of building the map will be reduced.

4.3 Relative Positioning Algorithm Accuracy Analysis

In the experiment, the position of the tracked object in global coordinates is mainly determined by the optical tracking system Vicon. Vicon realizes the high precision tracking, positioning and trajectory drawing of the moving object by six optical cameras, and the output accuracy can reach 0.01 mm, as shown in Fig. 9. The experiment uses the system as the basis for judging the accuracy of positioning and the excellence of trajectory.

The output accuracy of the EKF localization algorithm is analyzed in these experiments. Those two predetermined paths are set in the experiment, and the position pose of the vehicle is estimated by the fusion algorithm on the data of IMU and odometer. Then the motion of the next moment is planned by the control. Vicon tracks the position

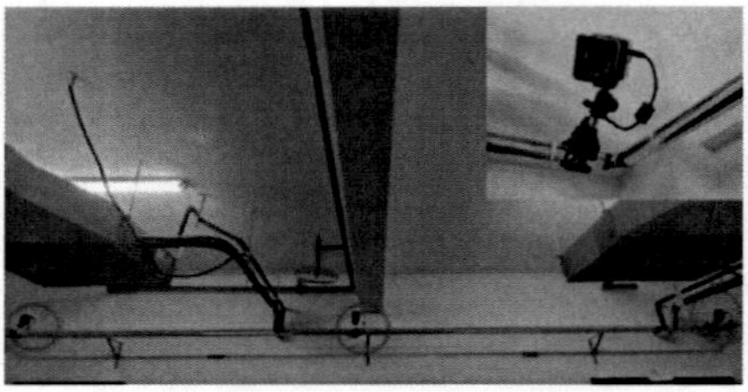

Fig. 9. Vicon optical systems

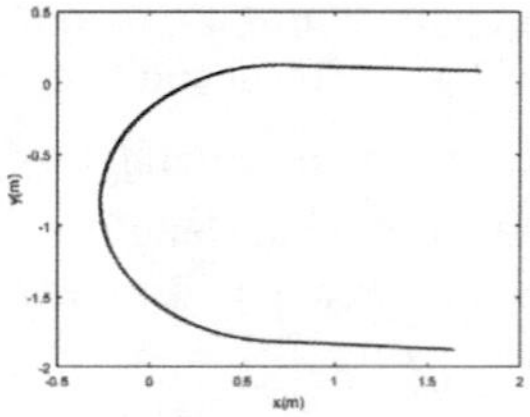

Fig. 10. Trajectory in plane coordinates

pose of the robot in the global coordinate system to determine the performance of the EKF fusion localization algorithm. The variation of the robot's poses in the coordinate system from Vicon is shown in Fig. 10.

In the odometer measurement with uncertain noise, the extended Kalman filter algorithm obtains the a priori positional estimate $\hat{x}_k$ and the error covariance matrix $\hat{p}_k$ through the state transfer matrix in the prediction phase. In the update phase, the Kalman gain value K_k is calculated to update the observation residual matrix through the data of IMU, so as to obtain a more accurate positional estimate $\overline{x}_k$ under the a posteriori estimate and the error The covariance matrix $\overline{p}_k$. In the motion, the robot obtains its own pose through the relative positioning method, and achieves next motion control, then measures. The whole process forms a closed loop. The output data in Fig. 10 shows that the deviation of the twice trajectory does not exceed 10 mm, thus it is indicated that the algorithm for this fusion localization has great performance.

5 Conclusion

In this paper, the EKF algorithm is used to fuse odometer and IMU information. These experiments are shown that the method can reduce the measurement noise and cumulative error of a single odometer and provide a more accurate positional estimation algorithm. At the same time, the relative positioning method is applied to a promenade with few features, and the accuracy analysis is shown that the method can avoid effectively the building failure caused by the similar environment in the front-end scan matching and describe the real environment map more accurately.

Acknowledgment. The authors would like to thank support from the Natural Science Foundation of Fujian Province, China (No. 2020J01272).

References

1. Xue-Meng, Y., Min-Ru, Y., Kai, C.: Overview on Issues and solutions of SLAM for mobile robot. Comput. Syst. Appl. **27**(7), 1–10 (2018)
2. Montemerlo, M., Thrun, S., Koller, D., et al.: FastSLAM: a factored solution to the simultaneous localization and mapping problem. In: Proceedings of the AAA I National Conference on Artificial Intelligence, pp. 593–598. AAA I, Menlo Park (2002)

3. Lin, S.-Y., Chen, Y.-C.: SLAM and navigation in indoor environments. In: Ho, Y.S. (eds.) Advances in Image and Video Technology. PSIVT 2011. LNCS, vol. 7087, pp. 48–60. Springer, Berlin (2011). https://doi.org/10.1007/978-3-642-25367-6_5
4. Hess, W., Kohler, D., Rapp, H., et al.: Real-time loop closure in 2D LIDAR SLAM. In: 2016 IEEE International Conference on Robotics and Automation (ICRA). IEEE, pp. 1271–1278 (2016)
5. Li, D., Zhang, K., Xu, R., et a1.: Feature extraction for map building based on laser navigation of automated guided robot without Refiecto. China Mech. Eng. **29**(22), 2733–2739 (2018)
6. Lin-zhi, Y., Hong-li, G., Xing-guo, S., et al.: The application of adaptive fading extended Kalman Filter in SLAM for mobile robot. Mach. Des. Manuf. **11**, 249–252 (2019)
7. Xu, B., Wang, D.: An improved EKF-SLAM algorithm. Manuf. Autom. **41**(12), 67–71, 94 (2019)
8. Jiawen, J., Mingxin, Y.: A indoor mapping and localization algorithm based on multi-sensor fusion. J. Chengdu Univ. Inf. Technol. **33**(4), 401–407 (2018)
9. Chen, C., Xu, J., Zhang W..: Mobile robot simultaneous localization and mapping based on multi-sensor fusion. Modern Electr. Tech. **43**(14), 164–169 (2020)
10. Li, J., et al.: Deep learning for 2D scan matching and loop closure. In: 2017 IEEE/RSJ International Conference on Intelligent Robots and Systems (IROS). IEEE (2017)

Predicting Concrete Compressive Strength Using Machine Learning

Polina Ladygina[1], Alexander Samochadin[1], Nikita Voinov[1](✉), Pavel Drobintsev[1], and Ilya Fedorov[2]

[1] Institute of Computer Science and Technology, Peter the Great St.Petersburg Polytechnic University, 195251 St.Petersburg, Russia
voinov@ics2.ecd.spbstu.ru
[2] Vedeneev VNIIG, JSC, 195220 St.Petersburg, Russia

Abstract. The paper is devoted to prediction of concrete compressive strength depending on its composition using machine learning methods. The use of machine learning improves the accuracy of the prediction of concrete strength and reduces the number of necessary experimental checks when selecting the composition.The approach is described which is based on training the models on the dataset of different concrete compositions with eight parameters describing the components of the concrete mix. Obtained results demonstrate that machine learning methods provide more accurate prediction than the traditional calculation method based on the Bolomey equation.

Keywords: Concrete compressive strength · Machine learning · Prediction models

1 Introduction

Concrete strength is an important characteristic used in the design of reinforced concrete structures for the construction of various objects. It is usually determined experimentally by measuring the minimum forces required for the destruction of a cylindrical concrete specimen under compression. To shorten the cycles of experimental verification the material strength is predicted using various calculation methods [1]. At first the engineer selects the composition, then roughly estimates the final strength. If the results obtained do not meet the requirements, the engineer changes the composition and performs the calculation again. This is followed by an experimental check and if the actual strength does not meet the criteria, the cycle starts again. The addition of new components to the mixture makes it difficult to perform calculation according to existing formulas which are based only on the water-cement ratio without taking into account superplasticizers, fine and coarse aggregates and so on.

For a more accurate forecast, methods can be used that take into account the nonlinear dependencies. Usage of machine learning methods (instead of calculation methods) for preliminary concrete strength prediction helps to reduce the number of composition

Y. Wang et al. (Eds.): IWAMA 2021, LNEE 880, pp. 450–457, 2022.
https://doi.org/10.1007/978-981-19-0572-8_57

selection cycles and allows design engineers to select the required concrete mix with less resource consumption.

This work is devoted to the study of various methods for predicting concrete compressive strength depending on its composition and using machine learning methods.

2 Existing Approaches to Prediction of Concrete Compressive Strength

Practical approaches to predicting strength are based on empirical relationships based on the analysis of experimental data and give only an approximate result. Therefore, when selecting compositions, a large number of experiments are currently being carried out to check the strength of the obtained material and are corrected based on the results. To obtain more accurate mathematical predictive models that take into account the nonlinear nature of the dependencies, modern methods based on machine learning can be used. Thus, the use of machine learning models on a sufficiently large dataset can improve the accuracy of the prediction of concrete strength and reduce the number of necessary experimental checks when selecting the composition.

Machine learning is a dynamically developing area, the achievements in which are now quite often used to solve various engineering problems. The idea of using it to predict the concrete strength has also been reflected in many scientific works.

Using machine learning for designing a concrete mixture was considered in [2]. In this research the author used an extensive dataset on which he applied two ML methods - linear regression and artificial neural networks (ANN), and concluded that the strength model based on the second approach is more accurate. However, this work did not take into account such features of the material as the period of concrete gaining strength up to 100% which equals 28 days. Another issue is that data with concrete ages of 3 and 14 days was also used, which could distort the results, because until the 28th day, concrete is not able to achieve maximum strength.

The aim of the study in [3] was the development of the I-PreConS (Intelligent PREdiction system of CONcrete Strength) with a modular network structure of 5 ANNs, which is able to predict the strength of concrete depending on its age: ANN-I predicts the strength within 24 h after pouring, the remaining ANNII - ANN-V modules predict strength from day 2 to day 28.

In [4] the use of a neural network expert system for predicting compressive strength is considered. The industrial application of this method is described in [5]. The authors are trying to get a more accurate prediction based on various parameters, which may not have an unambiguous effect on strength: in addition to the components and the period of concrete strength gain, the size and shape of the sample, the curing technique and environmental conditions are taken into account.

The study in [6] considers application of deep learning methods to solve the discussed problem. The authors conclude that predicting strength using a model based on a convolutional neural network (CNN) is the most accurate; however, CNN was trained on a small dataset (only 74 positions), which could lead to underfitting of the model and decrease in efficiency.

In [7] an attempt is made to compile a mathematical formula that allows estimating the concrete compressive strength using artificial neural networks (ANN). The resulting formula is suitable only for a rough estimate, since it calculates the strength based on only four parameters of the composition of the concrete mixture: the amount of water, cement, large and small aggregate. The influence of additives such as superplasticizers is not considered.

The analysis of modern methods of designing a concrete mixture made it possible to draw the following conclusion: the prediction of the concrete strength by machine learning methods is currently at the center of scientific discourse. A large amount of research has been carried out on this topic, but it does not lose its relevance due to the fact that the concrete mixture is a complex multicomponent system, and its composition can vary greatly depending on the purpose. It is rather difficult to predict the final characteristics of the material for each new composition using formulas, while machine learning allows to identify non-obvious dependencies and increase the accuracy of determining the strength of concrete without resorting to experimental methods so often [7].

Thus, in this paper various machine learning models are investigated to solve the problem of improving the accuracy of concrete strength prediction.

3 Proposed Approach Based on Machine Learning

To train the model, a dataset from the UCI machine learning data repository is used [8]. The dataset contains 1030 records of different concrete compositions with eight parameters describing the components of the concrete mix. The input of the model is a set of parameters of the concrete composition, as well as the period of strength gain.

Input parameters:

1. Cement.
2. Blast Furnace Slag.
3. Fly Ash.
4. Water.
5. Superplasticizer.
6. Coarse Aggregate.
7. Fine Aggregate.
8. Age (period of strength gain or curing time).

Output parameter is Concrete compressive strength (in Megapascal, MPa).

Two practical tasks solved within the study are the following:

1. Calculation of the predicted concrete strength according to the Bolomey method with averaged coefficients.
2. Prediction of concrete strength using machine learning methods with improved prediction accuracy compared to traditional calculation methods.

3.1 Preparation of the Dataset

The first step in training of machine learning models is to obtain information about the dataset, which includes plotting graphs against the input parameters and the target variable, checking for any missing or irrelevant values, etc. The dataset retrieved from the UCI repository contains 1,030 records of various concrete compositions and final strengths. The data has no missing values. Among the input variables there are 7 composition parameters (water, cement, coarse and fine aggregate, and others) and a parameter that determines the age of concrete at the time of assessing the strength of the material in MPa, which is the target variable. The recommended waiting time for testing a concrete sample (concrete age) is 28 days, however, the original dataset contains concrete compositions that are 1, 3, 7 and 14 days old, which is why a model trained on such a dataset can give inaccurate result. After deleting such data, 706 records remain in the set - this number is quite enough for training the models described above.

Further it is required to investigate the data in order to understand what parameters and how they affect the final strength of concrete. Consider a correlation matrix between the input parameters by calculating the Pearson correlation coefficient for each pair of parameters (Fig. 1).

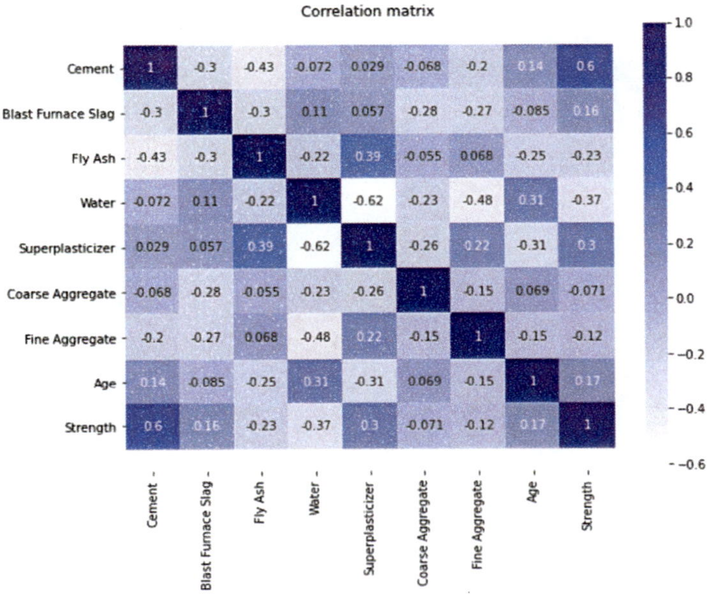

Fig. 1. Correlation matrix

Thus, a strong positive correlation was found between compressive strength and the amount of cement in the composition - this is expected, since the strength of concrete should really increase depending on the increase of this parameter. There is also a strong negative correlation between such components as superplasticizer and water, and at the same time a positive correlation between superplasticizer and fly ash. In addition, we see that the addition of superplasticizer and granular slag has a strong effect on strength, which confirms the advantage of forecasting methods that take into account the presence of these components in the composition over traditional formulas. Consider a scattering diagrams (Fig. 2) to visualize the relationship between the compressive strength and the input parameters: the amount of cement, the amount of water in the composition and the age of the concrete (Fig. 2, top) and the amount of fine aggregate, superplasticizer and ash (Fig. 2, bottom).

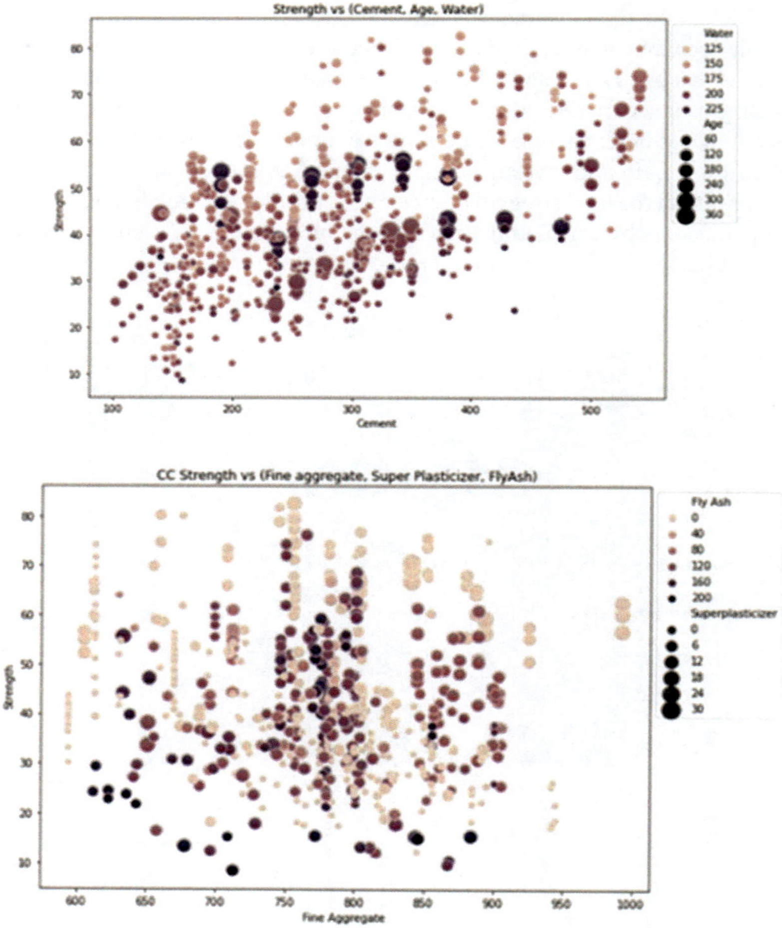

Fig. 2. Scattering diagrams of relationship between the concrete compressive strength and the input parameters

The following conclusions can be made here:

1. In some cases the compressive strength increases with the age of the concrete.
2. As the amount of cement in the composition increases, the strength also increases.
3. The composition of the concrete mix with a lower age requires more cement to achieve high strength.
4. The older the concrete, the more water it needs in the composition.
5. As the amount of fly ash increases, the final strength increases.
6. An increase in the amount of superplasticizer in the composition of the concrete mixture also has a positive effect on the strength of concrete.

Nevertheless it is rather difficult to trace all dependencies even using graphs - this confirms the need to use machine learning methods to solve this problem.

3.2 Results of the Models Training

Before training the data was divided into training and test parts (the test sample is 20% of all data). After preparing the dataset the machine learning models were trained, the work of the methods was evaluated and the most effective ones were selected.

Since the task of predicting the concrete strength is a regression task, for the assessment the RMSE - root-mean-square error (Fig. 3) and the coefficient of determination R^2 (Fig. 4) were used. Machine learning models from the Scikit-learn library were used for training.

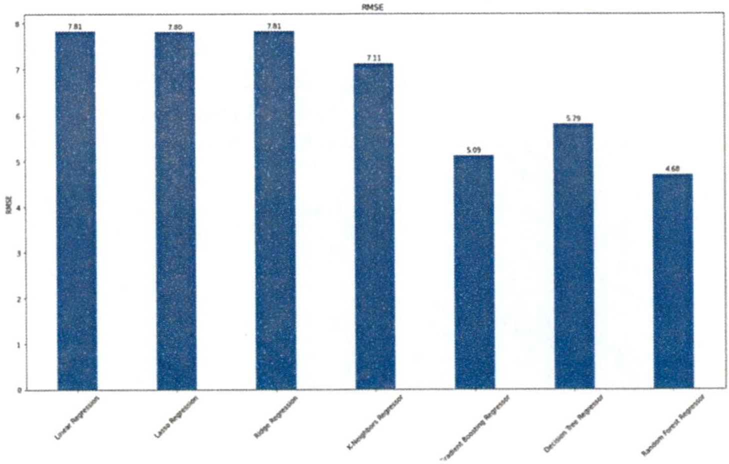

Fig. 3. Comparing RMSE for different methods

Figure 3 shows that the Random Forest Regressor model works for this task with the least root mean square error.

The calculated coefficient of determination (Fig. 4) also confirms the accuracy of the chosen method. The Gradient Boosting Regressor lags a little behind in accuracy, but also shows a good result. Linear models show the highest error.

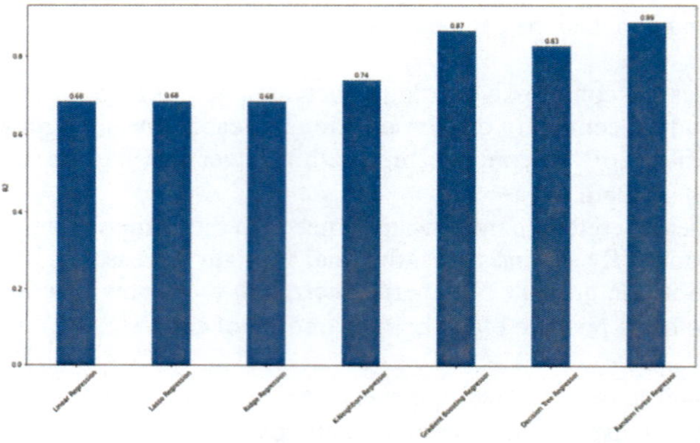

Fig. 4. Comparing R^2 for different methods

4 Results

Except for evaluation of the trained machine learning models, it is necessary to compare the prediction effectiveness by machine learning methods and the traditional calculation method. For the calculation according to the improved Bolomey equation, an indication of the compressive strength class of cement is required (32.5; 42.5; 52.5), while the machine learning models do not require an indication of the cement class, but take into account many additional parameters of the composition. Since the training dataset lacks an attribute of the cement class, it makes no sense to compare the standard deviation, but there is a visual comparison. Dot plots of concrete strength versus cement-water ratio were plotted for each machine learning model, for real data, as well as for the traditional calculation method (Bolomey's formulas with averaged coefficients) applied to each cement class. These graphs are shown in Fig. 5 and Fig. 6.

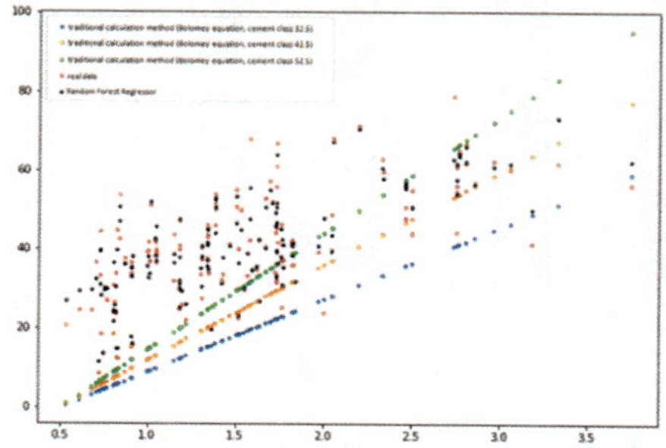

Fig. 5. Random forest regressor

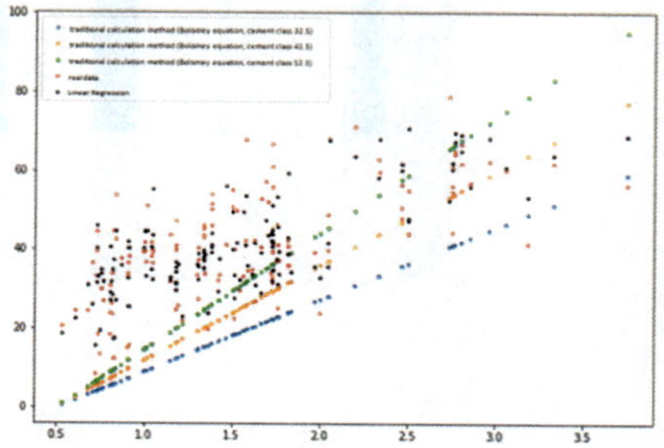

Fig. 6. Linear regression

The graphs show that even in the case of the least accurate model (Linear Regression) machine learning methods demonstrate a better prediction result than the traditional calculation method (the Bolomey equation in this case), despite the fact that there is no attribute of the cement class in the dataset used to train the models.

5 Conclusion

The paper considers an approach to predicting the concrete compressive strength depending on its composition using machine learning methods. Various approaches were considered, in particular, supervised learning, to which the prediction (regression) problem belongs.

Main machine learning models were studied to solve the problem: linear, nonlinear and ensemble algorithms, and various ways to assess the effectiveness of machine learning methods were considered. Various regression models are used to predict the concrete strength. The Scikit-Learn Python library is used to train the models. Random Forest Regressor demonstrated the best accuracy. A comparison of machine learning models and traditional calculation methods was made which showed the advantage of machine learning models for solving this problem.

References

1. Koval, S.B., Molodtsov, M.V.: Methods for calculating and predicting the concrete strength. Bulletin SUSU Series Construct. Eng. Architect. **16**(233), 25–29 (2011)
2. Yeh, I.-C.: Modeling of strength of high performance concrete using artificial neural networks. Cem. Concr. Res. **28**, 1797–1808 (1998)
3. Lee, S.C.: Prediction of concrete strength using artificial neural networks. Eng. Struct. **25**, 849–857 (2003)
4. Gupta, R., Kewalramani, M., Goel, A.: Prediction of concrete strength using neural-expert system. J. Mater. Civ. Eng. **18**(3), 462–466 (2006)
5. Bui, D.-K., Nguyen, T., Chou, J.-S., Nguyen-Xuan, H., Ngo, T.D.: A modified firefly algorithm-artificial neural network expert system for predicting compressive and tensile strength of high-performance concrete. Constr. Build. Mater. **180**, 320–333 (2018)
6. Deng, F., He, Y., Zhou, S., Yu, Y., Cheng, H., Wu, X.: Compressive strength prediction of recycled concrete based on deep learning. Constr. Build. Mater. **175**, 562–569 (2018)
7. Ziółkowski, P., Niedostatkiewicz, M.: Machine learning techniques in concrete mix design. Materials **12**, 1256 (2019)
8. Concrete Compressive Strength Data Set. https://archive.ics.uci.edu/ml/datasets/Concrete+Compressive+Strength

Research on Garbage Classification Based on Deep Learning

Jing Zhou[1], Jie Qian[1], Dongli Lu[2], Jun Guo[2], and Jinliang Zhang[3](✉)

[1] School of Economics and Management, Hubei University of Automotive Technology, Shiyan 442002, Hubei, China
[2] Zhejiang Hong Cheng Computer Systems Co., Ltd., HangZhou 311100, China
{ludl,guojun}@zjhcsoft.com
[3] College of Electrical and Information, Hubei University of Automotive Technology, Shiyan 442002, Hubei, China

Abstract. Many cities in our country are facing serious problems of garbage classification, with the rapid development of artificial intelligence and deep learning related technologies, it can provide a good and effective solution for garbage classification. In this paper, we combine the existing garbage classification standards, pre-processes and label the data set according to different garbage classification basis. Based on the VGG16 deep convolutional neural network structure, the activation function, feature extraction and selection are improved. Through training, verification and optimization of the model, the recognition and classification of the four categories of garbage images are implemented, and the effectiveness of improved VGG16 algorithm is also verified through experiments.

Keywords: Garbage classification · VGG16 · Activation function

1 Introduction

With the rapid development of society and economy, the standard of people's livelihood has gradually improved, and the amount of urban garbage increased sharply. The practices such as random disposal, incineration and landfill of garbage have brought great damage and pollution to people's living environment. Therefore, how to correctly classify and dispose of garbage has become an urgent problem to be solved worldwide. Many countries and cities have formulated corresponding garbage classification policies, while many people are vague about garbage classification concepts, and lack of subjective classification awareness, the actual situation is not optimistic. With the rapid development of high technology, how to use such as artificial intelligence, deep learning, and the Internet of Things to solve garbage classification and improve people's living environment has very important theoretical significance and practical value [1].

Research on deep learning has been carried out in foreign countries for a long time, most of the existing deep learning theoretical knowledge were put forward by foreign research scholars. After a long period of development, deep learning technology

Y. Wang et al. (Eds.): IWAMA 2021, LNEE 880, pp. 458–465, 2022.
https://doi.org/10.1007/978-981-19-0572-8_58

has obtained considerable scientific research results, which have being used in various application areas.

Some scholars have studied using the deep learning technology in the application field of garbage classification. In 2016, Mindy Yang of Stanford University and others constructed the public garbage classification data set (TrashNet Dataset), which contains 2527 images in six categories, and they use the SVM method on the data set to achieve an accuracy of 63% [2]. In 2019, Olugboja Adedeji and others built the ResNet network and achieved a good result of 87% on the TrashNet Dataset [3].

There are relatively few domestic researches in the field of garbage classification. Yang and others designed a new incremental learning framework GarbageNet to solve the problems of insufficient data, high-cost category increments and noisy labels faced by garbage classification [4]. Kang and others optimized the ResNet-34 algorithm from three aspects: the multi-feature fusion, residual unit feature reuse and activation function [5]. Although they some achievements have been made, further research is also needed.

In this work, based on actual application requirements, we combine the existing garbage classification standards, and label the data set according to different garbage classification basis for subsequent processing. The algorithm of VGG16 deep convolutional neural network is improved, and the data set is trained to realize the recognition and classification of four categories of garbage images, and the effectiveness of the algorithm is further verified through experiments.

The rest of this paper is organized as follows: Sect. 2 explains Selection of classification standards. Section 3 studies the algorithm improvement. Section 4 is the experiment and results analysis. The conclusions are summarized in the last section of this paper.

2 Classification Standards

The basic principle of the Garbage Classification Law is to classify and dispose of the waste generated in daily life. The garbage and other wasters that can be recycled need to be classified and collected, transported, and processed. This work is mainly based on the current industry development that is in line with the actual situation in our country. In 2019, the "Household Waste Classification Mark" officially issued by the Ministry of Housing and Urban-Rural Planning and Construction, which is selected as the main standard.

Compared with the standard in 2008, the new garbage classification standard was officially implemented and the scope of application has been further expanded. The original six main categories are replaced by four categories of garbage classification. Corresponding garbage classification labels have appeared in many cities, as shown in Fig. 1. It refers to recyclable garbage, hazardous waste, food waste and residual waste, including 11 simple and small categories, such as paper, plastic, glass, etc., the detailed classification is listed in Table 1.

Fig. 1. Garbage classification labels in China

Table 1. Garbage classification standard

Primary classification	Secondary classification
Recyclable garbage	Paper, plastic, metal, glass, fabric
Hazardous waste	Lamps, chemicals, batteries
Food waste	Vegetables, fruits, kitchen waste
Residual waste	Bricks, cigarette butts, tiles

3 VGG-16 Image Classification Model

VGGNet was first proposed by the Visual Geometry Group from the Department of Science and Engineering of Oxford University [6]. Its experiments proved that the depth of the neural network can affect the performance of the entire network to some extent. There two basic VGG model structures: VGG16 and VGG19, which have no essential difference except for the network depth. Due to the excellent performance of VGG16 in image classification, it can be used as a good feature extractor. Compared with VGG19, the VGG16 model is more simpler and lighter. Therefore, we choose the VGG16 model for garbage classification, and the network structure is shown in Fig. 2.

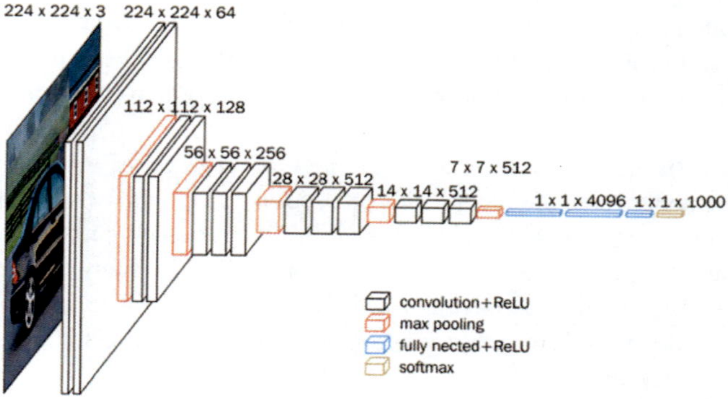

Fig. 2. The network structure of VGG16

3.1 Activation Function

In the traditional VGG16 network model, ReLU is used as the activation function, and the mathematical definition is

$$f(x) = \begin{cases} 0, \textit{if } x < 0 \\ x, \textit{if } x \geq 0 \end{cases} \tag{1}$$

As for the image data, after a series of convolution calculations in the network, the output will produce a large number of negative values, and these negative values will be directly set to zero through the ReLU activation unit, the network may face the phenomenon of neuron-death.

In order to better solve this problem, a new type of non-monotonic activation function named Mish is adopted [7], whose mathematical definition is

$$f(x) = x \tanh\left(\ln\left(1 + e^x\right)\right) \tag{2}$$

Compared with ReLU, this activation function helps the convolutional neural network to obtain more effective information from the image. At the same time, Mish does not have an upper boundary, thus it can avoid slowing down the network training when the gradient is close to zero in the training process. In addition, Mish is continuously differentiable, which well avoids the phenomenon of gradient explosion or gradient disappearance during gradient optimization.

3.2 Feature Selection

The image features are transformed after convolution and pooling operations, which will cause a lot of similar feature images and generate a lot of redundant information. Traditional feature selection methods are often time-consuming, so the maximum correlation minimum redundancy (MRMR) feature selection algorithm is adopted [8]. The MRMR algorithm eliminates irrelevant features in the garbage feature data set, calculates the similarity between features and feature tags, and obtains mutual information $M(x, y)$. Given two random variables x and y, their probability density functions (corresponding to continuous variables) are $p(x), p(y), p(x, y)$, the mutual information calculation formula is given as

$$M(x, y) = \iint p(x, y) \lg \frac{p(x, y)}{p(x)p(y)} dx\, dy \tag{3}$$

The correlation between garbage feature set A and category b is defined by the average of all mutual information values between garbage feature f_i and category b, it is given as

$$D(A, b) = \frac{1}{|A|} \sum\nolimits_{f_i} M(f_i, b) \tag{4}$$

Where $D(A, b)$ is the correlation function between garbage feature set A and category b, $M(f_i, b)$ is the mutual information between garbage feature f_i and category b.

The redundancy of all features in feature set is the average value of all mutual information values between each feature, and the calculation formula is

$$R(A) = \frac{1}{|A|^2} \sum_{f_i, f_j, \in A} M\left(f_i, f_j\right) \tag{5}$$

Where $R(A)$ is the redundancy of all the features in the garbage feature set A, $M\left(f_i, f_j\right)$ is the mutual information between feature f_i and f_j,

4 Experiment and Analysis

Model data sets selects about 4500 various types of garbage images as samples data, which are divided into three subsets, namely training set, validation set and test set, accounting for 60%, 20%, and 20% respectively. Figure 3 shows some of the most common garbage pictures in our daily life, which are included in the data sets. The workflow of data sets is shown in Fig. 4, the training set is used to train the model, the verification set is used to determine the network structure or control model complexity and other parameters, the test set is used to evaluate the prediction of the selected optimal model.

Fig. 3. The common household waste in daily life

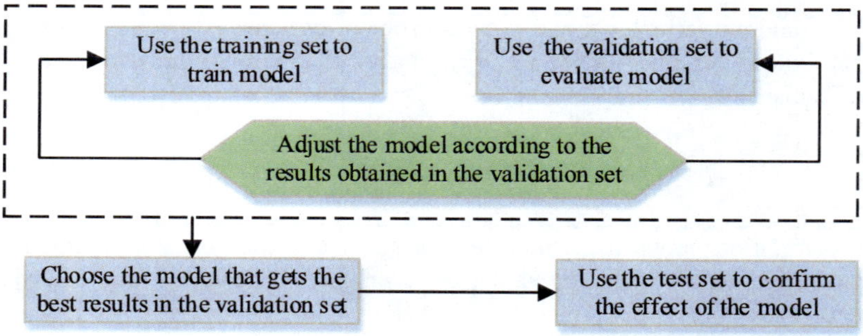

Fig. 4. The workflow of model data sets

The algorithm model training environment uses a cloud server with high computing power (RTX3090, 24G video memory), the neural network framework implementation is based on PyTorch1.8. Under the acceleration of CPU and GPU, the network model can be trained more quickly.

Convert the original image to 224 × 224 × 3 RGB image data as the input layer of the network; the hidden layer contains 13 convolutional layers, each of them is 3 × 3 convolution kernel, the step size is set to 1 in the convolution process. The Mish activation function is used for the small layers, and pooling with size 2 is used for the large layers, the parameters of the fully connected layer are modified, roughly divided into four categories, the learning rate is set to 1e−6.

Both the learning rate and epoch will affect the effect of model training. After many trials, the learning rate is finally determined to be 1e−6. The epoch is set to 100, and the single input batchsize is 128. By monitoring the changes of the error (Loss) and accuracy (Accuracy) curves, the training effect of the model is evaluated, and the training process is gradually improved and optimized.

During the training and verification process of improved VGG16 algorithm, the change curves of Loss and Accuracy are shown in Fig. 5 (a) and (b). In the early training stages, loss can quickly decline and converge. When the epoch exceeds 20, the loss curve gradually changes smoothly, and the accuracy has reached 0.9. In general, the validation loss curve converges faster than the training loss, and the accuracy curve is higher than the training accuracy, it is indicating that the pre-processing and labelling of the previous data set are effective, and the parameter selection is relatively reasonable, which has a good optimization on the training model.

In order to further verify the effect of the improved algorithm, the traditional VGG16 algorithm is selected for training and comparison of the same data sets. The loss and accuracy curves of the comparison are shown in Fig. 6 (a) and (b). From the figure, when the epoch is less than 10, the change trend of the loss curve is equivalent for the two algorithms. In the later stage, the improved VGG16 can quickly converge and basically stabilizes, while the traditional VGG16 has a slower convergence, and the curve platform is significantly higher, which requires more epochs to further reduce the loss. For the accuracy curve, the improved VGG16 is better than the traditional VGG16 in the entire training process.

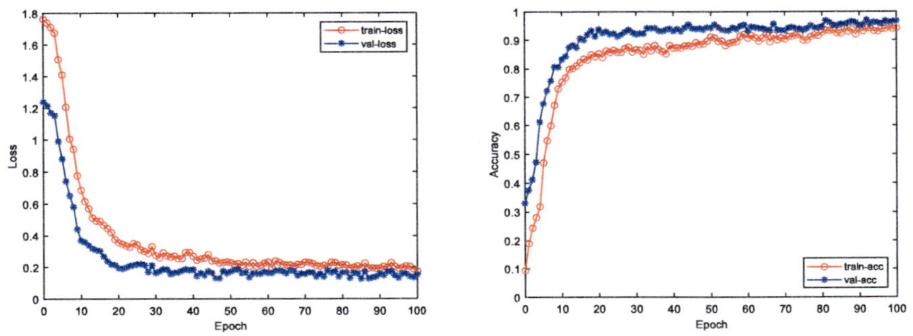

(a) (b)

Fig. 5. Loss and accuracy curve for improved VGG16: (a) the loss of training and validation loss; (b) the accuracy of training and validation.

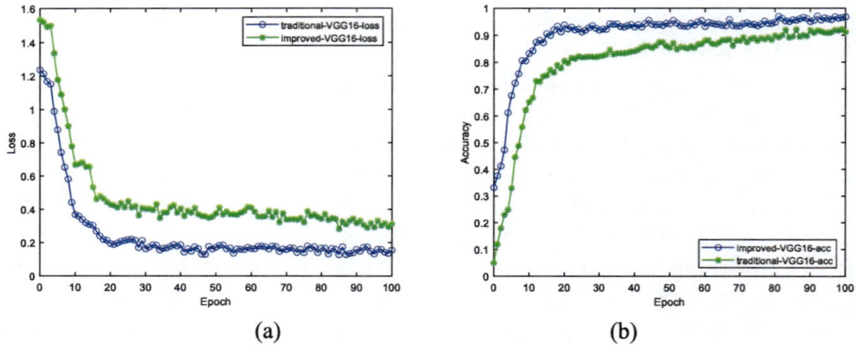

Fig. 6. Loss and accuracy curve for different algorithm: (a) the loss of improved VGG16 and traditional VGG16; (b) the accuracy of improved VGG16 and traditional VGG16

In addition, we use the camera to collect garbage images in real time for classification, four different types garbage classification results of the two algorithms are shown in the Table 2. It can be seen that the classification accuracy of the improved VGG16 is higher than traditional VGG16. Through actual tests, the improved algorithm also has a faster recognition and classification capabilities.

Table 2. Comparison of garbage classification accuracy

Categories	Traditional VGG16 (Accuracy)	Improved VGG16 (Accuracy)
Recyclable garbage	95.3%	96.5%
Hazardous waste	93.1%	95.3%
Food waste	92.7%	94.2%
Residual waste	92.5%	95.3%

5 Conclusion

Research on the application of artificial intelligence and deep learning related technologies to solve the garbage classification problem. Based on the commonly used image classification model VGG16 structure, the activation function, feature extraction and selection are improved. According to the current garbage classification standards, the four types of garbage are accurately classified. And the accuracy and effectiveness of the improved algorithm are verified through experiments.

However, there are still some limitations in the study. As for the VGG16 network, the structure is large, the parameters and characteristics are relatively complex. So the generated model is up to more than 500 M. It can run on a computer or laptop, but it is not suitable for embedded systems due to its limitations of the memory and chip processing capacity. Therefore, further research on the improvement of network structure and performance is needed to realize operation on embedded systems, which can promote the productization and industrialization of related systems or technologies.

Acknowledgment. The work is supported by Science Research and Technology Development Plan Project of Shiyan (2021K60), Innovation Training Project of Hubei University of Automotive Technology (DC2021022).

References

1. Li, J.Y., Chen, X.L., Zhang, A.H.: Survey of garbage classification methods based on deep learning. Comput. Eng. **1**, 1–11 (2021)
2. Yang, M., Thung, G.: Classification of trash for recyclability status. CS229 ProjectReport (2016)
3. Olugboja, A., Wang, Z.H.: Intelligent waste classification system using deep learning convolutional neural network. Proc. Manuf. 608–612 (2019)
4. Yang, J., Zeng, Z., Wang, K.: GarbageNet: a unified learning framework for robust garbage classification. IEEE Trans. Artif. Intell. (2021)
5. Kang, Z., Yang, J., Li, G., Zhang, Z.: An automatic garbage classification system based on deep learning. IEEE Access **8**, 140019–140029 (2020)
6. Simonyan, K., Zisserman, A.: Very deep convolutional networks for large-scale image recognition. ICLR 2015 (2014)
7. Misra, D.: Mish: A self regularized non-monotonic neural activation function. arXiv preprint arXiv:1908.0868142 (2019)
8. Kaur, K., Patil, N.: A fast and novel approach based on grouping and weighted mRMR for feature selection and classification of protein sequence data. Int. J. Data Min. Bioinform. **23**(1), 47–61 (2020)

Experimental Investigation of Coupled Hysteretic Thermo-Electro-Mechanical Properties of Piezo Stack Actuator

Yubo Meng(✉), Ke Zhao, and Peng Guo

College of Mechanical Engineering, Henan University of Engineering, Zhengzhou, China

Abstract. The piezoelectric injector works under the complex environment of high pressure and heavy load, and is easily coupled by various fields such as force field, electric field and temperature field. However, current understanding of their thermo-electro-mechanical performance is limited, The coupling effect will have a great influence on the output characteristics of piezoelectric actuator, and then further make the performance of piezoelectric injector worse. So in this paper, the thermo-electro-mechanical performance of piezo stack actuator is experimentally investigated over a temperature range from 25 °C to 80 °C,under the driving voltage of up to 150 V, the mechanical load of up to 1200N, and fit the piezo coupling coefficients with a polynomial. The results show that: in the range of [200 1200] N, the displacement increases and then decreases with the preload force; in the range of [25 80] °C, the displacement shows an increasing tendency, but with the increasing of temperature, the trend of growth shows down Through the way of testing, it providing experimental and theoretical basis for the optimal design of similar piezoelectric actuators.

Keywords: Thermo-electro-mechanical · Hysteretic · Optimal design

1 Introduction

Piezoelectric ceramics are widely used as actuators in many engineering applications because of their inverse piezoelectric effects. These devices are also used in dynamically loaded systems in that they have the attractive properties such as quick frequency response, small size, large force generation capabilities and low emission. The fuel injection system in internal combustion engines is one of importance example of such dynamic systems [1, 2]. When the piezoelectric ceramic is used in actuating applications [3], a linear description of electro-mechanical properties can be expressed as Eq. (1), and the description of thermo-electro-mechanical properties can be expressed as in Eq. (2) [4]:

$$S_3 = s_{33}^{E} T_3 + d_{33}^{T} E_3 \tag{1}$$

$$\varepsilon_3 = s_{33}^{(E,\theta)} + d_{33}^{(T,\theta)} E_3 + \alpha_3 \theta \tag{2}$$

Y. Wang et al. (Eds.): IWAMA 2021, LNEE 880, pp. 466–471, 2022.
https://doi.org/10.1007/978-981-19-0572-8_59

Where S_3, T_3, E_3, ε_3, θ are the strain,mechanical stress, electric field, the strain and the temperature, respectively, and s_{33}^E, d_{33}^T, $s_{33}^{(E,\theta)}$, $d_{33}^{(T,\theta)}$, α_3 are the compliance at constant electric field, the piezoelectric coefficient at fixed stress, the compliance at constant electric field and constant temperature, the piezoelectric coefficient at constant stress and constant temperature, the thermal-strain coefficients, respectively. Then the data given in the manufacturer data-sheets are based on these coefficients.

For the actuator, at small field levels the constitutive behavior is linear, which is governed by piezoelectric constitutive laws. At high field level, the ceramic shows strong nonlinearity and hysteresis [5]. Then the hysteretic behavior is the basic characteristic of the piezoelectric actuator, which means the above linear description is too simple. In addition, the input fields (i.e. mechanical, electric and thermal field) are both varying, so a coupled hysteretic modeling approach is needed for accurate description of the piezo material and actuator structure.

The effect of mechanical stress and temperature of piezoelectric actuator has been investigated [6]. In the literature [7] the couple effect of mechanical stress and electric has been analyzed, in the paper [8] the couple effect of temperature and electric has been studied, and the polynomials of piezoelectric coefficient under fixed preload can be expressed as in Eq. (3):

$$d_{33}^T(E,\theta) = a_0 + a_1 E + a_2 E^2 + a_3 \theta \tag{3}$$

Where E, θ are electric field and temperature, respectively, and a_0, a_1, a_2, a_3 are the coupling coefficient which obtained by the experiment.in paper [9], the increase of the actuator displacement is analyzed by the experiment method. Some literatures report about increase of stroke and explain that increase from the physical point of view [10]. But there are few studies on the coupling of piezoelectric actuators, so there is a lack of research on the accurate description of piezoelectric actuators.

The aim of this article is to investigate the quasi-static thermos-electro-mechanical coupling characteristics of piezoelectric actuator. Summarizing the main limitations of the linear method in describing the materials, mechanical stress and temperature and analyzing the negatively influence on the design process of piezo stack actuator. So the obtained data can provide the basis for optimizing the design of piezoelectric actuator from the theoretical and practical point of view.

2 Experimental Setup

Generally speaking, the displacement output of a single-wafer piezoelectric actuator is very small, so piezoelectric actuators are realized by mechanically layering or stacking hundreds of piezoelectric wafers in series and electronically connecting electrodes in parallel, as shown in the Fig. 1. So the displacement of piezo stack actuator is the sum of the displacement of the single-wafer, then some papers show that maximum strain of piezo stack actuators is 0.1–0.2% [7]. In this paper, a commercially available soft multilayer piezo stack actuator is used in for experimental study. Table 1 shows the geometric and properties parameters of the actuators used in this study.

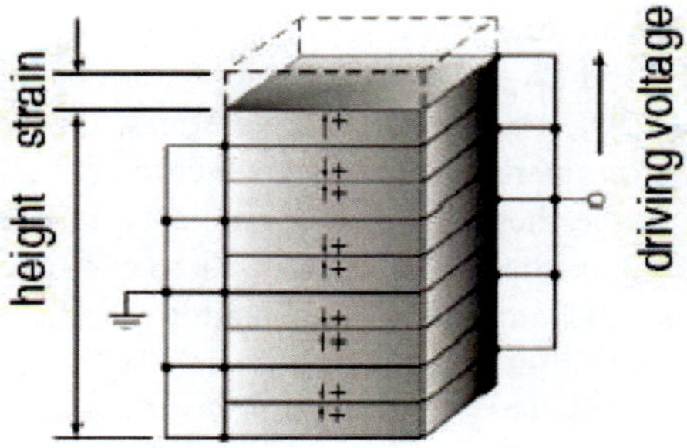

Fig. 1. Structure of piezo stack actuator

Table 1. The characteristics of piezo stack actuator

Length and width (mm)	7 × 7
Height (mm)	30
Capacity (μF)	6
Voltage (V)	150
Piezoelectric coefficient (m/V)	$\geq 650 \times 10^{-12}$
Density (kg/m^3)	7900

The schematic of the test setup for the piezo stack actuator is shown in Fig. 2. The preload is applied using a soft spring fixed by a screw nut and measured by a force sensor placed between the shaft and the moving loading head, which is in direct contact with the actuator. The other end of the actuator was placed on a fixed end. And the displacement is measured by a laser sensor which was placed on the rapid lifting mechanism which could adjust the sensor position as we desired. A heater jacket is mounted on the preload testing setup casing for the temperature measurements. The temperature is measured by a thermocouple and fed back to the heater through a PID controller for accurate temperature control.

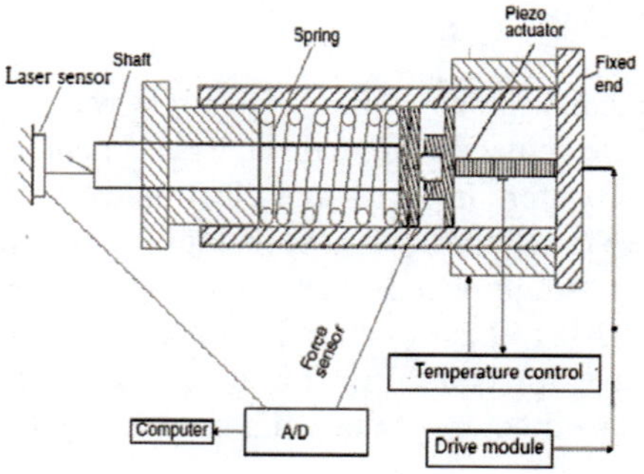

Fig. 2. Test setup

3 Fixed Mechanical Load Tests

The To analyze the effect of mechanical load on the characteristics of piezo stack actuator, a series of tests are carried out according to the following method:

(1) Mechanical load starts from a small value (i.e. 200 N) and gradually increases until a better preload value;
(2) The driving voltage is varied from 0 to 150 V (unipolar), the test cycle should be run at three times, and the test temperature is 25 °C;

The displacement - voltage curves under different mechanical loads are shown in Fig. 3. From the Fig. 3, the mechanical load do not only affect the peak of the displacement, but also affect the shape and area of hysteretic curve. As the energy represented by the area of the hysteresis loop is the main source of the self-heating in the actuator, therefore, the area of the hysteresis loop should be considered in the design of the mechanical load. In the Fig. 4, the peak displacement is recorded as a function of the mechanical load in the range 200 N–1200 N. Then we can see the maximum displacement appears near 800 N. It may not be the best value, but the optimum value should be around 800 N. Compared with the displacement output at 200 N, the displacement output increased by about 3.5 μm at 800 N. Meanwhile, as the data-sheet shows, the maximum displacement value is 29 μm, when the mechanical load is 0 N. So the mechanical load value can be used as the reference value for the optimization design of piezo stack actuator.

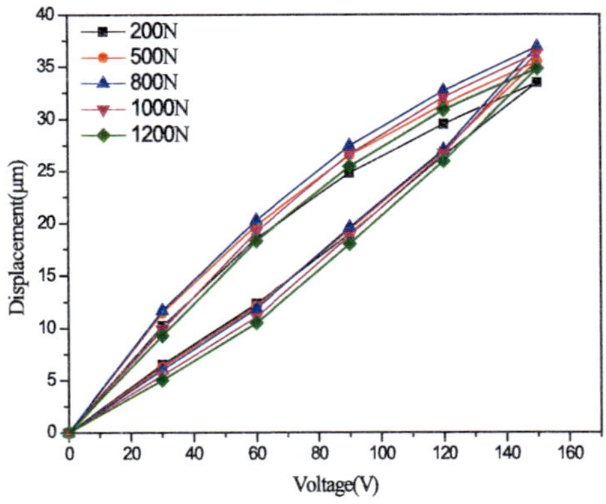

Fig. 3. The hysteresis curve under different mechanical load

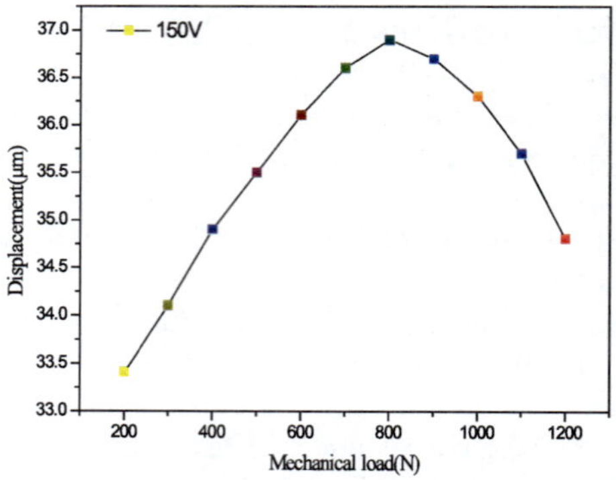

Fig. 4. The peak displacement under different mechanical load

4 Fixed Temperature Tests

In order to evaluate the effect of the temperature on the characteristics of piezo stack actuator, some tests are performed according to the following:

(1) The driving voltage is varied from 0 to 150 V (unipolar), and mechanical load is 800 N;
(2) The testing temperature starts from room temperature and gradually increases to 80 °C, and the test cycle should be run at three times;

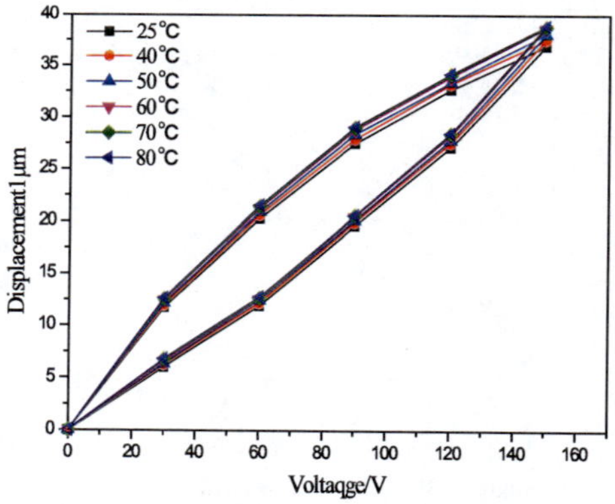

Fig. 5. Hysteresis curve under different temperature

As the Fig. 5 shows, the displacement-voltage curves at different temperature with the voltage from 0 to 150 V while the mechanical loads is 800 N. In the Fig. 5, in a certain temperature range, as the increase of temperature, the displacement of the actuator increases.

5 Conclusions

(1) In this paper, the quasi-static hysteresis characteristic of a commercial piezo stack actuator is measured and the displacement-voltage hysteresis curves are obtained under different mechanical load and temperature conditions;
(2) In the mechanical load range of [200, 1200] N, the displacement characteristic of piezo stack actuator increases first and then decreases with the increase of mechanical load. And the displacement reaches the maximum value near 800 N, which is about 3.5 μm higher than that of 200 N.
(3) In the temperature range of [25, 80] °C, the displacement characteristic of piezo stack actuator increases with the increase of temperature.

References

1. Biggio, M., Butcher, M., Giustiniani, A., Masi, A., Storace, M.: Memory characteristics of hysteresis and creep in multi-layer piezoelectric actuators: an experimental analysis. Phys. B **4**(35), 40–43 (2014)
2. Guo-Ying, G., Zhu, L.-M., Chun-Yi, S.: Modeling and compensation of asymmetric hysteresis nonlinearity for piezoceramic actuators with a modified Prandtl–Ishlinskii Model. IEEE Trans. Ind. Electron. **61**(3), 1583–1595 (2014). https://doi.org/10.1109/TIE.2013.2257153
3. Gu, G.Y., Yang, M.J., Li, F.X.: Real-time inverse hysteresis compensation of piezoelectric actuators with a modified prandtl-ishlinskii mode. Rev. Sci. Instrum. **83**(6), 065–106 (2012)
4. Gao, C.: New technology of piezoelectric effect and its application. Publishing house of electronics industry (2011)
5. Jiang, W., Zhang, R., Jiang, B., Cao, W.: Characterization of piezoelectric materials with large piezoelectric and electromechanical coupling coefficients. Ultrasonics **41**(2), 55–63 (2003). https://doi.org/10.1016/S0041-624X(02)00436-5
6. Han, Y.-M., Moon, B.K., Choi, S.-B.: Control performance investigation of piezoelectric actuators under variation of external heat environment. Trans. Korean Soc. Noise Vib. Eng. **25**(10), 707–713 (2015). https://doi.org/10.5050/KSNVE.2015.25.10.707
7. Meng, Y.B.: Characteristics analysis and optimization design of piezoelectric actuator for piezoelectric fuel injector. Shan Dong University (2018)
8. Senousy, M.S., Rajapakse, R.K.N.D., Gadala, M.: Experimental investigation and theoretical modeling of piezoelectric actuators used in fuel injectors. In: Kuna, M., Ricoeur, A. (eds.) IUTAM Symposium on Multiscale Modelling of Fatigue, Damage and Fracture in Smart Materials, pp. 219–227. Springer Netherlands, Dordrecht (2011). https://doi.org/10.1007/978-90-481-9887-0_21
9. Zsrezsan, T.G., Andersen, M.A., Zhang, Z., et al.: Preisach model of hysteresis for the piezoelectric actuator drive. In: 2015 41th IEEE Industrial Electronics Society Annual Conference, pp. 11–17 (2015)
10. Shao, S.B., Xu, M.L., Zhang, S.W., et al.: Stroke maximizing and high efficient hysteresis hybrid modeling for a rhombic piezoelectric actuator. Mech. Syst. Signal Process. **75**, 631–647 (2016)

Current Research State on Interface Dynamics of Spindle-Toolholder

Te Li[1](✉), Zhengya Xu[1], Jiju Guan[1], and Jinghua Song[2]

[1] School of Mechanical Engineering, Changshu Institute of Technology, Changshu 215500, China
lite126126@cslg.edu.cn

[2] Debenhengjia Precision Machinery (Kunshan) Co., Ltd., Shanghai, China

Abstract. Contact characteristics of joint surface of high-speed spindle-toolholder/toolholder-tool not only have important influences on dynamics of the spindle, also closely related to the stability of high-speed cutting. Accurate modeling and identification of the spindle-toolholder or toolholder-tool joint not only helps to improve the prediction accuracy of the dynamics of the spindle system in the design stage, also guide the assembly process and correct selection of cutting parameters in the cutting stage. In this paper, the dynamics modeling methods of the spindle-toolholder joint surface were reviewed.

Keywords: High-speed spindle · Toolholder · Joint surface · Modeling · Development · Review

1 Introduction

High speed spindle (HSS) is a core component in machine tool (MT), the dynamics of HSS deeply affecting the dynamic behavior and cutting stability of MT. With the increase of cutting requirement, dynamic stiffness of HSS is much important than ever [1]. The dynamic property of contact surface weight 60–80% of the HSS dynamics, whereas, the dynamics of spindle-holder (SH) is of 90% about the contact surface [2] Contact surfaces such as SH and holder-tool (HT) provide most basic functions that cutting requires. In chatter prediction, the calculation of stability region mainly relay on the tool point frequency response function (FRF). Therefore, accurately calculate and identify the static and dynamics of contact surfaces are the crucial point of precisely predicting cutting stability.

In this paper, the dynamics modeling methods were reviewed, the purposes are summarized the research methods and research directions at current, then forecast the research trends in the future.

2 Research on Dynamic Properties of SH Contact Surface

The research interests of SH interface mainly divided into 3 categories: (1) Combination behavior, which can be divide as static and dynamic law; (2) Dynamic modeling and

Y. Wang et al. (Eds.): IWAMA 2021, LNEE 880, pp. 472–478, 2022.
https://doi.org/10.1007/978-981-19-0572-8_60

computing of interface; (3) Parameter identification of interface. The last two are usually employed together to solve the dynamic problem. Three kinds of researches are contained in the above contents: (1) dynamics research of SH interface and (2) spindle; (3) cutting stability.

2.1 Dynamic Modeling of Spindle-Holder Interface

Interface dynamic stiffness is a combination of stiffness and damping under specific vibration model, which is the core content of interface properties, where damping is usually gained by experiment. The interfaces can only exist among components because they are not independent dynamic elements, in addition, the damping is not outstanding than other structural damping, therefore, the research of dynamic contact characteristics is extremely difficult [3].

On scale, the dynamic modeling of interface can be divided as macro and micro methods, where macro method adopt spring-damper to simulate the contact stiffness and damping, while ignoring surface morphology, roughness and geometrical factors. Based on microscopic appearance, parameters like surface roughness, microprotrusion distribution can be considered in the micro modeling approach, then the stiffness and damping can be obtained by statistics method under the deformation research of single microprotrusion. The spring-damper model can be divided as: (1) concentrate model; (2) distribution model. Where the distribution model was treated as ① uniform distribution model and ② non-uniform distribution model. In the non-uniform model, stiffness and damping coefficients are dissimilar in each element. At present, the first method is widely applied in studies.

Based on working status, the research also can be divided as 2 types, (1) non-rotating research and (2) rotating research. In SH interface, stiffness is more significant while damping effect is not obvious, hence, in most studies only the stiffness is considered [4].

2.2 Dynamic Modeling Under Non-rotating Status

(1) Virtual material method

Virtual material method is first used in modeling the static and dynamic characteristics of fixed joint. Supposing a layer of virtual material with tiny or zero thickness existing between the contact surfaces and rigidly connect to the components separately, hence the dynamic contact model of the joint can be modeled only by computing or identifying the material parameters such as elastic modulus, passion ratio, etc. [5].

Stiffness influential factor based method is also a general modeling and identification method for joint modelling [6], stiffness coefficient is a function including contact material, surface roughness and vertical pressure, which can be identified by frequency response experiment. Yang [7] built a bi-distributed joint interface model including spindle-chuck and chuck-tool based on Euler beam, where the contact surface was treated as zero-thickness elastic layer.

(2) **RCSA method**

On the research of spindle-holder joints, receptance coupling substructure analysis (RCSA) method is widely used in recent years. Only by theoretically calculation or experiment and then couple the substructures by RCSA [8], the receptacle of the whole assembly can be obtained, however, the model precision is heavily rely on the accurate identification of spindle-holder.

Based on the receptance theory, Schmitz [9] proposed a method of using RCSA to predict the (FRF) of tool point. The SH and the overhang segment of tool are combined with the translational stiffness/damping and rotational stiffness/damping. The real contact property of spindle-holder is included in the substructure, which is identified by hammer tests.

Identification precision is a bottleneck for further improvement of RCSA model. The stiffness-damping matrix of contact surface can be obtained by 2 ways: ① FE simulation, which can cause large error; ② experimental analysis model (EMA), by using some kind of algorithm to compare the results between modal experiment and FE simulation. Ertürk [5] built a dynamic model of spindle system including SH and HT substructures simultaneously, where the contact stiffness of substructures were written as:

$$K = \begin{bmatrix} k_{yf} + i\omega c_{yf} & 0 \\ 0 & k_{\theta m} + i\omega c_{\theta m} \end{bmatrix} \tag{1}$$

Where, k_{yf} translational stiffness caused by force, $k_{\theta m}$ rotational stiffness caused by torque, c_{yf}, $c_{\theta m}$ are the corresponding damping, i complex unit, ω is excitation frequency.

According to the coupling property, RCSA can be divided into 3 categories: ① lumped joint interface model including 2 substructures; ② lumped joint interface model including 3 substructures; ③ distributed joint interface model.

RCSA can better deal with the rotational response and deformations of holder/tool, it is also able to effectively calculate the contact stiffness and identify the contact damping, therefore, the RCSA is successfully applied in dynamic modeling and system response [10], but some drawbacks still exist: ① the acquisition of joint stiffness and damping are heavily rely on the precision of experimental identification, the dynamic parameters need to be re-identified once the MT has been changed; ② the rotational influence of spindle has been neglected, which means the identification were all based on static condition.

(3) **Finite element method**

Establishing dynamic model of joint interface by FEM follow such processes, that is, assemble each sub-matrix of stiffness to achieve a big matrix for whole system. The difficulty still exists in the calculation of node stiffness and damping.

The model built by Namazi [11] was easy for understanding, which was a uniform distribution spring model with translational spring and rotational spring were identified by measurement. But this method regarded the contact stiffness uniformly distributed, ignoring the nonuniform variation of compressive stress. Ahmadian [13] built a continuous contact model of HT system based on stiffness coefficient method, where the contact surface were treated as zero thickness elastic layer varying with contact length, then it was proved the predictive accuracy of spindle property was improved when the stiffness variation has been considered.

Based on above models, Ahmadian [14] built SH and HT model with varying stiffness by FE-RCSA method, where the dynamic property of tool can be obtained without experiment on holder or tool.

Overall, the stiffness models in regard with nonlinear properties and length factors mentioned above are all based on the contact status of continuous beam model, in short, the general stiffness equation was written as:

$$k(x) = (1 + i\eta)\sum_{q=0}^{Q} k_{X,q}x^q \tag{2}$$

Where, $k_{X,q}$ is polynomial of stiffness coefficient at a certain direction, η is structural damping coefficient, q is the DOF. Compare with spring-damping unit, this method defined a contact type based parametric stiffness distribution, it can better reflect the nonlinear contact condition, however, if a higher precision is needed, the experiment will become time consuming for the large stiffness matrix.

Although plenty of researches focus on the contact problem of interface with the help of FE software, however, there is still lack of suitable element type in FE software for establishing mechanical contact interface. Furthermore, most SH or HT models adopted beam elements for modeling, which are not appropriate for the solid FE modeling in some cases. Grossi [15] built a solid model by applying 3D element,which can better simulate the contact property of shrink TH if the mesh processes and boundary conditions are effectively controlled, furthermore, the disadvantage of low-precision owing to the sensitiveness between beam elements to contact interface can be avoided. Therefore, this model is able to applied in the RCSA to substitute the real experiment.

(4) **Fractal geometry method**

Because of the limitations of macro modeling method, micro modeling method has received huge development in recent years. In micro view, real contact generated between two rough surfaces containing convex and concave simultaneously, contact stiffness and damping are generated if elastic-plastic deformation produced between the convex. There are two ways for establishing the contact model in micro view [16]: ① statistical based method, which is based on traditional quantitative description of rough surface morphology, however, the result is not unique because of the difference of scale description; ② fractal geometric method.

Zhang [17] firstly presented the fractal geometric model of vertical stiffness for mechanical contact interface, demonstrated the contact stiffness is related to the fractal dimension and roughness amplitude, where the contact stiffness increase with the D. Generally, the vertical stiffness is written as:

$$K_n = \frac{2ED \cdot D/a_l^2}{\sqrt{c}(1-D)} \cdot (a_l^{\frac{1-D}{2}} - a_c^{\frac{1-D}{2}}) \tag{3}$$

Where, a_l is area of maximum contact point, a_c is critical contact area, E is elastic modulus. The model mentioned above is based on M-B fractal model [18], unfortunately, the M-B model is not appropriate for the rotary contact interfaces for it mainly adopted in the load capacity between two contact plane [19]. Later on, Zhang [20] built a SH model by 2D and 3D fractal theories, and simulated the contact status and contact stiffness under non-rotating and rotating conditions using FE software, respectively. Although the drawbacks are obvious from the micro perspective, it provided a new idea for the model built by combining the FEM and fractal approach together.

It was difficult to build contact model only by FEM especially if the machining error and surface roughness been taken into consideration, hence the simulation result has been deeply doubted [21], but the problem can be easily solved by combining the fractal method.

(5) **Artificial neural networks**

With the development of artificial neural networks (ANN), it was also adopted in the field of mechanical design and machining, where the ANN were used to predict the mechanical surface roughness. Usually the ANN has advantages in dealing with nonlinear inputs, and has higher computing accuracy while only need less experiment comparing with traditional ways [22]. Zhang [23] firstly proposed an ANN model for the static contact interface of MT. Mocahhedy [12] optimized the contact parameters by genetic algorithms. To improve the predict accuracy of stiffness, Yang [24] presented an algorithm that combine particle data group and Back propagation (BP) neural network.

It can be concluded by comparing the fractal method that, in the neural network based prediction researches, the final output is roughness amplitude Ra; referring to the fractal method, fractal dimension D and surface roughness amplitude G are the direct parameters for calculating the contact stiffness. Therefore, for the training process of ANN, the optimized D and G are exported from ANN model then substituted into the fractal model, the contact stiffness is finally obtained. Coincidently, Sonbaty [25] built an ANN model by this approach, where the cutting parameters were set as inputs and D and G were outputs.

3 Dynamic Modeling Under Rotating Condition

Except the contact condition of SH interface under high speed effect, the nonlinear and time varying properties of interface stiffness need more researches. This problem was a bottleneck for analyzing the spindle dynamics and cutting stability in the view of contact interface.

The dynamic properties vary during high rotating speed, especially the contact stiffness of SH system [26], hence the prediction result of cutting stability under static condition is not precise enough [27]. In order to build a SH model appropriate for the high speed, Jiang [28] studied the dynamic variation of contact interface and spindle dynamics considering the rotating speed and axial milling force. Xu's [29] study considered the drawbar force, while assumed the contact points are equally distributed and full contact. Zhao [30] calculated the vertical and tangential stiffness by fractal approach and FE method under rotating condition.

4 Discussion and Conclusion

In summary, the contact problems of static status have been resolved, currently, the problems are mainly related to dynamic status.

(1) FE simulation without considering morphology of interface is difficult to reflect the real contact condition; besides, the modeling accuracy via RCSA depends on the parameter identification accuracy of interface. However, the fractal method focusing on the surface morphology in micro view is able to reflect the stress variation of interface with high accuracy, where the variable stiffness characteristics can be well demonstrated. Furthermore, with the application of ANN, kinds of hybrid model has been established based on ANN or independently, for example: FE-RCSA method, FE-fractal method, et, al.
(2) Hybrid method such as ANN-fractal should receive more attention for it can reflect the interface property in micro view with high accuracy. Despite the ANN-fractal method, more coupled or hybrid modeling approach combine micro method and macro method should be developed as well.
(3) In rotating status, rotating effects should be considered when building the dynamic model of SH or TH contact interface for resolving the nonlinear variation property of stiffness during high speed rotation.

Acknowledgment. This work was financially supported by Open Research Fund by Jiangsu Key La-boratory of Recycling and Reuse Technology for Mechanical and Electronic Products (RPME-KF1609).

References

1. Olvera, D., Lacalle, L., Compeán, F., et al.: Analysis of the tool tip radial stiffness of turn-milling centers. Int. J. Adv. Manuf. Technol. **60**(9–12), 883–891 (2012)
2. Li, H., Shin, Y.: Integrated dynamic thermo-mechanical modeling of high speed spindles, part 1: model development. J. Manuf. Sci. Eng. **126**(1), 148–158 (2004)
3. Zhang Xueliang, X.U., Kening, W.S.: Review and prospect of the research on the static and dynamic characteristics of machine joint surfaces. J. Taiyuan Heavymach. Inst. **23**(3), 276–281 (2002)
4. Xu, C., Zhang, J., Wu, Z., et al.: Dynamic modeling and parameters identification of a spindle-holder taper joint. Int. J. Adv. Manuf. Technol. **67**(5–8), 1517–1525 (2013)
5. Tian, H., Li, B., Liu, H., et al.: A new method of virtual material hypothesis-based dynamic modeling on fixed joint interface in machine tools. Int. J. Mach. Tools Manuf **51**(3), 239–249 (2011)
6. Mao, K., Li, B., Wu, J., et al.: Stiffness influential factors-based dynamic modeling and its parameter identification method of fixed joints in machine tools. Int. J. Mach. Tools Manuf **50**(2), 156–164 (2010)
7. Yang, Y., Wan, M., Ma, Y.C., et al.: An improved method for tool point dynamics analysis using a bi-distributed joint interface model. Int. J. Mech. Sci. **105**, 239–252 (2016)
8. Namazi, M.: Mechanics and Dynamics of the Tool Holder-Spindle Interface, University of British Columbia (2006)

9. Schmitz, T., Donalson, R.: Predicting high-speed machining dynamics by substructure analysis. CIRP Ann. Manuf. Technol. **49**(1), 303–308 (2000)
10. Yang, Y., Zhang, W., Ma, Y., et al.: Generalized method for the analysis of bending, torsional and axial receptances of tool–holder–spindle assembly. Int. J. Mach. Tools Manuf **99**, 48–67 (2015)
11. Namazi, M., Altintas, Y., Abe, T., et al.: Modeling and identification of tool holder–spindle interface dynamics. Int. J. Mach. Tools Manuf **47**(9), 1333–1341 (2007)
12. Movahhedy, M., Gerami, J.: Prediction of spindle dynamics in milling by sub-structure coupling. Int. J. Mach. Tools Manuf **46**(3), 243–251 (2006)
13. Ahmadi, K., Ahmadian, H.: Modelling machine tool dynamics using a distributed parameter tool–holder joint interface. Int. J. Mach. Tools Manuf **47**(12), 1916–1928 (2007)
14. Ahmadian, H., Nourmohammadi, M.: Tool point dynamics prediction by a three-component model utilizing distributed joint interfaces. Int. J. Mach. Tools Manuf **50**(11), 998–1005 (2010)
15. Grossi, N., Montevecchi, F., Scippa, A., et al.: 3D finite element modeling of holder-tool assembly for stability prediction in milling. Procedia Cirp **31**, 527–532 (2015)
16. Wan, A.: Research on Experiment and Calculation Method of Dynamic Characteristics of Mechanical Joint Surface. Wuhan University of Technology (2012)
17. Xueliang, Z., Huang Yumei, F., Weiping, et al.: Fractal model of normal contact stiffness between roughness surfaces. Chinese J. Appl. Mech. **17**(2), 31–35 (2000)
18. Bhushan, B.: Introduction to Tribology. Wiley, New York (2002)
19. Han, Z., Qi, C., Kang, H.: Fractal model of normal contact stiffness between two cylinders' joint interfaces. J. Mech. Eng. **47**(7), 53–58 (2011)
20. Zhao, Y., Song, X., Cai, L., et al.: Surface fractal topography-based contact stiffness determination of spindle–toolholder joint. Proc. Inst. Mech. Eng. C J. Mech. Eng. Sci. **230**(4), 602–610 (2016)
21. Du, F., Li, B., Zhang, J., et al.: Ultrasonic measurement of contact stiffness and pressure distribution on spindle–holder taper interfaces. Int. J. Mach. Tools Manuf **97**, 18–28 (2015)
22. Erzurumlu, T., Oktem, H.: Comparison of response surface model with neural network in determining the surface quality of moulded parts. Mater. Des. **28**(2), 459–465 (2007)
23. Xueliang, Z., Yumei, H., Shuhua, W.: Modeling and application of static foundation characteristic parameters of machine tool joint surface. Manuf. Technol. Mach. Tool **11**, 8–10 (1997)
24. Yang Hongping, F.U., Weiping, S.B.: Modeling of machined joints normal stiffness using modified PSO-BP neural network algorithm. Trans. Chinese Soc. Agric. Mach. **42**(3), 119–223 (2011)
25. Sonbaty, I.A., Khashaba, U., Selmy, A., et al.: Prediction of surface roughness profiles for milled surfaces using an artificial neural network and fractal geometry approach. J. Mater. Process. Technol. **200**(1–3), 271–278 (2008)
26. Gagnol, V., L, et al.: Modal identification of spindle-tool unit in high-speed machining. Mech. Syst. Signal Process. **25**(7), 2388–2398 (2011)
27. Rantatalo, M., Aidanpää, J., Göransson, B., et al.: Milling machine spindle analysis using FEM and non-contact spindle excitation and response measurement. Int. J. Mach. Tools Manuf **47**(7), 1034–1045 (2007)
28. Jiang, Y., Liu, X., Wu, S., et al.: Dynamics characteristics of the spindle system with the interface and axial milling force. J. Mech. Eng. **51**(19), 66–74 (2015)
29. Xu, C., Zhang, J., Feng, P., et al.: Characteristics of stiffness and contact stress distribution of a spindle-holder taper joint under clamping and centrifugal forces. Int. J. Mach. Tools Manuf **82–83**(7), 21–28 (2014)
30. Zhao, Y., Xu, J., Cai, L., et al.: Contact stiffness analysis of spindle-toolholder system with high speed, J. Huazhong Univ. Sci. Technol. (Natural Science Edition), **44**(7), 91–95 (2016)

Predictive Maintenance System for Production Line Equipment Based on Digital Twin and Augmented Reality

Wentao Wei[1](✉), Lilan Liu[1], Muchen Yang[1], Jiaying Li[1], and Fang Wu[2]

[1] Shanghai Key Laboratory of Intelligent Manufacturing and Robotics, Shanghai University, Shanghai, China

[2] Huayu-Intelligent Equipment Technology Co., Ltd., Shanghai, China

Abstract. Synchronous monitoring and intelligent maintenance of production line equipment have always been two major problems in the field of intelligent manufacturing. To solve these problems, this paper proposes a predictive maintenance system for production lines based on digital twin and augmented reality (AR), which obtains real-time operation data of physical production lines through data processing module, drives synchronous movement of digital twin system of physical production lines, completes virtual and real synchronous mapping, and realizes visual monitoring of physical space; The AR predictive maintenance system is built based on digital twin. The equipment health threshold is obtained through the analysis of historical data, and the equipment health threshold is used in combination with real-time operation data to predict the fault of the production line. The equipment fault information and maintenance operation guidelines are presented through AR visualization effect, making the equipment maintenance work more efficient. This paper also shows the application of the system in the inspection of assembly line equipment, realizes the synchronous twin mapping and AR predictive maintenance of assembly line equipment, and verifies the feasibility of the system.

Keywords: Digital twin · Augmented Reality · Predictive maintenance · Intelligent inspection

1 Introduction

With the development of information technology, the traditional manufacturing industry is facing great changes. The development of 5G, artificial intelligence, big data and cloud computing technology makes it possible to connect everything and make intelligent decisions in the traditional manufacturing industry. More and more enterprises are integrating cutting-edge information science with the traditional manufacturing industry to pursue higher benefits. At the same time, in order to accelerate the industrial upgrading of the traditional manufacturing industry and occupy a dominant position in relevant fields, western developed countries have introduced their own advanced manufacturing development strategies, such as Germany's Industry 4.0 strategy and the United

Y. Wang et al. (Eds.): IWAMA 2021, LNEE 880, pp. 479–486, 2022.
https://doi.org/10.1007/978-981-19-0572-8_61

States' Industrial Internet strategy. Under the background of industrial intelligence and informatization, China has introduced a series of national development strategies for manufacturing, such as Made in China 2025, to promote the deep integration of the new generation of information technology and manufacturing, and vigorously develop intelligent manufacturing.

Digital twin is to create a virtual model of physical entity in a digital way, simulate the behavior of physical entity in the real environment with data, and add or expand new capabilities to physical entity through virtual interactive feedback, data fusion analysis, decision iteration optimization and other means [1]. In 2003, Professor Grieves of the University of Michigan first proposed the concept of digital twin in his PLM course, including physical products, virtual products and the connection between the two [2]. As a key technology in the field of intelligent manufacturing, digital twin has been widely concerned by researchers and related enterprises in recent years. Gartner, an IT research and consulting firm, ranked digital twin among the top 10 strategic technology trends for the third year in a row starting in 2017. The NASA is using digital twin to fuse physical and virtual information systems for failure prediction and health management of aircraft. Parameter Technology Corporation (PTC) uses digital twin technology to monitor physical systems in real time, enabling predictive maintenance of products.

Augmented reality (AR) technology combines computer-generated virtual information with real physical scenes to enhance and expand the information of physical systems. The enhanced scenes are often presented through AR glasses. The combination of AR and Digital Twin will better connect the virtual information system with the physical system, comprehensively and intuitively express the data of the virtual information system, expand the real-time monitoring ability of the physical information system by Digital Twin, and also enhance the interaction between users and the digital twin system.

The automobile rear axle assembly line in this study is an intelligent flexible assembly line with man-machine cooperation. The automobile rear axle assembly line is composed of industrial tightening robot, loading robot, conveyor belt, assembling jig, four wheel positioning adjustment device and other machine station and manual station. For the same series of products, it has a similar assembly process, only need to slightly adjust the assembly line on the fixture and worker. In this project, the auto rear axle assembly line is taken as the physical entity to build a mapping digital twin system, and the virtual twin model is driven synchronously by collecting twin data combined with the Internet of Things (IOT), so as to realize the virtual simulation and remote monitoring of the auto rear axle assembly line. The introduction of AR and fault prediction expands the functions of the digital twin system, making the real-time monitoring results of the digital twin system present directly in a visual way, and generating virtual maintenance guidelines to assist the worker to maintain the equipment.

2 Predictive Maintenance System Based on Digital Twinning and Augmented Reality

Based on the production line of automobile rear axle assembly, the digital twin technology collects the real-time working data of the production line equipment with sensors. After data fusion and data analysis, the synchronous mapping between the physical assembly production line and its digital twinning information system is constructed to realize the precise simulation and real-time monitoring of the production and assembly process in physical space. Based on the digital twin system of physical assembly line, we propose a predictive maintenance method of equipment failure combined with AR technology. The maintenance staff can maintain the equipment in advance according to the prediction results given by the fault prediction system, so as to avoid the production stagnation caused by equipment failure. The application of AR technology can assist the maintenance staff to quickly find the equipment fault location in the predicted results, and provide the maintenance method corresponding to the fault in the way of AR maintenance guidance animation. The predictive maintenance system based on digital twin and AR includes three parts: physical entity layer, data processing layer and simulation application layer. The simulation application layer also includes virtual digital twin system and AR predictive maintenance system. The whole system framework is shown in Fig. 1.

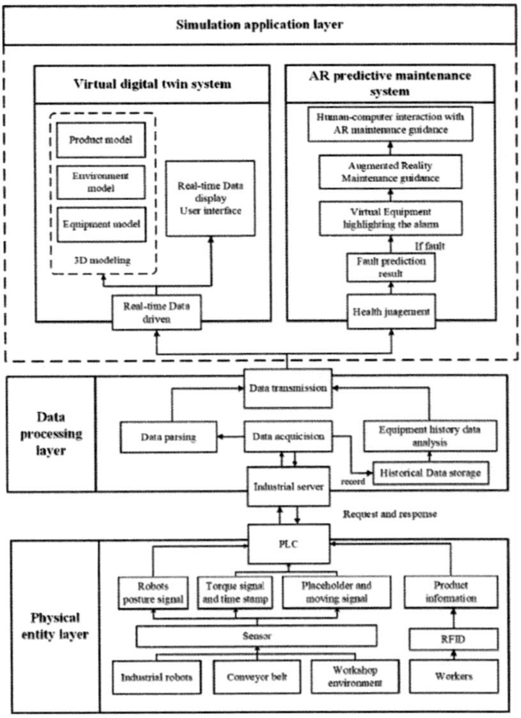

Fig. 1. The framework of predictive maintenance systems based on digital twin and AR

2.1 Physical Entity Layer

The physical layer is composed of industrial robots, conveyor belts, sensor network, programmable logic controller (PLC), industrial personal computer (IPC), assembly products, assembly workers and other elements in the automobile rear axle assembly production line. The sensor collects the real-time operation data of each equipment in the assembly line and transmits it to PLC through the PLC interface. PLC supports a variety of fieldbus and communication protocols, through the bus connected to the assembly production line LAN. The assembly workers scan the electronic tags on the assembly products with handheld Radio Frequency Identification devices, collect the data of the assembly products and transmit it to PLC. PLC interacted with industrial computer through assembly line Ethernet.

2.2 Data Analytics for Impending Failure Prediction

Based on the assembly line local area network established by the equipment communication protocol, the IPC sends the data acquisition instruction to the PLC, and the PLC responds to the instruction and transmits the robot position data, the robot torque and angle data, the conveyor belt placeholder signal and the moving signal, the product assembly information to the IPC for storage. In the data processing layer, the data acquisition module accesses the IP address of the data storage port of the IPC at fixed intervals to obtain the real-time operation data of the assembly line equipment stored in the IPC. On the one hand, the data format of the acquired real-time operation data of the assembly line is analyzed and transmitted to the virtual digital twin system of the simulation application layer. On the other hand, the acquired real-time operation data of the assembly line are stored in the time-series database InfluxDB, and the operation history data of the equipment stored in the time-series database are used for data analysis to extract the health state threshold that represents the health state parameters of the equipment on the assembly line. The health threshold will be transmitted to the AR predictive maintenance system in the simulation application layer together with the parsed real-time data.

The data in this project adopts JSON (Java Script Object Notation) data transmission format. JSON, as a lightweight data transmission format [3], can be used for data exchange between multiple languages. It is a simple data exchange format based on plain text that is easy to read and code. Figure 2 shows the real-time JSON format data obtained by accessing the IPC network port.

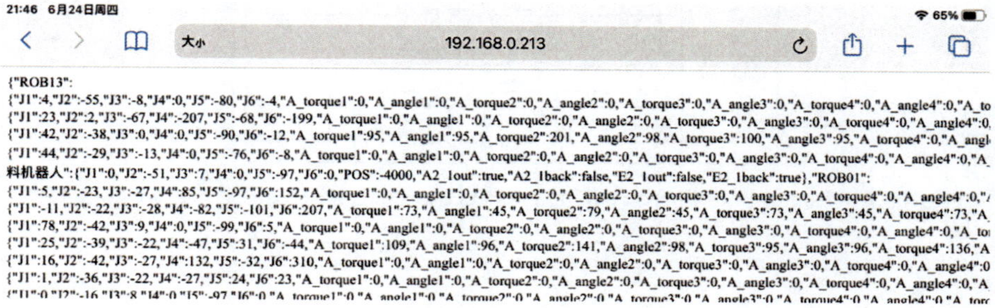

Fig. 2. Real-time JSON format data from Industrial Personal Computer

2.3 Simulation Application Layer

The simulation application layer is divided into virtual digital twin system and AR predictive maintenance system. The virtual digital twin system uses the real-time operation data of the physical assembly line equipment acquired by the data processing layer to drive the production line twin model to move synchronously and realize the remote monitoring of the physical assembly production line. AR predictive maintenance system is based on the assembly line extension of digital twin system, based on the assembly line running history data to establish a production line equipment failure prediction and evaluation of health system, real-time monitoring data of the assembly line equipment, fault forecast results will show in the form of AR on the mobile terminal equipment worn by the staff, According to the fault prediction information, AR maintenance instruction animation is triggered to assist the maintenance staff to maintain the equipment.

2.3.1 Virtual Digital Twin System

The digital twin system of the automobile rear axle assembly line includes the digital twin model of the assembly line equipment, the real-time operation data display of the equipment and the real-time data drive module. Digital twin model presents physical objects in virtual space in a digital way [4], including 3D model of production line and dynamic behavior rules of model. The virtual model of automobile rear axle assembly line includes product model, equipment model and environment model. In order to restore the real scene of the physical space as much as possible, the virtual models are twin modeled according to the geometric features related to the physical assembly line. In order to synchronously map the movement of assembly line equipment in physical space, according to the logic rules of assembly line equipment movement behavior, DoTween plug-in is used to construct the dynamic behavior rules corresponding to the digital twin model. The real-time data-driven module receives the real-time operation data of the assembly line equipment after parsing, and maps the data to the parameters of the digital twin model of the assembly line according to the corresponding mapping method, so as to realize the synchronous movement of the virtual digital twin system and the physical assembly line equipment. In order to better monitor the running state of each device in the physical assembly line through the digital twin system, a data display board was established to display the real-time running data of each device in the rear assembly line of the automobile, so as to facilitate the researchers to collect production data for analysis. Figure 3 shows the rear axle assembly line and its digital twin system.

2.3.2 AR Predictive Maintenance System

The AR predictive maintenance system judges the health of the real-time data of the operation of the equipment transmitted to the AR predictive maintenance system based on the health threshold of each device parameter of the automobile rear axle assembly line calculated by the digital processing layer,when actually affect the health status of the equipment parameter data is beyond the scope of corresponding health threshold, is judged to be related equipment of the risk of fault,system will present failure prediction results. The staffs wear AR device to scan the production line equipment and trigger marks, and the program can locate the relevant equipment components of the digital twin system of the assembly line based on the given failure prediction information and

Fig. 3. Automotive rear axle assembly line and its digital twin system

trigger the highlight flashing alarm. At the same time, the background program will follow the equipment failure of the assembly line The knowledge base and the fault prediction information match the corresponding AR maintenance guidance animation, and assist the maintenance staffs in the fault maintenance of the rear axle assembly line equipment in the form of AR. Figure 4 shows the animation effects of highlighting and maintenance guidance for the failure of tightening gun of assembly line tightening robot.

Fig. 4. Highlight flashing and maintenance instruction animation of assembly line tightening robot tightening gun failure

3 Application of Predictive Maintenance Systems

For a manufacturing facility, maintenance programs must prevent or reduce downtime, production decreases,delays and supply chain issues by minimizing equipment failure [5]. The current equipment maintenance strategy in manufacturing plants is usually based on the expert experience of maintenance staff and passive post-maintenance, which cannot avoid the production stagnation caused by equipment failure. The predictive maintenance system proposed in this paper can realize the predictive maintenance of equipment through the data analysis of the production line equipment, which fundamentally changes the maintenance strategy of equipment failure from passive treatment to active prevention, greatly reduces the failure rate of equipment, and realizes the smooth and efficient operation of the production line.

Equipment inspection is the key link of equipment maintenance. Patrol inspection methods mainly use manual inspection, remote video inspection, aircraft, robot inspection and other means. On the basis of traditional inspection, the current inspection system mostly utilizes the Internet of Things technology to improve the supervision of inspection and provide the function of analyzing and summarizing inspection data [6]. In this paper, the proposed predictive maintenance system based on digital twin and AR is applied to the device inspection, and the AR inspection application is released on the Android platform using C/S architecture. The maintenance staff carries out inspection on assembly line equipment by holding a tablet installed with the AR inspection application. The camera is used to scan the inspection equipment and identification images, triggering its corresponding AR digital twin model and its AR data display board. The AR digital twin model moves synchronously with the inspection equipment. The AR data display board displays the real-time operation data of the inspection equipment and the health status value of the equipment obtained from fault prediction. When the fault prediction result of the back-end fault prediction program is the equipment fault, the system will locate the fault parts of the equipment according to the fault information and trigger the highlighting alarm and the corresponding AR maintenance instruction animation, to assist the maintenance staff to select the machine and carry out maintenance on the inspection equipment. Figure 5 Maintenance staff is using Android tablet for AR inspection.

Fig. 5. AR inspection with Android tablet

4 Conclusion and Improvement

In this paper, we study the digital twin of the rear axle assembly line and the predictive maintenance of AR, establish the digital twin system of the rear axle assembly line, and realize the precise simulation and synchronous movement of the virtual and real assembly line. The AR predictive maintenance system based on intelligent mobile terminal devices is established for the assembly line equipment, which realizes the AR patrol inspection of the assembly line equipment, assists the maintenance staff to carry out preventive maintenance, and avoids the stagnation of the assembly line caused by equipment failure. In this project, In this project, the predictive maintenance system we built based on digital twins and AR has initially realized the AR predictive maintenance function on the rear axle assembly line of the automobile, but there are still the following problems: 1. For the assembly line equipment fault prediction accuracy is not high, the fault prediction model needs to be optimized. 2. The data types representing the health status of equipment are relatively single, which cannot comprehensively reflect the faults of each equipment component in the assembly line. 3. The animation effect of AR maintenance guidance needs to be further optimized.

Acknowledgment. The support of Shanghai University and Huayu Intelligent Equipment Technology Co., Ltd for author's research is greatly appreciated. The work described in this article has been conducted as part of the pillar program supported by Shanghai Science and Technology Committee of China (No. 19511105200).

References

1. Tao, F., Liu, W., Liu, J., et al.: Digital twin and its potential application exploration. Comput. Integr. Manuf. Syst. **24**(01), 1–18 (2018)
2. Grieves, M., Vickers, J.: Digital twin: mitigating unpredictable, undesirable emergent behavior in complex systems. In: Kahlen, F.-J., Flumerfelt, S., Alves, A. (eds.) Transdisciplinary Perspectives on Complex Systems, pp. 85–113. Springer, Cham (2017). https://doi.org/10.1007/978-3-319-38756-7_4
3. Gao, J., Duan, H.: Research on data transmission efficiency of JSON. Comput. Eng. Des. **32**(07), 2267–2270 (2011)
4. Fan, S.: Technical analysis of digital factory realization for intelligent manufacturing. Technol. Market **26**(10), 173 (2019)
5. O'Sullivan, J., O'Sullivan, D., Bruton, K.: A case-study in the introduction of a digital twin in a large-scale smart manufacturing facility. Procedia Manuf. **51**, 1523–1530 (2020)
6. Liu, C.: Research on Intelligent Inspection System of Industrial Equipment Based on Internet of Things. Yanshan University, China (2015)

Design of Intelligent Irrigation System Based on App Inventor and MCU

Xuehuan Jiang[1], Hang Tao[1], Xiue Gao[2](✉), Rui Tong[1], Yufeng Chen[1], and Bo Chen[2]

[1] School of Electrical and Information Engineering, Hubei University of Automotive Technology, Shiyan 442002, Hubei, China

[2] School of Information Engineering, Lingnan Normal University, Zhanjiang 524048, Guangdong, China

Abstract. This paper analyzes the research status and trend of potted irrigation system, combines the technical advantages of App Inventor's complete online development, and propose a design scheme of intelligent irrigation system based on MCU and App Inventor. Firstly, the overall system design, hardware design and software design scheme are explained in more detail; Secondly, a test environment for the intelligent irrigation system is set up, and the Bluetooth module is used to realize communication between MCU and mobile phone. System testing shows that the system can automatically complete watering tasks according to the soil status, and users can also understand the potted plants and the surrounding environment in real time, which improves the intelligent level and watering effect of the pot irrigation system.

Keywords: Automatic watering · Bluetooth · App inventor · MCU

1 Introduction

Green potted plants can purify indoor air, decorate life, and cultivate sentiments, so they are more and more popular. The growth trend and survival time of potted plants are affected by many factors such as soil, sunlight, moisture, fertilizers, etc. The amount of watering is the most important influencing factor. In particular, untimely watering or excessive watering can cause potted flowers to wilt or die. Therefore, it is necessary to design an automatic irrigation system to realize intelligent control of irrigation according to soil moisture.

Therefore, some automatic irrigation methods and systems based on potted environment monitoring have been proposed. The MCU mainly include STC89C51/52, STM32, etc.; the main interface methods with the host computer are serial port RS232, Bluetooth, WiFi, Zigbee and other several. The related work is elaborated as follows:

(1) Non-mobile application design. Such automatic watering methods and systems are not portable, and users usually cannot know the humidity of potted plants in real time. Ni Rui designed a temperature and humidity monitoring and control system based on AT89S51, using RS232 to communicate with the host computer, which

Y. Wang et al. (Eds.): IWAMA 2021, LNEE 880, pp. 487–494, 2022.
https://doi.org/10.1007/978-981-19-0572-8_62

can display the humidity of potted plants in real time [1]. Chen Yali designed an automatic watering system based on AT89C52 for the problem of unmanned watering of potted plants during holidays or business trips [2]. Tong Jinkai designed an automatic flowering control system based on STC89C52. The system consists of humidity acquisition, analogy-to-digital conversion, keyboard and relay control [3].

(2) Mobile application design. Such automatic irrigation methods and systems are generally flexible. Users can use the mobile phone APP to understand the soil moisture in real time and remotely control watering. Wang Hongmei designed a flower soil moisture data monitoring system based on AT89C51 + Bluetooth, and real-time monitoring of flower soil moisture through mobile APP [4]. Tao Zengjie and Li Huizhen designed an intelligent watering system based on STM32 + Bluetooth. The system collected soil moisture and environmental temperature data in real time, controlled the opening and closing of water pumps according to soil moisture, and controlled the heating and cooling of equipment according to environmental temperature [5, 6]. Li Shuolei designed an intelligent watering system based on STC89C52 + Zigbee. The system collects soil temperature, humidity and light intensity data in real time, which can realize all-weather supervision of plants [7].

In recent years, with the popularization and application of App Inventor mobile phone programming software with online development functions, the online development and application of the combination of MCU and App Inventor has gradually attracted widespread attention. Hexiao Huang constructed the teaching mode of applied embedded curriculum based on App Inventor computational thinking on the basis of App Inventor graphic programming course in primary and secondary schools [8]. Cui Chengyi designed an intelligent window control system based on MCU and App Inventor [9], Lukkarinen A. designed an event-driven programming tool based on App Inventor [10], and Prabhakar M. designed a robot based on App Inventor, the product handling procedure simplifies the task of picking and placing and improves the product handling speed [11]. This paper proposes a design scheme for an intelligent watering system based on MCU and App Inventor, which allows users to more conveniently and quickly understand the soil moisture, temperature and light intensity of potted plants in real time, reduces the requirements for user development capabilities, and enhances user interaction capabilities. And application flexibility.

2 System Overall Design

The system is mainly composed of power circuit, soil humidity, temperature, light intensity detection circuit, independent button circuit, LCD liquid crystal display circuit, Bluetooth and water pump control circuit. First, the system uses temperature, humidity, and light sensors to collect the corresponding environmental information; Second, it uses a analogy-to-digital conversion module to convert it into a digital signal for the MCU to read and display; Third, use the Bluetooth module to realize the upper and lower connection between the MCU and the mobile phone; Finally, use App Inventor to develop a mobile phone App for intelligent watering online to realize the remote automatic monitoring and intelligent watering of the potted soil environment by mobile phones (Fig. 1).

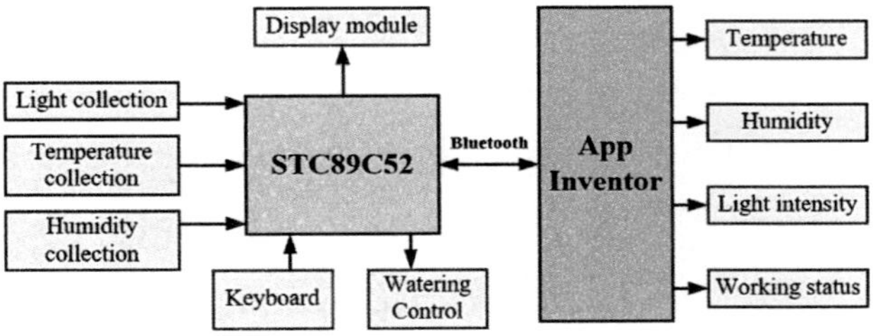

Fig. 1. Overall design

3 System Hardware Design

According to the overall design of the system, the schematic diagram of the hardware circuit is shown in Fig. 2. It mainly includes the MCU main control chip, power supply circuit, reset circuit, ISP serial port, Bluetooth module, analogy-to-digital conversion, LCD1602 display module, sensor module and other components.

3.1 MCU Interface and Display Circuit

The system uses the STC89C52, which has the advantages of simple operation of instructions, peripheral circuits, and I/O interfaces, which is convenient for programmers to modify the code. The display circuit adopts a liquid crystal display LCD1602, which can display letters, numbers and symbols at the same time, and can display up to two lines of 16 characters. In this system, it is necessary to consider the display of the actual soil temperature, humidity, light intensity, and pump working status. The temperature, humidity, light intensity, and pump working status are represented by T, S, G and State respectively, and the accuracy of the temperature is two decimals. Humidity and light intensity are expressed in a two-digit percentage system, so that the setting is more in line with the requirements of the LCD1602 display format.

3.2 Environmental Detection Circuit

The temperature sensor selects DS18b20, which uses single-channel communication to communicate with the MCU, and converts the collected data to obtain temperature data with two decimal places; the humidity sensor selects a humidity sensor, and changes the voltage value through the resistance value to detect the current humidity of the soil, The collected data is converted, and the humidity is expressed in a two-digit percentage system; the light intensity uses a photoresistor, and the voltage value of the photoresistor changes as the resistance of the photoresistor changes. The collected voltage is sent to MCU through analogy-to-digital conversion. Intensity is expressed using a two-digit percentage system.

The analogy-to-digital converter uses 8-bit conversion chip ADC0832, which has dual-channel antilog-to-digital conversion, which can meet the design requirements of this system, and has the characteristics of fast conversion speed and stable performance, small size, low power consumption, and strong compatibility.

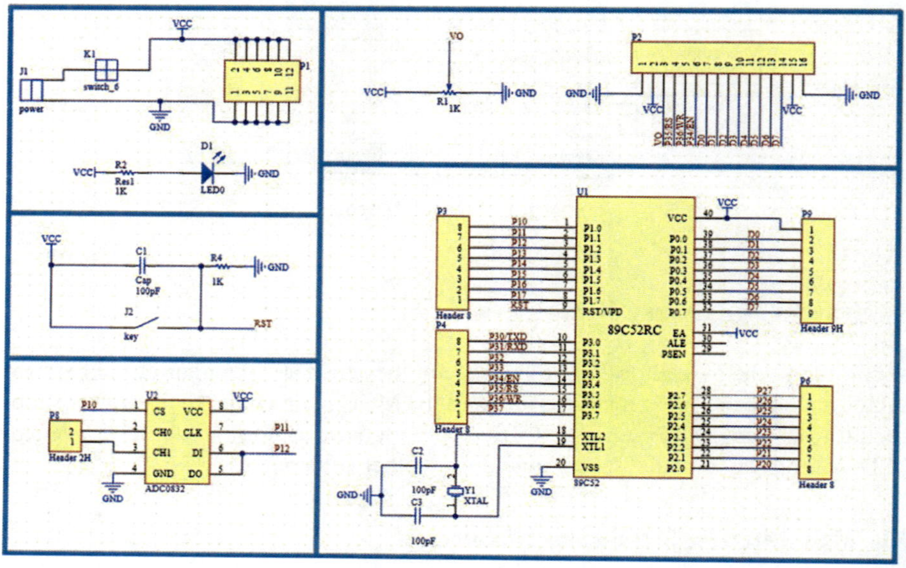

Fig. 2. Hardware circuit schematic diagram

3.3 Keyboard and Water Pump Drive Circuit

This part is mainly composed of keyboard and water pump drive circuit. Among them, the function of the keyboard is to control the threshold adjustment of the soil moisture, and the keys have the functions of switching the interface, adjusting up, and adjusting down respectively. While adjusting, the adjustment status will be displayed on the LCD display in real time.

Through the soil humidity information collected by the humidity sensor, the humidity information is compared with the threshold value, and finally the I/O port of the MCU outputs the high and low levels to reach the control relay pull-in state, so as to achieve the function of controlling the work of the water pump. When the soil humidity rises with the operation of the water pump, the threshold condition has been met. This information is transmitted to the MCU through the sensor, and the low level is pulled to release the relay and the water pump stops working. In this way, it is possible to dynamically adjust the watering time according to the humidity of the potted plants to meet the water demand of the potted plants.

4 Software Design

4.1 MCU Software Design

The MCU program of this system includes system initialization, display program, key detection program, soil moisture detection program, humidity threshold adjustment program, temperature and light detection program and Bluetooth driver. After power on, the system initializes, calls the display driver, enters the welcome interface, presses the button to enter the threshold setting interface, and presses the button to enter the detection interface. Then, the LCD displays the current humidity, temperature, humidity and relay working conditions.

By processing the soil humidity data detected by the humidity sensor received by the MCU, it is judged whether it is necessary to water the plants. Finally, the collected data is spliced, and the data is sent to the connected mobile phone via Bluetooth. The application receives the data in a fixed format, splits the data, and displays it in the corresponding display area.

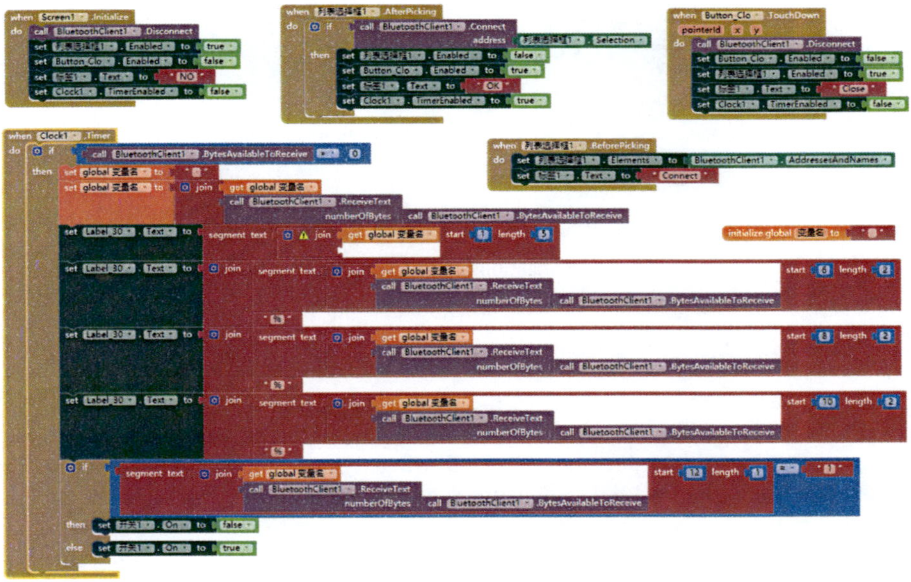

Fig. 3. Application logic design

4.2 Host Computer Application Program Design

For the more common Bluetooth assistants, only fixed format data receiving interface can be displayed. Therefore, this design utilizes the design advantages of App Inventor's fully online programming, and users only need to add service options to it according to their needs. The design framework is as follows: First, select the Bluetooth client to display Bluetooth connection, disconnection, and status changes; second, according to a linear layout, the temperature, humidity, light intensity, and water pump status are arranged on the interface. After finishing the layout of the interface, the logic of the application should be designed. First, initialize the button state and Bluetooth client, and enable the timer to continuously receive data from the MCU; second, split the data and display it in the corresponding position. The logic design of the upper computer application program is shown in Fig. 3.

After completing the design of the host computer application program, after continuous optimization and testing, its function meets the design requirements. Connect the Bluetooth of the mobile phone to the MCU device, and the Bluetooth connection status, temperature, humidity, and light intensity can be displayed on the mobile phone application. And the working condition information of the water pump, its application interface layout design is shown as in Fig. 4.

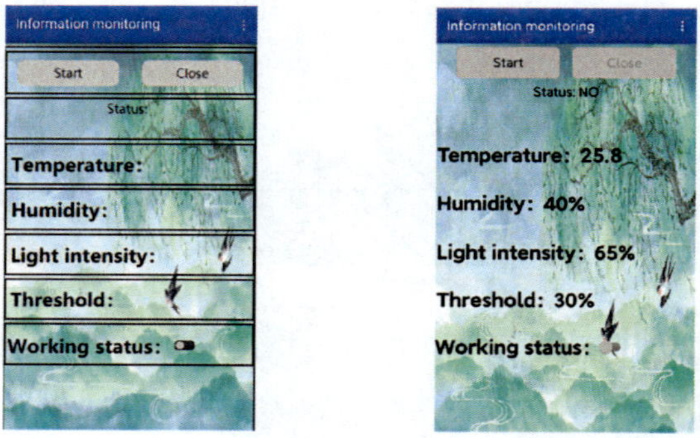

Fig. 4. Application interface layout design

5 System Test

In order to verify the effectiveness of the intelligent irrigation system, an intelligent irrigation test system was built, and the MCU was used to communicate with a mobile phone using Bluetooth connection. The following figure shows the change curve of temperature, humidity, and light intensity data monitored by the mobile phone within 210 s. It can be seen from the temperature change curve in Fig. 5 that the temperature changes drastically within 140–190 s, and the reason is that the temperature sensor is briefly touched. From the humidity change curve in Fig. 6, we can see that the humidity changes alternately (the lowest point of humidity in the figure is the humidity threshold). Here, sponge is used to simulate soil. Due to its strong water absorption capacity, the humidity sensor is placed on the sponge and a little artificially squeezed out. The water and humidity values also decrease, but when they fall below the threshold, the relay closes and the water pump works to increase the humidity to the sponge and close the relay at the same time. From Fig. 7, the light intensity change curve shows that due to the sensor's own problems, it is very sensitive to light changes, and a small amount of occlusion will cause major changes; but in actual use, the light sensor can play a certain role. The above test shows that the design requirements of the system have been met.

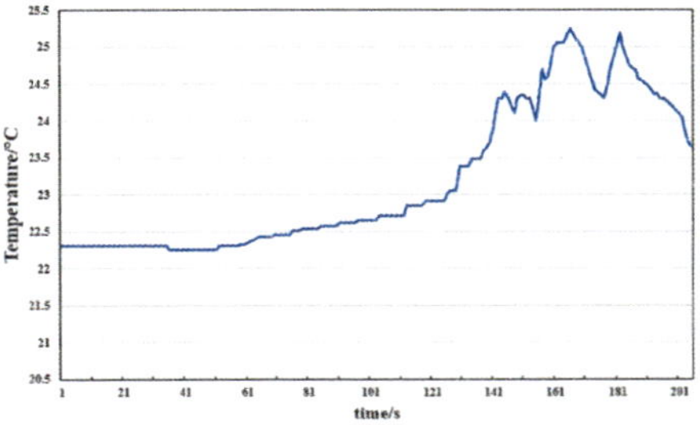

Fig. 5. Temperature curve

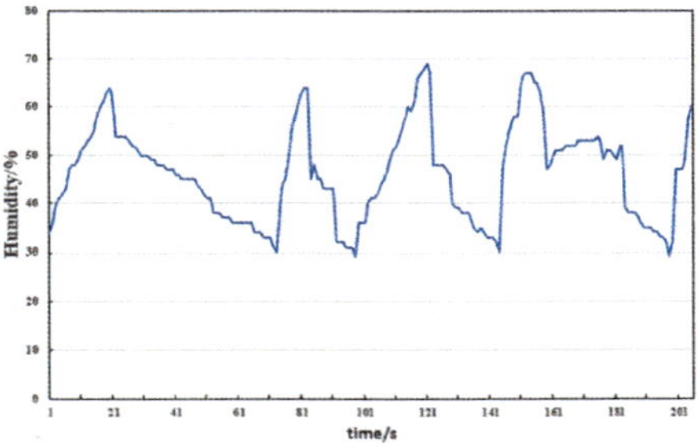

Fig. 6. Humidity curve

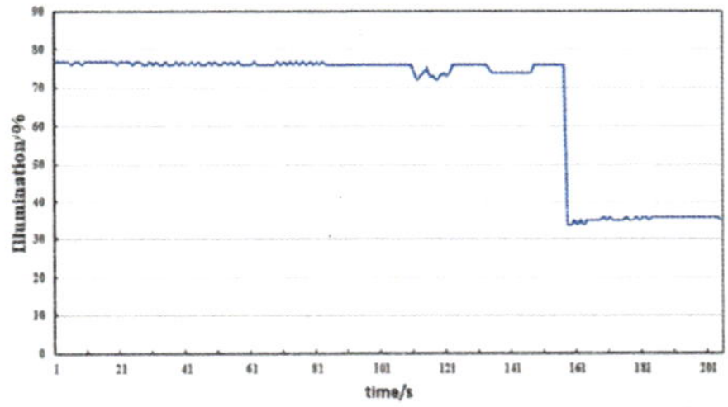

Fig. 7. Light intensity change curve

6 Conclusion

This paper designs an intelligent watering system based on STC89C52 and App Inventor. The system can quickly and accurately monitor the temperature, humidity and light intensity of the soil environment, and control the automatic start and stop of the water pump to achieve intelligent watering of potted soil. In addition, the system has low power consumption and stable performance, which meets the watering needs of potted plants in homes and offices.

Acknowledgment. This work is partially supported by Science and Technology Research Project of Hubei Provincial Department of Education under Grant Q20161805, Science Technology Research and Development Program of Shiyan under Grant 2021K60 and Teaching Team Project of Guangdong Province Department of Education under Grant T2019305.

References

1. Rui, N., Wanda, H.: Design of temperature and humidity monitoring and control system based on AT89S51 single-chip microcomputer. Autom. Instrument. **34**(05), 53–55 (2019)
2. Yali, C., Bing, Y.: Design on AT89C52 automatic watering system based on. Luohe Vocat. Tech. Coll. **19**(01), 23–25 (2020)
3. Tong, J., Xiao, P.: Design of automatic flowering control system based on STC89C52 single chip microcomputer. J. Liaoning Teach. Coll. **19**(03), 82–85 (2017)

4. Wang, H., Bai, L., Dong, M.: Flower soil moisture data monitoring system. Electron. World **10**, 135–136 (2020)
5. Tao, Z., Song, C., Huang, P., Cao, B.: Intelligent watering system based on STM32. J. Xihua Univ. (Nat. Sci. Edn.) **38**(05), 84–88 (2019)
6. Li, H., Zhang, P., Yang, F., Malingkun: Design of plant growth environment monitoring system based on mobile app. Sens. Microsyst. **39**(11), 57–59 (2020)
7. Li, S., Hu, N., Li, Z., Wang, J., Luo, Z.: Design and implementation of intelligent flower watering system based on Zigbee network. Electron. Des. Eng. **25**(15), 105–108 (2017)
8. Huang, H., Chen, L., Jin, S., Wang, Y., Hu, H.: Research on teaching practice of app inventor course embedded in computational thinking. Int. J. Embed. Syst. **14**(2), 117–126 (2021)
9. Cui, C., Chen, J., Gao, Q., Liu, X., Wang, K.: Intelligent window design in the comprehensive design experiment of single-chip microcomputer. Exp. Technol. Manag. **37**(9), 71–74 (2020)
10. Lukkarinen, A., Malmi, L., Haaranen, L.: Event-driven programming in programming education: a mapping review. ACM Trans. Comput. Educ. **21**(1), 31–32 (2021)
11. Prabhakar, M., Paulraj, V., Kannappan, D.A.K., Dhanraj, J.A., Ganapathy, D.: IOP Conference Series: Materials Science and Engineering. RIACT 2020, Chennai, India, 2–3 October 2020. 012003

How to Improve Conflict Management in Hospitals in the Healthcare Industry

Kareem Elmasry(✉) and Yi Wang(✉)

Plymouth Business School, University of Plymouth, 24 Sutton Road, Plymouth PL4 0HJ, UK
{Kareem.Elmasry,Yi.Wang}@plymouth.ac.uk

Abstract. The decision-making problem is the problems professionals face in hospitals in the health industry when managing conflict. Studies have shown that conflict is common [more specifically organisation conflict] in the hospitals in the industry and conflict resolution strategies are key. This report suggests the Thomas-Kilmann theory as a main solution on how to improve managing conflict in hospitals. This is since it has been applied and used for these conflict situations in the health industry in previous studies as an option.

Keywords: Conflict management · Healthcare industry · Thomas-Kilmann theory

1 Introduction

Conflicts can influence and happen at any of the stages of the decision-making process [1]. For example, conflict can happen when both parties disagree on whether the problem is or if they disagree which solution would be best etc.

Conflict is described as a disagreement or misunderstanding between one or more people/groups who have opposing views due to their personal beliefs or external influence towards an objective [2]. This does not necessarily mean that this is a negative situation that can cause a negative outcome. This definition can be seen as a means that challenges different people's opinions while learning and understanding each other's perspective. However, conflict is represented by Thakore [3] as a manifestation of animosity, negative behaviour, antagonism, violence, competition, and confusion. Although some may argue that these are similar definitions, Thakore's definition is best suited for the decision-making problem at hand as the problem can illustrate negative behaviour on both parties involved and animosity. However, Thakore [3] describes conflict as an expression of violence which can be seen as valid but is not relevant to the study below.

Conflict management is the method of resolving disagreements in such a way that negative outcomes are reduced, and positive outcomes are prioritised [4]. This important management ability entails a variety of strategies depending on the situation, as well as persuasion and innovative thinking. The importance of resolving or managing conflicts is of significance and it can directly affect all parties involved. Resolving conflict in a workplace can greatly increase the productivity and performance of all involved parties

Y. Wang et al. (Eds.): IWAMA 2021, LNEE 880, pp. 495–499, 2022.
https://doi.org/10.1007/978-981-19-0572-8_63

whether the problem lies between employees and superiors or other internal stakeholders [5]. Deciding how to resolve conflict can also be of importance especially in the health industry where conflict is common within organisations.

This paper first defines of conflicts, conflict management and the importance of resolving conflict. Then, a review of Thomaskilman mode as a solution is provided. After that, a critical analysis of the solution is carried out to further decipher if the solution is enough. By using multiple credible sources is used for a more accurate outcome. Finally a conclusion is made for the paper.

2 Literature Review

2.1 Overview of Thomas-Kilmann Conflict Model

The Thomas-Kilmann conflict model instrument is a credible solution that can be used as an option to solve this problem. The Thomas-Kilmann model is a tool used to understand and manage conflict situations [6]. The Thomas-Kilmann model is a tool used to understand and manage conflict situations. Jones [7] states that using these two dimensions as guidance, five techniques are created and used to manage conflict:

1. Collaborating
2. Competing
3. Avoiding
4. Accommodating
5. Compromising Conflict resolution using the Thomas-Kilmann model's five strategies has been comprehensively studied in the literature [8–11]. [12, 13] explores the different conflict-management strategies that are used through the application of the Thomas-Kilmann model using the model was successful and using the collaborating strategy was the most effective. Although, they concluded that using the strategy took the most time to resolve conflict.

Ogunyemi et al. [14] concluded that the collaborating strategy is the most effective, however, it also concluded that the competing strategy was equally effective and successful. Additionally, the study also concluded and mentions that the avoiding and accommodating strategy was the least effective and successful.

Pitsillidou, Farmakas, Noula and Roupa [15] discusses the types of conflict management that health professionals encounter in Cyprus hospitals on a daily basis. Through the application of the Thomas-Kilmann model using anonymous self-referral questionnaires from health professionals, the research concludes that the avoiding and compromising strategies were the most used and successful. Mahon [16] identifies that the compromising strategy is the favoured strategy when resolving conflict followed closely by the accommodating and avoiding strategies. Although both articles are almost 10 years apart and are used in different settings, similar research techniques have achieved near identical results. [8] examines the styles of conflict management derived from the Thomas-Kilmann model that is successfully used by first-year pharmacy students in their professional encounters. The level of success is measured through the level of empathy that students present when they confront and resolve conflicts. [9] identifies that the model is a quick and effective solution with successful outcomes when resolving conflicts. It does not mention a particular strategy of the five strategies as the most effective as it also concluded that each strategy is effective in different scenarios/situations.

3 Critical Analysis

The Thomas-Kilmann theory has many strategies for resolving conflicts. The five strategies provide insight into different techniques to use in different conflict situations. Thus, improving leadership skills [as an example] when managing a team as conflicts can occur and having the ability to resolve conflicts within a team using different methods is an added benefit. However, the Thomas-Kilmann theory does not advise or suggest which strategy is best and whether each strategy works in every situation [17]. Thus, making it harder for people to decide which strategy is best in immediate conflicts. Altmäe, Türk and Toomet [18] which also suggests that the theory although beneficial in some regards can be negative in terms of deciding what to use. This is a key factor to consider especially in hospitals where being able to think and decide in high-stress situations quickly is key for health professionals to be able to do [19]. The theory is more to do so with informing rather than teaching when to use in certain situations. However, it still gives a great insight into each strategy by presenting the positives and negatives of each one. Additionally, [20] describes the Thomas-Kilmann theory as a great tool. But more particularly that it is very beneficial to use because the strategies are easy to use by anyone. They are easy to adapt and anyone capable can use them effectively. And although this strategy is good in some cases, it is still a limitation to overuse a strategy as the person might not be as effective at resolving conflicts as someone who can use multiple strategies.

Another limitation of the theory is that all five strategies do not resolve conflict entirely. Meaning, all strategies are used to resolve the conflict between two parties however it can be argued that most strategies do not satisfy both parties' needs [21]. Schneider and Brown [22] describes the theory as not optimal in a high-stress situation. This means that the theory and its strategies are blind to stress and how stress can impact human nature. It depicts that the theory can be used when a person is level-headed and not swayed by their emotions. Results from studies by [23] and [24] have swayed towards the direction that human emotions can diverge from rational decision-making. And the theory does not account for this making it harder for humans to decide which strategy is best to use when resolving high-stress conflicts. This is significant as a hospitals environment can create high-stress environments which can lead to conflicts [25]. So, the theory may have some difficulties when using in high-stress conflict situations. Game theory is a viable, alternative strategy that can be applied to manage conflicts. Game theory is more of a rational approach where it considers the decision-makers and, unlike the Thomas-Kilmann theory, the people that are affected by the decision and the different outcomes that decision results in [26]. [27] investigates the theory and concluded that the theory was a favourable option to use when resolving conflict in a construction workplace. It also results in a more mature judgement when deciding how to resolve the conflict. However, the theory was limited as the results showed it was a more aggressive strategy and may have caused further conflicts. This would affect the hospital environment as hospital environments can already create stressful conflict environment and adding an aggressive resolution to this may not be the best.

The Thomas-Kilmann theory, with its limitations, can be improved to be better suited when managing conflict in hospitals. The theory does not account for managing conflict situations in stressful environments. The theory is for people who are always level-headed and stress-free. And in hospitals, stressful situations stem from conflicts [25]. The theory could be improved when applying it to stressful situations, as conflicts develop more aggressively when all parties are stressed [28]. Also, although the theory updates over the years and has recognised the difference when managing conflicts between different genders, races and cultures, the theory does not have a solution to dealing with these factors specifically [29]. In hospitals, the level of diversity is very high across the entire healthcare industry [30], and so having a conflict-management tool that can deal with situations that include multiple races, genders and cultures may arguably increase the effectiveness of the tool. This can ultimately improve the decision making process in hospitals.

4 Conclusion

The Thomas-Kilmann theory is a valid solution to the decision-making problem that hospitals in the health industry face when managing conflict. Although the theory has many limitations it is still suggested as an appropriate theory to use when resolving conflict in hospitals. This is because although the theory does not advise which strategy to be used specifically, the strategies the theory presents are easy to understand and use. And it can arguably be used in all conflict situations. Though the success of each strategy is dependent on the type of conflict, it is still viable, and studies have shown that health professionals have achieved success when applying the theory and its strategies in the work environment. Furthermore, although game theory [as discussed above] is a viable alternative, the Thomas-Kilmann, in my opinion, is better as it can provide alternative strategies to resolve conflict than the game theory. However, this can be a limitation as health professionals may not be able to decide which strategy is best and overuse a strategy. Acknowledgment.

References

1. Kuhn, T., Poole, M.: Do conflict management styles affect group decision making? Evidence from a longitudinal field study. Hum. Commun. Res. **26**(4), 558–590 (2000)
2. Tjosvold, D.: Defining conflict and making choices about its management. Int. J. Confl. Manag. **17**(2), 87–95 (2006)
3. Thakore, D.: Conflict and conflict management. J. Bus. Manag. **8**(6), 7–16 (2013)
4. Madalina, O.: Conflict management, a new challenge. Procedia Econ. Fin. **39**, 807–814 (2015)
5. McKibben, L.: Conflict management: importance and implications. Br. J. Nurs. **26**(2), 100–103 (2017)
6. Altmäe, S., Türk, K., Toomet, O.: Thomas-Kilmann's conflict management modes and their relationship to Fiedler's leadership styles [basing on Estonian organizations]. Balt. J. Manag. **8**(1), 45–65 (2013)
7. Jones, J.: Thomas-Kilmann conflict mode instrument. Group Org. Stud. **1**(2), 249–251 (1976)

8. Hastings, T., Kavookjian, J., Ekong, G.: Associations among student conflict management style and attitudes toward empathy. Curr. Pharm. Teach. Learn. **13**(12), 25–32 (2019)
9. Qadir, A.: Resolving conflicts at workplace – a discourse, using Thomas-Kilmann instrument mode framework. ZENITH Int. J. Multidisc. Res. **10**(6), 28–46 (2020)
10. Trippe, B., Baumoel, D.: Beyond the Thomas-Kilmann model: into extreme conflict. Negot. J. **31**(2), 89–103 (2015)
11. Atorough, P., Martin, A.: The politics of destination marketing. J. Place Manag. Dev. **5**(1), 35–55 (2012)
12. Islam, N., Rimi, N.: Conflict management technique in private commercial banks of Bangladesh: an application of Thomas-Kilmann conflict handling model. Eur. J. Bus. Manage. **9**(29), 91–99 (2017)
13. Lacity, M., Willcocks, L.: Conflict resolution in business services outsourcing relationships. J. Strat. Inf. Syst. **26**(2), 80–100 (2017)
14. Ogunyemi, D., Fong, S., Elmore, G., Korwin, D., Azziz, R.: The associations between residents' behavior and the Thomas-Kilmann conflict mode instrument. J. Grad. Med. Educ. **2**(1), 118–125 (2010)
15. Pitsillidou, M., Farmakas, A., Noula, M., Roupa, Z.: Conflict management among health professionals in hospitals of Cyprus. J. Nurs. Manag. **26**(8), 953–960 (2018)
16. Mahon, J.: Conflict style and cultural understanding among teachers in the western United States: exploring relationships. Int. J. Intercult. Relat. **33**(1), 46–56 (2009)
17. Sheridan, J.: A Study of Culture and Conflict Management Styles of Community College Employees. Education Masters, p. 120 (2007)
18. Dinçel, Y.: Causes of tension between physician-nurse in working environments and management of conflict. Sağlık ve Hemşirelik Yönetimi Dergisi **6**(3), 256–263 (2019)
19. Paans, W., Wijkamp, I., Wiltens, E., Wolvensberger, M.: What constitutes an excellent allied health care professional? A multidisciplinary focus group study. J. Multidiscip. Healthc. **6**, 347–356 (2013)
20. Womack, D.: Assessing the Thomas-Kilmann conflict mode survey. Manag. Commun. Q. **1**(3), 321–349 (1988)
21. Brown, J.: Empowering students to create and claim value through the Thomas-Kilmann conflict mode instrument. Negot. J. **28**(1), 79–91 (2012)
22. Schneider, A., Brown, J.: Negotiation barometry: a dynamic measure of conflict management style. Ohio St. J. Dispute Resol. **28**(3), 557–580 (2013)
23. Lerner, J., Li, Y., Valdesolo, P., Kassam, K.: Emotion and decision making. Annu. Rev. Psychol. **66**(1), 799–823 (2015)
24. Kirman, A., Livet, P., Teschl, M.: Rationality and emotions. Philos. Trans. Royal Soc. B Biol. Sci. **365**(1538), 215–219 (2010)
25. Moreland, J., Apker, J.: Conflict and stress in hospital nursing: improving communicative responses to enduring professional challenges. Health Commun. **31**(7), 815–823 (2015)
26. Burns, T., Meeker, L., Buckley, W.: Structural properties and resolutions of the prisoners' dilemma game. Game Theory Theory Confl. Resolut. **19**(5), 35–62 (1974)
27. Grzyl, B., Apollo, M., Kristowski, A.: Application of game theory to conflict management in a construction contract. Sustainability **11**(7), 1–12 (2019)
28. Giebels, E., Janssen, O.: Conflict stress and reduced wellbeing at work: the buffering effect of third-party help. Eur. J. Work Organ. Psy. **14**(2), 1–20 (2004)
29. Kilmann, R.: History of The Thomas-Kilmann Instrument: Developing the TKI Tool. Kilmann Diagnostics (2021). https://kilmanndiagnostics.com/a-brief-history-of-the-thomas-kilmann-conflict-mode-instrument/. Accessed 19 May 2021
30. Whelan, A., Weech-Maldonado, R., Dreachslin, J.: Diversity management in health: cross national organisational study. Int. J. Divers. Organ. Communities Nations **8**(3), 125–135 (2008)

Quick Response in Managing Volatile Demand in the Fashion Industry

Dale Wallington, Yi Wang(✉), and Kareem Elmasry

Plymouth Business School, University of Plymouth, 24 Sutton Road, Plymouth PL4 0HJ, UK
{Yi.Wang,Kareem.Elmasry}@plymouth.ac.uk

Abstract. Firms all over the globe are on a constant mission to speed up their supply chain in order to turn raw materials into finished products in stores as quickly as possible. With modern day supply chains often spanning several continents, and consumers demanding products at the click of a button, there has perhaps never been a more important time to speed this process up in order to gain a competitive advantage. The fashion industry is particularly dependent on having a fast supply chain. With demand and consumer tastes, changing so rapidly, top clothing companies need to be able to adapt almost instantly in order to meet this demand; a supply chain, which takes months from start to finish, simply won't be competitive, plus it is much easier to predict these changing consumer tastes a few weeks in advance compared to several months. A Quick Response (QR) strategy that ensures supply systems can quickly adapt whilst improving their own performance are crucial in a clothing companies' success in an increasingly competitive market.

Keywords: Quick response · Fashion industry · Fashion supply chain · Fast fashion

1 Introduction

In 2019, global retail sales of apparel and footwear reached $1.9 trillion, and is expected to rise to above $3 trillion by 2030 [1]. Wang (2, p. 10) suggests that the fashion industry is changing, with consumers demanding a 'higher quality of product' and 'greater transparency within the supply chain'. With this in mind, firms within the fashion industry must streamline their supply chains in order to remain competitive, especially in an industry, which is synonymous with such rapid change [3]. Sen (4, p. 571) states that 'short product life cycles' mean products are often only 'in-trend' for a short amount of time. Consequently, manufacturers of these fashion items must ensure they have planned in sufficient time to allow their products to be on shelves as these trends arise; though these trends are often impossible to predict, and being too late could have disastrous consequences.

Zara typically sells over 850 million units a year [5]. Compare that to the car manufacturing industry; VW sold 2.8 million units in 2020 [6]. Though these are obvious differences in these products, it is clear to see how a clothing manufacturer could struggle

Y. Wang et al. (Eds.): IWAMA 2021, LNEE 880, pp. 500–506, 2022.
https://doi.org/10.1007/978-981-19-0572-8_64

with such a huge inventory to manage. Traditionally, those in the fashion industry have had to deal with long and inflexible supply processes (Sen, 2008).

Berg et al. [7] agree, stating that firms who can shorten the often lengthy timespan between product development and launching the product can lower the risk of incorrectly guessing market trends. The new phenomenon of 'fast-fashion' means these times have been reduced significantly [8–11] Time-to-react focuses on how quickly the firm can increase/decrease demand based on volatility [3]. Though firms engage in expensive market research in order to predict what their customers will want, there is no better data than real data - what are consumers buying day-by-day. Consequently, new items can be on shelves within 3–4 weeks, as opposed to the norm of 16 weeks [12].

This paper will begin with a literature review on supply chain issues in the fashion industry, mentioning fast fashion. It will also review literature on Quick Response systems, before suggesting why these systems could be the best option to aid the typically slow supply chains within the fashion industry. Any negatives of the system will also be discussed, as well as if there are any other solutions to this problem. The essay will then finish with a brief conclusion.

2 Systematic Literature Review

The term fast fashion was coined as manufacturers sped up the time it took from items to be seen on the catwalk into the stores, and more recently refers to the 'readily available, inexpensively made fashion of today' [13]. More traditionally, the fashion season was split into two distinct parts; January to July as the spring/summer season and August through to December as the autumn/winter season, with huge sales to signify the end of each and to eliminate excess stock [14]. This is no longer the case, and those in the fast fashion industry must attempt to match supply with uncertain demand under increasingly short time frames [15]. Though Purvis et al. [16] states that fast fashion gives the consumer more choices, and consequently increases their shopping frequency, there are numerous articles with negative connotations towards fast fashion. Environmentally, a wealth of articles exist demonizing the industry [13, 17, 18], highlighting its huge carbon footprint and how it increases monocultures. Ethically, fast fashion has been shown to have links with slavery [19]. Whilst this sub-culture of the fashion industry has managed to massively reduce the time-to-market, it still has a long way to go before it is considered sustainable.

Over 20 years ago, [20] wrote how QR was initially developed in order to cut the lead times and adapt to market changes. QR is defined by [21] as a vertical strategy where manufacturers provide retailers with goods or services in the exact quantity required…resulting in minimum inventory levels through the pipeline'. [22] state that QR is most effective in industries where demand is highly volatile, replenishment lead time is long, and the product has a short life cycle. Consequently, and as QR allows supply systems to react quickly to changes whilst improving performance, it's particularly effective within the Fast Moving Consumer Goods (FMCG) and fashion industries [23]. [24] states that suppliers, manufacturers and distributors must communicate together to share information in order to forecast future demand, whilst detecting new opportunities.

An example of a firm who have successfully integrated QR into their practice already is Zara, who are famous for being one of the pioneering fashion firms to integrate QR successfully [25]. As opposed to their biggest rivals, like H&M and Gap, who outsource the majority of their production, Zara manufactures over 50% of its stock in its hometown of Galicia [26], with their 5 million sq.ft headquarters featuring underground 'tubes' that allow products to be sent between warehouses almost instantly [5]. Undoubtedly, Zara could achieve lower production costs overseas, but manufacturing in Spain means they benefit from faster time-to-market, reduced transportation costs, and lower exposure to changing tariffs and politics; their supply risk is lower [27].

Their QR system allows them to change the items in their stores every few weeks, as opposed to every 2–3 months, gaining them huge competitive advantage. Clothing prices are based on market demand as opposed to cost of manufacture. Though some economists may argue that this is a foolish method, Zara sells 85% of its items at full price compared to the industry average of 60% [28]. [5] states Zara have spent 30 years implementing a QR system that's so effective that almost 50% of the clothes they sell is manufactured on a 2–6 week demand forecast. This frees them from getting caught in the bullwhip effect mentioned above.

It should be mentioned that despite them achieving similar gross margins, World.co's net margins were just 2% compared to Zara's 10%, largely due to differing consumer tastes and an unprofitable contractor [29]. This shows that a successful QR strategy alone isn't enough to ensure competitive advantage.

3 Critical Analysis

When implemented correctly, there is a range of benefits available to those in the fashion industry. [30] state QR is particularly important in industries that require high inventory turnover ratio. [31] states managing inventory successfully is crucial for customer satisfaction; stockouts, or being 'out-of-stock', can be disastrous for a fashion firm. Equally, having a huge excess of inventory which could suddenly become off-trend or out of season can equal huge losses [32], highlighting the importance of QR further. By implementing a QR strategy, which is very similar to the Just In Time (JIT) method used by many manufacturers, those in the fashion industry can ensure inventory levels remain accurate and consistent whilst minimizing their 'working capital' - placing them in a stronger position strategically to use their resources.

Another benefit of implementing a QR strategy is its ability to avoid the bullwhip effect; a supply-chain phenomenon that occurs when orders sent to suppliers have a greater variability than the orders received from customers [33]. Firms then have to deal with a variance in larger stocks, extra production capacity, and the need for more storage space [16]. Additionally, QR can result in increased market share and customer loyalty as firms can respond to customers' changing needs almost instantly [24]. Firms who implement QR can often benefit from offering increased customer service, as they're able to tell a customer exactly where their product is in the product line [34]. Lastly, due to shorter order capture times, firms can often benefit from an increased cash flow due to QR as orders are fulfilled more quickly; this can only be beneficial for the firm [35]. An example of a firm who have successfully integrated QR into their practice already is Zara, who are famous for being one of the pioneering fashion firms to integrate QR successfully (Hines & Bruce, 2001). As opposed to their biggest rivals, like H&M and

Gap, who outsource the majority of their production, Zara manufactures over 50% of its stock in its hometown of Galicia [26], with their 5 million sq.ft headquarters featuring underground 'tubes' that allow products to be sent between warehouses almost instantly [5]. Undoubtedly, Zara could achieve lower production costs overseas, but manufacturing in Spain means they benefit from faster time-to-market, reduced transportation costs, and lower exposure to changing tariffs and politics; their supply risk is lower [27].

Their QR system allows them to change the items in their stores every few weeks, as opposed to every 2–3 months, gaining them huge competitive advantage. Clothing prices are based on market demand as opposed to cost of manufacture. Though some economists may argue that this is a foolish method, Zara sells 85% of its items at full price compared to the industry average of 60% [28]. [5] states Zara have spent 30 years implementing a QR system that is so effective that almost 50% of the clothes they sell is manufactured on a 2–6 week demand forecast. This frees them from being caught in the bullwhip effect mentioned above.

Although, it should be mentioned that despite them achieving similar gross margins, World.co's net margins were just 2% compared to Zara's 10%, largely due to differing consumer tastes and an unprofitable contractor [29]. This shows that a successful QR strategy alone is not enough to ensure competitive advantage. With this increased communication and technology between suppliers, some firms worry that sensitive information that would not normally be shared with those down the supply chain could fall into the wrong hands, exposing some of the firm's secrets [36]. Many UK retailers do not have strong links, and therefore do not trust their suppliers (Birtwistle et al., 2003). [37] agree, reinstating the importance of effective communication within QR strategies - something that not all those within a typical supply chain are capable of achieving and maintaining consistently. One weak link within the supply chain could be the difference between fulfilling a huge order or a stockout. Even with all the technology needed to implement QR, a study by [38] found that many firms still had to deal with human error within the supply chain.

Another negative of QR is the cost involved. Purchasing, installing and maintaining the additional technology needed is an expensive process for firms [22]. Additionally, staff need to be trained on often complicated and new systems, which can take long periods of time [39]. [40] argues that QR can actually damage a firm's profits. They state that QR could compromise retailer incentives to exert sales effort, equalling more sales of a competitor's product. [41] identifies extensively about the sustainability issues concerned with implementing a QR strategy. They state that given that production lead-time is so much shorter and that suppliers are under such pressure to meet short deadlines, it is highly unlikely that others in the supply chain will place any importance on their carbon footprint or greenhouse gases, in comparison to if they were not under this pressure.

[42] suggests one way for firms who invest in QR to eradicate this problem is to source their raw materials locally. Several large fashion firms have already committed to this; H&M have a list of guidelines and require all those within its supply chain to meet their environmental and social standards in order to work with them [43]. They also have invested in green technology elsewhere, such as solar panels on the roofs of their factories, in order to mitigate any negative environmental effects QR may bring [44].

[39] found that a major pitfall of implementing a QR strategy is that, from the view of the supplier, QR means that the suppliers carry the costs and the retailers gain the financial benefits. Additionally, [44] found that the long-set up times needed to implement a QR strategy could actually reduce the capacity for production.

4 Conclusions

This paper has conducted a literature review on the topics of fashion industry supply chains, fast fashion, and Quick Response. It then suggested a QR strategy could be effective in managing some of the typical issues found with fashion supply chains, given its benefits and limitations. With these limitations in mind, it then suggested that Agile Supply Chains could also be an efficient method of combating these issues. The fashion industry is particularly affected by slow supply chains for several reasons: high impulse purchasing by shoppers; low predictability; high volatility; and short life cycles. Consequently, firms are constantly trying to predict the latest trends and fashions for the upcoming season, but these lengthy supply chains, where product design to delivery can be 3–4 months, means predicting these patterns is an impossible task. This is where Quick Response can help. An effective Quick Response strategy significantly decreases the time-to-market for a fashion item, minimizing the guess-work of predicting future fashion trends and allowing those in the fashion industry to get a product on the shelves of stores in as little as a couple weeks. Additionally, the technological advancement in QR systems means they are able to accurately predict future demand; meaning firms who previously had to deal with either stock-outs or huge excess stock no longer have to. This, in turn, can increase turnover and customer loyalty, as consumers are more likely to visit and purchase items from stores with constantly changing stock, as opposed to those who only adapt seasonally. QR allows firms to reduce costs whilst improving service simultaneously; something previously considered impossible. However, consistent, effective communication combined with continuous monitoring of the technology involved is the only method to ensure QR remains effective.

References

1. Statista, U.S.: Apparel Market - Statistics and Facts (2020). https://www.statista.com/topics/965/apparel-market-in-the-us/#:~:text=In%202019%2C%20global%20retail%20sales,Asia%2DPacific%20and%20European%20regions. Accessed 26 April 2021
2. Lobina, E.: UK. In: Porcher, S., Saussier, S. (eds.) Facing the Challenges of Water Governance. PSWGPP, pp. 155–188. Springer, Cham (2019). https://doi.org/10.1007/978-3-319-98515-2_7
3. Christopher, M., et al.: Creating agile supply chains in the fashion industry. Int. J. Retail Distrib. Manag. **32**(8), 367–376 (2004)
4. Sen, A.: The US fashion industry: a supply chain review. Int. J. Prod. Econ. **114**(2), 571–593 (2008)
5. Hansen, S.: How Zara Grew Into the World's Largest Fashion Retailer (2012). https://www.nytimes.com/2012/11/11/magazine/how-zara-grew-into-the-worlds-largest-fashion-retailer.html?pagewanted=all. Accessed 26 April 2021

6. Statista. Sales figures for Volkswagen in FY 2020, by key brand (2021). https://www.statista.com/statistics/275868/sales-figures-for-volkswagen-by-brand/. Accessed 26 April 2021
7. Berg, A., et al.: Measuring the fashion world (2018). https://www.mckinsey.com/industries/retail/our-insights/measuring-the-fashion-world. Accessed 26 April 2021
8. Bhardwaj, V., Fairhurst, A.: Fast fashion: response to changes in the fashion industry. Int. Rev. Retail Distrib. Consumer Res. **20**(1), 165–173 (2010)
9. Sull, D., Turconi, S.: Fast fashion lessons. Bus. Strategy Rev. **19**(2), 4–11 (2008)
10. Van Belle, J., Valckenaers, P., Cattrysse, D.: Cross-docking: state of the art. Omega **40**(6), 827–846 (2012)
11. Bruce, M., Daly, L.: Buyer Behaviour for Fast Fashion. J. Fashion Market. Manag. **10**(3), 329–344 (2006)
12. Caro, F., Martínez-de-Albéniz, V.: Fast fashion: business model overview and research opportunities. In: Agrawal, N., Smith, S.A. (eds.) Retail Supply Chain Management. ISORMS, vol. 223, pp. 237–264. Springer, Boston, MA (2015). https://doi.org/10.1007/978-1-4899-7562-1_9
13. Nagurney, A., Yu, M., Masoumi, A.H., Nagurney, L.S.: Fast fashion apparel supply chains. In: networks against time. SpringerBriefs in Optimization. Springer, New York (2013). https://doi.org/10.1007/978-1-4614-6277-4_6
14. Bick, R., Halsey, E., Ekenga, C.: The global environmental injustice of fast fashion. Environ. Health **17**, 92 (2018)
15. Johnston, R., et al.: Cases in Operations Management, 3rd edn., pp. 82–90. Sage, Prentice Hall (2005)
16. Cachon, G., Swinney, R.: The value of fast fashion: quick response, enhanced design, and strategic consumer behavior. Manag. Sci. **57**(4), 778–795 (2011)
17. Purvis, L., et al.: Intermediation in agile global fashion supply chains. Int. J. Eng. Sci. Technol. **5**(2), 38–48 (2013)
18. Niinimäki, K., et al.: The environmental price of fast fashion. Nat. Rev. Earth Environ. **1**, 189–200 (2020)
19. Zamania, B., Sandin, G., Peters, G.: Life cycle assessment of clothing libraries: can collaborative consumption reduce the environmental impact of fast fashion? J. Cleaner Prod. **162**, 1368–1375 (2017)
20. Moulds, J.: Child labour in the fashion supply chain (2020). https://labs.theguardian.com/unicef-child-labour/. Accessed 27 April 2021
21. Iyer, A., Bergen, M.: Quick response in manufacturer-retailer channels. Manag. Sci. **43**, 559–570 (1997)
22. Fiorito, S., Giunipero, L., Yan, H.: Retail buyers' perceptions of quick response systems. Int. J. Retail Distrib. Manag. **26**(6), 237–246 (1998)
23. Choi, T.: Local sourcing and fashion quick response system: the impacts of carbon footprint tax. Transp. Res. Part E: Logistic. Transp. Rev. **55**, 43–54 (2013)
24. Lowson, B., et al.: Quick Response: Managing the Supply Chain to Meet Consumer Demand. Wiley, US (1999)
25. Hadjiconstantinou, E.: Quick Response in the Supply Chain. Springer, Berlin (1999)
26. Hansen, S.: How Zara Grew Into the World's Largest Fashion Retailer (2012). https://www.nytimes.com/2012/11/11/magazine/how-zara-grew-into-the-worlds-largest-fashion-retailer.html?_r=0. Accessed 28 April 2021
27. Hines, T., Bruce, M.: Fashion Marketing - Contemporary Issues. Butterworth Heinemann, London (2001)
28. Vitzthum, C.: Zara's 'Quick Response' Fashion Gives it an Edge Over Competitors (2001). https://www.wsj.com/articles/SB989517840687854661. Accessed 27 April 2021
29. Ivalua. Zara – a Masterclass in Supplier Relationships and Supply Chain Management (2020). https://www.ivalua.com/blog/supply-chain-management-zara/. Accessed 29 April 2020

30. Berfeild, S., Baigorri, M.: Zara's Fast-Fashion Edge (2013). https://www.bloomberg.com/news/articles/2013-11-14/2014-outlook-zaras-fashion-supply-chain-edge. Accessed 28 April 2021
31. Ghemawat, P., Nueno, J.: ZARA: Fast Fashion, Harvard Business School Strategic Management, pp. 1–35 (2006)
32. Golobovante, A., Ribeiro, P., Gomes, C.: Adoption of Quick Response and inventory management in fast fashion: two case studies in the state of Minas Gerais. Gestão Produção **25**(3), 557–567 (2018)
33. Muller, M.: Essentials of Inventory Management. AMACOM, New York (2011)
34. Coppini, M., et al.: Bullwhip effect and inventory oscillations analysis using the beer game model. Int. J. Prod. Res. **48**(13), 3943–3956 (2010)
35. Chatfield, D.C., Kim, J.G., Harrison, T.P., Hayya, J.C.: The bullwhip effect—impact of stochastic lead time, information quality, and information sharing: a simulation study. Prod. Oper. Manag. **13**, 340–353 (2004)
36. Forza, C., Vinelli, A.: Quick response in the textile-apparel industry and the support of information technologies. Integr. Manuf. Syst. **8**(3), 125–136 (1997)
37. Suri, R.: Quick Response Manufacturing: A Companywide Approach to Reducing Lead Times. CRC Press, Boca Raton (2020)
38. Songini, L.: Quick Response (2001). https://www.computerworld.com/article/2591779/quick-response.html. Accessed: 27 April 2021
39. Mazzillo, C.A., Anzanello, M.J.: Sistemática de seleção de variáveis para classificação de produtos em categorias de modelos de reposição. Gestão Produção **22**(1), 201–212 (2015)
40. Birtwistle, G., Siddiqui, N., Fiorito, S.: Quick response: perceptions of UK fashion retailers. Int. J. Retail Distrib. Manag. **31**(2), 118–128 (2003)
41. Birtwistle, G., Fiorito, S., Moore, C.: Supplier perceptions of quick response systems. J. Enterprise Inf. Manag. **19**(3), 334–345 (2006)
42. Krishnan, H., Kapuscinski, R., Butz, D.: Quick Response and Retailer Effort. Manag. Sci. **56**(6), 962–977 (2010)
43. Chan, H., Shen, B., Cai, Y.: Quick response strategy with cleaner technology in a supply chain: coordination and win-win situation analysis. Int. J. Prod. Res. **56**(10), 3397–3408 (2018)
44. Conscious Action Sustainability Report (2014). http://sustainability.hm.com/content/dam/hm/about/documents/en/CSR/reports/Conscious%20Actions%20Sustainability%20Report%202014_en.pdf. Accessed 29 April 2020
45. Shen, B., Zheng, J., Chow, P., Chow, K.: Perception of fashion sustainability in online community. J. Textile Ins. **105**(9), 971–979 (2014)
46. Lehtonen, J., Holmström, J., Slotte, J.: Constraints to quick response systems in the implosive industries. Supply Chain Manag. **4**(1), 51–57 (1999)

RFID Based Markable Passive Sensing System

Haitao Sang(✉), Shifeng Chen, Yongdi Huang, and Jihao Zeng

College of Information Engineering, Lingnan Normal University, Zhanjiang, China
sanght@lingnan.edu.cn

Abstract. With the further integration and evolution of Internet of things and artificial intelligence in theory, method and technology, human perception of the physical world has entered the ubiquitous intelligence stage. In this paper, the passive sensing technology of RFID as an example is analyzed to achieve more accurate and generalized sensing. We first use the passive sensing of backscatter communication. However, we find that the perception mechanism and method are not clear. Therefore, this paper proposes a hybrid sensing method based on RFID. Our idea is to combine the dual characteristics of binding and unbound sensing, which can not only directly sense the tag signal associated with the sensing object, but also use the influence of the sensing object on the tag reflection signal to sense the state change. Through the way of hybrid sensing, we can observe the spatiotemporal correlation of multi tag sensing signals in the tag array, which can realize the further understanding of the perception results. From the observations, we derived hypotheses for future validation.

Keywords: Passive device-free perception · Sensing algorithm · RFID · Influence of fluctuation

1 Introduction

Passive sensing is a new technology in recent years, with its unique perception mechanism and method, it has gradually become the core supporting technology in the field of ubiquitous perceptual computing. Compared with traditional active sensing, passive sensing mainly relies on the energy obtained from the environment to complete the calculation, sensing and communication, without power supply to the terminal node, therefore, it has potential advantages in endurance, deployment, maintenance and other aspects that traditional active sensing cannot match. However, most current passive sensing technologies are based on un taged reflected signals [1, 2], the source of multiple reflected signals cannot be distinguished, therefore, it is impossible to distinguish multiple perceptual objects at the same time, as a result, the scope of application is limited. The emergence of RFID technology provides a new opportunity for the realization of "markable" passive sensing. RFID system realizes the communication between RFID reader and tag based on backscatter mechanism. In the process of backscattering, the continuous wave signal transmitted by the reader antenna is modulated and reflected by the RFID tag, so that the reader can recognize the tag signal effectively. As shown in Fig. 1, using the environment sensitive characteristics of the backscatter communication

Y. Wang et al. (Eds.): IWAMA 2021, LNEE 880, pp. 507–514, 2022.
https://doi.org/10.1007/978-981-19-0572-8_65

mechanism, the RFID system can perceive the specified object based on the dynamic characteristics of the environmental factors carried in the tag reflection signal, such as body behavior recognition, breathing and heartbeat monitoring, etc.

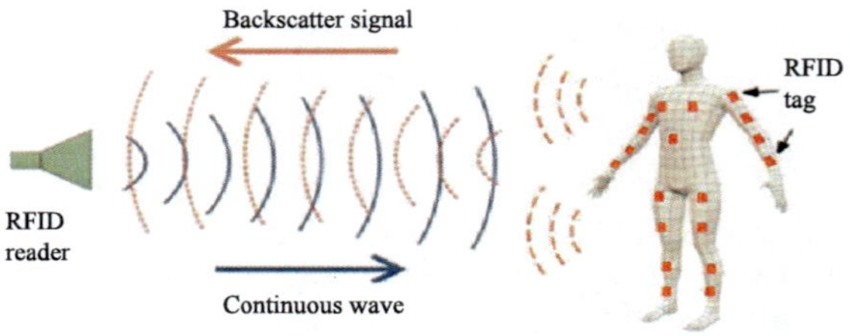

Fig. 1. RFID based markable passive sensing technology

Passive sensing based on RFID has the following challenging problems to be solved:

1. **Lack of theoretical model support for cross domain sensing:** there is still a lack of an effective theoretical perception model to describe the potential relevance of multiple sensing domains and RF signals in time, space, frequency and other signal domains. It is unable to measure the mathematical relationship between the relevant characteristic parameters by quantitative method, so it is unable to effectively guide cross domain perception.
2. **The sensing signal is easily interfered by many factors:** facing the dynamics, complexity and uncontrollability of the real complex environment, the radio frequency sensing signal obtained from the passive sensing RFID system is easily interfered by many factors in the environment, resulting in the relevant signal features being annihilated in the surrounding environmental noise and interference, It directly affects the sensing performance of passive sensing system, which challenges the generalization ability of passive sensing mechanism.
3. **Binding/unbound sensing lacks methodological guidance:** there is still a lack of a mature methodology to guide the core content of binding perception and unbound perception, such as perception model, deployment structure, perception method, collaborative way, etc.

2 RFID Passive Sensing Analysis

In view of the above challenging problems, it is difficult to achieve a comprehensive, comprehensive and thorough passive sensing of the sensing object only relying on the backscattered signal characteristics of a single RFID tag, and it is unable to eliminate the interference caused by environmental noise. We found that in the process of passive sensing in RFID system, multiple RFID tags can be effectively deployed around the sensing objects in a contact or non-contact way, forming a "RFID array" for sensing. Specifically, the new opportunities of RFID passive sensing include the following three aspects (as shown in Figs. 2 and 3):

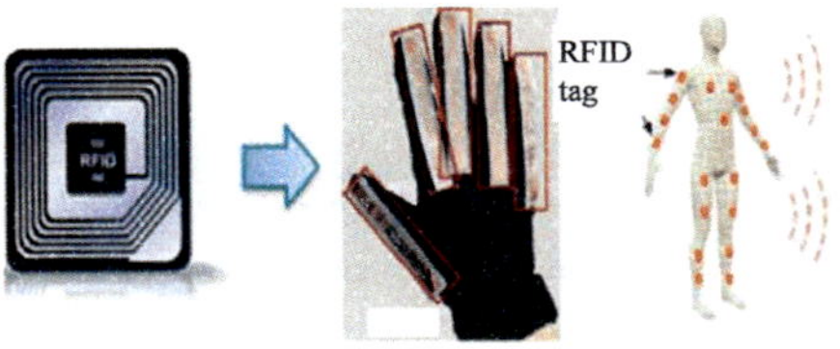

Fig. 2. Single tag to tag array sensing

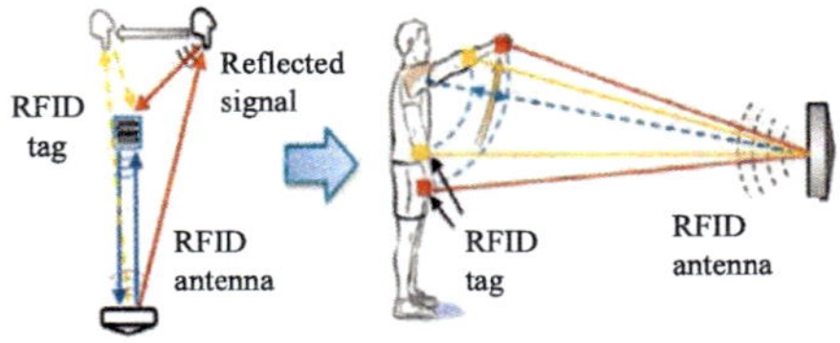

Fig. 3. "Absolute to relative" positioning

2.1 Expand the Perception Dimension

Early RFID passive sensing technology often uses more antennas to realize simple sensing such as positioning and tracking for "single tag", but single tag signal is vulnerable to interference, increasing the number of antennas will also greatly increase the hardware cost and deployment complexity [3]. The perception mechanism based on "tag array" can fully expand the perception dimension in the spatial category, by reasonably designing the topological relationship of multiple tags, the signal characteristics can be "differential" to eliminate the overall interference of external environmental factors on the tag signal [4–8], to combat the impact of dynamic environmental changes, and to improve the stability and reliability of the sensing signal.

2.2 Enhance Perception Sensitivity

The traditional "absolute positioning" method needs to give the absolute coordinates of the tag in space [9]. However, due to the influence of environmental factors, it is difficult to achieve the perception accuracy below centimeter level in the real complex environment. The "relative positioning" method can focus on the relative position relationship between tags in the RFID array, so as to enhance the sensitivity of perception. Based on the results of "relative positioning", for single tag individuals, it can more accurately determine the relative position relationship between tags in space [10], for tag array, by comparing and analyzing the relative position between tags and the existing topology, we can more sensitively capture the changes of tag array, so as to improve the perceptual sensitivity of perceptual objects.

2.3 Expand the Scope of Perception

RFID sensing system mainly relies on the change characteristics of backscatter signal to perceive the objects in the physical environment. This relatively single sensing method is difficult to comprehensively describe the state changes of different sensing domains. Therefore, the fusion of RFID and other modal perception to achieve multi-modal perception can more comprehensively perceive the target object from different perception categories. In addition, by fusing multi-modal signals, with the help of RFID tagging characteristics, it can complement the advantages of different modal sensing technologies, eliminate the "blind area" of sensing, and improve the robustness of sensing.

3 Hybrid Sensing Based on RFID

Binding perception and unbound perception are not opposite to each other. They have their own advantages and disadvantages. If they can be used together according to the specific application requirements, they are bound to play a complementary role.

3.1 Binding Sensing Method Based on Physical Model of Tag Signal

We can build physical models based on signal characteristics according to signal propagation characteristics to realize sensing, such as three-dimensional geometric model based on phase, signal propagation model based on signal strength and so on. The specific physical model can provide a mapping relationship between the reflected signal characteristics of tags and the related states of sensing objects. For example, after the tag array is deployed on the sensing object, a 2D or 3D physical model can be constructed based on the phase change of tags and the existing topological relationship between tags, so as to further decompose the corresponding relationship between target displacement, rotation angle and multi tag phase change.

3.2 Unbound Sensing Method Based on Tag Inductive Coupling

Researchers found that when the two tags are close, there will be inductive coupling phenomenon, and the change of the state of the sensing object in the external environment will further disturb the inductive coupling between tags, and the signal strength and reading speed of tags will be improved. If we use this feature properly, we can achieve accurate target perception. For example, by deploying a pair of double tags with close distance, the influence of human body on the inductive coupling of double tags can be used to determine whether there are abnormal walkers in the environment.

3.3 Unbound Sensing Method Based on Reflected Signal Model

Researchers found that in the process of backscattering, the human body, walls and other objects in the environment will reflect the RF signal, forming a multipath effect, resulting in the tag signal mixed with the characteristics of environmental changes. In this way, we can establish a reflection model to correlate the signal changes with the state changes of the sensing object. For example, by deploying a certain scale of regular tag array, we can sense the position change of human limbs or the trajectory of gestures.

3.4 Unbound Sensing Method Based on Signal Pattern Matching

When the relationship between the state change of the sensing object and the characteristics of the RFID signal is too complex to build an accurate physical model to describe, we can build a data-driven model based on deep learning method, and adopt the pattern matching method according to the signal change law of different actions, Association perception state and signal characteristics of tag array, so as to identify user action and behavior based on temporal and spatial correlation of tag array.

4 Case Studies

In the following, the author introduces some typical cases based on RFID binding perception, taking some related achievements of the research team in recent years as examples.

RFID based body behavior tracking system: The author's research team has developed a RFID based body behavior tracking system RF-Kinect, which uses RFID tags deployed on the body, the signal phase of the array can sense the body movement. In order to eliminate the interference caused by the multi-path reflection of human body, RF-Kinect system adopts the idea of "relative positioning" to extract the features of the phase signal of the tag, and realize the estimation and calibration of the three-dimensional limb angle. As shown in Fig. 4, RF-Kinect divides the limbs with the joints of the human body as nodes, binds two RFID tags on each limb, and uses two antennas facing the human body to scan these tags. First, RF-Kinect uses the phase difference between multiple tags on the same limb to preliminarily estimate the rotation angle of the limb, and then further uses the phase difference of the same tag on different antennas to estimate the AOA (angle of arrival) model to refine the previously estimated three-dimensional limb angle, and the limb posture meeting the constraint conditions will be retained. RF-Kinect further uses the limb angle filtering method based on the relative distance relationship to calculate the relative distance relationship between bones, remove the abnormal angle, and greatly reduce the search space. Finally, the independent postures are combined to recognize the movement behavior of the limbs.

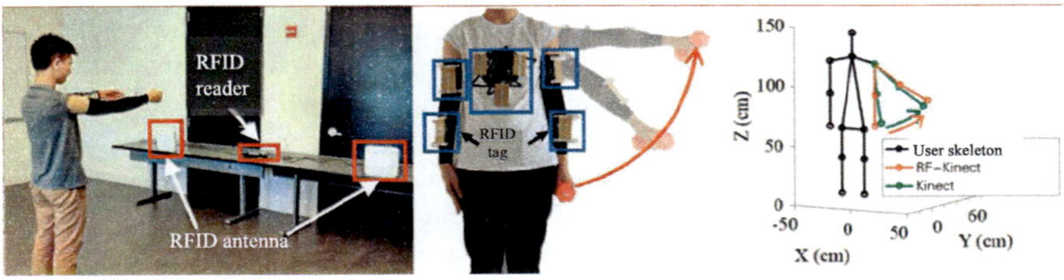

Fig. 4. Body behavior tracking system based on RFID

Gesture micro motion perception system based on RFID: The author's research team proposed a set of RFID based multi finger micro motion perception system RF-glove, which can realize the accurate perception of "micro motion gesture" with the accuracy of 1–2 cm. As shown in Fig. 5, RF-glove attaches passive tags to the five fingers of the glove respectively, and uses RFID antenna to continuously scan multiple tags to collect backscatter signal features. Considering that the relationship between signal feature change and gesture is relatively complex when multiple fingers perform "micro action", RF-glove perception is based on signal change pattern matching. When users perform "micro action" gesture with gloves, RFID reader can obtain a series of phase/signal strength curves. For each type of multi fingered "micro action", RF-glove learns the signal changes of the action through principal component analysis, and obtains a group of n × 5 size of the characteristic curve contour (n is the number of antennas).

Given a multi fingered micro action input, RF-glove compares the phase profile set corresponding to the action with each type of multi fingered micro action template to match the most similar micro action instruction.

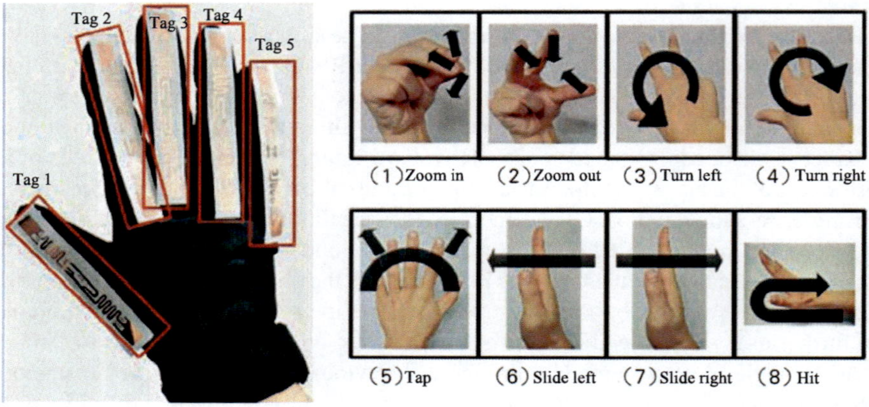

Fig. 5. Gesture micro motion perception system based on RFID

High precision non binding gesture perception system based on RFID: Considering the influence of fingers on the reflected signals of RFID tags, the author's research team has developed a set of RFID based high-precision unbound gesture sensing system RF-finger. RF-finger system is based on the grid tag array deployed in the environment, and uses the "disturbance effect" of fingers to sense the gesture of multiple tag signals. In order to achieve high-precision gesture perception, RF-finger system adopts the idea of "signal fusion", which fuses the different effects of fingers on multiple tags, constructs the signal reflection model, and realizes the accurate tracking of finger position. As shown in Fig. 6, RF-finger deploys a grid tag array in a space of A4 paper size, uses a single antenna to continuously scan tags behind the array, and the user performs gesture operation in front of the array. RF-finger firstly uses the reflection model of the signal to remove the free space signal from the received signal and extract the reflected signal, then, based on the reflection signals of multiple tags, the coarse-grained reflection signals are transformed into fine-grained finger position probability distribution by using the theoretical distribution characteristics of finger reflection signals. Finally, the fine-grained finger position probability distribution and machine learning algorithm are used to track the finger motion trajectory and recognize the common hand gestures. Generally speaking, RF-finger establishes a set of high-precision unbound reflective sensing model based on the grid tag array in the environment, and infers the finger position based on the idea of "signal fusion" combined with the sensing model, and finally realizes the fine-grained sensing of gesture and finger trajectory.

Human vital signs monitoring system based on RFID: The current heart rate sensing or the use of professional medical electrocardiogram (ECG) equipment, deployment is cumbersome and expensive; or using wearable devices or wireless devices for perception, only coarse-grained heart rate information can be measured, and it is difficult to detect premature beat, arrhythmia and other hidden dangers. Considering that human respiration and heartbeat have an impact on RFID signals, the author's research team has developed a set of RFID based human vital signs monitoring system RF-ECG. The RF-ECG system uses a set of RFID tag arrays deployed in the chest to sense respiration and heart rate. Among them, the chest changes of breathing change the position of tag, which is called "movement effect". The diastole and contraction of the heartbeat affect the reflex of the signal, which is called "reflex effect". In order to eliminate the significant interference caused by motion effect, RF-ECG system uses the idea of "signal separation" to sense heart rate in a fine-grained way. As shown in Fig. 7, the RF-ECG system deploys a set of tag arrays at the chest, and uses the antenna in front to continuously scan the tags for ECG sensing. RF-ECG first uses the spatial correlation of multiple tags to sense the changes of chest position in a binding way, and then removes the "motion effect" from the original signal; then, the unbound reflection signal model is used to extract the "reflection effect" caused by heartbeat, and the time correlation of multi tag is used to fuse the multi tag reflection signal to extract the heart rate signal. Finally, the fusion signal of heart rate is segmented by dynamic programming to achieve fine-grained heart rate perception.

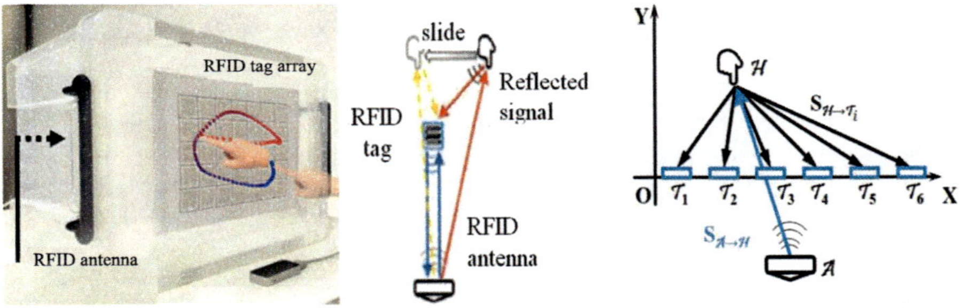

Fig. 6. High precision unbound gesture sensing system based on RFID

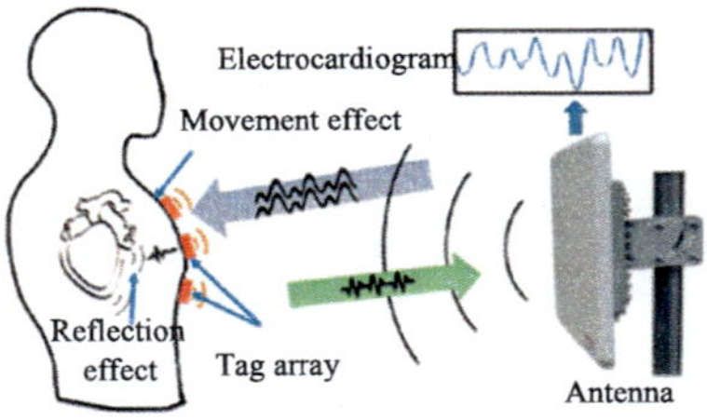

Fig. 7. Human vital signs monitoring based on RFID

5 Discussion and Conclusion

With the further integration and evolution of Internet of things and artificial intelligence in theory, method and technology, human perception of the physical world has entered the stage of "ubiquitous intelligence", and the ubiquitous intelligence framework of "AI + IOT" will further penetrate into various new environmental perception technologies. For the passive sensing technology taking RFID as an example, in order to achieve more accurate and generalized sensing, we need to complete the two tasks of "feature imaging" and "intelligent reasoning". Therefore, we need to further implement the "Data-Driven" reasoning process based on deep learning, reinforcement learning and other methods. Combined with the inspiration and paradigm brought by "model driven", we need to realize an accurate, generalized and robust ubiquitous intelligent perception mechanism based on a large number of taged perceptual data.

Acknowledgment. This work is supported by Zhanjiang Science and Technology Project (Grant no. 2019B01076), Lingnan Normal University Nature Science Research Project (No. ZL2004).

References

1. Wang, F., et al.: Person-in-WiFi: fine-grained person perception using WiFi. In: Proceedings of the IEEE International Conference on Computer Vision, pp. 5452–61 (2019)
2. Zheng, Y., et al.: Zero-effort cross-domain gesture recognition with Wi-Fi. In: Proceedings of the 17th Annual International Conference on Mobile Systems, Applications and Services, pp. 313–25 (2019)
3. Wang, J., et al.: RF-IDraw: virtual touch screen in the air using RF signals. ACM SIGCOMM Comput. Commun. Rev. **44**(4), 235–246 (2014)
4. Bu, Y., et al.: RF-3DScan: RFID-based 3D reconstruction on tagged packages. IEEE Trans. Mob. Comput. **20**(2), 722–738 (2019)
5. Bu, Y., et al.: RF-Dial: An RFID-based 2D human-computer interaction via tag array. In: IEEE INFOCOM 2018-IEEE Conference on Computer Communications, 837–45 (2018)
6. Wang, C,, et al.: RF-kinect: A wearable RFID-based approach towards 3D body movement tracking. In: Proceedings of the ACM on Interactive, Mobile, Wearable and Ubiquitous Technologies, vol. 2, no. 1, pp. 1–28 (2018)
7. Wang, C., et al.: Spin-Antenna: 3d motion tracking for tag array labeled objects via spinning antenna. In: IEEE INFOCOM IEEE Conference on Computer Communications, pp. 1–9 (2019)
8. Gong, Y., et al.: RF-brush: 3D human-computer interaction via linear tag array. In: 2018 IEEE 15th International Conference on Mobile Ad Hoc and Sensor Systems (MASS), pp. 290–8 (2018)
9. Yang, L., et al.: Tagoram: real-time tracking of mobile RFID tags to high precision using COTS devices. In: Proceedings of the 20th annual international conference on Mobile computing and networking, pp. 237–48 (2014)
10. Liu, X., et al.: Accurate localization of tagged objects using mobile RFID-augmented robots. IEEE Trans. Mob. Comput. **20**(4), 1273–1284 (2019)

Design of Remote Monitoring System for Greenhouse Environment

Yufeng Chen[1], Rui Tong[1], Xiue Gao[2](✉), Hang Tao[1], and Bo Chen[2]

[1] School of Electrical and Information Engineering, Hubei University of Automotive Technology, Shiyan 442002, Hubei, China

[2] School of Information Engineering, Lingnan Normal University, Zhanjiang 524048, Guangdong, China

Abstract. This paper analyzes the research status and trend of the greenhouse monitoring system, combined with the flexibility of Bluetooth and the technical advantages of the fully online development of App Inventor, and propose a design scheme for the greenhouse monitoring system based on Bluetooth and App Inventor. Firstly, the overall system design, hardware design and software design scheme are explained in more detail; Secondly, a test environment for the greenhouse monitoring system is set up, and use Bluetooth to realize the communication between the upper and lower computers. System testing shows that the system can realize data collection, display and remote monitoring of environmental parameters, and can automatically solve the situation when parameters are abnormal. It has the advantages of precise and efficient control, simple operation, etc., and improves agricultural production efficiency.

Keywords: Sensor · Data detection · App inventor · STC89C52

1 Introduction

With the development of modern agriculture, the application of greenhouse planting has become more and more extensive. The temperature, humidity, and light intensity of the greenhouse environment have a great impact on crop growth. It is necessary to monitor and control the operation of the corresponding greenhouse equipment to adjust the greenhouse environment. Therefore, higher requirements are placed on the greenhouse monitoring system in terms of data monitoring accuracy, automation and automatic control. There have been many relevant research and practical applications at home and abroad.

Han et al. [1] designed an intelligent drip irrigation monitoring system that can monitor soil moisture, due to the single environmental monitoring parameter, it is difficult to meet the needs of monitoring the diverse environmental parameters of the greenhouse. The greenhouse environment monitoring system designed by Wu [2] also only collects temperature and humidity data, which also has certain limitations. In the monitoring and management of multiple environmental parameters. The intelligent greenhouse control system designed by Jiang et al. [3] can achieve multi-point precise acquisition and control

Y. Wang et al. (Eds.): IWAMA 2021, LNEE 880, pp. 515–522, 2022.
https://doi.org/10.1007/978-981-19-0572-8_66

of parameters such as light intensity, temperature, humidity, and CO2 concentration. Zhao [4] designed the application of intelligent monitoring equipment based on PLC, which can realize the remote monitoring and control of environmental parameters such as temperature and humidity, CO2 concentration.

With the popularization and development of the Internet of Things, a greenhouse monitoring system based on the Internet of Things has also been applied in practice. In the monitoring system, the choice of communication technology is particularly important, including GPRS, WiFi, and ZigBee, etc. The greenhouse remote monitoring system designed by Zhang et al. [5] used ZigBee sensor network to realize remote monitoring and real-time monitoring. Zhang et al. [6] adopted the wireless sensor network (WANs) technology of the fusion of ZigBee and WiFi dual protocols to further realize the precise monitoring of the greenhouse. At the same time, with the popularization of smart phones, greenhouse management systems using mobile phones as the control terminal have also attracted more and more attention. Liu [7] designed a greenhouse remote management system based on Android and ZigBee, Sun et al. [8] designed a greenhouse environment monitoring system based on WeChat platform/WeChat official account, these systems have effectively improved the efficiency and intelligent level of greenhouse management.

As a mobile phone programming software, App inventor is more and more popular and widely used. Sun et al. [9] designed a voice-controlled bookshelf system based on APP Inventor, Huang et al. [10] designed a Bluetooth code lock based on APP Inventor, and Prabhakar M. et al. [11] designed a robot product handling system based on App Inventor. Therefore, this paper designed a greenhouse monitoring system based on Bluetooth and App Inventor, which not only realizes remote monitoring of greenhouses, but also enhances user interaction capabilities and application flexibility.

The rest of this paper is organized as follows: Sect. 2 introduces the system hardware design, Sect. 3 introduces the software design, and Sect. 4 tests the system functions. Discussion and conclusions are summarized in the last section of this paper.

2 System Overall Design

The greenhouse monitoring system is mainly composed of collection modules (temperature sensor, humidity sensor, light sensor, CO2 concentration sensor), LCD display module, Bluetooth module and control module. The block diagram of the greenhouse monitoring system is shown in Fig. 1. This system uses sensors to collect the temperature, humidity, and light data of the greenhouse environment, and transmits the collected data to the STC89C52 single-chip microcomputer after analog-to-digital conversion; After the single-chip data processing, the LCD module displays the current environmental data information; at the same time, In the single-chip microcomputer, the environmental parameter threshold can be set by buttons. In the control module, the collected environmental parameter data is compared with the set threshold to perform corresponding control operations; the Bluetooth module is used to complete the data transmission and reception, and realize the greenhouse Remote monitoring of the management system.

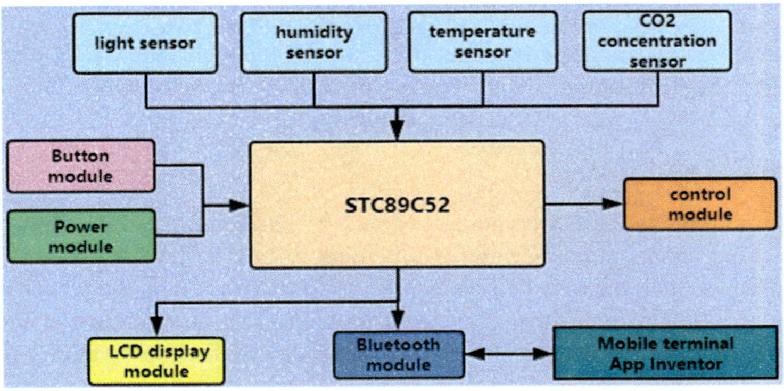

Fig. 1. System composition block diagram

3 System Hardware Design

3.1 Sensor Selection

The system needs to collect the temperature, humidity, light intensity and CO2 concentration of the greenhouse environment, and the sensor selection for collecting these environmental parameters needs to be considered. (1) The temperature sensor DS18B20 is selected for temperature collection. When it uses a single-bus interface to connect to the microprocessor, only one port is needed to realize the two-way communication between the microprocessor and DS18B20. The measurement temperature range is wide and the measurement accuracy is high. (2) The humidity sensor uses a fork-shaped sensor, which is easy to insert into the soil, has small size, low power consumption, fast response speed, strong anti-interference ability, simple control, and highcost performance. (3) The light sensor adopts T88/4-pin sensitive photoresistor sensor. The sensor is a comparator output, with a clean signal, good waveform and strong driving ability; it can be equipped with an adjustable potentiometer to adjust the detection light. (4) The CO2 concentration is collected by the RBY-CO2 sensor, and the analog quantity is output. The voltage value corresponding to the detected gas quantity is changed, and the single-chip microcomputer can directly use the conversion chip to connect. ADC0832 is selected as the digital-to-analog conversion chip, which is used as the peripheral circuit of the main control chip to perform A/D conversion on the collected data, convert the detected analog quantity into a digital quantity and then send it to the STC89C52 single-chip microcomputer.

3.2 Display Design

The display module selects LCD1602 for data display. LCD1602 character liquid crystal display module is a dot matrix liquid crystal display module specially used to display alphanumeric elements, symbols, etc. The so-called 1602 means that when it is displayed, there are 2 lines of content, and each line has 16 characters. The environmental parameters in this article are displayed as upper and lower lines, the temperature data is marked as T, accurate to two decimal places; the light intensity and humidity data are marked as G and S, and both are displayed as percentages, accurate to an integer; the CO2 concentration is displayed as C, The same is the percentage display, accurate to two decimal places.

3.3 Control Module Design

The control module is mainly composed of relays, water pumps, warm lights, etc., in order to achieve corresponding control functions. When the system detects that the humidity is lower than the threshold value, the relay will switch on the power supply of the water pump after receiving the signal from the single-chip microcomputer and control the pumping motor for watering. When the system detects that the humidity meets the conditions, the motor stops working; when the system detects that the temperature is too low, the warm light will turn on until the temperature meets the conditions; when the light intensity is insufficient, the system will power on the light to provide light, and turn off the light when the light is sufficient; when the detected CO2 concentration is lower than the threshold, The buzzer alarms, notifies the user to open the CO2 gas tank and replenish the CO2 in time.

4 Software Design

4.1 MCU Software Design

C language is used for programming and keil4 software is used for compiling. The program includes system initialization, key input program, display program, environment parameter detection program, control response program and Bluetooth driver. After power on, the system is initialized. Press button 1 to enter the threshold setting interface. Press button 3 to enter the parameter display interface. At this time, the display screen displays the current environmental parameters. The parameter values are processed to determine whether the corresponding control is needed, and the status and environmental parameters of the control terminal are uploaded to the mobile phone through Bluetooth, and finally return to the data collection stage to perform the same operation on the processed parameter values.

4.2 Mobile Terminal Application Design

This system uses a programming software called App Inventor to design mobile applications. App Inventor is a visual Android application production platform. Users can dynamically create components and general events, drag and drop components and logic blocks to complete the production of Android applications. In this design, the program design of the mobile terminal is mainly divided into two parts: component design and logic design. The first part is the component design, select the appropriate component to design the application interface, and set the display name, application icon and other attributes of the application installed on the phone. The page design is shown in Fig. 2. And the user can clearly see the environmental parameters of the greenhouse such as temperature, light, and Bluetooth connection status on the mobile phone.

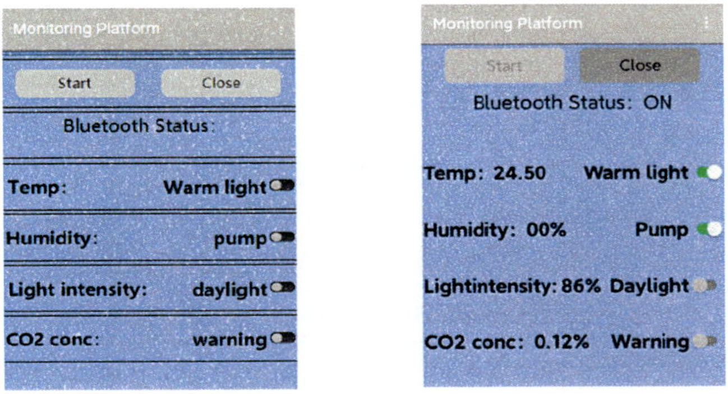

Fig. 2. Component design.

Next is the logic design, which realize of the event logic corresponding to the design component through the design logic block, splicing the functional logic. The logic design is shown in Fig. 3.

5 System Test

The system can complete the functions of real-time acquisition of various parameters, data processing, data display and data transmission, as well as module programming, joint debugging and joint testing. The data analysis of each monitoring parameter is shown in the following figures. Record the data once per second, collect 81s continuously and draw the change curve.

(1) The change curve of humidity is shown in Fig. 4. In order to simulate the soil environment, a system test was performed with a water-soaked paper towel, a humidity sensor was inserted into it. In order to verify the detection and control performance of the system, the water was artificially squeezed out at the 5th second and 54th second. At this time, the curve began to drop. When the humidity dropped to the set threshold of 30%, the water pump began to work. Due to the control is strong, and the humidity continues to decrease. After the water pump works, the humidity starts to rise and returns to the normal range relatively quickly, and the water pump stops working at the same time.

(2) The change curve of temperature is shown in Fig. 5. In order to verify the detection and control performance of the system, reduced the ambient temperature artificially at the 16th second, the curve begins to decline. When the temperature drops to a threshold of 28 °C, the warm light lamp works, the temperature starts to rise. Then the warm light lamp stops working, and the temperature slowly returns to the normal range.

(3) The change curve of Light intensity is shown in Fig. 6. In order to simulate the lack of light at sunset, the sensor is artificially blocked at the 1th second and 54th second. At the same time, the curve begins to decline. When the light intensity drops to the

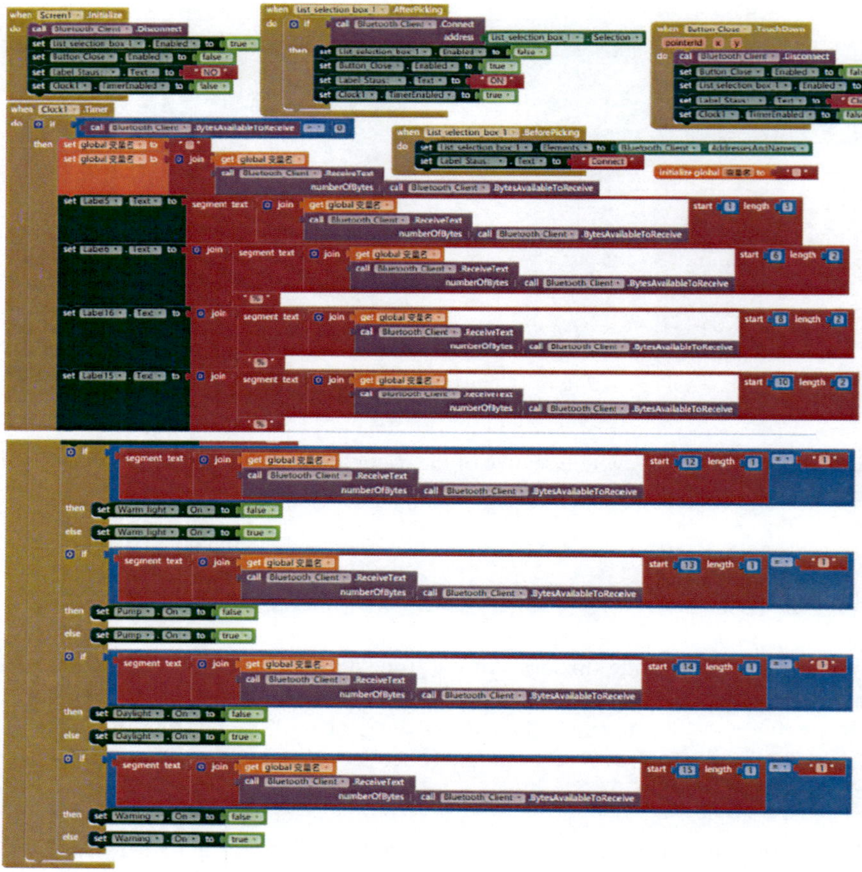

Fig. 3. Logic design.

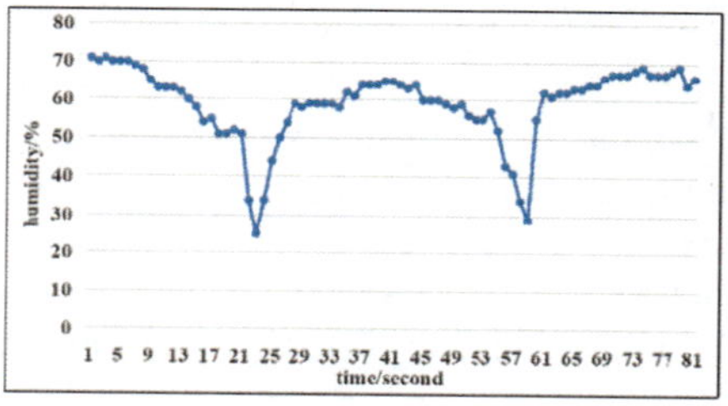

Fig. 4. Humidity curve.

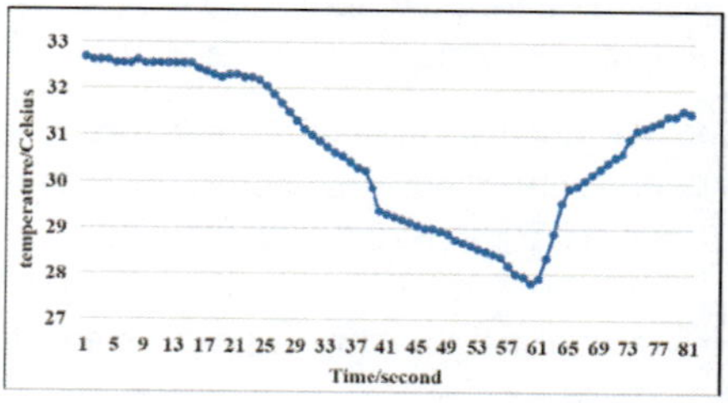

Fig. 5. Temperature curve.

set threshold of 25%, the relay closes, driving the fluorescent lamp to work. Due to the light is strong, and the curve rises rapidly. After removing the obstruction, the relay closes and the fluorescent lamp stops working. In the second experiment, changing the threshold to 30%, the same phenomenon can be observed.

(4) The change curve of CO2 concentration is shown in Fig. 7. Put the sensor into a small bottle containing CO2 to simulate a greenhouse environment, CO2 was released artificially at the 24th second and 54th second. When it fell to the threshold of 0.1%, the buzzer started to work. After a proper amount of CO2 was added, the CO2 concentration began to rise and the buzzer was turned off.

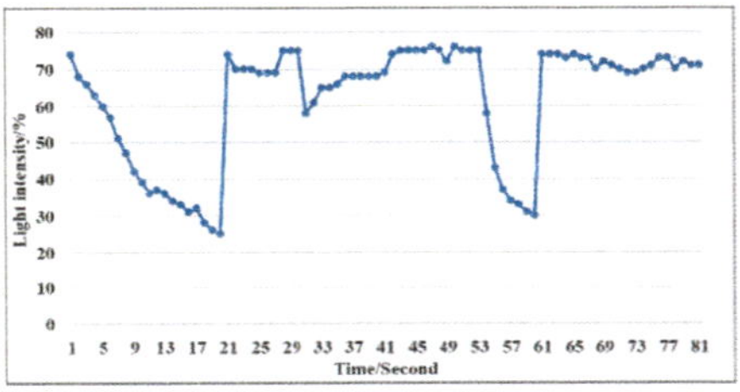

Fig. 6. Light intensity curve.

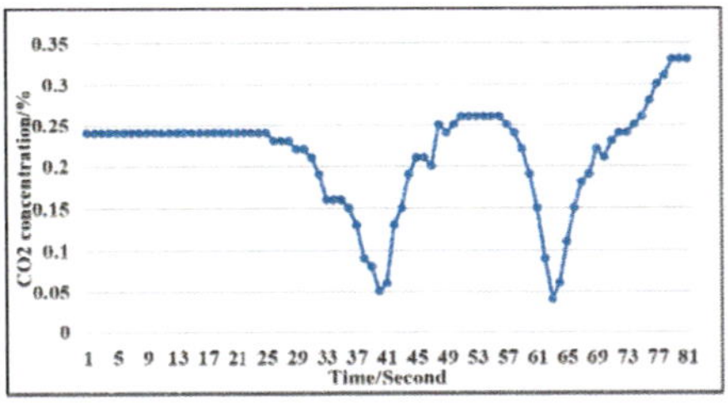

Fig. 7. CO2 concentration curve.

6 Conclusion

The system realizes real-time data collection, monitoring and control of the humidity, temperature, light intensity and CO2 concentration of the greenhouse soil. When the environmental parameters such as humidity and light do not meet the requirements, the system automatically controls the corresponding device to start working, and completes working on time and on demand. The system has the advantages of low cost, strong practicability, and strong expandability, and can be applied to data monitoring and control in other fields.

Acknowledgment. This work is partially supported by Science and Technology Research Project of Hubei Provincial Department of Education under Grant Q20161805.

References

1. Han, W.C., Huang, L.S., Pan, N.H.: Design of intelligent drip irrigation system based on STM32. J. Anhui Agri. **45**(18), 167–168 (2017)
2. Wu, B.: Design of greenhouse environment monitoring system based on single chip microcomputer. Comput. Knowl. Technol. **14**(24), 246–248 (2018)
3. Jiang, J., Yue, Y.D.: Design of intelligent greenhouse control system. Autom. Appl. **01**, 33–35 (2018)
4. Zhao, H.: Application of PLC-based intelligent monitoring equipment for agricultural greenhouses. Agri. Mech. Res. **43**(06), 214–218 (2021)
5. Zhang, T.H., Liu, X.F., Qu, B.H., Jia, Y.P.: Design on remote monitoring and control system for greenhouse based on ZigBee. J. Chongqing Univ.Technol. **34**(06), 200–204 (2020)
6. Zhang, H.Z., Cao, J.T., Shao, P.F., Zhang, M.Y.: Research on WANS-based Intelligent water saving Irrigation Control System for Greenhouse. Control Engineering of China. **26**(01), 108–113 (2019)
7. Liu, J.: Design and Implementation of Intelligent Greenhouse Environment Remote Control System Based on Android. Hunan University (2019)
8. Sun, H.Q., Chen, J.X., Cao, Y.X., Zhang, X.: (2020) Design of vegetable greenhouse environment monitoring system based on wechat. Electr. Test **3**, 74–75 (2020)
9. Sun, M., Lan, X.H., He, Z.L., Wei, Y.: Design and implementation of speech control bookshelf robot based on Bluetooth. Modern Electr. Tech. **43**(10), 179–183 (2020)
10. Huang, H.Y., Kang, S.Y., Fu, S.L., Wang, Y.H.: Design of bluetooth password lock for smart phone. Int. Things Technol. **10**(03), 108–111 (2020)
11. Prabhakar, M., Paulraj, V., Kannappan, D.A.K., Dhanraj, J.A., Ganapathy, D.: IOP conference series: materials science and engineering. In: RIACT 2020, Chennai, India, 2–3 October, p. 012003 (2020)

Simulation of Phase-Shift Full-Bridge Based on Dual-Loop Competitive Control Mode

Xuehuan Jiang, Lei Zhang, Jinliang Zhang(✉), Guosheng Peng, and Yufeng Chen

School of Electrical and Information Engineering, Hubei University of Automotive Technology, Shiyan 442002, Hubei, China

Abstract. This paper introduces the topology of the phase-shift full-bridge converter circuit, briefly analyzes the working principle of the circuit, and compares it with the buck converter. The mathematical model and small signal circuit model of the phase-shift full-bridge converter are derived by introducing the mathematical model of the buck converter. After analysis, the double-loop competitive control model based on the voltage closed-loop PI regulator and the current closed-loop PI regulator of the phase-shift full-bridge converter is established, and this control mode is added to the Simulink model of the phase-shift full-bridge converter. In the Simulink model, the output data of the closed-loop control system are observed by changing the load of the converter, and the data are compared and analyzed in a graphical way. It is verified that the double-loop competitive control mode improves the dynamic response speed and work efficiency of the phase-shift full-bridge converter system, which has an important reference value for the control of the phase-shift full-bridge converter system.

Keywords: Phase-shift full-bridge converter · Double-loop competitive control model · PI regulator · Simulink model

1 Introduction

Compared with other DC-DC converters, phase-shifted full-bridge (PSFB) converter has many advantages, such as fixed working frequency, simple control, low switching loss and high reliability [1]. Therefore, PSFB converter has been applied in many high power occasions. In these applications, the feedback control commonly used in the PSFB converter includes four control modes: single voltage loop feedback control, current loop feedback control, double closed-loop control and double-loop competition control [2, 3]. Double-loop competitive control mode combines the advantages of simple structure and fast dynamic response of the other three control methods [4].

In this paper, the voltage closed-loop feedback PI controller and the current closed-loop feedback PI controller are built using Simulink platform in MATLAB environment, and they are combined into a closed-loop control system with double-loop competitive control mode, which is applied to the specific PSFB converter simulation model, and two comparative tests are carried out. The experiment verifies that the PSFB converter model based on the dual-loop competitive control mode has fast response speed and small overshoot, which meets the control requirements of the PSFB converter.

Y. Wang et al. (Eds.): IWAMA 2021, LNEE 880, pp. 523–530, 2022.
https://doi.org/10.1007/978-981-19-0572-8_67

2 Small Signal Model Analysis of PSFB Converter

2.1 Topology Analysis of PSFB Converter

The circuit topology of the PSFB converter is shown in Fig. 1. S_1 and S_2 constitute the leading bridge arm, S_3 and S_4 constitute the lagging bridge arm. The PWM waveform of driving the leading bridge arm and the lagging bridge arm is different by a phase-shift angle [5]. This conduction mode makes the junction capacitance of the power switch resonant discharge, so that the anti-parallel diode of the power switch is conducted ahead of the power switch, so as to realize the zero voltage switching of he power switch [6, 7].

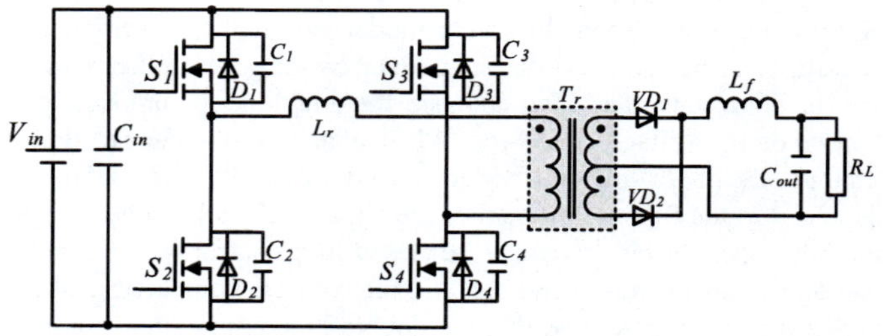

Fig. 1. Circuit topology of PSFB converter

2.2 Mathematical Model Analysis of PSFB Converter

The PSFB converter is essentially a Buck converter, but the leakage inductance of the transformer (T_r) of the PSFB converter will lead to the loss of duty cycle during the operation of the converter, which leads to the difference between the PSFB converter and the Buck converter [8, 9]. Therefore, the small signal model of PSFB converter is derived by introducing the small signal equivalent circuit of Buck converter. The ideal circuit structure of buck converter is shown in Fig. 2.

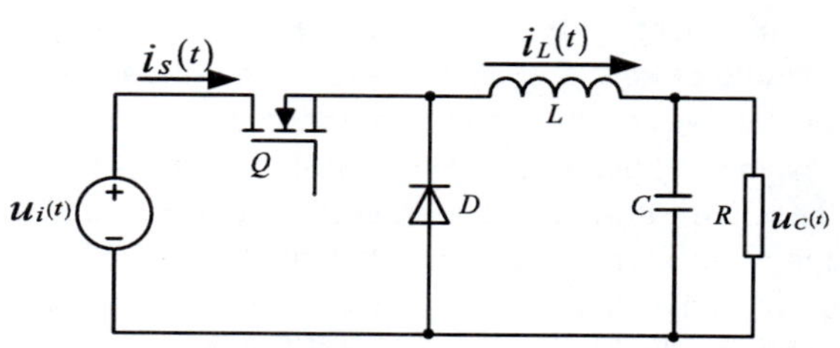

Fig. 2. The ideal circuit structure of buck converter

In the continuous mode of inductor current, the state equation of Buck converter in the state of switch (*Q*) on and diode (*D*) off as Eq. (1) and Eq. (2):

$$\begin{bmatrix} \dot{i}_L(t) \\ \dot{u}_C(t) \end{bmatrix} = \begin{bmatrix} 0 & -\frac{1}{L} \\ \frac{1}{C} & -\frac{1}{RC} \end{bmatrix} \begin{bmatrix} i_L(t) \\ u_C(t) \end{bmatrix} + \begin{bmatrix} \frac{1}{L} \\ 0 \end{bmatrix} [u_i(t)] \tag{1}$$

$$\begin{bmatrix} i_s(t) \\ u_c(t) \end{bmatrix} = \begin{bmatrix} 1 & 0 \\ 0 & 1 \end{bmatrix} \begin{bmatrix} i_L(t) \\ u_C(t) \end{bmatrix} + \begin{bmatrix} 0 \\ 0 \end{bmatrix} [u_i(t)] \tag{2}$$

The state equation of Buck converter under the condition of switch (*Q*) off and diode (*D*) conduction as Eq. (3) and Eq. (4):

$$\begin{bmatrix} \dot{i}_L(t) \\ \dot{u}_C(t) \end{bmatrix} = \begin{bmatrix} 0 & -\frac{1}{L} \\ \frac{1}{C} & -\frac{1}{RC} \end{bmatrix} \begin{bmatrix} i_L(t) \\ u_C(t) \end{bmatrix} + \begin{bmatrix} 0 \\ 0 \end{bmatrix} [u_i(t)] \tag{3}$$

$$\begin{bmatrix} i_s(t) \\ u_c(t) \end{bmatrix} = \begin{bmatrix} 0 & 0 \\ 0 & 1 \end{bmatrix} \begin{bmatrix} i_L(t) \\ u_C(t) \end{bmatrix} + \begin{bmatrix} 0 \\ 0 \end{bmatrix} [u_i(t)] \tag{4}$$

The Laplace transform of the state space average equation obtained from the above equation through the state space average method is shown in Eq. (5):

$$\begin{cases} \hat{i}_s(s) = D\hat{i}_L(s) + \frac{Du_i}{R}\hat{d}(s) \\ sL\hat{i}_L(s) = D\hat{u}_i(s) + u_i\hat{d}(s) - \hat{u}_o(s) \\ \hat{u}_o(s) = \hat{u}_c(s) = \frac{R}{1+sRC}\hat{i}_L(s) \end{cases} \tag{5}$$

The small-signal equivalent circuit of Buck converter with Laplace transform is shown in Fig. 3 below, and the open-loop transfer function of the converter is derived from the small-signal equivalent circuit as shown in Eq. (6):

$$G_{vd}(s) = \frac{\hat{v}(s)}{\hat{d}(s)} = \frac{U_i}{s^2LC + s\frac{L}{C} + 1} \tag{6}$$

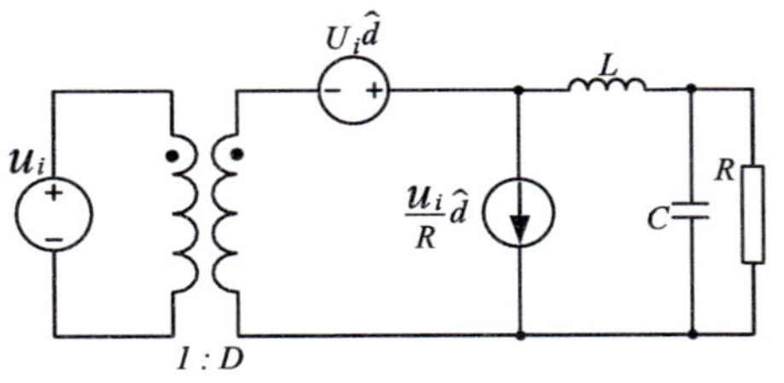

Fig. 3. Small signal equivalent circuit of buck converter

The mathematical models of the PSFB converter and the Buck converter are not exactly the same [10]. By analyzing the duty cycle loss of the PSFB converter, the effective duty cycle (D_{eff}) is calculated, as shown in Eq. (7), thus the mathematical model of the PSFB can be derived.

$$D_{eff} = D - \frac{2NL_r}{U_i T}\left[2I_L - \frac{U_o}{L_f}(1 - D)\frac{T}{2}\right] \tag{7}$$

N is the turn ratio of transformer, T is the period of PWM, D is the duty ratio of PWM. By analyzing Eq. (6) and Eq. (7), the small-signal equivalent circuit and open-loop transfer function of the PSFB converter can be derived [11]. The small signal equivalent circuit diagram of the PSFB converter is shown in Fig. 4, and the open-loop transfer function is shown in Eq. (8) and Eq. (9).

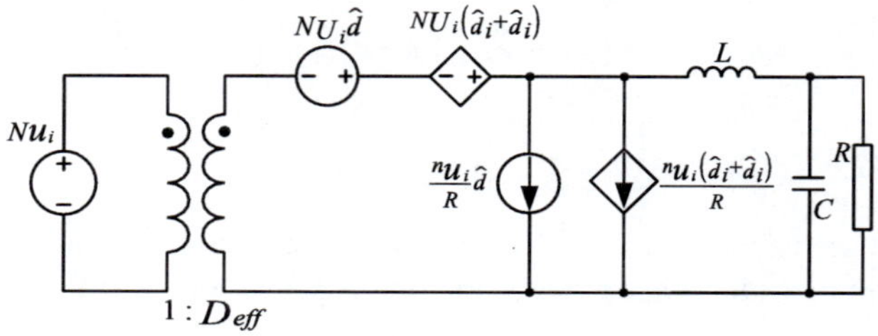

Fig. 4. Small signal equivalent circuit of PSFB converter

$$G_{vd}(s) = \frac{NU_i}{s^2LC + s\left(\frac{L}{R} + R_d\right) + \frac{R_d}{R} + 1} \tag{8}$$

$$G_{id}(s) = \frac{NU_i}{\frac{R(LCs^2 + s\frac{L}{R} + 1)}{1 + sRC} + R_d} \tag{9}$$

$R_d = 4N^2L_rf_s$, where f_s is the frequency of PWM.

3 Analysis of Dual-Loop Competition Control Model

Both the single voltage loop feedback control and the current loop feedback control can make the output of the PSFB relatively stable. In this paper, the current loop and the voltage loop are competitively controlled to control the PSFB converter. Compared with the former two-loop competitive control, the response speed is faster and the overshoot is smaller. The control mode of double-loop competition mode is shown in Fig. 5.

The control mode of dual-loop competition mode is essentially the single closed-loop control of the converter under different states. When the converter operates in steady state, only one loop works and the other is saturated. The voltage closed loop and current closed loop are calculated separately, and the results are compared. The PWM duty cycle is controlled by the output of the loop with small value. If the output voltage is greater than the given voltage, the output current is smaller than the given current, and the current

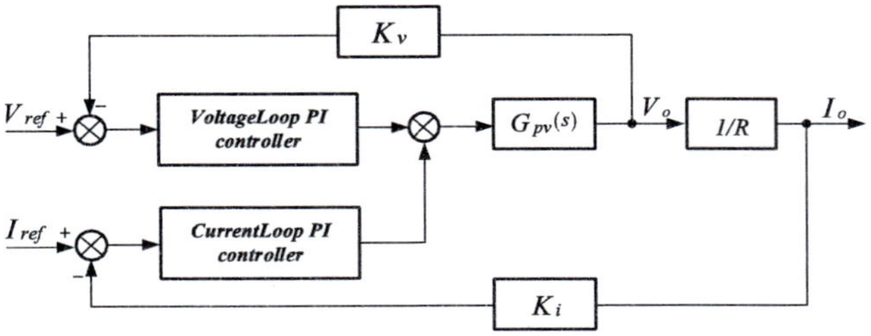

Fig. 5. Control mode block diagram of double-loop competition mode

loop achieves the maximum saturation output through integration, while the output of the voltage loop decreases. The voltage loop outputs the actual control output, and works at the constant voltage output. If the output voltage is less than the given voltage, the output current is greater than the given current, and works in the constant current loop. If the output voltage and output current are greater than the given value, then no matter which loop output is smaller, the output of the loop is smaller until the output voltage or output current is less than the given value, a small range of fluctuations near the given value [12, 13].

The simulation diagram of PSFB dual-loop competitive control mode is shown in Fig. 6 below. Through Eq. (8) and Eq. (9), the dynamic response of the open-loop state of the PSFB is analyzed, and the voltage closed-loop PI regulator and the current closed-loop PI regulator are designed. After the acquisition circuit collects the output voltage and current values, the corresponding error is obtained after the calculation of the set voltage reference value and current reference value. The smaller error is selected as the control variable to control the duty cycle transformation of PWM, so as to achieve the purpose of controlling the output stability of the converter.

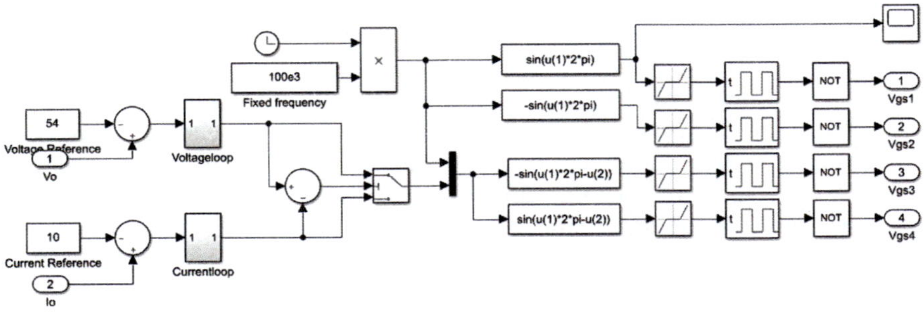

Fig. 6. Double-loop competition control model

4 Modeling of PSFB Converter with Dual-Loop Competitive Control Mode

After the analysis and derivation of the mathematical model of the PSFB converter and the modeling of the dual-loop competitive control mode, the simulation model of the PSFB converter with the dual-loop competitive control mode is built on the Simulink platform of MATLAB. The simulation model is shown in Fig. 7. The response of the converter under different conditions is simulated by changing the load of the converter. The model simulation time is 0.2 s, and the load is changed at 0.03 s.

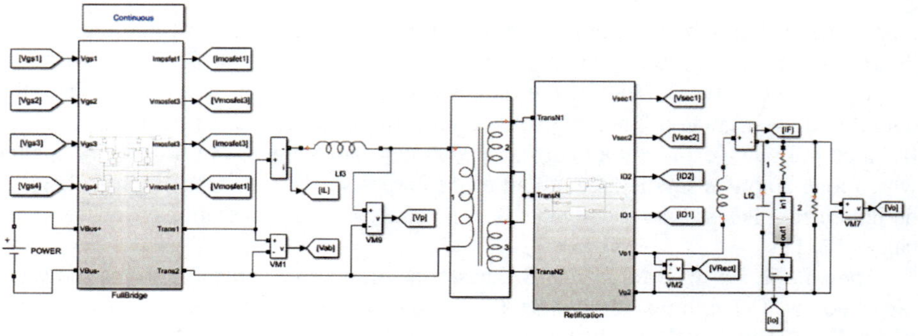

Fig. 7. Simulation model diagram of PSFB converter with double-loop competitive control mode

5 Simulation Test and Data Analysis

(1) Data analysis of closed-loop control mode without load change

The output voltage curve and inductance current of the PSFB converter model based on double-loop competitive control mode are shown in Fig. 8. The transverse axis is time, the longitudinal axis is voltage value and current value, the red curve is the output voltage curve, and the black curve is the filter inductance current curve. The output voltage of the converter rises to a given value at a relatively slow speed at the initial stage of operation, and the overshoot is small. After the inductance current increases at the maximum speed, it rapidly decreases to a given current value. The whole adjustment time is about 0.01 s, and the whole control process time is about 0.02 s, which meets the requirements of converter control.

(2) Data analysis of closed-loop control mode of load change

The output voltage curve and inductance current of PSFB converter model based on double-loop competitive control mode are shown in Fig. 9. In the figure, the horizontal axis is time, and the vertical axis is voltage value and current value. The simulation time is 0.2 s, and the load is changed at 0.03 s. The red curve is the output voltage curve, and the black curve is the filter inductor current curve. The output voltage of the converter rises to a given voltage at a relatively slow speed at

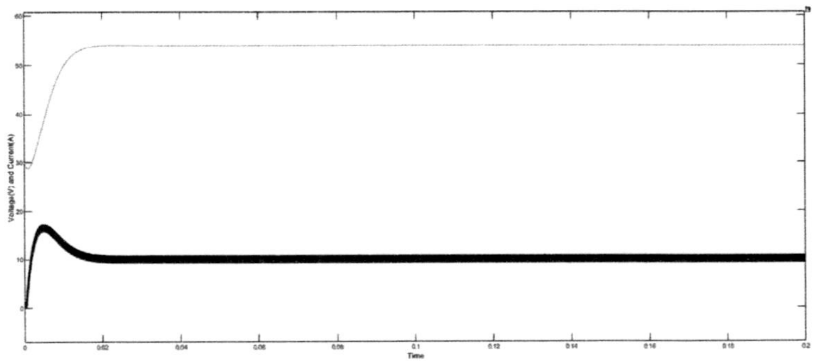

Fig. 8. Output curve without load change

the initial stage of operation, and the inductance current rises at the maximum speed and then decreases rapidly to a given current value. The load is switched at 0.03 s. At this time, the output voltage drops suddenly and quickly returns to the given value. After the inductor current increases, the maximum current is maintained to stabilize the output. The regulation time of double closed-loop competitive mode control is about 0.01 s, and the overshoot is small. The analysis shows that the dynamic response of the closed-loop control system is good. This control method has important reference value for the closed-loop control of PSFB converter.

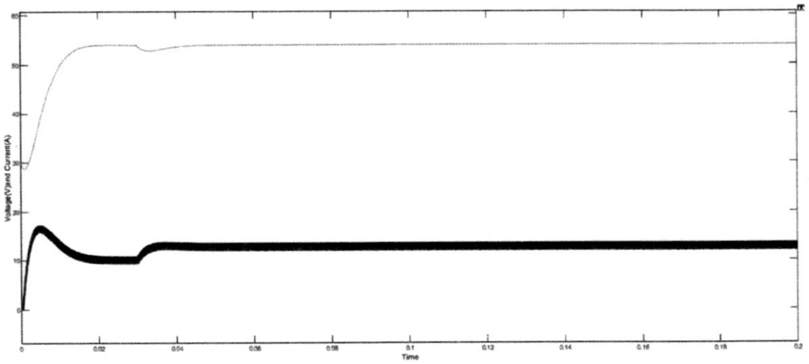

Fig. 9. Output curve of load change

6 Conclusion

In this paper, the topology of the PSFB converter is briefly analyzed and introduced, and the mathematical model of the PSFB converter is derived by introducing the Buck converter through reasoning method. On this basis, the small-signal circuit model of the PSFB converter and the open-loop transfer function of the system controlled to the output are obtained. The parameters of the closed-loop PI regulator are obtained by analyzing the system transfer function. Finally, the PSFB converter model of the double-loop competitive control mode is built on the Simulink platform, and it is verified that the dynamic response and efficiency of the system can be effectively improved by using the double-loop competitive control mode of the PSFB converter, which has an important reference value for the control of the PSFB converter suitable for low voltage and high current situations.

Acknowledgment. This work is partially supported by Science and Technology Research Project of Hubei Provincial Department of Education under Grant Q20161805, Local Science and Technology Development Project Guided by Central Government under Grant 2018ZYYD007.

References

1. Ghanbari, A.R., Raie, A.: General multi-mode control method for optimising the efficiency of full bridge converter based on prominent control modes and transition points calculation. IET Power Electr. **12**(8), 2038–2048 (2019)
2. Srivastava, M., Verma, A.K., Tomar, P.S.: Design and implementation of a novel auxiliary network based ZVS DC/DC converter topology with MPWM: an application to electric vehicle battery charging. IET Power Electr. **12**(13), 3340–3350 (2019)
3. Pellitteri, F., Miceli, R., Schettino, G., Viola, F., Schirone, L.: Design and realization of a bidirectional full bridge converter with improved modulation strategies. Electronics **9**(5), 724 (2020)
4. Divya Navamani, J., Vijayakumar, K., Lavanya, A.: FPGA-based digitally controlled isolated full-bridge DC-DC Converter with voltage doubler (IFBVD). Indian J. Sci. Technol. **9**, 16 (2016)
5. Huang, Z.Q., Song, J.G., Sun, M.J.: The design of phase shift soft switching full-bridge converter power supply based on UCC3895. Appl. Mech. Mater. **511–512**, 1141–1146 (2014). https://doi.org/10.4028/www.scientific.net/AMM.511-512.1141
6. Inba Rexy, A.,Seyezhai, R.: Design and simulation of active triple port full-bridge DC-DC converter for renewable energy source. Programm. Device Circ. Syst. **4**(4), 228–233 (2012)
7. Kim, J.-M., Lee, J., Ryu, K., Won, C.-Y.: Power device temperature-balancing control method for a phase-shift full-bridge converter. Energies **13**(7), 1623 (2020)
8. Escudero, M., Kutschak, M.-A., Meneses, D., Rodriguez, N., Morales, D.P.: A practical approach to the design of a highly efficient psfb DC-DC converter for server applications. Energies **12**(19), 3723 (2019)
9. Sevilay, C.: High efficiency design procedure of a second stage phase shifted full bridge converter for battery charge applications based on wide output voltage and load ranges. J. Power Electron. **18**(4), 975–984 (2018)
10. Le, T-.T., Park, M.-W., Yu, I.-K.: Current controller design of a phase shift full bridge converter for high current applications with inductive load. J. Korea Ind. Inf. Syst. Res. **23**(1), 43–52 (2018)
11. Tran, V.-L., Tran, D.-D., Doan, V.T., Kim, K.-Y., Choi, W.: A novel hybrid converter with wide range of soft-switching and no circulating current for on-board chargers of electric vehicles. J. Electr. Eng. Technol. **13**(1), 143–151 (2018)
12. Design of a Novel Integrated L-C-T for PSFB ZVS Converters. J. Power Electron. **17**(4), 905–913 (2017)
13. Zhao, L., Chuangyu, X., Zheng, X., Li, H.: A dual half-bridge converter with adaptive energy storage to achieve zvs over full range of operation conditions. Energies **10**(4), 444 (2017)

Conceptual Modelling and Topology Optimization Framework of Tower Crane Hook: A Case Study

Firankor T. Daba[1] and Hirpa G. Lemu[2(✉)]

[1] College of Engineering, Asosa University, Asosa, Ethiopia
[2] Faculty of Science and Technology, University of Stavanger, Stavanger, Norway
Hirpa.g.lemu@uis.no

Abstract. A crane a mechanical system operating under continuously varying load that exposes the crane components to fatigue failure, among others, the dynamic loads cause serious damages to the crane hook leading to accidents because most construction sites are very confined and close to the general public. Tower crane accidents not only hazard to workers in construction sites but also pedestrians within the vicinity. Therefore, a close study of the hook to enhance the strength and endurance requirements is found important. To develop the conceptual framework of design optimization and then apply topology optimization, designxplorer is used. Topology optimization technique is employed to optimize the material distribution in the crane hook to be designed. The objective is to find an optimized design of the crane hook without sacrificing the strength and durability. The topology optimization conducted as part of this paper reduced the mass by 6.69% (mass reduction, of crane hook from 15.75 kg to 13.678 kg). Simulation of the hook was done using the topological approach, where the model was created and then meshing was done in finite element analysis tool (ANSYS WB 19.2). The main contribution of this study is to investigate the possible methods of optimizing the strength and endurance required of the crane hook.

Keywords: Topology optimization · Structural optimization · Crane hook · Stress concentration area

1 Introduction

Due to its irregular curved beam shape, design development of a hook is a long process demanding a number of tests to validate the design and manufacturing variables. The continuous loading and unloading of a crane [1] subjects the hook to dynamics loads. Furthermore, one of the design methods of machine components with surface irregularities such as the crane hook structure is stress concentration factor approach, which is widely employed in durability evaluations and strength calculations [2].

Y. Wang et al. (Eds.): IWAMA 2021, LNEE 880, pp. 531–539, 2022.
https://doi.org/10.1007/978-981-19-0572-8_68

A hook block allows for a considerable amount of flexibility and safety in lifting operations as opposed to a direct connection. One of the most important functions of any hook block is facilitating a free turning or rotating hook arrangement. When loads are lifted, it is often necessary to turn the load to position it in a new location or to avoid striking obstructions. A crane hook attached directly to the hoist ropes would cause the ropes to twist if the load was turned from its original orientation. This would have several undesirable effects such as overstressing the ropes and boom pulleys, creating an unbalanced load, and causing the load to swing back in an uncontrolled fashion when released. A hook block allows loads to be freely rotated without changing the orientation of the hoist ropes. Minimizing the possible failure of a crane hook requires calculation of the induced stresses in the localized high stress areas.

Crane hooks can be exposed to premature failures owing to the accumulation of some local stresses that eventually lead to its failure. As a result, many industrial disasters are attributed to the catastrophic failure of crane hook. In most cases, cross-sectional changes, rough-machining marks and chatter marks are observed near the location of the hook failure.

Pavlović et al. [3] presented the analysis and optimization of T-cross section of the crane hook based on its geometric parameters. In this study, the optimization (objective) function is defined as the minimization of the cross-sectional area, while allowable stresses in the crane hook critical points of the structure are used as the constraint functions. In addition, some geometric constraints have been used.

Analytical calculations of the hook is often done by modifying the cross-section instead of the trapezoidal section. Thus, the analysis and the optimization work are based on this modified section, which lead to certain inaccuracies. In optimizing the design, in addition to changing the cross-sectional area, material is removed from the low-stress concentration area in the lifting hook and the design stress is compared [4].

Topology optimization (TO), as a mathematical technique is used to find a material distribution layout with even distribution of loading within the design space and hence attempts to reduce the mass of the structure for a given set of loads and boundary conditions. In design optimization of crane hook, the geometrical properties are used as optimization criteria [5]. Today, eventhough there are diverse commercial tools such as Hyper works optiStruct solver and Altair Inspire [6], Autodesks Fusion 360 [7], etc., there exists no general purpose tool that can address the general industrial problem.

To demonstrate the application TO technique, a crane hook of a high carbon steel AISI 4340 material was studied in [8] using finite element method under a loading of 2 tons whose model was developed CATIA V5. The study result showed the region of low stress that should be subjected to material removal material.

Topology optimization is a powerful approach for determining the best distribution of material within a defined design domain [9], which can be defined using shell or solid elements or both. The classical TO is set up to solve the minimum compliance problem, as well as the dual formulation with multiple constraints is available. Constraints on von-Mises stress and buckling factors are available with limitations [10]. TO is different from shape optimization because shape optimization methods work in a range of allowable shape which has fixed topological properties. TO can be implemented through the use of finite element methods for the analysis and optimization techniques based on the Homogenization method [11], level set [12, 13], optimality criteria methods [14, 15], etc. The objective of this article is to report the work to develop a conceptual framework for modelling and topology optimization of a tower crane on a case study conducted at Afro-Tsion construction [16] site.

2 Materials and Methods

2.1 Dimensions and Design Specifications of the Hook

The hook dimensions were determined for a load varying between 5 to 12.5 tons. Circular, rectangular and trapezoidal sections are often considered (Fig. 1 [17]). The design criteria to determine these dimensions is that the areas are kept the same for all cross-sections [18]. The proportional dimensions of a single shank hook are given in Table 1 and the cross-section parameters of the highly loaded region (Fig. 1 (b)) are given in Table 2. In the analytical stress analysis the height of the cross-section (h), inner width (b_i) and outer width b_o) of the cross-section were considered as dimensions of high stress concentration region in the bending stress (σ) calculation of the analytical method, which is done using the curved beam flexure formula given as

$$\sigma = \frac{M_z y}{Ae(r_n - y)}$$

where A = cross-section, e = the difference between the neutral axis and the centroidal axis. The other parameters in the equation are described in Fig. 1.

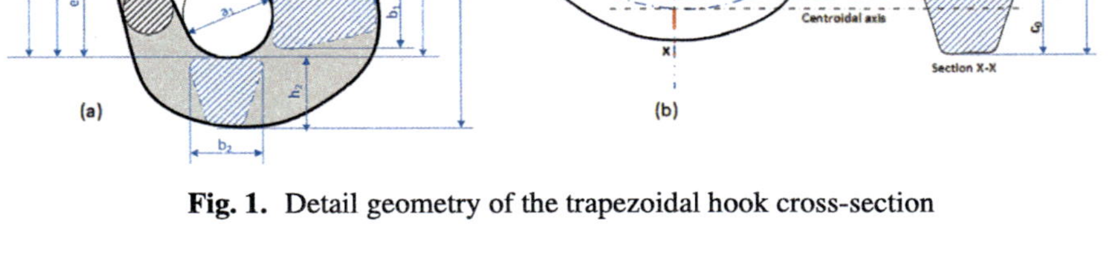

Fig. 1. Detail geometry of the trapezoidal hook cross-section

Table 1. Dimension values of the trapezoidal crane hook [17].

Crane hook parameters	Values (mm)
Distance from top to the loading surface (L_1)	318
Inner width of the cross-section at loading surface (b_2)	60
Distance from lock pin to loading surface (e_3)	165
Height of nose part of the crane hook (e_2)	90
A gap of the curvature (a_2)	63
Diameter of inner curvature of crane hook (a_1)	80
Shank length (B)	103
Hook total height (L)	393
Distance from the bottom of shank to loading surface (e_1)	215
Inner width of section with high-stress concentration (b_1)	71
Height of cross-section at loading surface (h_2)	75

Table 2. The measured dimension tower crane hook parameters.

Crane hook parameters	Values (mm)
Total height of crane hook (L)	338
Radius of outer surface (r_o)	140
Radius of inner surface (r_i)	42
Width of outer surface of cross-section (b_o)	38
Width of inner surface of cross-section (b_i)	76
Section depth (h)	90

2.2 Modelling and Optimization Methodology

The methodology used on how to use TO in the design process of a crane hook which is based on a specific topology optimization configuration. The methodology is tested and analyzed throughout the proposed component development process because the aim of this study is to establish a suitable topology optimization process starting from original design to the end of the optimal results design. The methodology is presented in a flow chart as shown in Fig. 2.

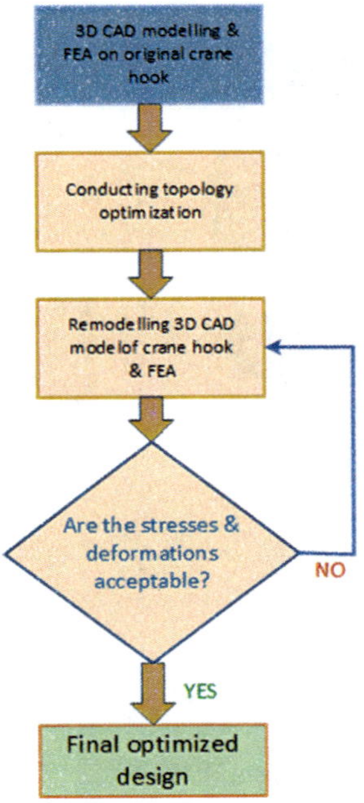

Fig. 2. The schematics of the conceptual modeling frameworks methodology

3 Geometrical Modelling and Analysis

SolidWorks was selected to generate the 3D model of the hook. Figure 3(a) is a picture of a tower crane hook from Afro-Tsion construction site [16] and Fig. 3 (b) is the 3D model developed in SolidWorks based on parameters and the dimensional specification of the original design from Afro-Tsion construction site, which has a carrying capacity of six tones.

The developed 3D model is then imported into ANSYS Workbench, a general-purpose finite element analysis (FEA) software package, for conducting numerical analysis. For the meshing of the crane hook, R-trias and Tetra mesh type meshes are used on the crane hook surface as these are more accurate for 3D parts. The average element size used is 2.5 mm. Element size was decided after meshing from 1 mm to 3 mm. High stress areas were meshed with fine mesh (1 mm). The meshed model has 179968 nodes and discretized into 105110 elements.

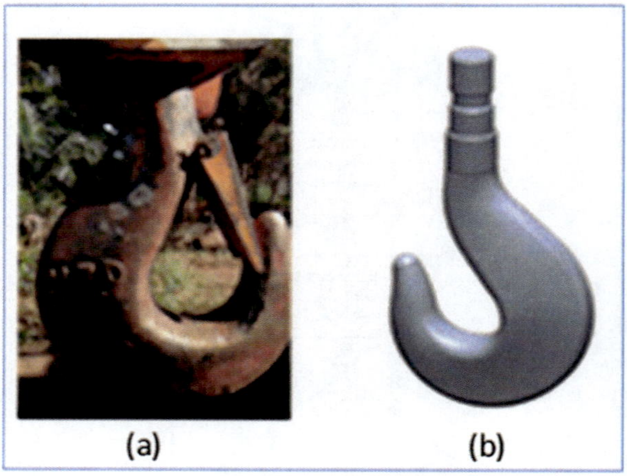

Fig. 3. (a) Crane hook photo captured from the site (29 Nov. 2019) and (b) 3D modelled hook

4 Discussion of Results

4.1 Structural Analysis of the Original Hook

The original hook is analyzed for static structural strength based on a trapezoidal cross-section. From the analysis, the equivalent stress (von Misses), total deformation, and the elastic strain were obtained by applying a load of 6 tons. The weight of the original hook design was 15.75 kg. The structural analysis shows that the highest equivalent stress result from numerical analysis is 92.216 MPa, while the highest total deformation and elastic strain results are 0.527 mm and 0.00046 respectively.

Based on the stress and displacement distribution, the TO is implemented to remove materials from areas that do not significantly contribute to carrying the applied loads. Then the topology optimization results guide the remodeling of the part. The new CAD model is then verified with FEA to carry the loads and to satisfy the design requirements. If the model satisfies the verification, physical model verification is done using any of the physical prototyping methods. If not, the remodeling is done again until verification is done. The final design is then prepared for the final design. The process is illustrated in Fig. 4.

4.2 Structural Analysis of the Optimal Crane Hook

According to the application of TO, the process removes materials from the part make inefficient contribution to load sharing according to the FEA result. This is done because the materials have a negligible effect on the performance of the hook. The material removal leads to the optimized shape that is remodeled and reanalyzed to check that the stress and deformation requirements are maintained. The whole process chain is illustrated in Fig. 4.

For this crane hook analysis, the highest value of the equivalent stress, total deformation, and equivalent elastic strain are 95.01 MPa, 0.318 mm, and 0.000475 m/m

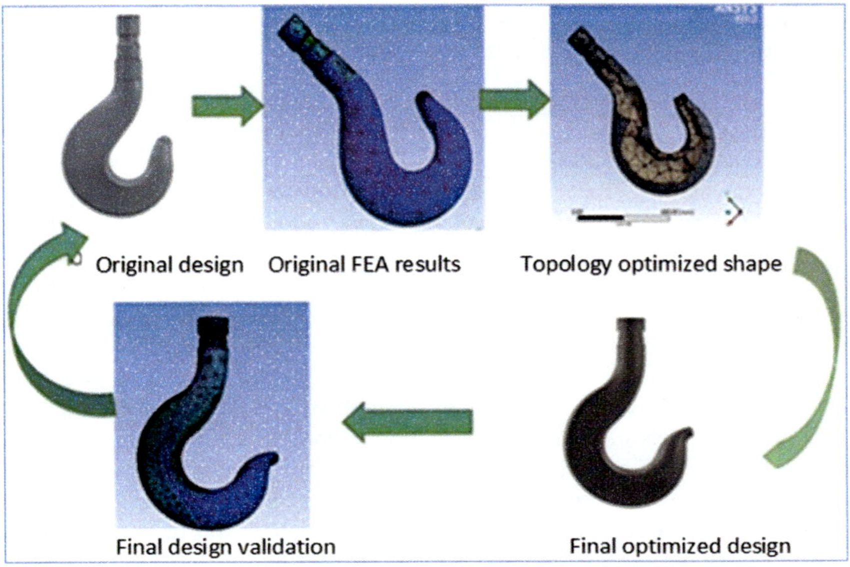

Fig. 4. Topology optimization process

respectively and the weight of the new model is 13.67 kg. Table 3 shows the comparison of the results for the original design and the optimized design. The model is obtained after material removal in the low-stress region up to a safe design limit and the weight of the 3D model is measured from the original weight and after material removal. The models before the topology optimized model with finite element analysis and after topology optimized model with ANSYS19.2 are 15.75 kg and 13.678 kg respectively. The redistribution of the material and topologically optimizing resulted in 6.69% reduction of weight.

Table 3. Comparison of original and optimized hook model

Crane hook stress analysis	Original hook	Optimized model
Load (tons)	6	6
Heighest von-Mises stress (MPa)	92.22	95.01
Maximum total deformation (mm)	0.311	0.318
Heighest elastic strain (m/m)	0.000464	0.000475
Weight (kg)	15.75	13.67

Safety factor based design is a commonly used method in mechanical design. In this study, the maximum and minimum safety factor of all the modelled crane hooks were investigated. The maximum value of safety factor obtained for both models of crane hook is 15 and almost equal minimum safety factor (1.057 for original and 1.055 for the optimized model). This means that nowhere on the crane hook is loaded (in terms of von Mises stress) higher than its yield stress when subjected to the 6 tons load.

5 Conclusions

From the conducted case study reported in this article, it can be concluded that topology optimization is a powerful design concept to reduce the weight of structural products. ANSYS 19.2 WB solver using the density approach is employed for this task. The analysis and simulation of the hook is done using the topological approach leading to minimum design cost and high flexibility of design iterations. The reduction of weight saves a huge amount of material without losing the design strength and processing energy. It also shows that the capability of topology optimization can be fully utilized and from the case study results, which is 6.69% weight reduction, it can also be concluded that topology optimized design can reduce a huge portion of the mass thus results in a lightweight design. The obtained safety factor after optimization are used to validate that the performance of the new design is as good as the original design, but improved material utilization and reduced material cost. Simulation based design and optimization approach is also a means of reducing design errors and time.

References

1. Bhasker, R.S., Prasad, R.K., Kumar, V.: Simulation of geometrical cross-section for practical purpose. Int. J. Mech. Res. Dev. **3**(1), 48–56 (2013)
2. Solanki, M., Bhatt, A., Rathour, A.: Design, analysis and weight optimization of crane hook: a review. Int. J. Sci. Res. Dev. **2**(9), 124–127 (2014)
3. Pavlović, G., Savković, M., Zdravković, N., Marković, G., Stanojković, J.: Analysis and optimization of T-cross section of crane hook considered as a curved beam. Res. Dev. Heavy Mach. **24**(2), 53–60 (2018)
4. Upendar, S.: Design and analysis of a crane hook. Int. J. Current Eng. Sci. Res. **5**(4), 10–15 (2018)
5. Desai, N., Zeytinoglu, N.: Design and optimization of the geometrical properties of a crane hook. World J. Eng. Technol. **4**(3), 391–397 (2016)
6. Altair University. https://altairuniversity.com/. Accessed 23 July 2021
7. Autodesk home. https://www.autodesk.com/products/fusion-360/overview. Accessed 23 July 2021
8. Ji, Y., Wang, H., Chen, H.Q., Gao, M.X., Wu, J.J.: Shape optimization of hook for marine crane. IOP Conf. Ser. J. Phy. **7**(11), 1–7 (2019)
9. Brackett, D., Ashcroft, I., Hague, R.: Topology optimization for additive manufacturing. In: Proceedings 6th International Conference Additive Manufacturing, Loughborough, July 2011, pp. 348–362 (2011)
10. Shrestha, B., Bhandari, A., Poudel, S., Rao, K.V.: Crane hook analysis for different cross-section using ANSYS. Int. J. Adv. Sci. Res. Eng. **5**(12), 67–73 (2019)
11. Allaire, G., Jouve, F., Maillot, H.: Topology optimization for minimum stress design with the homogenization method. Struct. Multidisc. Optim. **28**, 87–98 (2004)
12. Yulin, M., Xiaoming, Y.: A level set method for structural topology optimization and its applications. Adv. Eng. Softw. **35**(7), 415–441 (2004)

13. Yamada, T., Izui, K., Nishiwaki, S., Takezawa, A.: A topology optimization method based on the level set method incorporating a fictitious interface energy. Comput. Methods Appl. Mech. Eng. **199**(45–48), 2876–2891 (2010)
14. Hassani, B., Khanzadi, M., Tavakkol, S.-M.: An isogeometrical approach to structural topology optimization by optimality criteria. Struct. Multidisc. Optim. **45**, 223–233 (2012)
15. Yin, L., Yang, W.: Optimality criteria method for topology optimization under multiple constraints. Comp. Struc. **79**(20–21), 1839–1850 (2001)
16. Afro-Tsion official home page. https://www.afro-tsion.com/. Accessed 23 July 2021
17. Columbus. M. corpoaration Shank hooks specialty hooks and rigging products (2017). https://www.columbusmckinnon.com. Accessed 23 July 2021
18. Mehendale, S.A., Wankhade, S.R.: Design and analysis of EOT crane hook for various crosssections. Int. J. Curr. Eng. Sci. Res. **3**(12), 53–58 (2016)

Investigation of Static and Dynamic Loading Conditions on the Multi Jet 3D Printer Parts

Ramesh Chand[1], Vishal Santosh Sharma[2(✉)], and Rajeev Trehan[1]

[1] Department of Industrial and Production Engineering, Dr. B.R. Ambedkar National Institute of Technology, Jalandhar, Punjab, India

[2] School of Mechanical, Industrial and Aeronautical Engineering, University of the Witwatersrand, Johannesburg, South Africa

vishal.sharma@wits.ac.za

Abstract. Multi Jet Printing has gained popularity in the industry because of its ability to manufacture complex 3D printed parts. Also, it does not require any post-processing for improving dimensional accuracy and surface properties. So the fabricated parts can be directly used as functional parts. This article presents an investigation of the MultiJet Printing (MJP) based part fabricated for the point loading conditions and dynamic loading conditions. Used the primary materials VSIJET-M2RWT (Polymer) and support material VSIJET-M2R SUP (Wax) were in the liquid form. The results can conclude that how the material will behave under point load conditions and dynamic loading conditions. What will be the maximum load up to which material can withstand if the surrounding temperature will be increased. The results can conclude that how the functionality of part will affect actual environmental conditions.

Keywords: MJP · Flexural · Dynamic · Mechanical · Modulous · Storage modulus · Loss Modulus and Tan δ

1 Introduction

The Additive Manufacturing (AM) not only provides flexibility in the designing of the part. But also provide an effective end product for domestic and commercial applications. The Three Dimensional (3D) parts are firstly de-signed on the Computer Added Software (CAD). The standard interface (AM software) between the CAD and the 3D printer will convert the solid 3D geometry into the thin solid layers along the cross-section of the designed part.

Among the commercially available AM processes for the polymers that are extensively studied are Selective Laser Sintering (SLS), Stereolithography (SLA), Laminated Object Technology (LOM), Fused-Deposition Modelling (FDM), and MultiJet Printing (MJP). FDM, also known as Fused Filament Fabrication (FFF), is among the most common. Due to high thermal contraction, dimensional deviation, and high surface roughness, its marketplace in the 3D printing industry is replaced by the MJP. In MJP, the liquid material is used to fabricate the product in a layer-by-layer fashion. Due to

Y. Wang et al. (Eds.): IWAMA 2021, LNEE 880, pp. 540–546, 2022.
https://doi.org/10.1007/978-981-19-0572-8_69

better bonding between the layers and less thermal contraction than FDM, MJP offers immensely better surface and dimensional properties. It can work on a wide range of polypropylene (PP) and Acrylonitrile Butadiene Styrene ABS [1]. VSIJET-M2RWT material is used as the primary polymer material in the current study, and VSIJET-M2R SUP is used as a support material (wax) [1].

2 Background Work

An extensive study has been carried out to assess the mechanical characteristics of MJP components. Instead, fewer academics claimed time-temperature and mechanical characteristics depending on load, some of them as mentioned.

The findings of the finite element analysis (FEA)-based modelled analysis and the real-world results on the flexural fatigue behaviour of PC parts have been compared in the context of component build location on the FDM-based polycarbonate (PC) part. The experimental and virtual model demonstrate the excellent comparability of findings under the von Mises stress criterion [2]. Conducted experiments using FDM with various layer thickness The poly lactic acid (PLA) based component is under investigation for the viscoelastic characteristics that are dependent on time, frequency (0.1, 1, 5, and 10 Hz), and temperature (35 to 80 °C). Thus, in the case of the Maxwell model, the stiffness of the component is the greatest while it is being deposited in the direction of the filament. The stiffness decreases in the transverse directions [3]. Find out the results of the 3D FDM printing process settings on the behaviour of the ABS components under dynamic and cyclic loading. Therefore, it can be concluded that the I-optimal strategy is much superior than the standard experimental approach [4]. PETG components were created using FDM 3D printers that have a variety of print orientations (0°, 90°, and ± 45°). Ultimate tensile strength, modulus of elasticity, and fatigue life were tested on both flat and in-service stress conditions. Compared to other raster orientations, the tensile strength and fatigue characteristics determined to be at their maximum value at 90° [5]. Study the impact of infill density and printing pattern on FDM. A rise in infill density has resulted in a tenfold increase in tensile strength, Loss modulus, and damping; these attributes have diminished roughly and heat conductivity has been decreased [6]. Dynamic mechanical characteristics and tensile properties are studied, and PLA is considered according to four FDM process constraints: fill rate, nozzle temperature, layer thickness, and printing angle. According to the results, the interlayer fracture was observed when the printing angle was less than 45 °C. Layer thickness ensures the strength of the sample bond. When filler use increases, the manufactured components tend to become stronger. In the excellent dynamic mechanical properties, having a moderate nozzle temperature is critical [7]. With regards to the fatigue behaviour under various operating circumstances for the ACM produced by FDM 3D printers, notice the huge range of responses observed. A study proves that −45 °C/45 °C raster angle yields greater fatigue life. Higher ambient temperature in FDM has a negative effect on fatigue life [8]. To see how the newly developed composite Natural Nanorods Attapulgite (ATP) reinforced ABS, together with Nanopillars in different thicknesses, might be used to address the thermal and mechanical characteristics. While ABS/Organic-Attapulgite (OAT) just adds to tensile strength, coefficient, creep flexibility, and linear expansion, these results show that the two materials together (ABS/OAT) lead to greater tensile strength, greater coefficient, greater creep flexibility, and less linear expansion. The dimensional accuracy and thermal stability of the components are improved by using OAT [9]. Brought his experience and expertise to bear on designing and manufacturing a novel composite filament (Boron Nitride Nanotubes and Polystyrene) for the FDM 3D printer. Carried out thermal analyses of the composite glass transition temperature and thermal strength. According to the results, the technique of mixing the composite has a significant impact on the composite filament's mechanical characteristics [10]. In terms of the mechanical characteristics of the ABS-based component, examine the results of ultrasonic vibrational restrictions on the mechanical properties of the ABS-based part produced by FDM. Because to ultrasonic vibrations, the dynamic mechanical characteristics, tensile and bending, of non-crystalline polyamide-styrene (PA) and semi-crystalline polylactic acid (PLA) materials are greatly increased [11].

The main focus of current research is to study the three-point bending behaviour and the time-temperature-dependent viscoelastic characteristics Storage Modulus (E'), Loss Modulus (E"), Damping Parameter (Tan δ) and Glass Transition Temperature (Tg) of the MJP parts.

3 Material and Methods

In the current investigation for the 3-point bending and DMA all the samples were fabricated under the same temperature and pressure conditions. The parts were fabricated in the X-Y plane so that the maximum area of the samples remain in contact with the base plate and the chances of formation of voids can be reduced. The MJP based 3D printer is used. In the current study polymer material is used as main material and wax is used as a support material [1]. Material supported: wax is the only full cured support material for MJP printer. During the post processing at the 55–60 °C the support material is removed from the fabricated parts and then the mechanical testing is performed.

Three-point bending is performed on the UTM with standard attachments for 3-point bending test. Computerized Twin Screw UTM-NX was used. The UTM machine had load limit 1000 kg, travel limit 850 mm. The rate of loading 1 mm/min was used to perform the test. DMA was done by the "Mettler Toledo DMA "analyser under the simply supported loading positions. The DMA used in the current investigation is the multi-purpose dynamic mechanical analysers. It can be used for the liquid materials also.

4 Results and Discussions

4.1 Investigations on 3-Point Bending Test

Bend strength, flexural strength and modulus of rapture are the important property of the brittle material, that explain the ability of the material to resist the external load. If all the parametric and climate conditions of the part will remain the same the tensile strength and bending strength of the part will be same. So in other words it can be said that it is the alternative method to find out the elastic modulus of the part [12].

The ability of the 3D printing processes to fabricate the micro to the macro component make it more comfortable in the modern manufacturing sector. In the field of the fabrication of the supporting body parts of the electronic devices (touch panel of mobile and its body parts, different frame structures) to the macro mechanical parts (fixtures used in mobile phone sector). They regularly pass though the cyclic bending. So to understand the life of such 3D printed parts it is very important to understand its bending behaviour. The flexural strength in case of 3-point bending is calculated using the equation [12].

In the current study the ASTM D 790 standard for the flexural properties is considered. The dimension of the specimen is length 128, width 13 and thickness 3.2 mm considered. The average of all the five samples is considered in the investigation. Reflects that the maximum flexural strength was 17.272 MPa and Elastic modulus is 35.49 MPa. That is in the limit of the tensile strength claimed by the manufacturer of the main material used in the current investigation.

5 Investigations of Fabricated Parts Under Dynamic Loading Conditions

The investigation of the viscoelastic behaviour such as storage and loss modulus of part can be computed by Dynamic Mechanical Analysis (DMA). The specimens can be analysed for the complex modulus in different set of configuration such as single cantilever, double cantilever and three-point bending, using three different analysers. For the same set of material and same configuration the different DMA analysers give different results because each DMA machine has its own mathematical path way to find the complex modulus of the parts. The three-point bending mode in the DMA is found to be the less effective by the part geometry and the test parameters, if the clamping effect is avoided. That's why the three-point bending mode is selected to analyses in viscoelastic properties of the liquid polymer based part fabricated by MJP. During the fabrication of the standard ASTM-AD7028 is followed.

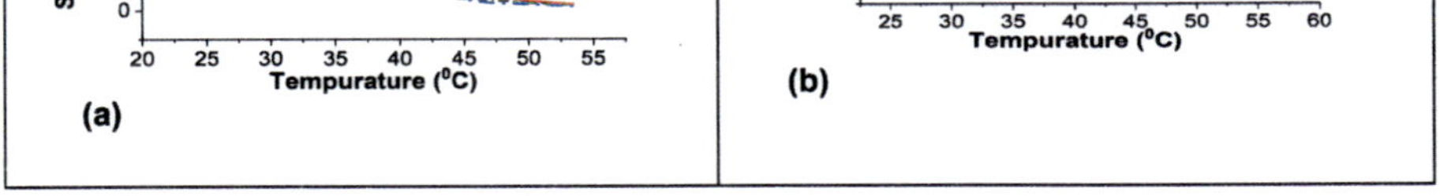

Fig. 1. Storage modulus as per the first set (a) and second set of parameters (b)

DMA was performed on five samples by the "Mettler Toledo DMA-1. DMA based four parameters Loss Modulus (E"), Storage Modulus(E'), Damping Parameter (Tan δ) and Glass Transition Temperature (T_g) were analyzed. Storage Modulus (E') represents the elastic behavior of the component whereas Loss Modulus (E") shows the viscous response of the material in terms of the energy dissipation by the material during each cycle under the loading conditions with respect to temperature [13]. Damping Parameter (Tan δ) is the ratio of the E"/E' and it provides the balance in between the viscous phase and elastic phase. T_g is the temperature on which the material starts losing its mechanical properties from the initial hard state it gets deformed in to the rubbery state. T_g is the function of degree of polymerization of the polymer based material. In the present study the T_g is calculated on the basics of very first rise in the Tan δ curve.

Intended for analyzing the behavior MJP based 3D printing part at a first set of parameters (frequency-1 Hz, temperature range −22–55 °C, Rate of temperature increment 2 °C/minute and Load of 0.5 N) and second set of by varying the Frequency from 0.1,1,10 and 100 Hz, keeping other parameters same. Figure 1 reflects the Storage modulus as per the first set (a) and second set of parameters (b). In case of first set of parameters the value of storage modulus is changing in range (7000–9000) MPa. With rise in temperature there is reduction in storage modulus and the values become stagnated after 50 °C. Whereas under the second set of parameters the range of values is lying in between (7000–11000) MPa. The elastic response of the martial with rise in temperature is different with respect to frequency. It is found that the E' of the material is continuously increasing as the frequency of oscillation is increasing. But it will suddenly decrease at 40 °C as the frequency of vibration increases from 10 Hz to 100 Hz. So above 40 °C the ability of the 3D printed part to store the energy is decreasing. So it can be said that at 40 °C and the 1 Hz the material can absorb maximum amount of energy.

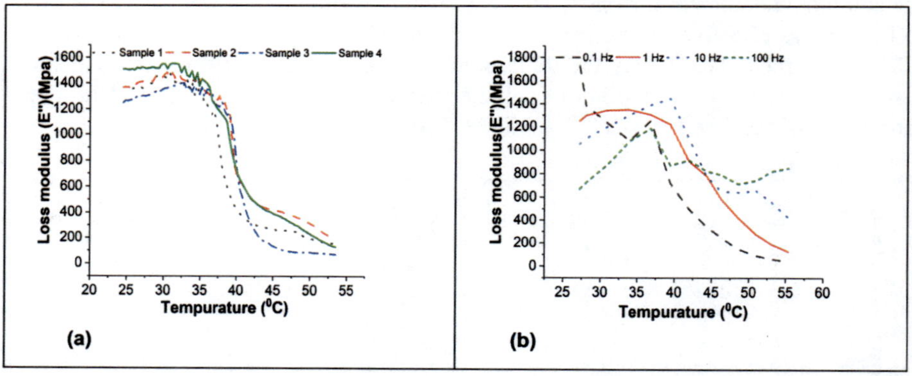

Fig. 2. Loss modulus as per the first set (a) and second set of parameters (b)

As shown in Fig. 2 (a) under the first set of parameters the ability of the material to loss the energy in the form of heat and temperature is following a similar trend but the rate of loss of all the four samples is different. That means the material is not behaving homogeneously to the external loading conditions. Whereas under the second set of parameters as the rate of oscillations is increased from 0.1 Hz to 100 Hz the loss modulus is decreasing. So the material is behaving like an elastic material as the frequency of vibration is increased.

Considering Fig. 3 as per the first set of parameters damping behavior of the material is not following a similar trend even the geometric factor of all the four samples is same. The very first rise in the Tan δ curve is considered as the its T_g peak. The T_g for the first set of parameters is lying in between 39.685–51.516 °C. Under the second set of parameters the trend of the various graphs in different frequency is not similar as the frequency of the oscillation is increasing from the 0.1,1,10 and 100 Hz the damping capacity of the material is increased (refer Fig. 3(b)). So that it can be said that the material behaves well under the high frequency oscillations. The T_g for the second of parameters is lying in between 45.431–55.361 °C.

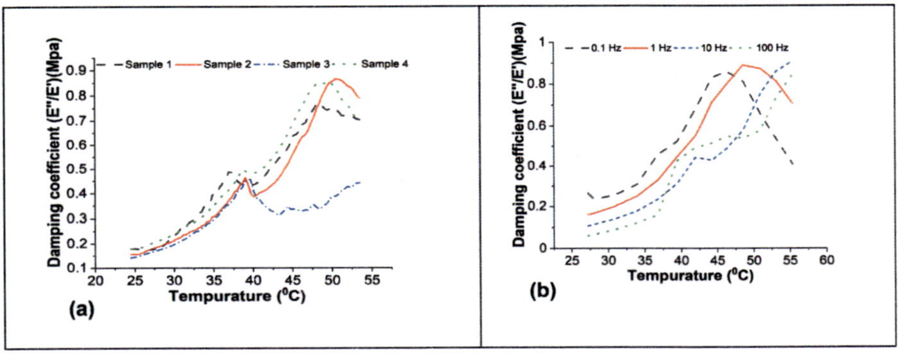

Fig. 3. Damping coefficient as per (a) the first set and (b) second set of parameters

6 Conclusions

The current study emphasized on the static and dynamic behavior of the MJP based part under different temperature and loading conditions. The following conclusions are made from the results.

- 3-point bending test can be used to find the modulus of Elasticity of the material.
- In the case of the first set of values (frequency-1 Hz, temperature range –22–55 °C, Rate of temperature increment 2 °C/minute and Load of 0.5 N), the storage modulus varies in the range (7000–9000) MPa. The storage modulus decreases as temperature rises, and around 50 °C, the values become stagnant.
- The range of storage modulus values for the second set of parameters (Frequency from 0.1,1,10, and 100 Hz, temperature range −22–55 °C, Rate of temperature increase 2 °C/minute, and Load of 0.5 N) is between (7000–11000) MPa. The elastic response of the martial to temperature rise differs according on frequency. It has been discovered that the material's E' is constantly growing as the frequency increases.

- Under the first set of parameters, the material's capacity to lose energy in the form of heat and temperature follows a similar trend, although the rate of loss differs amongst the four samples. This indicates that the material is not responding uniformly to the external loading conditions.
- With regard to the second set of parameters, the loss modulus decreases as the rate of oscillations increases from 0.1 Hz to 100 Hz. As the frequency of vibration increases, the material behaves like an elastic material.
- The T_g for the first set of parameters is lying in between 39.685_51.516 °C. The T_g for the second of parameters is lying in between 45.431–55.361 °C.

References

1. 3D Systems. MultiJet Plastic Printers, pp. 2–3 (2018)
2. Puigoriol-Forcada, J.M., Alsina, A., Salazar-Martín, A.G., Gomez-Gras, G., Pérez, M.A.: Flexural fatigue properties of polycarbonate fused-deposition modelling specimens. Mater. Des. **155**, 414–421 (2018). https://doi.org/10.1016/j.matdes.2018.06.018
3. Anoop, M.S., Senthil, P., Sooraj, V.S.: An investigation on viscoelastic characteristics of 3D-printed FDM components using RVE numerical analysis. J. Braz. Soc. Mech. Sci. Eng. **43**(1), 1–13 (2021). https://doi.org/10.1007/s40430-020-02724-5
4. Mohamed, O.A., Masood, S.H., Bhowmik, J.L.: Characterization and dynamic mechanical analysis of PC-ABS material processed by fused deposition modelling: An investigation through I-optimal response surface methodology. Measurement **107**, 128–141 (2017). https://doi.org/10.1016/j.measurement.2017.05.019
5. Dolzyk, G., Jung, S.: Tensile and fatigue analysis of 3D-printed polyethylene terephthalate glycol. J. Fail. Anal. Prev. **19**(2), 511–518 (2019). https://doi.org/10.1007/s11668-019-00631-z
6. Aw, Y., Yeoh, C., Idris, M., Teh, P., Hamzah, K., Sazali, S.: Effect of printing parameters on tensile, dynamic mechanical, and thermoelectric properties of FDM 3D printed CABS/ZnO composites. Materials **11**(4), 466 (2018). https://doi.org/10.3390/ma11040466
7. Wang, S., Ma, Y., Deng, Z., Zhang, S., Cai, J.: Effects of fused deposition modeling process parameters on tensile, dynamic mechanical properties of 3D printed polylactic acid materials. Polym. Testing **86**, 106483 (2020)
8. Shanmugam, V., et al.: Fatigue behaviour of FDM-3D printed polymers, polymeric composites and architected cellular materials. Int. J. Fatigue **143**, 106007 (2021)
9. Wang, L., et al.: Properties of abs/organic-attapulgite nanocomposites parts fabricated by fused deposition modeling. J. Renew. Mater. **8**(11), 1505–1518 (2020). https://doi.org/10.32604/jrm.2020.010544
10. Akintola, T.M., Tran, P., Sweat, R.D., Dickens, T.: Thermomechanical multifunctionality in 3d-printed polystyrene-boron nitride nanotubes (Bnnt) composites. J. Comp. Sci. **5**(2), 61 (2021). https://doi.org/10.3390/jcs5020061
11. Li, G., et al.: Effect of ultrasonic vibration on mechanical properties of 3D printing non-crystalline and semi-crystalline polymers. Materials **11**(5), 826 (2018). https://doi.org/10.3390/ma11050826
12. J. Instruments. Flexural Strength Testing
13. Toledo, M.: DMA 1 - Dynamic Mechanical Analyzer Swiss Quality. https://www.mt.com/in/en/home/products/Laboratory_Analytics_Browse/TA_Family_Browse/DMA/DMA1.html

Multi-parameter Identification of PMSM Based on IGWO Algorithm

Xianwei Ke, Jinliang Zhang(✉), Wei Jian, Guosheng Peng, and Yufeng Chen

School of Electrical and Information Engineering, Hubei University of Automotive Technology, Shiyan 442002, Hubei, China
{201911029,zhangjl_dy,jianwei,20180010,chenyf_dy}@huat.edu.cn

Abstract. To solve the multi-parameter identification problem of permanent magnet synchronous motor (PMSM), an improved Grey Wolf Optimizer (IGWO) algorithm is proposed. Firstly, the nonlinear convergence strategy is introduced to overcome the shortcoming of the linear convergence of Grey Wolf Optimizer (GWO) algorithm. Secondly, the coefficient weights of wolf α, β and δ are adjusted in the location update to highlight the impact of wolf α. Finally, dimension-learning-based hunting (DLH) optimization strategy is added to improve the diversity of the optimization process. The simulation results show that the proposed IGWO has higher precision and faster convergence speed for multi-parameter identification of PMSM.

Keywords: Grey wolf optimization algorithm · Permanent magnet synchronous motor · Multi-parameter identification · Dimension-learning-based hunting

1 Introduction

Permanent magnet synchronous motor (PMSM) plays an important role in industry because of its excellent performance. In motor control systems, PMSM parameters usually change with the temperature of working condition, which are difficult to be obtained directly [1]. In double closed-loop control system, the parameters of inner-loop PI controller are designed by using the values of d-q axis inductance and stator resistance, and the parameters of outer-loop PI controller are designed by using the values of permanent magnet flux and moment of inertia [2]. Therefore, the accurate identification of PMSM parameters is very important to improve motor control system performance [3]. In reference [4], the finite element model was constructed to obtain the calculation formula of motor iron loss resistance and d-q axis inductance, but the solution of the model was complicated.In reference [5], as for the data saturation problem in the later stage of traditional recursive least squares iteration, a discount recursive least square identification algorithm is proposed, which reduces the effect of data saturation to some extent by introducing discount factor, but the error of identification result is still large. In reference [6], the extended kalman filter is used to identify the moment of inertia effectively, but only single parameter is identified. In reference [7], an adaptive update formula of inertia

Y. Wang et al. (Eds.): IWAMA 2021, LNEE 880, pp. 547–555, 2022.
https://doi.org/10.1007/978-981-19-0572-8_70

weight was proposed to improve the particle swarm optimization algorithm. However, two parameters were self-identified and the error was large.

In order to improve the performance of PMSM parameter identification, an improved Grey Wolf Optimizer (IGWO) algorithm is proposed. Firstly, a nonlinear convergence strategy is introduced, which makes the global search prominent in the early stage and local search prominent in the later stage, it also accelerates the algorithm convergence. Secondly, in view of the different effects of wolf α, β and δ in search optimization, the location update was adjusted to highlight the dominant effect of α wolf. Then, dimension-learning-based hunting (DLH) search strategy is introduced to improve the diversity of the search algorithm. Based on the double closed-loop control of PMSM, a 4-order identification model is constructed based on its mathematical model, and the parameters of PMSM are identified by using IGWO.

2 Mathematical Model of Permanent Magnet Synchronous Motor

PMSM is a nonlinear and strong-coupling complex control object. For convenience of study, the motor core saturation effect and permanent magnet conductivity are neglected; the back electromotive force is regarded as standard sine signal. The stator voltage equation of PMSM in d-q axis can be written as

$$\begin{cases} u_d = R_s i_d + L_d \frac{di_d}{d_t} - \omega_e L_q i_q \\ u_q = R_s i_q + L_q \frac{di_q}{d_t} + \omega_e L_d i_d + \omega_e \psi_f \end{cases} \tag{1}$$

Where u_d and u_q are the d-q axis component of the stator voltage, i_d and i_q are the d-q axis components of the stator current, R_s is the stator resistance, L_d and L_q are d-q axis inductors, and ψ_f is the permanent magnet flux. When PMSM runs stably, the change of i_d and i_q is very small, then (1) can be further reduced to

$$\begin{cases} u_d = R_s i_d - \omega_e L_q i_q \\ u_q = R_s i_q + \omega_e L_d i_d + \omega_e \psi_f \end{cases} \tag{2}$$

Equation (2) R_s, L_q, L_q and ψ_f are four parameters which need to be identified, but the equation is only a 2-order model. Normally, the 2-order model cannot identify the four parameters at the same time.

$$\begin{cases} \hat{u}_{d0}(k) = -\omega_{e0}(k) i_{q0}(k) \widehat{L}_q \\ \hat{u}_{q0}(k) = i_{q0}(k) \widehat{R}_s + \omega_{e0} \widehat{\psi}_f \\ \hat{u}_{d1}(k) = i_{d1}(k) \widehat{R}_s - \omega_{e1}(k) i_{q1}(k) \widehat{L}_q \\ \hat{u}_{q1}(k) = i_{q1}(k) \widehat{R}_s + \omega_{e1}(k) i_{d1}(k) \widehat{L}_d + \omega_{e1}(k) \widehat{\psi}_f \end{cases} \tag{3}$$

In order to obtain the 4-order model, the motor is firstly sampled under the vector control strategy of id = 0, and then a negative-sequence d-axis current is injected for a short time, sample in the same way. Thus, a fourth-order discrete model as shown in (3) can be obtained. where$\omega_{e0}(k)$,$i_{q0}(k)$ are the *K-th* sampling data under control condition of $i_d = 0$, and $\omega_{e1}(k)$, $i_{d1}(k)$, $i_{q1}(k)$ are the *K-th* sampling data under control condition of $i_d \neq 0$.

3 Grey Wolf Optimizer Algorithm

GWO algorithm is a new algorithm based on the grading mechanism and hunting process of gray wolves in nature [8]. The algorithm was first proposed in 2014 by Mirjalili et al., it divides wolves into groups of α, β, δ and ω, with social status descending in a sort of pyramid. In the algorithm optimization, the first three wolfs lead the search, and the wolf ω follows. The whole process of the gray wolf hunting prey can be divided into: surrounding, hunting, attacking and searching.

First, the mathematical formula for surrounding prey is

$$\begin{cases} \boldsymbol{D} = |\boldsymbol{C} \cdot \boldsymbol{X}_P(t) - \boldsymbol{X}(t)| \\ \boldsymbol{X}(t+1) = \boldsymbol{X}_P(t) - \boldsymbol{A} \cdot \boldsymbol{D} \end{cases} \tag{4}$$

where: $\boldsymbol{X}_P(t)$ and $\boldsymbol{X}(t)$ are defined as the current location of target prey and grey wolf respectively; $\boldsymbol{A}$ and $\boldsymbol{C}$ are coordinate coefficient vectors; $\boldsymbol{D}$ is expressed as the distance between the prey and individual gray wolf; t is the current number of iterations.

Vectors $\boldsymbol{A}$ and $\boldsymbol{C}$ change as the algorithm optimizes, and can be described as

$$\begin{cases} \boldsymbol{A} = 2\boldsymbol{a} \cdot \boldsymbol{r}_1 - \boldsymbol{a} \\ \boldsymbol{C} = 2 \cdot \boldsymbol{r}_2 \end{cases} \tag{5}$$

where: $\boldsymbol{r}_1$ and $\boldsymbol{r}_2$ are random vectors of [0,1] and the components of $\boldsymbol{a}$ change linearly from 2 to 0 as the number of iterations increases. As shown in (6)

$$a = 2 - 2(t/t_{max}) \tag{6}$$

where: $t_{\max}$ is the maximum number of iterations.

After the pack encircled the prey, it went on a hunting, whose position update formula is as follows

$$\begin{cases} \boldsymbol{D}_a = |\boldsymbol{C}_1 \cdot \boldsymbol{X}_\alpha - \boldsymbol{X}| \\ \boldsymbol{D}_\beta = |\boldsymbol{C}_2 \cdot \boldsymbol{X}_\beta - \boldsymbol{X}| \\ \boldsymbol{D}_\delta = |\boldsymbol{C}_3 \cdot \boldsymbol{X}_\delta - \boldsymbol{X}| \\ \boldsymbol{X}_1 = |\boldsymbol{X}_\alpha - \boldsymbol{A}_1 \cdot (\boldsymbol{D}_\alpha)| \\ \boldsymbol{X}_2 = |\boldsymbol{X}_\beta - \boldsymbol{A}_2 \cdot (\boldsymbol{D}_\beta)| \\ \boldsymbol{X}_3 = |\boldsymbol{X}_\delta - \boldsymbol{A}_3 \cdot (\boldsymbol{D}_\delta)| \end{cases} \tag{7}$$

where: $\boldsymbol{A}_1, \boldsymbol{A}_2$ and $\boldsymbol{A}_3$ are the coefficient vectors; $\boldsymbol{X}_\alpha$, $\boldsymbol{X}_\beta$ and $\boldsymbol{X}_\delta$ represent the current positions of Wolf α, β and δ, respectively; $\boldsymbol{C}_1$, $\boldsymbol{C}_2$ and $\boldsymbol{C}_3$ are random vectors; $\boldsymbol{D}_a$, $\boldsymbol{D}_\beta$ and $\boldsymbol{D}_\delta$ are the distances between the current participating gray wolf in position updating and the gray wolf α, β and δ.

$$\boldsymbol{X}(t+1) = (\boldsymbol{X}_1 + \boldsymbol{X}_2 + \boldsymbol{X}_3)/3 \tag{8}$$

When the pack gets close to its prey and the prey does not move, the grey wolf starts to attacks. In (6), the value of coefficient vector $\boldsymbol{A}$ will also change with the linear decrease of "a". When $|\boldsymbol{A}| \le 1$, wolves attack prey and tend to search locally. While $|\boldsymbol{A}| > 1$ the pack will break off and spread out, tending to global search.

4 Improved Grey Wolf Optimizer Algorithm

4.1 Convergence Factor for Nonlinear Variation

For swarm intelligence algorithm, it is very important to give consideration to both global and local search, improper treatment will make the algorithm slow convergence or easy to fall into local optimal. In the optimization process of GWO algorithm, as shown in (6), the change of convergence factor "a" is linear, and the linear change strategy of "a" cannot well reflect the algorithm actual convergence process. Therefore, the nonlinear change strategy is introduced

$$a = 2 - 2\left((1/e - 1) \cdot \left(e^{t/t_{max}} - 1\right)\right) \tag{9}$$

4.2 Location Update Adjustment Policy

In GWO algorithm, its position update is shown in (8). It treats the current position of wolf α, β and δ equally to the subsequent optimization direction of the algorithm, and does not highlight the special influence of the current optimal solution of wolf α. Based on this, a new update strategy is introduced as follows

$$\begin{cases} \xi_1 = (f_\alpha + f_\beta + f_\delta)/f_\alpha \\ \xi_2 = (f_\alpha + f_\beta + f_\delta)/f_\beta \\ \xi_3 = (f_\alpha + f_\beta + f_\delta)/f_\delta \end{cases} \tag{10}$$

$$\boldsymbol{X}(t+1) = (\xi_1 \boldsymbol{X}_1 + \xi_2 \boldsymbol{X}_2 + \xi_3 \boldsymbol{X}_3)/(\xi_1 + \xi_2 + \xi_3) \tag{11}$$

where: f_α, f_β and f_δ are the fitness values of wolf α, β and δ under the current iteration number, and ξ_1, ξ_2 and ξ_3 are the weight coefficients of different positions influenced by wolf α, β and δ in the next iteration under the current iteration number.

4.3 Dimension-Learning-Based Hunting (DLH) Optimization Strategy

In GWO algorithm, each wolf in the pack creates a new position under the influence of three wolves (wolf α, β, and δ). This approach leads to slow convergence of GWO and premature loss of diversity of the population, it is also easy to fall into local optimum. Based on this, this paper adopts (10) and (11) to adjust the position update strategy of GWO, and then introduces the search strategy based on DLH.

In DLH strategy, the hunting direction of an individual wolf pack is influenced by the presence of gray wolves in a neighborhood surrounding its location. First, the neighborhood radius is calculated by (12). The formula is as follows

$$\boldsymbol{R}_i(t) = \|\boldsymbol{X}_i(t) - \boldsymbol{X}_{i\_GWO}(t+1)\| \tag{12}$$

Then, a neighborhood is constructed using the radius calculated in (12). The formula is as follows

$$\boldsymbol{N}_i(t) = \left\{\boldsymbol{X}_j(t) | D_i\left(\boldsymbol{X}_i(t), \boldsymbol{X}_j(t)\right) \le \boldsymbol{R}_i(t), \boldsymbol{X}_j(t) \in \boldsymbol{Pop}\right\} \tag{13}$$

Then, the candidate location of the grey wolf under the DLH search strategy is calculated as follows

$$X_{i\_DLH,d}(t+1) = X_{i,d}(t) + rand \cdot (X_{n,d}(t) - X_{r,d}(t)) \tag{14}$$

Where: subscript d represents the dimension of the optimization problem; $X_{r,d}(t)$ is the location of randomly selected gray wolves in the group. $X_{n,d}(t)$ is the position of the gray wolf randomly selected in the grey wolf set.

Finally, the final position is determined according to the size of the fitness value, and the formula is as follows

$$X_i(t+1) = \begin{cases} X_{i\_GWO}(t+1), f(X_{i\_GWO}) < f(X_{i\_DLH}) \\ X_{i\_DLH}(t+1), f(X_{i\_GWO}) \geq f(X_{i\_DLH}) \end{cases} \tag{15}$$

where: $f(X_{i\_GWO})$ is the fitness value of candidate position $X_{i\_GWO}(t+1)$; $f(X_{i\_DLH})$ is the fitness value of the candidate position $X_{i\_DLH}(t+1)$.

4.4 Improved Algorithm Performance Testing

In order to evaluate the optimization performance of IGWO algorithm, this paper introduces the GWO as a reference experiment. The two algorithms optimize the benchmark functions F_1, F_2 and F_3, respectively. Where, F_1 is a single peak function, which is used to evaluate the algorithm performance of local optimization; F_2 and F_3 are multi-peak functions, which are used to evaluate the algorithm performance of global optimization. F_1, F_2 and F_3 are shown in Table 1.

Table 1. Three test functions

Function	F_{min}
$F_1(x) = \sum_{i=1}^{n} x_i^2$	0
$F_2(x) = \sum_{i=1}^{n} \left[a_i - \frac{x_1(b_i^2 + b_i x_2)}{b_i^2 + b_i x_3 + x_4} \right]^2$	0.0003
$F_3(x) = \left(\frac{1}{500} + \sum_{j=1}^{25} \frac{1}{j + \sum_{i=1}^{2} (x_i - a_{ij})^6} \right)^{-1}$	1

Experimental parameters are set as follows: wolf pack size is 30, and t_{max} is 500. The optimization curves of the two algorithms for the three test functions are shown in Fig. 1. It can be seen that the optimization accuracy and convergence speed of IGWO algorithm for F_1, F_2 and F_3 are better than that of GWO algorithm.

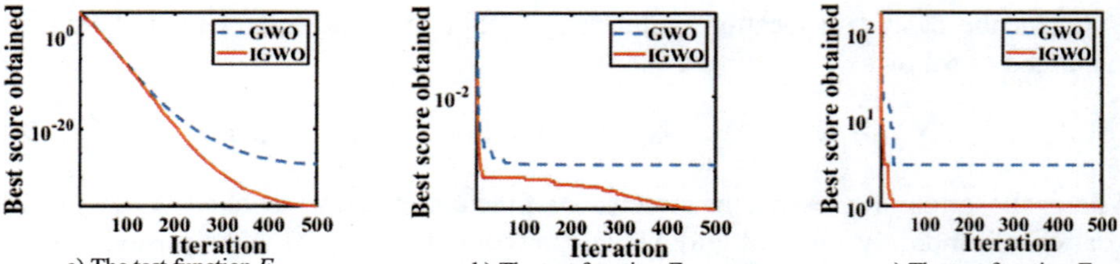

a) The test function F_1 b) The test function F_2 c) The test function F_3

Fig. 1. The optimization curve of the test function

5 PMSM Multi-parameter Identification Base on IGWO Algorithm

Its basic idea is to compare the sampling value of the motor control system with the theoretical calculation value of the algorithm, continuously optimize to make the two infinitely approximate, and finally make the parameter with the minimum value of objective function f be the optimal solution. The objective function is shown in (16).

$$\begin{cases} f = \mu_1(m_1)^2 + \mu_2(m_2)^2 + \mu_3(m_3)^2 + \mu_4(m_4)^2 \\ m_1 = u_{d0}(k) - \hat{u}_{d0}(k), m_2 = u_{q0}(k) - \hat{u}_{q0}(k) \\ m_3 = u_{d1}(k) - \hat{u}_{d1}(k), m_4 = u_{q1}(k) - \hat{u}_{q1}(k) \end{cases} \tag{16}$$

Where the weight coefficients μ_1, μ_2, μ_3 and μ_4 are 0.25 respectively, $u_{d0}(k)$, $u_{d1}(k)$, $u_{q0}(k)$ and $u_{q1}(k)$ are the sampling values, and $\hat{u}_{d0}(k)$, $\hat{u}_{d1}(k)$, $\hat{u}_{q0}(k)$ and $\hat{u}_{q1}(k)$ are the values calculated by substituting the identification results into the motor voltage equation.

6 The Experimental Simulation

6.1 The Experiment Design

In order to verify the feasibility and effectiveness of IGWO algorithm in identifying PMSM parameters, a simulation experimental model was built in matlab/simulink. The specific block diagram is shown in Fig. 2.

The simulation parameters are set as follows. $R_s = 1.5\,\text{ohm}$, $L_d = 5.25\,\text{mH}$, $L_q = 5.25\,\text{mH}$, $\psi_f = 0.263\,\text{Wb}$, and number of pole pairs $P_n = 4$. When the motor is running steadily, sampling is carried out under the vector control strategy of $i_d = 0$ and $i_d = -2$ respectively, and the single sampling time is $1\mu s$. The standard GWO algorithm is introduced to carry out for comparative experiment. The wolf pack size is 150 and the maximum number of iterations is 200.

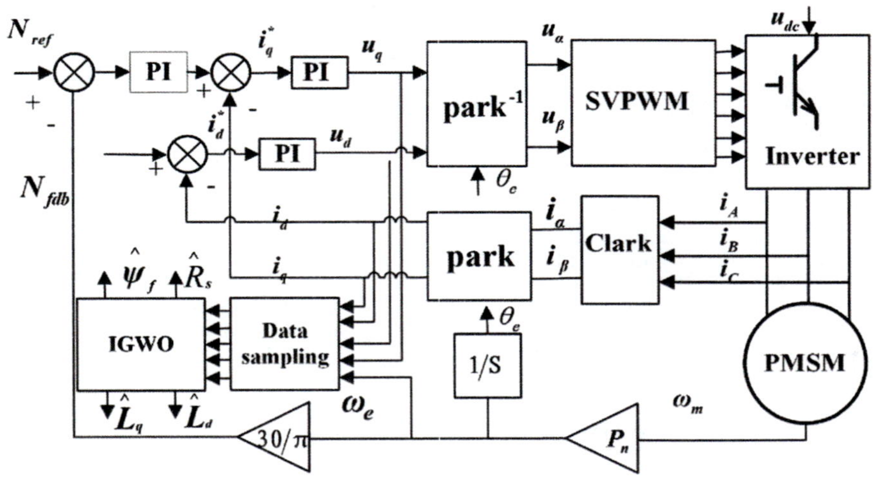

Fig. 2. Block diagram of PMSM parameter identification based on IGWO algorithm

6.2 Experimental Analysis

Figure 3 shows the fitness value of objective function f.The smaller the fitness value, the higher the identification accuracy. Figure 4 shows the curve graph of parameter identification results. It can be seen that, for R_s, L_d, L_q and ψ_f, the overall identification accuracy and convergence speed of IGWO algorithm are better than that of GWO algorithm, which is also consistent with the reaction results of the curve shown in Fig. 3.

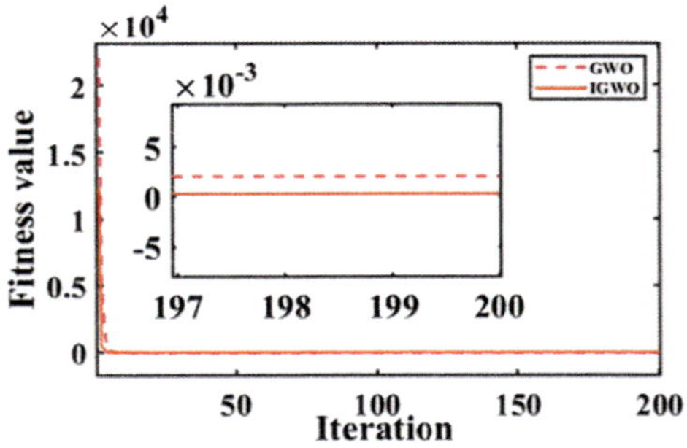

Fig. 3. Fitness change curve

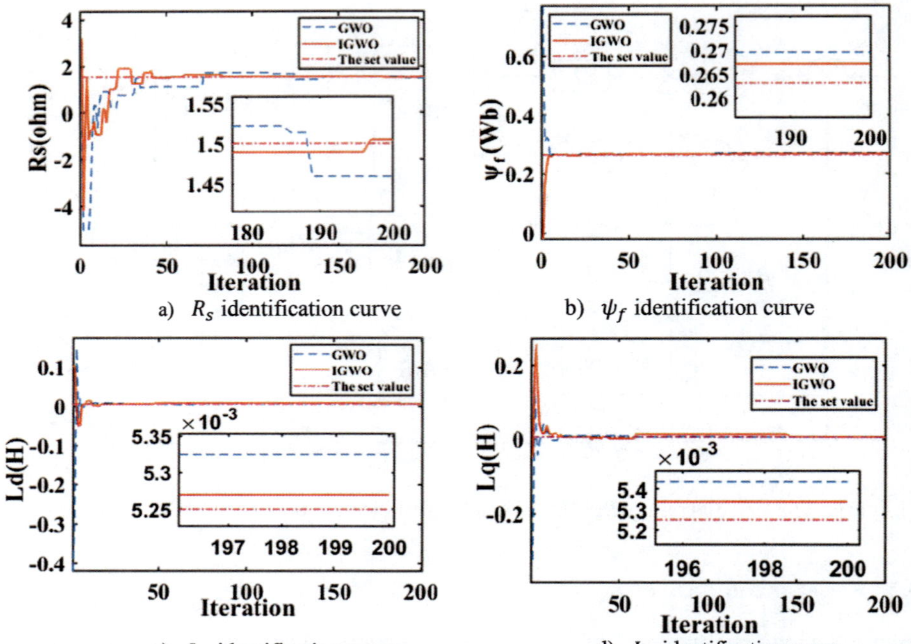

a) R_s identification curve

b) ψ_f identification curve

c) L_d identification curve

d) L_q identification curve

Fig. 4. Identification result curve

7 Conclusion

To solve the problem of slow convergence and low precision in PMSM parameter identification, an IGWO algorithm based on GWO algorithm is proposed in this paper. Firstly, for the lack of linear convergence of GWO algorithm, a nonlinear convergence strategy is introduced. Secondly, the coefficients of wolf α, β and δ were adjusted to highlight the influence of wolf α. Then, a hunter-search strategy based on dimensionality learning is introduced to improve the diversity in the optimization process. For three benchmark function test results, the IGWO algorithm show much better optimization performance than GWO. The motor parameter identification results also show that the IGWO has higher identification accuracy and faster convergence speed than GWO.

Acknowledgment. The work is supported by Local Science and Technology Development Project Guided by Central Government (Grant No. 2018ZYYD007).

References

1. Wang, S.: Parameter Identification and Control Strategy Research of Permanent Magnet Synchronous Motor. Beijing Jiao tong University, Beijing (2011)
2. Yan, H.F.: Research on Parameter Identification Algorithm of Permanent Magnet Synchronous Motor. Harbin Institute of Technology, Harbin (2015)

3. Zhou, Hu.: Research on Parameter Identification Algorithm of Permanent Magnet Synchronous Motor. University of Electronic Science and Technology of China, Chengdu (2012)
4. Lee, J.-Y., Lee, S.-H., Lee, G.-H., Hong, J.-P., Hur, J.: Determination of parameters considering magnetic nonlinearity in an interior permanent magnet synchronous motor. IEEE Trans. Magn. **42**(4), 1303–1306 (2006). https://doi.org/10.1109/TMAG.2006.871951
5. Shi, J.F., Ge, B.J.: Research on on-line identification method of permanent magnet synchronous motor. Electric Mach. Control **22**(3) (2018)
6. Huang, S.R., He, D.L.: Research on identification of moment of inertial of permanent magnet synchronous motor based on extended Kalman filter. Electric Mach. Control Appl. **42**(12), 7–11 (2015)
7. Yuan, Y.M.: Parameter identification of permanent magnet synchronous motor based on adaptive particle swarm optimization algorithm. Measure. Control Technol. **37**(7), 42–45, 13 (2018)
8. Mirjalili, S., Mirjalili, S.M., Lewis, A.: Grey wolf optimizer. Adv. Eng. Softw. **69**, 46–61 (2014)

Application and Development of Artificial Intelligence in Fault Diagnosis

Xiang Zhao[1] and Yi Wang[2(✉)]

[1] Henan Polytechnic, Zhengzhou, Henan, China
[2] Department of International Shipping, Logistics, and Operations Management, Plymouth Business School, Plymouth University, Plymouth PL4 8AA, UK
YiWang@plymouth.ac.uk

Abstract. Intelligent manufacturing 2025 initiative provides new research directions in mechanical equipment operation reliability, key techniques within the initiative, such as equipment state monitoring and fault diagnosis is an important part of the preventive maintenance system. To ensure the safety of the mechanical equipment running and the stability is of great significance. In recent years, with the continuous development of computer science and AI technology, the development prospect of artificial intelligence for fault diagnosis has been fruitful in many industries. In this paper, the common fault diagnosis model is taken as an example, and the application of artificial intelligence in industrial field in the new era is analyzed, and the development prospect and potential challenges on this basis are pointed out.

Keywords: Artificial intelligence · Fault diagnosis · Design development prospect

1 Introduction

With the continuous advancement of "intelligent manufacturing", the intelligent degree of mechanical equipment is increasing day by day, and the reliability of equipment is put forward higher requirements. When early equipment fails and needs to be maintained, human experts are generally used to observe the operating status of the equipment, test the abnormal changes in the parameters such as noise, running track, temperature and vibration, compare them with the normal state, and make fault diagnosis based on long-term accumulated maintenance experience [1]. However, this method is time-consuming and labor-intensive, and requires high level of maintenance personnel. Therefore, with the continuous development of intelligent manufacturing, the timely and accurate diagnosis and analysis of equipment failure has become a top priority. Traditional fault diagnosis models are based on the working principle of equipment to establish a forward mathematical model, through real-time analysis of the abnormal conditions between each link to analyze the working status of equipment. However, with the emergence of multimode, uncertainty, high density and other characteristics of industrial field, it is very

Y. Wang et al. (Eds.): IWAMA 2021, LNEE 880, pp. 556–562, 2022.
https://doi.org/10.1007/978-981-19-0572-8_71

difficult to establish accurate mathematical model. In order to further improve the diagnosis efficiency and meet the industrial field with big data characteristics in the new era, new requirements for higher level fault diagnosis methods are put forward. In this paper artificial intelligence-based fault diagnosis are evaluated after the introduction, then an artificial intelligence diagnostic model for energy classification is dicussed, afterwards a critical evalution of artificial intelligence diagnostic model is presented. The paper ends with a conclusion.

2 Artificial Intelligence-Based Fault Diagnosis

In recent years, due to the increasing complexity, intelligence and mechatronic integration of industrial equipment, it is difficult to use traditional fault diagnosis methods to meet the needs of modern system equipment fault diagnosis. Thus, artificial intelligence based fault diagnosis technology is becoming more and more important. Compared with the traditional fault diagnosis method, the artificial intelligence based fault diagnosis model is changed from principles-driven to data-driven. It does not need to understand the working principle and fault cause of the equipment in detail. By collecting historical data of the normal working condition and fault state of the equipment, To build artificial intelligence network model to simulate human thinking mechanism to learn, explain and analyze the input data, gain the ability to interpret data knowledge, at the same time, according to the characteristics of the input data automatically adjust and update the network weights, improve the ability of feature extraction, or learning new knowledge, so as to establish the right diagnosis model [2].

2.1 Diagnosis Model Based on Neural Network

As a typical artificial intelligence method, neural network is widely used in automatic control and pattern recognition. With its multi-dimensional and non-linear dynamic characteristics, it has the ability to imitate human's intuitive association and memory function, which makes it have great advantages in the fault diagnosis of non-linear systems [3]. The neural network model consists of three layers: input layer, hidden layer and output layer. Each layer has a corresponding neural network connected to the next layer, as shown in Fig. 1 below.

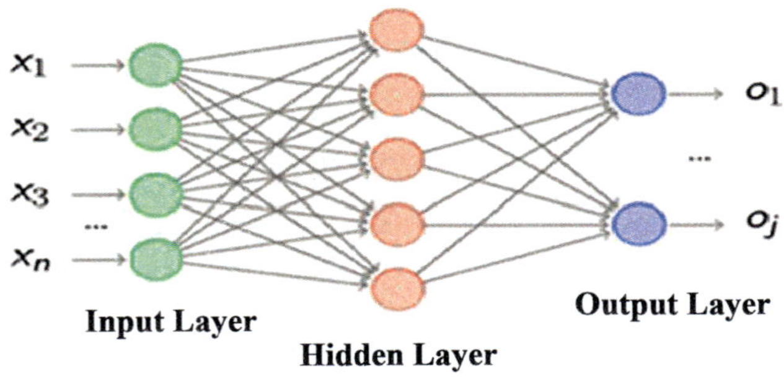

Fig. 1. Neural network model

The input layer is directly connected with the input data to obtain historical data information, which is standardized and passed to the next layer. Hidden layer is used for feature extraction, can let hidden layers by adjusting the weights of neural unit to respond to some sort of pattern formation, step by step to extract the topological structure characteristics of the hidden in the data, with the layers of network structure, the extracted features also gradually becomes abstraction, finally obtain input data translation, rotation and scale invariant features; The output layer is used to connect the hidden layer and output the results of the model. By adjusting the weight to form correct responses to different hidden layer neuron stimuli, the operation results of the model are output.

For fault diagnosis model, the input data is generally equipment running status, historical data, using neural network to study data in different working conditions, the corresponding equipment status, fitting system performance parameter and nonlinear function relationship between the fault type, the neural network input node corresponding to the output node fault symptoms correspond to the cause of the problem, Later, when the model encounters similar abnormal data again, the corresponding fault detection and conclusion can be obtained.

2.2 Diagnosis Model Based on Expert System

With its powerful ability to solve complex and experiential problems, expert system has become the most widely developed type of intelligent diagnosis model. It is based on specific areas within knowledge and experience of the diagnosis system, according to experts, rich practical experience, the thinking of experts to analyze and solve problems [4], establish a fault diagnosis knowledge base, rule base and reasoning machine, the knowledge base required for solving problems in knowledge, according to the knowledge base knowledge, rule base through the rules of reasoning mechanism and reasoning machine, Make a reasonable explanation for the fault phenomenon, as shown in Fig. 2 below:

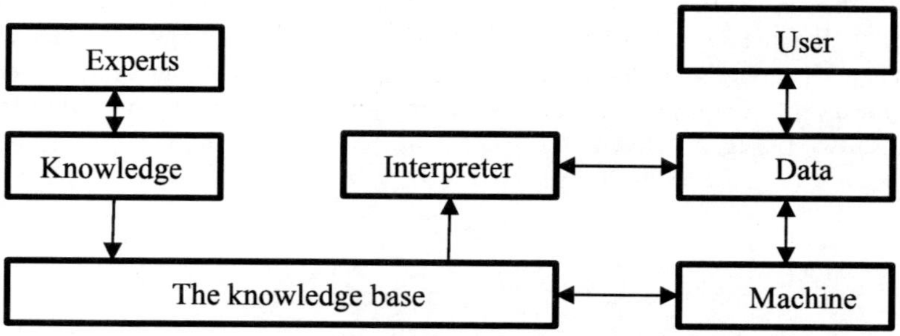

Fig. 2. Structure diagram of expert system

From the point of system framework, knowledge base contains diagnostic object model of knowledge and experience knowledge of human experts, covering the system of various fault conditions, and set the appropriate algorithm and rule by the engineer, after the reasoning machine read to the abnormal data model can fault monitoring, analysis and processing, such as process, help users to make decisions, Malfunction the system as if it were an expert on the spot.

2.3 Diagnosis Model Based on Fault Tree

As a kind of graph deductive method, the fault tree analysis method shows the logical and causal relationship between each part of the system [5]. It analyzes many factors of the system failure, from the whole to part according to the inverted tree step by step refinement analysis, through the expression of the fault chart image, intuitively reflects the relationship between the fault, system, component, fault cause. The modeling process based on the fault tree is shown in Fig. 3 below:

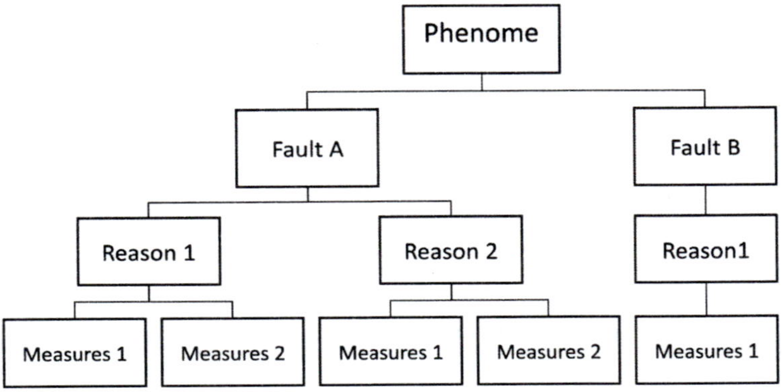

Fig. 3. Fault tree structure diagram

In fault tree analysis, the least expected fault state of the system is called the top event, and other fault states that may lead to the top event are intermediate events. The prevention and diagnosis of the top events as the primary goal of the model, searching out may induce the top in the middle of the events, drill-down qualitative analysis was carried out on the various events, find out the fault tree of all of the minimum cut sets, combination to make people understand what events can cause damage to the system security failure, and thus calculate the failure probability, Then diagnose, detect and repair the fault in time. For the fault tree model, the longer the running time, the richer the rules will be, and the higher the accuracy and rate of fault diagnosis will be.

2.4 Diagnosis Model Based on Fuzzy Theory

As the most successful diagnostic model in the field of fault diagnosis, fuzzy diagnosis method collects system data and expert experience, uses fuzzy theory to carry on fuzzy modeling and fuzzy clustering processing to obtain diagnosis results [6]. It has the simple model, convenient application, concept clear and direct diagnosis are shown in Fig. 4 below, has strong structural knowledge representation ability, a fuzzy diagnosis does not

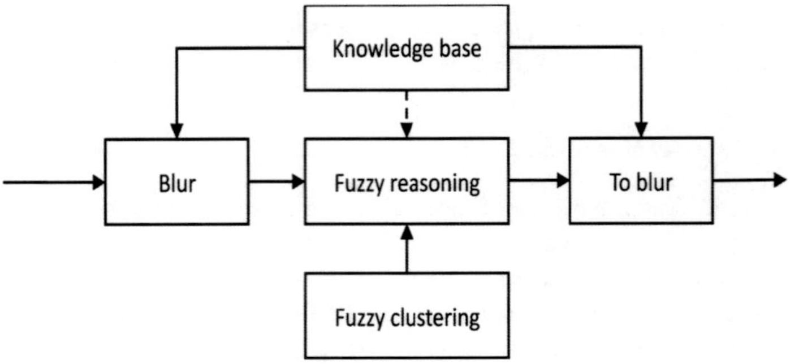

Fig. 4. Fuzzy theoretical structure diagram

need to establish accurate mathematical model of the system, but the way by simulating the human carries on the analysis, for complex systems also have stronger generalization.

Face real industrial field failure to quantitative analysis of a large, fuzzy principle by the status of data using fuzzy equivalence matrix standardization, fuzzy clustering by knowledge base and inference, the first-order matrix is obtained by judging, by solving the blur again after the output fault belongs to which subset, to diagnose the fault type,This diagnostic method shows high reliability for systems where accurate data and models cannot be obtained.

3 Development of an Artificial Intelligence Diagnostic Model for Energy Classification

Complexity and unpredictability of system fault prediction methods are also put forward higher request, single artificial intelligence fault diagnosis model is difficult to obtain accurate, complete and effective diagnosis knowledge, so in view of the complex system of industrial field, and artificial intelligent fault diagnosis model also spawned new features.

3.1 Multi-Information Fusion Diagnosis

The fault diagnosis method based on artificial intelligence is often used to model the vibration signal collected by acceleration sensor. With the constant improvement of the electronic technology and signal detection technology, especially the development of The Times, 5 g to reproduce more categories of sensor, using multi-dimensional data information for fault diagnosis analysis also become more easy, for example, the equipment vibration signal and noise signal, temperature, humidity, pressure, torque, etc., all-round information represents the working condition of equipment more detailed, It also provides more training variables for the fault diagnosis model, which can effectively improve the training accuracy and accuracy of the model, reduce the influence of external factors, and enhance the robustness of the fault diagnosis model.

3.2 Compound Fault Diagnosis

Both the traditional analytical model method and the diagnosis method based on artificial intelligence can only analyze and diagnose a single fault type, but the actual industrial site will not be so ideal, often a variety of faults occur at the same time. For composite fault diagnosis problem, some scholars put forward a tabbed K nearest neighbor algorithm classification method, by mining and the relationship between different fault types and fault data correlation strength, extract the node characteristics of social structure classifier to label different dimensions of data classification, in the case of equipment complex fault accurate diagnosis analysis.

3.3 Potential Failure Warning

Industrial field devices of continuous work for a long time, make the system running inevitably occur in the process of fatigue damage, appear early failure, these early damage and failure often has the potential and weak dynamic response, through the model data cannot be analyzed a exception, system can still function properly, and even some intermittent failures can disappear by oneself, But by the time the model was able to diagnose the fault, it was too late. Therefore, the model is required to be able to monitor the system based on big data and timely detect and warn early faults and anomalies of the system, carry out maintenance and adopt targeted maintenance strategies according to the system operation rules or data change trends, so as to eliminate potential risks of the system.

3.4 Hybrid Diagnostic Model

Since the concept of fault diagnosis was put forward, various diagnosis methods have emerged in an endless stream. However, there is still no diagnosis method suitable for all scenarios. Therefore, multi - model fusion collaborative fault diagnosis has entered people's sight. The hybrid diagnosis model includes the combination of traditional signal processing methods and artificial intelligence models, as well as the hybrid diagnosis of multiple artificial intelligence models. The signal processing method can process and analyze the massive data of industrial field, and then submit it to artificial intelligence modeling and diagnosis, which can effectively improve the learning speed and accuracy of the model. According to the characteristics of different models, it is very important to realize the mixed diagnosis of different models in multi-mode, big data and uncertain industrial field.

4 Challenges of Artificial Intelligence Diagnostic Model

Artificial intelligence tries to find the internal structure of the data and discover the essential relationship between variables by analyzing the historical data of the equipment and learning the way of thinking of people.Relevant studies have shown that the way of data representation has a great influence on the success of training and learning. Good representation can eliminate the influence of changes in input data unrelated to learning tasks on learning performance, while retaining useful information for learning tasks.In recent years, artificial intelligence has developed rapidly in fault diagnosis, but there are also some challenging problems.

(1) In essence, the diagnostic mode method of artificial intelligence is to model by learning massive data of industrial site. However, rapid processing and feature recognition of industrial big data have always been a problem for engineers. How to obtain and analyze accurate and effective system data is the primary challenge of artificial intelligence modeling.
(2) The establishment of the model is based on the analysis of historical fault data, so artificial intelligence can only learn the types of faults that have been shown before, and for faults that have never occurred, the model may still fail to identify even if the data has been abnormal.
(3) The artificial intelligence diagnostic model is different from the mathematical model of forward analysis, but is constructed by multiple deep learning intelligent networks. The understanding of the model is more like a "black box", which can only obtain the prediction results of the model but cannot be theoretically explained, so it may be difficult to explain the cause of system failure.of the supply-chains must be created.

5 Conclusions

With the new round of industry 5.0 proposed and intelligent manufacturing requirements, advanced industrial equipment and stable operation requirements for system fault diagnosis into a new stage, fault diagnosis method based on artificial intelligence for the efficient and stable operation of the industrial site to provide a guarantee. Based on the development of artificial intelligence, the principle and application of common artificial intelligence fault diagnosis models are described in this paper. The future development direction and potential challenges are proposed, which provides some help for the development of intelligent fault diagnosis of industrial equipment.

References

1. Shilin, W., Yuguang, N., Xiaoming, L., Zhongwei, L.: Application of Multivariable State Estimation Fault Early Warning in Industrial Process vol. 6 (2014)
2. Baoguo, S.: Application of artificial intelligence technology in electrical automation control. Electr. World **11**, 180–181 (2021)
3. Zou, J., Gao, Z., Wang, N.: Application of artificial neural network in pattern recognition. In: China Command and Control Society: China Command and Control Society, vol. 7 (2021)
4. Shi, J., Huo, Z., Zhu, R.: Research on Fault Diagnosis Expert System of Mechatronics System, vol. 3 (2014)
5. Chen, H., Zhao, A., Li, T., Cai, C., Cheng, S., Xu, C.: Fuzzy Bayesian network reasoning fault diagnosis of complex equipment based on fault tree. Syst. Eng. Electr. Technol. **43**(05), 1248–1261 (2021)
6. Wang, X.: Research on Fault Diagnosis Method of Industrial Equipment Based on Fuzzy Multi-Attribute Decision Making. Qilu University of Technology (2020)

Power Convex Operator-Based Multiple-Criteria Decision Making for Hesitant Multiplicative Fuzzy Information

Ye Mei[1](✉), Bo Chen[2], Junjie Yang[3], and Yufeng Chen[1]

[1] School of Electrical and Information Engineering, Hubei University of Automotive Technology, Shiyan 442002, China
{20070015,chenyfdy}@huat.edu.cn

[2] School of Electronic and Electrical Engineering of Lingnan Normal University, Zhanjiang 524048, China
chenbo@lingnan.edu.cn

[3] School of Computer Science and Intelligence Education of Lingnan Normal University, Zhanjiang 524048, China
yangjunjie@lingnan.edu.cn

Abstract. Multi-hesitant fuzzy set (MHFS) is an extended hesitant fuzzy set, which allows the values to be repeated more than once. Here, we introduce a new convex combination of MHFS values. The weighted MHF power average operator is established and the corresponding properties are explained. Next, a novel aggregation operator-based multiple-criteria decision making approach is designed for the problem of ranking alternatives. Lastly, we provide an example to validate the proposed method and evaluate its feasibility and robustness.

Keywords: Convex operation · Aggregation operator · Multiple-criteria decision making · Multi-hesitant fuzzy sets

1 Introduction

Hesitant fuzzy (HF) and multi-hesitant fuzzy (MHF) sets have been initially proposed by Torra and Narukawa [1] and Torra [2], which are the extended fuzzy sets based on Zadeh's principle [3]. The degrees of "membership" from zero to one are often used to replace the possible truth values. HF sets is commonly employed to solve the problem of multiple-criteria decision making (MCDM), and has received increasing attention in recent years. Several studies have been focused on the aggregation operators of HF sets for solving MCDM problems [4, 5]. In addition, HF sets-based distance and correlation coefficient measures have also been established [6–8]. For certain values that are repeated more than once in HF sets, MHF sets has been developed to address such constraints.

In this paper, the novel MCDM approach is developed based on power convex aggregation operators of MHF sets. First, both HF sets and MHF sets are defined in details, and their comparison methods are presented. Next, the corresponding aggregation operator is introduced. Then, a novel convex aggregation operator-based MCDM approach

Y. Wang et al. (Eds.): IWAMA 2021, LNEE 880, pp. 563–569, 2022.
https://doi.org/10.1007/978-981-19-0572-8_72

is developed. After that, we provide an example to validate the proposed method by performing sensitivity and comparative analyses. Lastly, few conclusions are made.

2 Hesitant and Multi-hesitant Fuzzy Sets

Both HF sets and MHF sets are defined in this section. The comparison methods for HF sets subsequent analysis are also described.

Definition 1 [6, 9]. Let X be a universal set, and a HF sets E on X is a function that returns a subset of [0, 1] when it is implemented to X.

The simplified formula is [5] shown as follows:

$$E = \{\langle x, h_E(x)\rangle | x \in X\} \tag{1}$$

Definition 2 [2]. X represent a universal set, MHF sets is described as E_M based on a function H_A that will return a multi-subset of [0, 1] when it is implemented to X.

According to Definition 1, MHF sets can be described as follows:

$$E_M = \left\{\langle x, H_{E_M}(x)\rangle \middle| x \in X\right\} \tag{2}$$

Definition 3 [5]. Let H_A and H_B be two hesitant fuzzy numbers on X, the convex combination of MHF numbers is presented as follows:

(i) if $H_A \prec H_B$, only if $s(H_A) = s(H_B)$;
(ii) if $s(H_A) < s(H_B)$, then:

if $H_A \sim H_B$, only if $f(H_A) = f(H_B)$;
if $H_A \succ H_B$, only if $f(H_A) < f(H_B)$;
if $H_A \prec H_B$, only if $f(H_A) > f(H_B)$.

3 The Convex Combination Operation and the Weighted Multi-hesitant Fuzzy Power Average Aggregation Operator

Next, we introduce the convex combinations of MHF numbers, and demonstrate the corresponding aggregation operators.

Definition 4 [10]. Let H_1 and H_2 be 2 MHF numbers. A convex combination of H_1 and H_2 is defined as

$$C^2(w_1, \mathrm{H}_1, w_2, H_2) = w_1 \otimes H_1 \oplus w_2 \otimes H_2 = \left\{ \left(w_1\gamma_1^\lambda + w_2\gamma_2^\lambda\right)^{1/\lambda} \middle| \gamma_1 \in H_1, \gamma_2 \in H_2 \right\}, \lambda > 0 \tag{3}$$

where $w_1 \geq 0$, $w_2 \geq 0$ and $w_1 + w_2 = 1$. The nonlinear weighted-average aggregation tool used in this study is power average operator (PAO).

Definition 5. The map PA : $R^n \to R$, is used in the PAO, which can be described as:

$$PA(a_1, a_2, \cdots, a_n) = \frac{\sum_{i=1}^{n}(1 + S(a_i))a_i}{\sum_{i=1}^{n}(1 + S(a_i))} \quad (4)$$

Here $S(a_i) = \sum_{i=1, j \neq i}^{n} Supp(a_i, a_j)$, and $Supp(a_i, a_j)$ represents the support for a_i from a_j, which satisfy the following properties:

(i) $Supp(a_i, a_j) \geq Supp(a_p, a_q)$ *iff* $|a_i - a_j| < |a_p - a_q|$;
(ii) $Supp(a_i, a_j) = Supp(a_j, a_i)$;
(iii) $Supp(a_i, a_j) \in [0, 1]$.

Obviously, the greater the support from each other, the closer the two values obtained.

Definition 6. Let $H_i (i = 1, 2, \ldots, n)$ be a subset of MHF numbers. The weighted MHF PAO (WMHFPAO) maps $WMHFPA : MHFN^n \to MHFN$, which contains a weight vector $w = (w_1, w_2, \ldots, w_n)$ with $w_i \geq 0 (i = 1, 2, \ldots, n)$ and $\sum_{i=1}^{n} w_i = 1$, and

$$\begin{aligned} & WMHFA(H_1, H_2, \ldots, H_n) \\ & = C^n\left(w_k, H_{\sigma(k)}, k = 1, 2, \ldots, n\right) \\ & = w_1 \otimes H_{\sigma(1)} \oplus (1 - w_1) \otimes C^{n-1}\left\{w_i \Big/ \sum_{k=2}^{n} w_k, H_{\sigma(i)}, i = 2, 3, \ldots, n\right\}. \end{aligned} \quad (5)$$

where $(\sigma(1), \sigma(2), \ldots, \sigma(n))$ is meant by a permutation of $(1, 2, \ldots, n)$, thereby $H_{\sigma(1)} \leq H_{\sigma(2)} \leq \cdots \leq H_{\sigma(n)}$.

Theorem 1. Let $H_i (i = 1, 2, \ldots, n)$ using the WMHFPAO is also a MHFN, and

$$\begin{aligned} & WMHFPA(H_1, H_2, \ldots, H_n) \\ & = w_1 \otimes \frac{\mathrm{T}(H_1)H_1}{\sum_{i=1}^{n} w_i(HT(H_{\theta 1}))} \oplus (1 - w_1) \otimes C^{n-1}\left\{\frac{w_1}{\sum_{k=2}^{n} w_i}, H_i\right\} \\ & = \left\{\frac{1 + S(H_1)}{\sum_{i=1}^{n} w_i(1 + S(H_1)}\gamma_1 + \frac{1 + S(H_2)}{\sum_{i=1}^{n} w_i(1 + S(H_2)}\gamma_2\right. \\ & \left. + \cdots \frac{1 + S(H_n)}{\sum_{i=1}^{n} w_i(1 + S(H_n)}\gamma_n \,\middle|\, \gamma_i \in H_i, i = 1, 2, \cdots n\right\} \end{aligned} \quad (6)$$

Here, $S(H_j) = \sum_{i=1, i \neq j}^{n} w_j Supp(H_{ij}, H_i)$.

Proposition 2. Let $H_i(i = 1, 2, \ldots, n)$ be a subset of MHF numbers, the following properties are true.

(1) (Monotonicity) Let $H_i'(i = 1, 2, \ldots, n)$ be a subset of MHF numbers. If for all i, $H_i \leq H_i'$, then

$$\text{WMHFPA}(H_1, H_2, \ldots, H_n) \leq \text{WMHFPA}\left(H_1', H_2', \ldots, H_n'\right) \tag{7}$$

(2) (Commutativity) If $H_1^*, \ldots, H_n^*$ is a permutation of $H_1, \ldots, H_n$, thus

$$\text{WMHFPA}\left(H_1^*, H_2^*, \ldots, H\right)_n^* = \text{WMHFPA}(H_1, H_2, \ldots, H_n) \tag{8}$$

(3) (Boundedness) If $H^- = \left\{\gamma_1^-, \gamma_2^-, \ldots, \gamma_n^-\right\}$ and $H^+ = \left\{\gamma_1^+, \gamma_2^+, \ldots, \gamma_n^+\right\}$, where $\gamma_i^- = \underset{\gamma_i \in H_i}{min}\ \gamma_i$ and $\gamma_i^+ = \underset{\gamma_i \in H_i}{max}\ \gamma_i$, then

$$H^- \leq \text{WMHFPA}(H_1, H_2, \ldots, H_n) \leq H^+ \tag{9}$$

4 The Aggregation Operator-Based MCDM Approach with Numbers

The MCDM selection/ranking problem with MFH data contains a series of alternatives, as delineated by $A = \{a_1, a_2, \ldots, a_n\}$. The alternatives are determined according to $C = \{c_1, c_2, \ldots, c_m\}$ a_{ij} indicates the value of the alternative a_i based on the criterionc_j, and $a_{ij} = \left\{\gamma_{ij}^k, k = 1, 2, \ldots, l\left(a_{ij}\right)\right\}(i = 1, \ldots, n; j = 1, \ldots, m)$ consists of MHF numbers, which can be generated by various decision makers. The number of elements in a_{ij} is represented by $l\left(a_{ij}\right)$ and $w = (w_1, w_2, \ldots, w_m)$. This technique is useful when only a small number of decision makers is available. The alternatives can be evaluated by decision makers according to the decision criteria, and a decision maker can give numerous evaluation values. Particularly, when a similar value is generated by two or more decision makers, it is considered as a repetition. a_{ij} represents a group of evaluation values made by all decision makers.

In this approach, aggregation operators and MHF numbers are integrated to address the above-mentioned MCDM problems. A sequence of steps is given below:

Step 1. Normalization of the decision matrix.

The major criteria of MCDM problem belong to minimizing and maximizing types. The evaluation values should be normalized in order to unify all criteria, except for the criteria with maximizing type and similar measurement values. Assuming that the matrix $R = \left(a_{ij}\right)_{n \times m}$, where the MHF numbers are indicated by $a_{ij} = \left\{\gamma_{ij}^1, \gamma_{ij}^2, \ldots, \gamma_{ij}^k\right\}$ $(i = 1, 2, \ldots, n;\ j = 1, 2, \ldots, m;\ k = 1, 2, \ldots, l\left(a_{ij}\right))$, the normalized matrix is $\tilde{R} = \left(\tilde{a}_{ij}\right)_{n \times m}$. Where $\tilde{a}_{ij} = \left\{\overset{\sim 1}{\gamma}_{ij}, \overset{\sim 2}{\gamma}_{ij}, \ldots, \overset{\sim k}{\gamma}_{ij}\right\}$ $(i = 1, 2, \ldots, n;\ j = 1, 2, \ldots, m;$ $k = 1, 2, \ldots, l\left(a_{ij}\right))$. The number of elements in a_{ij} is represented by $l\left(a_{ij}\right)$.

The normalization equation for the criteria with maximizing type is as follows:

$$\tilde{\gamma}_{ij}^{k} = \gamma_{ij}^{k}, k = 1, 2, \ldots, l(a_{ij}) \tag{10}$$

The normalization equation for the criteria with minimizing type is as follows:

$$\tilde{\gamma}_{ij}^{k} = 1 - \gamma_{ij}^{k}, k = 1, 2, \ldots, l(a_{ij}) \tag{11}$$

Seemingly, the normalization values $\tilde{a}_{ij} = \left\{\tilde{\gamma}_{ij}^{1}, \tilde{\gamma}_{ij}^{2}, \ldots, \tilde{\gamma}_{ij}^{k}\right\}$ $\{i = 1, 2, \ldots, n; j = 1, 2, \ldots, m\}$ are also MHF numbers.

Step 2: Calculation of the support.

All supports are calculated as follows:

$$Supp(a_{ij}, a_{ik}) = 1 - d(a_{ij}, a_{ik})\ i = 1, 2, \ldots, n; j, k = 1, 2, \ldots, m; j \neq k \tag{12}$$

Here $Supp(a_{ij}, a_{ik})$ is the support for a_{ij} from a_{ik}, which meets the requirements described in Eq. (12).

$$d(a_{ij}, a_{ik}) = \frac{1}{2}\left(\max_{\gamma_{ij} \in a_{ij}} \min_{\gamma_{ik} \in a_{ik}} |\gamma_{ij} - \gamma_{ik}|\right) + \max_{\gamma_{ik} \in a_{ik}} \min_{\gamma_{ij} \in a_{ij}} |\gamma_{ik} - \gamma_{ij}| \tag{13}$$

Step 3: Calculation of the weighted support.

The weighted support $S(a_{ij})$ of the multi-valued neutrosophic (MVN) number a_{ij} by the other MVN numbers is determined based on the weight w_j of c_j $(j = 1, 2, \cdots, m)$.

$$S(a_{ij}) = \sum\nolimits_{k=1, k \neq j}^{m} w_k Supp(a_{ij}, a_{jk})\ (k = 1, 2, \cdots m) \tag{14}$$

Step 4: Obtain the weights associated with each evaluation value. Based on Step 3, the weight τ_{ij} associated with a_{ij} can be obtained as follows:

$$\tau_{ij} = \frac{w_j(1 + S(a_{ij}))}{\sum\limits_{j=1}^{n} w_j(1 + S(a_{ij}))}\ (j = 1, 2, \cdots m) \tag{15}$$

Here $\tau_{ij} \geq 0$ and $\sum_{j=1}^{m} \tau_{ij} = 1$.

Step 5: Aggregate the MVNNs.

Utilize the WMHFPA operator to aggregate the MVNNs a_{ij}, and the aggregate value y_i of the alternative a_i $(i = 1, 2, \ldots, n)$ is determined..

Step 6: Calculation of the accuracy function $a(y_i)$ and score function $s(y_i)$ values of $y_i (i = 1, 2, \ldots, m)$ according to Definition 3.

Step 7: The ranking of each alternative.

5 An Illustrative Example

To validate the feasibility and robustness of the proposed method, an example is adapted from a previous report [11]. A company wishes to invest in the projects of car (α_1); food

(α_2); computer (α_3); arms (α_4); and TV (α_5) companies. The company makes a decision based on 4 criteria: the environmental impact (c_1), risk (c_2); growth prospects (c_3); and social-political impact (c_4). The associated weight is $w = (0.33, 0.18, 0.37, 0.12)$. The five possible alternatives $\alpha_i (i = 1, 2, \ldots, 5)$ are analyzed based on the MHF data of 2 decision makers (Table 1). The decision makers provide the evaluation value $\alpha_{ij} (i = 1, 2, 3, 4, 5; j = 1, 2, 3, 4)$ in the form of MFH numbers based on their experience and knowledge. When a similar value is generated by these decision makers, it is regarded as a repetition.

Table 1. MHF decision matrix

	c_1	c_2	c_3	c_4
α_1	{0.5,0.6,0.9}	{0.3,0.3,0.6}	{0.3,0.3,0.8}	{0.2,0.7}
α_2	{0.3,0.4,0.6}	{0.7,0.8}	{0.3,0.4,0.1}	{0.4,0.6}
α_3	{0.5,0.3}	{0.2,0.4,0.6}	{0.7,0.8}	{0.3,0.4}
α_4	{0.1,0.3,0.5}	{0.3,0.6}	{0.7,0.3}	{0.4,0.5,0.6}
α_5	{0.3,0.5}	{0.6,0.4}	{0.3,0.6}	{0.2,0.2,0.5}

When considering the numbered positions of the criteria's MHF values, WMHFPAO will be employed. A sequence of steps is given below:

Step 1. Normalization of the data in Table 1.

Given that all criteria are of maximizing type with similar measurement values, data normalization is not required and $\tilde{R} = \left(\tilde{a}_{ij}\right)_{5\times4} = \left(a_{ij}\right)_{5\times4}$.

Step 2. According to Eq. (11), the supports $Supp(a_{ij}, a_{ik})$ $(i = 1, 2, \ldots, n;\ j, k = 1, 2, \ldots, m;\ j \neq k)$ are obtained as follows

$$Supp(a_{i1}, a_{i2}) = Supp(a_{i2}, a_{i1}) = \begin{bmatrix} 0.0333 \\ 0.0167 \\ 0.0500 \\ 0.0833 \end{bmatrix} Supp(a_{i1}, a_{i3}) = Supp(a_{i3}, a_{i1}) = \begin{bmatrix} 0.3833 \\ 0.4333 \\ 0.4000 \\ 0.4500 \end{bmatrix}$$

$$Supp(a_{i2}, a_{i3}) = Supp(a_{i3}, a_{i2}) = \begin{bmatrix} 0.3833 \\ 0.4833 \\ 0.3500 \\ 0.4000 \end{bmatrix}$$

Steps 3–5. Calculation of the weighted supports.

Utilize Eq. (9) to aggregate the MVN numbers of all alternatives, and the aggregate value of $\alpha_i (i = 1, 2, \cdots, n,\ j = 1, 2, \cdots, m)$ is obtained.

$$s(y_1) = 0.5810;\ s(y_2) = 0.5782;\ s(y_3) = 0.5240;\ s(y_4) = 0.5247;\ s(y_5) = 0.5485$$

The accuracy function values are not computed due to the varying score function values.

Step 6. The ranking of each alternative.

According to Step 3, the final ranking is $a_3 \prec a_4 \prec a_5 \prec a_2 \prec a_1$, which may be attributed to $s(y_3) < s(y_4) < s(y_5) < s(y_2) < s(y_1)$. Therefore, a_1 and a_3 are the best and worst alternatives, respectively.

6 Conclusion

In this study, the convex combinations of MHF numbers are introduced and the corresponding power aggregation operator is established. A novel aggregation operator-based MCDM approach is proposed to solve the problems associated with MHF numbers. Moreover, an example is provided to validate the developed method. The main feature of this method is that the aggregation operator can improve the decision-making process by offering more choices in terms of actual environments. Nevertheless, more investigations are needed to optimize the values of criteria through a specific MFH approach in a given environment.

References

1. Torra, V., Narukawa, Y.: On hesitant fuzzy sets and decision. In: The 18th IEEE International Conference on Fuzzy Systems, Jeju Island, Korea, pp. 1378–138 (2009)
2. Torra, V.: Hesitant fuzzy sets. Int. J. Intell. Syst. **25**(6), 529–539 (2010)
3. Zadeh, L.A.: Fuzzy sets. Inf. Control **8**, 338–356 (1965)
4. Mahmood, T.: Power aggregation operators and similarity measures based on improved intuitionistic hesitant fuzzy sets and their applications to multiple attribute decision making. Comput. Model. Eng. Sci. **126**, 1165–1187 (2021). https://doi.org/10.32604/cmes.2021.014393
5. Xu, Z.S., Xia, M.M.: Distance and similarity measures for hesitant fuzzy sets. Inf. Sci. **181**(11), 2128–2138 (2011)
6. Singha, B., Sen, M., Sinha, N.: Modified distance measure on hesitant fuzzy sets and its application in multi-criteria decision making problem. Opsearch **57**(2), 584–602 (2019). https://doi.org/10.1007/s12597-019-00431-x
7. Liu, X., Wang, Z., Zhang, S., Garg, H.: Novel correlation coefficient between hesitant fuzzy sets with application to medical diagnosis. Expert Syst. Appl. **183**, 115393 (2021). https://doi.org/10.1016/j.eswa.2021.115393
8. Meng, F., Xu, Y., Wang, N.: Correlation coefficients of dual hesitant fuzzy sets and their application in engineering management. J. Ambient. Intell. Humaniz. Comput. **11**(7), 2943–2961 (2019). https://doi.org/10.1007/s12652-019-01435-7
9. Na, C., Xu, Z.: Hesitant fuzzy ELECTRE II approach: a new way to handle multi-criteria decision making problems. Inf. Sci. **292**, 175–197 (2015)
10. Mei, Y., Peng, J.J., Yang, J.J.: Convex aggregation operators and their applications to multi-hesitant fuzzy multi-criteria decision-making. Information **9**, 207 (2018). https://doi.org/10.3390/info9090207
11. Schmeidler, D.: Subjective probability and expected utility without additivity. Econometrica **57**, 517–587 (1989)

Check for updates

Research on SOC and SOP Co-simulation Estimation of Lithium-Ion Battery for Vehicle

Mingjie Dai and Xuehuan Jiang(✉)

School of Electrical and Information Engineering, Hubei University of Automotive Technology, Shiyan 442002, Hubei, China

Abstract. The state of charge (SOC) and continuous peak power (SOP) of lithium-ion battery are important criteria to evaluate the stable operation of electric vehicles. In the paper, a battery model was built based on the second-order Thevenin equivalent circuit. In UDDS condition, an extended Kalman filter algorithm was used to estimate the SOC. The battery SOC, terminal voltage and internal parameters were taken as the constraints to estimate the SOP. Finally, the simulation is implemented in MATLAB Simulink environment and analyzed in detail. The results show that the proposed method has better estimation accuracy and more stable static performance, which has a certain reference value for the research of lithium battery SOP estimation.

Keywords: Second-order Thevenin model · SOC estimation · SOP estimation

1 Introduction

Accurate estimation of SOC and SOP is an important criterion for evaluating the performance of electric vehicle battery. There are many methods for SOC estimation at home and abroad. The open-circuit voltage method is relatively simple, but it is not suitable for the real-time estimation environment of electric vehicles because it requires a long standing time [1]. The emphasis of neural network method is not the battery internal structure, which focus on the prediction of current, voltage and other data, data set collection requires a lot of experimentation, the process is relatively complicated [2, 3]. The method of amper hour intergal is based on the value of the previous moment to obtain the value of the current moment. The advantage is that the calculation method is simple, while the cumulative error will increase due to its iteration [4]. Extended Kalman filter (EKF) is suitable to nonlinear system. Even in the case of inaccurate initial value, it can also observe the state variables, and the error is calculated separately [5, 6]. The estimation methods of SOP mainly include interpolation method, data driven method and model method. The interpolation method based on HPPC cycle discharge experiment requires a large number of experiments to calculate the power value at the charging-discharge cut-off voltage. The method is relatively simple, but the polarization phenomenon inside the battery and its own dynamic characteristics are ignored. The model method is mainly divided into terminal voltage based method and SOC based method. The terminal voltage method can improve the accuracy to a certain extent through the selection of model,

Y. Wang et al. (Eds.): IWAMA 2021, LNEE 880, pp. 570–577, 2022.
https://doi.org/10.1007/978-981-19-0572-8_73

while the SOC method is easy to cause the problem of excessive peak current, which has safety risks. Data-driven method is mainly a driving method based on BP neural network, which requires a large amount of training data to obtain the value of SOP, resulting in a large amount of calculation and difficulty. In this paper, the second-order Thevenin model is established, and the EKF algorithm is used to estimate SOC [7, 8]. The SOC estimation results are taken as one key constraint of SOP estimation to estimate SOP under multiple constraints, and the error analysis and verification of the algorithm are carried out.

2 Battery Modeling and SOC Estimation

2.1 Construction of the Second-Order Thevenin Equivalent Circuit Model

An appropriate battery model is the basis for estimating SOC and SOP. Considering the complexity, accuracy and model compatibility, the second-order Thevenin model is selected to build the equivalent circuit, whiceh is shown in Fig. 1.

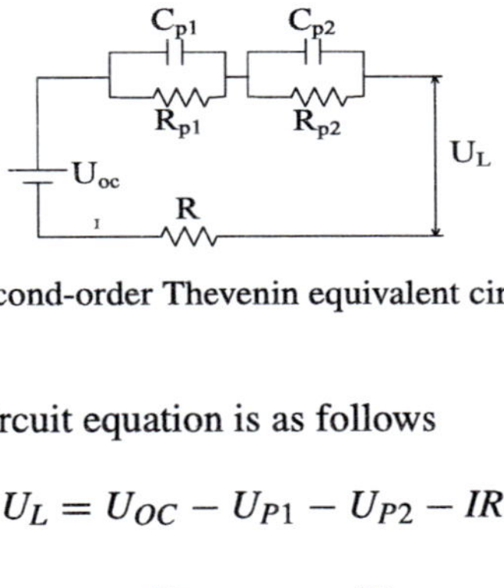

Fig. 1. Second-order Thevenin equivalent circuit model

The equivalent model circuit equation is as follows

$$U_L = U_{OC} - U_{P1} - U_{P2} - IR \tag{1}$$

$$I = \frac{U_{P1}}{R_{P1}} + C_{P1}\frac{dU_{P1}}{dt} \tag{2}$$

U_{P1}, U_{P2} represents the respective voltage in the two RC loops. I is the current in the circuit, R_{P1} and R_{P2} is the two polarization internal resistance, C_{P1} and C_{P2} is the polarization capacitance, U_{OC} is the open-circuit voltage, U_L is the terminal voltage, t is the operating time, R is the ohm internal resistance [9].

According to the second-order RC model constructed above, the parameters to be identified include U_{OC}, R_{P1} and R_{P2}, C_{P1} and C_{P2}, and R. In the paper, ternary lithium-ion square shell battery (3–4.2 V) with a nominal capacity of 32 Ah was used to identify its parameters under HPPC cycle discharge test. The identification results are shown in Table 1.

Table 1. Battery parameters identification value

SOC	U_{oc} (v)	R (mΩ)	R_{p1} (mΩ)	C_{p1} (F)	R_{p2} (mΩ)	C_{p2} (F)
0.0	3.423	2.5833	2.046	50000.69	1.073	65826.42
0.1	3.521	2.2188	1.095	69256.99	1.080	40240.71
0.2	3.596	2.0313	1.190	12808.68	1.100	92736.85
0.3	3.644	2.0156	1.140	10729.94	1.078	83004.39
0.4	3.696	2.0000	1.050	84639.03	1.135	30248.95
0.5	3.784	1.9531	1.014	80169.67	1.194	19311.70
0.6	3.880	1.9375	1.106	11549.97	1.138	121686.4
0.7	3.948	1.9688	1.140	45213.41	1.357	80606.25
0.8	4.020	1.9687	1.061	45006.78	1.357	110572.4
0.9	4.075	1.9688	0.988	30860.00	1.148	80140.93
1.0	4.181	1.9531	0.723	40869.57	1.064	78675.78

2.2 SOC Estimation Method and Analysis

The Kalman filter method uses the previous state estimation and the current output to observe the current state estimate value. The greatest advantage of this method is that the error is calculated separately and does not apply to the system. Suppose that the system output equation and state equation are given as

$$\boldsymbol{x}_{k+1} = \boldsymbol{A}\boldsymbol{x}_k + \boldsymbol{B}\boldsymbol{u}_k + \boldsymbol{\Gamma}\boldsymbol{w}_k \tag{3}$$

$$\boldsymbol{y}_k = \boldsymbol{C}\boldsymbol{x}_k + \boldsymbol{D}\boldsymbol{u}_k + \boldsymbol{v}_k \tag{4}$$

x_k is the value of the variable at time k, u_k is the input quantity of the system, y_k is the output value of the system at time k, $\boldsymbol{A}$ is the transfer matrix of the system, $\boldsymbol{B}$ is the input matrix of the system, $\boldsymbol{C}$ is the output matrix of the system, $\boldsymbol{D}$ is the feedforward matrix of the system, $\boldsymbol{w_K}$ is the process noise of the system, $\boldsymbol{v_k}$ is the measurement noise of the system, $\boldsymbol{\Gamma}$ is the noise matrix. Kalman filtering basic step for the first prior state estimation is as shown in Eq. (5), and then calculate the error covariance matrix and the kalman gain through Eq. (6), Eq. (7), Under this premise, the system state estimator and the error covariance matrix at the next moment are updated, use Eq. (8), Eq. (9), Eq. (10).

$$\hat{\boldsymbol{x}}_{k+1/k} = \boldsymbol{A}\hat{\boldsymbol{x}}_k + \boldsymbol{B}\boldsymbol{u}_k \tag{5}$$

$$\boldsymbol{P}_{k+1/k} = \boldsymbol{A}\boldsymbol{P}_k\boldsymbol{A}^T + \boldsymbol{\Gamma}\boldsymbol{Q}_k\boldsymbol{\Gamma}^T \tag{6}$$

$$\boldsymbol{L}_k = \boldsymbol{P}_{k+1/k}\boldsymbol{C}^T(\boldsymbol{C}\boldsymbol{P}_{k+1/k}\boldsymbol{C}^T + \boldsymbol{R}_k)^{-1} \tag{7}$$

$$\hat{\boldsymbol{x}}_{k+1} = \hat{\boldsymbol{x}}_{k+1/k} + \boldsymbol{L}_k(\boldsymbol{y}_{k+1} - \boldsymbol{C}\boldsymbol{x}_{k+1/k} - \boldsymbol{D}\boldsymbol{u}_{k+1}) \tag{8}$$

$$\boldsymbol{Z} = \boldsymbol{y}_{k+1} - \boldsymbol{C}\boldsymbol{x}_{k+1/k} - \boldsymbol{D}\boldsymbol{u}_{k+1} \tag{9}$$

$\boldsymbol{Z}$ is the error between the measured output value and the predicted output value, $\boldsymbol{L_K}$ is the weighted value of a non-zero term caused by system inaccuracies. The larger the error, the larger the update range.

$$\boldsymbol{P}_{k+1} = (\boldsymbol{E} - \boldsymbol{L}_k\boldsymbol{C})\boldsymbol{P}_{k+1/k} \tag{10}$$

$\boldsymbol{E}$ is the identity matrix. According to the above equation, it can be shown that the Kalman filter algorithm is applicable to the multi-input multi-output system. Compar-ed with other algorithms, this algorithm reduces the burden of system calculation and is easy to implement [10, 11].

3 Multi-constraint SOP Estimation

The battery is a highly nonlinear system. In fact, the peak power is restricted by temperature, available capacity, SOC and other aspects. If a singal constraint, it is difficult to estimate the peak power accurately, which will cause irreversible damage to the battery. Therefore, the peak battery power must be limited to a safe range, which is deremined by the battery cut-off protection voltage,maximum pulse current and SOC. At the same time period, SOP estimation based on multi-constraint conditions can maximize the performance of batteries on the basis of protecting battery safety [12].

3.1 Maximum Current Estimation

Model-based maximum current estimation is given as

$$\begin{cases} i_{\max,k+L}^{dis,v} = \dfrac{u_{ocv}(z_k, c_N) - \left(e^{-\frac{T_S}{\tau_1}}\right)^L U_{P1,K} - \left(e^{-\frac{T_S}{\tau_2}}\right)^L U_{P2,K} - U_{L,\min}}{L\eta T_S / C_N \frac{\partial U_{OCV}}{\partial Z}\Big|_{Z=Z_K} + R_{P1}\left(1 - e^{-\frac{T_S}{\tau_1}}\right)\sum\limits_{J=0}^{L-1}\left(e^{-\frac{T_S}{\tau_1}}\right)^{L-1-j} + R_{P1}\left(1 - e^{-\frac{T_S}{\tau_2}}\right)\sum\limits_{J=0}^{L-1}\left(e^{-\frac{T_S}{\tau_2}}\right)^{L-1-j} + R} \\ i_{\min,k+L}^{chg,v} = \dfrac{u_{ocv}(z_k, c_N) - \left(e^{-\frac{T_S}{\tau_1}}\right)^L U_{P1,K} - \left(e^{-\frac{T_S}{\tau_2}}\right)^L U_{P2,K} - U_{L,\max}}{L\eta T_S / C_N \frac{\partial U_{OCV}}{\partial Z}\Big|_{Z=Z_K} + R_{P1}\left(1 - e^{-\frac{T_S}{\tau_1}}\right)\sum\limits_{J=0}^{L-1}\left(e^{-\frac{T_S}{\tau_1}}\right)^{L-1-j} + R_{P1}\left(1 - e^{-\frac{T_S}{\tau_2}}\right)\sum\limits_{J=0}^{L-1}\left(e^{-\frac{T_S}{\tau_2}}\right)^{L-1-j} + R} \end{cases} \tag{11}$$

$i_{\min,k+L}^{chg,v}$ and $i_{\max,k+L}^{dis,v}$ are the minimum charging current maximum discharge current within continuous sampling time intervas L based on the maximum cutoff voltage respectively. $U_{L,\max}$ and $U_{L,\min}$ are the maximum and minimum cut-off voltages respectively. C_N is the rated capacity of the battery, and η is the charging-discharge efficiency. And τ_1, τ_2 represent the time constants of the two RC loops.

In addition to the limitations of the battery model itself, SOC is another important factor in the calculation of maximum current. When the actual SOC value is close to the preset SOC value during battery use, the current must be limited to protect the battery itself.

The maximum current calculation method is given as follows

$$i_{\max,k+L}^{dis,z} = \frac{Z_k - Z_{\min}}{L\eta T_S/C_N},\ i_{\min,k+L}^{chg,z} = \frac{Z_k - Z_{\max}}{L\eta T_S/C_N} \tag{12}$$

$i_{\max,k+L}^{dis,z}$ and $i_{\min,k+L}^{chg,z}$ are the SOC-based maximum discharge current and minimum charging current within consecutive sampling time L respectively, $Z_{\min}$ is the preset minimum SOC, and $Z_{\max}$ is the preset maximum SOC [13].

3.2 Principle of SOP Estimation Under Multiple Constraints

Under the condition of single constraint, the estimation error of peak power is often large, which causes some damage to the battery. Therefore, it is necessary to use a multi-constraint model for the SOP estimation, the calculation method of the maximum current under the condition of multiple constraints is shown in Eq. (13) [14].

$$i_{\max}^{dis} = \min\left\{i_{\max}, i_{\max,k+L}^{dis,v}, i_{\max,k+L}^{dis,z}\right\},\ i_{\min}^{chg} = \max\left\{i_{\min}, i_{\min,k+L}^{chg,v}, i_{\min,k+L}^{chg,z}\right\} \tag{13}$$

$i_{\max}$ and $i_{\min}$ respectively represent the preset maximum discharge current and the preset minimum charge current, $i_{\max}^{dis}$ and $i_{\min}^{chg}$ are the maximum discharge current and the minimum charge current under multiple constraints. Therefore, the continuous time peak power estimation equation is shown in Eq. (14).

$$P_{\max}^{dis} = U_{L,K+L} i_{\max,k+L}^{dis},\ P_{\min}^{chg} = U_{L,K+L} i_{\min,k+L}^{chg} \tag{14}$$

$P_{\max}^{dis}$ and $P_{\min}^{chg}$ represents the maximum peak discharge power and the minimum peak charge power. The estimated value of the terminal voltage at any time is shown in Eq. (15).

$$\begin{aligned} U_{L,K+L} = {} & U_{ocv}(z_k, C_N) - i\eta LT_S/C_N \frac{\partial U_{OCV}}{\partial Z}|_{Z=Z_K} - \left[\left(e^{-\frac{T_S}{\tau_1}}\right)^L U_{P1,K}\right. \\ & \left. + i_k R_{P1}\left(1 - e^{-\frac{T_S}{\tau_1}}\right)\sum_{j=0}^{L-1}\left(e^{-\frac{T_S}{\tau_1}}\right)^{L-1-j} + \left(e^{-\frac{T_S}{\tau_2}}\right)^L U_{P2,K} + i_k R_{P2}\left(1 - e^{-\frac{T_S}{\tau_2}}\right)\sum_{j=0}^{L-1}\left(e^{-\frac{T_S}{\tau_2}}\right)^{L-1-j}\right] - R_l i_k \end{aligned} \tag{15}$$

Substitute Eq. (15) into Eq. (14) to obtain the real-time peak power expression.

$$\begin{aligned} P_{\max}^{dis} &= i_{\max,k+L}^{dis}\left(\begin{aligned} & U_{ocv}(z_k, c_N) - \left(e^{-\frac{T_S}{\tau_1}}\right)^L U_{P1,K} - \left(e^{-\frac{T_S}{\tau_2}}\right)^L U_{P2,K} \\ & - i_{\max,k+L}^{dis}\left(R_l + \frac{\eta LT_S}{C_N}\frac{\partial U_{OCV}}{\partial Z}|_{Z=Z_K} + R_{P1}\left(1 - e^{-\frac{T_S}{\tau_1}}\right)\sum_{j=0}^{L-1}\left(e^{-\frac{T_S}{\tau_1}}\right)^{L-1-j} + R_{P2}\left(1 - e^{-\frac{T_S}{\tau_2}}\right)\sum_{j=0}^{L-1}\left(e^{-\frac{T_S}{\tau_2}}\right)^{L-1-j}\right)\end{aligned}\right), \\ P_{\min}^{chg} &= i_{\min,k+L}^{chg}\left(\begin{aligned} & U_{ocv}(z_k, c_N) - \left(e^{-\frac{T_S}{\tau_1}}\right)^L U_{P1,K} - \left(e^{-\frac{T_S}{\tau_2}}\right)^L U_{P2,K} \\ & - i_{\min,k+L}^{chg}\left(R_l + \frac{\eta LT_S}{C_N}\frac{\partial U_{OCV}}{\partial Z}|_{Z=Z_K} + R_{P1}\left(1 - e^{-\frac{T_S}{\tau_1}}\right)\sum_{j=0}^{L-1}\left(e^{-\frac{T_S}{\tau_1}}\right)^{L-1-j} + R_{P2}\left(1 - e^{-\frac{T_S}{\tau_2}}\right)\sum_{j=0}^{L-1}\left(e^{-\frac{T_S}{\tau_2}}\right)^{L-1-j}\right)\end{aligned}\right) \end{aligned} \tag{16}$$

4 Simulation Results Are Tested and Analyzed

4.1 Analysis of SOC Estimation Results

Figure 2 is the curve of SOC model estimate and real value under UDDS condition. Figure 3 shows the errors of the two curves under this working condition. It can be seen that the errors of the model remain within 0.05 and show periodic fluctuations. The reason for this phenomenon is that the EKF algorithm adopts a fixed model, battery parameters will change, and the algorithm itself will ignore the high level term of Taylor expansion, which will also cause some errors.

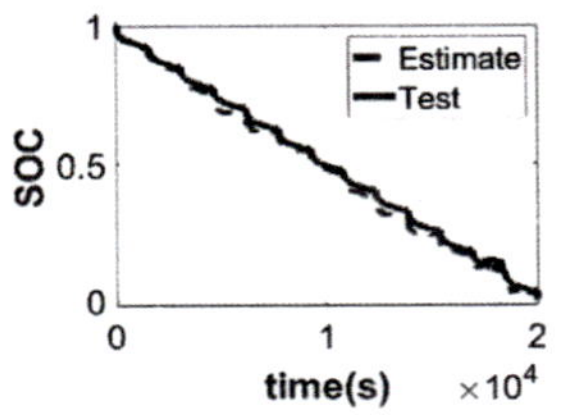

Fig. 2. True and estimated values of SOC

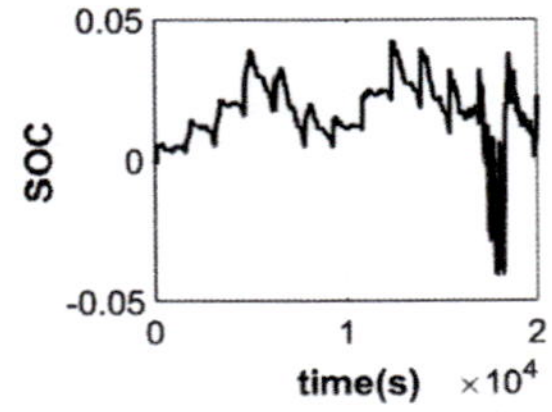

Fig. 3. Estimation error of SOC

4.2 Analysis of Continuous Peak Discharge Power

Figure 4 shows the continuous peak discharge power curves at 30 s, 2 min and 5 min. In the same time period, the discharge power decreases gradually with the discharge time. when the SOC of the battery is high, the continuous peak discharge power is mainly related to the current limit of the battery itself. Until the end of discharge, the terminal voltage will fall to the lowest value. When the SOC value is below 50%, in order to ensure that there is no excessive discharge phenomenon, the continuous peak discharge current decreases and the continuous peak discharge power also decreases. Compared with 2 min and 5 min, the change of continuous peak discharge power in 30 s is relatively gentle, because in a short time, the battery model parameters are almost constant and the error of EKF algorithm cause less interference to it, and do not produce a large number of iterations.

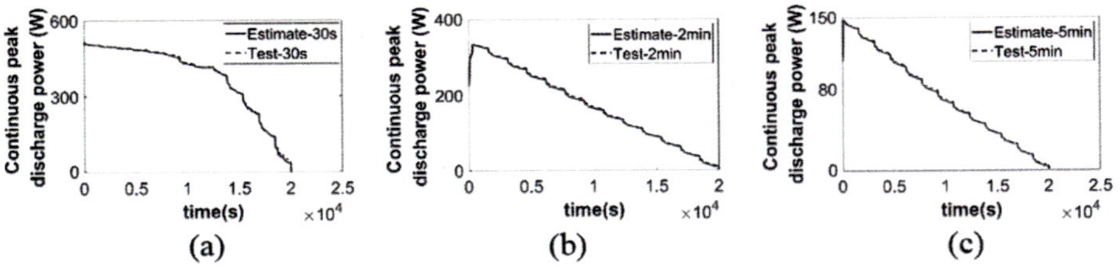

Fig. 4. True values and errors of persistent peak discharge powers

Figure 5 shows the error curve between the true value and the estimated value of peak discharge power at 30 s, 2 min and 5 min. The figure shows that in the initial stage, the continuous power error in 30 s, 2 min, 5 min duration compared with smaller, this is because the error of the initial time affected by the initial value of EKF algorithm, in the 30 s of time, the role of a relatively short time, is not very obvious error, with the increase of time, iteration algorithm is cumulative, due to the influence of this algorithm, the error is large. When the estimation of SOC approximates to the true value, the error of continuous peak discharge power becomes smaller. SOC is an important constraint of SOP estimation, so the error tends to converge. The error of the algorithm is maintained within 15 W in the duration of 30 s, and within 4 W in the duration of 2 min and 5 min. The accuracy of the algorithm is within the acceptable range.

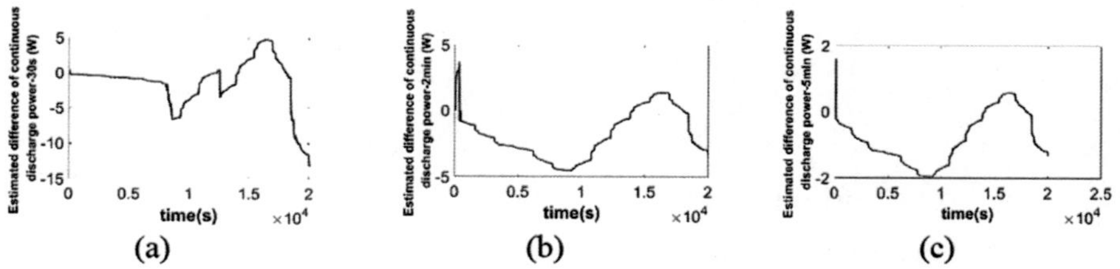

Fig. 5. Estimation error of persistent discharge power

5 Conclusion

In this paper, a lithium battery model was built based on the second-order RC equivalent circuit. In UDDS condition, SOC estimation was carried out based on EKF algorithm. The results of SOC estimation, terminal voltage and battery design limitations are taken as important limitations of SOP estimation. The error curves of each parameter are analyzed and compared. The results show that the error fluctuation of SOP estimation method under multiple constraints is relatively stable, which can provide reference value for the research of lithium battery. At the same time, the variation of internal characteristic parameters of lithium battery with time and the omission of high-order Taylor expansion in SOC estimation algorithm are important sources of errors, which are also important improvement directions to improve the accuracy of SOC estimation in the future.

Acknowledgment. This work ispartially supported by Science and Thchology Research Project of Education under Grant Q20161805, Local Science and Techology Development Project Guided by Central Government under Grant 2018ZYYD007.

References

1. Liu, N., Lv, T., Ye, W., Zuo, Y., Zhang, X.: Online parameter identification and SOC joint estimation of lithium-ion supercapacitors. Chin. J. Power Sources **44**(05), 736–739+748 (2020)

2. Lee, S., et al.: State of charge and capacity estimation of lithium-ion battery using a new open circuit voltage versus state of charge. J. Power Sources **185**(2), 1367–1373 (2008)
3. Sbarufatti, C., et al.: Adaptive prognosis of lithium ion batteries based on the combination of particle filters and radial basis function neural networks. J. Power Sources **344**, 128–140 (2017)
4. Kim, W.-Y., et al.: A nonlinear model based observer for a state of charge estimation of a lithium ion battery in electric vehicles. Energies **12**(17), 3383 (2019)
5. Xiong, R., et al.: Model based state of charge and peak power capability joint estimation of lithium-ion battery in plug in hybrid electric vehicles. J. Power Sources **229**, 159–169 (2013)
6. Yanke, X., Xingming, F., Xin, Z., Linlin, G.: The estimation method of lithium ion battery SOC based on EKF algorithm. J. Guilin Univ. Electron. Technol. **38**(03), 189–193 (2018)
7. Xinna, J., Qimeng, G., Yuwei, P., Yang, H.: Online state of power estimation methods for lithium-ion batteries in EV. Chin. J. Power Sources **43**(09), 1448–1452 (2019)
8. Xinbo, Y., Yuejiu, Z., Wenkai, G., Dongsheng, R., Xuebing, H., Minggao, O.: Power state estimation of high specific energy storage lithium battery system based on extended equivalent circuit model. Power Syst. Technol. **45**(01), 57–66 (2021)
9. Zhang, S.: Lithium battery SOC estimation method and implementation based on second-order Kalman filter. North China University of Technology (2018)
10. Ying, H.: Estimating the state of charge of a battery system by Kalman filter method. Autom. Appl. Technol. **46**(11), 6–9 (2021)
11. Zhang, Y.: Parameter identification and SOC estimation of power battery for electric vehicle. Jilin University (2014)
12. Hao, Z., Wenbo, Z., Yuanwang, D., Meng, L., Xiang, J.: Peak power estimation of power battery discharge based on SA+ BP hybrid algorithm. J. Jiangsu Univ. (Nat. Sci. Ed.) **41**(02), 192–198 (2020)
13. Yang, Z.: Research on multi-scale joint estimation of state of charge and state of power of vehicles's lithium ion power battery. Nanchang University (2019)
14. She, L.: Joint estimation of SOC and SOP for traction power battery in electric vehicles. Wuhan University of Technology (2018)

Check for updates

Startup Performance of Dry Gas Seals with Different Types of Grooves Considering the Slip Flow Effect

Qiangguo Deng[1], Pengyun Song[2](✉), Xiangping Hu[3](✉), Hengjie Xu[2], Xuejian Sun[1], and Wenyuan Mao[2]

[1] Department of Mechanical and Electrical Engineering, Kunming University of Science and Technology, Kunming 650504, Yunnan, China

[2] Department of Chemical Engineering, Kunming University of Science and Technology, Kunming 650504, Yunnan, China

songpengyunkm@sina.com

[3] Department of Energy and Process Engineering, Norwegian University of Science and Technology, 7491 Trondheim, Norway

Xiangping.Hu@ntnu.no

Abstract. Finite difference method is used to solve modified Reynolds equation based on FK slip flow model. The startup performance of spiral groove dry gas seal (S-DGS), T-groove dry gas seal (T-DGS) and linear groove dry gas seal (L-DGS) in startup stage are analyzed. The results show that in the startup stage, the slip flow effect reduces the opening force, and the influence of slip flow effect on the opening force is the most obvious for S-DGS, while it is the weakest for L-DGS. Considering the slip flow effect, the hydrodynamic pressure effect of S-DGS is the largest and that of L-DGS is the smallest. Under the same conditions, the shaft speed range of S-DGS with obvious slip flow effect with the Knudsen number being 0.001 is the smallest, and that of T-DGS is the largest. The research provides theoretical guidance for the reasonable evaluation of the influence of slip flow effect on the startup performance of dry gas seals in the startup stage.

Keywords: Dry gas seal · Startup performance · Slip flow effect · Hydrodynamic pressure effect

1 Introduction

Dry gas seal is a non-contact mechanical seal, and it is widely used because of its excellent sealing performance. In the startup stage, the gap (gas film thickness) of the seal rings gradually increases from zero to several microns, and there is obvious slip flow effect on the seal ring faces. Therefore, it is necessary to consider the influence of slip flow effect when one analyzes the performance of dry gas seal in the startup stage.

Hydrodynamic pressure groove is arranged on the rotating ring of dry gas seal, which can pressurize the gas film between the sealing rings and help improve the opening performance. There are many researches on the structural design and sealing performance

Y. Wang et al. (Eds.): IWAMA 2021, LNEE 880, pp. 578–586, 2022.
https://doi.org/10.1007/978-981-19-0572-8_74

analyses of hydrodynamic pressure groove. Li [1] analyzed the steady-state characteristics of T-groove dry gas seal (T-DGS). Bai [2] obtained the gas film dynamic pressure and temperature distributions of T-DGS under different vibration frequencies.Hu [3] used finite element method (FEM) calculate pressure distribution of linear groove dry gas seal (L-DGS), which shows that L-DGS with proper structure can improve the opening force. Song [4] analyzed the sealing performance of spiral groove dry gas seal (S-DGS) in steady state by analytical method. Ruan [5] used FEM to solve the modified Reynolds equation considering the slip flow effect, and verified that the S-DGS has obvious slip flow effect under low pressure and low speed.

At present, most researchers mainly focus on the optimization of groove structures, the analysis of steady-state characteristics and the performance analysis considering the slip flow effect in the stable operation stage. There are few literatures on the influence of the slip flow effect on the sealing opening force, leakage rate and opening speed during in the startup stage. In this paper, the opening force, the leakage rate and the opening speed of S-DGS, T-DGS and L-DGS in the startup stage considering the slip flow effect are analyzed by a numerical analysis method. The research provides reference for the reasonable evaluation of the influence of slip flow effect on the startup performance of dry gas seals in the startup stage.

2 Slip Flow Models

Fukui and Kaneko (FK) [6] proposed a modified Reynolds equation to study the slip flow effect by introducing flow factors derived from linearized Boltzmann equation. For an isothermal, compressible lubricant, ideal gas behavior, the modified Reynolds equation in steady-state condition is given by:

$$\frac{1}{r}\frac{\partial}{\partial r}\left(Qrh^3\frac{\partial p^2}{\partial r}\right)+\frac{1}{r^2}\frac{\partial}{\partial \theta}\left(Qh^3\frac{\partial p^2}{\partial \theta}\right)=12\eta\omega\frac{\partial (ph)}{\partial \theta} \tag{1}$$

where r is radius, p is pressure, h is gas film thickness, ω is angular velocity, η is viscosity, Q is the relative flow rate coefficient, which is defined as $Q = Q_p/Q_c$, Q_p is flow rate of the Poiseuille flow, and Q_c is flow rate of the continuum Poiseuille flow. The results are shown as follows:

$$\begin{cases} Q_c = D/6 \\ \begin{cases} Q_p = D/6 + 1.0162 + 1.0653/D - 2.1354/D^2 (D > 5) \\ Q_p = 0.13852D + 1.25087 + 0.15653/D - 0.00969/D^2 (0.15 < D < 5) \\ Q_p = -2.22919D + 2.10673 + 0.01653/D - 0.0000694/D^2 (0.01 < D < 0.15) \end{cases} \end{cases} \tag{2}$$

where Kn is Knudsen number, which represents the degree of slip flow of the gas, the larger the value, the more obvious the slip flow effect. D is inverse Knudsen number, which can be defined as $D = D_0 ph$. D_0 is characteristic inverse Knudsen number, which related to th atmospheric pressure p_a and it is defined as:

$$D_0 = \frac{p_a h_0}{\mu\sqrt{2R_c T_0}} \tag{3}$$

3 The Sealing Performance Parameters

The pressure at the iner sealing ring (i.e. $r = r_i$) is the atmospheric pressure p_a and the pressure at the outer sealing ring (i.e. $r = r_o$) is the medium pressure p_o. The cyclic boundary condition can be described by the following:

$$p(r,\theta) = p(r,\theta + 2\pi/N_g) \quad (4)$$

The opening force, leakage rate and closing force are respectively defined as:

$$F_o = \int_0^{2\pi}\int_{r_i}^{r_o} p(r)r\mathrm{d}r\mathrm{d}\theta \quad (5)$$

$$F_{close} = p_i\pi(r_b^2 - r_i^2) + p_o\pi(r_o^2 - r_b^2) + p_{sp}\pi(r_o^2 - r_i^2) \quad (6)$$

$$Q_L = \int \frac{Mp(r)rh^3}{12\mu R_r T}\frac{\partial p}{\partial r}\mathrm{d}\theta \quad (7)$$

where r_i, r_o, r_b are respectively inner, outer and balance radius, p_{sp} is spring pressure.

4 Verification of Dry Gas Seal Pressure Governing Equation

The governing equation of gas film pressure in Eq. (1) is discretized by finite difference method (FDM), and it is numerically calculated by MATLAB software. To verify the correctness of the present calculation program, the present numerical results are compared with Gabriel [7], Cai [8] and Yin [9], which are shown in Table 1. The maximum opening force related-errors between the present and Gabriel, Cai and Yin are respectively 12.58%, 8.13% and 0.77%. The calculation results in this paper are closer to that of Yin. These may be caused by different calculation methods. Therefore, the present calculation program is reasonable.

Table 1. Opening force of different literatures

Gas film thickness h_0/μm	Opening force F_o/N			
	Gabriel	Cai	Yin	The present calculation program
2.03	40711.8	34000	35691.2	35589.9
3.05	33168.7	30000	31684.5	31469.3
5.08	29569.2	27000	29420.8	29194.6

5 Results and Discussion

5.1 Dry Gas Seal Groove Structures

S-DGS, T-DGS and L-DGS are selected to analyze the opening characteristics. The groove structure of three types dry gas seals have been given respectively in Fig. 1. Geometry and sealing parameters based on Gabriel. To compare the startup performance of different groove structures, the inner and outer diameters of three types dry gas seals are the same, and the groove-to-table ratio of each groove is 1. That is, the circumferential length and area of the groove and the table are equal. For T-DGS, it means that the sum of the circumferential length of the large groove and the small groove is equal to the sum of the circumferential length of the large table and the small table.

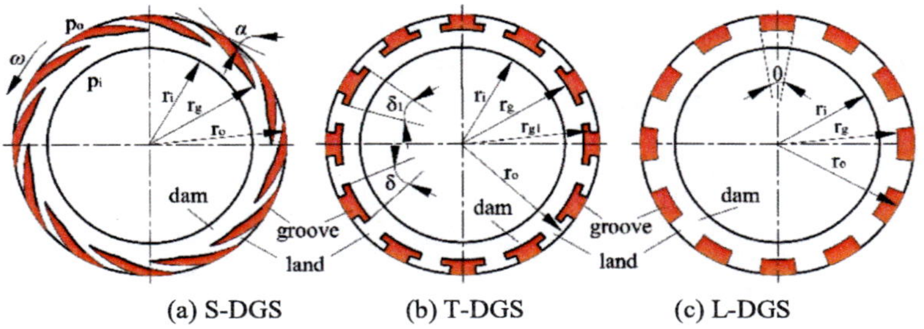

Fig. 1. Dry gas seal groove structures

where groove middle radius r_{g1} is 73.39 mm, L-DGS central angle θ is 15°, T-DGS large central angle δ is 20°, T-DGS small central angleδ_1 is 10.3°.

5.2 Relative Errors

The sealing parameters such as opening force, leakage rate and opening shaft speed of seal end faces are calculated in the modified Reynolds equation based on FK slip flow model. In order to directly reflect the influence degree of the slip flow effect on the startup performance, the relative errors of the opening parameters of each type of seal with or without slip flow effect are defined:

E_1 = [(Opening speed of slip flow − Opening speed of non-slip flow)/Opening speed of non-slip flow] × 100%

E_2 = [(Equilibrium film thickness of non-slip flow − Equilibrium film thickness of slip flow)/Equilibrium film thickness of non-slip flow] × 100%

E_3 = [(Equilibrium speed of slip flow − Equilibrium speed of non-slip flow)/ Equilibrium speed of non-slip flow] × 100%

E_4 = [(Leakage rate of non-slip flow − Leakage rate of slip flow)/Leakage rate of non-slip flow] × 100%

5.3 Startup Performance Analysis

In the startup stage of dry gas seal, the seal end faces go through the process of contact to non-contact to stable working film thickness, and the gas film thickness is dynamically increasing. The opening film thickness is taken as 0.52 μm. Table 2 shows the startup performance parameters of the three types dry gas seals. Compared with T-DGS and L-DGS, the opening speed of S-DGS is the lowest. So, S-DGS is the easiest to open, and L-DGS needs the highest shaft speed to open. It proves that S-DGS has the best hydrodynamic pressure effect while L-DGS has the worst.

Considering the slip flow effect, the opening speed N of the same-type of seal are larger than those of the non-slip flow. Because the gas film thickness is very small in the startup stage, and the slip flow effect is obvious, or the effective viscosity of the gas film is reduced, and the viscous shear force is weakened, which leads to the reduction of the hydrodynamic pressure effect of the gas film and the opening force. The opening speed related-errors of S-DGS, T-DGS and L-DGS is respectively 17.89%. 17.05% and 17.34%. It means that the influence of slip flow effect on S-DGS is stronger than those of T-DGS and L-DGS in the startup stage.

Table 2. Startup performance when gas film thickness $h_0 = 0.52$ μm

Parameters	S-DGS		T-DGS		L-DGS	
	Non- slip flow	Slip flow	Non- slip flow	Slip flow	Non- slip flow	Slip flow
N/(r/min)	780	950	3600	4340	4158	5030
Q_L /(10^{-6} kg/s)	3.32	3.81	3.27	3.7	3.27	3.71

According to the quasi-steady state concept, when the opening force is equal to the closing force, the seal ring is the equilibrium state in the non-contact period. At this time, the shaft speed is called the equilibrium speed, and the seal end faces gap is called the equilibrium film thickness. The opening gas film ($h_0 = 0.52$ μm) is the first one of the equilibrium film thicknesses.

Figure 2 shows the startup performance parameters of the three types dry gas seals with increasing the shaft speed. It can be seen that for the same type of seal and the same speed, the slip flow effect reduces the equilibrium film thickness and results in small leakage rate. For example, for S-DGS with shaft speed of 6000 r/min, without considering the slip flow effect, the equilibrium film thickness h_0 is 1.9 μm, and the leakage rate Q_L is 2.82×10^{-4} kg/s. Moreover, when the slip flow effect is considered, the equilibrium film thickness h_0 is 1.86 μm, and the leakage rate Q_L is 2.73×10^{-4} kg/s. The equilibrium film thickness related-errors E_2 is 2.1%, and the leakage rate related-errors E_4 is 3.2%. In order to maintain the equilibrium state where the opening force is equal to the closing force, the equilibrium film thickness considering the slip flow effect will automatically decreases to supplement the opening force loss caused by the slip flow effect. At the same time, the equilibrium film thickness considering the slip flow effect is smaller than the equilibrium film thickness without the slip flow effect, and the leakage rate corresponding to the small film thickness is naturally smaller than that of the large film thickness.

When the shaft speed is 950 r/min, the equilibrium film thickness of S-DGS h_0 is 0.67 μm without considering slip flow effect and h_0 is 0.52 μm under slip flow condition, and the equilibrium film thickness related-errors E_2 is 22.39%. Comparing the equilibrium film thickness related-errors caused by slip flow effect at 950 r/min (E_2 = 22.39%) and 6000 r/min (E_2 = 2.1%). It found that with the increasing of the shaft speed, the film thickness gradually increases, and the slip flow effect becomes smaller and smaller, so the influence of slip flow effect on startup performance of dry gas seal becomes lower and lower.

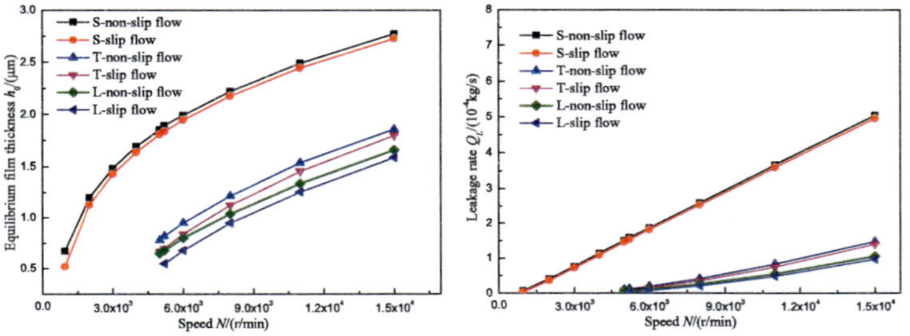

(a) h_0 *vs N* (b) Q_L *vs N*

Fig. 2. Startup parameters of different grooves on quasi-steady state

Figure 3 shows the opening force and leakage rate with shaft speed when the opening film thickness h_0 is 0.52 μm. Results show that the same type of seal and the same speed, the slip flow effect decreases the gas film opening force and increases leakage rate. Under the same shaft speed, the opening force and leakage rate of S-DGS are larger than those of T-DGS and L-DGS. S-DGS can open the sealing ring faster, followed by T-DGS, while L-DGS is the slowest. The T–DGS and L-DGS need a larger speed to improve the opening force.

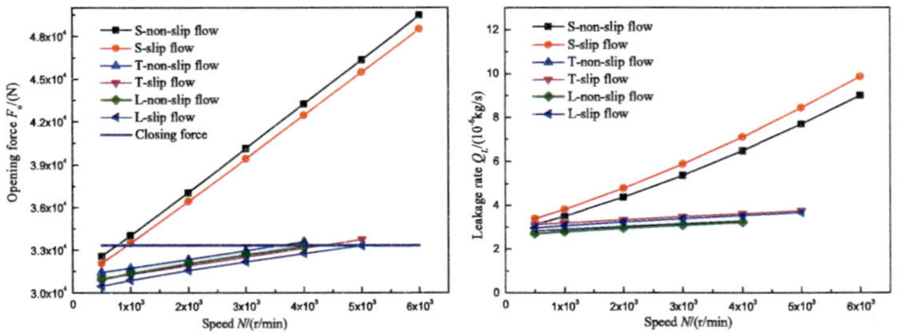

(a) F_o *vs N* (b) Q_L *vs N*

Fig. 3. Performance parameters of different grooves (h0 = 0.52 μm)

5.4 Judgment of Influence Interval of Slip Flow

The startup performance of the dry gas seal is affected by the slip flow effect, but with the increasing of the shaft speed, the film thickness gradually increases, and the influence of the slip flow effect on the startup performance gradually decreases. Therefore,it is necessary to define the scope of the slip flow phenomenon, that is, to clarify the relationship between the sealing parameters and the Knudsen number. To define whether slip flow effect should be considered for calculation in the startup stage.

Knudsen defined the obvious slip flow effect in the range of $0.001 \leq Kn \leq 0.1$. In this paper, the working condition parameters of dry gas seal are from the Gabriel.The influence interval of slip flow effect of different types dry gas seals are analyzed. The average Knudsen number of a certain instantaneous equilibrium film thickness is taken as the analysis object, and this value is the average value of the Knudsen number of each calculation node in the FDM.

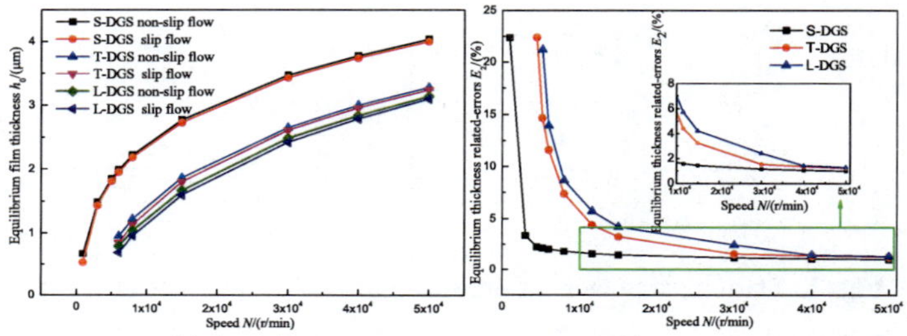

(a) h_o *vs N* (b) E_2 *vs N*

Fig. 4. The equilibrium film thickness of different types of seals

Figure 4(a) shows the equilibrium film thickness of S-DGS, T-DGS and L-DGS with the shaft speed. It can be seen that as the shaft speed increases, the equilibrium film thickness of the three types of seals gradually increases. The equilibrium film thickness related-errors of the three types seals with influence of the slip flow effect are shown in Fig. 4(b), which rapidly reduce with the increasing of the shaft speed. It shows that the slip flow effect is more obvious at the low-speed period.

Table 3. The equilibrium film thickness related-errors of three types seals

Parameters	S-DGS		T-DGS		L-DGS	
	Opening speed	Equilibrium speed	Opening speed	Equilibrium speed	Opening speed	Equilibrium speed
N/(r/min)	950	12150	4340	43800	5030	37250
Kn	0.0049	0.001	0.0059	0.001	0.0051	0.001
E_2/(%)	22.39	1.55	22.39	1.29	22.21	1.82

Table 3 shows the Knudsen number and the equilibrium film thickness related-errors E_2 of three types seals at the opening speed and an equilibrium speed with $Kn = 0.001$. It can be seen that two parameters are relatively large at the opening speed, indicating that the influence of slip flow effect is more obvious at the opening speed. However, the influence of the slip flow effect gradually reduces with the increasing of the shaft speed, it results in the equilibrium film thickness related-errors E_2 gradually decreases. When $Kn \leq 0.001$, the equilibrium film thickness related-errors E_2 of three types seals are all very small, which means that $Kn = 0.001$ being an indicator for considering the slip-flow effect is reasonable. Knudsen defined the slip flow effect as being very small in this situation, which can be ignored. Further, results shows that S-DGS is most affected compared to T-DGS and L-DGS at low-speed period. The equilibrium film thickness related-errors rapidly decrease with the increasing of the shaft speed, and the descending speed of S-DGS is greater than that of the T-DGS and L-DGS. This indicates that under the same conditions, the slip flow effect significantly affects S-DGS in the startup stage, and the slip flow speed range of S-DGS is the narrowest, and that of T-DGS the widest.

6 Conclusions

With the analyses and results, the following conclusions can be reached:

(1) In the startup stage of dry gas seal, the slip flow effect reduces the opening force of the same seal. The influence of slip flow effect on the opening force is the most obvious for S-DGS, while it is the weakest for L-DGS.
(2) Considering the slip flow effect, the hydrodynamic pressure effect of S-DGS is the largest, and that of L-DGS is the smallest.
(3) For the same type of seal and the same speed, the slip flow effect reduces the equilibrium film thickness and results in small leakage rate. With the increasing of the shaft speed, the slip flow effect becomes smaller and smaller. S-DGS is significantly affected by the slip flow effect in the startup stage, and the slip flow speed range of S-DGS is the narrowest, and that of T-DGS the widest.

Acknowledgment. The research is supported by National Natural Foundation of China (granted numbers: 51465026) and the Basic Research Programs of Yunnan Province (granted numbers: 202101AU070019).

References

1. Li, T.Z., Zhang, Q.X., Cai, J.N., et al.: Steady-state performance analysis of T-shape groove dry gas seals by a finite element method. J. Beijing Univ. Chem. Technol. **30**(2), 58–62 (2003)
2. Xian, B.S., Jia, W., Zu, D.L., et al.: Thermoelastohydrodynamic gas lubrication of T-groove face seals: stability of sealing film. Tribology **39**(2), 131–139 (2019)
3. Hu, D.M., Wu, Z.X.: Analysis and calculation of the gas seal on the end face of the linear groove. Fluid Mach. **24**(9), 16–22 (1996)

4. Song, P.Y., Zhang, S.: An approximately analytical method of characteristics of spiral groove dry gas seals under slip flow conditions. J. Drain. Irrig. Mach. Eng. **32**(10), 877–882 (2014)
5. Bo, R.: Finite element analysis of the spiral groove gas face seal at the slow speed and the low pressure conditions-slip flow consideration. Tribol. Trans. **43**(3), 411–418 (2000)
6. Fukui, S., Kaneko, R.: A database for interpolation of poiseuille flow rates for high Knudsen number lubrication problems. ASME J. Tribol. **112**(1), 78–83 (1990)
7. Gabriel, R.P.: Fundamentals of spiral groove noncontacting face seals. Lubr. Eng. **50**(3), 215–224 (1994)
8. Cai, W.X.: Theory research of spiral-groove gas seal. J. Wuhan Univ. Technol. **16**(4), 112–118 (1994)
9. Xiao Ni, Y., Xu Dong, P.: Selection of a shape function in finite element analysis for a spiral groove dry gas seal. Lubr. Eng. **3**, 13–14 (2006)

Research on Workspace of 6-DOF Rope Traction Parallel Mechanism with Spring Passive Branch Chain

Yuguang Chen(✉) and Haodi Wang

College of Aeronautical Engineering, Civil Aviation University of China, Tianjin 2898, China

Abstract. Because the rope can only be pulled but not pressed, this paper introduces the spring branch chain which can both be pulled and pressed to study the rope traction parallel mechanism. According to the shortcomings of previous studies on the workspace of rope traction parallel mechanism, a new workspace evaluation index is proposed to evaluate two different configurations of rope traction parallel robot. It is concluded that the introduction of spring passive branch chain can increase the workspace volume and posture space volume of the mechanism. Due to the introduction of spring mechanism, shape defects appear in the workspace. In order to quantify the severity of this problem, a shape evaluation function is established for evaluation. Several different spring branch chain parameters are selected to solve the value of its evaluation function and the shape of workspace. The obtained workspace shape map is compared with the value of the evaluation function to demonstrate the scientificity and sensitivity of the function.

Keyword: Parallel mechanism · Workspace · Evaluation index · Defect function

1 Introduction

Rope traction parallel robot has the advantages of simple structure, large workspace, fast movement speed and easy disassembly [1, 2], which is applied to tactile device [3], lower limb rehabilitation mechanism [4–6] and high-speed assembly [7, 8]. However, the driving rope of the rope traction parallel robot can only be pulled but not pressed. Therefore, this paper proposes to introduce a spring into the rope traction parallel mechanism to improve the performance of the rope traction parallel robot.

The spring, an elastic element, is connected between the frame and the end moving platform as a branch chain of the rope traction parallel mechanism, which will exert a passive force affected by its position and attitude on the moving platform. This passive force can help the tension of the mechanism to achieve complete constraint positioning and improve the performance of the mechanism [9]. The workspace reflects the motion performance of the end effector, which has been studied by scholars at home and abroad. Bosscher et al. [10] equivalent the end moving platform to a particle through the geometric vector method, and obtained the workspace boundary of the under constrained configuration rope traction parallel mechanism in analytical form, but it is limited to the

Y. Wang et al. (Eds.): IWAMA 2021, LNEE 880, pp. 587–595, 2022.
https://doi.org/10.1007/978-981-19-0572-8_75

specific attitude angle and is not suitable for the case where the size of the end actuator needs to be considered. Gao et al. [11] studied the inverse kinematics and workspace of parallel humanoid neck robot with spring skeleton.

By summarizing the previous research results on workspace, this paper establishes a new workspace evaluation index, analyzes the workspace of rope traction parallel robot with spring branch chain, and summarizes the influence of spring parameters on workspace. Then the shape defect caused by the introduction of spring branch chain is studied and discussed.

2 Static Modeling

The structural diagram of the rope traction parallel structure with spring branch chain is shown in Fig. 1, in which the ground coordinate system is *OXYZ* and the local coordinate system is $O'X'Y'Z'$. Both ends of the rope and spring are hinged with the frame and the moving platform respectively, and the position vectors of the hinge joint with the frame in the ground coordinate system are $\boldsymbol{A}_i$ and $\boldsymbol{A}_{Si}$; The position vectors of the hinge point with the moving platform in the local coordinate system are $\boldsymbol{B}_i$ and $\boldsymbol{B}_{Si}$.

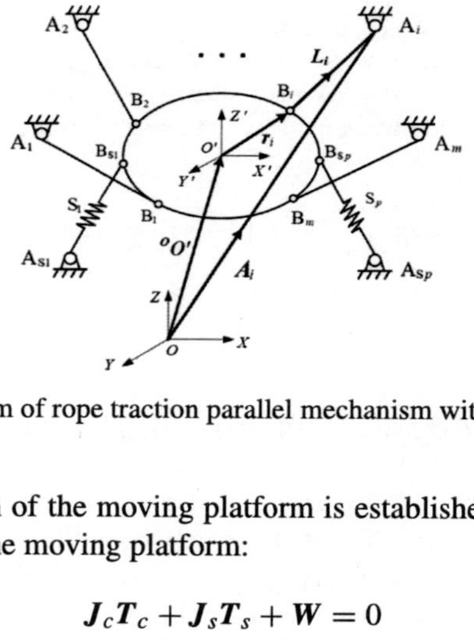

Fig. 1. Model diagram of rope traction parallel mechanism with spring branch chain

The balance equation of the moving platform is established according to the force balance relationship of the moving platform:

$$\boldsymbol{J}_c\boldsymbol{T}_c + \boldsymbol{J}_s\boldsymbol{T}_s + \boldsymbol{W} = 0 \tag{1}$$

Where: $\boldsymbol{J}_c$ is the Jacobian matrix of rope branch chain force and $\boldsymbol{J}_s$ is the Jacobian matrix of spring branch chain force; $\boldsymbol{W} = \begin{bmatrix} \boldsymbol{F}_e\ \boldsymbol{M}_e \end{bmatrix}^{\mathrm{T}}_{n\times 1}$. $\boldsymbol{F}_e$ is the external force acting on the moving platform, $\boldsymbol{M}_e$ is the external torque acting on the moving platform, $\boldsymbol{T}_c = \begin{bmatrix} t_1\ t_2\ \ldots\ t_m \end{bmatrix}^{\mathrm{T}}$, $\boldsymbol{T}_s = \begin{bmatrix} t_{s1}\ t_{s2}\ \ldots\ t_{sp} \end{bmatrix}^{\mathrm{T}}$.

$$\boldsymbol{J}_C = \begin{bmatrix} \boldsymbol{u}_1 & \boldsymbol{u}_2 & \cdots & \boldsymbol{u}_m \\ \boldsymbol{r}_1 \times \boldsymbol{u}_1 & \boldsymbol{r}_2 \times \boldsymbol{u}_2 & \cdots & \boldsymbol{u}_m \times \boldsymbol{u}_m \end{bmatrix}_{n\times m} \tag{2}$$

$$J_S = \begin{bmatrix} \boldsymbol{u}_{s1} & \boldsymbol{u}_{s2} & \cdots & \boldsymbol{u}_{sp} \\ \boldsymbol{r}_{s1} \times \boldsymbol{u}_{s1} & \boldsymbol{r}_{s2} \times \boldsymbol{u}_{s2} & \cdots & \boldsymbol{r}_{sp} \times \boldsymbol{u}_{sp} \end{bmatrix}_{n\times p} \tag{3}$$

Where $\boldsymbol{u}_i = \frac{\boldsymbol{L}_i}{|\boldsymbol{L}_i|}$ is the unit direction vector of the i rope branch chain, and $\boldsymbol{r}_i = {}^{O}\boldsymbol{R}_{O'}\boldsymbol{B}_i$ is the position vector of the hinge point $\boldsymbol{B}_i$ in the ground coordinate system. In Eq. (3), vectors $\boldsymbol{u}_{si}$ and $\boldsymbol{r}_{si}$ have the same form, and ${}^{O}\boldsymbol{R}_{O'}$ is the rotation transformation matrix from local coordinate system to ground coordinate system,

$${}^{O}\boldsymbol{R}_{O'} = \begin{bmatrix} c\gamma c\beta & c\gamma s\beta s\alpha - s\gamma c\alpha & c\gamma s\beta c\alpha + s\gamma s\alpha \\ s\gamma c\beta & s\gamma s\beta s\alpha + c\gamma c\alpha & s\gamma s\beta c\alpha - c\gamma s\alpha \\ -s\beta & c\beta s\alpha & c\beta c\alpha \end{bmatrix} \tag{4}$$

Where α, β and γ are the rotation angles of the moving platform around the X axis, Y axis and Z axis relative to the ground coordinate system respectively, $c\alpha$ and $s\alpha$ represent $\cos\alpha$ and $\sin\alpha$ respectively, and other symbols have the same meaning.

The tensile force on the branch chain of the *i* rope is t_i; According to Hooke's law, the driven force on the *i* spring branch chain is:

$$t_{Si} = k(|\boldsymbol{L}_{Si}| - l_{0i}) \tag{5}$$

Where: k is the stiffness coefficient of the spring branch chain, $|\boldsymbol{L}_{Si}|$ is the length of the first spring branch chain, and $\boldsymbol{L}_{Oi}$ is the initial length of the *ith* spring branch chain.

3 Workspace Analysis

The workspace of the rope traction parallel robot is the collection of all pose points that can meet the force balance conditions of the end actuator and the allowable tension conditions of the rope. General work requirements require a wide range of workspace and regular shape. The spring introduced in this paper will change the workspace of the rope traction parallel robot, so it is necessary to analyze and study the spring mechanism. The parameters of the spring include the spring stiffness coefficient, the initial length and the position of the hinge joint of the spring branch chain. In this chapter, a new workspace evaluation index is proposed for the rope traction parallel robot with spring branch chain. The differences of workspace before and after the introduction of spring are compared, and the effects of different spring parameters on the performance of rope traction parallel robot are analyzed and studied.

3.1 Workspace Evaluation Index

When studying the workspace of the rope traction parallel robot, the position space under a certain attitude is usually used to reflect the size of the attitude space of the mechanism. However, this method can not accurately reflect the change of the attitude space when the position and attitude of the moving platform change. The other method is to solve all the position points that meet the attitude range, the calculation process is very complex and the amount of calculation is huge, and the results are difficult to be expressed intuitively. Therefore, it is necessary to propose a new evaluation method and index that can effectively and accurately reflect the performance of workspace.

Mark any position of the moving platform in the ground coordinate system as $X_i = \left[o_{x_i}\ o_{y_i}\ o_{z_i}\right]^{\mathrm{T}}$, Any attitude point of the platform in the ground coordinate system is recorded as $X_{ij} = \left[o_{\alpha_j}\ o_{\beta_j}\ o_{\gamma_j}\right]^{\mathrm{T}}$. Under the condition of the set attitude angle step n, the number of different values of the angle α_j around the *x-axis* in all the reachable attitude points at the position point X_i is recorded as $\eta_{j\alpha}$, The number of location points satisfying condition $\eta_{i\alpha} \geq q(q \in \mathrm{N}_{+})$ in the whole workspace is recorded as $\eta_{\alpha q}$. $\eta_{\alpha q}$ represents the number of position points where the rotation attitude angle of the moving platform around the x-axis reaches q or more under the step condition, and $\eta_{\beta q}$ and $\eta_{\gamma q}$ represent the corresponding indexes of the rotation attitude of the mechanism around the y-axis and z-axis respectively. The larger the values of $\eta_{\alpha q}$, $\eta_{\beta q}$ and $\eta_{\gamma q}$, the more position points of the moving platform rotating around the x-axis, y-axis and z-axis to reach the range span Q.

3.2 Influence of Introducing Spring Passive Branch Chain

When studying the workspace of the rope traction parallel robot with spring passive branch chain, the parameters of the rope and spring are shown in Table 1, and the solution parameters of the workspace and hinge point position are shown in Table 2, Table 3 and Table 4.

Table 1. Spring and rope parameter settings

Parameter	Spring stiffness coefficient k_1/(N/m)	Spring stiffness coefficient k_2/(N/m)	Hinge point position Z_{S1}/m	Hinge point position Z_{S2}/mm	Initial spring length l_{01}/m	Initial spring length l_{02}/m	Rope tension range/N
Numerical value	52.56	49.72	0.72	1.95	3.28	5.2	[10, 150]

Table 2. Workspace solution parameter settings

Parameter	Horizontal range r/m	Vertical range h/m	Attitude search range/°	Position search step/m	Attitude search step/°	Moving platform mass /N
Numerical value	[0, 0.5]	[0, 0.7]	[−80, 80]	0.05	16	200

Table 3. Hinge joint position of rope, spring and frame(m)

Frame hinge joint	Position
A_1	(0.6cos165°, 0.6sin165°, 0.7)
A_2	(0.6cos-75°, 0.6sin-75°, 0.7)
A_3	(0.6cos-75°, 0.6sin-75°, 0.7)
A_4	(0.6cos45°, 0.6sin45°, 0.7)
A_5	(0.6cos45°, 0.6sin45°, 0.7)
A_6	(0.6cos165°, 0.6sin165°, 0.7)
A_{S1}	$(0, 0, Z_{S1})$
A_{S2}	$(0, 0, -Z_{S2})$

Table 4. Position of hinge point between rope, spring and moving platform (m)

Articulated point of moving platform	Position
B_1	(0.2cos-135°, 0.2sin-135°, 0)
B_2	(0.2cos-135°, 0.2sin-135°, 0)
B_3	(0.2cos-15°, 0.2sin-15°, 0)
B_4	(0.2cos-15°, 0.2sin-15°,0)
B_5	(0.2cos105°, 0.2sin105°, 0)
B_6	(0.2cos105°, 0.2sin105°, 0)
B_{S1}	(0, 0, 0)
B_{S2}	(0, 0, 0)

According to the data in Table 1, Table 2, Table 3 and Table 4, the distribution of three attitude indexes $\eta_{\alpha q}$, $\eta_{\beta q}$ and $\eta_{\gamma q}$ of 6C configuration rope traction parallel mechanism in its workspace is solved, as shown in (a)–(c) in Fig. 2, The distribution of the three attitude indexes $\eta_{\alpha q}$, $\eta_{\beta q}$ and $\eta_{\gamma q}$ of the 6c2s configuration rope traction parallel mechanism in its workspace is shown in (a)–(c) in Fig. 3.

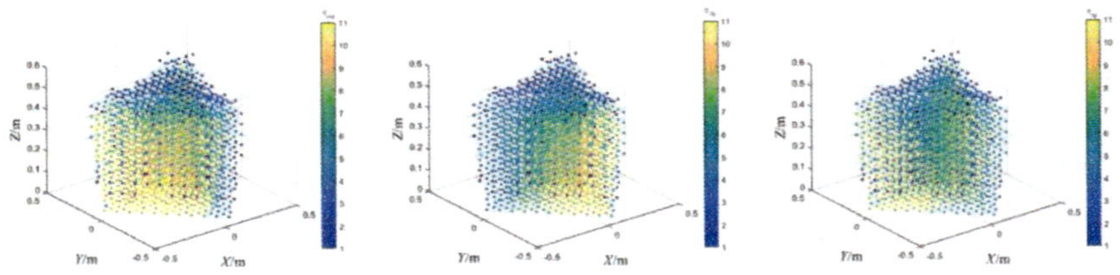

(a)Parameter $\eta_{\alpha q}$ (b))Parameter $\eta_{\beta q}$ (c)Parameter $\eta_{\gamma q}$

Fig. 2. 6C configuration workspace and index distribution

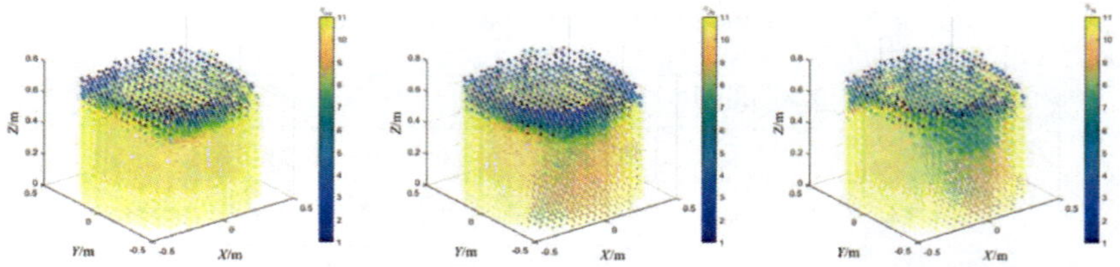

(a)Parameter $\eta_{\alpha q}$ (b))Parameter $\eta_{\beta q}$ (c)Parameter $\eta_{\gamma q}$

Fig. 3. 6c2s configuration workspace and index distribution

By comparing Fig. 2 and Fig. 3, it can be seen that after the spring passive branch chain is introduced into the rope traction parallel mechanism, the working space of the moving platform has been significantly improved, and the reachable attitude range has also been significantly increased. The results show that the introduction of spring into the rope traction parallel mechanism can increase the volume of workspace and attitude space.

4 Workspace Defect Analysis

When the spring branch chain is introduced into the rope traction parallel mechanism, the unreasonable value of each branch chain parameter will cause the shape defect problem in the working space of the mechanism as shown in Fig. 4, and the occurrence of this cavity will affect the motion performance of the moving platform.

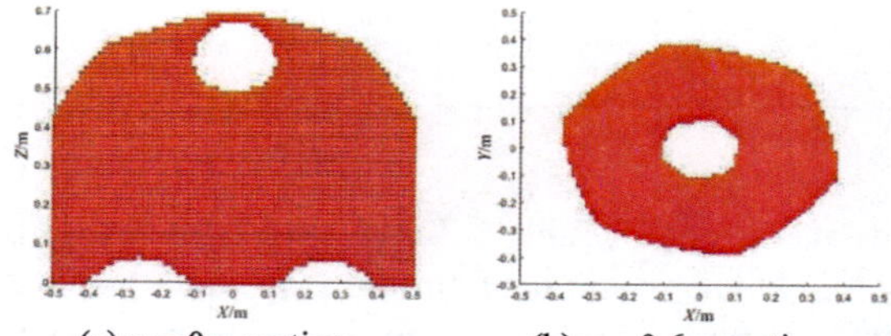

(a) y = 0m section (b) z = 0.6m section

Fig. 4. Schematic diagram of shape defects

Since the mechanism arrangement in this paper basically presents a symmetrical state, the defect shape formed basically presents a central symmetrical distribution on the horizontal section. Therefore, a model as shown in Fig. 5 can be established in this paper to evaluate the severity of the defect problem.

The defect evaluation function P constructed in this paper is established by mixing the horizontal position factor $\mu(d_{-})$, vertical position factor $\lambda(d_{\perp})$ and defect range factor D in the section from the defect part to the vertical central axis of the workspace, as follows:

$$\mu(d_{-}) = C_{-}(r_{max} - d_{-}), d_{-} \in (0, r_{max}] \tag{6}$$

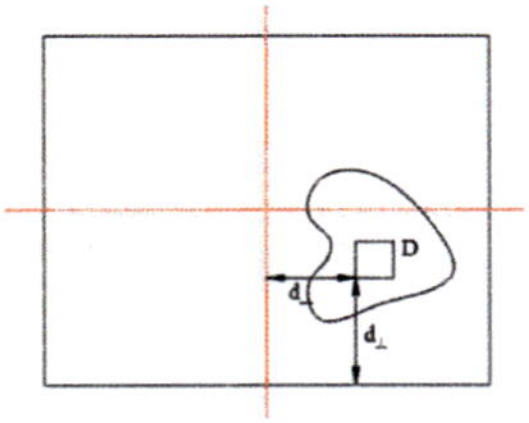

Fig. 5. Schematic diagram of evaluation model

$$\lambda(d_\perp) = \begin{cases} C_{\perp t} \cdot p_t(h_{max} - d_\perp), d_\perp \in [h^*, h_{max}) \\ C_{\perp b} \cdot p_b \cdot d_\perp, d_\perp \in [0, h^*) \end{cases} \tag{7}$$

$$P = \int \int_D \mu(d_-)\lambda(d_\perp)\mathrm{d}d_-\mathrm{d}d_\perp \tag{8}$$

Where $\mu(d_-)$ and $\lambda(d_\perp)$ respectively reflect the influence of the horizontal distance d_- and the vertical position $d_\perp$ on the evaluation function P. Considering the different possibility of defects in the upper and lower workspaces of height h^*, $\lambda(d_\perp)$ in Eq. (2) is set as a piecewise function with h^* as the dividing point, and p_t and p_b are the weight coefficients of piecewise evaluation. The coefficients C_-, $C_{\perp t}$ and $C_{\perp b}$ in Eqs. (1) and (2) are constants, which are determined by the integral interval of the two factors. The function is to make $P = p_t + p_b$ when the range D reaches the maximum. In this paper, several different spring branch chain parameters as shown in Table 5 are selected to solve the value of the evaluation function and the shape of the workspace, as shown in Fig. 6.

Table 5. Parameter value and evaluation function value of spring branch chain

	k/(N/m)	p_k	Z_{S1}/m	Z_{S2}/m	p_1	p_2	P
(a)	76	1.5	0.40	0.2	2.52	1.41	0.4247
(b)	53	1.0	0.71	1.84	3.85	1.86	0.0680
(c)	52	0.95	0.72	1.95	3.76	1.93	0.0607

Comparing (a), (b) and (c) in Fig. 6, it can be seen that the value of the evaluation function p becomes smaller and smaller as the defect shape decreases, and comparing (b) and (c), it can be seen that even if the defect shape is only a little different, the value of P is different, which proves the sensitivity of the constructed work space defect evaluation function. In this way, even if different configurations are adopted, it shows that the method of modeling shape defects and constructing evaluation function through mixed factors is still applicable.

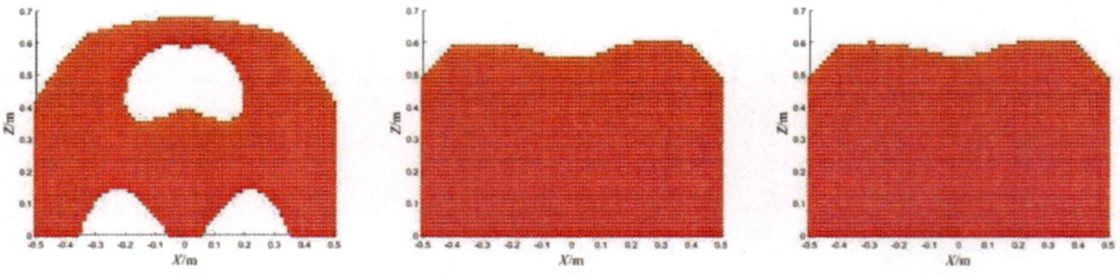

(a) (b) (c)

Fig. 6. 3 example workspace y = 0 section

5 Discussion and Conclusion

Aiming at the shortcomings of the existing workspace evaluation indexes, this paper proposes an evaluation index based on numerical method, which can intuitively and comprehensively reflect the attitude space of the mechanism moving platform, and compares the workspace of 6C configuration and 6c2s configuration based on the proposed index. The results show that the introduction of spring branch chain can increase the position space and attitude space of the mechanism at the same time.

The shape defect of the mechanism workspace caused by the spring branch chain is discussed. Taking the configuration studied in this paper as an example, a hybrid evaluation function for evaluating the severity of shape defects is established, and several workspace shape defects are selected for evaluation. The sensitivity of the evaluation function is proved, which provides a research basis for eliminating this phenomenon or reducing its impact.

Acknowledgment. This paper is supported by the graduate science and technology innovation fund of Civil Aviation University of China.

References

1. Hiller, M., Fang, S., Mielczarek, S., et al.: Design, analysis and realization of tendon-based parallel manipulators. Mech. Mach. Theory **40**(4), 429–445 (2005)
2. Diao, X., Ou, M.: A method of verifying force-closure condition for general cable manipulators with seven cables. Mech. Mach. Theory **42**(12), 1563–1576 (2007)
3. Gallina, P., Rosati, G., Rossi, A.: 3-d.o.f. wire driven planar haptic interface. J. Intell. Robot. Syst. **32**(1), 23–36 (2001). https://doi.org/10.1023/A:1012095609866
4. Surdilovic, D., Cojbasic, Z.: Robust robot compliant motion control using intelligent adaptive impedance approach. IEEE (1999)
5. Surdilovic, D., Bernhardt, R.: STRING-MAN: a new wire robot for gait rehabilitation. In: 2004 IEEE International Conference on Robotics and Automation, 2004. Proceedings. ICRA'04. IEEE (2004)
6. Surdilovic, D., Radojicic, J.: Robust control of interaction with haptic interfaces. In: IEEE International Conference on Robotics & Automation. IEEE (2007)
7. Kawamura, S., Won, C., Tanaka, S., et al.: Development of an ultrahigh speed robot FALCON using wire drive system. In: IEEE International Conference on Robotics & Automation (1995)

8. Kawamura, S., Kino, H., Won, C.: High-speed manipulation by using parallel wire-driven robots. Robotica **18**(1), 13–21 (2000)
9. Duan, Q., Vashista, V., Agrawal, S.K.: Effect on wrench-feasible workspace of cable-driven parallel robots by adding springs. Mech. Mach. Theory **86**, 201–210 (2015)
10. Bosscher, P., Riechel, A.T., Ebert-Upho, F.I.: Wrench-feasible workspace generation for cable-driven robots. IEEE Trans. Rob. **22**(5), 890–902 (2006)
11. Gao, B., Song, H., Zhao, J., et al.: Inverse kinematics and workspace analysis of a cable-driven parallel robot with a spring spine. Mech. Mach. Theory **76**, 56–69 (2014)

Simulation Research on Working Device of Small Tonnage Forklift

ZhiBin Wang(✉), Xinyong Li, Jian Wu, and Te Li

School of Mechanical Engineering, Changshu Institute of Technology, Suzhou, China

Abstract. With the development of logistics machinery, small-tonnage forklifts have become the most widely used type of forklifts. The working device is the most frequently moving part of the forklift, and its stability directly affects the performance of the forklift. This article first determines the critical state of the work, analyzes the force of each part, uses ANSYS to impose constraints and loads on the fork and mast of the small tonnage forklift, and obtains the strength and stiffness diagrams of the parts through simulation. The results show that under full load, the strength of the parts does not exceed the yield limit of 345 MPa, and the strength meets the design requirements; the maximum deformation of the parts does not exceed the allowable deformation, and the stiffness meets the structural design requirements.

Keywords: Small-tonnage forklift · Working device · Simulation · Strength · Stiffness

1 Introduction

Forklifts are a kind of handling machinery, which has been rapidly developed in recent years, especially small-tonnage forklifts, which are widely used due to their flexibility and compactness. A forklift is mainly composed of an engine, a chassis and a working device, and the working device mainly includes a fork and an inner and outer mast. The performance of the forklift directly affects the efficiency and life of the whole vehicle. Therefore, the fork and the inner and outer mast must have sufficient strength and stiffness.

As the most important part of the forklift, the work device is indispensable for its research. Zhang Quanyu of the Department of Automotive Engineering of Chengde Petroleum Technical College applied CATIA software to build a three-dimensional model of a crawler forklift working device, and used ADAMS to optimize the simulation analysis with the minimum thrust of the hydraulic cylinder in the working process of the working device as the optimization objective. The performance of the working device is a good reference; Luo Li of Shandong University of Technology uses the parameterized modeling and optimization analysis functions of the ADAMS/View module to obtain the best hinge point position that minimizes the force on the tilt cylinder. The structural design of the working device of the medium tonnage forklift provides ideas; He Shunli from Zhejiang University used ANSYS to simulate and analyze the mast of the medium

Y. Wang et al. (Eds.): IWAMA 2021, LNEE 880, pp. 596–602, 2022.
https://doi.org/10.1007/978-981-19-0572-8_76

tonnage forklift, and combined the analysis results to optimize the structure of the mast to improve the work performance of the mast; Zuo Yu from Taiyuan University of Science and Technology Fei uses the Ansys Workbench platform to perform modal analysis on the internal and external mast models of the 3t forklift, draws the corresponding mode diagrams, and obtains the main stress concentration of the mast. By avoiding the phenomenon of resonance, the fatigue life of the door frame is improved. These articles have carried out detailed research on a certain part of the medium-tonnage forklift, and have achieved the expected optimization results, but have not studied the complete working device of the most widely used small-tonnage forklift, which has certain limitations.

This paper takes a small tonnage forklift with a lifting load of 1.6t as the research object, and chooses ANSYS to simulate and analyze the complete working device of the forklift. The most dangerous working conditions were selected during the research, and Ansys was used to perform statics analysis on the most dangerous working conditions of the forklift to obtain the stress cloud diagram and deformation curve of the working device, and check whether the strength and rigidity meet the requirements according to relevant standards.

2 Working Principle

The forklift is mainly composed of four parts: power unit, chassis, working device and electrical equipment. The assembly diagram of the forklift is shown in Fig. 1 below. Among them, the working device is a characteristic structure that distinguishes forklifts from other vehicles, and is mainly used for picking, placing, lifting and lowering goods. The working device is mainly composed of an inner mast, an outer mast, a lifting cylinder, a chain and a fork, as shown in Fig. 2. The lower end of the outer door frame is hinged on the frame, and the inner door frame is provided with rollers and is embedded in the outer door frame. When the inner door frame rises, it can partially extend out of the outer door frame. The fork has a roller, which is embedded in the inner mast and can move up and down. The bottom of the lifting cylinder is fixed at the lower part of the outer mast, and the top is connected with the top of the inner mast through a sprocket. One end of the lifting chain is fixed on the top of the piston cylinder, and the other end is connected to the fork frame by bypassing the sprocket. When working, the top of the piston rod lifts with the sprocket, and the chain lifts the fork. At the beginning of lifting, only the fork lifts up, and the inner main frame can only be driven up after the piston rod hits the inner main frame, and the rising speed of the inner main frame is 1/2 of the fork.

This article mainly focuses on the simulation research of inner mast, outer mast and fork. The lower end of the outer door frame is fixed, the rollers and the door frame guide rail are in frictional contact, the friction coefficient is set to 0.5, and the contact between the remaining parts is set to bind.

As the most frequently moving part of the forklift, the working device bears a variety of loads and is extremely prone to failue, causing losses to the enterprise. The simulation study of the working device has a certain reference value for the design of the forklift, and can ensure the structural stability of the forklift.

Fig. 1. Forklift assembly drawing

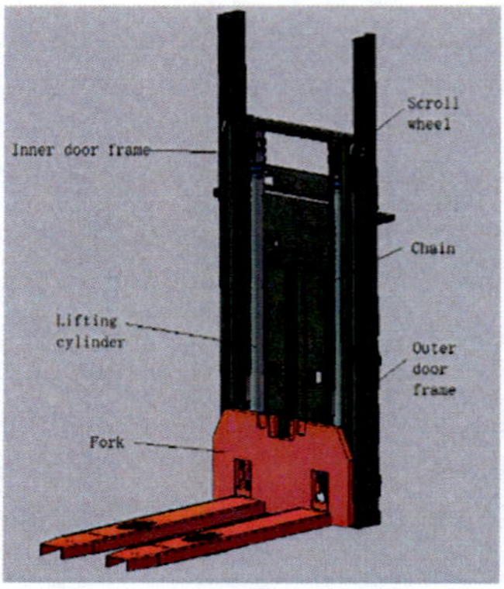

Fig. 2. Working device structure diagram

3 Simulation Design of Working Device

3.1 Overall Analysis

First, the overall simulation analysis of the working device. The forklift has no forward tilt, so the most dangerous working condition is when the forklift is lifted to the maximum height of 1.5 m. The forks are made of carbon structural steel Q345 plate welding parts, and the mast is made of Q345. There is frictional contact between the roller and the gantry rail, the friction coefficient is set to 0.5, and the contact between the remaining parts can be bound according to the automatic setting of the system. Before meshing, the rounded corners and threaded holes that have little effect on the result are simplified to improve the solution speed, and the hybrid meshing is adopted as a whole to improve the comprehensive division effect. Apply a downward cargo gravity of 16000 N to the forks and fix the lower end of the mast. The operating results are as follows:

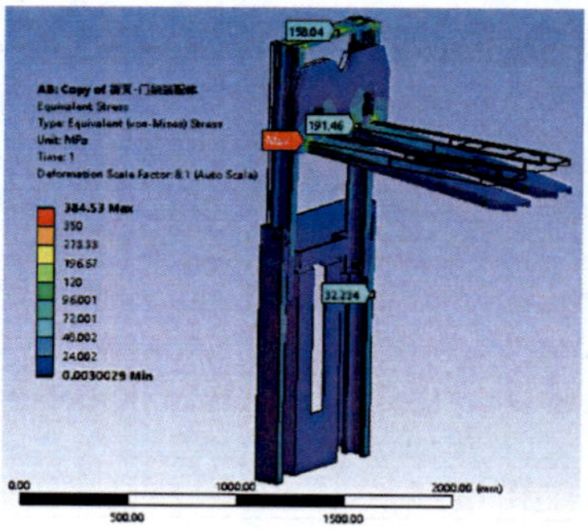

Fig. 3. Overall stress distribution of mast

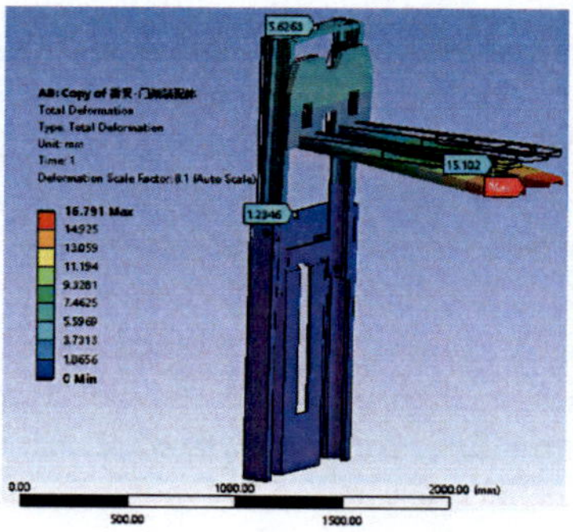

Fig. 4. Overall deformation of the mast

It can be seen from Fig. 3 that the overall maximum stress of the mast is 384.53 MPa, which is located at the welding place of the fork and the baffle plate, where the stress is singular and the stress increases, and the stresses in other parts are far less than the yield limit of the material, so Meet the requirements. It can be seen from Fig. 4 that the overall maximum deformation is at the tip of the fork, and the maximum deformation is 16.791 mm, because the fork is approximately a cantilever beam, and the maximum deformation is at the end of the beam.

3.2 Local Analysis

The fork used in this machine is an integral structure. The rated lifting weight of the forklift is 1600 kg, and the forks are connected to the inner main frame through rollers. Therefore, in the analysis, the displacement in the x, y, and z directions between the roller seat plate and the roller connecting hole is restricted. Under the full load condition, a force of 16000 N is applied to the fork panel to simplify the threaded hole on the fork and increase the settlement speed. The available fork strength and deformation results are as follows:

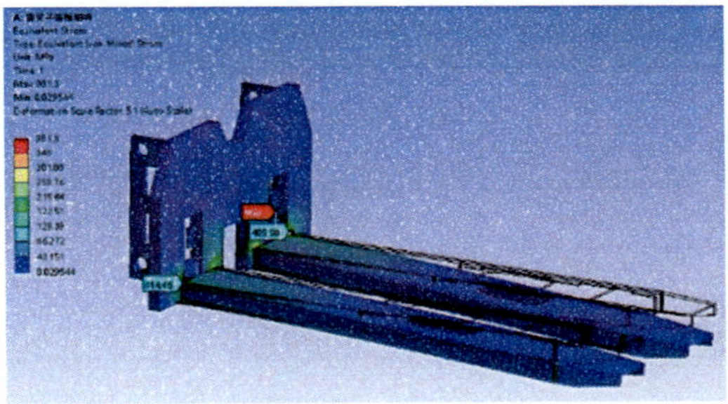

Fig. 5. Fork stress cloud chart

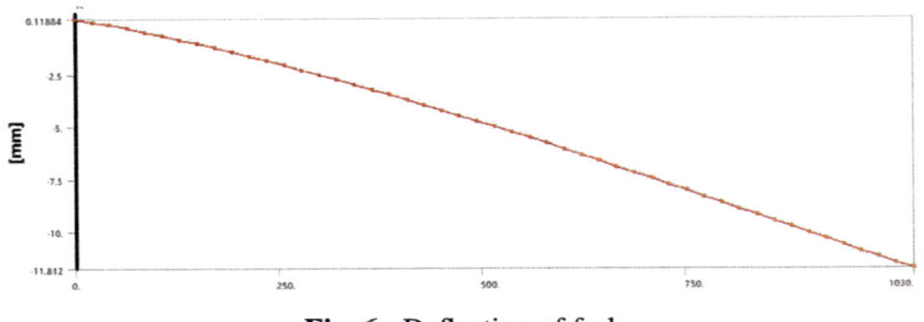

Fig. 6. Deflection of fork

It can be seen from Fig. 5 that the maximum stress is at the welding place of the fork and the baffle plate, where the stress is singular and the stress increases. Except for this, the stress in other parts is lower than 345 MPa, which does not exceed the yield limit of the material, and the strength meets the requirements. The rigidity is mainly measured by the longitudinal flexural deformation of the fork. The allowable deformation of the fork is: fork length/50 = 1030/50 = 20.6 mm. It can be seen from Fig. 6 that the deflection deformation is approximate to a linear function, which increases with the increase of the distance to the vertical section of the fork. It is similar to the cantilever beam structure. The maximum deformation is at the end of the fork, which is $11.8 - 0.118 = 11.682$ mm < 20.6 mm, so The design of the fork meets the requirements.

The stage of adding loads and constraints to the three-dimensional simulation of the mast must be considered in accordance with the force situation when the secondary mast is fully loaded, fully extended, kept vertical, and the cargo is lifted to the highest level. The upper part of the inner main frame is under the pressure of the fork rollers, and the calculated size is 21007 N. The lower part is connected to the outer main frame by the rollers. The simulation results are as follows:

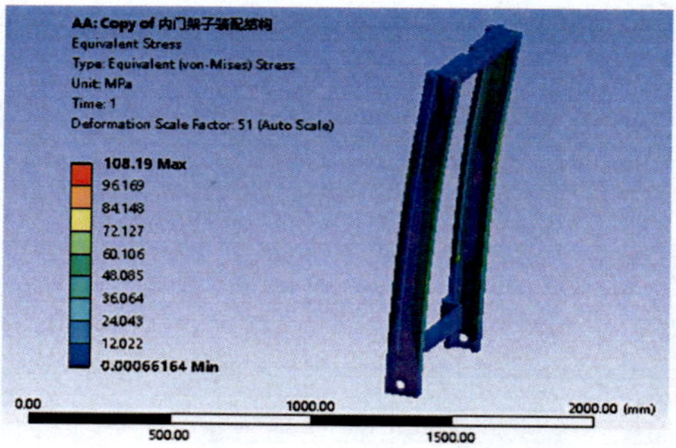

Fig. 7. Stress cloud diagram of inner door frame

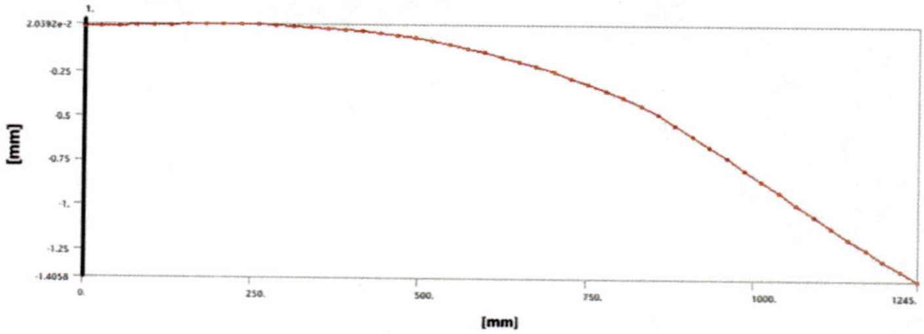

Fig. 8. Deflection and deformation of the inner door frame

It can be seen from Fig. 7 that the maximum stress of the inner door frame is 108.19 MPa, which is far below the yield limit of the material, and the inner door frame is safe. It can be seen from Fig. 8 that the deformation of the inner door frame is not obvious at the bottom of the door frame, but from 300 mm away from the bottom of the door frame, the deformation increases with the increase in height. The maximum deformation is located at the top of the door frame and the size is 1.4 mm. The moment generated by the load makes the upper part of the inner mast deformed in the horizontal direction. The standard lifting height of the secondary mast is 1600 mm, and the allowable horizontal offset is $[f] = H/100 = 1600/100 = 16$ mm > 1.4mm, the design meets the requirements.

The bottom of the outer door frame is welded to the bottom plate to impose a fixed constraint on the bottom of the door frame. The force of the door frame mainly comes from the lateral pressure of the upper end of the outer door frame of the inner door frame. A load of 21007 N is added. The simulation results are as follows:

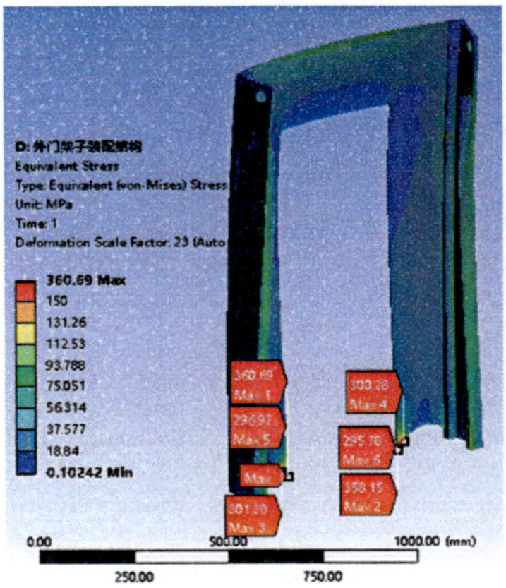

Fig. 9. Stress cloud diagram of outer portal frame

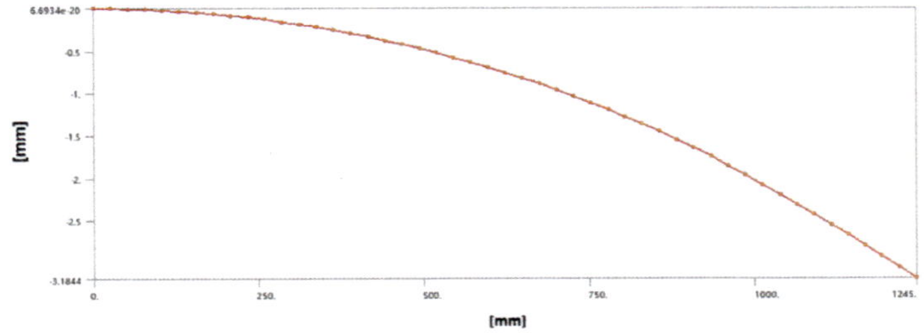

Fig. 10. Deflection and deformation of the outer door frame

It can be seen from Fig. 9 that due to the influence of the stress concentration at the welding place of the back plate and the floor, the global stress peak value is higher, the maximum stress is 360.69 Mpa, which exceeds the yield strength of Q345, but the analysis of the stress cloud chart shows that the stress is excessive except for the local stress concentration The stresses of other parts are around 100 MPa, and the distribution is reasonable. So choose Q345 to meet the requirements. It can be seen from Fig. 10 that the deflection deformation is approximately a parabolic function, and the deformation increases with the increase of the distance. The maximum deformation is 3.2 mm, which is located on the top of the outer door frame. This is due to the concentrated force and the couple of forces that make the upper part of the outer door frame horizontal. The direction deformation is large, and the allowable deformation of the outer door frame is: H/200 = 1600/200 = 8 mm > 3.2 mm, and the design meets the requirements.

4 Conclusion

In this paper, through static analysis of the working device of small-tonnage forklift, the stress cloud diagram and deflection deformation curve of the inner and outer mast and fork are obtained. Through the analysis, the maximum stress of the inner mast is 108.19 MPa, and the fork and outer mast are in addition to stress concentration. The outside does not exceed the material yield limit of 345 MPa, and the strength meets the requirements; the maximum deformation of the fork is 11.682 mm, the maximum deformation of the inner mast is 1.4 mm, and the maximum deformation of the outer mast is 3.2 mm, which are all less than the allowable deformation of the parts, and the rigidity meets the requirements. This article has a good reference for the design of small-tonnage forklift working devices. On the premise that the strength and rigidity meet the requirements, it lays the foundation for the later optimized design of the forklift.

References

1. He, S.: Forklift mast dynamic and static co-simulation and structural optimization. Zhejiang University (2020)
2. Qiu, W.: Finite element analysis of the strength and stiffness of the forklift working device. Qual. Tech. Supervision Res. **2018**(02), 30–32+44 (2018)
3. Zuo, Y.: Research on the static and dynamic characteristics of the three-stage gantry system of a forklift. Taiyuan University of Science and Technology (2016)
4. Li, L.: Simulation research on forklift working device. Shandong University of Technology (2015)
5. Quanyu, Z.: Optimization simulation analysis of crawler forklift working device. J. Chengde Petrol. Coll. **17**(01), 46–51 (2015)
6. Emin, L., Bin, W., Chen, Z.: Finite element analysis of forklift frame based on ANSYS. Sci. Technol. Eng. **11**(09), 2078–2081 (2011)
7. Xu, J., Yang, F., Li, G., Liu, M.: Dynamic simulation and optimization design of forklift mast. Eng. Mach. **39**(12), 25–29+129 (2008)
8. Lifang, S.: Computer-aided analysis of forklift mast structure based on ANSYS. Comput. Knowl. Technol. **20**, 383–385 (2008)
9. Yashan, S.: Finite element calculation and stress analysis of forklift forks. J. Tianjin Inst. Urban Constr. **1**, 43–47 (1995)

The Lubrication Performance of mm-scale Specimen Based on Magnetic Fluid

Jian Wu[1], Jiejie Cao[1](✉), and Yangyang Chen[2]

[1] College of Mechanical and Electrical Engineering, Changshu Institute of Technology, Changshu 215500, China
wujian@cslg.edu.cn
[2] Jiangsu Offshore Longyuan Wind Power Co., Ltd., Nantong, China

Abstract. The friction of the mm-scale is different from the macro due to the high surface-to-volume ratio. In this paper, the lubrication performance of mm-scale specimens based on magnetic fluid has been studied. The lubrication experiments of mm-scale under three magnetic fields (0 mT, 20 mT, 40 mT) have been carried out. The coefficient of friction, the thickness of the film were measured, and the wear surface was analyzed. The results show that the magnetic fluid without a magnetic field exhibits the lubricating properties of a normal fluid. When a magnetic field is applied, the viscosity of the magnetic fluid increases, the friction coefficient is lower with low speed and the film thickness is higher, which indicates that the lubrication effect is improved with magnetic fluid under a given magnetic field. The best lubrication has shown when the magnetic fluid is saturated. The wear surface under the action of two magnetic fields also supports.

Keywords: mm-scale Specimen · Lubrication performance · Magnetic fluid

1 Introduction

MEMS (micro electro mechanical systems) has grown dramatically in the last decade, which is the core components of sensors, actuators, and actuators. The friction is significantly different from the macro because of the high surface-to-volume ratio [1]. To reduce the friction in mm-scale, the various method has been developed such as surface treatment [2], organic monolayers [3] and so on.

As a new type of functional material, the magnetic fluid provides new solutions for lubrication [4]. Patel [5] analyze the performance of a hydrodynamic short journal bearing under the presence of a magnetic fluid lubricant. Jian [6] used an abrasion testing machine and a vertical four-ball testing machine to study the friction performance of magnetic liquid as a lubricant, and analyzed the influence of magnetic particle concentration on its friction and wear performance. Wang [7] measured the bearing capacity and friction coefficient of magnetic liquid under a magnetic field. However, all of this researches focus on the macro scale, and there is no research has reported based on magnetic liquids in mm-scale.

Y. Wang et al. (Eds.): IWAMA 2021, LNEE 880, pp. 603–610, 2022.
https://doi.org/10.1007/978-981-19-0572-8_77

In this paper, the lubrication performance of mm-scale specimens based on magnetic fluid has been studied. A measuring system for magnetic fluid lubrication has been developed. The lubrication performance under three magnetic fields (0 mT, 20 mT, 40 mT) has been tested. The coefficient of friction, the thickness of film was measured, and the wear surface was analyzed.

2 Experimental Details

2.1 Experimental Apparatus

In reference [8], the principle of measurement system has been described, it contains a pin-on-disk tribometer (Rotating disk and an mm-scale pin specimen), a displacement sensor for normal force, and a laser-based measurement system for friction which generated by the two contact surfaces. The lubrication mechanism has also been described, a quantitative pump has been used to deliver the lubricant to the friction surface. To study the lubrication performance of the magnetic fluid, in this paper, the lubrication system has been modified, as shown in Fig. 1.

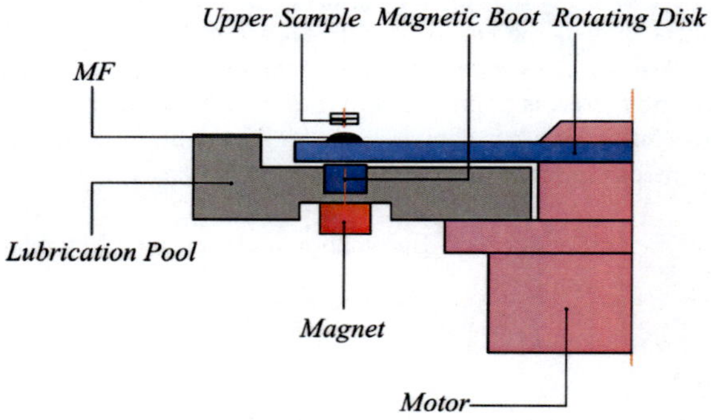

Fig. 1. The design of magnetic field in lubrication bath

A blind hole at the bottom of the lubrication pool has been processed, which has been used to hold a magnetic boot (with a diameter of $\varphi 10$ mm and a height of 2 mm). Permanent magnets are placed behind the lubrication pool. The magnetic field generated by the permanent magnet can reach the friction surface through the magnetic boot. For the magnetic fluid is saturated and unsaturated, a series of permanent magnets have been used to obtain different magnetic fields. Table 1 shows the relationship between magnets and the magnetic field on the friction surface.

Table 1. The relationship between the quantity of magnet and friction

Diameter (mm) height (mm)	ϕ6 * 3	ϕ8 * 3	ϕ8 * 6	ϕ8 * 9	ϕ12 * 3	ϕ10 * 12
Magnetic field strength (mT)	10	20	30	40	50	60

2.2 Magnetic Fluid

Magnetic fluid (Magnetic fluid), also known as magnetic fluid or ferrofluid, is a colloidal solution formed by stably dispersing nano-scale magnetic particles in a carrier liquid after being treated with a surfactant [4]. In this experiment, the selected nano-magnetic liquid is a double lipid-based magnetic fluid, and its related properties are shown in Table 2. Figure 2 shown the magnetization characteristic curve which is measured by the vibrating sample magnetometer (VSM). It can be seen that the external magnetic field intensity H reaches to 3500Oe (35 mT), the magnetic fluid tends to be saturated. 20 mT and 40 mT have been used in these tests.

Table 2. Characteristic parameter of the magnetic fluid

Property	Density (g/cm^3)	Viscosity (mpa·s)	Saturation magnetization M_s(Gs)	Volume component
Parameter	1.13	14.5	300	0.1%

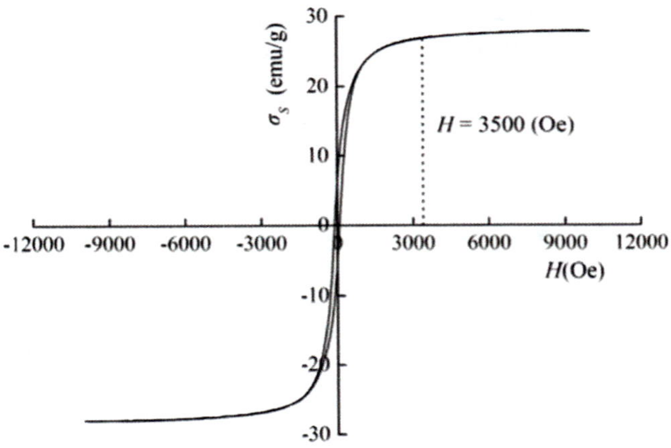

Fig. 2. The hysteresis curve of diesters magnetic fluid

2.3 Testing Method

The testing method of lubricant on mm-scale Specimen has been described in Ref [8]. In this test, a double lipid-based nano-magnetic fluid has been used. To get the lubrication performance with or without a magnetic field, the following experimental procedure is used:

1) According to Ref [9], all mechanical and electrical components should be installed accurately.

2) The rotating disk and specimen are cleaned before the test, and it's installed correctly. The magnetic boot has been installed in the bottom of the lubrication pool.

3) Adjust the motor to apply a normal force on the specimen surface. A given magnetic liquid (5 mL) has been drop on the surface of a rotating disk.

4) Running the measurement system, the monitor records the friction coefficient and the film thickness. When the speed reaches 1.1 m/s, the system turns off automatic, and uninstalled the specimen,

5) Start measurement the procedure with a new specimen, and 20 mT and 40 mT magnetic fields have been applied respectively.

Figure 3 shown the photograph of lubricated by magnetic fluid under a 40 mT magnetic field. The magnetic liquid forms circular droplets on the friction disk with the action of a magnetic field. When the installation is completed, the specimen is completely immersed in the magnetic liquid, and stable lubrication has been formed between the friction pairs during operation.

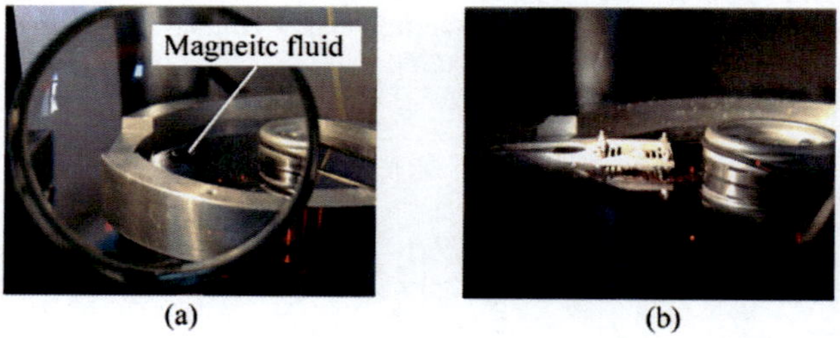

Fig. 3. The photograph of specimen installation under magnetic field (a) upper sample unassembled; (b) upper sample assembled

3 Results and Discussion

The lubrication performance of the magnetic liquid on the mm scale has been measured using the modified test system. Three states of the magnetic field were measured respectively, without, 20 mT and 40 mT magnetic field. And the friction coefficient and the film thickness were measured.

3.1 Measurement of Friction

Figure 4 shown the friction coefficient of three magnetic fields. In this chart, a logarithmic scale is used on the abscissa to show the friction at a low speed more clearly.

It can be seen that the three friction curves are different. When no magnetic field is applied, the magnetic liquid exhibits the properties of ordinary fluids. The friction coefficient curve has three regions. With a low speed (<0.05 m/s), the friction coefficient

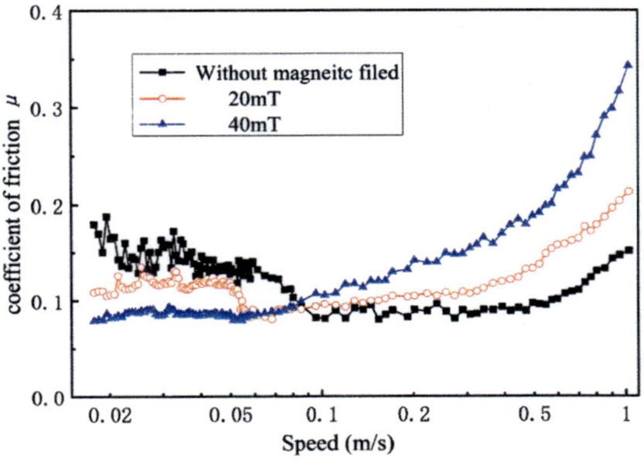

Fig. 4. The friction coefficient of magnetic fluid under three magnetic fields

is high (0.12–0.2), and there is some scattering of the results. When the sliding speed increases to 0.05 m/s, the friction decreases sharply. Increasing the speed further, the friction increases slightly. When a given magnetic field is applied, the viscosity of the magnetic fluid has changed. At low speeds, the friction coefficient is low and relatively stable. As the speed increases, the friction coefficient gradually increases. The lubrication characteristics of magnetic fluids under two magnetic field strengths are also different. The friction coefficient of low-viscosity magnetic fluids is relatively large and there is some scattering of the results. At high speed, high-viscosity magnetic fluids are more obvious.

3.2 Measurement of Film Thickness

Figure 5 shown the film thickness of three magnetic fields. In this chart, a logarithmic scale has also been used on the abscissa.

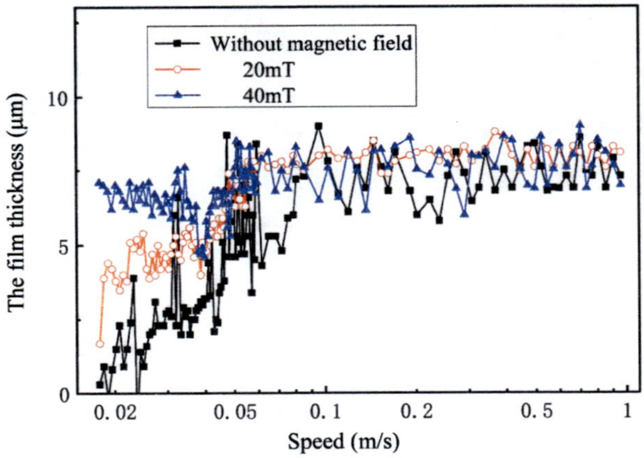

Fig. 5. The thickness of film under three magnetic fields

Three curves are similar, at low speed, there is some scattering of the results. Without and with weak (20 mT) magnetic fields, the film thickness is increasing. And with a strong magnetic field (20 mT), the thickness of the film is relatively stable and has less scatter. Because without a magnetic field, the magnetic fluid exhibits the lubricating properties of a normal fluid. Low-viscosity magnetic fluids also have this feature. When the magnetic fluid reaches saturation, the viscosity is higher, and the lubricating properties at this time show strong lubricating properties. When the speed reaches to 0.05 m/s, the surface of the friction surface forms a stable lubricating film, and the thickness of the film is relatively stable, the three curves have a relatively stable film thickness of 7 um.

3.3 Wear

To analyze the lubrication of the magnetic fluid, scanning electron microscopy (SEM) analysis has been used to perform on the surface of the specimen. Figure 6 shown the wear of specimen surface with (20 mT) and without magnetic field applicated. The sliding speed of the two specimen is 0.05 m/s, the applied load is 1.5 N, and the running time is 8 h.

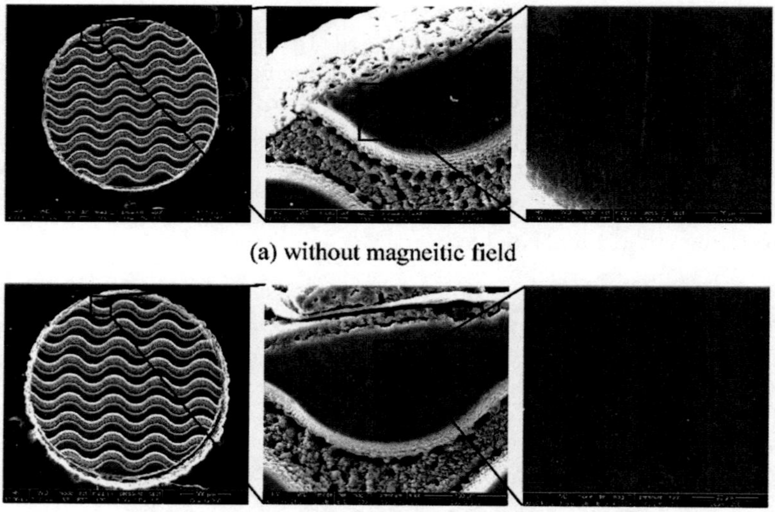

(a) without magneitic field

(b) 20mT magnetic field

Fig. 6. The SEM photography of wear surface under different magnetic field

From the analysis in the above figure, when the magnetic field is not applied, there are obvious wear marks on the surface of the specimen; and when a 20 mT magnetic field is applied, the microscopic surface is not abraded. The reason is that the magnetic field can increase the viscosity of the nano-magnetic liquid, thereby increasing its carrying capacity, so that the two test pieces are completely separated during the mixing and lubrication phase.

To further analyze the lubrication performance of the magnetic liquid on the surface of the specimen. X-ray photoelectron spectroscopy (XPS) was used to characterize the surface of specimens lubricated by the magnetic fluid. Figure 7 shown the elemental analysis results of specimen surface with (20 mT) and without magnetic field applicated.

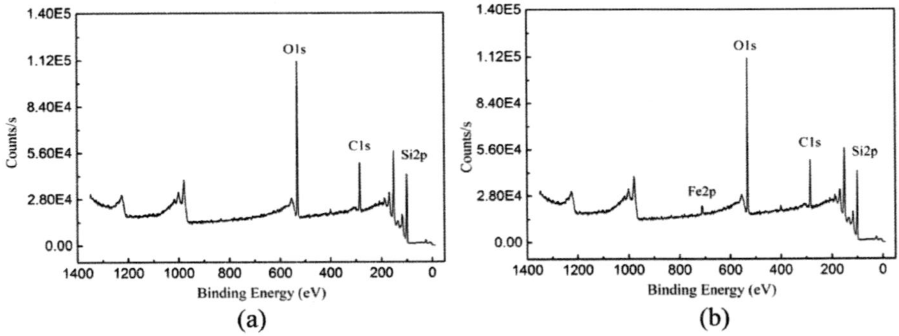

Fig. 7. The XPS analysis of the surface of the specimen (a) without magnetic field; (b) 20 mT magnetic field applied

It can be seen from the above figures that the main elements on the specimen surface are: O, N, C, and Si (material). In addition to these elements, the surface of the specimen after lubricated with the magnetic fluid has Fe at 711 eV, and the content is relatively high. During the whole test process, the Fe only contains in the magnetic fluid. The magnetic particles in the magnetic fluid stay on the surface of the specimen during the test process. The reason may be analyzed as follows: due to the influence of the shear stress of the asperities and the furrow effect, a certain amount of frictional heat will be generated on the specimen surface, which can soften the nanomagnetic particles adsorbed on the surface of the specimen. A certain physical and chemical reaction occurs on the surface of the base material and the specimen adheres to the surface of the test piece, a permeable layer has been formed, and fills the pits and cracks on the surface of the friction pair under the action of squeezing force.

4 Conclusion

In this paper, the lubrication performance of mm-scale specimens based on the magnetic fluid has been studied. A measuring system for magnetic fluid lubrication has been developed. The lubrication experiments of mm-scale under three magnetic fields (0 mT, 20 mT, 40 mT) have been carried out. The coefficient of friction, the thickness of the film was measured, and the wear surface was analyzed. The results show that there are obvious differences in the lubrication performance of magnetic fluids under a magnetic field. Without a magnetic field, the magnetic fluid shows the lubricating properties of a normal liquid. When a given magnetic field is applied, the friction coefficient is lower with low speed and the film thickness is higher, which indicates that the lubrication effect is improved with magnetic fluid under a given magnetic field. The best lubrication has performed when the magnetic fluid is saturated.

Acknowledgment. The work Open Project of Jiangsu Elevator Intelligent Safety Key Construction Laboratory (JSKLESS201703), and The Doctoral Science Foundation of Changshu Institute of Technology (No. KYZ2015054Z).

References

1. Ku, I.S.Y., Reddyhoff, T., Holmes, A.S., et al.: Wear of silicon surfaces in MEMS. Wear **271**(7–8), 1050–1058 (2011)
2. Tayebi, N., Polycarpou, A.A.: Reducing the effects of adhesion and friction in microelectromechanical systems (MEMSs) through surface roughening: comparison between theory and experiments. J. Appl. Phys. **98**(7), 073528 (2005)
3. Ashurst, W.R., Yau, C., Carraro, C., et al.: Alkene based monolayer films as anti-stiction coatings for polysilicon MEMS. Sens. Actuators A **91**(3), 239–248 (2001)
4. Rosensweig, R.E.: Ferrohydrodynamics. Courier Corporation (2013)
5. Patel, N.S., Vakharia, D.P., Deheri, G.M.: A study on the performance of a magnetic-fluid-based hydrodynamic short journal bearing. Int. Sch. Res. Not. **2012**, 603460 (2012)
6. Jian, L., Yunfeng, T.: Lubrication theoretical and experimental study of magnetic fluid applied to sliding friction. Electr. Eng. Mater. **3**, 29–31 (2004)
7. Wang, L., Guo, C., Yang, Z., Zhang, D.: Experiment research on friction coefficient of Fe3O4 ferrofluid. Lubr. Eng. **179**(7), 142–143+157 (2006)
8. Wu, J., Liu, T., Yu, N., et al.: A pin-on-disk tribometer for friction and lubricating performance in mm-scale. Tribol. Lett. **69**(2), 1–6 (2021)
9. Wu, J., Sørby, K.: Effect of lubricant viscosity on the friction behavior of the mm-scale Specimen. In: Wang, Y., Martinsen, K., Yu, T., Wang, K. (eds.) Advanced Manufacturing and Automation X. IWAMA 2020. LNEE, vol. 737, pp. 339–346. Springer, Singapore (2020). https://doi.org/10.1007/978-981-33-6318-2_42

Development and Test of Navigation System for Unmanned Carrier Platform in Greenhouse

Hao Li[1], Lin Liu[2], Jin Yuan[2](✉), Yan Zhang[1], and Kai Tao[1]

[1] College of Mechanical and Electronic Engineering, Shandong Agricultural University, Tai'an 271018, China

[2] Shandong Agricultural Equipment Intelligent Engineering Laboratory, Tai'an 271018, China

jyuan@sdau.edu.cn

Abstract. In order to solve the navigation and positioning problem of unmanned carrier platform in facility greenhouse environment, aiming at the problems and shortcomings of existing schemes, this paper proposes a facility greenhouse unmanned carrier platform navigation system combining electromagnetic line and encoder. The system uses electromagnetic line arranged above the travel path of facility greenhouse for navigation, The position data of the electromagnetic line node mark and the encoder of the vehicle body are fused for accurate positioning, which realizes the autonomous navigation, real-time positioning and autonomous operation of the unmanned carrier platform in the facility greenhouse environment. The prototype is tested in the laboratory environment to test the navigation deviation and positioning accuracy at different speeds. The test shows that when the prototype runs at the speeds of 0.5 m/s, 1.0 m/s and 1.5 m/s respectively, the maximum absolute value of the horizontal deviation between the actual running track and the ideal track is 5cm, and the maximum positioning accuracy error is 7 cm. Meet the requirements of trajectory deviation and positioning accuracy.

Keywords: Greenhouse facilities · Independent entry · Electromagnetic wire · Encoder

1 Introduction

Facility agriculture plays an important role in China's agricultural field. Especially in recent years, China's facility agriculture has made great development, and the planting area of facility greenhouse has increased year by year [1, 2]. Compared with field agriculture, the internal space of facility agriculture is narrow and the scene is complex. The existing operation equipment can not carry out mechanized and intelligent operation. Restricted by the low overall level of equipment manufacturing industry, there is a large gap between China's facility horticultural equipment and developed countries such as the Netherlands, the United States and Japan in terms of operation efficiency, completeness, stability and intelligence, especially the lack of an independent operation basic platform of facility greenhouse based on autonomous navigation control [3, 4].

Y. Wang et al. (Eds.): IWAMA 2021, LNEE 880, pp. 611–619, 2022.
https://doi.org/10.1007/978-981-19-0572-8_78

At present, there are four main types of navigation methods for unmanned delivery platforms in facility greenhouses: ridge type, guide rail type, tracking type and multi-source data fusion type [5–7]. The ridge and ditch navigation mode mainly relies on the mechanical structure of the carrier platform for independent navigation, and the operation is stable. However, it has strong dependence on the greenhouse environment, and has strong destructiveness to the ground, and the transformation process is more complex. The guideway navigation mode is mainly to lay the track in the greenhouse, so that the carrier platform can conduct autonomous navigation with high reliability and good stability. However, this mode limits the activity range of the carrier platform. Generally, it can only operate in the main line of the facility greenhouse, and can not enter a single line. The tracking navigation mode is mainly to lay electromagnetic wires or cables on the ground as identification marks to guide the platform for autonomous navigation, which has high reliability and good stability. However, the cables laid below the surface need to be excavated and re laid when the operation line changes, and the process is more complex. The multi-source data fusion navigation mode mainly uses industrial camera and lidar as the main data source, and uses sensor data such as gyroscope and encoder for auxiliary fusion. This mode has low requirements for facility greenhouse hardware and high flexibility. However, the industrial camera and lidar as the main data source have high cost and are sensitive to external light. The requirements for temperature, humidity and other conditions are high, so it is impossible to carry out all-weather operation.

According to the characteristics of the internal environment of the facility greenhouse and the advantages and disadvantages of the existing navigation methods, this paper designs a navigation system of the facility greenhouse unmanned carrier platform based on the combination of electromagnetic line and encoder, which uses the integration of electromagnetic line node mark and the position data of the encoder of the vehicle body for accurate positioning. The electromagnetic line is used as the navigation mode, and the electromagnetic line is arranged above the operation line to guide the vehicle forward. When the object line is changed, it is more convenient to change the electromagnetic line in the air.

2 Structure and Working Principle of the Whole Machine

2.1 Complete Machine Structure

The structure of the whole machine is shown in Fig. 1. The developed facility greenhouse intelligent unmanned carrier platform mainly consists of platform chassis, scissor lifting platform, power drive device, steering drive device and operation control system. The platform can realize the basic operations of forward, backward and steering in the solar greenhouse environment, autonomous navigation operation along the operation route, and operation of crops at different heights.

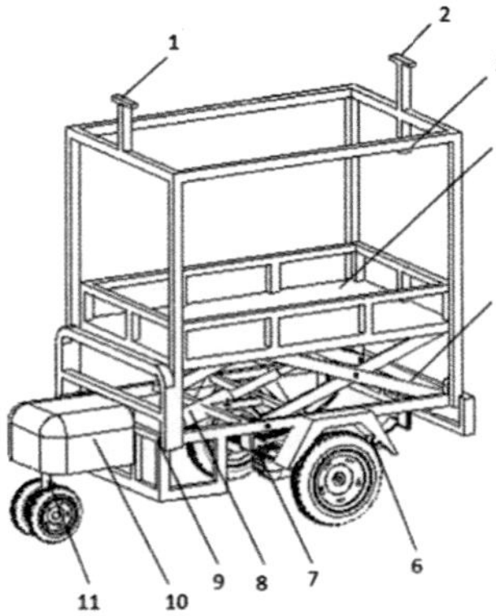

1. Front electromagnetic sensor 2. Rear electromagnetic sensor 3. Pose sensor 4.Work platform 5. Shear arm 6. Encoder 7. Drive motor 8. Push rod motor 9. Frame 10. Control box 11. Steering shaft

Fig. 1. Three dimensional model of facility greenhouse unmanned carrier platform

2.2 Working Principle

During the autonomous navigation operation of the platform, the electromagnetic sensor detects the deviation of magnetic flux on both sides, adjusts the position of the platform relative to the magnetic line through the front steering steering gear, keeps the platform in the central position of the route, and detects the right angle mark positions at the bottom and head of the line. When the mark is detected, the encoder is calibrated, and the corresponding mark positions are incremented, The specific position of the current platform is known by calculating the value of the flag bit and the value of the encoder. When the platform moves in and out autonomously, the electronic compass records the angle before turning, the steering gear rotates a certain angle, and the front wheel guides the platform to turn. In this process, the electronic compass feeds back the steering angle in real time. When the body turns 90°, the steering gear angle returns to the right, and the platform completes the independent action, enters a single line, and goes straight along the magnetic line, When traveling, the platform performs actions similar to those of entering the bank, so as to complete the independent access of the platform.

3 Design of Autonomous Navigation System

3.1 Electromagnetic Navigation Device

The copper conductor is arranged above the passage path of the facility greenhouse. The copper conductor is about 2 m above the ground. The signal generator at both ends generates 20 kHz AC signal to generate a magnetic field near the copper wire. When the current in the conductor changes according to a certain law, the magnetic field around the conductor will also change, and a certain electromotive force will be induced in the coil. According to Faraday's law, the relationship between the internal induced voltage e of the coil magnetic field sensor and the magnetic field *B (T)*, the number of turns n of the electromagnetic coil and the cross-sectional area a is as follows:

$$E = (NA) \times (\mu_0 \mu_r) \frac{dB(t)}{dt} = -\frac{d\phi(t)}{dt} \tag{1}$$

Since the current frequency passing through the conductor in this design is 20 kHz and the coil is small, the distance from the coil center to the conductor is r, it is considered that the magnetic field distribution is uniform in a small range. Then the induced electromotive force in the coil can be approximately:

$$E = -\frac{d\Phi(t)}{dt} = \frac{k}{r}\frac{dI}{dt} = \frac{k}{r} \tag{2}$$

That is, the magnitude of the induced electromotive force in the coil is directly proportional to the change rate of the current and inversely proportional to the distance from the center of the coil to the conductor. The constant k is a quantity related to coil placement method, coil area and some physical constants. The specific induced electromotive force constant must be determined by actual measurement.

As shown in Fig. 2, two inductors perpendicular and parallel to the electromagnetic line are symmetrically distributed with the electromagnetic line as the center, which are respectively used to detect the electromagnetic line signals perpendicular to the electromagnetic sensor and parallel to the electromagnetic sensor, and to ensure the acquisition of the signal of the device moving forward and turning along the laying direction of the electromagnetic line. The vertical inductance is used to detect the magnetic flux difference on both sides of the electromagnetic line, so that the platform runs directly below the magnetic line. The parallel inductance is used to detect the electromagnetic line perpendicular to the forward direction of the platform, so as to ensure that the platform can timely sense the right angle magnetic line mark during the moving process.

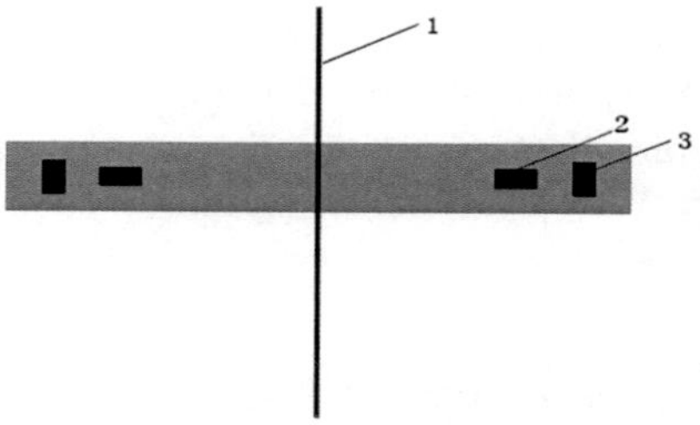

1. Electromagnetic wire 2. Vertical inductance 3. Parallel inductance

Fig. 2. Sensor inductance distribution

3.2 Research on Location Method

The positioning method used in this paper is mainly composed of encoder and magnetic line right angle node mark. The encoder measures the driving distance of the platform in real time, and calibrates the encoder value when encountering the magnetic line right

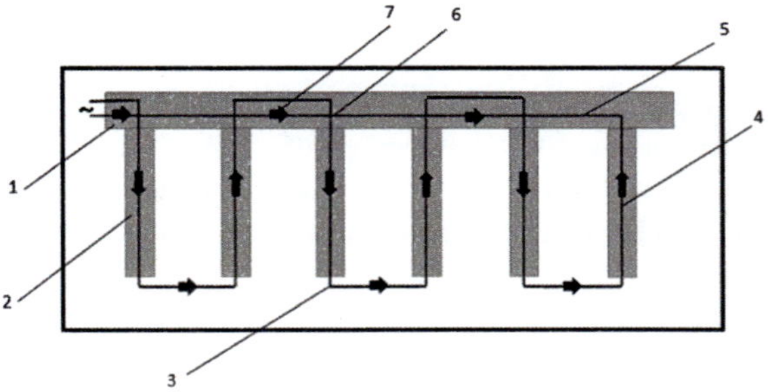

1. Main line 2. Single line 3. Line bottom sign 4. Single line magnetic line 5. Main line magnetic line 6. Line beginning sign 7. Running direction

Fig. 3. Magnetic wire layout

angle mark node. This method has high positioning accuracy, and the running distance can be returned in real time.

As shown in Fig. 3, copper wires are arranged above the traffic path. The right angle nodes of the magnetic line are used as signs at the beginning and bottom of the line. When the platform runs in the main line and one line, the encoder records the driving mileage and determines the position of the current platform. When the platform runs to the bottom or head of the line, because the magnetic lines are arranged at right angles, The electromagnetic sensor will collect the sudden change of magnetic flux as shown in Fig. 4, so the controller can judge that the platform runs to the bottom or beginning of the line, add one to the corresponding position mark in the program, and calibrate the encoder at the same time.

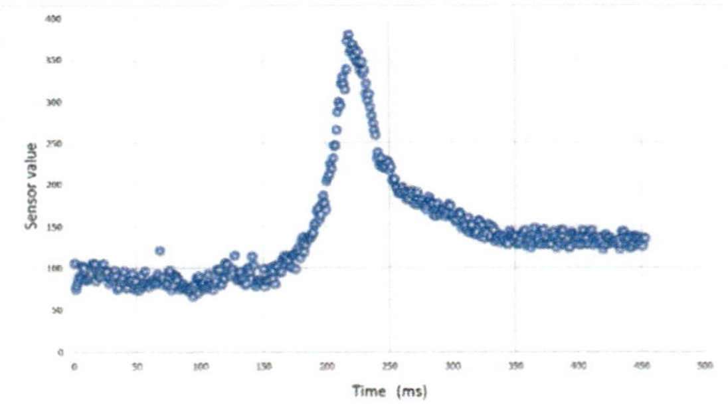

Fig. 4. Variation of magnetic flux under right angle sign

According to the positioning mode and principle, formula (3) can be summarized to calculate the current position of the platform:

$$C = R + P \tag{3}$$

Where, C is the current position of the platform; R is the distance between the mark and the initial position when the electromagnetic sensor detects the right angle mark the latest time; P is the moving distance recorded by the encoder within the current right angle mark interval.

Where, R is the distance between the current right angle sign and the initial position, and R_i is the distance between the platform and the initial position when R_i is the ith sign. Since the length A and interval B of the row in the facility greenhouse are fixed, when i is an odd number, $R = R_{i-1} + A$, when i is an even number, $R = R_{i-1} + B$, that is:

$$R_i = \begin{cases} \sum_{i=0}^{n} R_{i-1} + A(i = 2n) \\ \sum_{i=0}^{n} R_{i-1} + B(i = 2n + 1) \end{cases} \tag{4}$$

P is the moving distance recorded by the encoder in the current right angle mark interval. In this paper, the encoder used has a resolution of 600. By reading the encoder signal through the controller, the current number of rotations and pulses less than one cycle of the encoder can be obtained. The moving distance within the right angle sign interval can be expressed as:

$$P = \left(N + \frac{n}{600}360^\circ\right)2\pi r \tag{5}$$

Where, N is the number of revolutions of the encoder, n is the number of pulses less than one revolution, and r is the radius of the wheel.

3.3 Independent Access Mode

In the facility greenhouse, whether the operation platform can go out independently is an important evaluation index for the intelligence of the operation platform. In this paper, the platform navigation mainly depends on the electromagnetic line. In the steering process, due to the circular arc of the vehicle track, the electromagnetic sensor may be disconnected during the steering process. In order to ensure that the platform is always closed-loop control during operation, Therefore, the electronic compass is used as the feedback source in the steering process.

When the platform detects the steering sign, the electronic compass records the current angle, the steering gear rotates at a fixed angle, and the operation platform performs the steering operation. When the compass value changes by 90°, the access process ends. As shown in Fig. 5, according to the platform motion track when the steering gear rotates at different angles.

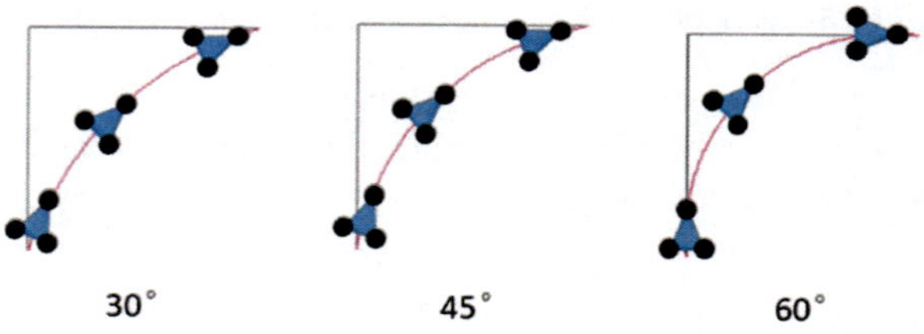

Fig. 5. Trajectory under different steering angles

As shown in Table 1, with the increase of the angle of the corner, the turning radius of the platform gradually decreases, and the off-line time becomes shorter. When the turning area is fixed, the included angle between the platform and the tangent decreases, and the part where the track coincides with the magnetic lines on both sides increases.

Table 1. Parameter values under different steering angles

Steering angle (°)	Turning radius (m)	Off line duration (s)	Included angle (°)
30	2.78	4.3	25
45	1.62	2.4	17
60	0.85	1.5	0

When the angle is greater than 60°, because the steering angle is too large, there will be dead center in the steering process, resulting in the platform unable to move. Therefore, the fixed angle for steering gear rotation is 60°.

4 Prototype Verification Test

In order to verify the navigation trajectory and positioning accuracy of the developed and processed prototype, the route scene of the simulated facility greenhouse is built in the corridor outside the laboratory. The horizontal deviation value and positioning accuracy error between the actual trajectory and the ideal trajectory in the navigation process of the test platform (Fig. 6).

Fig. 6. Laboratory test

Firstly, the offset between the test platform and the specified track at different travel speeds. The test results are shown in the figure. As can be seen from Fig. 7, with the increase of platform running speed, the offset of platform running track will increase.

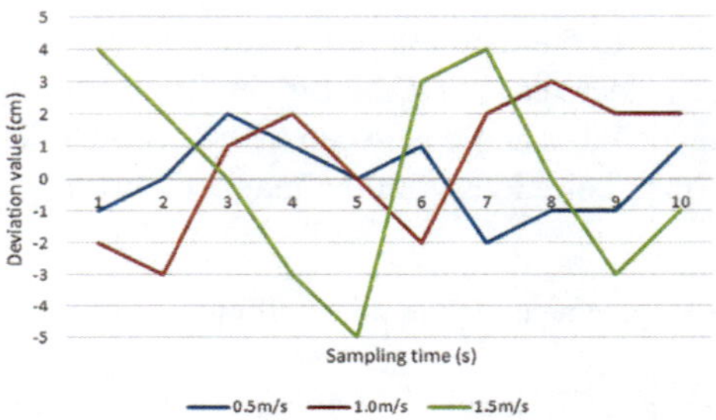

Fig. 7. Track offset at different speeds

Secondly, six measured position stops are preset on the platform running path. When the platform runs to the stop, the real-time position data fed back by the system is recorded. The data are recorded as follows (Table 2):

Table 2. Platform positioning data

Grade	Actual location	System data	Error value
1	235	232	3
2	385	380	5
3	535	530	5
4	685	684	1
5	835	828	7
6	985	982	3

5 Conclusions

According to the internal environmental characteristics and operation requirements of facility greenhouse, this paper designs a facility greenhouse transportation platform navigation system combining electromagnetic wire and encoder, which can realize independent navigation, real-time positioning and operation in facility greenhouse, and provide a basic operation platform for cargo transportation, plant protection, information collection and fruit harvest in facility greenhouse, It greatly reduces the labor demand of greenhouse facilities. The processed prototype is tested in the laboratory. The test results show that the maximum trajectory error of the platform is 5cm and the maximum positioning accuracy error is 7 cm within the design speed range. Meet the basic design requirements.

Acknowledgements. This work was supported by the National Natural Science Foundation of China (51675317) and the Major Agricultural Application Technology Innovation Project of Shandong Province (2019JZZY010715).

References

1. Qi, F., Li, K., Li, S.: Study on the enlightenment of the development of intelligent equipment for facility horticulture in the world to China. J. Agric. Eng. **35**(02), 183–195 (2019)
2. Qi, F., Wei, X., Zhang, Y.: Development status and future research direction of facility horticulture equipment technology in China. J. Agric. Eng. **33**(24), 1–9 (2017)
3. Su, J., Gao, H., Tian, J.: Development status and trend of protected horticulture in China. J. Nanjing Agric. Univ. **42**(04), 594–604 (2019)
4. Li, Z., Sun, S., Ding, X.: Current situation and evaluation of facility horticulture mechanization in China. J. Xinjiang Agric. Sci. **51**(06), 1143–1148 (2014)
5. Zhao, Y., Wu, C., Hu, X.: Research progress and existing problems of agricultural robots. J. Agric. Eng. **19**(01), 20–24 (2003)
6. Yu, Y., Wu, P., Feng, Q.: Research and application status and development countermeasures of vegetable seedling raising equipment in China. J. Res. Agric. Mechanization **36**(06), 1–6 (2017)
7. Liu, N., Jiang, X., Cheng, J.: Current situation of foreign organic facility horticulture and its enlightenment to the sustainable development of facility agriculture in China. J. Agric. Eng. **34**(15), 1–9 (2018)

Check for updates

Dynamic Porosity Estimation Model Based on Wind Speed Variation in Cotton Canopy

Yichong Liu[1], Huiyuan Cui[1], Xinghua Liu[2], and Xuemei Liu[2](✉)

[1] College of Mechanical and Electronic Engineering, Shandong Agricultural University, Tai'an 271018, China

[2] Shandong Agricultural Equipment Intelligent Engineering Laboratory, Tai'an 271018, China

Abstract. Air-assisted spray technology can increase the droplet deposition rate and reduce droplet drift. In order to broaden the transport channel of droplets, the effect of wind speed on the dynamic change of static porosity of crop canopy was studied. Taking cotton as an example, we established four simulated cotton plants and measured the wind speed in the canopy. Through the numerical calculation of the windward area, airflow speed and leaf parameters, the change of leaf inclination in the stratified zoning is determined when the leaves are affected by different airflow speeds. Finally, numerical results of static porosity changes in different regions are obtained. The results show that the static porosity model can be used to calculate the dynamic porosity.

Keywords: Dynamic porosity · Zonal porosity · Cotton morphology · Leaf angle change

1 Introduction

Plant canopy is the main place where plants interact with the external environment, and its structure affects the light transmittance and air permeability of plants. In spraying operations, droplet deposition also occurs inside the canopy. At present, airflow assisted spraying is one of the effective methods to prevent and control pests and diseases in real time. In order to broaden the droplet transport channel and explain the attenuation of air velocity in the canopy, it is particularly important to study the interaction process between the airflow and the canopy leaves.

For the moment, relevant scholars have carried out research on the interaction of crop canopy and airflow field, crop canopy and droplet deposition. Duga et al. Used four kinds of fruit trees with different canopy characteristics as the test objects, and carried out three kinds of wind deposition models, and combined with the characteristics of the canopy, explained the deposition process of three spray models [1]. Miranda-Fuentes et al. Studied the effect of spraying dosage and wind speed on the spray effect of dense planting olive trees [2]. Li et al. analyzed the distribution of the UAV rotor wind field in the rice canopy, and found that the lower the canopy height, the greater the attenuation rate of the wind field equivalent area [3]. Hong et al. Predicted the air velocity distribution inside and around the tree crown by integrating the CFD model of sliding grid technology,

Y. Wang et al. (Eds.): IWAMA 2021, LNEE 880, pp. 620–628, 2022.
https://doi.org/10.1007/978-981-19-0572-8_79

and defined the tree crown as a virtual porous medium. The simulation results can better predict the air distribution [4, 5].

In previous work [6, 7], we proposed a layered porosity that describes the distribution of cotton leaves based on morphological characteristics, and then used the Gaussian process model to predict the distribution of spray deposits on the cotton canopy. The above research is a quantitative analysis of static porosity. The change of porosity is an immeasurable dynamic process. The dynamic change of static porosity affected by air flow has not been reported.

Taking cotton as the research object, using the numerical analysis method, this paper compiles the porosity calculation model of canopy plant stratification, and analyzes the numerical changes of dynamic porosity of cotton plants under the action of auxiliary air flow at full flowering stage.

2 Simulation of Cotton Plant Construction

The measured cotton was located in the experimental field of the South Campus of Shandong Agricultural University. Taking the four cotton plants of the smallest unit in the cotton population as the research object, the cotton height at the full flowering stage is about 1.2 m at the same time, according to the row spacing and plant spacing (80 × 30 cm) required by planting agronomy, as shown in Fig. 1(a).

The distance between the nodes of the same fruit branch is set to the same value. The spiral angle of adjacent fruit branches in the spiral distribution of main stem is 135°. The included angle of fruit branch leaf and fruit branch stem, fruit branch and main stem, fruit branch and main stem leaf is uniformly set at 45°. The leaf shape presents palmately split, generally 3-split or 5-split. Artificial simulated leaves with similar leaf length and width are used to replace cotton leaves as the main material of simulated cotton leaves. According to the measured phenotypic parameters, select appropriate main stem and fruit branch materials to complete the construction of four cotton plants with the same size, as shown in Fig. 1(b).

(a) Row spacing and plant spacing (b) Simulation model of cotton

Fig. 1. Simulated cotton plant

3 Measurement of Wind Speed Inside Canopy

3.1 Measurement Process of Wind Speed

The experimental site is the spray performance laboratory of Shandong Agricultural University. The test bench is based on 4 cotton plants. The white board is used to represent the ground, and the outlet is placed above the cotton. During the test, the influence of natural wind is not considered.

According to the construction of simulation plants, considering the agronomic requirements of actual cotton dense planting, the rectangular area surrounded by the main stems of four cotton plants is used as the target area for wind speed measurement (80 × 30 cm). The target area is divided into 8 areas (15 × 20 cm), as shown in Fig. 2(a), One measuring point is set at an interval of 40 cm from bottom to top in each small area. We use the centrifugal fan of the test bench as the wind source, use the profile to make a tripod suitable for different heights to fix the anemometer. The air velocity measurement tool is a handheld hot-wire anemometer (405i, Testo). The mobile phone is equipped with a Smart Probe APP. The hot-wire anemometer is connected to the mobile phone via Bluetooth to read data. As shown in Fig. 2(b).

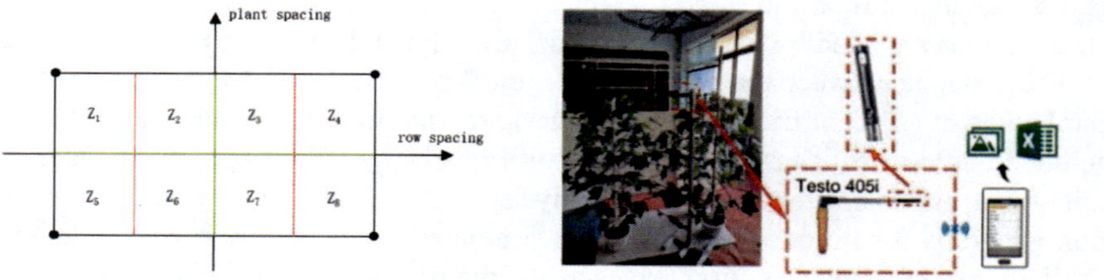

(a) Schematic diagram of regional division (b) Cotton canopy wind speed measurement tool

Fig. 2. Measurement of wind speed inside cotton canopy

An air outlet is set 50 cm above the cotton canopy. By adjusting the working frequency of the frequency converter, the speed of the air bag outlet is adjusted to 15 m/s and 20 m/s. when the fan reaches a stable state, the wind speed at the measuring point in the target area is measured.

3.2 Measurement Results of Wind Speed

The measurement time of each sampling point is 60 s. set the anemometer for 2.0 s to save one piece of data, and obtain 30 pieces of data at each sampling point. In order to ensure the accuracy of measurement, calculate the average wind speed within 60 s at each measuring point as the statistical wind speed at that point. The measurement results are shown in Table 1.

Table 1. (a) Wind speed measurements at different positions of canopy under wind speed of 15 m/s.

Area	Distance from ground					
	40 (cm)		80 (cm)		120 (cm)	
	Velocity (m/s)	Standard deviation (m/s)	Velocity (m/s)	Standard deviation (m/s)	Velocity (m/s)	Standard deviation (m/s)
Z_1	0.47	0.48	3.74	0.38	8.91	0.72
Z_2	1.12	1.69	8.40	1.63	12.13	1.58
Z_3	1.20	1.15	8.41	1.01	12.08	1.46
Z_4	0.60	0.42	4.70	0.91	9.39	0.95
Z_5	0.45	0.29	2.86	0.53	7.92	0.66
Z_6	1.00	1.18	6.12	1.75	10.74	1.76
Z_7	1.67	0.96	6.42	1.17	9.44	1.69
Z_8	0.49	0.60	3.47	0.93	9.81	0.99

(b) Wind speed measurements at different positions of canopy under wind speed of 20 m/s

Area	Distance from ground					
	40 (cm)		80 (cm)		120 (cm)	
	Velocity (m/s)	Standard deviation (m/s)	Velocity (m/s)	Standard deviation (m/s)	Velocity (m/s)	Standard deviation (m/s)
Z_1	2.13	0.67	7.68	0.15	13.68	0.11
Z_2	6.56	1.39	11.34	1.96	14.55	1.80
Z_3	6.10	1.02	11.16	1.89	15.32	1.83
Z_4	3.36	1.51	8.47	0.75	13.14	1.47
Z_5	1.59	0.6	6.61	1.05	12.73	0.92
Z_6	5.13	1.4	8.53	1.92	14.00	1.54
Z_7	5.49	1.63	8.92	1.86	12.93	1.57
Z_8	2.92	1.53	8.25	0.55	13.57	0.84

4 Calculation Model of Dynamic Porosity

4.1 Calculation Model of Static Porosity

The static porosity calculation model is based on the layered projection plane, and the layered porosity is determined by calculating the total leaf area of visible leaves on the projection plane. A single cotton plant is divided into lower, middle and upper layers along the vertical direction of the plant. The lower layer is 0–0.4 m, the middle layer is 0.4–0.8 m, and the upper layer is 0.8–1.2 m. And the number of cotton plants is 8.

Using the MATLAB program software and the matrix method, the static porosity calculation model of the three-layer projection surface of the initial position of the leaf is obtained, as shown in Fig. 3. Among them, the red circle represents the position of the main stem of the 8 cotton plants, the projections of the main stem leaves and fruit branches and leaves are shown as yellow circles, and the red rectangular wire frame represents the cross-sectional area of the analysis domain.

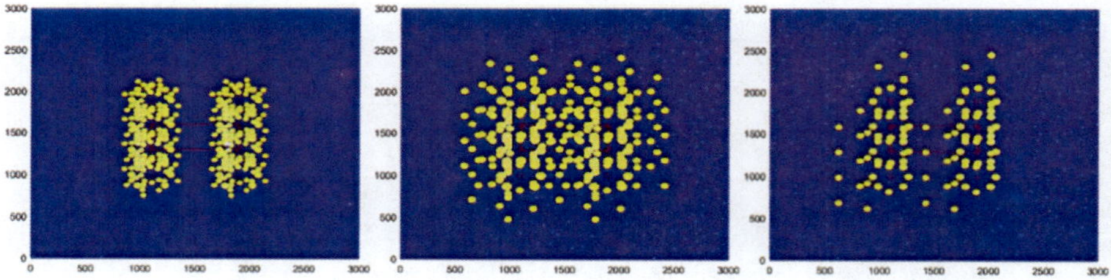

(a) Upper projection plane (b) Middle projection plane (c) Lower projection plane

Fig. 3. Schematic diagram of the three-layer static projection surface

4.2 Calculation of Projected Leaf Area

A schematic diagram of the coordinate axis of the projected image of a single leaf is shown in Fig. 4(a). θ represents the leaf pitch angle. The projected image of the circular leaf presents an elliptical shape. In order to simplify the programming of the program, we use the equivalent leaf area method to convert the elliptical area of the projected leaf into a circular leaf with the equivalent leaf area. The conversion process is shown in Fig. 4(b).

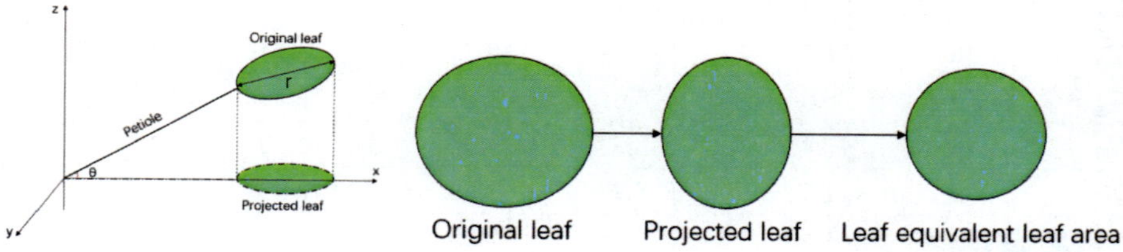

(a) Projected image of a single leaf (b) Equivalent leaf area of a single leaf

Fig. 4. Projection image of single leaf and area processing of equivalent leaf

Calculate the leaf area of the equivalent circular leaf of a single projection leaf according to formula (1), and calculate the radius value of the equivalent circular leaf according to formula (2):

$$S_B = S_A = \pi r^2 \cos\theta \tag{1}$$

$$R = \sqrt{S_B/\pi} \tag{2}$$

Where
S_A—Projected image area of a single leaf (cm^2)
S_B—The leaf area of a round leaf with equal leaf area (cm^2)
r—Initial circular leaf radius value (cm)
R—Equal leaf area round leaf radius value (cm)

4.3 The Amount of Change in Leaf Inclination

In the spraying operation, the cotton branches and leaves are disturbed by the airflow, and the inclination angle of the leaves changes, so the static porosity also changes. In order to study the dynamic process of porosity change, it is necessary to consider the change of leaf inclination angle under different wind speeds in different layers. The change of the leaf inclination angle of the plant population is a complicated process, and we only consider the simplified round leaves here. The change of leaf inclination angle belongs to the elastic deformation of the leaf affected by wind [9], which can be estimated by the deflection curve equation.

$$w_l(x) = \frac{\rho \cdot S}{12k_7 \cdot E \cdot I} \boldsymbol{U} \cdot cos(\boldsymbol{\beta}) \cdot \boldsymbol{L} \tag{3}$$

$$\Delta\beta = arc\,cos\left(\frac{2 \cdot \Delta x^2 - (w_l(x_2) - w_l(x_1))^2}{2 \cdot \Delta x^2}\right)\frac{180}{\pi} \tag{4}$$

Where
$w_l(x)$—The deflection value of the main leaf vein on the abscissa x value of the leaf surface (mm)
ρ—Air density (1.293 kg/m^3)
S—Leaf area of cotton (mm^2)
E—Elastic modulus of material (MPa)
I—Moment of inertia of cantilever beam (mm^4)
$\mathbf{U} = k_1 \cdot u^2 + k_2 \cdot u + k_3$—Quadratic polynomial of wind speed (m/s), u—Velocity of auxiliary air flow (m/s)
$\boldsymbol{\beta} = \beta - e^{k_4 \cdot u + k_5}$—Correction value of leaf inclination angle considering leaf deformation (°), β—Inclination of leaf (°)
$\mathbf{L} = 2L + L \cdot e^{k_6 \cdot (\frac{x}{L})^2}$—Petiole length variable considering nonlinear large deformation (mm), L—Length of cantilever beam (mm), x—The abscissa of the deflection value analysis point (mm)
$\boldsymbol{\theta}_l = [k_1, k_2, k_3, k_4, k_5, k_6, k_7]$ = [1.0181, –2.8919, 0.9873, 1.8483, 3.5920, 5.94, –1.1582]—Identification parameter vector
Δβ—Variation of leaf inclination under wind load (°)
Δx—Distance from leaf centroid to leaf tip (mm)
$w_l(x_2)$, $w_l(x_1)$—They are the deflection values of leaf centroid and leaf tip respectively (mm)

The distance from the leaf centroid to the leaf tip is defined as the radius value of the layered circular leaf. In addition to considering the leaf elastic modulus and air density, the calculation of deflection value also needs to obtain the layered wind speed and the windward area of layered leaves. The windward area refers to the leaf area of the initial leaf, and the windward area of the three layers of leaves is also different. Different layers are divided into two parts: the windward area of the main stem leaf and the windward area of the fruit branch leaf, as shown in Table 2.

Table 2. Windward area of layered leaves

Types of leaves	Windward area (cm^2)		
	Upper level	Middle level	Lower level
Main stem and leaf	58.09	91.61	63.62
Fruit branches and leaves	36.32	62.18	50.27

After statistics and calculations, the changes in the inclination angle of the leaves in each area under the wind speed measurement results are obtained. The change in the inclination angle of the leaves is shown in Table 3.

Table 3. (a) Leaf inclination angle change in each area under 15 m/s wind speed

Area	Change of leaf inclination on main stem (°)			Change of leaf inclination on fruit branch (°)		
	Upper plane	Middle plane	Lower plane	Upper plane	Middle plane	Lower plane
Z_1	39.96	27.04	8.45	32.03	18.18	7.49
Z_2	71.82	53.11	12.50	58.27	39.26	11.14
Z_3	70.63	53.29	13.13	57.25	39.50	11.59
Z_4	35.82	32.73	9.53	35.16	22.62	9.10
Z_5	45.60	21.00	8.25	29.00	14.91	7.07
Z_6	54.38	39.56	11.82	42.42	29.66	10.67
Z_7	46.45	41.08	14.04	36.43	30.28	12.86
Z_8	48.65	25.94	8.64	38.47	16.07	7.89

(b) Leaf inclination angle change in each area under 20 m/s wind speed

Area	Change of leaf inclination on main stem (°)			Change of leaf inclination on fruit branch (°)		
	Upper plane	Middle plane	Lower plane	Upper plane	Middle plane	Lower plane
Z_1	80.90	47.83	16.70	71.04	33.49	13.89
Z_2	84.66	87.68	41.19	77.70	62.10	33.18
Z_3	108.60	84.63	37.07	90.60	61.28	30.22
Z_4	79.46	54.97	20.16	67.80	39.42	16.77
Z_5	72.05	41.92	14.65	64.73	31.93	12.12

(continued)

Table 3. (*continued*)

(b) Leaf inclination angle change in each area under 20 m/s wind speed						
Area	Change of leaf inclination on main stem (°)			Change of leaf inclination on fruit branch (°)		
	Upper plane	Middle plane	Lower plane	Upper plane	Middle plane	Lower plane
Z_6	82.76	55.64	27.96	73.73	39.88	20.62
Z_7	76.09	60.82	29.42	66.40	40.36	23.98
Z_8	80.04	52.16	19.83	70.68	37.74	15.34

4.4 Dynamic Porosity

Dynamic porosity refers to the process that after the air flow enters the plant canopy, the interaction between the air flow and the leaves leads to the change of leaf inclination, the change of leaf shape and leaf area in the layered projection plane, and the dynamic change

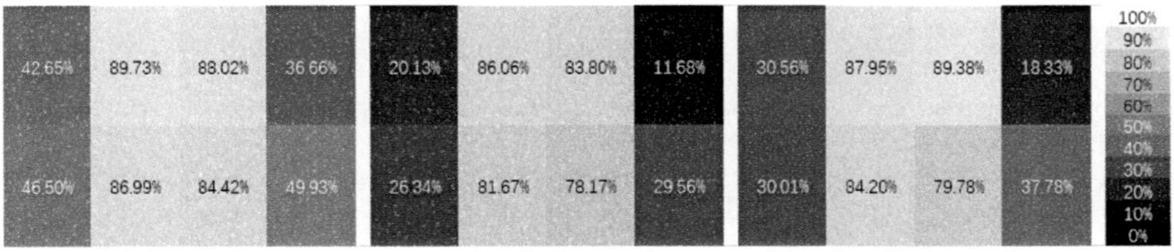

Initial position 15 m/s wind speed 20 m/s wind speed

(a) Porosity change of upper projection plane

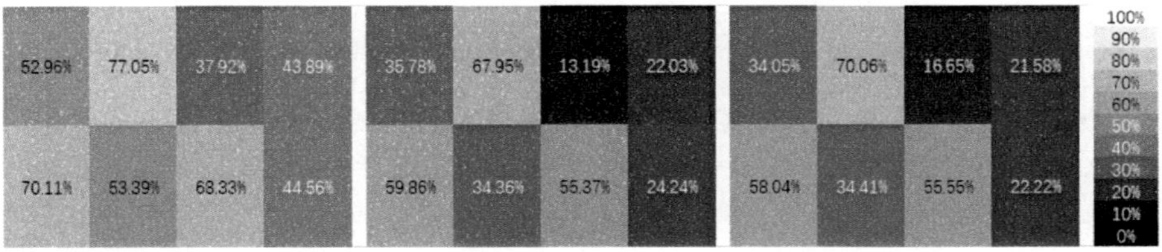

Initial position 15 m/s wind speed 20 m/s wind speed

(b) Porosity change of middle projection plane

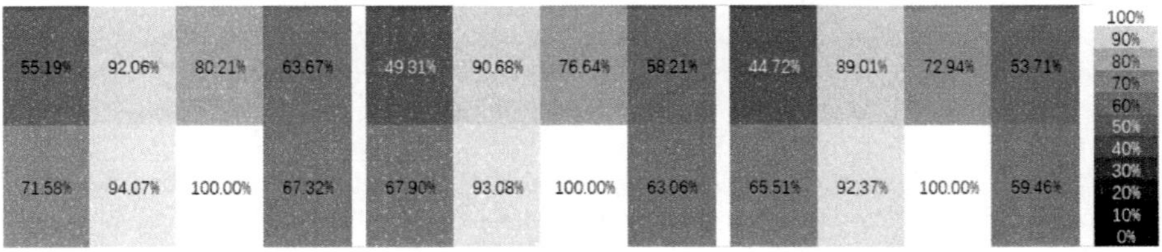

Initial position 15 m/s wind speed 20 m/s wind speed

(c) Porosity change of lower projection plane

Fig. 5. Porosity variation of three-layer projection plane of static porosity in the layered projection plane. Due to the difference of windward area between main stem leaf and fruit branch leaf, the change of Leaf Inclination under the same wind speed is inconsistent. The layered porosity model generated by MATLAB can obtain the zonal porosity of the layered projection surface, the initial position and the layered zonal porosity under two wind speeds, as shown in Fig. 5. Therefore, we believe that the initial wind speed has a significant effect on the change of porosity. Under the conditions of planting technology and growth period, 20 m/s wind speed is more conducive to broaden the air transport channel and the effective deposition of droplets.

5 Conclusion

In this paper, a calculation model of static porosity is designed. Based on this model, the numerical variation of dynamic porosity is studied. Taking cotton as an example, four simulated cotton plants were established and the wind speed in the canopy was measured. Through the numerical calculation of relevant parameters, the change of leaf inclination in different layers and zones is determined when the leaves are affected by different air velocities. Finally, the numerical results of static porosity change in layered and zoning are obtained. The results show that the static porosity model can be used to calculate the dynamic porosity.

Acknowledgements. This work was supported by the National Natural Science Foundation of China (52075308).

References

1. Duga, A.T., Dekeyser, D., Ruysen, K., et al.: Numerical analysis of the effects of wind and sprayer type on spray distribution in different orchard training systems. Bound. Layer Meteorol. **157**(3), 517–535 (2015)
2. Miranda-Fuentes, A., Rodríguez-Lizana, A., Gil, E., et al.: Influence of liquid-volume and airflow rates on spray application quality and homogeneity in super-intensive olive tree canopies. Sci. Total Environ. **537**, 250–259 (2015)
3. Li, J., Shi, Y., Lan, Y., et al.: Vertical distribution and vortex structure of rotor wind field under the influence of rice canopy. Comput. Electron. Agric. **159**, 140–146 (2019)
4. Hong, S., Zhao, L., Zhu, H.: CFD simulation of airflow inside tree canopies discharged from air-assisted sprayers. Comput. Electron. Agric. **149**, 121–132 (2018)
5. Hong, S., Zhao, L., Zhu, H.: CFD simulation of pesticide spray from air-assisted sprayers in an apple orchard: tree deposition and off-target losses. Atmos. Environ. **175**, 109–119 (2018)
6. Liu, X., Liu, X., Li, Y., et al.: Estimation model of canopy stratification porosity based on morphological characteristics: a case study of cotton. Biosys. Eng. **193**, 174–186 (2020)
7. Liu, X., Liu, X., Li, Y., et al.: Predicting spray deposit distribution within a cotton plant canopy based on canopy stratification porosity and Gaussian process models. Biosys. Eng. **204**, 1–14 (2021)
8. Liu, X., Yuan, J., Li, Y., et al.: Deformation modelling of cotton leaf under assisted airflow using bidirectional fluid-structure coupling method. Trans. CSAE **37**(5), 68–76 (2021)

Design and Test of Integrated Air Supply and Fine Mist System in Solar Greenhouse

Yan Zhang[1], Huiyuan Cui[1], Xinghua Liu[2], Xuemei Liu[2](✉), Hao Li[1], and Xianfeng Du[1]

[1] College of Mechanical and Electronic Engineering, Shandong Agricultural University, Tai'an 271018, China

[2] Shandong Agricultural Equipment Intelligent Engineering Laboratory, Tai'an 271018, China

Abstract. Aiming at the problems of low pesticide utilization, poor atomization uniformity of the whole shed, and uneven droplet deposition during the use of the existing fixed ceiling high-pressure atomization spray system in the solar greenhouse, combined air duct air delivery and air-assisted precision. A kind of integrated wind-supplied misting system in the solar greenhouse is designed to increase the pesticide application effect of hanging vine crops in the greenhouse. Through the spraying test of the ridge-growing simulation cucumber vine in the solar greenhouse environment, the air outlet angle of the air outlet was used as the test factor, and the leaf droplet deposition density was used as the test index. The effect of droplet deposition and the optimal structural parameters of the air duct. Four sets of control angles were set for the direction of the wind from the openings, which were 45°, 90°, 135°, and 180°. Experiments show that the installation of an air delivery system can significantly improve the uniformity of droplet deposition in the fixed ceiling high-pressure atomization spray system; when the air outlet wind speed is 10 m/s, the best air duct opening angle is 135°, compared with the no air supply mode, the deposition of the upper, middle, and lower three layers of the canopy of the hanging vine plant has the best uniformity of droplet deposition when the angle of the air outlet direction is 135°, and the coefficient of variation of the upper layer It is 34.55%, the coefficient of variation of the middle layer is 47.68%, and the coefficient of variation of the lower layer is 49.66%. The canopy deposition density of the upper, middle, and lower layers is reduced by 64.20%, 113.29%, and 49.81%, respectively, compared with the windless conditions, greatly increasing the droplets. Deposition uniformity.

Keywords: Solar greenhouse · Plant protection machinery · Wind-assisted pesticide application · Mist · Droplet deposition

1 Introduction

As the main body of my country's facility agriculture industry, the solar greenhouse industry has problems such as high temperature and high humidity, continuous soil cropping, soil-borne diseases, etc., which are very easy to cause pests and diseases, which seriously affect the yield and quality of vegetables. Therefore, the task of pest control in

Y. Wang et al. (Eds.): IWAMA 2021, LNEE 880, pp. 629–636, 2022.
https://doi.org/10.1007/978-981-19-0572-8_80

the greenhouse every season Heavy. In accordance with the development requirements of precision agriculture, greenhouse crop production must not only effectively control pests and diseases, but also pay attention to the protection of the ecological environment and reduce the residual amount of agricultural products [1].

At present, in order to improve the convenience and plant protection effect of plant protection operations in facilities, some domestic scholars have carried out a lot of research and development on plant protection machinery and equipment. Chen Baolin [2] and Li Ming [3] developed the 3M-50 autonomous walking type the greenhouse mist machine completes the precise spraying operation through the air delivery system. Li Hui et al. [4] analyzed the three-dimensional simulation and experiment of the airflow field and droplet deposition based on the hanging rail type normal temperature fog machine. Most of the above studies are applied from one side of the planting area or within the planting row, and there are few studies on the combination of high-pressure spraying and air-sending application methods above the planting area. Some foreign researchers have also done a lot of research on greenhouse plant protection equipment. For example, Rowe et al. [5] designed an automatic spraying system for greenhouses. The S55 series spraying robot developed by Shamshiri R. R. of Holland Green Machine Company in Holland [6] can be equipped with a dual-fluid nozzle to provide a smaller droplet and coverage area. However, some foreign researches are mostly based on large multi-span greenhouses such as Venlo-style glass greenhouses. The cost of the facilities is high and the maintenance is complicated, which is not suitable for the current production status of my country's facilities.

In this paper, the existing fixed ceiling high-pressure atomization spray system is improved in the solar greenhouse, and a set of solar greenhouse integrated air delivery mist system is designed, and the atomization of this system and the fixed ceiling high-pressure atomization spray system is analyzed and compared through comparative experiments. Performance and operational effects, and analyze the influence of the canopy of hanging vines in the facility on the droplet deposition and distribution of the system.

2 Whole Machine Structure and Working Principle

2.1 Whole Machine Structure

The integrated air delivery and mist mitigation system of the solar greenhouse is mainly composed of a high-pressure plunger pump, an axial flow fan, a thin-film air duct, a nozzle, a high-pressure nylon tube, a suspension bracket for the air duct, a wire, a water pipe, a filter and a chemical tank. Its structure is schematic. As shown in Fig. 1.

2.2 Working Principle

The high-pressure plunger pump sucks the liquid medicine, and then transports the pressurized water through the high-pressure nylon tube, which is atomized by the high-pressure atomizing nozzle, and at the same time, the air flow generated by the axial flow fan passes through the small circle of the film duct The hole escapes, accelerates the diffusion, disturbs the droplets, and settles to adhere to the surface of the crop.

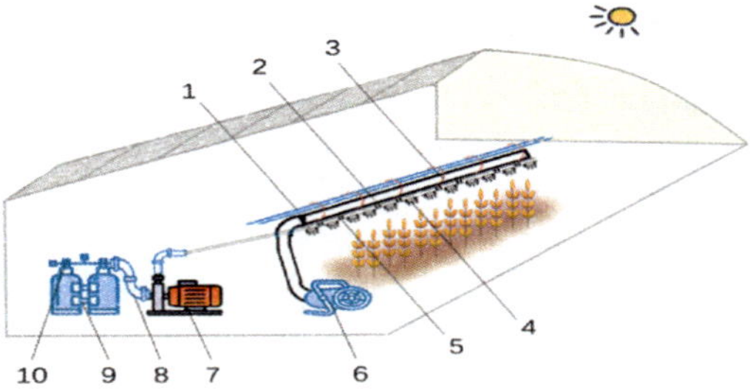

1. Iron wire 2. Film duct 3. Air duct suspension bracket 4. Sprinkler 5. High pressure nylon tube 6. Axial fan 7. High pressure plunger pump 8. Water pipe 9. Medicine tank 10. Water filter

Fig. 1. Schematic diagram of the whole machine structure

3 Selection and Design of Key Components and Parameters

3.1 High Pressure Spray System Design

The current installation of the equipment adopts a fixed type, and the water pump adopts the "Fujun" brand high-performance plunger pump. This pump adopts a high-efficiency three-phase asynchronous motor, with a working voltage of 380 V, a rated power of 4 KW, and a rotation speed of 1440 r/min. It is directly connected to the plunger pump through a bearing. The working pressure is 0.4–10.0 MPa, which can be adjusted manually. The nozzle adopts the "Qihong" brand three-section extended filter nozzle, the model is QH3010 three-section extended filter nozzle, which contains a washable 10 μm filter, the pore size is 0.30 mm, the spray volume is 0.06–0.12 L/min, the spray volume The corresponding relationship with water pressure is shown in Table 1.

Table 1. Nozzle flow rate under different pressures

Water pressure (MPa)	Flow (L/min)
1.0	0.06
1.5	0.07
1.8	0.08
3.5	0.09
5.0	0.10
7.0	0.11
8.0	0.12

A 500-square-meter greenhouse is equipped with a high-pressure plunger pump, 50 nozzles and 50-m hoses. When installing, place the high-pressure plunger pump on the side of the greenhouse, and the spray pipe is fixed in the center of the greenhouse shelf, and the 50 nozzles are divided into 25 pairs are evenly distributed on the top of the greenhouse with an interval of 2 m.

3.2 Design of Thin Film Duct

The shape of the film air duct has two types: cylindrical and trumpet. There are double rows of holes on the outer circumferential surface of the membrane air duct. The aperture size and the hole spacing can be combined with each other, including the equal aperture and equal hole spacing type (a), the equal aperture hole spacing first sparse and then dense type (b), the hole diameter is first large and then small and the hole spacing Sparse first and then dense (c), as shown in Fig. 3. [8]. According to the actual test environment, in order to facilitate the calculation of aperture size and fan piezoresistance and other parameters, this test uses a cylindrical membrane air duct, and the aperture type adopts the equal aperture and equal aperture spacing type (Fig. 2).

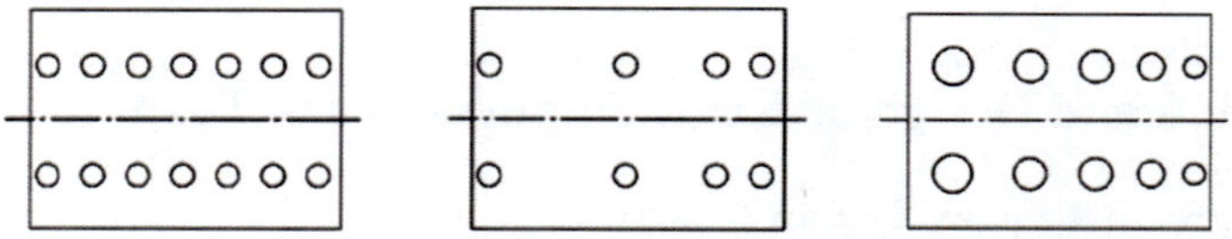

(a) Type One (b) Type two (c) Type three

Fig. 2. Different aperture types on the membrane duct

According to the actual size of the greenhouse and the hanging vine plant, cucumber is used as an example. Cucumber planting generally adopts large and small rows, with a small row spacing of 40 cm and a large row spacing of 80 cm. Open a pair of holes every 0.8 m, and set the angles of the air outlet directions of the openings to set three sets of contrasts, respectively 45°, 90°, 135°, and 180°. The size of the holes on the holes can be replaced according to the principle of equal replacement of the outlet air volume of the fan. Calculate according to the following formula (1):

$$\pi r_1^2 = 2n\pi r_2^2 \tag{1}$$

Where
r_1—The radius of the film duct (mm)
r_2—The radius of the opening on the air duct (mm)
n—Number of holes on the air duct
Taking into account the hub diameter of the fan, the shading degree of the thin film air duct and the ease of installation of the thin film air duct, here r1 is taken as 110 mm and n = 60. It is calculated that r2 = 10.04 mm. After rounding, the diameter of the hole on the air duct is about 2.0 cm.

3.3 Selection of Axial Fan

The wind speed and air volume in the pipeline satisfy formula (2).

$$\nu = \frac{Q}{3600s} = \frac{Q}{3600\pi r_1^2} \tag{2}$$

Where

ν—The wind speed of the airflow in the duct (m/s)

S—Cross-sectional area of film duct (m^2)

Q—Amount of wind in the wind tube (m^3/h)

The value of the airflow velocity ν in the duct can refer to the technical parameters of the fog machine at room temperature [20], and $\nu = 7–8$ m/s.

Considering the air volume loss, Eq. (3) is used to calculate the fan air volume.

$$Qq = 1.2Q \tag{3}$$

Substituting various parameters, after calculation, Qq = 957.5–1094 m^3/h.

The calculation of the wind pressure H can refer to the calculation method of the resistance of the ventilation duct. The wind pressure of the film duct mainly considers the frictional resistance along the way, and the calculation formula (4) is as follows.

$$H = R \cdot L = \left(\frac{2\lambda}{r_1} \cdot \frac{\rho\nu^2}{2}\right) \cdot L \tag{4}$$

Where

L—The length of the air duct (m)

R—Friction resistance per unit length along the inner surface of the air duct (Pa/m)

ρ—Air density (kg/m^3)

λ—Friction coefficient along the duct

The friction coefficient λ along the duct can be calculated according to the empirical formula (5).

$$\lambda = 0.0125 + \frac{0.0022}{r_1} \tag{5}$$

Taking into account the other pressure loss in the air duct, the air pressure Hh of the fan is calculated according to the formula (6).

$$H_h = 1.2H \tag{6}$$

Substituting in the various parameters, $H_h = 173.73–226.91$ Pa can be obtained by calculation.

Therefore, this article chooses the "Zheshen" brand SFB-3-2 axial flow fan. The specific technical parameters are: the impeller diameter is 30 cm, the speed is 2900 r/min, the air volume is 4500 m^3/h, and the total pressure is 230 Pa.

4 Experimental Design and Methods

4.1 Test Site and Equipment

The experiment was carried out in the No. 2 solar greenhouse (50 m × 10 m × 3.5 m) in the east-west direction of the Science and Technology Innovation Park of Shandong Agricultural University in July 2021. During the test, the indoor ambient temperature in the shed was about 27.6 °C, and the relative humidity was about 58%. The equipment used in the test is a solar greenhouse integrated air delivery and mist system, as shown in Fig. 3, mainly including a high-pressure plunger pump, the working pressure can be manually adjusted within 0.4–10.0 MPa; SFB-3-2 type shaft 1 flow fan; 1 medicine box, 1 plastic film tube, total length 50 m, tube diameter 220 mm, with a pair of small holes every 0.8 m, with a diameter of 2.0 cm, suspended in the middle of the shed at a distance of 2.5 m from the ground, The angles of the air outlet direction of the air duct openings are set for four groups of controls, respectively 45°, 90°, 135°, and 180°.

The air-sending mist mitigation system is mainly composed of two parts, the high-pressure spray system and the air-assisted system that integrate the control system. The high-pressure spray system and the air-assisted system host are placed in the greenhouse near the side wall, the film air duct is located directly above the high-pressure spray pipe, and the pipeline is installed in the middle of the shed at a distance of 2.5 m from the ground, with a total of 50 nozzles. The nozzle spacing is 2 m, and the two-hole direct joint is reversely installed. The wind speed at the opening of the test film duct is 8 m/s, the hydraulic pressure of the high-pressure plunger pump is 8 MPa, and the spraying flow rate of the single nozzle is 0.12 L/min. The medicine time is 5 min. The chemical liquid used in the test is a simulated chemical liquid from tap water.

Fig. 3. Integral wind delivery and mist mitigation system for solar greenhouse

4.2 Sampling Method and Acquisition Card Layout

I arrange three rows of simulated cucumber vines as spray objects in the solar greenhouse. The density of the leaves is similar to the actual cucumber plants. The row spacing is 0.5 m, the plant height is 2 m, and the plant spacing is 30 cm. Starting from 0.3 m, one simulated cucumber is selected every 1.5 m. The vine is used as the test target. Choose a leaf from the bottom up to the height of 0.4 m, 1 m, 1.6 m from the ground and randomly select a leaf on the west side of the bottom, middle, and upper layers of the plant, and fix it to the top and bottom of the selected simulated cucumber vine leaves with a small clip. On both sides, as shown in Fig. 4(a), set up sampling points. The water-sensitive paper is a 3 cm × 11 cm high-precision test water-sensitive paper provided by Chongqing Liuliushanxia Plant Protection Technology Co., Ltd. Use tap water as the spray medium to spray simulated cucumber vines. Collect the water-sensitive paper. After each spray, remove the water-sensitive paper from the sampling point in sequence, dry it, and put it into a sealed bag and number the position, as shown in Fig. 4(b). Water-sensitive paper processing, using a portable scanner to scan each piece of water-sensitive paper with a resolution of 600 × 600 in grayscale and save it as a png format picture. Use the fog drop analysis software provided by Chongqing Liuliushanxia Plant Protection Technology Co., Ltd. Picture analysis. Figure 4(c) shows the results of the droplet analysis software for sampling water-sensitive paper from the front of the leaf at 3.3 m in the middle layer under no wind conditions when the nozzle pressure is 8 MPa.

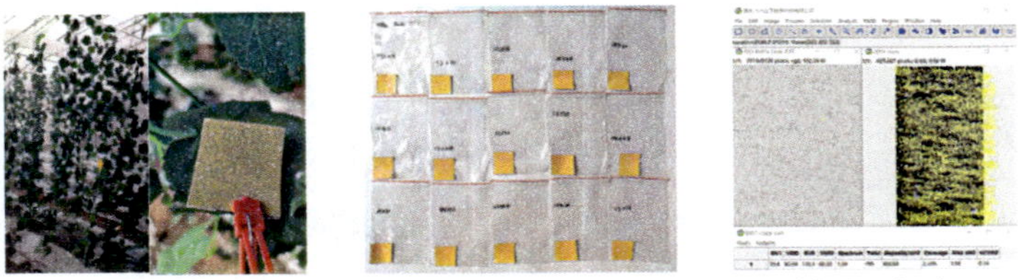

(a) Placement (b) Collection (c) Analysis

Fig. 4. Test water sensitive paper placement, collection and analysis

5 Test Results and Analysis

5.1 Test Results

Under normal circumstances, the deposition density of fog droplets for pest control should not be less than 20/cm^2. Comparison of droplet deposition density between the upper, middle, and lower layers of the plant when the wind speed of the air outlet is 10 m/s, and the angles of the air outlet directions are 45°, 90°, 135°, and 180°, respectively, and those under no wind conditions, As shown in Fig. 5.

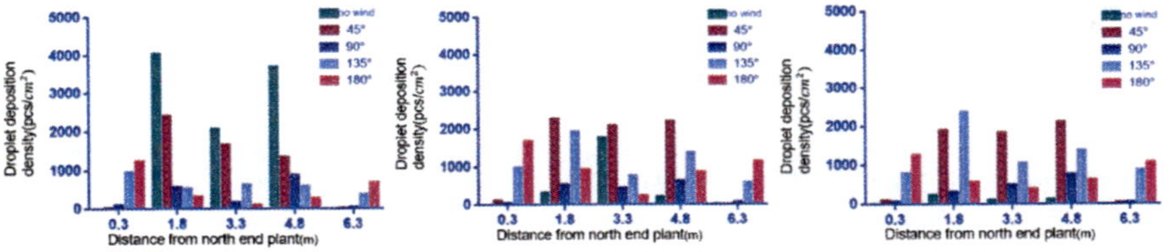

(a) Upper layer middle (b) Middle layer (c) Lower layer

Fig. 5. Comparison chart of droplet deposition density under different conditions

From the droplet deposition density statistics table, it is found that under the same flow rate, the uniformity of the droplet deposition density in each interval under the condition of air delivery is significantly higher than that of the uniformity of the droplet deposition density in each interval under the condition of no air delivery. The overall droplet deposition volume under delivery conditions is higher than that under no wind conditions. The highest coefficient of variation of the upper layer is 100.91% when the angle of the air outlet direction of the air duct opening is 90°, and the smallest is 34.55% when the angle is 135°. The largest coefficient of variation of the middle layer is the air duct opening. When the angle of the wind direction is 90°, it is 100.91%, the smallest is when the angle is 135°, it is 47.68%. The coefficient of variation of the lower layer is the largest when there is no wind, which is 99.47%, and the smallest is the angle of 180° When there is no wind, it is 46.36%; under no wind conditions, the average droplet deposition density of each layer is 842.87, 1212.60, 347.53, 1017.87, and 761.40 pcs/cm^2. When the angles of the wind outlet direction are 45°, 90°, 135°, and 180°.

6 Conclusion

Through the experiment, it is concluded that the integrated air delivery mist system of the solar greenhouse has a uniform droplet deposition density than the ceiling-mounted high-pressure atomization spray system. The uniformity of the canopy distribution of vine plants. The uniformity of droplet deposition is best when the angle of the air outlet direction of the duct opening is 135°, the upper layer variation coefficient is 34.55%, the middle layer variation coefficient is 47.68%, and the lower layer variation coefficient is 49.66%. The canopy deposition density was reduced by 64.20%, 113.29%, and 49.81% respectively compared with that under no wind conditions, which greatly improved the uniformity of droplet deposition.

Acknowledgements. This work was supported by the National Natural Science Foundation of China (52075308), the Major Agricultural Application Technology Innovation Project of Shandong Province (2019JZZY010715) and the Natural Science Foundation of Shandong Province (ZR2019PC024).

References

1. Cao, Z., Zhang, J., Geng, C., et al.: Greenhouse target spray robot control system. Trans. Chinese Soc. Agricult. Eng. **26**(2), 228–233 (2010)
2. Chen, B.: Development and test of the control system of 3M-50 self-operating greenhouse mist dispersing machine. Shandong Agricultural University (2018)
3. Li, M.: Development and experimental research of 3M-50 autonomous walking greenhouse fog mist machine. Shandong Agricultural University (2017)
4. Li, H., Qi, L., Wang, P.: Three-dimensional simulation and experiment of airflow field and droplet deposition of hanging rail type normal temperature fog machine. Trans. Chinese Soc. Agricult. Mach. **45**(04), 103–109+122 (2014)
5. Rowe, D.E., Malone, S., Yates, Q.L.: Automated greenhouse spray system for increased safety and flexibility. Crop. Sci. **40**(4), 1176–1179 (2000)
6. Shamshiri, R.R., Weltzien, C., Hameed, I.A., et al.: Research and development in agricultural robotics: a perspective of digital farming. Int. J. Agricult. Biol. Eng. **11**(4), 1–11 (2018)
7. Zhou, C.: Dr. Zhou inspected the vine-hanging and vine-releasing technique for the production of high seedling crops in solar greenhouse. Agricult. Eng. Technol. (Greenhouse Horticulture) **2012**(10), 24+26+28+30 (2012)
8. Guan, C., Hu, H., Han, B., Chen, Y., Yang, Y.: Research on the air duct fog machine for greenhouse. Chinese J. Agricult. Mach. Chem. **37**(01), 55–58 (2016)

A Survey of Few-Shot Learning and Its Application in Industrial Object Detection Tasks

Xiufeng Zhang, Chen Wang(✉), Yu Tang, Zhixiao Zhou, and Xuxiang Lu

Department of Mechanical Engineering, Hubei University of Automotive Technology, Shiyan 442002, China
20090011@huat.edu.cn

Abstract. Few-shot learning aims to obtain a better deep learning model through training with a smaller amount of data. In recent years, with the development and application of deep learning, deep learning has faced the problem of few-shot in many scenarios. Therefore, the study of few-shot learning will promote the application and deployment of deep learning in the industrial field. First of all, this survey specifically introduces two few-shot learning implementation methods: based on data augmentation and based on transfer learning. Then, this survey lists actual cases to illustrate the application of few-shot learning in industry. One case is about Few-shot parts surface defect detection based on metric learning. Another case is about poultry egg detection in few-shot based on meta learning. Finally, this survey summarizes the research achievements and research value of few-shot learning. Survey summarizes the problems and challenges faced by few-shot learning and the future development direction of few-shot learning.

Keywords: Few-shot learning · Data augmentation · Transfer learning · Metric learning · Meta learning

1 Introduction

In recent years, object detection algorithms based on deep learning have been widely used in many industrial scenarios [1]. These algorithms [2] have achieved great success in vision tasks and have solved some practical problems in the fields of industrial manufacturing, intelligent transportation, and optical character recognition [3]. However, deep learning needs on a large amount of well-annotated data. To make a model with better performance, a large amount of data must be owned and made. At the same time, the distribution characteristics of the training data is similar to the distribution of the task target data. In real application scenarios, only a small amount of data can be obtained. And these data need take a lot of time and labor costs to label task data. Sometimes, it is difficult to label accurately. For example, in the task of industrial defect detection, the occurrence of industrial product of defects is a small probability event. It is difficult to obtain a large number of data samples from the same type of defect. In autonomous driving tasks, data of emergencies is also difficult to simulate and obtain. In text recognition tasks, some text content involves patent protection or privacy protection. In such

Y. Wang et al. (Eds.): IWAMA 2021, LNEE 880, pp. 637–647, 2022.
https://doi.org/10.1007/978-981-19-0572-8_81

difficult object detection tasks, the generalization ability of the model is poor trained by a small amount of data. So the effect of target detection is also poor. How to solve the problem of detection accuracy of small samples has become an urgent problem to be studied.

Object detection [4] was proposed as a basic vision task in the 1960s, which requires to classify and locate objects in the image with computing equipment. At present, the object detection algorithm has developed from the method of geometric principles and low-level features of the target to the method of deep learning based on big data. In comparison, humans only need very little data to learn a good detection model. Inspired by this, in order to make the learning algorithm's ability closer to the level of the human brain, the concept of few-shot learning [5, 10] was proposed. The goal of few-shot learning is to learn a task-solving model from a small number of samples. In summary, few-shot learning is not only of practical value in actual detection tasks, but also is a direction worthy of research in theoretical research.

At present, there have been many research results on few-shot learning under deep learning at home and abroad, and their research angles and implementation methods are different [11, 14]. The classification of few-shot learning methods is also slightly different, and there is no unified specification and conclusion. For example, Zhu J.T, Yao G.L et al. [15] divided the few-shot learning methods into four categories in their review. There are data augmentation, metric learning, external memory, and parameter optimization. Liu H.Y and Wang X.J [16] divided the few-shot learning method into three categories: data, model, and algorithm. This survey, First of all, collected and summarized the methods of few-shot learning. Then, the survey specifically introduced the implementation and research progress of three types of few-shot learning methods:

1. Few-shot learning based on data augmentation.
2. Few-shot learning based on transfer learning.

This survey also introduced the case of few-shot learning method in solving practical tasks. Finally, this survey summarizes the research status and main challenges of few-shot learning. In the end, we take about the future development trend and existing problems of few-shot learning.

2 Few-Shot Learning Based on Data Augmentation

In a general deep learning method, the training process is realized by minimizing the loss function of the object detection under a given training set (D). The loss function can represent applications such as image classification and detection box regression. The training algorithm is the process of searching for the parameter of the model in the hypothetical model space (P). The training process of deep learning is actually the process of fitting the true distribution in the actual test task through the training set. The data distribution of the model obtained by training is often inconsistent with the distribution of real data. Therefore, training the object detection model requires a large amount of samples. In the task of few-shot learning, the number of training samples is limited. If the general deep learning method is still used, it will be difficult to train a model with good detection effect.

At this time, the most direct method is to enhance the data. Without adding new data, we can use image transformation to increase the number of data samples. Data augmentation is also called data expansion, which refers to adding new data to the original small sample data set with the help of auxiliary data or transform rules. Then, we can get a larger data set. Commonly used methods include image flipping, image cropping, image scaling, image rotation, brightness adjustment, hue and saturation conversion, and so on. Zoph et al. [17] proposed Auto Augment image augmentation. It uses 22 kinds of data augmentation strategies to create a search space. And it used a search algorithm to select a suitable data augmentation strategy for the characteristics of a specific application task. 1%-2% improvement in detection accuracy has been achieved on the data set. On the other hand, we can also operate on multiple samples at the same time. Such as SMOTE [18], Sample Pairing [19] and Mixup [20] and other algorithms, these methods are very simple. The changes to the original image are very small, too. And a large amount of new data can be obtained quickly. The data obtained through the data augmentation method is new data for the algorithm network (Fig. 1).

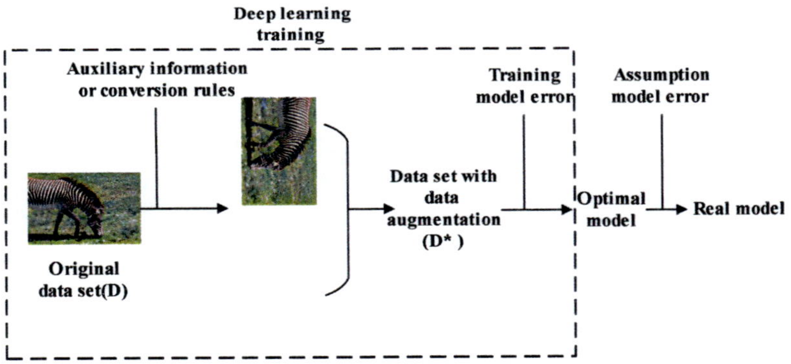

Fig. 1. Structure diagram of few-shot learning method based on data augmentation

2.1 Generative Adversarial Nets

However, the data augmentation method has a limited effect on the improvement of the model. Sometimes, data augmentation methods have better performance on specific data sets. And the same method does not improve on other task data sets. Therefore, it is necessary to use more advanced methods to expand data. One of the most representative is the use of neural networks to generate new data. For example, use Generative Adversarial Networks (GAN) [21, 22] to synthesize new data from the original few-shot data. The basic idea of this method is to train the GAN with a original data. Then the trained network was used to generate new data as a larger new data set. The GAN framework network consists of a network called Generator and a network called a Discriminator. The Generator generates random simulation samples based on the input random noise. The task of the Discriminator is to judge whether the input sample is real and output the probability that it is a real sample. The goal of Generator is to generate real samples as much as possible to deceive Discriminator. And Discriminator should correctly distinguish between real samples and simulated samples. The two networks constitute a dynamic game process. And their own technology were constantly improved in the process of confrontation. Finally, reaching a balance point, the generated simulation samples can achieve good effect. The objective function is defined as follows.

$$min_G\, max_D V(D, G) = E_{x\sim P(x)}[IbD(x)] + E_{z\sim P(z)}[Ib(1 - D(G(z)))] \quad (1)$$

Mehotra et al. [23] applied GAN to few-shot learning and proposed generative adversarial residual pairwise network to solve the single-sample learning problem. The algorithm used a GAN-based generator network to provide an effective regular representation of the invisible data distribution and used the residual pairwise network as a discriminator to measure the similarity of pairwise sample. Liu et al. [24] proposed a few-shot learning method combining W-GAN and DNN. They used W-GAN to enhance the new category data set and then used the enhanced data set to train the DNN classifier. The W-GAN algorithm in this paper used the Wasserstein distance to optimize the JS distance. This method can better evaluate the distribution difference between real samples and simulated samples. At the same time, the training time of the algorithm network was shortened and the model became more robust. The indicative figure of the network is shown in Fig. 2.

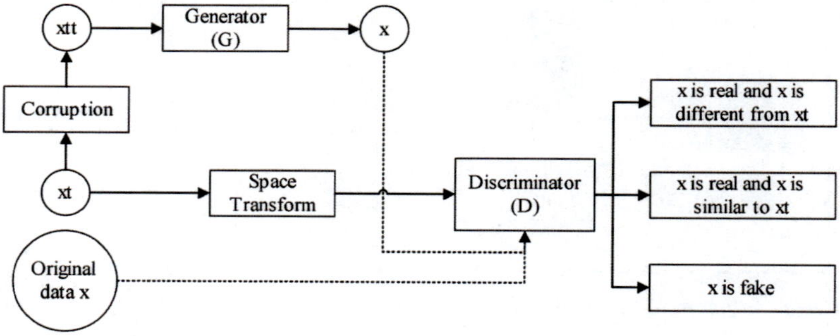

Fig. 2. An indicative figure of generative adversarial residual pairwise network

The difference between different samples in the same category can be generalized to other categories. By learning this difference, Hariharan et al. [25] proposed a new method for synthesizing data for few-shot learning. The method includes two stages: representation learning and few-shot learning. In representation learning stage, a large number of training samples of the base class data set were used to train a good feature extractor parameter. In the few-shot learning stage, both the new category data set and the base category data set were used to fine-tune the model. In order to solve the problem of too little new category data in the few-shot learning stage, it is proposed to use a feature generator to learn the change relationship between two different sample feature vectors in the same category. And it can be applied in generation data for the new category.

However, the existing data generation method has the following disadvantages. The first point is that the method of data augmentation does not capture the complex data distribution. And the ways of solving the few-shot learning problem had many limitations.

The augmentation strategies were usually customized for each data set. It is not easy to migrate to other data sets, especially data sets from other domains. The second point is that generative method has insufficient generalization ability for few-shot categories, and the generated data cannot simulate real data well.

3 Few-Shot Learning Based on Transfer Learning

In order to solve the problem of few-shot learning, we hope that the algorithm can realize the ability of using past knowledge to learn new knowledge. So the existing prior knowledge was supposed to add to the algorithm network. Therefore, transfer learning [26, 27] came into being. Transfer learning refers to learning from the source domain and source task with sufficient training data and transferring knowledge to the target domain and target task with few-shot data. Transfer learning only requires a certain correlation between the source field and the target field. So the knowledge and features learned in the source field and data can help train the classification model in the target field. The transfer learning realized the transfer of knowledge between different fields. Generally speaking, the stronger the correlation between the source domain and the target domain, the better the effect of transfer learning (Fig. 3).

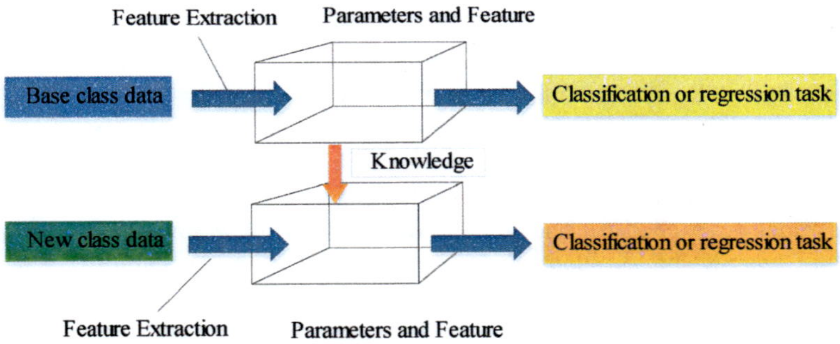

Fig. 3. The basic idea of transfer learning

The based idea of transfer learning is that the learning between similar tasks can be found in the same law. Given a source domain data as a base class and a learning task T1. Then we can determine a target domain data as a new class and a learning task T2. The purpose of transfer learning is that the knowledge about the base class was used to help improve the learning of the prediction function in the new class and have better perform in the task T2. The specific transfer methods can be divided into three categories: feature-based transfer, shared parameter-based transfer and relationship-based transfer. If the source base data class and target data in transfer learning are different but related, then further processing was needed. Transfer learning divides the data set into training set, support set and query set. The training set is a source data set that contains a large amount of annotated data. The support set refers to the target task training set containing a small amount of annotated data. Query set refers to the test set of the target task.

3.1 Meta-learning

Meta-learning is a method to realize transfer learning. The goal of meta-learning is to let the machine learn to learn. The idea of meta-learning was proposed as early as the 1990s [28]. Meta-learning fits the distribution of a series of similar tasks. The meta-learner was used to synthesize the parameters of each learning task. In this way, good initialization parameters can be obtained. Meta-learning is essentially to find model parameters that were better for all tasks. When updating parameters, meta-learning combines the loss of all tasks on the test set as prior knowledge to guide the gradient update. Therefore, the meta-learning method can be used as a good way to initialize the parameters. The value of parameter was already a better value for the new task. And meta-learning only needs fewer iterations to update for reaching the optimal hypothesis. Specifically, the meta-learning strategy is to make the original neural network have the ability to learn new tasks with few-shot. the meta-learner can get a large number of different learning tasks. These tasks were used to train a basic learner (neural network) to achieve few-shot learning. The training process is divided into two stages: meta-training and meta-testing. The meta-learner will randomly select C numbers categories from the basic category and K numbers samples from this category as a support set. The support set will be input to the basic learner as training samples. Then, randomly select M numbers samples from the remaining samples of C numbers categories as the query set for the test of the basic learner. By analogy, the above process is continuously repeated to obtain multiple independent learning tasks T. And the number of data sets for each task were set to be the same. Use these learning tasks to train the basic learner, so that it can better deal with the differences between tasks and ignore the characteristics of specific tasks. So it can have a good effect when encountering new tasks with few-shot. In the meta-testing stage, the data set of the new task was set the same as the task during training, and the categories in the new task were not seen in the previously learned tasks. In meta-learning, the value of K usually does not exceed 20. The problem of generation of learning task T was usually called the C-way K-shot problem (Fig. 4).

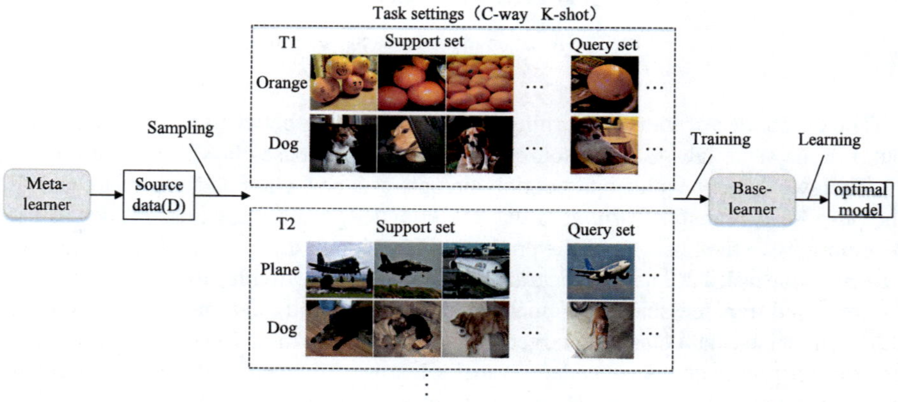

Fig. 4. Meta-learning training strategies

Fu et al. [29] combined meta-learning and SSD target detection algorithms. The Meta-SSD model was proposed as a small sample model with better generalization performance. Ravi et al. [30] used the meta-learning optimizer to classify few-shot images. This method uses LSTM's meta-learner to learn parameter optimization algorithms. Using this method to update the network parameters of the classifier enables make the algorithm model achieve better classification results on few-shot data. At the same time, the optimal initialization parameters of the network model are obtained, which speeds up the training convergence speed. Gidaris et al. [31] proposed a method that includes a weight generator based on the attention mechanism. At the same time, a CNN was redesigned as a cosine similarity function between the feature representation and the category weight vector. First the feature extractor was trained by the training set. Then a meta-learner was used to generate the corresponding parameter weights by the new few-shot data. When there are multiple samples, the attention mechanism was used to select the corresponding initial weights. Yu et al. [32] proposed a meta-learning method for multi-task clustering in 2018. This method considers the differences between multiple tasks. The model clusters all tasks and divides different tasks into different clusters. The tasks in the same cluster are relatively similar, and they share a set of classifier parameters. For a new task, the classifier of each cluster is calculated at the current level. The adaptation effect on the sample data was represented by the adaptation parameter. Then, the model parameters of the current classification task are obtained by linear combination of the parameters under all clusters and adaptation parameter. Finally, the classification result was output.

Meta-learning was a high-level cross-task learning strategy that it can be applied to various few-shot learning models, not only a specific few-shot learning model. The implementation methods of meta-learning were vary. Their applicable fields were also relatively wide ranging from images to text and voice fields. The method of meta-learning can well realize the task of few-shot learning. In the future, researchers can design meta-learners to achieve better and faster learning with few-shot.

4 Application in Industrial Tasks

At present, there have been many few-shot learning work and results in the field of industrial detection. And there are many excellent works and practical applications in the fields of fault diagnosis and image classification and detection. The problem of studying in few-shot learning was more suitable for industrial production and practical application work.

In 2020, Yu et al. [33] used the few-shot learning method based on metric learning to realize the few-shot defect detection of parts. This work proposed a variability feature pyramid network (DC-FPN) based on the characteristics of different defect sizes and geometric variability of defects. First of all, this method used FPN as the basic network to complete feature extraction. In order to overcome the geometric deformation problem that cannot be solved by traditional convolution operations, DC [34] was introduced into the basic network ResNet50. Then the idea of transfer learning was used. The large-scale data set MSCOCO was used for training. And its trained structural parameters were transferred and learned into the model proposed in the article as the defect feature extraction part. Then the positioning data set of the public data set ImageNet was used to fine-tune the model parameters of the improved FPN pre-training. DC-FPN output the ROI characteristics of the defect map of part data. Finally, detection network was improved based on the RepMet network. The multi-modal network structure was used to extract ROI feature information to obtain different feature vectors. The distance between the feature vector proposed by each mode and the feature vector obtained by the feature vector module was measured. The algorithm structure diagram of surface defect identification and detection was shown as the figure below (Fig. 5).

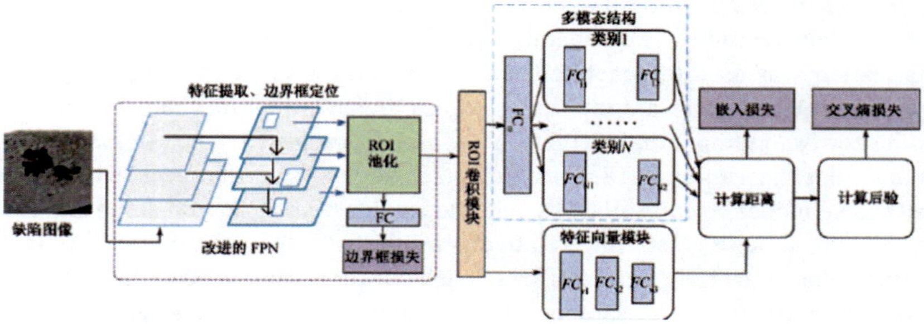

Fig. 5. Few-shot parts surface defect detection model structure based on metric learning [33]

The results show that the algorithm model has a maximum average accuracy of 35.76% on the surface defect data set established on the 3 types and 5 samples, which improved the performance of the RepMet model by nearly 70%. The model does not require large-scale data sets for training. The model has learned the ability to learn detect new classes after training on a few-shot surface defect data set. It had a well accuracy and generalization ability. However, the defect detection accuracy needs to be further improved.

Moreover, Li and Wang [35] proposed a prototypical network adapted to the detection of poultry egg images in few-shot. This method used metric-based meta-learning to construct a poultry egg image detection model under the condition of few-shot. Based on the prototype network, a prototypical-eggnetwork for egg image detection was designed, and a convolutional neural network was built with an inverse residual structure introducing an attention mechanism. The network was used to map the egg image to the embedding space. This way realized the classification of poultry egg images. This paper uses the network to verify the classification and detection effects of fertilized eggs and infertile eggs, double yolk eggs and single yolk eggs, cracked eggs and normal eggs under few-shot conditions. The detection accuracy respectively was 95%, 98%, and 88%. The paper realizes the non-destructive detection of poultry egg images under few-shot.

5 Discussion and Conclusion

Due to constraints in the real world, the situation of few-shot generally will always appear. The sample annotation work always consumed a lot of time and manpower. Therefore, research on few-shot learning is a very necessary and extremely important. Especially for applications in the industrial manufacturing. Few-shot learning can help the industry solve practical problems. Such as detection, fault diagnosis and prediction. At present, there have been many achievements in the study of few-shot learning. Its applications in the industrial field were also very extensive. The case mentioned above showed that the application of few-shot learning in the industrial field has strong practicability and extensiveness. Few-shot learning can produce higher research value and economic value.

In general, the study of few-shot learning has made great progress. But there is still a big gap compared with the accuracy of human classification. The method based on data augmentation was a way to perform data expansion or feature enhancement on small sample data sets. This method can avoid parameter adjustments. And it is easy to introduce noisy data. The method based on transfer learning was a way to transfer the knowledge learned in the old field to the new field. This way did not require a strong correlation between the base class and new class. However, if there were the stronger correlation between their, there was better the migration effect. Since the deep learning model was a black box model, it is difficult for people to understand the features parameter. What features need be retained during training and parameter migration? It was more difficult to adjust parameters. Improving the interpretability of deep learning can help understand feature transfer and find suitable transfer features between the source domain and the target domain.

In conclusion, the few-shot learning model helps to reduce the burden of artificial intelligence in the industrial deployment. It reduced the cost of artificial intelligence about collecting a large number of annotation costs. On the other hand, it also was possible to deploy deep learning on rare cases and achieve the goals of the industrial field. Applications such as detection, fault diagnosis and intelligent decision-making will be realized based on few-shot learning,

References

1. Luo, H.L., Yuan, P., Tong, K.: Review of the methods for salient object detection based on deep learning. Acta Electronica Sin. **49**(07), 1417–1427 (2021)
2. Chen, H.D., Ding, X.Y., Liu, Y.X.: Review of target detection algorithm based on deep learning. J. Beijing Union Univ. **35**(03), 39–46 (2021)
3. Yang, C.L., Zhang, X.Q.: Survey of material defect detection applications based on machine vision and deep learning. Mater. Rep. **2022**(16), 1–19 (2021)
4. Fan, L.L., Zhao, H.W., Zhap, H.Y., et al.: Survey of target detection based on deep convolutional neural networks. Opt. Precis. Eng. **28**(05), 1152–1164 (2020)
5. Jankowski, N., Duch, W., Grąbczewski, K.: Meta-learning in Computational Intelligence. SCI. Springer, Heidelberg, pp. 97–115 (2011). https://doi.org/10.1007/978-3-642-20980-2
6. Lake, B., Salakhutdinov, R.: One-shot learning by inverting a compositional causal process. In: Advances in Neural Information Processing Systems, pp. 2526–2534 (2013)
7. Li, F.F., et al.: A Bayesian approach to unsupervised one-shot learning of object categories. In: Proceedings of the 9th IEEE International Conference on Computer Vision (2003)
8. Feifei, L., Fergus, R., Perona, P.: One-shot learning of object categories. IEEE Trans. Pattern Anal. Mach. Intell. **28**(4), 594–611 (2006)
9. Wang, Y.X., Girshick, R., Hebert, M., et al.: Low-shot learning from imaginary data. In: Proceedings of the IEEE Conference on Computer Vision and Pattern Recognition, pp. 7278–7286 (2018)

10. Fu, Y., Xiang, T., Jiang, Y.G., et al.: Recent advances in zero-shot recognition: toward data-efficient understanding of visual content. IEEE Sig. Process. Mag. **35**(1), 112–125 (2018)
11. Zhao, K.L., Jin, X.L., Wang, Y.Z.: Survey on few-shot learning. J. Softw. **32**(2), 349–369 (2021)
12. Pan, X.J., Zhang, X.L., Dong, W.M., et al.: A survey of few-shot object detection. J. Nanjing Univ. Inf. Sci. Technol. (Nat. Sci. Edn.) **11**(6), 698–770 (2019)
13. Liu, Y., Lei, Y.B., Fan, J.L., et al.: Survey on image classification technology based on few-shot learning. Acta Automatica Sin. **47**(2), 297–315 (2021)
14. Pan, C.Y., Huang, J., Hao, J.G., et al.: Survey of weakly supervised learning integrating zero-shot and few-shot learning. Syst. Eng. Electron. **42**(10), 2246–2256 (2020)
15. Zhu, J.T., Yao, G.L., Zhang, G.X., et al.: Survey of few shot learning of deep neural network. Comput. Eng. Appl. **57**(7), 22–33 (2021)
16. Liu, H.Y., Wang, X.J.: A survey of few-shot object detection. Navig. Control. **20**(01), 1–14 (2021)
17. Zoph, B., Cubuk, E.D., Ghiasi, G., Lin, T.-Y., Shlens, J., Le, Q.V.: Learning data augmentation strategies for object detection. In: Vedaldi, A., Bischof, H., Brox, T., Frahm, J.-M. (eds.) ECCV 2020. LNCS, vol. 12372, pp. 566–583. Springer, Cham (2020). https://doi.org/10.1007/978-3-030-58583-9_34
18. Chawla, N.V., Bowyer, K.W.: Smote: synthetic minority over-sampling technique. J. Artif. Intell. Res. **16**(1), 321–357 (2002)
19. Inoue, H.: Data augmentation by pairing samples for images classification. arXiv:1801.02929 (2018)
20. Zhang, H., Cisse, M., Dauphin, Y.N., et al.: Mixup: beyond empirical risk minimization. In: Proceedings of International Conference on Learning Representations (2018)
21. Goodfellow, I., Pouget-Abasie, J., Mirza, M., et al.: Generative adversarial nets. In: Advances in Neural Information Processing Systems, pp. 2672–2680 (2014)
22. Goodfellow, I., Bengio, Y., Courville, A.: Deep Learning. MIT Press, Cambridge (2016)
23. Mehrotra, A., Dukkipati, A.: Generative adversarial residual pairwise networks for one shot learning. arXiv preprint arXiv:1703.08033 (2017)
24. Liu, Y.F., Zhou, Y., Liu, X., et al.: A new generation of artificial intelligence small sample data augmentation method based on Wasserstein Gan: a case study of cancer staging data in biological field. Engineering **5**(1), 338–354 (2019)
25. Hariharah, B., Girshick, R.: Low-shot visual recognition by shrinking and hallucinating features. In: Proceedings of IEEE Conference on Computer Vision and Pattern Recognition (2017)
26. Aytar, Y., Zisserman, A.: Tabula rasa: model transfer for object category detection. In: Proceedings of the 2011 International Conference on Computer Vision, pp. 2252–2259. IEEE, Barcelona (2011)
27. Tommasi, T., Orabona, F., Caputo, B.: Safety in numbers: learning categories from few examples with multi model knowledge transfer. In: Proceedings of the 2010 IEEE Computer Society Conference on Computer Vision and Pattern Recognition, pp. 3081–3088. IEEE, San Francisco (2010)
28. Naik, D.K., Mammone, R.: Meta-neural networks that learn by learning. In: Proceedings of International Joint Conference on Neural Net (1992)
29. Fu, K., Zhang, T.F., Zhang, Y., et al.: Meta-SSD: towards fast adaptation for few-shot object detection with Meta-learning. IEEE Access **7**, 77597–77606 (2019)
30. Ravi, S., Larochelle, H.: Optimization as a model for few-shot learning. In: Proceedings of the ICLR (2016)
31. Gidaris, S., Komodakis, N.: Dynamic few-shot visual learning without forgetting. In: Proceedings of the IEEE Conference on Computer Vision and Pattern Recognition, pp. 4367–4375 (2018)

32. Yu, M., Guo, X., Yi, J., et al.: Diverse few-shot text classification with multiple metrics. In: Proceedings of the NAACL-HLT, pp. 1206−1215 (2018)
33. Yu, C.C., Sa, L.B., Ma, X.Q., et al.: Few-shot parts surface defect detection based on the metric learning. Chin. J. Sci. Instrum. **41**(07), 214–223 (2020)
34. Dai, J.F., Qi, H.Z.H., Xiong, Y.W., et al.: Deformable convolutional networks. In: IEEE International Conference on Computer Vision, pp. 764–773 (2017)
35. Li, Q.X., Wang, Q.H.: Feature detection method of small sample egg image based on meta-learning. Trans. Chin. Soc. Agric. Mach. 1–8 (2021)

A Review of Neural Architecture Search

Tang Yu, Chen Wang(✉), Zhang Xiufeng, Liu Chao, Zhou Zhixiao, and Lu Xuxiang

Department of Mechanical Engineering, Hubei University of Automotive Technology, Shiyan 442002, China
{202011053,20090011}@huat.edu.cn

Abstract. In recent years, deep neural networks (DNN) have achieved great success in many applications. Deep neural networks are usually designed manually by professionals, but this work depends on the experience of designers and is time-consuming and laborious. Neural architecture search (NAS) is a technology that can automatically design deep neural networks. This paper studies the development of neural architecture search technology, and analyzes its basic technical route.

Keywords: Deep learning · Neural architecture search · Convolution Neural Network

1 Introduction

As the basis of artificial intelligence, deep neural network plays a more and more important role in various fields. Including computer vision [1], natural language processing [2], speech recognition [3], and so on. DNN can be used as a feature learner that is, learning meaningful features directly from a large amount of data. However, according to No free Lunch Theorem (NFL Theorem) [4], the expectation of any algorithm is not higher than that of random search. This implies that our algorithm must be designed for specific problems. In most cases, the better DNN structure is designed manually by personnel with relevant professional competence. Such as VGG [5], ResNet [6] and DenseNet [7]. However, the manual design of DNN not only has high cost, but also has high requirements for professional knowledge. This means that the neural architecture search (NAS) technology which can automatically generate high-performance networks has become an inevitable requirement.

The performance of DNN mainly depends on the network structure and weight. Therefore, DNN optimization problem is essentially a double-layer optimization problem. First, we should find the appropriate network structure; the second is to optimize the parameters of the network. The principle of NAS is to give a set of candidate neural network structures (called search space), find some network structures with a certain strategy (search strategy), use a certain index to measure the advantages and disadvantages of the network structure (performance evaluation), and finally find the optimal network structure. It can be seen that search space, search strategy and performance evaluation are the core elements of NAS. This optimization problem is modelled by mathematical expression, as shown in Eq. 1:

Y. Wang et al. (Eds.): IWAMA 2021, LNEE 880, pp. 648–652, 2022.
https://doi.org/10.1007/978-981-19-0572-8_82

Where $\mathcal{A}$ denotes the search space of the potential neural architectures, measures the performance of the architecture A on the fitness evaluation dataset $\mathcal{D}_{fitness}$ after being trained on the training dataset $\mathcal{D}_{train}$. The is usually non-convex and non-differentiable [8].

NAS has faced many problems since its birth. For example: high computational ovredhead, double-layer optimization structure, discrete optimization problems, complex constraints and so on. The NAS [9] published by Google in ArXir in 2016 is widely regarded as the earliest NAS algorithm.

2 Search Spaces

Search space is a crucial part of NAS. It defines how to describe the structure of neural network. In a well-designed search space, even if the random search strategy is used, it is possible to find a network structure with excellent performance. In addition, the design of search space determines the upper limit of the final network performance. Excellent search strategy only makes the final result close to or equal to this upper limit. When designing the search space, we often need a balance between the degree of freedom of structural search and the search efficiency. These two aspects are contradictory.

The search space of neural network can be divided into three scales according to different levels: topology, operation type and weight. In terms of topology, the structure of neural network can be abstracted into a directed acyclic graph (DAG) without isolated nodes. Each node in the DAG represents an operation. Each node receives the data of the precursor node, calculates it and then transmits it to the post node (except the head and tail nodes). A node can have more than one precursor node or post node to form a multi branch or jump connection structure (Fig. 1).

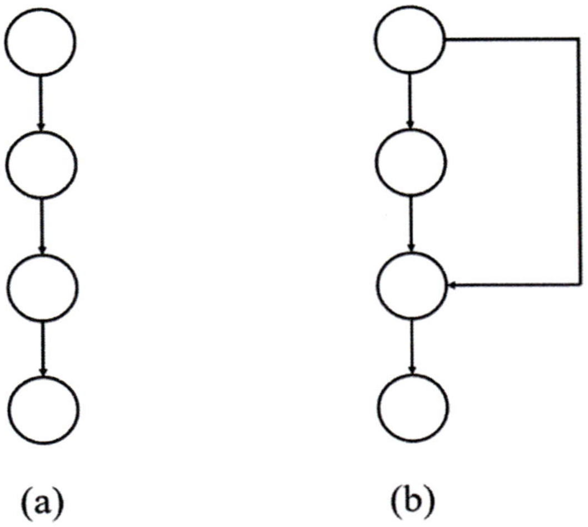

Fig. 1. Schematic diagram of chain structure and branch structure. (a) Chain structure. (b) Branching structure

Operation type refers to the calculation method used by each node in DAG. Although some methods use very basic mathematical operations, such as AutoML-Zero [10], most algorithms still use cell-based operations. Cell refers to some basic modules widely considered to have excellent performance in the networks manually designed by predecessors, such as deny block. Except that the types of the first and last layers must be input and output, the operations of the middle layers are obtained through some search strategy, but the combined use of these layers sometimes must obey certain constraints.

3 Search Strategy

The search strategy defines the iterative optimization process of neural network structure. Common search methods include random search, Bayesian optimization, evolutionary algorithms [11], reinforcement learning [12] (RL) and gradient based algorhythms. Evolutionary algorithm, and gradient based algorithm are the mainstream algorithms, which are also the main methods discussed in this paper.

3.1 Search Strategy Based on Reinforcement Learning

Reinforcement learning is used to describe and solve the problem that agents use learning strategies to maximize reporting or achieve specific goals in the process of interacting with the environment. RL has four basic elements: agent, action, environment and reward. Barret zoph et al. [13] applied reinforcement learning to NAS, which has become one of the paradigms of search strategy based on reinforcement learning (Fig. 2).

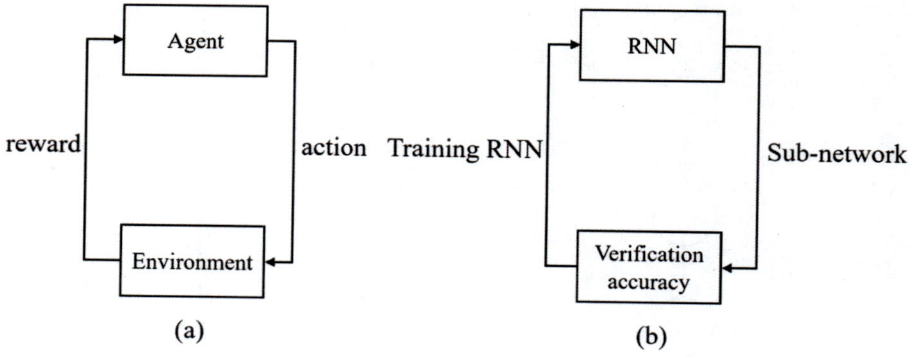

Fig. 2. Schematic diagram of reinforcement learning and NAS based on Reinforcement Learning reinforcement learning. (b) NAS based on Reinforcement Learning

3.2 Search Strategy Based on Evolutionary Algorithms

Evolutionary algorithm is a global optimization method with high robustness and wide applicability. When used in NAS, the network structure is generally coded so that each individual in the evolutionary algorithm population represents a neural network structure. The performance of an individual on a single check data set is called fitness. Evolutionary algorithm uses the way of survival of the fittest to iterate continuously, and finally obtains a better neural network structure.

3.3 Search Strategy Based on Gradient

Most NAS based on RL and evolutionary algorithm regard the network topology as discrete space. Although this approach is intuitive, the black box optimization problem in discrete space often produces huge computational overhead. DARTS [14] algorithm creates an idea of continuous discrete optimization problems. The network search space is relaxed and transformed into continuous variables. This algorithm can be solved by gradient descent method, and optimize the network structure, weight and other parameters.

4 Performance Evaluation Strategy

The purpose of neural network optimization must be to maximize some performance metrics. Training and verifying the network in a conventional way is the most intuitive performance evaluation strategy, but the process time cost is very high. Therefore, sacrificing accuracy to improve speed is the main idea of performance evaluation strategy. The most commonly used methods are reducing the number of iterations, reducing training samples, training on low resolution images and so on.

A more complex approach is to predict the performance of the neural network, that is, infer its final performance from the performance of the previous iterations during training, or predict the performance of the whole network with the characteristics of the searched units. In addition, weight sharing is also a common means, which is, using the previously trained weight as the initial weight of the current network to be evaluated, which speeds up the convergence speed.

5 Summary

This paper gives a comprehensive overview of NAS. We mainly introduce three aspects: search space, search strategy and performance evaluation. NAS automates the most labor-intensive part of machine learning, greatly improves the development speed of intelligent technology and reduces the threshold of using machine learning model. However, the current NAS still has great limitations. The search space of many NAS algorithms is still based on manually designed modules. We believe that wider search space is one of the most important directions for NAS development, which also puts forward higher requirements for search strategy and performance evaluation.

References

1. He, K., Zhang, X., Ren, S., Sun, J.: Deep residual learning for image recognition. In: Proceedings of the IEEE Conference on Computer Vision and Pattern Recognition, pp. 770–778 (2016)
2. Devlin, J., Chang, M.-W., Lee, K., Toutanova, K.: BERT: pre-training of deep bidirectional transformers for language understanding. arXivpreprint arXiv:1810.04805 (2018)
3. Zhang, Y., Chan, W., Jaitly, N.: Very deep convolutional networks for end-to-end speech recognition. In: 2017 IEEE International Conference on Acoustics, Speech and Signal Processing (ICASSP), pp. 4845–4849. IEEE (2017)
4. Wolpert, D.H., Macready, W.G.: No free lunch theorems for optimization. IEEE Trans. Evol. Comput. **1**(1), 67–82 (1997)
5. Simonyan, K., Zisserman, A.: Very deep convolutional networks for large-scale image recognition. Computer Science (2014)
6. He, K., Zhang, X., Ren, S., et al.: Deep residual learning for image recognition. In: 2016 IEEE Conference on Computer Vision and Pattern Recognition (CVPR). IEEE (2016)
7. Huang, G., Liu, Z., Laurens, V., et al.: Densely Connected Convolutional Networks. IEEE Computer Society (2016)
8. Sun, Y., Xue, B., Zhang, M., Yen, G.G.: Completely automated CNN architecture design based on blocks. IEEE Trans. Neural Networks Learn. Syst. (2019)
9. Zoph, B., Le, Q.V.: Neural architecture search with reinforcement learning. arXiv preprint arXiv:1611.01578 (2016)
10. Real, E., Liang, C., So, D.R., et al.: AutoML-Zero: Evolving Machine Learning Algorithms from Scratch (2020)
11. Bäck, T., Fogel, D.B., Michalewicz, Z.: Handbook of evolutionary computation. Release **97**(1), B1 (1997)
12. Kaelbling, L.P., Littman, M.L., Moore, A.W.: Reinforcement learning: a survey. J. Artif. Intell. Res. **4**, 237–285 (1996)
13. Zoph, B., Le, Q.V.: Neural architecture search with reinforcement learning. In: International Conference on Learning Representations (2016)
14. Liu, H., Simonyan, K., Yang, Y.: DARTS: differentiable architecture search. In: International Conference on Learning Representations (2018)

Applying Decision Tree in Fast Fashion Process

Emmanuel Aldovino and Yi Wang(✉)

Department of International Shipping, Logistics, and Operations Management, Plymouth Business School, Plymouth University, Plymouth PL4 8AA, UK
{e.aldovino,yi.wang}@plymouth.ac.uk

Abstract. This paper will explore how pivotal decision making is within the role of management, through the exemplification of theories surrounding decision making and the common practice of decision making within different industries. Decision making is prevalent within various instances as the role of management utilises and make decisions to achieve business objectives. Many businesses understand the importance of decision making due to the consideration of resources and assets being used when businesses make these decisions as it can affect the direction of the business and also a business's overall success. This paper will analyse the business H&M, investigating the industries direction into fast fashion and the harmful effects it creates surrounding this industry. Many reports in relation to this issue provide different types of literature, models and theories that can be applied to a business-like H&M and the ties their production has within fast fashion. In this case, many aspects such as any moral implications regarding the case are also being considered. Decision-making trees will be applied and examined where many changes such as a company's logistics and businesses reputation. As well as a critical analysis of decision-making trees being a beneficial use in reducing the amount of risk a company may face. This paper will identify various types of literature to assess both the benefits and drawbacks of decision-making trees, as well as how useful the tool decision trees are for a company like H&M to consider.

Keywords: Fast fashion · Corporate social responsibility [CSR] · Decision tree · Fashion and textile industry

1 Introduction

Firstly, the assimilation of corporate social responsibility within businesses has grown since before the creation of modern-day companies an example being the Quaker boycott of slave-grown sugar and rum in the UK during the 1700s. Many industries and business alike encourage the direction of being socially responsible due to the adverse effects it has on different communities and also people's lives. We can see decades of example that the decision of being more socially responsible has long been a requirement of many businesses [1]. Elkington's triple bottom line [TBL] concept expresses the fact that companies and many different types of industries simultaneously have the ability to both create and destroy economic, social, and environmental value. In other words, with regard to TBL's concept we can see the different knock-on effects it may have and thus

Y. Wang et al. (Eds.): IWAMA 2021, LNEE 880, pp. 653–660, 2022.
https://doi.org/10.1007/978-981-19-0572-8_83

within many companies may reinforce the importance of CSR [2]. CSR can be simply defined by many to be a business's decision to be transparent and make more ethical decisions regarding the of respect many other stakeholders [3]. Furthermore, being more socially responsible; meaning organisations making these more ethical decisions proved to be an underlying reason for many business successes [4]. Therefore, the power to manage the direction and make crucial decisions for a business are those within senior positions. Moreover, it has become imperative that the duty management faces are 'issues beyond the narrow and economic, technical and legal requirement of the firm'. In other words, the responsibilities and the role management has now developed to more than just the coordination, planning, organisation of employees and departments within a business, but also the further consequences of those decisions as well [5]. It can further be said that the ethical ideals these individuals may have can play a pivotal role within the direction of a business [6]. This can be applied to the case of H&M where the production of fast fashion product promotes waste and also contributes to the contamination of freshwater due to the toxic chemicals used within the production of their products. This also further contributes to the large and ever-growing carbon footprint left behind by the fashion industry and shows a disregard for the conservation of the earth's natural resources. This paper starts with a brief introduction of the background, then it reviews the decision tree applied in Fashion industry, afterwards a critical evaluation of decision tree is made. The paper ends with a conclusion.

2 Literature Review

2.1 Fast Fashion Industry

The fashion industry has further developed within recent decades as the boundaries of fashion started to expand and grow [7]. The fashion industry has grown from the deterioration of mass production to the increased amount of fashion seasons. As well as due to change within the supply chains has left many retailers to want low cost and low-quality products. Which further promotes harmful ideals and also that many organisations have already assumed this stance within their supply chain [8]. The term fast fashion can be defined to be an expression at which low-cost clothing are produced rapidly by mass-market retailers to keep up with the newest trends [9]. Moreover, contemporary's society's stance in fashion has been highly competitive with the desire of fashion products and accessories needing a constant refresh. Therefore, many retailers increase the number of seasons they may have to then encourage consumers to go within their stores and 'Buy now'. Thus, demonstrating how much many fashion retailers prioritise an increase in profit margins through the sale of fast and quick merchandise compared to the affects those products would have within the environment [10]. Retailers such as Zara and H&M have used many pricing strategies to encourage consumers to buy from them due to the low price [11]. In addition, looking in the past key inspirations for many fast-fashion retailers often look at runway models and catwalks to see the latest trends. Often many retailers like H&M, Zara, Mango and Topshop attempt to copy the styles of clothing presented within the high-end fashion industry and they rapidly try to adopt these styles to attract more consumers and introduce them to stores within 3–5 weeks [12]. Furthermore, other influences and sources that retailers may use are fashion trends within social media. As in many cases due to social media, influencers, and many celebrities, they have transformed the world of fashion to a great extent affecting the operational environment of fashion. Social media is an important tool for retailers as it allows them to communicate with their consumers and also promote any product or seasonal sales that may happen. As well as, because social media is such an important channel for marketing to use, many retailers would often rely on word of mouth and viral marketing to promote their fast fashion products [13].

Modern business has shown an interest in more sustainable supply chain management [SSCM]. There has been a further increase in ways of increasing attention among the matter of fast fashion where many brands often attempt to integrate more sustainable practices within the company. Fast fashion retailers like H&M have attempted to address their effect on the fashion industry and the production of fast fashion [14]. On the other hand, there have been areas in which H&M participated in the adverse effects of fast fashion. H&M among many other retailers have been reportedly accused of greenwashing which is the practice of making false claims about how environmentally friendly an organisations product is [15]. The general conception that H&M has is that they are not known for their sustainable practices [16]. The majority of H&M's supply of fast fashion products come from Bangladesh with approximately 163 suppliers working for the organisation [Shen, 2014]. In a paper in April 2005, around 64 garment workers were killed and 80 injured when the spectrum factory producing these fast-fashion clothing collapsed. It was identified that the employees were working in poor working conditions. However, although the Spectrum tragedy has happened. Safety issues faced by many garment working has been led to believe that many of the owners of these factories or the Bangladeshi government have not made any systematic change within the improvement of these working conditions. Implying that both organisations and governments haven't taken a steps to prevent any tragedies to happen in the future [17].

With reference to the US director of Global Labour Justice discloses that "H&M's fast fashion supply chain model creates unreasonable production targets… which results in women working unpaid overtime and working very fast under extreme pressure" [18]. Further, reinforcing the issue within the fast fashion industry and how unsustainable H&M's supply chain management is. Essentially contributing to the mass unhealthy environmental rhetoric, they originally did promise to fix these issues [19]. As well as a paper has said that the fashion industry's is responsible for more than 10% of the annual global emissions, proving to be more harmful international flights and maritime shipping combined. Emphasising the amount of harm fast fashion products contribute to the world's environmental footprint [20]. In the last decade, there have been many accidents that have happened due to the effects of fast fashion. One being the Rana Plaza factory that collapsed on the 24th of April 2013 where a 9-story factory in the city of northwest Dhaka collapsed. Over 1100 people lost their lives and approximately 2400 people were injured that day. This is one of the most recorded deadliest accidents to happen, but only one of many other accidents that have happened within countries like Bangladesh, Pakistan, and other garment-producing factories [21].

2.2 Decision Tree

Decision trees allow a decision-maker to view all important and decisive aspects of the problem at once. It can help with any decision alternatives, uncertain outcomes, also their probabilities, any economic consequences that may occur and also the chronological order of events [22]. Furthermore, decision-making problems have three common elements and once these elements are identified, the decision-maker can decide on a decision that suitably fits the criteria chosen. Therefore, decision trees adding clarity to these uncertain decisions. This can aid users that use this tool when attempting to summarise and understand many parts of their data and also helps these users transpire their data a lot easier. This emphasises why decision trees are a popular tool to use as it further provides a graphical representation on any possible statistical solutions, which when used to find out a certain criterion and also applied to a process that guides analysis [23]. Decision trees further split data into subdivisions which allow users to see the probability of any variables being met [24]. Moreover, when calculating the expected value [EMV]can be regarded to be the most preferred approach in most decisions problems and is an approach that can interpret probabilities for each outcome of each decision to calculate an expected payoff [25]. With this, decision-makers will choose the decision which gives the most EMV – otherwise known as playing averages.

Any branch on a decision tree that may lead to a node means it has already occurred whereas if branches lead out of a node, then it means it is yet to occur [26]. The root nodes which are normally found at the beginning of the decision tree are the foundation and do not have any oncoming branches going into them and represent decisions with an uncertain outcome. These will then further branch out to an internal node that possesses incoming and outgoing branches. These internal nodes are what signify any conditional probabilities of the target value [27]. This further extenuate the validity of the decision tree outcome, which further shows the link between the criteria and analysis. Likewise, it branches to the terminal node where only one branch is incoming and no outgoing. During the end node, the algorithm calculates the dependant variables with the EMV being outlined on the right side [28]. Once the decision tree has been calculated to the terminal node all costs and problems have been solved and should be noted down by the decision-maker. Every different path has been considered, as well as uncertainty should be determined between all nodes and branches [29].

3 Discussion

A decision tree is a representation of data and allows a decision-maker to view all data simultaneously, further allowing for said data to be compared to the different branches. Decision trees allow the decision-maker to interpret the data clearly and further analyse the correlation between all different types of input and output variable, but also the process too. DT has the ability to sort out small or mass amounts of data and can display the information easily for the decision-maker to use [30].

The benefits of implementing decision trees are that due to the tool being presented as a linear diagram, it allows users to easily interpret the tool. Thus, allowing for easy comparison between data which can be crucial when trying to find out any alternative business decisions. In other words, having a linear diagram allows for the user to easily distinguish between the data with no further complications, the transparency decision trees make can allow for complex data to be presented in a much easier format [31]. Furthermore, other benefits of using decision trees are how flexible decision trees are. With the advantage to handle what is known to be continuous data which is data that can take any value and discrete data which is where only certain types of data values can be taken [32]. This further allows for decision-makers to apply certain values to reach a certain criterion and also allow it to apply to many situations. In addition, DT can allow the user to identify any non-linear relations within the decision trees [33] as well as can further interpret any non-binary labels [34] which are trees that allow for more than 2 branches to link to either a node or a root within the DT [35].

On the other hand, the drawback of decision trees is that as the DT expands and becomes larger, the data accumulated can be extremely complex to analyse, because of the increase in size. Thus, the more data and subsequent depth you add to the tree the more likely the decision tree is going to be more ineffective and become a redundant tool. Due to the decision tree being strictly incompressible [36]. In other words, it can affect how accurate the DT is and increase how inaccurate the data may also be. In the same vein, decision trees can also be very costly when implementing. This is because the decision-maker must possess advanced skills within that of quantitative data and also the ability to analyse statistical data. In other words, for the implementation of decision trees to be effective the organisation will need an individual that possesses these key skills to interpret and analyse data, which could be expensive since it is a very specialist type of job. Therefore, not being effective if the individual does not have the subsequent experience [37]. To summarise, the main disadvantage of decision trees is that the more data that is used, the more complex it becomes. Thus, making it harder to interpret and lastly give inaccurate results.

Conversely, this risk can be minimised by limiting the amount of data being added to the decision tree and also by stopping the growth of the tree before when trying to find out the EMV. With reference [38] reducing the amount of data being added could be considered a 'pruning phase' which allows for the accuracy of data to not get too complicated and limits the amount of error that may occur. This prevents the overfitting of messy data by adherently reducing it. However, pruning can still take a considerable amount of time as it is imperative that the decision tree gives the most reliable result. On the other hand, over-pruning, where the decision tree is cut down too much, can still have adverse effects on the data. As it may increase the risk of uncertainty, thus decreasing the accuracy of the overall results [39]. However, it can also be argued that there would need to be a balance as the oversimplification of decision trees as previously said may prove the DT to be ineffective.

4 Conclusions

Many decision trees can be very effective in use when used in the right way. If the right amount of data is used where the depth of the tree does not overcomplicate the overall results and pruning is also used, can prove to be a very useful tool for H&M to use. Decision trees can help an organisation like H&M to make decisions that follow a more CSR focussed. Which is reinforced by the validity and proficiency this decision-making tool can bring to a business. This paper has highlighted H&M and their involvement within the fast fashion industry, in relation to decision making techniques within management and limiting the amount of error that may happen. Furthermore, a critical review on decision trees has exemplified how effective the tool can be in assisting in making complex decisions, thus being a valuable tool for a company like H&M when making decisive decisions. However, although it can be seen as a useful tool for many organisations, it is important to place limits on the amount of complex data that is collected through the process of pruning. This makes the decision tree more comprehensible and transparent, making it easier to read. This can be applied to managers within H&M to make better managerial choices following their environmental motives. Having the ability to evaluate these decisions that follow their CSR model can allow for a business-like H&M to build up their reputation and promote more sustainable supply chain practices and make decisions that provide safe places employees can work in. To restate, CSR plays an important role within many industries and organisations. It is extremely important for a company like H&M to improve on its practices and put forward communities and people's welfare.

The literature review has also identified the impact management has within an organisation. The role of management will need to be improved for an ethical solution to be implemented successfully. Therefore, the decision between the choice of whether H&M continues their current practice of greenwashing and harmful fast fashion or switch to a much greener alternative which in doing so is much better for the environment and also any stakeholders that may be involved. Thus, allowing them to take the right direction into remedying their previous actions and gain the public's trust. The practice H&M participates in requires change as the long-term adverse effects it may have on their profit margins, reputation and overall brand will degrade. A good direction that H&M may take is a decision or direction that can give them the highest probability of having A] establishing a good reputation B] increase their sales and lastly C] clearing up any greenwashing issues. Therefore, it is imperative to discuss the need for decision-making techniques for them to increase the rate of success and implementation.

References

1. Lovell, A.M.C., Valero-Silva, N.: Business Ethics and Values: Individual, Corporate and International Perspectives. Pearson Education, Corp., Harlow (2013)
2. Alhaddi, H.: Triple bottom line and sustainability: a literature review. Bus. Manag. Stud. **1**(2), 6 (2015)
3. Liang, H., Renneboog, L.: On the foundations of corporate social responsibility. J. Financ. **72**(2), 853–910 (2017)
4. Carroll, A.B.: Corporate social responsibility. Organ. Dyn. **44**(2), 87–96 (2015)
5. Davis, K.: The Case for and against business assumption of social responsibilities. Acad. Manag. J. **16**(2), 312–322 (1973)
6. Duarte, F.: Working with corporate social responsibility in Brazilian companies: the role of managers' values in the maintenance of CSR cultures. J. Bus. Ethics **96**(3), 355–368 (2010)
7. Djelic, M.-L., Ainamo, A.: The coevolution of new organizational forms in the fashion industry: a historical and comparative study of France, Italy, and the United States. Organ. Sci. **10**(5), 622–637 (1999)

8. Doyle, S.A., Moore, C.M., Morgan, L.: Supplier management in fast moving fashion retailing. J. Fashion Mark. Manage. Int. J. **10**(3), 272–281 (2006)
9. Fernie, J., Sparks, L.: Logistics and retail management. Int. J. Retail Distrib. Manage. **27**(8), 328–330 (1998)
10. Sydney Loves Fashion: Fast Fashion is not a Trend. Sydney Loves Fashion (2008). https://www.sydneylovesfashion.com/2008/12/fast-fashion-is-trend.html
11. Mo, Z.: Internationalization process of fast fashion retailers: evidence of H&M and Zara. Int. J. Bus. Manage. **10**(3), 217 (2015)
12. Barnes, L., Lea-Greenwood, G.: Fast fashioning the supply chain: shaping the research agenda. J. Fashion Mark. Manage. Int. J. **10**(3), 259–271 (2006). https://www.emerald.com/insight/content/doi/10.1108/13612020610679259/full/html
13. Wang, Y.: Fashion Supply Chain and Logistics Management. Routledge, New York (2019)
14. Turker, D., Altuntas, C.: Sustainable supply chain management in the fast fashion industry: an analysis of corporate reports. Eur. Manage. J. **32**(5), 837–849 (2014). https://www.sciencedirect.com/science/article/pii/S026323731400022X
15. Dahl, R.: Green washing. Environ. Health Perspect. **118**(6), 246 (2010)
16. Dach, L., Allmendinger, K.: Sustainability in corporate communications and its influence on consumer awareness and perceptions: a study of H&M and primark. Procedia Soc. Behav. Sci. **130**, 409–418 (2014). https://www.sciencedirect.com/science/article/pii/S1877042814029589
17. Campaign, C.C.: Hazardous workplaces: Making the Bangladesh Garment industry safe (2012). https://ecommons.cornell.edu/bitstream/handle/1813/102069/CCC_2012_Report_Hazardous_Workplaces.pdf?sequence=1
18. Hodal, K.: Abuse is daily reality for female garment workers for Gap and H&M, says report. The Guardian (2018). https://www.theguardian.com/global-development/2018/jun/05/female-garment-workers-gap-hm-south-asia
19. Gelici, N.: The grass is always greener on the other side – but who says green takes it all? An exploratory study on how greenwashing affects purchase intention, word-of-mouth and brand credibility. essay.utwente.nl (2019). http://purl.utwente.nl/essays/78267. Accessed 19 May 2021
20. The World bank: How Much Do Our Wardrobes Cost to the Environment? World Bank (2019). https://www.worldbank.org/en/news/feature/2019/09/23/costo-moda-medio-ambiente#:~:text=The%20fashion%20industry%20is%20responsible
21. Becker-Leifhold, C., Heuer, M.: Eco-friendly and Fair: Fast Fashion and Consumer Behaviour. Routledge, New York (2018)
22. Myles, A.J., Feudale, R.N., Liu, Y., Woody, N.A., Brown, S.D.: An introduction to decision tree modeling. J. Chemom. **18**(6), 275–285 (2004)
23. Wang, Y., Li, Y., Song, Y., Rong, X., Zhang, S.: Improvement of ID3 algorithm based on simplified information entropy and coordination degree. Algorithms **10**(4), 124 (2017)
24. Syed Nor, S.H., Ismail, S., Yap, B.W.: Personal bankruptcy prediction using decision tree model. J. Econ. Finan. Admin. Sci. **24**(47), 157–170 (2019)
25. Kingsford, C., Salzberg, S.L. What are decision trees? Nat. Biotechnol. **26**(9), 1011–1013 (2008). https://doi.org/10.1038/2Fnbt0908-1011. Accessed 13 June 2021
26. Anderson, D.R.: An Introduction to Management Science: Quantitative Approaches to Decision Making. Thomson/South-Western, Mason (2008)
27. Last, M., Maimon, O., Minkov, E.: Improving stability of decision trees. Int. J. Pattern Recogn. Artif. Intell. **16**(02), 145–159 (2002)
28. Peters, J.F., Skowron, A. (eds.): Transactions on Rough Sets III. LNCS, vol. 3400. Springer, Heidelberg (2005). https://doi.org/10.1007/b136502

29. Chiu, M.-H., Yu, Y.-R., Hao, L.: The Use of Facial Micro-Expression State and Tree-Forest Model for Predicting Conceptual-Conflict Based Conceptual Change Facial micro-expressions in science education View project Models and modelling in science learning View project (2016)
30. Osei-Bryson, K.-M.: Evaluation of decision trees: a multi-criteria approach. Comput. Oper. Res. **31**(11), 1933–1945 (2004)
31. Brodley, C.E., Utgoff, P.E.: Multivariate decision trees. Mach. Learn. **19**(1), 45–77 (1995)
32. Ross, B.C.: Mutual Information between discrete and continuous data sets. PLoS One **9**(2), e87357 (2014)
33. Kim, H.M., et al.: Synthesis, enzymatic inhibition, and cancer cell growth inhibition of novel δ-lactam-based histone deacetylase (HDAC) inhibitors. Bioorg. Med. Chem. Lett. **16**(15), 4068–4070 (2006)
34. Quinlan, J.R.: Learning decision tree classifiers. ACM Comput. Surv. **28**(1), 71–72 (1996)
35. Shaier, S.: ML Algorithms: One SD (σ)-Decision Trees Algorithms. Medium (2019). https://towardsdatascience.com/ml-algorithms-one-sd-%CF%83-decision-trees-algorithms-746e866ac3f. Accessed 20 May 2021
36. Song, Y.-Y., Lu, Y.: Decision tree methods: applications for classification and prediction. Shanghai Arch. Psychiatry **27**(2), 130–135 (2015). https://www.ncbi.nlm.nih.gov/pmc/articles/PMC4466856/
37. Shen, B.: Sustainable fashion supply chain: lessons from H&M. Sustainability **6**(9), 6236–6249 (2014). https://www.mdpi.com/2071-1050/6/9/6236
38. Patil, D.D., Wadhai, V.M., Gokhale, J.A.: Evaluation of decision tree pruning algorithms for complexity and classification accuracy. Int. J. Comput. Appl. **11**(2), 23–30 (2010)
39. Quinlan, J.R.: Simplifying decision trees. Int. J. Hum. Comput. Stud. **51**(2), 497–510 (1999)

Application of Virtual Simulation Technology in Welding Training

Yiqing Zhang(✉) and Yanyan Ren

Henan Polytechnic, Zhengzhou, China

Abstract. This paper studies the application of virtual welding simulation technology in welding training. As an important processing method, welding plays a very important role in the manufacturing industry. Due to the obvious disadvantages of high training cost, environmental pollution and health hazards in traditional welding training, the combination of welding training and virtual simulation technology can be used for welding teaching, which can train welding personnel in an economic, efficient, environmental friendly and safe way, obviously improve the shortcomings of traditional welding teaching.

Keywords: Virtual · Simulation · Welding · Training

1 Introduction

Welding refers to a process in which two or more materials are integrated by diffusion or combination between atoms or molecules, and the materials are combined by heating or pressurization [1]. In industry, welding can perfectly combine different materials or connect materials with great thickness difference. Welding manufacturing can significantly shorten the processing time and save materials. In the manufacturing of metal structure, the ratio of welded parts to riveted parts can save 15%–20% of metal materials. The self weight of the transportation equipment manufactured by welding is reduced, so as to improve the transportation efficiency. When manufacturing complex large-scale mechanical parts, it can be realized by assembly and welding. The composite process composed of welding, casting and forging can also be used to reduce the workload of casting and forging and reduce the cost. In addition, welding can realize the welding of different materials, optimize the design and save valuable materials. According to statistics, welding is a very important mechanical manufacturing process. More than 50% of all kinds of industrial products adopt one or more welding methods and connection technologies. Welding is called "industrial tailor". It plays an increasingly important role in the manufacturing industry. Welding plays a very important role in modern industrial products such as vehicles, ships, aerospace, petrochemical industry, electronics and appliances, atomic energy reactors and power stations, as well as major engineering construction such as buildings, bridges, urban subways, natural gas or petroleum pipelines

At present, with the rapid development of science and technology, welding robots have developed rapidly, their applications are more and more, and various welding

Y. Wang et al. (Eds.): IWAMA 2021, LNEE 880, pp. 661–666, 2022.
https://doi.org/10.1007/978-981-19-0572-8_84

automation methods are relatively mature. However, the traditional manual welding methods still occupy an important position in welding methods because of their convenience and flexibility, and the basic position of manual welding in welding training is still irreplaceable. During the welding process, welders should contact with electricity, combustible and explosive gases, flammable liquids and pressure vessels. Sometimes they need to work at heights or underwater, sometimes they need to work in special spaces such as closed containers, boilers, cabins, trenches and pipelines, and operate in high temperature and smoke environment. Not only the working environment is difficult, but also there is a certain risk, which may lead to explosion, fire Accidents such as electric shock, scalding, falling and drowning. Smoke, toxic gas, arc radiation, noise and other harmful factors in the welding process will damage the welder's eyes, skin and nervous system; When operating in a narrow working space for a long time with poor ventilation, the respiratory system will be damaged, which will not only endanger the safety and health of welders and production personnel, but also affect the smooth progress of production in case of welding explosion and fire, resulting in serious losses to national property. At present, most manual welding training adopts traditional methods. Practitioners have to face the adverse effects of various harmful factors, and will consume a lot of welding materials, solder and energy at the same time. Due to the above problems, it is urgent to use the combination of virtual simulation technology and traditional welding teaching to train welders.

Virtual simulation technology is the product of the combination of virtual reality technology and simulation technology. It is an advanced simulation technology based on the rapid development of network information technology, virtual reality technology, multimedia technology and other related technologies [2, 3]. Virtual simulation technology realizes a large number of entities, which are integrated and controlled through the virtual environment, so as to build a complete and unified virtual environment. As a typical feature, entities include digital models, simulators or other virtual simulation systems. In the virtual environment, the characteristics of the real and objective world are reflected through the mutual cooperation of entities or interaction with the virtual environment. Its integration, networking and virtualization meet the needs of the development of current simulation technology [4]. With the strong support of national policies and the rapid development of information technology, with the transformation and upgrading of manufacturing industry, welding technology is developing towards digitization, networking and intelligence, and virtual simulation welding technology has also been developed and applied rapidly.

Virtual welding simulation can simulate the welding process of Weldments by different welding methods according to different welding process parameters (workpiece size, welding method, welding current, voltage, speed, etc.) to produce different weld shapes, so as to make the weld forming process. Virtual welding simulation has the advantages of saving a lot of training costs and training time, enhancing training management, improving training effect and improving training environment [5]. The welding simulator system is generally equipped with the simulation training system of common welding methods such as gas shielded welding, electrode arc welding and argon arc welding. The simulation environment provided by the system has a high degree of simulation, provides a highly realistic environment for the practitioner, and allows the practitioner to feel the immersive welding environment and welding process, so as to effectively interact with the real environment, And high investment in the training process to achieve efficient training. For trainers who have welding ability and want to improve their skills, the virtual simulation system can provide an efficient training platform. After completing the weld, the operation information can be accurately measured, and the shortcomings, deficiencies and how to improve of the operator can be displayed, so that the operator can quickly understand their own shortcomings, master the improvement methods, and achieve simple and efficient training results, And translate the learned skills into the actual welding operation.

The welding virtual simulation simulator system has the following advantages [6, 7]: first, green, environmental protection and safety. The equipment is free from chemical pollution, strong light injury, high temperature injury and strong electromagnetic radiation, and the training environment is safe; (2) Three shots, one shot. The simulator is equipped with physical welding gun and welding wire, including gas shielded welding gun, electrode arc welding clamp, argon arc welding gun and welding wire; (3) The equipment has a good sense of touch and feedback. The weldments equipped with the simulator have touch force perception, and cooperate with the system's accurate spatial positioning technology to provide trainers with a feeling consistent with the real space; (4) 3D stereo vision welding simulation. The trainer can recognize the spatial position from the gun to the weldment in the three-dimensional picture by wearing the three-dimensional helmet, which is the same as that recognized by the naked eye, which greatly improves the authenticity of training; (5) "One to one" guided practice mode. Before welding training, practitioners can conduct "one-to-one" guided practice mode. In the virtual welding scene, practitioners only need to follow the "guided welding gun" to carry out relevant welding operations, which is similar to the "one-to-one" guidance given by teachers; (6) Adopt broadcast teaching mode. In teaching, teachers can synchronize the welding picture to the student machine to facilitate students' observation and learning. Students can also synchronize the content of the teacher machine to their own student machine to learn welding techniques in real time; (7) 3D welding video can be shared. The welding videos of practice can be shared through the network, and good welding videos can be downloaded or imported into the simulator for training comparison; (8) With synchronous auxiliary display function. During welding training, the system can provide graphical prompts of welding data during training, such as walking angle, speed, arc length, etc. students can adjust their posture in time according to the prompts to achieve the best training effect; (9) Reduce training costs. Virtual simulation training can save space occupation, no consumption of weldments and welding materials, no smoke, noise and visible light pollution, save cleaning cost and training investment cost. (10) The simulator can provide multi weldment and multi position welding. (11) Teacher side remote control equipment. The teacher side has the power to remotely manage any student training machine, such as assigning tasks, setting examination time period and other functions. (12) Welding knowledge can be queried at any time. The simulator provides the data of welding related knowledge, and can update and delete the data. (13) Tasks can be dynamically set, and students can conduct hierarchical training. (14) The training results can be evaluated. The teacher management terminal can synchronously manage multiple simulators and dynamically analyze training tasks. The assessment results are transmitted to the database, and the system automatically evaluates, records and prints them. Teachers can track and evaluate the complete welding training process of trainees.

The welding virtual simulation teacher management system can remotely realize the management of multiple training equipment, including the functions of practice, assessment and video recording of multiple training equipment, monitor the data and training status of training equipment, present the welding information on any student machine on the management system, and teachers can explain the teaching content to multiple students with the help of projection equipment, Students' data and operation contents can also be uploaded to the network.

At present, many colleges, welding majors and welder training institutions have chosen the welding training method of combining virtual simulation technology with practical operation [8]. For example, Jiangsu Jingjiang secondary vocational school uses VR technology in the teaching of welding specialty to effectively improve teachers' teaching quality and students' learning enthusiasm, so that students can change passive learning into active learning [9]. Wang Yizhong of Tianjin University proposed an arc welding training system based on virtual reality tactile technology. The system uses a tactile feedback device to further improve the coordination of operators [10]. In terms of numerical simulation, in 2014, Yu Xiangdong of Shanghai Maritime University analyzed the division of stress field and temperature field in the welding process of Q690 high-strength steel, and obtained the thermal cycle simulation result by using SYSWELD software, which is in line with the actual effect [11]. In the training of power grid steel structure welding operators, the simulation training system is applied to the MIG (MAG) welding skill training of power grid steel structure. By constantly exploring new welding training modes, the deficiencies in the traditional welding training are improved and the overall training effect of welding operators is improved [12].

Higher vocational colleges mainly cultivate practical talents, but from the teaching situation of welding specialty, welding training belongs to money burning specialty, which requires a large number of weldments and welding materials for repeated practice for many times, so as to effectively improve students' welding level. Due to the limitations of teaching conditions, poor welding working environment and boring welding practice, It often affects students' practical ability and operation level [13]. In the welding training, the students should first understand the main operation steps of welding and the environmental characteristics of the training site through the virtual simulation technology, and then carry out the training of welding parameters and welding techniques through the simulated welding operation. After repeated training for many times, the students will help to improve the operation quality in the practical operation [14]. After reaching the standard through the virtual simulation, they can carry out the practical operation, It can avoid the loss of personnel and property caused by initial welding and unfamiliar with the welding training site [15]. In addition, the equipment model in virtual simulation can be updated in time through the development of the industry, so as to closely connect teaching with the industry. Students in virtual simulation can also make bold attempts without worrying about the danger caused by equipment damage or improper operation, so as to cultivate students' creative thinking [16].

Using virtual reality technology to carry out the application research of welding training can improve the learning efficiency of employees, reduce the learning cost and reduce the risks brought by the welding learning process [17]. Taking the training of welders in schools or enterprises with 50 people as an example, the annual practical training is 4 weeks, and the training using virtual welding simulation equipment can save 80000 yuan of consumables and tools, and reduce environmental pollution. Enterprises can use welding simulation technology to apply virtual welding simulation to product optimization and production, reduce the use of welding operation, reduce enterprise production costs and improve enterprise safety from the aspects of materials, construction period and personnel.

With the increasing progress of society and the rapid development of economy, there is an increasing shortage of welding workers. Whether welding professional training in Higher Vocational Colleges or welding personnel training in enterprises, the combination of virtual simulation technology and welding training has obvious advantages. It can train welding personnel in an economic, efficient, environmental and safe teaching method, which has realistic economic, social and environmental benefits.

References

1. Chen, Z.: Welding engineer's Manual, p. 97. China Machine Press (2002)
2. Ayan, A.: Research on virtual teaching application of VR technology. Donghua University (2017)
3. Xiao, S.A.: Application of VR technology in exhibition design. Shandong Academy of Arts and crafts (2016)
4. Gengyi, C.: Social road of virtual reality. China Informatization **6**, 16–17 (2015)
5. Shu, Z., Weihong, L.: Application of virtual reality welding training simulation system in welder training. Welding **1**, 64–66 (2015)
6. Shengchang, L., Yimin, Z., Hai, J.: Research on manual arc welding operation simulation training system. Sci. Technol. Innov. Guide **13**, 89 (2009)
7. Jun, H., Liu, X., Xu, S.: Research progress of virtual welding simulation system. Comput. Modernization **6**, 40–44 (2011)
8. Yi, L., Tao, J.: Development of virtual laboratory teaching platform for metal welding course in petroleum colleges and universities. Pet. Educ. **03**, 37–40 (2013)
9. Yun, P.: Application of VR technology in welding teaching classroom. New Media Appl. **11**, 105 (2018)
10. Gong, Z.: Application and research of welding simulation based on virtual reality. Master's thesis of University of Electronic Science and technology (2018)
11. Wang, Y., Nan, Z., Chen, Y.: Study on welder training by means of haptic guidance and virtual reality for arc welding, pp. 954–958 (2006)
12. Li, Y., Qu, B., Han, M., et al.: Application of welding simulation training system. J. State Grid Inst. Technol. **20**(6), 79–81 (2017)
13. Wang, C.: Safe operation technology for welding and thermal cutting. China Labor and Social Security Press (2011)
14. Xianwei, M., Yulong, X., Yutong, T., et al.: Research status and application of welding intelligence. Electr. Weld. Mach. **49**(09), 84–87 (2019)
15. Wang, L.: Introduction of simulation technology into welding practice teaching. Today Keyuan, **20**, 279–279 (2009)

16. Chen, Y., Yifeng, X., Jinwen, Q.: Teaching reform of virtual welding experiment under the background of new engineering. Electr. Weld. Mach. **49**(12), 117–120 (2019)
17. Min, B., Jie, B., Yang, Q.: Research on the construction of industrial robot virtual simulation training center. Vocat. Technol. **16**(8), 121–122 (2017)

Mobile Robot System Based on Intelligent Inspection of Substation

Shiying Xue(✉)

Henan Polytechnic, Zhengzhou, China

Abstract. Substation inspection is an important way to ensure the normal operation of substation equipment. The use of mobile robots to replace or assist labor is to improve the efficiency and safety of substation inspections. The completion of inspection tasks is of great significance to the realization of unattended substations suction line in a laboratory, which proved the feasibility of the method.

Keywords: Mobile robot · Substation · Unmanned inspection · Automatic control

1 Introduction

The safe and stable operation of electric power has important and far-reaching significance to guarantee our daily life and the stability of the country and society. In recent years, the electric power industry has developed rapidly in the direction of large capacity and ultra-high voltage to ensure the stable and normal operation of power equipment and the safety of the entire power system. Operation has become a core issue in power management. As the core of the entire power supply network, substations play an important role in power dispatch and continuous power supply. Ensuring the safe and stable operation of the substation system has become the key to ensuring continuous power supply [1].

Traditional substation inspections use manual inspections. There are high labor intensity, low inspection efficiency, and inconsistent inspection standards in the inspection process. In addition, they encounter harsh environments such as rain and snow, which increase labor. Difficulties in inspections [2, 3]. In addition, relying on the experience and perception of inspectors to make simple judgments, incomplete and inaccurate judgments are prone to appear, laying hidden dangers to the normal and stable operation of substation equipment. Therefore, robot inspections instead of manual inspections came into being [4, 5]. It avoids all kinds of inspection difficulties, inspection efficiency, and inspection standards during manual inspections, and is of positive significance for ensuring the safe and stable operation of substation equipment.

2 Substation Inspection Robot

In February 2012, the Shenyang Institute of Automation of the Chinese Academy of Sciences developed the country's first rail-mounted substation inspection robot in two

Y. Wang et al. (Eds.): IWAMA 2021, LNEE 880, pp. 667–674, 2022.
https://doi.org/10.1007/978-981-19-0572-8_85

years. It was successfully commissioned in the 220 kV Wangtie substation in Anshan, Liaoning. All-weather, all-regional equipment inspection [6].

In November 2012, Huituo Intelligent Machine Co., Ltd. officially put the self-developed substation intelligent inspection robot into use at the 110 kV Niuzhai substation in Zhengzhou. The robot can analyze images and videos of outdoor switch positions and meter readings. Autonomously judge the normal operation status and abnormal status of the substation equipment and give an alarm for abnormal conditions [7]. As shown in Fig. 1.

Fig. 1. Huituo intelligent robot

In January 2014, the substation inspection robot developed by Zhejiang Guozi Robot Technology Co., Ltd. was put into operation in Ruian Substation [9] Substation personnel only need to remotely operate on a computer or tablet, and the robot can inspect the substation equipment according to the well-equipped inspection strategy, and stop taking pictures and uploading in the area that needs to be photographed or inspected. As shown in Fig 2.

Fig. 2. Zhejiang Guozi robot

In October 2014, with its technical strength, Langchi Robotics successfully developed a 1000 kV UHV substation inspection robot, and successfully operated it in the country's third UHV line 1000 kV Zhebei-Fuzhou UHV AC transmission and transformation project [8]. It has realized the safe and stable operation in all-weather, all-climate, and high-pressure environments of all levels, and escorted the safe operation and maintenance of substations. As shown in Fig. 3.

In China, with the clarification of the power robot market, more and more manufacturers have invested in the development of substation inspection robots, which has greatly promoted the advancement of substation inspection robots for autonomous movement, intelligent detection, analysis and early warning. According to the National Grid Corporation of China's 2015–2020 smart grid planning guidelines, in 2018, we will carry out pilot applications of miniaturized and tooled robots, and build a substation intelligent robot inspection information management platform; in 2020, we will comprehensively promote miniaturized and tooled robots. All the system transformation operation and maintenance teams are configured [10].

3 The Overall Structure of the Mobile Robot for Substation Inspection

The substation mobile robot uses its own sensors to detect substation equipment, and sends the equipment detection data to the substation personnel. If it encounters equipment abnormalities, it will promptly alarm and report the situation to the substation personnel. No matter how severe the cold or the heat, the robot can tolerate the harsh environment, and the robot can stick to the post of equipment inspection, thus reducing the labor intensity of substation personnel, increasing the reliability of equipment inspection, and realizing the inspection of substation equipment Intelligent and automated.

According to the inspection requirements for the inspection of substation equipment in the Regulations of the Regulations for the Condition Inspection and Test of Power Transmission and Transformation Equipment, it can be seen that the inspection personnel of the substation need to check the appearance, switch position, The inspection cycle of meter readings, equipment temperature, is relatively short compared to full inspection [11].

It can be seen from Table 1 that the overall technical requirements for daily inspections of outdoor equipment in substations are not very high. However, due to the wide distribution and large number of equipment in substations, it is more suitable to use intelligent mobile robots to replace or assist manual inspections.

The substation inspection mobile robot integrates infrared detection devices, camera devices, pickups and other sensors, and integrates navigation technology, electromagnetic compatibility technology, robot vision technology, wireless transmission technology and other cutting-edge intelligent technologies, so that mobile robots can pass The detection data obtained by its own sensors analyzes the appearance, temperature, sound, etc. of the power equipment, and determines whether the equipment has internal thermal defects, external mechanical damage or electrical problems and other safety hazards and abnormal conditions, and timely reports the hidden hazards and failure precursors The data is transmitted to the substation staff, and the inspection robot replaces the operation and maintenance personnel to conduct on-site inspections of the substation equipment, which can not only reduce the work intensity of the on-site inspectors and reduce the risk of personal injury, but also timely feedback the equipment information to the management system. Improve the reliability, safety and stability of substation management.

Fig. 3. Langchi robot

4 Substation Inspection Mobile Robot System Function

As shown in Fig. 4, the intelligent substation inspection robot integrates robots, power equipment non-contact detection, multi-sensor fusion, pattern recognition and other technologies.

It can perform round-the-clock, all-round, and fully autonomous inspection and monitoring of substation equipment, which can be effective Improve the efficiency of inspection work and the intelligent level of substation management. The advanced detection technology applied by intelligent inspection robots ensures the safe and stable operation of the substation system and promotes the process of unattended substations.

According to its tasks and functions, the substation inspection robot system has at least the following functions:

(1) The robot can perceive the information of its environment, can determine its position in the substation by reading the information of the positioning module, and can determine its own status and the status of tasks performed.
(2) By processing navigation information, it can identify the distance and angle relationship between the robot body and the desired path, intersection turning information, etc.; through efficient path planning algorithms and accurate and effective navigation control, etc., to ensure that the least consumption in the shortest time The resource completes the inspection task.
(3) It can control the inspection robot to complete the action, track the route, maintain a straight line, turn at the intersection, stop at a specific location, perform specific inspection tasks, etc.; it can detect the execution of the robot's movement in real time, including whether the specified action is completed, Whether there is an error or accident during the execution of the action, etc.

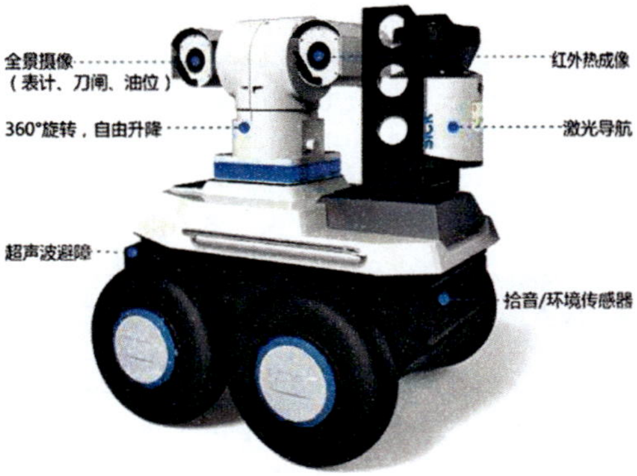

Fig. 4. Mobile robot for substation inspection

(4) It can obtain the information of detection devices such as pan-tilt, camera, and supplementary light equipment, and realize the action control of these detection devices to achieve the desired posture and capture the clearer image of the target device.
(5) With remote control function, the inspection robot can receive commands from the monitoring center and be remotely controlled by the monitoring center.
(6) Preliminary judgments can be made on the test results, and emergency measures can be taken when the equipment has a serious failure. During the inspection process, if the robot finds that the temperature of a certain substation equipment exceeds the set maximum temperature, the robot should have emergency alarm capability, and can suspend the current inspection task to monitor the faulty equipment at a fixed point.

5 The Overall Structure of the Mobile Robot for Substation Inspection

The substation inspection staff pre-set the scope of inspection equipment and inspection content according to the "Technical Specifications of Intelligent Robot Inspection System for Substations", and then send the equipment inspection task through the base station monitoring center of the mobile robot system, and the mobile robot receives the base station monitoring After the inspection task information issued by the center, it walks out of the charging room and starts to inspect the equipment; the mobile robot automatically plans an inspection route suitable for the inspection task according to the inspection task involving the equipment inspection stop.

When inspecting equipment, the mobile inspection robot first needs to accurately position the inspection equipment, and control the pan-tilt to adjust to a proper posture through the pan-tilt control system, and then perform visible light or infrared imaging detection on the inspected equipment. In this process, the mobile inspection robot saves and uploads the acquired visible light image and infrared image in real time. In the process of equipment inspection, if it is detected that the temperature of the equipment being inspected is too high and exceeds the maximum temperature value preset by the system, the mobile inspection robot should send an alarm message to the staff of the base station monitoring center. After the inspection of one equipment inspection stop is completed, the base station monitoring center sends the "equipment inspection completed" instruction to the mobile inspection robot, and then the mobile inspection robot continues to perform inspections at the next equipment inspection stop.

5.1 Mobile Robot Body Module

According to the mechanical structure, the mobile station inspection robot body module can be divided into several parts such as the driving mechanism, the chassis of the vehicle body, the charging mechanism and the outer shell.

(1) Drive mechanism

In view of the substation environment where the patrol robot performs patrol inspection, the patrol robot is designed as a four-wheeled mobile trolley. The front two wheels are driving wheels, which are driven by two servo DC motors. The driving wheels are used to support and assist driving and steering. The car body of this structure has many advantages: the structure is very simple, the machinability is good; the movement ability is good, it can smoothly perform straight and turn, and it can also perform 360° rotation around the middle driving wheel, which is strong in the complex environment of the substation.

(2) Car body chassis

The chassis of the car body is essentially a steel plate that carries the weight of the car body components. The bottom of the car body chassis is equipped with a fan and an opening, which solves the problem of heat dissipation when the inspection robot performs inspection tasks.

(3) Charging mechanism

The inspection robot can realize autonomous charging of the inspection robot through the charging mechanism. During autonomous charging, the DC motor of the inspection robot drives the rack to reciprocate back and forth to drive the charging electrode to move. When the charging electrode moves to a certain posture, the inspection robot stops at the cabin. There is a charging pole plate with a certain height from the ground inside, which can realize the charging of the inspection robot.

(4) Shell

The outer shell of the inspection robot can be divided into three parts: the front decorative panel, the middle main shell and the rear shell. The decorative panel can be easily removed. The middle main shell is used to protect the various mechanisms and components installed inside the robot. The rear shell can be easily removed from the rear of the robot to provide staff for the installation and maintenance of the internal components of the robot. convenient.

5.2 Mobile Robot Control and Detection Module

The control detection module can be divided into 4 parts: power management module, navigation and positioning module, motion control module and equipment detection module.

(1) Power management module

The power module is the power source of the entire inspection robot system, including two parts: power supply battery and charging system. Commonly used batteries include lithium batteries, lead storage batteries, nickel-cadmium batteries, etc. Considering that the robot has a long working time and needs to be recharged, and taking into account the cost of various batteries, the inspection robot chooses lead storage battery as the power supply of the system. The inspection robot is normal. When working, the power supply battery provides energy, and the working current can reach 5–10 A.

(2) Navigation and positioning module

The function of the navigation and positioning module is to realize the navigation and positioning of the inspection robot based on the environmental information of the location of the robot collected by each sensor, and provide information for the motion control of the inspection robot. The information collected by various sensors in the navigation and positioning module includes self-positioning, turning range measurement, climbing measurement, robot inclination measurement, and so on. Therefore, the inspection robot is equipped with sensors such as ultrasonic sensors, gyroscopes, photoelectric encoders, etc., which are used to detect the distance from the front-end obstacles, the posture of the vehicle body, and the speed of the motor.

(3) Motion control module

The motion control module is the mechanism that controls the inspection robot to achieve the expected motion, which mainly includes CPU controller, motor driver, photoelectric encoder, memory chip, sensor, interface conversion circuit and so on.

(4) Equipment detection module

The equipment detection module is composed of visible light CCD, infrared imager, pan-tilt and other devices. After the inspection robot arrives at the destination, it will automatically adjust the angle of the moving pan-tilt to the best, and then adjust the focal length of the visible light camera. Finally, the inspection robot will start to capture the visible light and infrared images of the equipment we pay attention to, and will capture The acquired image is transmitted to the base station system.

6 Concluding

The mobile robot system based on intelligent substation inspection includes the main body and control module. It uses its own sensors to obtain equipment inspection information and realizes the interaction between the robot and the on-duty management platform. It is more helpful while ensuring the completion of the substation inspection task. Real-time understanding of the operation of the substation, and immediate measures to be taken in the face of failures and emergencies, are of far-reaching significance for the realization of substation automation, intelligent management and the stable operation of substation equipment.

References

1. Zhang, W., Liang, J., Qin, J.: Multi-agent based substation robot inspection remote centralized control system. Electric Power Inf. Commun. Technol. **2020**(12) (2020)
2. Yang, X., Wang, Z.: Line power distribution. Hubei Electric Power **2021**(01) (2021)
3. Li, Y.: Research on path planning of substation inspection robot. Decis. Explor. (Part 2) **2020**(02) (2020)
4. Li, J., Chen, X.: Discussion on the technical specifications of the inspection of the substation wheeled inspection robot. Autom. Appl. **2020**(02) (2020)
5. Zhang, Y.: The application of intelligent inspection robot in 500kV substation. Agric. Staff **2020**(08) (2020)
6. Wang, G.: Functional design and application of remote control module based on intelligent inspection robot for substation. Technol. Innov. Appl. **2020**(21) (2020)
7. Hu, L., Yuan, S., Jing, K.: Discussion on the technical specifications for inspection of substation wheeled inspection robots. Inf. Record. Mater. **2020**(08) (2020)
8. Yang, X., Gong, X., Li, M., Liang, X.: Substation inspection robot instead of manual evaluation plan. Robot Ind. **2020**(06) (2020)
9. Zhong, Z., Zhang, C., Liu, J.: A design of an intelligent lighting system that cooperates with substation inspection robots. Sci. Technol. Wind **2019**(31) (2019)
10. Wang, K.: Research and Implementation of Key Technologies and Systems for Substation Inspection Robots. University of Electronic Science and Technology of China (2016)
11. Tan, F.: Research and Application of Inspection Robots for Unattended Substations. Changsha University of Science and Technology (2016)

Application of Mechatronics Technology in Intelligent Manufacturing

Kunyang Cao(✉)

Department of Mechanical and Electrical Engineering, Henan Polytechnic, Zhengzhou, China

Abstract. With the development and progress of science and technology of the times, the application of mechatronics technology in various industries has become more and more extensive. The application of mechanical and electrical integration technology into intelligent manufacturing can effectively improve production efficiency, reduce labor consumption, lower production costs, improve processing accuracy, enhance enterprise competitiveness, and promote the healthy development of intelligent manufacturing industry.

Keywords: Mechatronics · Intelligent manufacturing · Automated production · Robotics

1 Introduction

At present, the application of mechatronics technology in the world is more and more extensive, and many countries apply this technology to all areas of life, particularly in the manufacturing sector and application control automation is more significant [1]. At the same time, as the world's information industry continues to deepen, our country's traditional industries have also been effectively promoted, and the traditional manufacturing industry has gradually developed in the direction of intelligence, systemization, and standardization.

2 Mechatronics Technology Analysis

Mechatronics technology is mainly an advanced technology that integrates sensor technology, intelligent technology, and information technology. Mechatronics technology has better control capabilities, higher sensitivity, and stronger data processing capabilities. In the application process, it realizes the comprehensive processing of various data information, which not only ensures the safety and reliability of data processing, but also has high processing efficiency. The application of mechatronics technology in intelligent manufacturing can not only break through the bottleneck of traditional information processing, strengthen the exchange of information, but also discover the hidden safety hazards and manufacturing problems in intelligent manufacturing in time, and make targeted improvements and perfections. In addition, mechatronics technology has more

Y. Wang et al. (Eds.): IWAMA 2021, LNEE 880, pp. 675–678, 2022.
https://doi.org/10.1007/978-981-19-0572-8_86

precise control of linear and non-linearity, which is the key to ensuring the stable operation of manufacturing systems and control systems [2]. Mechatronics technology has strong model advantages, especially its modular design allows its functions to be more comprehensive, and it has strong transmission and processing capabilities. Applying it to intelligent manufacturing can be designed in combination with specific application conditions, which has a significant effect on improving the efficiency of intelligent manufacturing.

3 Overview of Intelligent Manufacturing

As far as industrial development is concerned, the field of intelligent manufacturing is mainly based on intelligent manufacturing technology, focusing on flexible and efficient processing of the actions and contents of smart devices. With the continuous advancement of science and technology, intelligent robots have been widely used in the field of industrial intelligent manufacturing. This is undoubtedly a major technological breakthrough for mechatronics technology. The current industrial production and manufacturing mainly concentrate on emerging technologies such as artificial intelligence technology and mechatronics technology, and major breakthroughs have been made in product production quality and production efficiency [3]. Judging from the current development status, intelligent industrial production robots have many application advantages, mainly in terms of information and materials, and engaged in production, which can basically achieve the expected production effect.

Intelligent manufacturing is embodied in the application of automatic production lines and intelligent manufacturing systems. Among them, the automated production line is mainly based on mechatronics technology, through the combination of programmable control devices, to achieve man-machine interface control and photoelectric control functions. The intelligent manufacturing system is mainly aimed at the human-machine intelligent integration system. Taking the basic production line as an example, the printing and packaging part is mainly composed of human experts and intelligent robots; in terms of tool selection, the computer system is the main focus; in the original production process, the production is mainly made by the decision-making function of the computer intelligent system. In the actual application process of intelligent manufacturing technology, operators are mainly used to complete the analysis and decision-making of the manufacturing process with the aid of a computer simulation system [4]. In this process, the staff can use the computer system equipment to complete the analysis of the production process under the premise of meeting the daily production volume.

4 Research on the Application of Mechatronics Technology in Intelligent Manufacturing

4.1 Application of Industrial Intelligent Robot

In intelligent manufacturing, industrial intelligent robots are a key component of highly automated production by replacing labor with machines. Intelligent robot is a kind of automated mechanical equipment, which realizes automated production by imitating certain parts of the human body and through the control of the control system. The research and development of intelligent robots is inseparable from the application of mechanical technology, microelectronic technology, computer technology, sensor detection technology and automatic control and other mechatronics technologies. The main body, control system and drive system are the three main components of the industrial robot. The main body mainly contains the actuator and the base; the realization of the control system requires the input of the program to issue instruction signals to control the actuator and drive system; The drive system is the part that executes the action, which is composed of transmission and power devices. The control system plays the role of the brain in the industrial intelligent robot, which is the key to its automatic production, and mechatronics technology plays a key role in it. Through computer programming, the robot can realize automatic and intelligent operations such as welding, painting, assembling, cutting, packaging and palletizing, and handling, which brings great convenience to the development of production [5].

4.2 Application of Sensing Technology

Sensing technology is the core of mechatronics technology. The application of mechatronics technology in intelligent manufacturing can combine sensor technology with intelligent manufacturing technology. In order to improve the sensitivity of sensor technology in intelligent manufacturing, it is also necessary to establish a sensor network system, so that the sensor technology is not interfered by external signals, realize the accurate collection and transmission of the target signal, and send the information to the computer system, so as to issue instructions and form intelligent and automated control of industrial production. In the application of sensor technology in smart manufacturing, smart devices may be affected by the interference of sensor technology information. Therefore, it is necessary to effectively combine sensor network systems in smart manufacturing systems and connect computer systems to achieve rapid information acquisition and analysis. At present, under the current situation of the combined application of mechatronics technology and intelligent manufacturing, fiber optic cables are mostly used for sensing equipment, the sensitivity and accuracy of sensing technology can be better guaranteed, and the design is more scientific and reasonable, which is conducive to improving the level of intelligent manufacturing technology and systems, and providing more development advantages for intelligent manufacturing.

4.3 Application of Electronic Numerical Control Technology

During the construction of intelligent mechatronics, the application of numerical control technology played an important role in the construction of the entire industry, furthering the modernization of the industry, enhancing the production efficiency of the entire industry, and improving product quality. Numerical control technology can create the entire manufacturing process into a proportional model, and realize the overall process of model construction through the control of electronic equipment. Through the application of electronic information and electronic induction technology, the equipment that has problems in the manufacturing process can be restored to the original state through the control of the program. This also has important guiding significance for the maintenance of intelligent machinery and equipment.

4.4 Application of Automated Production Technology

Automated production is the performance of mechatronics technology in specific production activities, which can effectively save costs and help improve overall production efficiency and product quality. At present, the more common automated production lines, such as printing and packaging, electronic product production and assembly, etc., have greatly improved production efficiency, while also effectively reducing production costs, helping enterprises to obtain better production and economic benefits [6]. The application of automated production technology in the intelligent manufacturing process can combine relevant elements such as computers, production lines, numerical control technology and warehouse management to form a complete production system in combination with actual production. On this basis, the integrated remote intelligent control is carried out, and scientific adjustment and optimization are carried out according to the production needs, and finally a good production effect of intelligent manufacturing is achieved.

5 Conclusion

In summary, the application of mechatronics technology overcomes a series of drawbacks in the traditional industrial production process, and can provide technical advantages for intelligent manufacturing automated production line technology, industrial robot production and research and development, etc. It has been widely used in large-scale production in manufacturing enterprises, and has achieved good application effects. On the basis of understanding the application background of mechatronics technology, the organic combination of multiple technologies and the scientific promotion of specialization is an important content for realizing the efficient development of machinery manufacturing enterprises and an important means to promote the innovation and upgrading of industrial production technology.

References

1. Yun, P.: Application of VR technology in welding teaching classroom. New Media Appl. **11**, 105 (2018)
2. Yi, L., Tao, J.: Development of virtual laboratory teaching platform for metal welding course in petroleum universities. Petrol. Educ. **03**, 37–40 (2013)
3. Chen, Y., Yifeng, X., Jinwen, Q.: Teaching reform of virtual welding experiment under the background of new engineering. Electric Weld. Mach. **49**(12), 117–120 (2019)
4. Bao, M., Bao, J., Qiao, Y.: Research on the construction of industrial robot virtual simulation training center. Vocat. Technol. **16**(8), 121–122 (2017)
5. Li, Y., Qu, B., Han, M., et al.: Application of welding simulation training system. J. State Grid Inst. Technol. **20**(6), 79–81 (2017)
6. Xianwei, M., Yulong, X., Yutong, T., et al.: Research status and application of intelligent welding. Electric Weld. Mach. **49**(09), 84–87 (2019)

Maintenance Method of Spindle Non Rotation Fault of FANUC NC Machine Tool

Xueping Chu and Yiqing Zhang(✉)

Henan Polytechnic, Zhengzhou, China

Abstract. Aiming at the non rotation fault of FANUC 0id serial spindle without alarm, this paper starts with the spindle control principle, analyzes the conditions to be met for spindle rotation, puts forward two maintenance methods: key parameter maintenance method and control principle maintenance method, and illustrates how to use these two maintenance methods with examples. It provides a simple, fast and effective fault maintenance method for maintenance personnel.

Keywords: Serial spindle · Spindle does not rotate · Fault repair

1 Introduction

The spindle of the CNC machine tool drives the tool or the workpiece to rotate, forming the main movement in the cutting motion, and is one of the important parts of the machine tool. The CNC system has a self-diagnosis function. When a fault occurs, the system will give a corresponding alarm. This is actually the CNC system pre-estimated some failures. However, in actual production, there are often failures without alarms, and it is difficult to repair such failures. For the spindle of a CNC machine tool, it often encounters a failure that does not rotate but does not give an alarm. To repair this fault, on the one hand, the maintenance personnel are required to be very clear about the spindle control principle, on the other hand, the maintenance method is required to be appropriate. There are many maintenance methods, such as the maintenance method of "one look, two questions and three thoughts" [1], but each method is based on the spindle control principle and the application of the control principle. Therefore, starting from the FANUC 0id spindle control principle, this paper analyzes the maintenance method of spindle non rotation without alarm fault.

2 The Principle of Fanuc 0id Serial Spindle Control

FANUC 0iD has many spindle control functions, mainly including steering control, speed control, orientation control, positioning control, CS axis control, gear shift control, etc. The realization of each control function requires the spindle to rotate normally, so the spindle rotation failure is of obvious importance. Due to its many functions, the FANUC 0iD spindle control principle is somewhat complicated. The control principle is shown in Fig. 1.

Y. Wang et al. (Eds.): IWAMA 2021, LNEE 880, pp. 679–683, 2022.
https://doi.org/10.1007/978-981-19-0572-8_87

As shown in Fig. 1, the operator inputs program commands such as "S1000 M03" to the NC system through the system panel, where s command specifies the speed of the spindle and M command specifies the steering, stop, orientation and other auxiliary functions of the spindle. Therefore, in Fig. 1, the M command realizes the spindle steering control and the s command realizes the spindle speed control.

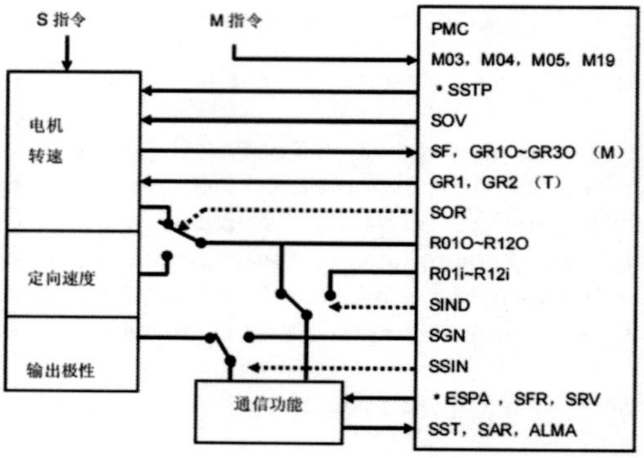

Fig. 1. The principle of FANUC 0iD spindle control [2]

2.1 Steering Control Principle

When the system reads the M instruction, such as M03, CNC outputs the data 03 after M to PMC specific code register F10 in the form of binary code. Because PMC can only process signals, the M code must be output to PMC in the form of signal MF. Of course, this requires the M code delay time set by the system parameters. When PMC receives this signal, it means that at this time, the M command is input to PMC, then the M command decoding control is executed, and the decoding result is input to CNC, and then the spindle motor is driven through the spindle amplifier through the serial digital spindle interface of CNC, and the forward rotation of the main shaft is realized through the spindle transmission system. The spindle encoder detects and feeds back the spindle speed to the CNC. When the fed back spindle speed is the same as the command set speed, the CNC judges that the execution of the M command is over, notifies the PMC, and the PMC executes the M command end control, feeds back the result to the CNC, and the CNC automatically executes the next command.

2.2 Speed Control Principle

There are two control methods for the speed of FANUC 0id serial spindle: one is that the CNC directly executes the command, but the serial communication needs to be carried out between the spindle amplifier and the CNC to exchange speed and control signals [2]. One is to control the spindle speed with 12 bit binary code through PMC.

As shown in Fig. 1, when the system reads the s command, when the spindle stop signal * SSTP is 1 and receives the magnification signal input by PMC, the system automatically calculates the actual spindle speed according to the spindle speed parameter setting value and magnification value. This speed value is passed through the serial spindle interface and drives the spindle motor through the amplifier. At the same time, CNC will output the actual speed value of the motor to PMC, which is convenient for users to query.The execution process of S instruction is shown in Fig. 2.

If PMC inputs shift or orientation request to CNC, that is, SOR signal in Fig. 1 is 1, the speed of spindle motor is determined by the setting value of orientation speed parameter in Fig. 1. When the SOR signal is 0 after the shift or directional execution, the spindle will automatically return to the previous speed. The execution process of spindle orientation or shift speed is shown in Fig. 3.

If PMC controls the spindle speed and outputs the control signal SIND is 1, the speed of the spindle motor is controlled by the 12 bit binary value input by PMC.

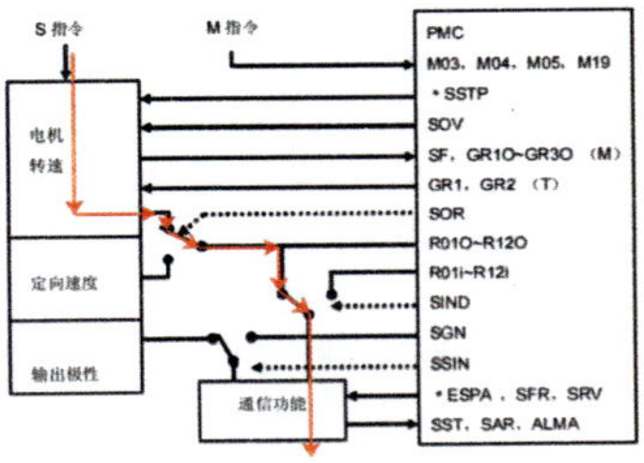

Fig. 2. The execution process of S instruction

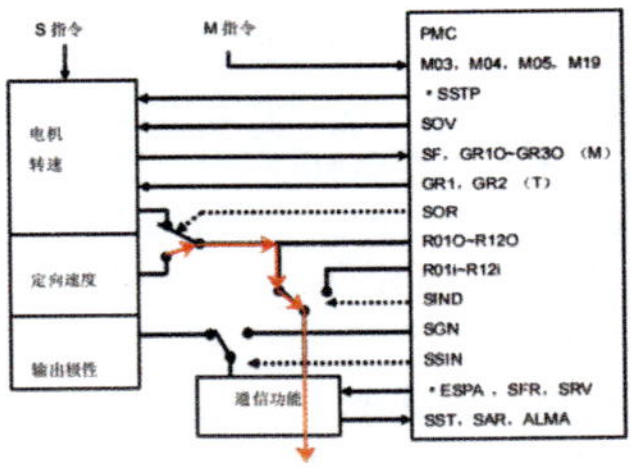

Fig. 3. The execution process of spindle orientation

3 Maintenance Method of Spindle Non Rotation Fault

It can be seen from the above that the spindle rotates normally, the correct spindle speed parameters must be set, the CNC sends the correct spindle control signal and the hardware part is normal. The spindle control signal needs to be exchanged between CNC and PMC, which is determined by the correct PMC control program. The hardware part requires that the system, spindle interface, amplifier, encoder and motor are correctly connected according to the instructions, and the mechanical transmission part and spindle itself are normal. Therefore, the spindle control is complex, so a fast and effective maintenance method is particularly needed.

3.1 Key Signal Method

The key signals of spindle control include serial spindle ready signal g70.7, spindle emergency stop signal g71.1, spindle reverse signal g70.4, forward rotation signal g70.5 and spindle stop signal G29.6. Using these key signals for fault maintenance is a good method. If the command is issued, but the spindle does not rotate, you can enter the PMC signal viewing interface to view the status of these signals. The fault point is the abnormal signal status. If the signal status is abnormal, open the PMC control program to check whether the program has errors. After modification, restart the system to eliminate the fault.

FANUC system spindle amplifier LED lamp adopts different display results to indicate which of the above key signals is abnormal.

If the LED light shows [01] and the yellow light is on, it means that the spindle is not ready. Check g70.7 and g71.1; If the LED light displays [--], indicating that the spindle steering signal is not sent, check g70.4 and g70.5; If the LED light shows [00], it indicates that the turn signal has been sent, but the spindle is not in the start state, check g29.6. [3].

Using the key signal method to repair faults is actually not that simple.

Example 1 for a FANUC 0id three-axis machining center, the spindle cannot rotate after the command is issued.

Using the above key signal method, observe the LED display [01] and check the signal g71.1. Check that g71.1 is 0, indicating that the machine tool spindle is in emergency stop state. The cause of the emergency stop of the main shaft shall be determined in combination with the electrical diagram. Different machine tool manufacturers have different electrical diagrams. Pay attention to the corresponding relationship between the actual electrical components in the electrical diagram and the corresponding signal state in the PMC control program. Therefore, the key signal method provides a method for maintainers to quickly determine the fault point, and the maintenance should be completed in combination with the electrical diagram.

Example 2 for a FANUC 0id three-axis machining center, the spindle cannot rotate after the command is issued.

In a numerical control maintenance skill competition, a contestant's equipment spindle did not rotate. It is observed that the above key signals are normal, but the spindle does not rotate, and the machine tool cannot do any action. Checked the electrical diagram, no problem. In case of such failure, pay attention to the phenomenon that all actions of the machine tool cannot be executed. It is observed that the machine tool does not have any alarm, but when you see the running program, after pressing the cycle start button, the cursor of the program always stays at the beginning of the program. It indicates that the system is in reset state. After checking the system reset signal g8.7, it is found that the signal is 1, which indicates that the PMC control program is wrong. After modifying and restarting PMC, the machine tool spindle can rotate normally.

3.2 Maintenance Method Using Control Principle

According to the spindle control principle, after the spindle steering is determined, the spindle must have a speed before it can rotate. According to the analysis in Fig. 1, there are two methods to control the spindle speed, but both need to set reasonable parameter values. CNC will notify PMC of the actual speed of the spindle with a 12 bit binary value. If the spindle does not rotate, check this value first. If this value is 0, it means that either the speed value is not entered or the value of the spindle parameter is not set. The specific reason depends on whether the machine tool runs m and s commands for the first time after startup. Because even if the parameter setting is correct, when the M command is run for the first time after startup, if the s value is not entered, the spindle speed is 0.

Example 3 for a FANUC 0id three-axis machining center, in the manual continuous feeding mode, press the "positive" button on the machine panel.

When the "rotate" button is pressed, the spindle does not rotate and there is no alarm information. After inquiry, the operator appears when manually operating the spindle forward rotation after startup. Since the value of S is not entered after startup, the value of S is 0 by default. Manual operation when the SIND signal is 0, the spindle speed is executed by CNC, so the spindle speed cannot be input in this case, so the spindle cannot rotate.

4 Conclusion

According to the spindle control principle, this paper puts forward two maintenance methods: key signal method and using control principle method. The key signal method requires the maintainer to be very familiar with each key signal, and then combined with the LED status display of the spindle to determine the fault. Using the control principle requires the maintainer to fully understand the spindle control principle. In fact, the application of these two methods is not simple and requires high maintenance personnel. However, these two methods are generally two feasible rapid maintenance methods.

References

1. Xuepingchu: Maintenance method of spindle non rotation fault of FANUC NC machine tool. Hebei Agric. Mach. **3**, 66–67 (2013)
2. Beijing FANUC Mechanical & Electrical Co., Ltd, Fanuc connection debugging tutorial (0i-d), p. 546 (2011)
3. Beijing FANUC Mechanical & Electrical Co., Ltd, FANUC CNC maintenance and adjustment(0i-D), p. 198 (2011)

Implementation of a High-Performance Multi-threaded Media Player Module for a Cross-Platform Game Engine Based on OpenGL

Hua Li[1](✉) and Hai-yan Chen[2]

[1] School of Information Engineering, Lingnan Normal University, Zhanjiang 524048, Guangdong, China

[2] Northeastern University, Qinhuangdao Branch, Qinhuangdao 066004, China

Abstract. Game engines based on OpenGL and its extensions occupies an important share in game development. With the development of the game industry, the sound and picture rendering effect and picture fluency in large-scale games have increasingly higher requirements. Inserting media playback into the main thread may cause the screen to freeze and the game experience to drop due to blocking. This paper proposes a high-performance media player module based on multi-threading, which aims to optimize the sound and picture scenes of the game, improve the stability of the game operation, and create higher-quality games to create greater economic benefits.

Keywords: Game engine · Multithreading · Media player · OpenGL · DMA transfer

With the rapid development of 5G technology in China and the popularity of WiFi6 protocols, the demand for electronic entertainment products will multiply. While the number of personal computers in China will increase further and performance will continue to increase. Games based on OpenGL and its related expansion of the game engine have large amount of security and wide coverage, and with the development of image technology, the degree of improvement it brings to the experiencer in the senses decreases year by year, which requires game developers to pay more attention to the interaction between the game plot and the player's emotions. In-game video, music and other media resources can bridge the plot, create an atmosphere and give players an immersive feeling. So the game engine media resources modular optimization has a high commercial value.

1 Media Playback in the Game

After years of development, the media resources in the game have gone through a process from nothing to nothing. Noww the "Witch 3: The Hunt" "Assassin's Creed: Origin" as the representative of the large-scale click game soundtrack, and video accounted for a

Y. Wang et al. (Eds.): IWAMA 2021, LNEE 880, pp. 684–690, 2022.
https://doi.org/10.1007/978-981-19-0572-8_88

higher and higher proportion. Game developers are increasingly paying attention to the global immersion they bring. At present, the media system in the game engine and the overall coupling is too high, while the media playback directly inserted into the main thread may be due to the loss of media files, audio and video decoding a large number of CPU resources caused by the game unresponsive and screen Caton. In addition, media files inserted into different parts of the game after different scenes, it is difficult to make holistic modifications. Therefore, how to achieve more efficient and perfect media playback has become an important topic in the game industry.

2 The Media System is Modular

2.1 Introduction to Modular Ideas

Modular design is the combination of certain elements of the entire system to form a subsysyscies with specific functions. Program design in functional blocks can effectively reduce program complexity and make program design, debugging and maintenance operations simple. In the game engine, the separation of media resources from the engine itself for modular processing, can effectively overcome the original coupling of the two defects, at the same time can establish a standardized workflow for unified production, maintenance, so as to improve work efficiency, reduce costs, produce more economic benefits.

2.2 The Design of the Module

2.2.1 Multithreaded Audio and Video Decoding Policy

FFmpeg is an open source computer program that can record, convert and convert digital audio and video. It uses an LGPL or GPL license to provide a complete solution for recording, converting, and streaming audio and video. FFmpeg was developed under the Linux platform, but it can also be compiled and run in other operating system environments, including Windows, Mac OS X, and so on. It is a complete cross-platform open source solution for audio and video development for recording conversions and steaming audio and video, which can be efficiently coded, decoded and deprecation.

The module uses a multi-threaded, Golang-based and Goroutine-based synth programming mode that avoids the thread jam when audio and video is decoded and played. In C++, the same process can consist of multiple threads, run in parallel, and perform different tasks, while assigning load to multiple CPU cores improves the smoothness of the game when multimedia is playing. On this basis, Goruntine, as a Co-program of the Go language, can run on multiple C++ threads at the same time, and the Go language implements memory sharing between these Goruntines, while executing Goruntine requires very little stack memory and is easy to use efficient and light weight.

2.2.2 The DMA Transport Strategy is Used in the Pixel Buffer

Pixel Cache (PBO) can quickly pass pixel data over the graphics card through DMA without affecting the CPU's time period (i.e., without interruption), while it also has asynchronous DMA transmission, the original image can be loaded directly into the PBO and the PBO is controlled by OpenGL, although the central processor is involved in loading textures to PBO, but does not involve pixel data transfer from the PBO to the texture object work, but by the GPU is responsible for PBO to texture object data transfer, in addition, OpenGL can also schedule later asynchronous DMA transmission, that is, the CPU does not have to wait for the pixel data transfer, can continue other work at the same time.

2.2.3 Audio and Video Queue Synchronization Policy

Queues, as a special linear table, are only allowed to be deleted at the front end of the table, inserted at the back end of the table, in this module, audio and video data is stored separately in two buffer queues and inserted into both queues at the end of the queue after de-re-use, when the game When the engine calls a media resource, only the header data in the video queue is bound to the texture, while the audio team list header data is synchronized with the video by the trigger of the listener.

3 The Implementation of the Module

3.1 Media System

The implementation of the media system is shown in Fig. 1:

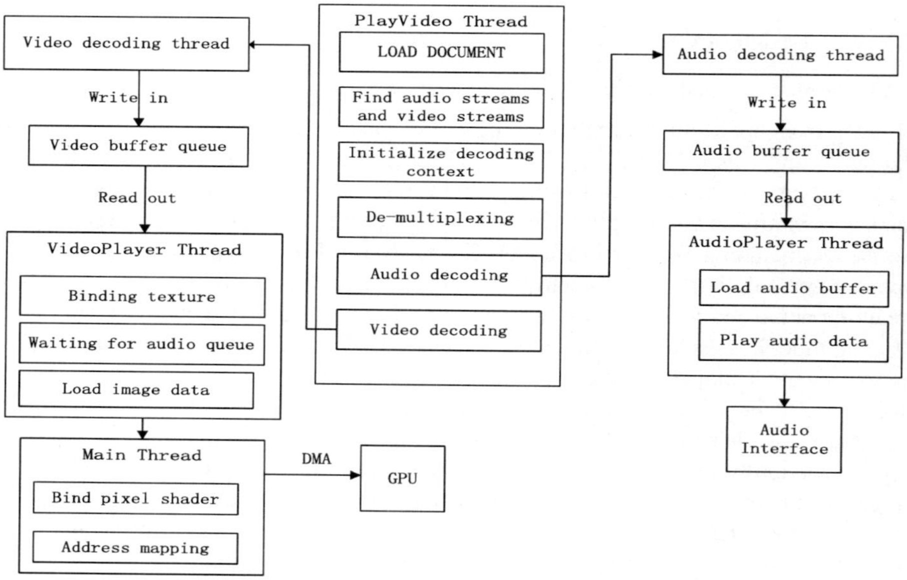

Fig. 1. Media system working model

When you play media resources using the game engine, Media first turns on the PlayVideo thread, where media files are loaded, audio and video streams are found, media files are deprecation, audio streams, and video streams are decoded. The thread separates the heavy decoding task from the main thread, and the on-demand loading decoding of audio and video can be achieved well with multiple threads.

Decoded data is immediately written to audio and video queues, which are designed to decode audio and video on demand, and to re-use memory space to save memory space, while the queue enables synchronization and mutual exclusion of thread reads and writes.

Audio playback threads and video playback threads read data from their queues, data writes and reads are implemented through the queue and thread-safe, audio data reading is implemented by playback functions encapsulated by each operating system, and video data reading is synchronized in the video thread.

After the video data is read, the data is saved to the texture, which is passed in through the pointer, the texture data main thread is shared with the playback thread, and the texture is updated using the DMA transport using OpenGL's pixel buffer through the main thread's Update module.

3.2 Buffer Queue

Thread model for media system buffer queues:

- Pictured 2 Shown: In order to make efficient use of memory, Free queues and active queues are used inside the buffer queue to re-use memory, and when data is written, the buffer queue takes free blocks out of the idle queue, writes data into the active block, and when the data is read, the buffer queue inserts the number of free blocks from the active block. According to the block, the efficient use of memory is realized (Fig. 2).

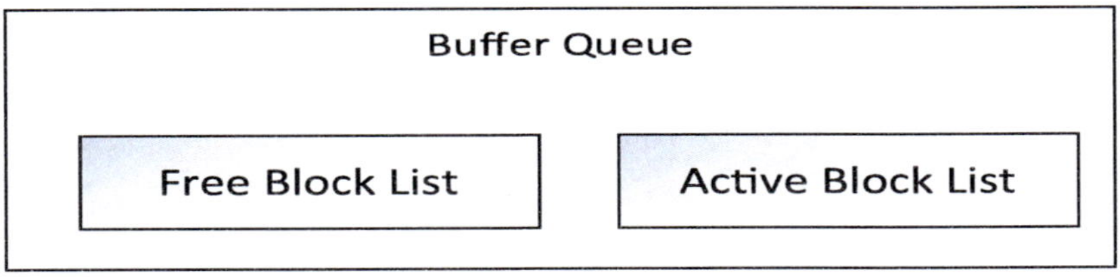

Fig. 2. Buffering the queue structure

- Pictured 3 Shown: The game engine sets the maximum capacity that can be accommodated in the buffer and the maximum capacity that triggers writes, and when data is written, the queue deposits the data and determines whether the total capacity of the current buffer exceeds the maximum capacity that can be accommodated if the data in the queue exceeds the buffer setting Maximum total capacity, blocking the decoding thread. When the playback thread reads the data, the queue determines if there is data and reads if there is data, otherwise the playback thread is blocked. When data is written, the buffer attempts to wake up the playback thread. When reading data, if the total data in the buffer is less than the one that triggered the write Maximum total capacity, wake up the decoding thread to start writing data, and determine whether the data is finished playing, and if the playback is complete, wait for the data to finish reading (Fig. 3).

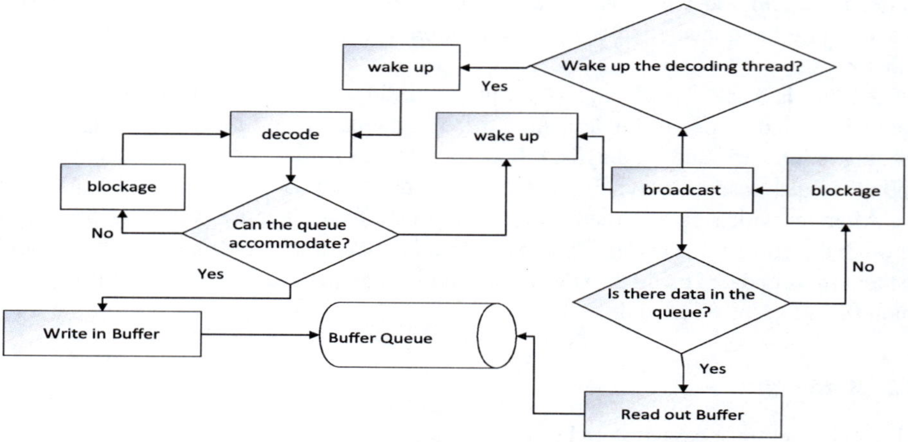

Fig. 3. Buffer queue thread model

3.3 The Engine Uses the Module

The game engine's call to the module looks like Fig. 4:

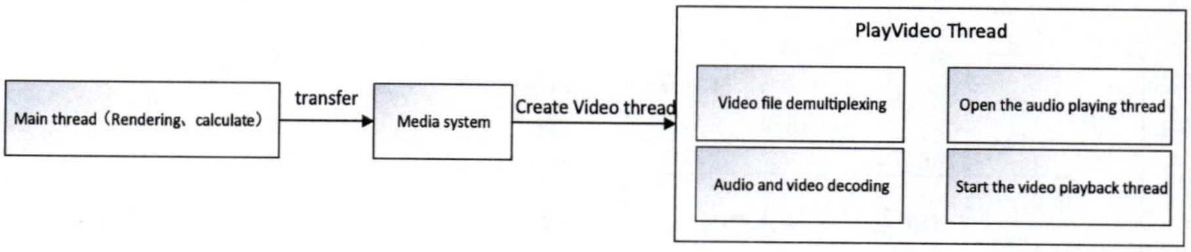

Fig. 4. The use of the game engine for modules

The media system encapsulates the internal implementation through the Media class, and the game engine can create different Media subsyses to play media resources throughout the game through simple class inheritance, as shown in Fig. 5, where each specific media playback creates a corresponding sub-class and Media to form a generalization relationship, while the implementation of the BindTextture() function in the sub-class requires the use of texture type as a method parameter by interface. Figure 6 is an example of a game engine successfully using a multimedia playback module to enable simultaneous playback of video on a cube in a game scene by binding the same texture on multiple sides.

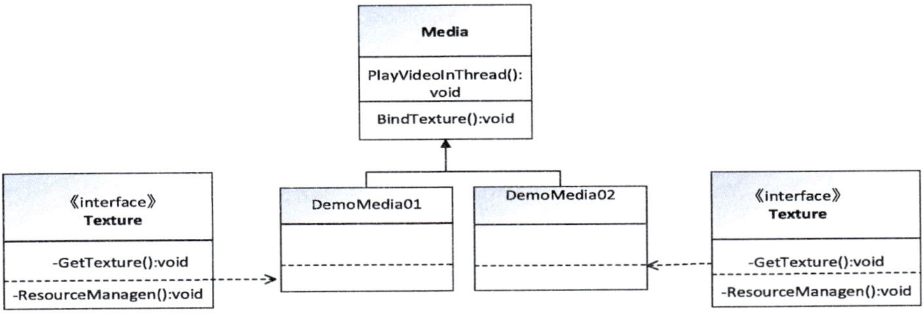

Fig. 5. Through the media class implements module functionality

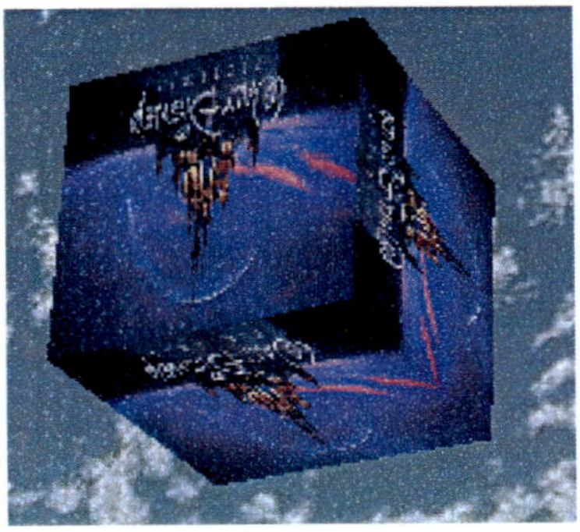

Fig. 6. Video playback in a test scene

4 Conclusion

Based on multithreaded audio-video decoding strategy, DMA transmission strategy in OpenGL pixel buffer, audio-video queue synchronization strategy, this paper proposes and implements an efficient audio-video playback module. Firstly, the framework and development environment used in the system design are introduced, and then the implementation of the core module of the media system is discussed, including: media system, the package of buffer queue, the use of the engine for the module, and finally, the playback is tested. The results show that the system can greatly improve the efficiency of media playback and screen smoothness, free up CPU computing resources, and realize the function of more efficient media file playback in the game. Due to the continuous innovation and development of game technology, the multimedia module in the game engine proposed in this paper can be further expanded and perfected to cope with more complex game scenes.

References

1. Chen, K.: Design and Implementation of 3D Game Engines. Zhejiang University (2007)

2. Cong, H., Zhou, S., Tang, L.: FFMPEG-based cross-platform video codec research. J. Wuhan Univ. Technol. **33**(11), 139–142 (2011)
3. Zhangshun, W., Zhang, X.: FFmpeg-based video encoding storage research and implementation. J. Hangzhou Univ. Electron. Sci. Technol. **03**, 30–34 (2006)
4. Xu, T.: Research on The Problem of Co-issues in Go Language. Beijing University of Posts and Telecommunications (2019)
5. Yan, J., Ning, W.: DMA high-speed data transmission system based on PCI bus. J. Univ. Electron. Sci. Technol. **05**, 858–861 (2007)
6. Dong, C.: Audio-visual Synchronization Research and Realization. Jilin University (2007)
7. Liu, L., Bian, J., Zhang, L., Sen, M.: Audio-visual synchronization based on FFPEG decoding. Comput. Eng. Des. **34**(06), 2087–2092 (2013)
8. Zhu, Y., Chen, Y.: Talk about OpenGL-based texture mapping techniques. **23**(03), 70–71 (2009)
9. Chen, L.: 3D Technology - Research and Implementation of Real-Time Graphic Rendering of Large-Scale Scenes. University of Electronic Science and Technology (2005)
10. Wang, E., Pei, Y., He, Y.: 3D video enables multiple video simultaneous playback. Comput. Syst. Appl. **20**(08), 208–211 (2011)

A Review of Vision-Based Robot Gripping System Application Research

Zhixiao Zhou, Chen Wang(✉), Chao Liu, Xiufeng Zhang, Yu Tang, and Xuxiang Lu

Department of Mechanical Engineering, Hubei University of Automotive Technology, Shiyan 442002, China
20090011@huat.edu.cn

Abstract. In response to the current problem that industrial robots are not intelligent enough to adapt to complex production environments, the introduction of the increasingly developed machine vision technology into the field of industrial robots provides favourable conditions for the development of intelligent manufacturing technology. The research results at this stage are reviewed for the three main steps of camera calibration, identification and matching of target work pieces, and optimization of the robot's grasping trajectory for the target work pieces in the industrial robot grasping system on the production line. The research results show that the vision grasping system is technically mature and can meet the real-time requirements in industrial applications, and the research and improvement of the algorithms of each part have certain reference value for the development of industry and related research.

Keywords: Part sorting · Machine vision

1 Introduction

Vision is an important source for people to obtain information from the outside world, and statistics surface that 80% of the amount of information that humans obtain from the outside world comes from vision [1]. Machine vision is a branch of artificial intelligence is developing rapidly, generally refers to the use of machines to replace the human eye, the formation of visual capabilities, the world that the human eye can see into the world that the machine can perceive, after this conversion, the machine can be well aware of the objective world, measurement, transformation, etc.

In traditional sorting, manual sorting is generally used, because the worker's operation will have the problem of fatigue, for long time operation, obviously the efficiency is not satisfied. Even if an industrial robot is used for sorting, because traditional industrial robots are usually taught or programmed offline, the robot's position at the target point is fixed and each action is repetitive, so it has poor adaptability to changes in the processing object and working environment. Combining machine vision with sorting robots can greatly reduce the disadvantages of poor stability and subjectivity caused by enterprises' reliance on manual operation, reduce labour costs, and effectively improve the automation and intelligence of production equipment [2].

Y. Wang et al. (Eds.): IWAMA 2021, LNEE 880, pp. 691–696, 2022.
https://doi.org/10.1007/978-981-19-0572-8_89

2 Status of Research

As the manufacturing level of various sensing devices continues to improve and vision algorithms continue to mature, a lot of research work has been carried out by domestic and foreign scholars for vision robot grasping technology.

2.1 Status of Foreign Research

Foreign industrial technology developed earlier, and the United States, Japan, and European countries conducted research on vision robots in the 1970s. Vision-based robots were first studied by Shirai et al. at the University of the Ryukyus in Japan, who proposed an open-loop-based vision robot control system in 1973. The position of the target was first obtained through machine vision, and then the robot was controlled to grasp it, forming the earliest vision robot system [3]. With the continuous development of science and technology, the computing power of computers has been significantly improved, especially in industrial operations, such as the vision-based potato grasping system developed by Suez Canal University [4], and the robot handling system developed by German VMT that can position and adjust the position of the target [5], etc., which improves the accuracy rate while reducing the labor intensity, especially for some manual operation of dangerous complex work, which is of great importance.

2.2 Current Status of Domestic Research

With the rise of concepts such as "Industry 4.0" and "Made in China 2025", domestic research on machine vision technology has developed rapidly and is gradually narrowing the gap with foreign countries. Liling GAO et al. [6] achieved object localization and measurement in 3D space using binocular vision model by improving the colour similarity-based segmentation method in HSV space. Genzu YAN et al. [7] developed an efficient and intelligent food sorting system based on the professional image processing software Sherlock combined with a high-speed parallel robot. Haifeng FANG et al. [8] used a DELTA parallel robot and an industrial camera as the based on the DELTA parallel robot and industrial camera, combined with robot motion control technology and composite detection technology to design a parallel robot-based recyclable waste sorting system. The process is shown in Fig. 1.

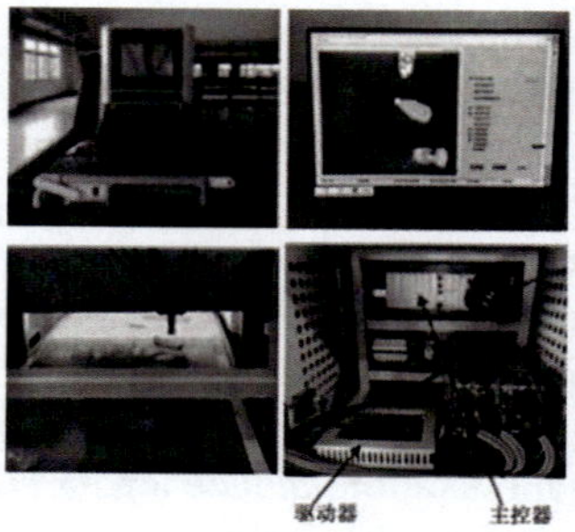

Fig. 1. Waste sorting system

3 Key Technologies for Vision Robot Positioning and Grasping

3.1 Camera Calibration

The design goal of a computer vision system is to estimate the position, shape, etc. of an object in three-dimensional space from the images captured by the camera. Camera calibration is to obtain camera parameters through experiments and calculations to establish a mapping relationship between the pixels coordinates of an object point in the camera and the three-dimensional coordinates of that point in the world coordinates [9]. The methods of camera calibration are: self-calibration method, active vision method and Zhang Zhengyou calibration method. With the development of manufacturing technology, the manufacture of bionic compound eyes that simulate the structure of insect compound eyes has become possible, and bionic compound eye imaging systems [10] have come into being, and some calibration methods based on bionic compound eye cameras have been continuously updated and the calibration accuracy has been improved.

Recently, Zeqiang YUAN et al. [11] designed and fabricated a 17-eye bionic surface compound eye prototype with a total viewing angle of more than 180°, and completed the but target localization of 17 words and 38 pairs of inter-adjacent subeyes in the compound eye system by the Zhang Zhengyou calibration method, based on which, based on the nature of a large overlapping field between the adjacent subeyes of the compound eye, an external reference calibration method for non-adjacent subeyes was proposed Based on this, a non-adjacent subeye calibration method is proposed based on the nature of the large overlapping field between the adjacent subeyes, which can achieve high precision large field of view target localization (Fig. 2).

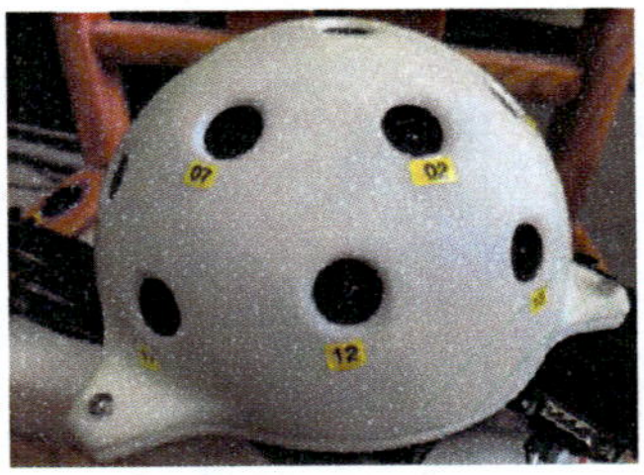

Fig. 2. Ocular bionic curved compound eye prototype

3.2 Target Identification Matching

The effect of image content analysis depends heavily on the effect of image segmentation, and its accuracy directly determines the accuracy of image recognition and the robustness of the system at a later stage. In recent years, as vision robots work with higher and higher accuracy, the working conditions are also gradually complicated. Therefore, the requirements for image processing are getting higher and higher, and the traditional algorithms are subject to poor segmentation results. Many scholars began to study hybrid image segmentation techniques to obtain the optimal image analysis effect by optimizing the algorithm. Chengpeng ZHU et al. [12] proposed a workpiece image recognition method based on ORB algorithm. The ORB algorithm is combined with the Surf descriptor to obtain the image corner points with rotational scale invariance to improve the efficiency and accuracy of image recognition.

3.3 Deep Learning Based Target Recognition

With the development of artificial intelligence, the current deep learning-based image segmentation and recognition methods are gradually becoming the mainstream image processing methods. The task of multi-target image segmentation and recognition can be solved effectively in a large neural network model. Convolution neural network is mainly a neural network used to process images. The difference with other network structures is that convolution neural network adds convolution and pooling layers on top of the most basic neural network structure. Therefore, the structure of convolution neural network can be roughly divided into four parts: input layer, convolution layer, pooling layer, and output layer. Input layer: The input layer is mainly responsible for inputting the target information, which specifically contains the information of the image, the information of the pixel, and so on (Fig. 3).

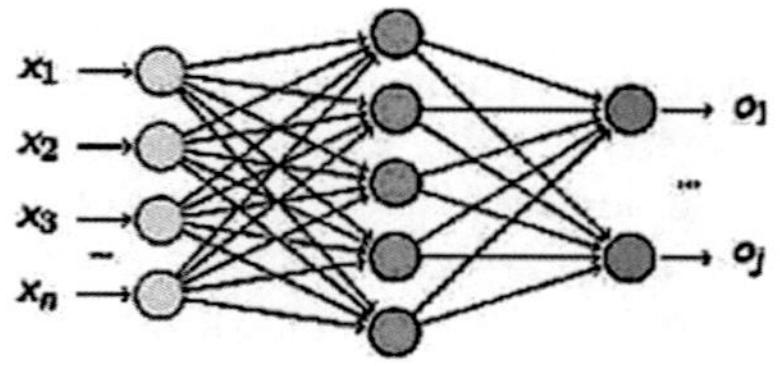

Fig. 3. Convolutional neural networks

With the continuous improvement of computer performance, the hardware for high performance computing has been improved, which has improved the computing power and computing speed to a certain extent, and the existing high performance computers can complete the large-scale matrix operations in deep learning. Therefore, deep learning has been developed rapidly during this period, and various algorithmic models for deep learning have been proposed. Jiang ZHU [13] et al. introduced an attention mechanism in the feature extraction network of Faster R-CNN to map the weights of different positions of the crankshaft shingle cover image to the feature channels, so that the deep learning model can pay more attention to the edge and center semantic information, and successfully apply this algorithm to the visual localization and target detection of the loading robot.

3.4 Crawl Track Planning

After completing the part identification and positioning, the robot needs to plan the path of grasping parts to achieve fast sorting of the robot. Let f be a set of moving sequences, $v_1, v_2, \ldots, v_n$ be the n target parts, and T_d be the robot sorting path under this moving sequence:

$$f = (v_1, v_2, \ldots, v_n) \tag{1}$$

$$T_d = \sum_{i=1}^{n-1} d(v_i, v_{i+1}) + d(v_n, v_1) \quad (2)$$

Where $d\ (v_i, v_{i+1})$ is the distance from part v_i to part v_{i+1}. It can be seen that this model can be reduced to the travel sale problem. The travel merchant problem can be described as a merchant who has to go to several cities and can only pass through each city once and eventually return to the city of origin. How the route should be chosen so that the total route length is the shortest. Many scholars have studied the TSP solution algorithm, and because of the limitations of a single intelligent optimization algorithm, such algorithms mostly use a combination of several algorithms or the improvement of the operator of the original algorithm. Changyou WU et al. [14] designed a new crossover operator to solve the TSP based on the brainstorming algorithm, which has high solution accuracy and is not easy to fall into local optimality. Wenzhuang LU et al. [15] designed an improved ant colony algorithm to solve the TSP, which uses the particle swarm algorithm to crossover and mutate the better solutions generated by the ant colony algorithm to obtain a higher quality population.

4 Concluding Remarks

This paper mainly analyzes the current research hotspots of vision industrial robots and gives some examples related to machine vision sorting technology. At present, vision industrial robots are developing very rapidly, and the corresponding hardware and software technologies are becoming more and more mature, driving the technological progress of the industry and bringing more development potential and opportunities for related application fields. In the future, high-precision and high-speed industrial robots will run to the assembly line, restricted working space and other scenarios, driving traditional industries to stride towards digitalization and intelligence, providing newer and stronger impetus for the improvement of production efficiency and the change of production methods.

References

1. Szpytko, J., Hyla, P.: Workspace supervising system for material handling devices with machine vision assistance. J. Konbin **11–12**(1), 7–16 (2009)
2. Wang, L., Pan, R.R., Zhou, J., Gao, W.: Advances in the application of machine vision in textile intelligence. Cotton Textile Technol. 1–3 (2021)
3. Wang, C.: Research on robot arm grasping system based on machine vision. China University of Mining and Technology (2019)
4. Elmasry, G., Cubero, S., Molto, E., et al.: In-line sorting of irregular potatoes by using automated computer-based machine vision system. J. Food Eng. **112**(1–2), 60–68 (2012)
5. Li, J.Y., Yang, C., Wang, J., et al.: Vision-based positioning of robot handling system. Manuf. Autom. **033**(004), 40–42 (2011)
6. Gao, L.: Binocular stereo vision for object recognition and localization. Beijing Jiaotong University (2012)

7. Yan, Z., Li, M., Xu, K., et al.: Research on machine vision technology for high-speed robot sorting system. Pack. Food Mach. **1748**(001), 28–31 (2014)
8. Fang, H., Xu, K., Wan, X., Kai, W.: Design of parallel robot-based recyclable waste sorting system. Comb. Mach. Tools Autom. Process. Technol **2020**(11), 24–27+32 (2020)
9. Chen, W.Y., Xu, J., Yang, F.: A camera calibration method using dual neural networks. Infr. Laser Eng. 1–10 (2021)
10. Luo, J.: Research on multifocal bionic compound eye. Chongqing University (2018)
11. Yuan, Z., Gu, Y., Qiu, S., Zhang, X.: Calibration and target localization of multi-camera type bionic surface compound eye. J. Photon. 1–13 (2021)
12. Zhong, P.C., Li, W., Liu, J.H.: A workpiece image recognition method based on improved ORB algorithm. Mach. Tools Hydraul. **48**(21), 12–16 (2020)
13. Zhu, J., Du, R., Li, J., Cai, M., Xu, H.: An attention mechanism-based vision localization and detection method for crankshaft shingle cover loading robot. J. Instrum. **42**(05), 140–150 (2021)
14. Wu, C., Fu, X.: An agglomerative greedy brain storm optimization algorithm for solving the TSP. IEEE Access **8**, 201606–201621 (2020)
15. Lv, W.Z., Cao, J.Y., Dang, M.Z., Pei, Y.X.: Application of improved ant colony algorithm in path planning of welding robots. Mech. Des. Res. **35**(06), 32–36 (2019)

Research on Tool Wear State Monitoring Method Based on Feature Processing

Liu Chao[1](✉), Wang Chen[1,2](✉), Zhang Xiufeng[1], Lu Xuxiang[1], and Tang Yu[1]

[1] Hubei University of Automotive Technology, Shiyan 442002, China
{201911038,20090011}@huat.edu.cn
[2] Shanghai Key Laboratory of Intelligent Manufacturing and Robotics, Shanghai 200072, China

Abstract. In order to accurately monitor the tool wear degree in milling process, the tool wear state is monitored based on feature processing. Firstly, the force, vibration and acoustic emission signals were collected in the processing process, and the multi-dimensional information of the serial signals was obtained by analyzing the time domain, frequency domain and wavelet packet. Then, the Spearman coefficient was used to obtain the feature weight coefficient and extract the features strongly correlated with the average rear tool surface wear to carry out feature dimension reduction. Finally, KNN and ANN are respectively used to identify the tool wear state. By comparing the loss function, accuracy and confusion matrix, it is found that ANN model can accurately identify the tool wear state, and the recognition accuracy and generalization degree have been improved to some extent.

Keywords: Tool · Feature processing · Spearman coefficient · KNN · ANN

1 Introduction

Tool state monitoring is an important part of machine tool state monitoring in machining process. With the continuous processing, tool wear and even fracture will increase the cost of tool, affect the surface quality of the workpiece, reduce production efficiency and even affect the operation of machining system, resulting in significant losses [1]. Therefore, it is of great practical significance to monitor tool condition.

According to different measurement methods, tool condition monitoring methods can be divided into direct measurement method and indirect measurement method. The direct method usually uses the resistance measurement method, the optical measurement method, the ray measurement method and so on, obtains the tool wear state directly [2]; The indirect measurement method, the vibration signal, cutting force signal, power signal, acoustic emission signal and multi-information fusion signal are obtained by the sensor installed on the machine tool. After data from pretreatment and feature extraction, the classification identification model is input to identify the tool wear state [3]. In contrast, indirect measurement method is widely used because of its low cost, convenient measurement and less interference to the machining process.

Y. Wang et al. (Eds.): IWAMA 2021, LNEE 880, pp. 697–702, 2022.
https://doi.org/10.1007/978-981-19-0572-8_90

Traditional data classification models include K-nearest Neighbor (KNN) [4], Random Forests (RF) [5], Support Vector Machine (SVM) [6] and other Machine learning models. These machine learning models are limited by the quality of features extracted and require a certain amount of experience and skill of the researcher.

In recent years, with the continuous development of deep learning technology, its application in the field of tool condition monitoring is more and more extensive, and corresponding achievements have been achieved [7–9]. Since the essence of tool information is timing signal, without data preprocessing, deep learning cannot achieve good learning effect. Tool signals are small sample data, so too many layers of deep learning classification network will cause overfitting.

In this paper, the tool wear state is monitored and studied based on feature processing. The second section is the data pretreatment part, the third section uses KNN and ANN respectively to identify the tool state, and the fourth section summarizes the content of this paper and prospects for the future tool state recognition technology.

2 Data Preprocessing

2.1 Data Set Description and Partitioning

PHM2010 public data set is used for experimental analysis, and model training and validation of the proposed method are carried out based on the data set. PHM2010 data set is derived from the open data of the 2010 High Speed CNC Tool Health Prediction Competition held by the Society for Forecasting and Health Management (PHM) in New York, USA [10]. The experimental conditions are shown in Table 1.

Table 1. The experimental conditions of PHM2010

The hardware conditions	Model parameters	Cutting condition	Parameter
CNC milling machine	RÖders Tech RFM 760	Spindle speed (r/min)	10400
Force sensor	Kistler three way dynamometer	Feed (mm/min)	1555
Vibration sensor	Kistler triaxial vibration sensor	Depth of axial cutting (mm)	0.2
Acoustic emission sensor	Kistler acoustic emission sensor	Radial cutting width (mm)	0.125
Amplifier	Kistler charge amplifier	Sampling frequency (kHz)	50
Number of mining equipment	NI DAQ PCI 1200	Milling mode	Down milling
Wear measuring equipment	LEICA MZ12	Cooling mode	Dry cutting
Tool	Carbide knife with ball head	The workpiece material	Inconel718

The test was repeated for 6 times under the above cutting conditions. The shape of end milling is square, the length of the end milling is 108 mm each time, and the time of each tool running is equal. After each tool running, the wear amount of the tool's rear tool surface is measured. The milling force signals in three directions (X, Y, Z) and the root mean square (AE-RMS) values of vibration signals and acoustic emission signals in three directions (X, Y, Z) were collected for each cutter. The wear of three milling cutter blades was measured off-line with a microscope in three experiments. At the last stage of the experiment, the experiment was terminated when the tool wear was so serious that it could not work. Finally, 315 complete cutting times were obtained for each tool, and 315 effective records were obtained.

PHM2010 data set contains 6 sets of data from c1-c6, among which c1, c4 and c6 are training sets, and c2, c3 and c5 are test sets. In this paper, c1, c4 and c6 experimental data sets with wear values are selected, and the one-leave-one method is adopted to cross-verify the model. c1 and c6 are taken as training sets and c4 as verification sets, and new data use cases are created and represented as C1, C6 and C4 respectively.

2.2 Data Set Description and Partitioning

This paper to extract the 12 characteristics in the time domain and 4 characteristics in the frequency domain,at the same time, the original data is decomposed by 4-layer wavelet transform in the time-frequency domain, and the signal is decomposed into 8 frequency bands with different center frequencies, and the energy of each band (E1–E8) is calculated. Therefore, 24 features of different dimensions can be extracted from each sensor channel.

Spearman correlation coefficient [10] was used to determine the features of a high degree of correlation with the mean tool wear before model training. Spearman correlation coefficient is a statistical method which is used to evaluate two variables linear or nonlinear correlation,so it can reflect the monotonic relationship between two continuous variables, and it was usually as the method of feature selection in machine learning, Correlation analysis was made between multidimensional feature data and tool wear values, and correlation degree threshold was set to 0.95. The multidimensional features with higher mean value of tool wear were screened out respectively, as shown in Table 2.

Table 2. The experimental conditions of PHM2010

Signal types		The time domain	Time-frequency domain
The force signal	X	Mean, RMS, RA	E_5, E_7
	Y	Mean, RMS, RA	E_5, E_6
	Z	–	E_4, E_5, E_7
Vibration signal	X	Mean, RA	E_4
	Y	Mean, RA	E_4
	Z	–	E_4

Mean, RMS and RA represent the Mean value, root Mean square and root amplitude of the characteristic signal respectively. The tool wear state is divided into three stages: initial wear, normal wear and sharp wear, as shown in Table 3. In order to facilitate the training and testing of the model, the one-Hot coding of the wear value of the rear tool surface was transformed into the tool wear state.

Table 3. Tool wear status labels

The wear state	Number of milling	Tags
The early wear and tear	1–30	0
Normal wear and tear	31–210	1
Severe wear	211–315	2

3 Tool State Recognition Based on KNN and ANN Models

The core of KNN algorithm idea is that the category of the sample to be classified is determined by the category of the K neighbor samples nearest to the sample, that is, according to the principle of majority voting, and the classification with the largest number of K neighbors is set as the classification of the sample to be tested. Assuming that T is the sample to be classified and A is the sample set of labeled categories, the steps of KNN algorithm are as follows:

Step 1: Calculate the distance between t and each sample in A by using A certain distance calculation method.
Step 2: Take the K points closest to all the samples in t and A, and mark these K points as B.
Step 3: Calculate the number of samples for each category of B.
Step 4: The classification with the largest number of samples in B is denoted as the prediction category of t.

ANN algorithm is a basic neural network model built in this paper. It composes of input layer, Dropout layer and classification layer. The network activation function is relu and the optimizer is Adam. The experiment verified the feedback of the results through the model and selected the optimal hyperparameter. Each training iteration had 400 generations and the Dropout factor was 0.3.

In this paper, the sample C4 data set was selected as an example, and the K value was set to 3. All the comparison programs were run on high-performance workstations of Windows10 operating system, which was configured with inter 3.10 Ghz 64-bit core CPU and 128 GB memory. The software environment: based on 64-bit Win10 operating system, TensorFlow framework, CUDA11.0 and Spyder integrated development environment.

The loss function, accuracy and confusion proof results of KNN and ANN algorithms are shown in Figs. 1, 2 and 3. Where (a) and (b) are the results of KNN and ANN respectively.

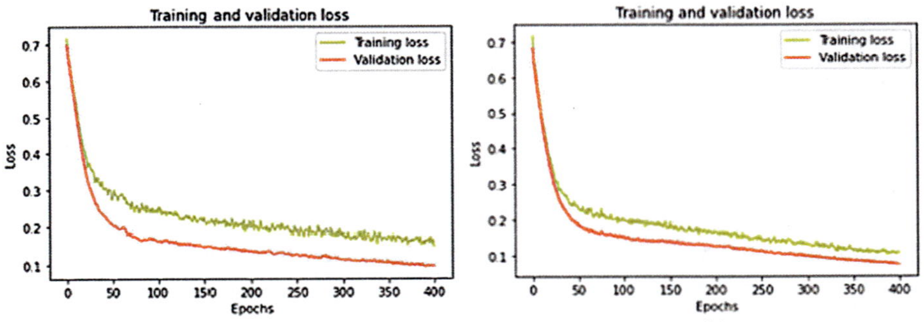

(a) (b)

Fig. 1. Algorithm loss function graph

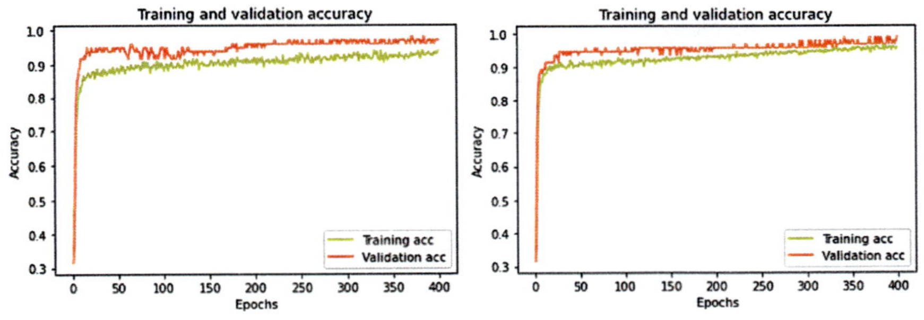

(a) (b)

Fig. 2. Algorithm accuracy graph

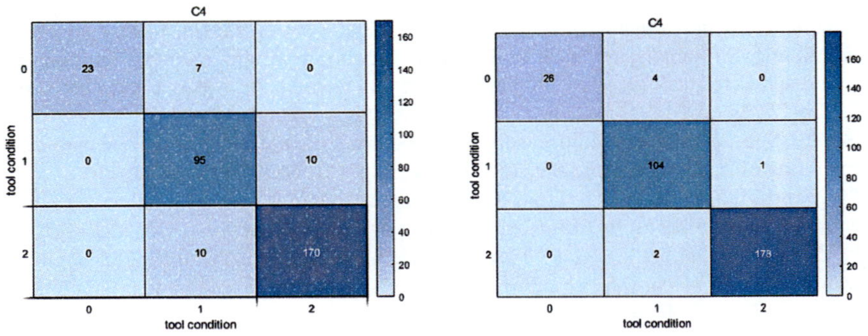

(a) (b)

Fig. 3. Confusion matrix graph

Through the above results can be analyzed out, KNN as the classical algorithm, it did well in the classification problem, but as a result of KNN parameter adjustment need special professional researchers, Therefore, the accuracy of the KNN algorithm in this paper is not as high as that of the basic ANN neural network. Of course, the ANN algorithm in this paper is still a basic deep learning network, and its performance still has a lot of room for improvement in the future.

4 Conclusion

In order to accurately monitor the tool wear degree in the milling process, KNN and ANN algorithms are used to identify the tool state of PHM2010 data set under the premise of feature processing. By comparing the results of loss function, accuracy and confusion matrix, the superiority of ANN is demonstrated. It should be noted that due to manual feature extraction, the result of recognition is affected by subjective factors. At the same time, ANN is a basic Neural Network, and its feature extraction ability has much room for improvement compared with Convolutional Neural Network (CNN) and Recurrent Neural Network (RNN). Therefore, it is planned to use a deeper neural network to extract features for tool state recognition in the future. At the same time, attention mechanism and residual structure can be added to the network for network optimization to further improve the accuracy of tool state recognition.

References

1. He, Z.J., Zhou, Z.X., Huang, X.M.: Tool wear state monitoring based on variational mode decomposition and correlation dimension and relevance vector machine. Acta Metrol. Sinica **39**(02), 182–186 (2018)
2. Zhu, J.M., Zhan, H., Zhang, T.C., et al.: Tool wear status identification based on acoustic emission signal measurement in cutting. Acta Metrol. Sinica **36**(03), 268–272 (2015)
3. Liu, X.L., Li, X.B., Ding, M.N., et al.: Intelligent management and control technology of cutting tool life–cycle for intelligent manufacturing. J. Mech. Eng. **57**(10), 196–219 (2021)
4. Chen, R.X., Wu, Z.Y., Hu, X.L., et al.: Wear state recognition for different tools based on the joint matching of depth characteristics. Chinese J. Sci. Instrum. **41**(12), 138–145 (2020)
5. Ravikumar, S., Ramachandran, K.I.: Tool wear monitoring of multipoint cutting tool using sound signal features signals with machine learning techniques. Mater. Today: Proc. **5**(11), 25720–25729 (2018)
6. Niu, B., Sun, J., Yang, B.: Multisensory based tool wear monitoring for practical applications in milling of titanium alloy. Mater. Today: Proc. **22**(Pt 3), 1209–1217 (2020)
7. Liu, H., Liu, Z.Y., Jia, W.Q., et al.: Current research and challenges of deep learning for equipment remaining useful life prediction. Comput. Integ. Manuf. Syst. **27**(01), 34–52 (2021)
8. Li, M.Q., Miao, H.B., Wang, T.: Tool wear rate prediction in BTA drilling base on fuzzy neural network. Mach. Des. Res. **36**(01), 134–137 (2020)
9. Ma, M., Wang, T.: Research on monitoring method of aeroengine lubricating oil based on CNN-MSLSTM. Acta Metrol. Sinica **42**(02), 232–238 (2021)
10. Agogina, A., Goebel, K.: Mill Data Set [EB/OL]. http://tiarc.nasa.gov/project/prognostic-data-repository
11. Fu, R.R., Hou, P.G., Shi, P.M., et al.: Automatic EOG artifact elimination of eeg based on subspace decomposition. Acta Metrol. Sinica **38**(06), 749–753 (2017)

Tool Wear Monitoring and Life Prediction Analysis

Xuxiang Lu, Chen Wang(✉), Chao Liu, Xiufeng Zhang, Yu Tang, and Zhixiao Zhou

Department of Mechanical Engineering, Hubei University of Automotive Technology, Shiyan 442002, China
20090011@huat.edu.cn

Abstract. Tool wear monitoring can effectively guarantee the machining quality of work pieces and improve productivity. As a basis and prerequisite for tool life prediction, the common monitoring methods are briefly described and research progress on force, power, vibration and acoustic emission signals are introduced. Multi-signal fusion for tool life prediction is a hot topic of research at home and abroad, and common tool life prediction models are listed in this paper. Finally, an outlook on tool wear detection and life prediction is given, looking forward to the advancement and general application of tool fault diagnosis technology.

Keywords: Tool wear · Monitoring methods · Tool life

1 Introduction

In the 21st century, the emergence of technologies such as big data, artificial intelligence and cloud computing promises a new phase of convergence and testing of research paradigms. The research paradigm determines the efficiency and new starting point of science and technology innovation. With the advent of Smart Manufacturing 2025 and Industry 4.0, the demands on the manufacturing industry are becoming higher and higher, especially the intelligence of the manufacturing industry. Today, the rapid development of aerospace, transportation, automotive and new energy fields has put forward higher requirements for manufacturing intelligence [1]. In the process of work piece processing and manufacturing, machine tool processing plays an important role. The tool is an important part of the machine tool and plays a decisive role in the quality of the work piece.

The methods for monitoring tool wear are divided into direct and indirect measurement methods. The direct measurement method obtains the tool wear on the rear face of the tool by some specific direct measurement; the indirect method measures the signal associated with the tool wear and obtains the tool wear indirectly by mapping the relationship [2]. At present, scholars mostly optimize the tool breakage prediction model algorithmically to improve the accuracy of tool breakage prediction. Scholars at home and abroad have summarized the general rules of tool failure and developed a variety of tool condition detection means and methods [3].

Y. Wang et al. (Eds.): IWAMA 2021, LNEE 880, pp. 703–709, 2022.
https://doi.org/10.1007/978-981-19-0572-8_91

The wear of milling cutters is determined based on the wear of milling cutters to a predetermined value, the standard predetermined value recommended by the national standard (GB /T 16460–2016) is the rear face wear width value (VB). KASIM et al. [4]. And JAHARAH S A [5] and other scholars adopted the milling cutter milling failure standard of ISO 8688–2:1989. Their recommended values for tool wear failure in the abovementioned ISO standard and national standards are 0.5 mm for maximum wear and 0.3 mm for uniform wear.

2 Current Research

General fault diagnosis includes 3 links: acquisition of signal; extraction of features; fault identification and prediction. Scholars at home and abroad have conducted a lot of research on the 3 stages of fault diagnosis, and the results are remarkable [6]. The three links of tool fault diagnosis and the common methods are shown in Table 1 below.

Table 1. Common method

Signal acquisition	Force signals Acoustic signals Vibration signals Power signals	Acquisition of information during tool operation through the installation of force sensors, acoustic emission sensors, vibration and power sensors
Feature extraction	Time domain Frequency domain Time-frequency domain	Statistical features, chaos and entropy, Fourier and wavelettransforms, etc.
Projections	Life expectancy Fault identification	Neural networks support vector machines, hidden Markov models, etc.

2.1 Force Signals

Milling force is an extremely important signal in milling and it is an accurate reflection of tool wear. Under the same conditions, a worn tool will generate more milling force than a new tool. Gao D et al. [7] and Freyer B. H. et al. [8] investigated the data-driven prediction method of the wavelet transform of cutting forces from Hidden Markov Trees. Chen X. Q et al. [9] found that the mean values of the force components in all three directions increased gradually with increasing cutting length and tool wear and it was found that there was a correlation between the force signal and the tool wear width.

Liu et al. [10] proposed a Beetle Antennae Search Algorithm (BAS) to improve the BPneural network, optimize the threshold and initial weights, and establish prediction model. Sui et al. [11] established an experimental cutting force for machining creeping graphite cast iron and showed that the isotropic cutting force tends to decrease and increase with the increase of coating thickness. Wei et al. [12] proposed an ultrasonic milling force modeling method considering axial vibration, combining chip formation force model, friction force model and axial impact force model to obtain an axial ultrasonic vibration-assisted milling force model, which has a high prediction accuracy and a consistent change trend, with an error range of 7. 53%–17. 35% at the maximum peak.

2.2 Power Signals

Lin X [13] proposed a tool condition monitoring method based on sequential spindle current signals and applied support vector machines for tool breakage identification. Pal S et al. [14] used wavelet packet tree and principal component analysis to extract tool wear features from motor currents, correlated cutting conditions with tool wear features, and predicted tool wear based on cutting conditions. Chen B. et al. [15] designed a spindle The Hilbert-Huang transforms and wavelet transform were applied to combine the tool wear state tool wear coefficients for online tool monitoring. Dong W et al. [16] Collected spindle power signals for different working conditions (spindle speed, milling depth) through machine tool power sensor san dosed the different working conditions as input and tool wear as output to establish the data set. The monitoring model was trained by a neural network and the tool wear was monitored under different milling conditions.

2.3 Vibration Signals

Antic' A et al. [17] illustrated the process of acquiring real-time high-quality vibration state information, emphasizing the vibration signal acquisition and processing module, and tested the model in both lateral and longitudinal directions during machining operations. Hsieh W. H. et al. [18] detected changes in tool wear through vibration analysis and found that vibration is sensitive to direction and feature selection is important for neural network classification. Bajaj Naman S. et al. [19] built a data acquisition module and used Bayesian optimization to search for optimal model parameters with a model accuracy of 93.3%. Chen D. X. et al. [20] illustrated that ultrasonic vibration-assisted Machining (UVAM) technology can improve cutting performance, established UVAM process for titanium alloy Ti6A14 V, and provided a theory for the study of the formation mechanism of serrated cuts. Literature [21] pointed out the force-thermal agglomeration problem of high-speed milling of titanium alloy, and solved the force-thermal agglomeration problem by high pressure cooling-assisted ultrasonic vibration machining method, and ensured the surface quality of work piece Ra $\leq$ 0. 4 μm by test.

The vibration signal will be affected by the installation position, usually installed on the surface of the work piece to be machined, when the number of mi1ling increases, the height of the work piece decreases with the increase of the milling thickness, the distance between the tool and the sensor becomes closer, then the vibration signal will increase; milling process, there is a periodic air cutting process, the vibration signal collected by the sensor is divided into the milling process and the air cutting process, then the signal processing of the two processes is The processing of the signals from both processes is complex.

2.4 Acoustic Emission Signals (AE)

The sources of acoustic emission signals are generally from: work piece shaping deformation, friction between cutting chips and back tool surface, tool breakages. Ying et al.

[22] studied the relationship between acoustic emission signals and cutting process parameters during high-speed milling of different coated milling tools and hardened steel with different stiffness. The results showed that the variation of milling depth of different coated milling tools did not affect the acoustic emission signal significantly, and the AE wave form could reflect the state of saw tooth chip generation well. In the paper [23], a tool wear monitoring device was designed and tested on 7075 Al plates to analyzed the acoustic emission signals after wavelet decomposition and to monitor the wear status in real time. Traditionally, the tool wear condition predicted by digging deeper into the mathematical model and extracting the information from the sensor signal. The literature [24] proposed wavelet transform at multi-resolution level applying multi-threshold counting feature extraction method, turning 19NiMoCr6 steel test, compared with support vector machine artificial neural network, decision tree and other ML algorithms, its method root mean square prediction error is reduced by 36.53%. The applied approach using a decision tree was found to have a prediction error of less than 6% for a 10% Al-ceramic composite, diamond-coated carbide milling tool based on AE signals.

Acoustic emission signals are not suitable for complex environments, the signal has more sources, and the tool wear is not accurately reflected from the AE signal; the processing and analysis of AE signals is the core of the technology, and modern industrial development is changing rapidly, with higher requirements for the analysis and processing of signals.

3 Tool Life Model

Tool wear is studied in order to optimum cutting parameters, improve tool life and ensure the machining accuracy of the part [26]. The tool wear curve consists of three processes, initial wear, normal wear and rapid wear stages.

A lot of scholars have proposed different models for the factors affecting tool life as show in Table 2.

Table 2. Different lifetime models

Lifetime models	Author
$T = \frac{D^q}{VC^{a^x p a_e^y f} Z^p}$	Fang X F\Shi H Y [27, 28]
$T = \frac{C_1}{C^t \theta^v C^2 f^{C^3} d^{C^4}}$	Kovac P [29]
$\ln NC = (\mathrm{n} + 1) \ln \sigma f + \ln C$	Cheng Y N [30]
$T_{\mathrm{model}} = T_1 + T_2 + T_3 + T_4 + T_5$	Daniel J [31]
$T_r = R^{-1}(\mathrm{r})$	Qian D C [32]
$C = vT^n$	Taylor F [33]

Tool life prediction is influenced by various factors, and if only a single factor influence is considered, then errors are bound to arise. Li Peng et al. [34] used a combination of vibration and acoustic emission signals from spindle and table to achieve complementary advantages and compensate for a single one, and finally evaluated the tool by fully connected layers and classifiers with an identification accuracy of 95.3%; He Yan et al. [35] used vibration, force and acoustic emission signals, considered the sequential and multidimensional characteristics of the data, and used a long and short term memory convolution neural network (LSTM-CNN) for The model was compared with a bidirectional gated recurrent unit (BD-GRU) and a deep heterogeneous gated recurrent unit (DH-GRU),and the accuracy of the model was verified through experiments. The stacked sparse auto encoder network was trained by unsupervised learning and supervised learning, a deep learning milling The stacked sparse auto encoder network was trained by unsupervised learning and supervised learning, a deep learning milling tool wear status recognition model was established), and its multi-signal fusion deep learning model had a high recognition rate; Li Heng et al. [37] also used vibration, force and acoustic emission signals, divided the features into independent samples by type, and made a weighted integrated decision on the tool wear amount predicted by each independent sample.

4 Concluding Remarks

There are various methods for tool wear monitoring and life prediction, but most of them are at the theoretical stage and cannot be put into practice. Deep learning methods have shown initial success, but further research is needed.

CNC machine tools are increasingly integrated, consider the sensor inside the machine tool, in order to facilitate and accurately extract the sensor signal, compared to sensors installed outside the machine tool, reducing unnecessary signal acquisition errors, but also to save money to buy equipment.

Industrial Big Data is character by massive amounts of normal operating data and small samples of fault data. Building deep learning fault diagnosis models for industrial big data is a challenging task. The combination of "small samples" and "big data" is a new way of thinking about the use of large amounts of industrial big data.

References

1. Guo, J.C., Li, A.H.: Advances in monitoring technology of tool wear condition. Tool Eng. **53**(05), 3–13 (2019)
2. Wang, L.Q., Li, S.P., Lv, Z.J.: A review of tool failure condition monitoring research. Mech. Elect. Technol. **04**, 110–114 (2019)
3. Luo, H., Zhang, D.H., Luo, M.: The tool wear and remaining useful life estimation of difficult-to-machine aerospace alloys: a review. China Mech. Eng. (2021)
4. Kasim, M.S., et al.: Wear mechanism and notch wear location prediction model in ball nose end milling of Inconel 718. Wear, Elsevier **302**(1–2), 1171–1179 (2013)
5. Jaharah, S.A., et al.: Prediction of tool life when end milling of Ti6Al4V alloy using hybrid learning system. Arab. J. Sci. Eng. **39**(6), 5095–5111 (2014)
6. Lei, Y.G., Kong, D.T.: Opportunities and challenges of machinery intelligent fault diagnosis in big data era. J. Mech. Eng. **54**(5), 94 (2018). https://doi.org/10.3901/JME.2018.05.094

7. Gao, D., Liao, Z., Lv, Z., Lu, Y.: Multi-scale statistical signal processing of cutting force in cutting tool condition monitoring. Int. J. Adv. Manuf. Technol. **80**(9–12), 1843–1853 (2015). https://doi.org/10.1007/s00170-015-7116-0
8. Freyer, B.H., Heyns, P.S., Theron, N.J.: Comparing orthogonal force and unidirectional strain component processing for tool condition monitoring. J. Intell. Manuf. **25**(3), 473–487 (2012). https://doi.org/10.1007/s10845-012-0698-6
9. Chen, X.Q., Li, H.Z.: Development of a tool wear observer model for online tool condition monitoring and control in machining nickel-based alloys. Int. J. Adv. Manuf. Technol. **45**(7–8), 786–800 (2009)
10. Liu, X.F., Miao, H.B., Wen, J.Y.: Prediction of drilling force based on BAS-BP neural network. Mod. Mach. Tool Autom. Manuf. Techn. **08**, 49–52 (2019)
11. Sui, J.B., Ding, F., Wang, C.Y.: Cutting force experiment and prediction model for machining compacted graphite iron. Tool Technol. 26–32 (2020)
12. Wei, X.T., Yue, C.X., Liu, X.L.: Modeling and experimental study of axial ultrasonic vibration-assisted milling force. Mech. Sci. Technol. Aerospace Eng. (2020)
13. Lin, X., Zhou, B., Zhu, L.: Sequential spindle current-based tool condition monitoring with support vector classifier for milling process. Int. J. Adv. Manuf. Technol. **92**(9–12), 3319–3328 (2017). https://doi.org/10.1007/s00170-017-0396-9
14. Pal, S., et al.: Tool wear monitoring and selection of optimum cutting conditions with progressive tool wear effect and input uncertainties. J. Intell. Manuf. **22**(4), 491–504 (2011)
15. Chen, B., Lin, Z.C., Guo, G.Q.: A new method of real time monitoring of machining status in CPS system. Mach. Des. Manuf. **S2**, 62–64+68 (2018)
16. Dong, W.H., Hu, Y., Tian, G.J.: Woodworking tool wear condition monitoring based on discrete wavelet transformation and genetic algorithm-BP neural network. J. Central. South. Univ. Forestry. Technol. **41**(06), 157–166 (2021)
17. Antic, A., et al.: A model of tool wear monitoring system for turning. Techn. Gazette **20**(2), 247–254 (2013)
18. Hsieh, W.H., Lu, M.C.: Application of back propagation neural network for spindle vibration-based tool wear monitoring in micro-milling. Int. J. Adv. Manuf. Technol. **61**(1–4), 53–61 (2012)
19. Bajaj Naman, S., et al.: A Bayesian optimized discriminant analysis model for condition monitoring of face milling cutter using vibration datasets. J. Nondestruct. Eval. Diagnost. Prognost. Eng. Syst. **5**, 1–12 (2021)
20. Chen, D., Chen, J., Zhou, H.: The finite element analysis of machining characteristics of titanium alloy in ultrasonic vibration assisted machining. J. Mech. Sci. Technol. **35**(8), 3601–3618 (2021). https://doi.org/10.1007/s12206-021-0731-9
21. Zhang, X.Y., et al.: High quality and efficient ultrasonic vibration cutting of titanium alloys. J. Mech. Eng. **57**(05), 133–147 (2021)
22. Xie, Y.X., Wang, C.Y., Zheng, L.J.: AE signal characteristics of high-speed milling hardened steels with coated tools. China Mech. Eng. **30**(17), 2030–2039 (2019)
23. Zhang, X.C., Liu, H.Y., Shi, Y.C.: Experimental study on vibration drilling tool wear monitoring based on AE signal. Mach. Tool Hydraulics **48**(13), 189–192 (2020)
24. Chacón, J.L.F., Fernández, T., de Barrena, A., García, M.S., de Buruaga, X., Badiola, J.V.: A novel machine learning-based methodology for tool wear prediction using acoustic emission signals. Sensors **21**(17), 5984 (2021). https://doi.org/10.3390/s21175984
25. Twardowski, P., et al.: Identification of tool wear using acoustic emission signal and machine learning methods. Precis. Eng. **72**, 738–744 (2021). https://doi.org/10.1016/j.precisioneng.2021.07.019
26. Karanuzelu, B.: Investigating effects of milling conditions on cutting temperatures through analytical and experimental methods. J. Mater. Process. Technol. **262**, 532–540 (2018)

27. Fang, X.F., et al.: Research on tool life prediction model based on BP-RBF neural networks. Tool Eng. **54**(12), 69–73 (2020)
28. Shi, H.Y., Wu, X.Q., Zhang, C.Y.: Prediction of cutting tool life based on support vector regression. Tool Eng. **49**(11), 47–50 (2015)
29. Kovac, P., et al.: Using the temperature method for the prediction of tool life in sustainable production. Measurement **133**, 320–327 (2019)
30. Cheng, Y.N., et al.: Research on damage life reliability of heavy-duty milling carbide tool. Manuf. Technol. Mach. Tool. **01**, 84–88+95 (2019)
31. Johansson, D., et al.: Selecting cutting data tests for cutting data modeling using the colding tool life model. Procedia CIRP **72**, 197–201 (2018)
32. Qian, D.H., et al.: Research of tool reliability and lifetime distribution model based on statistical analysis. Mod. Manuf. Eng. **02**, 134–138 (2016)
33. Taylor, F.W.: On the art of cutting metals. American Society of Mechanical Engineers (1906)
34. Li, P., Huang, Y.X., Xia, P.C.: Multi-signal fusion assessment of tool wear based on 1DCNN-LSTM. Mach. Electron. **39**(05), 8–14 (2021)
35. He, Y., Ling, J.J., Wang, Y.L.: In-process tool wear monitoring model based on LSTM-CNN. China Mech. Eng. **16**, 1959–1967 (2020)
36. Mu, D.F., Liu, X.L., Yue, C.X.: Research on wear status recognition of milling cutter by deep learning and multi-signal fusion. Mechanical Science and Technology for Aerospace Engineering (2020)
37. Li, H., Ye, Z.K., Cha, W.B.: Tool Wear Online Monitoring Based on Multi-sensor Information Decision-making Level Fusion Acta Armamentarii (2021)

Remaining Useful Life Prediction Based on Multi-source Sensor Data Fusion Under Multi Working Conditions

Yang Ge[1(✉)], Jian Wu[1], Jiancong Qin[2], Lingyun Ma[2], and Jianxin Ding[2]

[1] School of Mechanical Engineering, Changshu Institute of Technology, Changshu 215500, China
geyang@cslg.edu.cn
[2] Dongnan Elevator Co., Ltd., Suzhou 215006, China

Abstract. Remaining useful life prediction (RUL) is an important prerequisite for preventive maintenance. The feature fluctuation problem of multi working condition equipment brings challenges to the accuracy of RUL prediction. In this paper, a RUL prediction method for multi working condition equipment is proposed. Firstly, the standardized method of multi working condition equipment health features is proposed to reduce the fluctuation of the features. Secondly, a multi-source data fusion method of equipment health indicators is proposed, which fuses multi-dimensional health features into one-dimensional health indicators, which can reflect the deterioration situation of equipment. Then the sliding window method is used to construct the training and testing samples by time series. In order to reduce the over fitting problem caused by unbalanced training samples, a KNN sample weight assignment method is proposed. Finally, a multilayer BLSTM RUL pre-diction network is established. In order to enhance the prediction accuracy of the network, an improved Huber loss function is added into the prediction network. The experimental results of C-MAPSS data show that the RUL prediction method proposed in this paper has good prediction effect.

Keywords: Multiple working conditions · Remaining useful life prediction · Unbalanced data · Multilayer neural network · Feature fusion

1 Introduction

Remaining useful life (RUL) prediction can provide time basis for health manage-ment of equipment with high reliability requirements. Especially for aviation, aerospace, nuclear energy and other equipment, if the RUL can be accurately predicted, it is helpful for timely preventive maintenance. the occurrence of safety accidents can be avoided and the cost of maintenance can be reduced. At present, the common RUL prediction methods can be generally divided into model-based method and data-driven method [1, 2]. The model-based method is to establish a mathematical model to describe the relationship between the RUL and the decay features of the equipment, and the observation data is used to update the model parameters in real time, so as to predict the RUL. This method

Y. Wang et al. (Eds.): IWAMA 2021, LNEE 880, pp. 710–718, 2022.
https://doi.org/10.1007/978-981-19-0572-8_92

often needs some experience knowledge and clear degradation mechanism. The common model-based methods include particle filter (PF) [3], Markov equation [4], exponential degradation model [5], Wiener model [6] and so on. The method based on da-ta-driven is to establish the black box relationship between operation data and degrada-tion state through machine learning, which does not need to study the degradation mechanism and a lot of prior knowledge. For example, deep neural network based on convolution long and short-term memory (CLSTM) network [7], adaptive Bayesian algo-rithm [8], single-stage and multi-stage iterative prediction model of RUL based on long short memory (LSTM) neural network [9].

Due to the good prediction effect, deep learning is a hot topic in data-driven methods, and many studies have applied it to the RUL prediction of equipment. An autoencod-er-based gate recurrent unit (AE-GRU) model is proposed in [10]. A deep bi-directional long-term and short-term memory (DBLSTM) neural network is proposed in [11], which can perform forward and backward prediction and smoothing on the sequence data re-spectively, and realize the coding of time information and the recognition of long-term correlation. An ensemble GRU network is designed to extend PF for the case where ana-lytic measurement equations are not available [12]. An intelligent prediction method of RUL of synchronous belt based on vibration is proposed in [13].

In fact, many equipment often have multiple working conditions. The prediction effect of classical deep learning method is worse than that of the same working condition. Therefore, a new RUL prediction method is proposed in this paper. The main innovations are as follows.

1. A feature fusion method which can reflect the trend of equipment deterioration is proposed. In order to eliminate the influence of working conditions on feature information, a K-means based method is proposed to identify working conditions and standardize sensing data. A weighted linear model is established to fuse multiple features into one feature.
2. Unbalanced data processing. A weight assignment method based on k-Nearest Neighbor (kNN) is proposed, which can reduce the training over fitting problem to a cer-tain extent and improve the prediction accuracy.
3. Regression loss function constructing. An improved Huber loss function is pro-posed to enhance the robustness of the training model when outliers appear.

2 Methodology

2.1 Working Condition Identification

Let $X = \{X_1, X_2, \cdots, X_n\}$ be the known data set, where $X_1, X_2, \cdots, X_n$ are n data, and every data is N dimension, namely $X_i = (x_{i1}, x_{i2}, \cdots, x_{iN})$. K-means clustering algorithm is to find the set with K cluster centers, namely $C = \{C_1, C_2, \cdots, C_K\} = \{(c_{11}, c_{12}, \cdots, c_{1N}), (c_{21}, c_{22}, \cdots, c_{2N}), \cdots, (c_{K1}, c_{K2}, \cdots, c_{KN})\}$, which satisfies the following objective function.

$$G(X, C) = \min\left(\sum_{i=1}^{K}\sum_{j=1}^{n_i} d(C_i, X_j)\right) \quad (1)$$

where, n_i is the number of data objects classified as class C_i,$d\left(C_i, X_j\right)$ is the Euclidean distance between the cluster center and the data object, and its calculation formula is as follows.

$$d\left(C_i, X_j\right) = \sqrt{\left(c_{i1} - x_{j1}\right)^2 + \left(c_{i2} - x_{j2}\right)^2 + \cdots + \left(c_{iN} - x_{jN}\right)^2} \tag{2}$$

The core idea of K-means clustering algorithm is to divide the data set into K classes to minimize the objective function. Firstly, the clustering center and radius are obtained by clustering the training data set. Then, the distance between the real-time working condition parameters and each cluster center is calculated with the test data, and the closer one is the working condition.

2.2 Feature Fusion

Let $\mathrm{X}_k^i = \left(x_{k,1}^i, x_{k,2}^i, \cdots, x_{k,N}^i\right)$ denote the feature eigenvector of the ith equipment at time k. The linear weighted model as shown in Eq. (3) is established to obtain the fusion health index of the ith equipment at time k.

$$\mathrm{HI}_k^i\left(\mathrm{X}_k^i, \lambda\right) = \lambda_0 + \sum\nolimits_{n=1}^{N} \lambda_n x_{k,n}^i \tag{3}$$

where $\lambda = (\lambda_0, \lambda_1, \cdots, \lambda_N)$ is the weight vector. If we have historical data, we can use the least square method to solve the value of λ. In order to make the health index *HI* have obvious degradation trend, the value of *HI* is defined as Eq. (4).

$$\widehat{\mathrm{HI}}_k^i\left(\mathrm{X}_k^i, \lambda\right) = \begin{cases} 0, k \le T_i 1\% \\ \exp\left(\frac{\log(0.01)}{0.99T_i}(T_i - k)\right), T_i 1\% < k < T_i 99\% \\ 1, k \ge T_i 99\% \end{cases} \tag{4}$$

where T_i is the life of the *i* th equipment. In the whole life process, the *HI* value of the equipment changes from 0 to 1. When the equipment is a new product, HI = 0. With the increase of service time, *HI* value increases until the unit fails, HI = 1.

2.3 Bidirectional Long Short Memory Network

LSTM is a special recurrent neural network (RNN).LSTM can extract long-term dependent information and avoid gradient disappearance. LSTM adds a memory unit in the RNN hidden layer to record historical information, and adds three gates (input, forge, output) to control the use of historical information.

The degradation of equipment performance is strongly dependent on time. Data transformation is not only related to the previous time, but also reflected in the subsequent time. BLSTM is optimized on the basis of LSTM. It has bi-directional transmission ability and can extract the information of time before and after. The BLSTM structure established in this paper is shown in Fig. 1.

2.4 Regression Loss Function

(1) Mean Square Error (MSE)

The MSE is the mean of the square of the distance between the predicted value $f(x)$ of the model and the true value y of the sample. The calculation formula of MSE is shown in Eq. (5).

$$MSE = \frac{1}{m}\sum_{i=1}^{m}(y_i - f(x_i))^2 \tag{5}$$

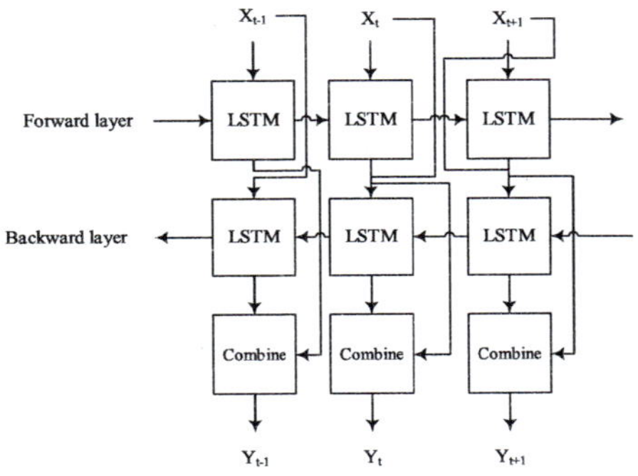

Fig. 1. BLSTM structure

Where y_i and $f(x_i)$ represent the real value and predicted value of the i-th sample respectively, and m is the number of samples.

(2) Mean Absolute Error (MAE)

The calculation formula of MAE is shown in Eq. (6).

$$MAE = \frac{1}{m}\sum_{i=1}^{m}|y_i - f(x_i)| \tag{6}$$

(3) Huber loss (smooth MAE)

Compared with the square loss, Huber loss is insensitive to outliers, but it also keeps differentiable. It is based on absolute error, but it becomes square error when the error is small. We can use the super parameter δ to adjust the threshold of this error. When δ tends to 0, it degenerates to MAE, and then it degenerates to MSE when δ tends to infinity. Its expression is shown in Eq. (7), which is a continuous differentiable piecewise function.

$$L_\delta^i = \begin{cases} \frac{1}{2}(y_i - f(x_i))^2 & if\ |y_i - f(x_i)| \le \delta \\ \delta|y_i - f(x_i)| - \frac{1}{2}\delta^2 & \text{otherwise} \end{cases} \tag{7}$$

MSE is simple to calculate, but MAE has better robustness to outliers. One of the major problems of using MAE training neural network is that its gradient is always large, which may lead to the loss of minimum value at the end of gradient descent training model. In this paper, the regression loss function of Eq. (8) is established based on Huber loss function.

$$L = \sum_{k=1}^{K} \frac{\omega_k}{m} \sum_{i=1}^{m} L_{\delta}^{i} \tag{8}$$

where ω_k is the weight of the k-th sample. After many tests, the δ value in this paper is determined as 1.35. In this paper, KNN method is used to determine the weight of training samples, the specific steps are as follows.

KNN algorithm includes the following four steps:

①Prepare the data and preprocess the data.

②Calculate the distance from the test sample point to each training sample point.

③Sort each distance and select k samples with the smallest distance.

④Assign weight for the selected k samples according to the distance, k for the smallest distance and 1 for the largest distance.

3 Experimental Verification

The experimental data comes from the C-MAPSS turbofan engine simulation data set provided by NASA. The data set records 24 kinds of feature data of aero-engine each time, including 21 sensor data and operation state parameters combined by different flight altitude, Mach number and throttle angle. The data set consists of four sub data sets (FD001-FD004). Each sub data set has a test set and a training set. The training set records the complete process data of the engine from normal operation to failure, and the test set data is several cycle data before engine failure.

3.1 Fault Feature Analysis

The first 10 observation cycles features and the last 10 observation cycles features of 260 training samples in the training data set are selected as the comparison objects. The data of the first 10 cycles can be regarded as normal state feature, and the data of the last 10 cycles can be regarded as fault state feature. The 21 dimensional sensor data of each cycle are compressed into 2 dimensional features by t-SNE, and the results are shown in Fig. 2, which shows that the features of engine fault state and normal state are obviously different.

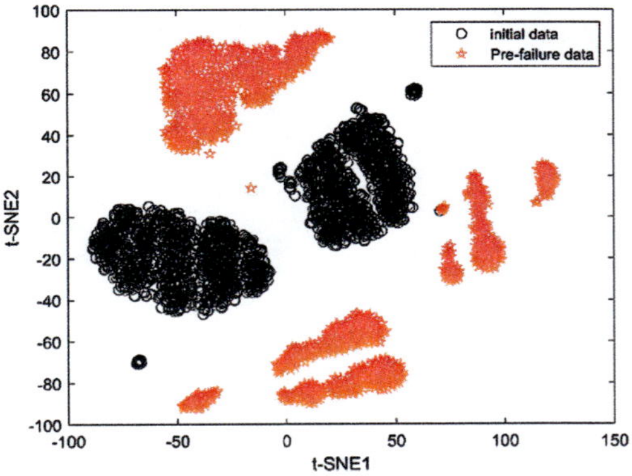

Fig. 2. Normal state and fault state features reduced by t-SNE

3.2 Health Indicators

According to the data fusion method, 21 standardized features of 260 training samples and 259 test samples are fused into one health indicator. The cubic curve fitting method is used to smooth the fused health indicators, and the results are shown in Fig. 3.

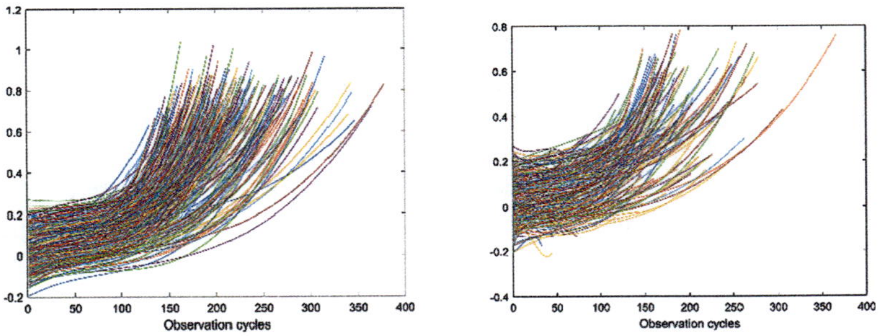

Fig. 3. (a) The cubic curve fitting results of health indicators for training samples. (b) The cubic curve fitting results of health indicators for test samples.

3.3 Evaluation Index

To compare the performance of the proposed RUL prediction method with some existing methods, the mean absolute error, mean square error and root mean square error are used to measure.

Mean absolute error

$$MAE = \frac{1}{n}\sum_{i=1}^{n}|PT_i - RT_i| \tag{9}$$

Mean square error

$$MSE = \frac{1}{n}\sum_{i=1}^{n}(PT_i - RT_i)^2 \tag{10}$$

Root mean square error

$$RMSE = \sqrt{\frac{1}{n}\sum_{i=1}^{n}(PT_i - RT_i)^2} \tag{11}$$

where, *PT* is the predicted value of RUL, *RT* is the true value of RUL. The smaller the above three evaluation indexes are, the higher the prediction accuracy is.

3.4 Predicted Results

The experiment is carried out in two ways. In the first way, 21 original sensor data of training sample set are directly used as feature vectors, and the corresponding RUL is used as labels to form training data set. The 21 sensing data of the last observation cycle of the test sample are taken as the test data set. The network structure is built in Kears environment by using multi-layer LSTM and multi-layer BLSTM networks respectively. Three loss functions, MSE, MAE and Huber, are added to the network for RUL prediction. The prediction results are shown in Table 2.

In the second way, the time series of health indicators proposed are used for RUL prediction. The network structure of prediction is the same as the first way. The prediction results of RUL are shown in Table 1. It should be noted that the training samples of methods 1 to 6 in Table 2 are not assigned weights, while the training samples of method 7 are weighted by KNN method. The prediction method proposed in this paper has the best accuracy.

Table 1. RUL prediction results of the first way

Number	Method	MAE	MSE	RMSE
1	LSTM-MSE	22.4555	933.2257	30.5487
2	LSTM-MAE	22.7667	932.9309	30.5439
3	LSTM-Huber	**21.9668**	**927.9551**	**30.4624**
4	BLSTM-MSE	23.6130	954.9986	30.9031
5	BLSTM-MAE	24.0734	993.7834	31.5243
6	BLSTM-Huber	22.6206	928.4385	30.4703

Table 2. RUL prediction results of the second way

Number	Method	MAE	MSE	RMSE
1	LSTM-MSE	20.545	824.992	28.723
2	LSTM-MAE	20.641	815.211	28.552
3	LSTM-Huber	19.839	778.254	27.897
4	BLSTM-MSE	21.318	782.485	27.973
5	BLSTM-MAE	20.936	779.924	27.927
6	BLSTM-Huber	19.330	693.593	26.336
7	Proposed method	**19.173**	**669.194**	**25.869**

4 Conclusion and Future Research

A RUL prediction method is presented for the equipment under multiple working conditions. Firstly, a standardized method of sensing data under multiple working conditions is proposed to reduce the influence of the change of working conditions on the fluctuation of sensing data. Secondly, a health indicator fusion method is proposed, in which multiple sensing features ate fused into a health indicator, which can reflect the deterioration trend in the whole life cycle of the equipment. Then, a weight assignment method of training samples based on KNN is proposed to reduce the influence of over-fitting caused by the imbalance of training samples. Finally, a multi-layer BLSTM network is built to predict the RUL of the test samples, and an improved Huber loss function is added to the training network. The C-MAPSS data is used to verify the proposed method, and the experimental results show that the proposed RUL prediction method has a good prediction effect for the equipment under multiple working conditions.

Acknowledgments. This work was supported in part by the Suzhou Science and Technology Foundation of China under Grant SYG202021.

References

1. Li, Z., Wang, Y., Wang, K.S.: A deep learning driven method for fault classification and degradation assessment in mechanical equipment. Comput. Ind. **104**, 1–10 (2019)
2. Soualhi, M., Nguyen, K.T.P., Soualhi, A., Medjaher, K., Hemsas, K.E.: Health monitoring of bearing and gear faults by using a new health indicator extracted from current signals. Measurement **141**, 37–51 (2019)
3. Cai, H., Feng, J., Li, W., Hsu, Y.-M., Lee, J.: Similarity-based particle filter for remaining useful life prediction with enhanced performance. Appl. Soft Comput. **94**, 106474 (2020). https://doi.org/10.1016/j.asoc.2020.106474
4. Ramezani, S., Moini, A., Riahi, M., Marquez, A.C.: A model to determining the remaining useful life of rotating equipment, based on a new approach to determining state of degradation. J. Central South Univ. **27**(8), 2291–2310 (2020). https://doi.org/10.1007/s11771-020-4450-7

5. Wang, P., Long, Z., Wang, G.: A hybrid prognostics approach for estimating remaining useful life of wind turbine bearings. Energy Rep. **6**, 173–182 (2020)
6. Liao, G., Yin, H., Chen, M., Lin, Z.: Remaining useful life prediction for multi-phase deteriorating process based on Wiener process. Reliab. Eng. Syst. Safety. **207**, 107361 (2021)
7. Ma, M., Mao, Z.: Deep-convolution-based LSTM network for remaining useful life prediction. IEEE Trans. Industr. Inf. **17**(3), 1658–1667 (2021)
8. Rezamand, M., Kordestani, M., Orchard, M.E., Carriveau, R., Ting, D.S.K., Saif, M.: Improved remaining useful life estimation of wind turbine drivetrain bearings under varying operating conditions. IEEE Trans. Industr. Inf. **17**(3), 1742–1752 (2021)
9. Ding, N., Li, H., Yin, Z., Zhong, N., Zhang, L.: Journal bearing seizure degradation assessment and remaining useful life prediction based on long short-term memory neural network. Measurement **166**, 108215 (2020). https://doi.org/10.1016/j.measurement.2020.108215
10. Lu, Y.-W., Hsu, C.-Y., Huang, K.-C.: An autoencoder gated recurrent unit for remaining useful life prediction. Processes **8**(9), 1155 (2020). https://doi.org/10.3390/pr8091155
11. Li, H., Wang, W., Li, Z.W., Dong, L.Y., Li, Q.Z.: A novel approach for predicting tool remaining useful life using limited data. Mech. Syst. Signal Process. **143**, 22 (2020)
12. Deng, Y., Du, S., Jia, S., Zhao, C., Xie, Z.: Prognostic study of ball screws by ensemble data-driven particle filters. J. Manuf. Syst. **56**, 359–372 (2020)
13. Khazaee, M., Banakar, A., Ghobadian, B., Mirsalim, M.A., Minaei, S.: Remaining useful life (RUL) prediction of internal combustion engine timing belt based on vibration signals and artificial neural network. Neural Comput. Appl. **33**(13), 7785–7801 (2020). https://doi.org/10.1007/s00521-020-05520-3

Check for updates

An Industrial Application of Cellular Manufacturing Using African Buffalo Optimization

Tamal Ghosh(✉)

IVB, NTNU i Gjøvik, Teknologivegen 21, 2815 Gjøvik, Norway
tamal.ghosh@ntnu.no

Abstract. An industrial case is presented in this paper where the existing job shop layout is converted to a cellular structure using a typical percentage utilization based bi-objective cell formation technique. The aim is the total cell movements. The African Buffalo Optimization (ABO) algorithm is used to obtain the stated cellular structure. The proposed algorithm obtains an initial solution based on random machine cells and part families using percentage utilization. Thereafter, it improves the solutions based on performance metric. The obtained solution shows improved parts movement.

Keywords: Cellular manufacturing · African buffalo optimization · Industrial data

1 Introduction

Group technology and its application in Cellular Manufacturing Systems (CMS) presents a combined setup of jobshop (diversity) and flowshop (high production) and demonstrates an alternative form of manufacturing [1]. The objective is to manufacture parts and classify them into families based on similarities in processing requirements, time, or sequences, etc. Application of CMS could be a way to reduce the throughput time, the work-in-process, tool requirements, enhance the product quality, and overall control of operations. CMS could possibly make the organisation highly flexible and responsive to customer needs and facilitate low-cost manufacturing of complex production of multiple parts with variable volumes. According to Ref. [2] the aim of CMS is to decompose the production system into several tiny systems (blocks or cells). These are independent and could exploit the processing resemblances of parts on machines. The procedure of allocating part families to the machine cells is termed as the cell formation problem (CFP) and one part family is expected to be assigned to a machine cell. CFP, also termed as machine-part grouping problem, exploits the machine-part incidence matrix and attains block diagonal cellular structure to generate cells. Data in CFPs could be classified as, binary data, ratio data, and sequential data. An incidence matrix, packed with '0' and '1' depending upon the machining requirement of parts, is known as 'binary data'. If the '1' elements are replaced with processing time (percentage utilization) of

Y. Wang et al. (Eds.): IWAMA 2021, LNEE 880, pp. 719–725, 2022.
https://doi.org/10.1007/978-981-19-0572-8_93

parts, it is termed as 'ratio data', and if the elements present sequential integer numbers, it is called 'sequential data' [3]. An example of simple cell formation based on ratio data could be depicted in Fig. 1. The rows represent machines and columns represent parts. Elements (real valued figures) represent percentage utilization for each part over each machine. After cell formation, block diagonal cells are achieved with part families and respective machines. Elements outside diagonal blocks are known as bottleneck machines (exceptional elements) and empty places inside cells are known as voids.

	P1	P2	P3	P4	P5
M1		0.23	0.11		0.36
M2	0.32			0.43	0.11
M3			0.66		0.28

	P3	P5	P2	P1	P4
M1	0.11	0.36	0.23		
M3	0.66	0.28			
M2		0.11		0.32	0.43

Fig. 1. Incidence matrix for 3 × 5 example before (left) and after (right)

Binary data are mostly used in the past few decades [4–8]. In real world, many other production factors are important. Percentage utilization is one such factor, which is considered as the 'ratio data' in literature.

Ref. [9] first recommended ratio data in a most logical way. Many researchers have considered ratio data subsequently [3, 10, 11]. Ref. [10] quantified that the workload data and ratio data are identical which is further transformed into an incidence matrix. The total processing time on a machine/work station for any part could be obtained by multiplying its production quantity and its unit processing time. All the '1's in the incidence matrix are then converted to workload values. These would take any value in the ratio scale (0–1) and termed as the ratio level data. However, rarely any past article actually shows the systematic way to obtain the workload value from total processing time. Recently ref. [12] has pointed this out and portrayed a way to attain that goal.

To solve CFPs a sizable number of solution approaches are available in literature of CMS since early 80s. These are exact methods, graph theoretic approaches, mathematical programming, similarity coefficient techniques, clustering algorithms, soft-computing techniques such as neural network, metaheuristics, and fuzzy methods [13]. Particularly the direction of the research indicates towards soft-computing techniques due to their robust convergence properties and ability of handling combinatorial optimization problems with ease. Since the CFPs are NP-Hard (Non-Ploynomially Hard) in nature, obtaining global optimal solution is difficult [14]. These problems could be efficiently solved using meta-heuristic techniques. Therefore, many meta-heuristic techniques are being applied to CFPs for better solutions during past two decades [15]. Ref. [15] demonstrates the usage of hybrid metaheuristics for CFPs since the past decade. Hybrid techniques are improved and prompt while attaining global best solutions.

African Buffalo Optimization (ABO) is a recent addition in the class of metaheuristic algorithms [16], which is highly efficient in solving various optimization problems (details of ABO are portrayed in Sect. 3), and it has never been studied for CFPs before. Most of the articles, published in the domain of CMS, have considered theoretical data for evaluation of cell formation models and techniques. However, practical or real-world data-oriented study is rarely available. This research work seals that gap. In this work, an attempt is made to develop the cells using ABO technique considering the percentage utilization of machines. To solve this problem a real-valued matrix is considered from past literature. Further, a novel mathematical formula is proposed which successfully obtains part families minimizing total percentage utilization induced by exceptional elements (EE) or bottleneck machines. Subsequently an industrial case study is presented which is obtained from MSME Tool Room, Kolkata, India. The collected data are used to develop manufacturing cells realistically using the proposed technique.

2 Industrial Data and Problem Formulation

A case study is conducted in MSME (Ministry of Micro, Small and Medium Enterprises) Tool Room (previously known as Indo-German Tool Room), India. This tool room is a government aided company which develops all the special types of tools required by governmental, semi-governmental or quasi-governmental organizations. MSME Tool Room has an existing production floor (Fig. 2) with various machines arranged in jobshop formation (Fig. 3).

Fig. 2. Production floor of MSME tool room, Kolkata

Production flow data are received for the period of one year (2013–2014). In this period 368 different types of parts or products are manufactured over 10 different kinds of machines. This company maintains the data in consolidation sheets. An example sheet (scanned) is depicted in Fig. 4. Data sheets are further refined and gathered for historical data acquisition.

Reference [12] has recently defined Utilization based CFP which is popularly stated as “operational time” or “ratio data” in past literature. This is also named as Machine Part Utilization Matrix (MPUM). After pre-processing and proper transformation, the MPUM is obtained of size 10×368 for the industrial case, which is substantially large.

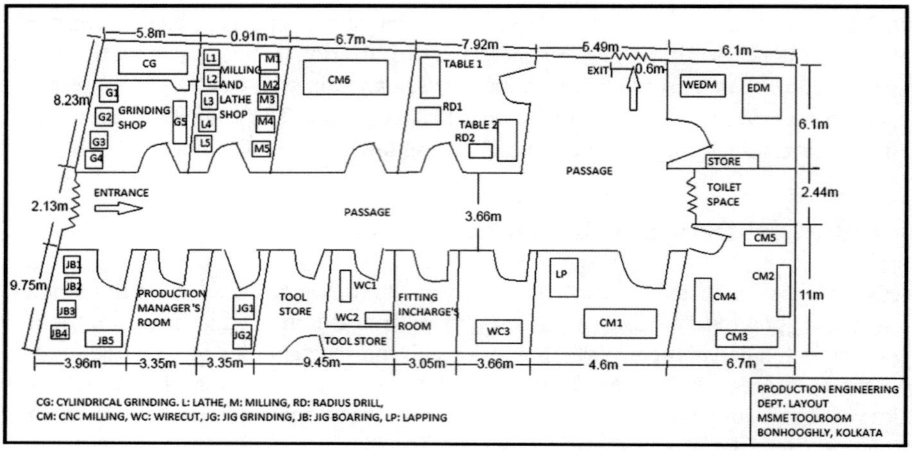

Fig. 3. Arrangement of machines of MSME tool room in jobshop structure

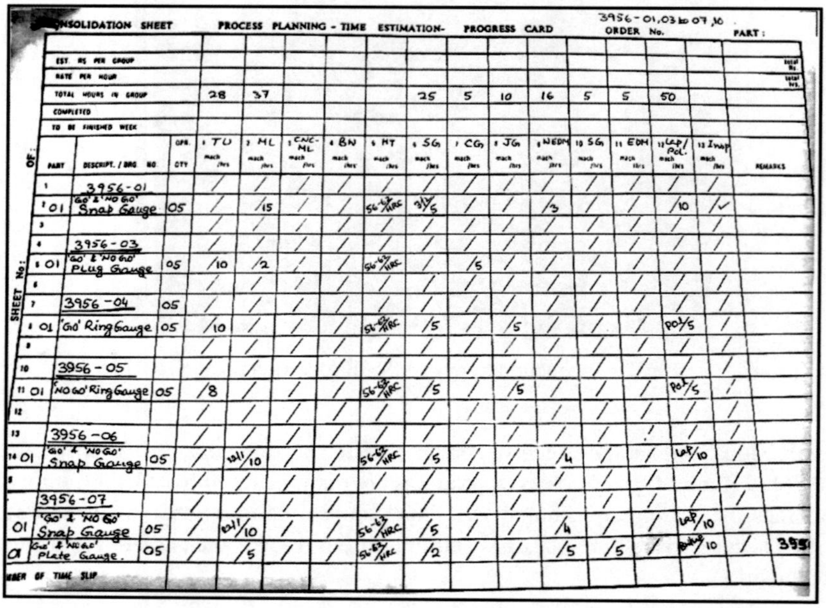

Fig. 4. Example consolidation sheets for production data of MSME tool room

3 African Buffalo Optimization (ABO)

ABO is a metaheuristic optimization algorithm recently developed [16]. This technique mimics the migratory behaviour of the large herd of African buffalos foraging for food across the plains and forests. They migrate in search of green fields and travel across vast

African territory. Depending upon the seasonal differences the buffalos travel continuously across the plains, mountains, and forests in quest of green land. ABO advances with two natural inter-communication strategies among the herd members, (i) the roaring *waaa*, which indicates the sense of risks or scarcity of greenery in present location and directing the herd to move to next location. (ii) The roaring *maaa*, which confirms promising green meadows and suggesting the herd to stay [17]. ABO is being practiced in different problem domains by researchers, such as Travelling Salesman Problem (TSP) and parameter optimization of PID controllers. ABO advances with the ith buffalo and its two natural inter-communication methods among the herd members ($i = 1,2,3,\ldots,n$), (i) the roaring *waaa*, denoted by w_i, which helps in exploring the solutions in global search. (ii) The roaring *maaa*, denoted by m_i, which helps in exploiting in the local search space. The values of m and w for $(i + 1)^{th}$ buffalo are updated using the following equations,

$$m_{i+1} = m_i + lp_1 \times (bg - w_i) + lp_2 \times (bp_i - w_i) \quad (1)$$

$$w_{i+1} = \frac{(w_i + m_i)}{\lambda} \quad (2)$$

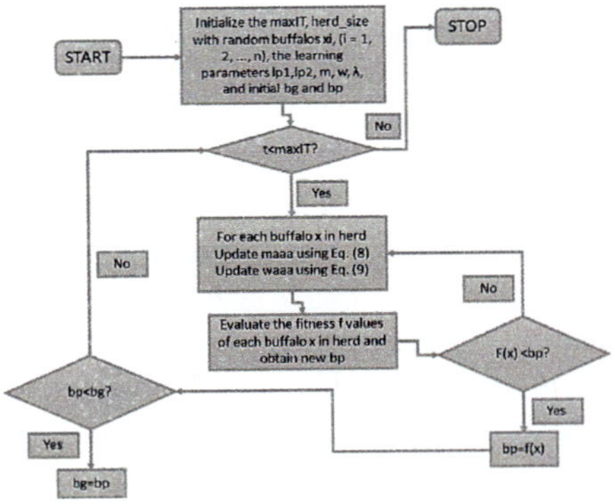

Fig. 5. ABO flowchart

Where lp_1 and lp_2 are learning parameters, bg and bp_i are the best position of the herd and the ith buffalo's best known position respectively, and λ is a prefixed random number. The artificial herd of buffalos is generated with initial herd $x_i = \{x_1, x_2, \ldots, x_n\}_{i \in [1,n]}$ representing the n coordinates for a continuous domain or n vectors for discrete domain. The flowchart of ABO is portrayed in Fig. 5.

4 Results and Discussions

A large dataset (10 × 368) eventually increases the complexity to find the near-best solution. The ABO is applied on this data and three cells are formed. The result obtained is, UGE: 53.039, TEU 886, Voids: 657 and CPU Time: 2.33×10^3 s. Further the new cellular structure is made with three distinct cells (Fig. 6). It could further be observed that the material handling is improved in the obtained cell formation. For example, a trimming die would travel from the CNC milling shop to benchwork to jig grinding shop to WEDM shop to benchwork and to surface grinding shop in order to be completely processed. In the existing jobshop formation it would travel 7.3 + 17.67 + 27.737 + 12.49 + 17.37 = 82.57 m, but in the proposed cellular shop-floor it would travel around 7.3 + 12.49 + 15.85 + 13.41 + 15.54 = 64.59 m. Therefore, material handling cost is reduced by 21.77% approximately.

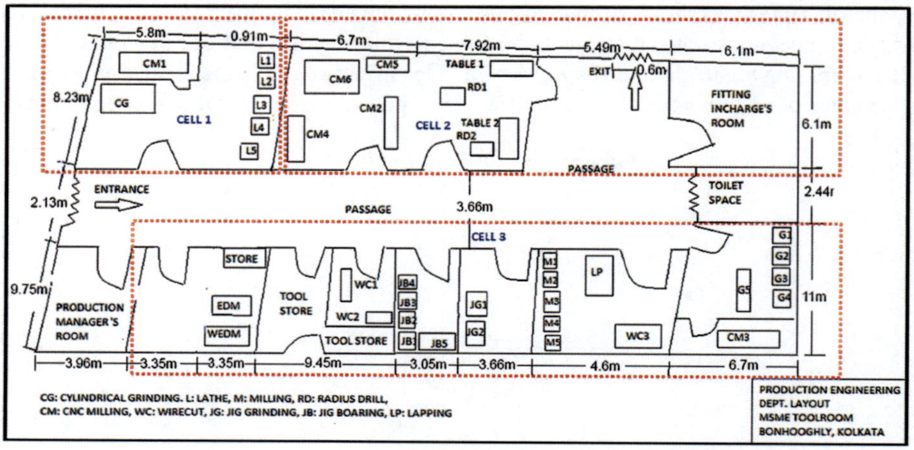

Fig. 6. Proposed arrangement of machines with Cellular configuration

In another example a length gauge would travel from turning to cylindrical grinding to obtain complete processing. In the existing shop-floor it would travel 1.83 + 2.74 = 4.57 m, whereas in the proposed cellular shop-floor it is an intra-cell movement in cell 1 which is approximately 1.5 m, which is 65.6% better than existing arrangement. Therefore, the proposed solution improves the material movement which in turn reduces material handling cost.

5 Conclusions

In this paper the Utilization-based cellular structure is implemented in production toolroom of MSME, India. This is considered as one of the latest NP-Hard problems with an aim to minimize the total cell movements. A latest ABO technique is utilized as a methodology. This proposed technique is shown to obtain promising result. The obtained cellular structure is shown to be more efficient than the existing jobshop formation for the company. This solution could be useful to the operational personnel of the company, who could further decide on material handling costs reduction and production time improvement.

Acknowledgement. This work is supported by the SFI Manufacturing (Project No. 237900) and funded by the Norwegian Research Council.

References

1. Goldengorin, B., Krushinsky, D., Pardalos, P.M.: Cell Formation in Industrial Engineering: Theory Algorithms and Experiments. Springer Optimization and Its Applications. Springer, Heidelberg (2013). https://doi.org/10.1007/978-1-4614-8002-0
2. Selim, M.S., Askin, R.G., Vakharia, A.J.: Cell formation in group technology: review evaluation and directions for future research. Comput. Ind. Eng. **34**(1), 3–20 (1998)
3. Sengupta, S., Ghosh, T., Dan, P.K.: FAKMCT: fuzzy ART K-means clustering technique: a hybrid neural network approach to cellular manufacturing systems. Int. J. Comput. Integr. Manuf. **24**(10), 927–938 (2011)
4. Dimopoulos, C., Mort, N.: A hierarchical clustering methodology based on genetic programming for the solution of simple cell-formation problems. Int. J. Prod. Res. **39**(1), 1–19 (2001)
5. Srinivasan, G.: A clustering algorithm for machine cell formation in group technology using minimum spanning trees. Int. J. Prod. Res. **32**(9), 2149–2158 (1994)
6. Xambre, R., Vilarinho, P.M.: A simulated annealing approach for manufacturing cell formation with multiple identical machines. Eur. J. Oper. Res. **151**(2), 434–446 (2003)
7. Žilinskas, J., Goldengorin, B., Pardalos, P.M.: Pareto-optimal front of cell formation problem in group technology. J. Global Optim. **61**(1), 91–108 (2014). https://doi.org/10.1007/s10898-014-0154-6
8. Olivares, R., Soto, R., Crawford, B.: Performance evaluation of the parameterless bat algorithm to solve the manufacturing cell design problem. In: 13th Iberian Conference on Information Systems and Technologies (CISTI) (2018)
9. Venugopal, V., Narendran, T.T.: Cell formation in manufacturing systems through simulated annealing: an experimental evaluation. Eur. J. Oper. Res. **63**(3), 409–422 (1992)
10. George, A.P., Rajendran, C., Ghosh, S.: An analytical-iterative clustering algorithm for cell formation in cellular manufacturing systems with ordinal-level and ratio-level data. Int. J. Adv. Manuf. Technol. **22**(1), 125–133 (2003)
11. Mahapatra, S.S., Pandian, R.S.: Genetic cell formation using ratio level data in cellular manufacturing systems. Int. J. Adv. Manuf. Technol. **38**(5), 630–640 (2008)
12. Ghosh, T., Doloi, B., Dan, P.K.: Utilization-based grouping efficiency and multi-criteria decision approach in designing of manufacturing cells. Proc. Inst. Mech. Eng. Part B J. Eng. Manuf. **231**(3), 505–522 (2017)
13. Papaioannou, G., Wilson, J.M.: The evolution of cell formation problem methodologies based on recent studies (1997–2008): review and directions for future research. Eur. J. Oper. Res. **209**(3), 509–521 (2010)
14. Nair, G.J., Narendran, T.T.: ACCORD: a bicriterion algorithm for cell formation using ordinal and ratio-level data. Int. J. Prod. Res. **37**(3), 539–556 (1999)
15. Ghosh, T., Sengupta, S., Chattopadhyay, M., Dan, P.K.: Meta-heuristics in cellular manufacturing: a state-of-the-art review. Int. J. Indust. Eng. Comput. **2**(1), 87–122 (2011)
16. Odili, J.B., Kahar, M.N.M.: Solving the traveling salesman's problem using the African buffalo optimization. Comput. Intell. Neurosci. **2016**, 1–12 (2016). https://doi.org/10.1155/2016/1510256
17. Odili, J.B., Kahar, M.N.M., Noraziah, A.: Parameters-tuning of PID controller for automatic voltage regulators using the African buffalo optimization. PLOS ONE **12**(4), e0175901 (2017). https://doi.org/10.1371/journal.pone.0175901

An Outlier Detection Algorithm Based on Spectral Clustering

Li Xiaoqiang[1](✉) and Liu Tingfeng[2]

[1] Henan Polytechnic, Zhengzhou 450046, China
[2] School of Information Science and Engineering, Henan University of Technology, Zhengzhou 450001, China

Abstract. Outliers refer to data that is far away from conventional data objects, resulting from different mechanisms, and there is a clear gap from conventional data. Outlier detection algorithms include statistics-based, distance-based, density-based, and cluster-based algorithms. Spectral clustering has become a research hotspot in the field of clustering because it can cluster on arbitrary-shaped data samples.Based on Self-Tuning spectral clustering, optimize the scale parameters, improve the method of determining the initial centroid and the optimal clustering center in the K-means algorithm, calculate the outlier index for each data, and sort the outliers according to the size of the outlier index.Group points.Through the verification of artificial data and real data sets, the experiments show that the outlier detection algorithm based on spectral clustering is superior to the traditional outlier detection algorithm.

Keywords: Spectral clustering · Self-Tuning · Scale parameters · Factor of outlier

1 Introduction

Outlier point detection is an important branch of data mining. With the rapid development of data mining technology, people pay more and more attention to the overall data trends and the outlier data points that significantly deviate from the overall data trend, because these data points tend to often contain more important information. D.Outlier points proposed by Hawkins [1] Definition: Outliers are observations that vary so far from other observations that it is suspected by different mechanisms.The purpose of outlier detection is to effectively identify anomalous data in the dataset and mine significant potential information in the dataset.

The Spectral clustering algorithm is a structural map-bIn recent years, outlier detection has been widely used in network fault detection, financial fraud, health insurance, and bad weather forecasting [2] and so on. The methods for outlier detection mainly include statistical, distance-based, density-based, cluster-based methods [3, 4]. Cluster analysis is used to discover groups of cluster objects strongly correlated locally, while clustering-based outlier detection is used to detect and discover objects that are not strongly associated with other clustering objects. Thus, clustering analysis can be applied

Y. Wang et al. (Eds.): IWAMA 2021, LNEE 880, pp. 726–738, 2022.
https://doi.org/10.1007/978-981-19-0572-8_94

to outlier detection. After years of research and development, clustering algorithms have appeared a large number of classical algorithms such as: K-means, DBSCAN, EM algorithm. However, many traditional clusters are only suitable for certain specific problems, with certain limitations, such as the sensitivity to the input parameters and the shape of the spatial distribution of the dataset. Spectral clustering algorithm [5] Overcoming the shortcomings of traditional algorithms, convergence to the global optimal solution, more adaptable to data distribution and better clustering, thus get wide attention.

The Spectral clustering algorithm is a structural mapping-based clustering method. In the dataset containing the outlier points, Spectral clustering can be used for the description of structural features of the dataset and to detect the presence of outliers. The essence of spectral clustering is to transform the clustering problem into the optimal partition of the graph, which has minimum cut criterion, ratio cut criterion, average cut criterion, maximum minimum cut criterion, normalized cut criterion, etc. Spectral clustering algorithm is generally divided in two steps, first establishing the graph based on the similarity measure of the data sample points, and then constructing the algorithm to cluster the graph. This paper improves in two aspects: on one hand, the Self-Tuning algorithm proposed by Zelnik-Manor and Perona [6]. To recompute the scale parameters, and improve on the k-means algorithm on the other hand. The proposed algorithm is compared with the traditional outlier detection algorithm, and the effectiveness analysis and results show that the proposed algorithm has more advantageous in detecting outlier points.

2 Related Theory

2.1 Similarity Matrix

The similarity matrix is a measure of distance between any two points in the sample point, The closer the distance, the higher the point similarity. For a graph G, we generally describe it by the set V of points and the set E of edges. That is, the G (V, E). Where V is all the points in our dataset (v1,v2,...vn). For any two points in V, there can be edges connected, or no edges connected.wij is the weight between point viAnd point v_j. Since it is an undirected graph, therefore. The adjacency matrix W of the graph is composed of weight values between all vertices. The degree di of any vertex vi in the figure is defined as the sum of the weights of all the edges connected to it,

$$d_i = \sum_{j=1}^{n} w_{ij} \tag{1}$$

The degree matrix D is the n * n diagonal matrix composed of d_i.

The spectral clustering algorithm requires in put is a graph that contains similarity between all samples and samples, which is a matrix whose size is n * n. The adjacency matrix W, similarity matrix is equal to the adjacency matrix can be obtained by using the similarity matrix S of the sample point distance measure. There are three ways to construct the adjacency matrix W:ε-close neighbor method (ε-neighborhood graph), the k close neighbor method (k-nearest nerghbor graph), and the full connectivity method (fully connected graph). The full connection method is the most commonly used, you can choose different kernel functions to define the edge weights, commonly used are polynomial kernel function, Gaussian kernel function, Sigmoid kernel function, Gaussian kernel function here:

$$W_{ij} = \exp(-\frac{||x_i - x_j||^2}{2\delta^2}) \tag{2}$$

2.2 The Laplacian Matrix (Graph Laplacians)

There are usually three methods for Laplace matrix construction, with non-standardized Laplace matrices:

$$L = D - W \tag{3}$$

The Standardized Laplacian includes the symmetric-normalized Laplacian matrix Lsys And the random walk (Random Walk) of the normalized Laplace matrix Lrw.

$$L_{sys} = D^{\frac{1}{2}} L D^{\frac{1}{2}} = E - D^{\frac{1}{2}} W D^{\frac{1}{2}} \tag{4}$$

$$L_{rw} = D^{-1} L = E - D^{-1} W \tag{5}$$

D is the degree matrix and W is the similarity matrix, and its element value is wij.

2.3 Spectral Clustering Algorithm

Spectral clustering algorithm [6]. The theoretical basis is the graph theory, mapping the data points into the weighted undirected graph, the sample data points as the nodes of the graph, the weights of the edges represent the similarity between the data point characteristics, and then use some segmentation criterion to obtain the best two-dimensional partition of the data. The standard spectral clustering algorithm deals with a dichotomy, divided using only one eigenvector with poor and unstable effects. The K-way spectral clustering algorithm [7]. The emergence of overcoming the disadvantages of the dichotomy clustering algorithm while using multiple feature vectors can not only approximate the optimal k-way partition, but also avoid instability due to loss of information.

Spectral clustering works well in some fields, but there are also urgent problems to be solved. The algorithm requires to utilize a Gaussian function to construct similarity matrix matrices, which will involve scaling parameters without any range and regularity at selection, which makes spectral clustering sensitive to parameters and decreases sharply if improperly chosen. Many scholars optimize the spectral clustering algorithm to make it work better. Tao Tong et al. [8]. We propose a one-step spectral clustering algorithm based on self-velocity learning. This method uses missing value mapping matrix to handle missing data, one-step clustering method to reduce error accumulation due to multiple steps of traditional spectral clustering, and uses self-step rule term to avoid the influence of noise and outliers; Zelnik-Manor proposes we propose an adaptive spectral clustering algorithm to automatically calculate scale and group number, and can handle multiscale data with previous methods; Afzalan and Jazizade [9]. A heuristic for iterative feature search multiscale, high-dimensional feature space is proposed and evaluated. The proposed heuristic algorithm does not require a priori assumptions on the number of clusters, or on the scaling parameters for affinity measures containing a series of values of the number of clusters; Xie Juanying et al. [10]. We propose unsupervised feature selection idea based on spectral clustering algorithm, spectral clustering all features, cluster features with high similarity into one, define feature discrimination and feature independence; Tan Malone et al. [11]. A spectral clustering algorithm based on mutual KNN and standardized matrix is proposed to effectively improve the execution efficiency and accuracy of clustering classification.

To sum up, the main problems of spectral clustering are: (1) the number of categories is difficult to automatically determine when clustering with the spectral clustering algorithm; (2) the spectral clustering algorithm has a great dependence on the parameters, and the selection of parameters has a great influence on the clustering results.

3 Improved the Adaptive Spectral Clustering Algorithm

3.1 Optimize the Scaling Parameters

For a given sample set, X = {x1, x2, ..., xn}, Each sample x has m dimension features, namely xi = (xi1, xi2,...xi m),(i = 1,2,...,n)。A local scaling parameter σ is defined for each data point in the self-turning algorithmi,

$$\sigma_{\mathrm{i}} = d(x_i, x_k) \tag{6}$$

That is, parameter σiTake point x_i. To the k th point x in its immediate neighborhood k. The Euclidean distance of, but the σ_i It may be affected by the outlier point. Thus this paper scales ing parameter σ_i Take point x_i Weighted distance of k points in the k close neighborhood of.

$$\sigma_{\mathrm{i}} = \sum_{j=1}^{K} w_{ij} d_{ij} \tag{7}$$

$w_{ij} = 1 - \ln(d_{ij} / \sum_{k=1}^{K} d_{jk})$ $0 < d_{ij} / \sum_{k=1}^{K} d_{jk} < 1$ Among them, since, so the $w_{ij} > 1$.

3.2 Optimize the K-means Algorithm

Since the K-means clustering is relatively sensitive to the selection of the initial clustering center, the K-means algorithm is unstable, and hence the spectral clustering algorithm is unstable. In this paper, we propose the dichotomy K-means algorithm that at the beginning of the algorithm, the two farthest objects go as the initial center, forming two initial clusters that continue to divide until the k clusters are generated. This approach

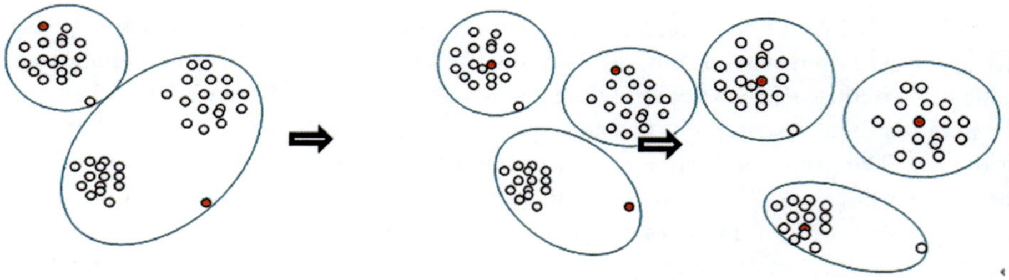

Fig. 1. A modified K-means simple clustering process

makes the algorithm less susceptible by initialization as it has only two centroids per step. In this way, objects in the same class can guarantee great similarity in subsequent processing, and very low similarity between different classes.

We introduce the Davies-Bouldin index (DBI), an indicator for assessing the merits of the clustering algorithms. The smaller the DBI value, the smaller the DBI value indicates that the clustering result is close within the same cluster, and the separation of different clusters is far away. That is, the smaller the inner class distance, the greater the interclass distance.

$DBI = \frac{1}{N}\sum_{i=1}^{N}\max(R_{ij})$ $R_{ij} = \frac{S_i+S_j}{M_{ij}}$ Among them, measure the similarity of class i to class j. S_i The average distance between the within-class data and the cluster center of mass was calculated with the formula of, X $S_i = \left\{\frac{1}{T}\sum_{j=1}^{T_i}|X_j - A_i|^q\right\}^{1/q}$ j Represents ote the j data point in class i; A_i Represents ote the center of class i; T_i Represents the number of data points in class i; q takes 1 represents the mean of the distance from each point to the center, and q takes 2 represents the standard deviation of the distance from each point to the center, and they can be used to measure the degree of dispersion, a $M_{ij} = \left\{\sum_{k=1}^{N}|a_{ki} - a_{kj}|^p\right\}^{1/p}$ kiRepresents the value of the K th attribute of the central point of class i, M_{ij}. This is the distance between class i and the center of class j.

The core of the improvement algorithm is as follows: (1) Calculate the two farthest objects in the dataset: x_1, x_2. (2) The distance of the remaining data objects to the x_1,x_2 was calculated separately, Divit into the class at the center with a small distance, marking while recording the minimum distance. (3) After the division is completed, the cluster center is recalculated to obtain the c_1, c_2 $\max\{\min(d(c_1, j), d(c_2, j))\}$. (4) Using the maximum and minimum distance method, j = 1,2 ,……n, gets x_j, DBI uses the formula to determine whether the best clustering center; (5) go to (2), the dataset repartition; (6) DBI is calculated according to the DBI formula$_{new}$And compare with the last calculated DBI_{old}Compcomparison, if $DBI_{new} < DBI_{old}$, To find a reasonable x_j, The k value was added by 1 to the previous one, and the new cluster center meeting the conditions cannot be found otherwise, and the cluster ends. (7) Class in turn until new clustering centers satisfying the conditions cannot be found, and finally output the clustering results.

3.3 Outlier Index

After clustering, outlier points need to be mined for which the outlier index is introduced. First, we have the following definitions:

$C = \{C_1, ..., C_k\}$ $|C_1| \geq ... \geq |C_k|$ $|C_i|$ $C_i(i = 1, ...k)$ Definition 1 is assumed to be a set of cluster sequences, and, where representing the number of objects in the cluster, k is the number of clusters.Given two positive numbers α and β, if the following equation holds, we define b as the boundary of size clustering.

$$|C_1| + ... + |C_b| \geq |D|\alpha \tag{8}$$

$$|C_b|/|C_{b+1}| \geq \beta \tag{9}$$

Thus, the collection of large clusters can be defined as $LC = \{Ci|i \leq b\}$, while the collection of small clusters can be defined as $SC = \{Cj|j > b\}$.

Definition 1 gives a method for quantifying size clustering. Equation (8) indicates the proportion α. of data for large clusters larger than the entire data. Therefore, clusters containing a large number of data objects are large clusters. While formula (9) indicates that large clusters are β times of small clusters, indicating that large and small clusters must be enough gap in scale. In this way, the effect of small clustering on the dataset can be ignored.

For any object, the x, outlier exponential factor of outlier (FOO) in the dataset:

$$FOO(x) = \frac{1}{|C_i|} dist(x, C_i) \tag{10}$$

Among these, dist (x, C_i) is the Euclidean distance between the object x and the cluster C_i center, |Ci| is the cluster CiNumber of objects in.

3.4 Improve the Spectral Clustering Algorithm Steps

Input: A dataset with n data, the parameter α,β

Output: m outlier candidates.

(1) Build the elements in the A, matrix of the similarity matrix

$$A_{ij} = \exp(-\frac{||x_i - x_j||^2}{\sigma_i \sigma_j}) \tag{11}$$

Among them, σi = dm(xi,xk) Is the weighted Eucan distance from the sample to the k nearest neighbor.

(2) The element D (i, j) on the main diagonal of the construction degree matrix D, degree matrix D is the sum of line i elements of the similarity matrix W, and the other elements are 0;

(3) Construct Laplace matrix L = D-W is $L = D^{-\frac{1}{2}} W D^{-\frac{1}{2}}$; then compute normative Rapp RMatmatrix:

$$L_{sys} = D^{-\frac{1}{2}} L D^{-\frac{1}{2}} = E - D^{-\frac{1}{2}} W D^{-\frac{1}{2}} \tag{12}$$

(4) The order is the L $\lambda_1 \leq \lambda_2 \leq ...\lambda_K$ sys. The first k minimum eigenvalues, v1,v2,...,vk. Is the corresponding eigenvector, then construct the matrix V = [v1,v2,...,vk $\in R^{n\times k}$], vi Is the column vector, i = 1,2,...K;

(5) The V was normalized by row to obtain the matrix U,

$$U = \frac{V_{ij}}{(\sum_j V_{ij}^2)^{\frac{1}{2}}} \tag{13}$$

(6) Each row in U was treated as a point within a sample space R^k and clustered with an improved K-means as the k class.

(7) The original sample x_i, Assign to the j class cluster if and only if the i row of the matrix U belongs to the j class cluster.

(8) The k class is divided into the large cluster set Cj, small cluster set Ci. according to the parameter α, β and the definition of large and small clusters.

(9) The outlier inde x FOO, was calculated for each data x in the cluster.

(10) Data points were sorted by outlier index size,and returns the first m outlier point.

4 Experimental Results and Analysis

4.1 The Artificial Dataset

We will conduct experiments with different datasets to evaluate our proposed algorithm and compare it with two outlier detection algorithms (KNN algorithm, LOF algorithm). We use artificial datasets to demonstrate that the proposed spectral clustering-based outlier detection method efficiently identifies outliers. The clustering effect of the artificial dataset Data1 is shown in Fig. 1, where A,B,C,D are four normal clusters of different sizes and E is the only outlier class, with some outlier data distributed around it.Specific information is shown in Table 1. Experiments were performed with the outlier clustering based detection algorithm, KNN algorithm, LOF algorithm, whose parameters α and β were set to 0.75 and 5, the parameter MinPts of LOF algorithm to 30, while the nearest neighbor parameter k of the detection algorithm was set to 4, and the results are shown in the following Fig. 2.

Table 1. Data1 information for the artificial dataset

Data1	A	B	C	D	E
Number of data	50	56	116	66	6
Number of outlier points	4	0	8	7	6

Figure 3 shows that the KNN algorithm can detect most outliers (indicated by ×) with detection errors; From Fig. 4, the LOF algorithm identified the outlier points slightly better than the KNN. It is seen from Fig. 5 that outlier points based on spectral clustering detection algorithms can correctly detect all outliers.

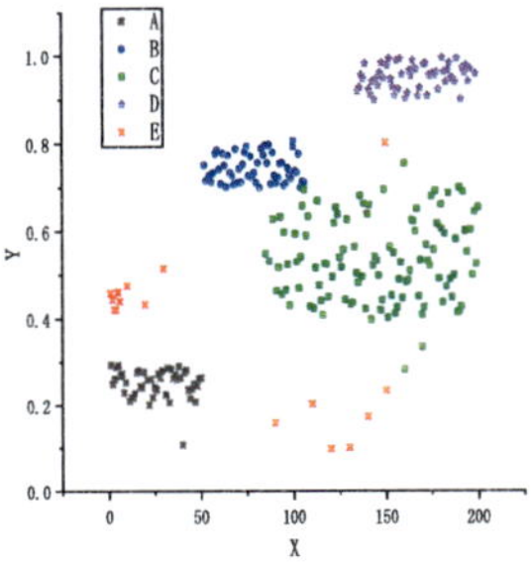

Fig. 2. Artificial dataset

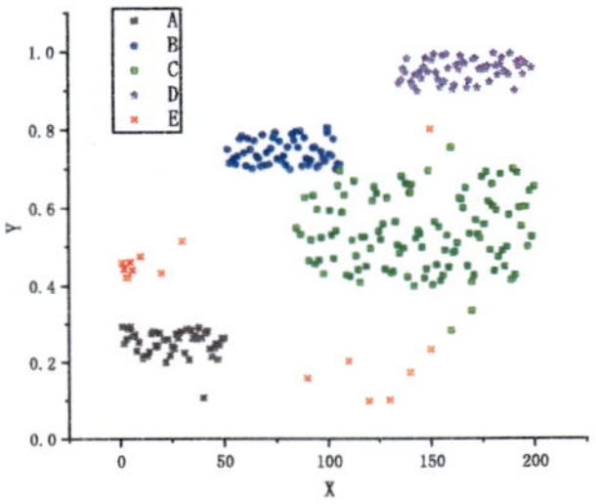

Fig. 3. KNN outlier detection algorithm

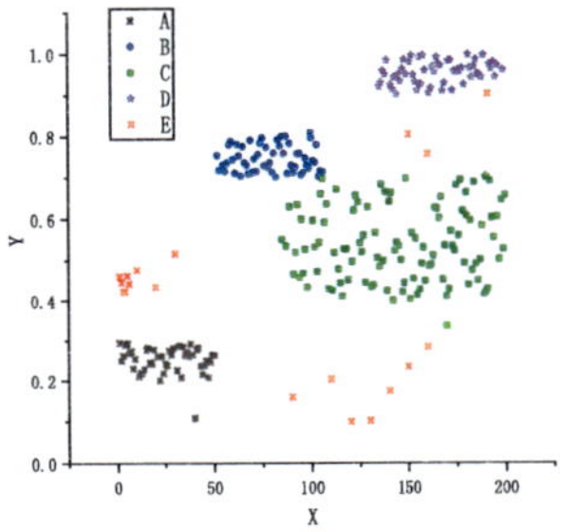

Fig. 4. LOF outlier detection algorithm

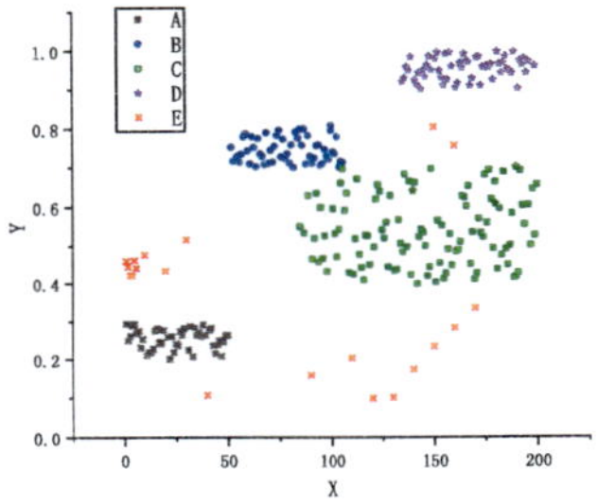

Fig. 5. Outlier detection algorithm based on spectral clustering

The above outlier detection effect were analyzed using precision rate (Precision) and false detection rate (false positive) indicators, indicating the accuracy of the algorithm, and the closer the value to 1, the better the detection effect.The false detection rate is a measure of the degree of misdetected irrelevant information. The closer the value is to 0, the better the effect.

Table 2. Detection and comparison of the three algorithms

Algorithm name	KNN algorithm	LOF algorithm	An algorithm based on spectral clustering
Precision rate (Precision)	0.801	0.875	0.996
Misdetection rate (false positive)	0.28	0.22	0.04

As can be seen from the Table 2, the accuracy rate of the KNN algorithm and LOF algorithm is lower than the outlier detection algorithm based on spectral clustering algorithm is significantly lower than the KNN algorithm and LOF algorithm, which shows that the spectral clustering-based outlier detection algorithm is significantly better than other algorithms.

The above outlier detection effect were analyzed using precision rate (Precision) and false detection rate (false positive) indicators, indicating the accuracy of the algorithm, and the closer the value to 1, the better the detection effect.The false detection rate is a measure of the degree of misdetected irrelevant information. The closer the value is to 0, the better the effect.

4.2 UCI Real Dataset

To test the validity of the algorithm on real data, we selected Breast Cancer Wisconsin (W B C), Arrhythmia, Glass, Wine, Yeast, Cardiotocogrpahy from the UCI standard dataset as the experimental dataset.Taxonomic properties in mixed data often can not be directly processed in outlier detection, and some datasets with semantically meaningful outliers are required to follow the knowledge background and expert knowledge of the data set, in data preprocessing to adapt to the outlier detection variant topic dataset.In the experiment, we preprocessed the dataset first.

(1) Processing of missing values: Since the standard outlier detection cannot handle the missing values, the specific number of missing values in each dataset is first counted. All missing data in the same property were filled with the sample mean in this property.
(2) Data type transformation: When we process numerical data, if we encounter non-numerical data categories that can take the form of converting it into numbers, we use LabelEncoder. in sklearn
(3) Downsampling: For larger data sets, outliers are less abundant in the data.The purpose of downsampling is to select a part of the data from most sets and recombine it with a few sets into a new data set to solve the data distribution imbalance.

The WBC dataset, Wisconsin-Breast Cancer, records the measurements of breast cancer cases in two categories: benign and malignant. This dataset was downsampled to 21 points, and the malignancies were treated as an outlier. Ionosphere is a binclassified dataset with the ionolayer raw data with dimension 34, removing all properties with zero values, so the total dimension is 33. “bad” is an outlier cluster, and “good” is a normal cluster. The Wine dataset is the result of a chemical analysis of wine produced from three different varieties, with the first category subsampled to 10 points and treated as an outlier, with the remainder being normal points. Glass contains several types of glass, of which class 6 is a distinct minority class considered outlier. Yeast is the yeast dataset that predicts cellular localization sites for proteins containing 64 outliers. The cardiac angiography (Cardio) dataset includes measurements of the fetal heart rate (FHR) and uterine contractions (UC) features of cardiac angiography classified by obstetric specialists. This is a taxonomic dataset classified as normal, suspicious, and pathological. For outliers detection, normal classes formed inliers, while pathological classes were downsampled to 176 points and were considered as outlier classes. Suspected categories were discarded. Basic information of the data set is provided in Table 3.

It can be seen from the table that the number of attributes of the dataset, namely the dimension of the dataset, is relatively large, and it is difficult to realize the high-dimensional space for data visualization, so the dimension is necessary to reduce to

Table 3. Basic information on the experimental dataset

Data set	Number of samples	Number of attributes	Number of outliers	The fraction of outlier points
WBC	278	30	21	5.6%
Ionosphere	351	33	126	36%
Wine	129	13	10	7.7%
Glass	214	9	9	4.2%
Yeast	1364	8	64	4.7%
Cardiotocogrpahy	1831	21	176	9.6%

directly reflect the distribution of the dataset. We conduct two-dimensional display on the principal component analysis.As shown in Fig. 6.

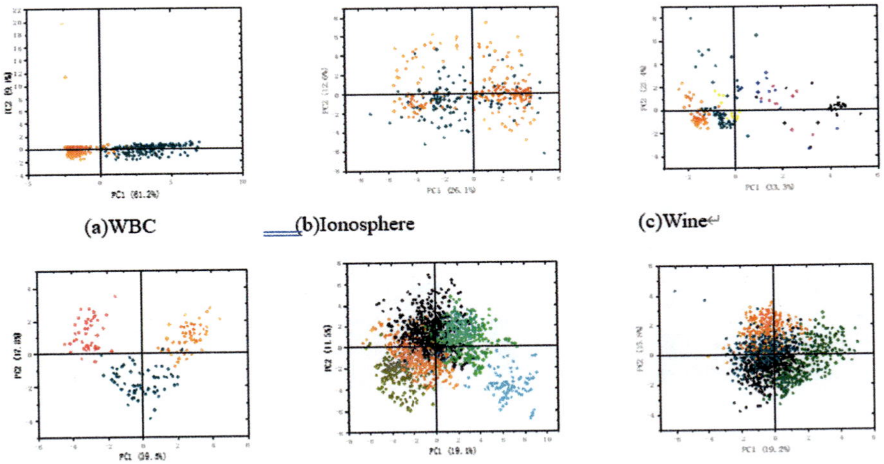

Fig. 6. Distribution of the datasets

In Fig, it is clear that the distribution of the dataset is shown.To visually show the distribution or density of the data, we show the relevant properties of the WBC, Arrhythmia, Glass, Wine, Yeast, Cardiotocogrpahy dataset with Heatmap.The dataset case shown in Fig. 7 provides a direct reference for selecting properties representing all variables for the main component analysis.The heat map shows in the form of the map, the more concentrated the number of attributes in a region, the darker the data identification color on the graph, which can be understood as little attribute difference; if the data in the area is relatively small, the lighter the color, the data distribution in the area.Heatmaps not only better understand the characteristic status of each attribute in the observed dataset, but also which region similar features of object properties are concentrated in the dataset.

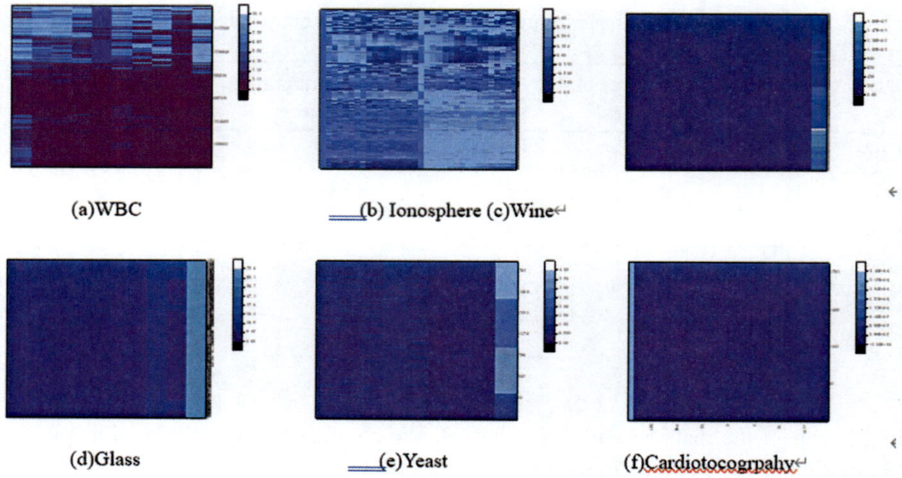

(a)WBC (b) Ionosphere (c)Wine

(d)Glass (e)Yeast (f)Cardiotocogrpahy

Fig. 7. The Heatmap of the dataset

The practical results of the three algorithms on these six datasets were evaluated by KNN, LOF, spectral clustering-based algorithms with precision P and recall R analysis.

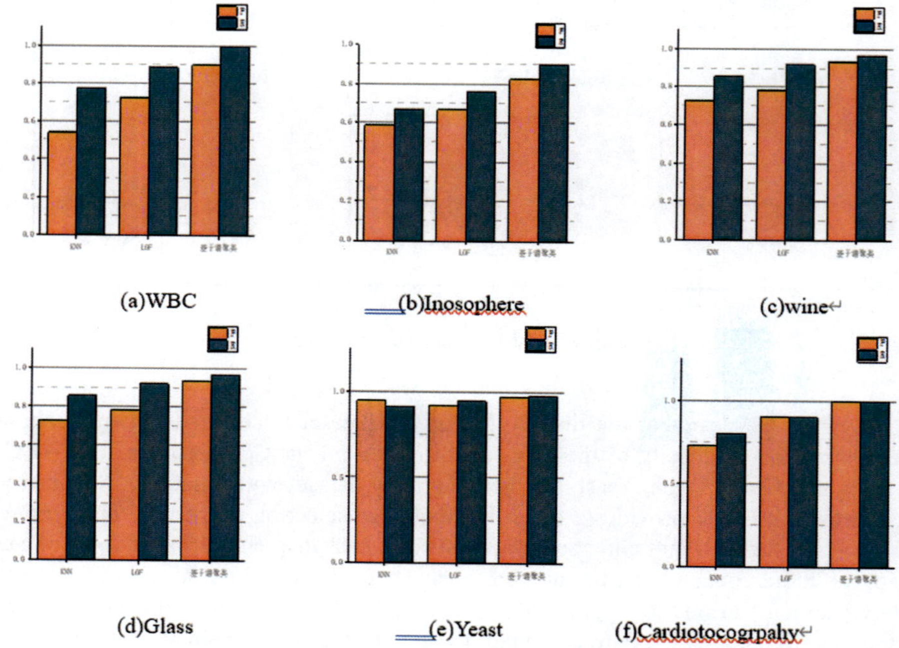

(a)WBC (b)Inosophere (c)wine

(d)Glass (e)Yeast (f)Cardiotocogrpahy

Fig. 8. Detection effect of the three algorithms on different datasets

As can be seen from Fig. 8, the KNN and LOF algorithms are greatly affected by the data dimension, and when the data dimension is high, the algorithm effect is not ideal.The spectral clustering-based outlier detection algorithm performs accuracy and recall over the other two algorithms on these six datasets and is sensitive to outlier points and can identify most of the outliers.

In the outlier detection algorithm, the dataset is quite different, and the index evaluation algorithm such as precision rate and error detection rate or recall rate could not be expressed very accurately, so we chose F1-score as the evaluation index. F1-score is the harmonic average of precision and recall, indicating that the algorithm is more efficient when F1 is high.

$$F1 = \frac{2 \times P \times R}{P + R} = \frac{2 \times TP}{M + TP - TN} \tag{14}$$

In the formula: P is exact rate (Precision), R is recall rate (Recall), TP is true rate (True Positive), TN is true negative rate (True Negative), and M is the total number of samples. The accuracy is $P = TP/(TP + FP)$, and the recall rate is $R = TP/(TP + FN)$. The confusion matrix includes the TP, TN, false positive rate FP (False Positive), the false negative rate FN (False Negative), the toal number of semples calculated is $M = TP + TN + FP + FN$. As shown in Fig. 9 are the curves detected F1 by three algorithms for six data sets.

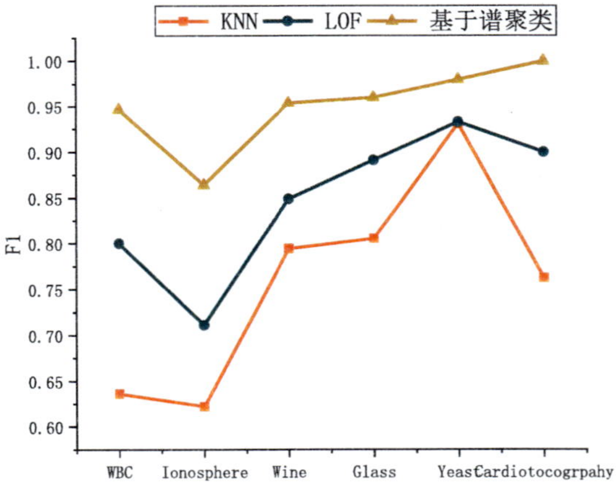

Fig. 9. The F1 curve of the dataset

As can be seen from Fig. 9, the outlier detection algorithm based on spectral clustering is obvious and relatively stable, and the KNN algorithm and LOF algorithm differ greatly on the outlier detection of different datasets and has poor effect.

5 Summary

In this paper we propose an effective outlier detection method based on adaptive spectral clustering that can detect both global and local outliers, solving the dependence of spectral dependence on the parameter and difficult number of clusters. This algorithm uses the Self-Tuning spectral clustering algorithm to reimprove the scaling parameters, while improving the K-means algorithm to make the clustering stable. The size cluster definition and the outlier exponent are used to identify the outliers. The feasibility and stability of the algorithm were verified by performing experiments with artificial and real datasets.

Acknowledgement. The project is funded by the National Major Science and Technology Special Project funded "New Technology Research and Equipment Development of Grain Measurement and Control" (No: 2017YFD0401004); "Research and Development and Demonstration of Grain Container Information Tracing Technology and Demonstration" (No: 2018YFD0401404); Open Project of Key Laboratory of Grain Information Processing and Control of the Ministry of Education (KFJJ-2016-103); Henan Provincial Science and Technology Research Project "Remote Internet of Things Experimental and Verification System Based on Cloud Platform" (No: 212102210170).

References

1. Hawkins, D.M.: Identification of Outliers. Springer Netherlands, Dordrecht (1980). https://doi.org/10.1007/978-94-015-3994-4
2. Huang, Q., et al.: Study of outlier identification methods. Softw. Guide **6**, 35–41 (2019)
3. Aggarwal, C.C.: Outlier analysis. In: Aggarwal, C.C. (ed.) Data Mining, pp. 237–263. Springer International Publishing, Cham (2015). https://doi.org/10.1007/978-3-319-14142-8_8
4. Domingues, R., Filippone, M., Michiardi, P., Zouaoui, J.: A comparative evaluation of outlier detection algorithms: experiments and analyses. Pattern. Recogn. **74**, 406–421 (2018)
5. Luxburg, U.V.: A tutorial on spectral clustering. Stats. Comput. **17**(4), 395–416 (2007)
6. Zelnik-Manor, L., Perona, P.: Self-Tuning Spectral Clustering. Advances in Neural Information Processing Systems (NIPS). MIT Press, Cambridge, MA (2004)
7. He, L., Xue, A.: Background outlier detection based on K-way spectral clustering. Comput. Eng. **39**(3), 197–202 (2013)
8. Tong, T., Gan, J., Wen, G., Li, Y.: One-step spectral clustering based on self-paced learning. Pattern. Recogn. Lett. **135**, 8–14 (2020)
9. Afzalan, M., Jazizadeh, F.: An automated spectral clustering for multi-scale data. Neurocomputing. **347**, 94–108 (2019)
10. Xie, J., Ding, L., Wang, M.: Unsupervised feature selection algorithm based on spectral clustering. Softw. J. **31**(04), 1009–1024 (2020)
11. Tan, M., Wen, G., Tong, T., Wu, L., Du, T.: The mutual KNN-based and normalized spectral clustering algorithm. Comput. Eng. Des. **40**(07), 1878–1884 (2019)

Measurement of Flexoelectric Response in Polyvinylidene Fluoride Beam

Jianfeng Lu[1,2](✉), Jian Wu[1,2], and Xinyong Li[1,2]

[1] School of Mechanical Engineering, Changshu Institute of Technology, Changshu 215500, People's Republic of China
Jianfeng.lu@cslg.edu.cn

[2] Jiangsu Key Laboratory of Recycling and Reuse Technology for Mechanical and Electronic Products, Changshu 215500, People's Republic of China

Abstract. Flexoelectricity describes the polarization response to the deformation gradient that exists in all dielectric materials, and is of great importance for understanding the gradient-induced physical phenomenon in micro/nano scale. In this paper, we explore the flexoelectric response of α-phase polyvinylidene fluoride (PVDF) cantilever beam undergoing quasi-static loading conditions. The flexoelectric coefficient of the PVDF is in the order of 10^{-7} C/m at room temperature and the value is 3–4 orders of magnitude higher than theoretical predication. Based on the experiment results, we analyzed the physical mechanism of the flexoelectricity in PVDF and then excavated the influence factors of effective electromechanical coupling coefficient. This work offers a very helpful experimental and theoretical basis for the design of the cantilever-based structural device with piezoelectric effect by using the flexoelectric effect.

Keywords: Flexoelectric coefficient · Experiment · Cantilever beam · PVDF

1 Introduction

Flexoelectricity, which describes the polarization response to the strain gradient in dielectric materials. The attributes of flexoelectricity has critically influenced academic dialogue on electromechanical coupling research field both from theoretical and experimental. The mathematical formulation of the flexoelectricity can be expressed as [1]

$$P_i = \mu_{ijkl} \frac{\partial \varepsilon_{kl}}{\partial x_j} \tag{1}$$

Here P_i is the induced electric polarization, ε_{kl} is the elastic strain, $\partial/\partial x_j$ denotes the gradient operator respect to x_j, and μ_{ijkl} refers to the fourth rank tensor of flexoelectric coefficient.

Compared with peizoelectricity, flexoelectricity has its own distinguishing features and gained much importance in recent years. Firstly, the flexoelectricity has no symmetry constraint, and exists in all solid dielectric materials, this characteristic broden the

Y. Wang et al. (Eds.): IWAMA 2021, LNEE 880, pp. 739–746, 2022.
https://doi.org/10.1007/978-981-19-0572-8_95

selection of flexoelectric materials. Secondly, the strain gradient scales inversely with the material diemension, this allows the flexoelectric effect to match and even domaint over piezoelectric in micro and nano scale.

The theoretical model of flexoelectricity was first proposed by Kogan to describe the energy coupling between induced polarization and mechanical strain gradient [2]. Accroding to Tagantsev's work [3], the flexoelectric coefficient of crystalline dielectric is estimated to be in the order of 10^{-10}~10^{-11} C/m. On the theoretical front, Shen and coworkers developed a mathematical formulation of flexoelectricity based on the variational principle, and this works provide the physical fundamentals and computational method for flexoelectricity [4]. On the experimental front, Cross et al. did many pioneering work in the study of flexoelectricity in ferroelectric materials, and the results indicate that the materials with high permittivity may have larger flexoelectric coefficient (3–4 orders magnitude) than theoretical prediction. However, this principle is not applicable to polymer materials. Fu and coworkers observed giant flexoelectric coefficient(78 μC/m) in polyvinylidene fluoride films, which the permittivity is less than 1/1000 of the ferroelectric material [5]. Poddar and coworker investigated the flexoelectric response of ferroelectric and relaxor of polyvinylidene fluoride polymers by using a simple cantilever measurement technique [6], and the temperature dependence of flexoelectric response in films of ferroelectric and relaxor of polyvinylidene fluoride polymer [7]. Chu et al. investigated the flexoelectric effect of several thermoplastic and thermosetting polymers of cantilevers with the flexoelectric coefficients in the order of 10^{-9}~10^{-8} C/m [8]. In our previous work [9], we derived an improved model to verify the existence of longitudinal flexoelectricity in α-phase PVDF truncated cone bar. These studies indicate that the flexoelectricity in polymer might be more complicated than that of ferroelectric material.

In this article, in light of the charge measurement, the transverse flexoelectric effect in α-phase PVDF polymer cantilever under bending is investigated. We will first give the relationship between the induced polarization charge and the effective flexoelectric coefficient, then quantitative measurement of flexoelectric coefficient. A summary of flexoelectric coefficients of materials at room temperature will be presented and discussed, and the effective electromechanical coupling coefficient of PVDF polymer is calculated.

2 Experimental Descrption and Anslysis

The flexoelectric measurement was performed on cantilever by using experimental setup as show in Fig. 1. The α-phase PVDF polymer cantilever (40 mm length, 15 mm width, and 1.5 mm thickness) was sputtered with silver as electrode both on the bottom and upper surface, and the length of the electrode was 20 mm. The input sinusoidal signal was generated by Tektronix AFG 3022C and send to the voice coil motor by an NF HSA4014 power amplifier. The deflection of the cantilever was measured by Keyence LK-H025 laser displacement sensor and the induced polarization charge was monitored and collected by B&K-2962 charge amplifier, both of them were displayed and stored in the THS-3014 oscilloscope. The material parameters of PVDF are presented in Table1.

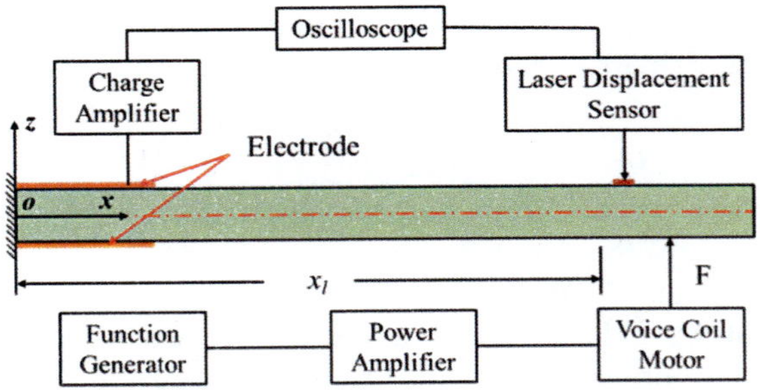

Fig. 1. Experimental setup for the measurement of the flexoelectric coefficient μ12.

Table. 1. Material parameters of PVDF cantilever

Item	Mass density ρ (kg/m^3)	Young's odulus E(GPa)	Poisson's ratio ν	Permittivity ε_0	Natural frequency(Hz)
Value	1.8×10^3	2.1	0.28	11.1	163

The Cartesian coordinated is set to the neutral layer of the cantilever, and x axis along the length of the beam while z axis along the thickness. The mode shape of natural vibration of a cantilever can be expressed as

$$\begin{aligned} W_i(x) = A_r[(sin\beta_i l - sinh\beta_i l)(sin\beta_i x - sinh\beta_i x)+ \\ +(cos\beta_i l + cosh\beta_i l)(cos\beta_i x - cosh\beta_i x)] \end{aligned} \tag{2}$$

Here $A_i = C_1/sin\beta_i l - sinh\beta_i l, i = 1, 2 \ldots$ =1,2, C_1 is determined by the electrode position. For simplicity, we only consider the fundamental mode where $i = 1$ and $\beta_i l = 1.875$. The strain gradient of the cantilever along the thickness can be written as

$$\varepsilon_{x,z} = \frac{\partial \varepsilon_x}{\partial z} = -\frac{\partial^2 W(x)}{\partial x^2} \tag{3}$$

$W(x)$ is actually a superposition of different deformation modes.

For the centrosymmetric crystal cantilever, when suffering from bending, the flexoelectric effect can be described as [1]

$$P_3 = \mu_{11}\frac{\partial \varepsilon_{33}}{\partial x_3} + \mu_{13}\left(\frac{\partial \varepsilon_{11}}{\partial x_3} + \frac{\partial \varepsilon_{22}}{\partial x_3}\right) + \mu_{44}\left(\frac{\partial \varepsilon_{13}}{\partial x_1} + \frac{\partial \varepsilon_{23}}{\partial x_2}\right) \tag{4}$$

To simplify the above expression, the contributions induced by shear strain ε_{13} and ε_{23} are neglected. The polarization can be simplified as

$$P_3 = \mu_{11}\frac{\partial \varepsilon_{33}}{\partial x_3} + \mu_{13}\left(\frac{\partial \varepsilon_{11}}{\partial x_3} + \frac{\partial \varepsilon_{22}}{\partial x_3}\right) = [\nu\mu_{11} + (1+\nu)\mu_{13}]\frac{\partial \varepsilon_{11}}{\partial x_3} \tag{5}$$

Here $\mu_{13}^{eff} = \nu\mu_{11} + (1+\nu)\mu_{13}$ is effective transerve flexoelectric coefficient and ν is the Poisson's ratio. Thus the induced charge can be written as

$$Q = \int_{A_e} P_3 dA_e = \mu_{13}^{eff} \int_{A_e} \frac{\partial \varepsilon_{11}}{\partial x_3} dA_e \tag{6}$$

Here A_e is the area of the electrode.

In the experiment, the PVDF cantilever is fixed at one end and the free end is driven by sinusoidal signal. The induced charge and the displacement under different frequency were recorded. The strain gradient can be calculated by Eq. (3) with the assistance of Eq. (2). The effective flexoelectric coefficient can be calculated from Eq. (6). When the frequency is 1.0 Hz, the relationship between induced charge and the strain gradient is shown in Fig. 2, it can be perceived that the strain gradient and the induced charge fit quite well with linear relationship. From which the effective flexoelectric coefficient $\mu_{13}^{eff} = 2.8 \times 10^{-7}$ C/m is calculated from the slope, and the effective flexo-coupling coefficient calculated from $f = \mu_{13}^{eff}/\varepsilon_0\varepsilon_r$ is 2849 V. The result is 3 orders of magnitude larger than theoretical prediction [10].

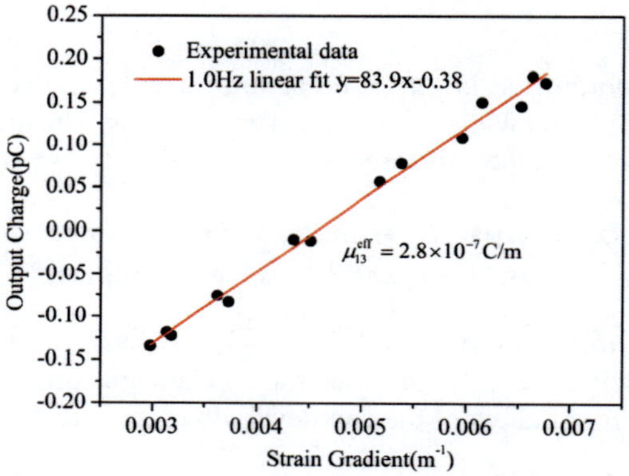

Fig. 2. The induced flexoelectric charge Q as a function of strain gradient for PVDF beam under the frequency of 1.0 Hz.

The flexoelectric coefficient and flexocoupling coefficient at different frequencies are shown in Table 2. It can be detected that minimum of the effective transverse flexoelectric coefficient is 2.61×10^{-7} C/m at the frequency of 0.5 Hz, the maximum is 2.87×10^{-7} C/m at 1.5 Hz, and the variation of which is 10%. It can be inferred that the flexoelectric coeffcients are 2 order smaller than that of the ferroelectric material [11–14].

Table. 2. Effective flexoelectric coefficient of PVDF material

Frequency (Hz)	0.5	1.0	1.5	2.0	2.5
Flexoelectric coefficient ($\times 10^{-7}$C/m)	2.61	2.80	2.87	2.80	2.82
Flexocoupling coefficient ($\times 10^{3}$V)	2.66	2.85	2.92	2.85	2.87

The comparison of flexocoupling coefficient in PVDF material with ferroelectric is shown in Table 3. It can be clearly seen that the flexoelectric coefficient in PVDF material is 2 order smaller that ferroelectric. As suggested by Tagantsev [15], the flexoelectric coefficient should be linear to with the permittivity in dielectric crystal. The realtive permittivity of the PVDF material is 11.1, while the ferroelectric is in the order of 10^4 [16–19], the difference is as high as 10^3. However, the flexocoupling coefficient of PVDF is 100 higher than that of ferroelectric. Hence, with the consideration of dielectric properties of the materials, the PVDF polymer appeared better electromechanical coupling effect. Besides, the PVDF ploymer has its special characteristic for the flexibility and machinability, and will play an important role in the future research of flexoelectricity.

Table. 3. Summary of flexoelectric coefficients of materials at room temperature

Material	Transverse flexoelectric coefficient	Relative permittivity	Flexocoupling coefficient
	$\mu_{12}(\times 10^{-6}$ C/m)	χ/ε_0	$f = \mu_{13}^{eff}/\varepsilon_0\varepsilon_r$
ST [12]	100	≈20, 000	564
PZT [18]	2	≈2, 200	102.7
BT [20]	5	≈10, 000	56.5
PMN($PbMg_{1/3}NB_{2/3}O_3$) [17]	4	≈13, 000	34.8
PVDF Cantilever beam	0.29	≈11.1	2849

It is well known that the effective electromechanical coupling (EMC) coefficient is an important parameter to evalute the capability of energy conversion, and can be obtained by measuring the energy stored in the piezoelectric strunture [21]. Previous study has shown that the effective electromechanical coupling coefficient in non piezoelectric material can be determined as [22]

$$k_{EMC}^2 = \frac{12\mu_{13}\mu_{13}}{Ea_{33}h^2 + 12\mu_{13}\mu_{13}} \tag{7}$$

Here $a_{33}= \varepsilon_0\varepsilon_r$ is the permittivity. Together with parameters in Table 1, the permittivity $a_{33} = 9.83 \times 10^{-11}C^2/N \bullet m^2$, and the frequency is set at 1.0 Hz where $\mu_{13} = 2.8\times 10^{-11}C/m$. The reported document has shown that the effective electromechanical coupling coefficient $k_{31} = 0.12$ [23]. The EMC coefficient in α-phase PVDF material is displayed in Fig. 3.

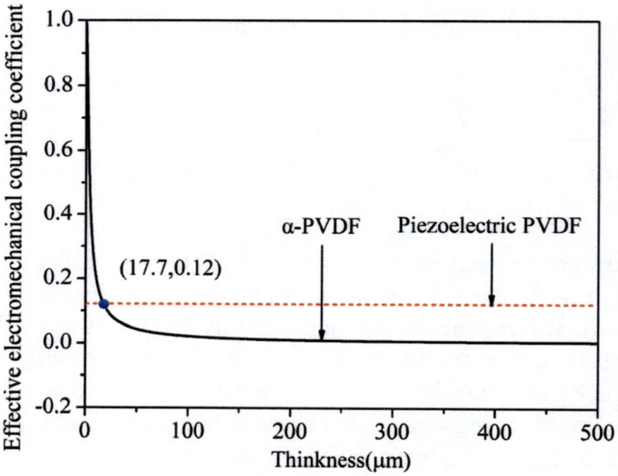

Fig. 3. The EMC coefficient as a function of thickness of the cantilever beam.

It is clearly seen that the EMC coefficient is dependent on the thickness of the cantilever, k_{EMC} increases with decrease of the thickness h; It must be noticed that when the thickness is less than 50 μm, the k_{EMC} increase sharply; And when the thickness reduced to 17.7 μm, the EMC coefficient of α-phase PVDF material is equal to the value of the piezoelectric PVDF material; Besides, the EMC coefficient of α-phase PVDF material is greater than that of the piezoelectric PVDF material when the thickness of cantilever less than 17.7 μm, and the flexoelectric response increases sharply. The above observed event manifested that the α-phase PVDF material can replace the piezoelectric PVDF material in the cantilever structure when the thickness less than 17.7 μm. This finding suggests a theoretical basis for the fabrication of cantilever structural device with piezoelectric effect using non-piezoelectric material.

3 Conclusions

In conclusion, the transverse flexoelectric effect of α-phase PVDF material was investigated by charge measurement. The strain gradient was generated by bending of the cantilever structure and the charge was collected to calculate the induced polarization. The transverse flexoelectric coefficient was calculated in the order of 10^{-7} C/m and the flexocoupling coefficient was 1 or 2 orders of magnitude larger than ferroelectric material. The EMC coefficient was dependent on the thickness of the cantilever structure, and a fairly strong flexoelectric response was observed when the thickness was less than 17.7 μm. This study provided an alternative way for the fabrication of piezoelectric cantilever with non-piezoelectric constituent.

Acknowledgment. This work is supported by Doctoral Science Foundation of Changshu Institute of Technology (No. KYZ2018036Q).

References

1. Ma, W.: A study of flexoelectric coupling associated internal electric field and stress in thin film ferroelectrics. Phys. Status Solidi (b) **245**(4), 761–768 (2008). https://doi.org/10.1002/pssb.200743514
2. Kogan, S.M.: Piezoelectric effect during inhomogeneous deformation and acoustic scattering of carriers in crystals. Soviet Phys. Solid State **5**(10), 2069–2070 (1964)
3. Tagantsev, A.: Theory of flexoelectric effect in crystals. Zh. Eksp. Teor. Fiz. **88**(6), 2108–2122 (1985)
4. Shen, S., Hu, S.: A theory of flexoelectricity with surface effect for elastic dielectrics. J. Mech. Phys. Solids **58**(5), 665–677 (2010)
5. Baskaran, S., et al.: Giant flexoelectricity in polyvinylidene fluoride films. Phys. Lett. A **375**(20), 2082–2084 (2011)
6. Poddar, S., Ducharme, S.: Measurement of the flexoelectric response in ferroelectric and relaxor polymer thin films. Appl. Phys. Lett. **103**(20), 202901 (2013). https://doi.org/10.1063/1.4829622
7. Poddar, S., Ducharme, S.: Temperature dependence of flexoelectric response in ferroelectric and relaxor polymer thin films. J. Appl. Phys. **116**(11), 114105 (2014). https://doi.org/10.1063/1.4895988
8. Chu, B., Salem, D.: Flexoelectricity in several thermoplastic and thermosetting polymers. Appl. Phys. Lett. **101**(10), 103905 (2012)
9. Lu, J., et al.: Improved approach to measure the direct flexoelectric coefficient of bulk polyvinylidene fluoride. J. Appl. Phys. **119**(9), 094104 (2016)
10. Zubko, P., Catalan, G., Tagantsev, A.K.: Flexoelectric effect in solids. Annu. Rev. Mater. Res. **43**, 387–421 (2013)
11. Ma, W., Cross, L.E.: Large flexoelectric polarization in ceramic lead magnesium niobate. Appl. Phys. Lett. **79**(26), 4420–4422 (2001)
12. Wenhui Ma, L., Cross, E.: Flexoelectric polarization of barium strontium titanate in the paraelectric state. Appl. Phys. Lett. **81**(18), 3440–3442 (2002). https://doi.org/10.1063/1.1518559
13. Shu, L., Wei, X., Jin, L., Li, Y., Wang, H., Yao, X.: Enhanced direct flexoelectricity in paraelectric phase of Ba(Ti0.87Sn0.13)O3 ceramics. Appl. Phys. Lett. **102**(15), 152904 (2013). https://doi.org/10.1063/1.4802450
14. Li, Y., Shu, L., Huang, W., Jiang, X., Wang, H.: Giant flexoelecticity in Ba0. 6Sr). 4TiO3/Ni0. 8Zn0. 2Fe2O4 composite. Appl. Phys. Lett. **105**(16), 162906 (2014). https://doi.org/10.1063/1.4899060
15. Yudin, P., Tagantsev, A.: Fundamentals of flexoelectricity in solids. Nanotechnology. **24**(43), 432001 (2013)
16. Hana, P.: Study of flexoelectric phenomenon from direct and from inverse flexoelectric behavior of PMNT ceramic. Ferroelectrics **351**(1), 196–203 (2007)
17. Ma, W., Cross, L.E.: Observation of the flexoelectric effect in relaxor Pb (Mg1/3Nb2/3) O3 ceramics. Appl. Phys. Lett. **78**(19), 2920–2921 (2001)
18. Wenhui Ma, L., Cross, E.: Strain-gradient-induced electric polarization in lead zirconate titanate ceramics. Appl. Phys. Lett. **82**(19), 3293–3295 (2003). https://doi.org/10.1063/1.1570517
19. Ma, W., Cross, L.E.: Flexoelectric effect in ceramic lead zirconate titanate. Appl. Phys. Lett. **86**(7), 072905 (2005)
20. Ma, L.W., Cross, E.: Flexoelectricity of barium titanate. Appl. Phys. Lett. **88**(23), 232902 (2006). https://doi.org/10.1063/1.2211309

21. Chen, Q.M., Wang, Q.M.: The effective electromechanical coupling coefficient of piezoelectric thin-film resonators. Appl. Phys. Lett. **86**(2), 3 (2005)
22. Xu, L., Shen, S.: Size-dependent piezoelectricity and elasticity due to the electric field-strain gradient coupling and strain gradient elasticity. Int. J. Appl. Mech. **05**(02), 1350015 (2013). https://doi.org/10.1142/S1758825113500154
23. Song, J., et al.: Design optimization of PVDF-based piezoelectric energy harvesters. Heliyon **3**(9), e00377–e00377 (2017)

Research on the Starwheel Loading Performance of the Roadheader

Tianjiao Wu, Kuidong Gao(✉), Lisong Lin, Yuanjin Zhang, and Sheng Chen

College of Mechanical and Electronic Engineering, Shandong University of Science and Technology, Qingdao 266590, China

Abstract. The cantilever roadheader is widely used in roadway excavation. The loading mechanism in roadheader directly affects the working efficiency of roadheader roadway excavation. As the core component of the whole loading mechanism, the star wheel plays a vital role. In order to study the loading performance of the star wheel, the loading process of the star wheel is simulated by discrete element software, and the particle motion variation law and particle loading efficiency of the star wheel under different speed parameters and teeth number parameters are studied. The results show that the loading capacity of particles can be improved by increasing the star wheel speed and the number of star wheel teeth in a certain range. The research results can provide technical guidance for the structural design of the star wheel and the selection of speed parameters.

Keywords: Boom-type roadheader · Material loading · Star wheel · Particle

1 Introduction

Cantilever roadheader is the core equipment of coal mine roadway heading face. Its working performance and equipment reliability are the key factors affecting the safe and efficient production of coal mines [1]. The loading mechanism is mainly composed of shovel plate and star wheel, which is responsible for collecting and loading the broken rock broken by roadheader [2, 3].

There are a lot of reports on the research of roadheader loading in the existing literature. Wang Shaowu measured the corresponding turntable speed and loading data through the loading experiments of different star wheels, and summarized the relationship between star wheel speed, chain speed and shipping capacity [4]. Liang analyzed the force situation of the star wheel model according to the coal dumping situation when the star wheel loading mechanism raked and transported the coal, and determined the minimum speed of the star wheel under this working condition [5]. Zhang compared the advantages and disadvantages of the overall and split structure of the shovel plate, and gave a calculation method to determine the size of the star wheel, the speed range, and the loading power [6]. Gao studied the performance characteristics of the loading mechanism of the roadheader, analyzed the motion state of the material pushed by the star wheel mechanism, determined various loading mechanism parameters related to it [7]. Xu analyzed the parameter theory of the star wheel loading mechanism, studied

Y. Wang et al. (Eds.): IWAMA 2021, LNEE 880, pp. 747–758, 2022.
https://doi.org/10.1007/978-981-19-0572-8_96

the method of determining the size of the star wheel structure, and gave a specific size of the star wheel loading mechanism [8]. Xiaohuo studied the dynamic behavior of the helical three-star wheel loading mechanism, used the Lagrangian method to establish the mechanical model of the helical three-star wheel, and analyzed the equivalent stress cloud diagram of the star wheel [9]. Linhui applied the finite element analysis software to analyze the influence of the opening of the loading mechanism on the structural strength and the influence of the pile depth on the loading efficiency [10]. Fernadez used the discrete element method to simulate the conveying process of spherical materials by a horizontal screw conveyor, and obtained the influence of different screw conveying rod structures and parameters on the movement of materials in the feeding hopper [11]. Shimizu and Cundall used the discrete element method to simulate the movement of particles in a horizontal screw conveyor and a vertical screw conveyor for the first time [12]. Moysey and Thompson used the discrete element method to simulate a single-head screw conveyor. The results show that the discrete element method can well represent the movement behavior of the particle group during the conveying process [13]. Mcbridge and Cleary used the discrete element method to influence the internal speed of the vertical screw conveyor, the pitch of the screw blades and the material parameters on the conveying performance [14].

Most of the existing studies have studied the body structure of the loading mechanism by finite element method, and there are few reports on the material loading effect and material particle distribution. Therefore, this paper uses the discrete element method to study the influence law of the number of star wheel teeth and rotating speed on the loading effect of roadheader and the distribution of gravel. The research results provide a reference for the reliability design and working parameter matching of roadheader loader.

2 Materials and Methods

2.1 Contact Model of the Particle

Herts-Mindlin contact model is the most commonly used in discrete element simulation. It has the characteristics of high accuracy and high calculation efficiency. The schematic diagram of the model is shown in the Fig. 1. The dynamic system of normal contact force and tangential contact force is a spring damping system.

In the model, the normal force in the contact between particles and particles and between particles and wall is as follows:

$$F_n = \frac{4}{3} E^* \sqrt{R^*} \delta_n^{\frac{3}{2}} \tag{1}$$

$$\frac{1}{E^*} = \frac{(1-\mu_i^2)}{E_i} + \frac{(1-\mu_j^2)}{E_j} \tag{2}$$

$$\frac{1}{R^*} = \frac{1}{R_i} + \frac{1}{R_j} \tag{3}$$

where E^* is the equivalent Young's modulus, GPa; R^* is the equivalent radius, m; δ_n is the normal overlap, m; μ is the Poisson's ratio; R is the contact curvature radiuses, m.

The tangential force in the contact between particles and particles and between particles and the wall in the model can be given by the formula:

$$F_t = -S_t\delta_t \tag{4}$$

$$S_t = 8G^*\sqrt{R^*\delta_n} \tag{5}$$

where δ_t is the tangential overlap, m; S_t is the tangential stiffness, N/m; G^* is the equivalent shear modulus (Fig. 2).

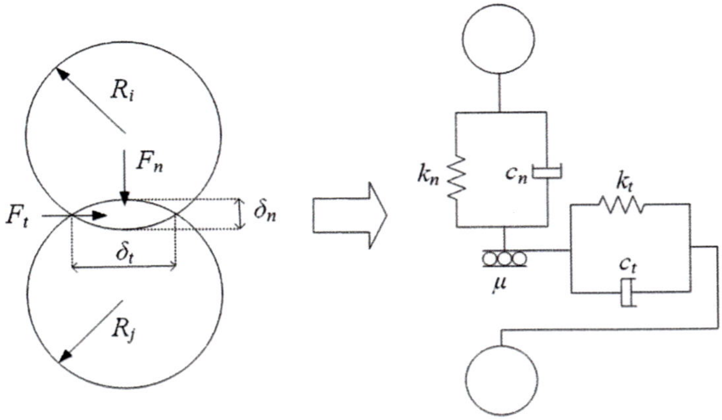

Fig. 1. Contact model of the particle.

2.2 Geometric Model of Particles

In the simulation process, spherical particles are widely used in many fields because of their simple shape, few particle contact surfaces and high calculation efficiency of computer equipment. However, spherical particles have a disadvantage that can not be ignored. The contact between particles belongs to point contact, and sliding friction has little effect on the movement of particles. The shape of particles has an important influence on the motion behavior, and the loaded rock particles can not guarantee the consistency of size and shape. Therefore, according to the geometry of rock particles, four typical particle models are used for modeling in discrete element, as shown in Table 1. The four typical particle shapes are single spherical, triangular, pyramid and flake particles.

In order to study the characteristics of these four typical particles, the repose angle stacking calibration test was carried out. In the simulation process, the more spherical surfaces make up the particle shape, the greater the calculation time and cost, and the lower the calculation efficiency. Therefore, based on the calculation time T_1 of spherical particles with the simplest shape and the highest calculation efficiency, T_i is

Table 1. Four typical shapes of particle in DEM

sphere	triangle	tetrahedron	flake

the calculation time of other particles, and T_1/T_i is the calculation efficiency standard, the relationship between different particle stacking angles and calculation efficiency is obtained.

The results show that the repose angle of spherical, triangular, tetrahedral and flake particles is larger and larger, and the stacking time is longer and longer. In other words, the more spherical particles, the larger the stacking angle of particles, and the simulation calculation will increase. Therefore, within the acceptable error range, tetrahedral particles can be used as the particle shape in the simulation to obtain accurate results.

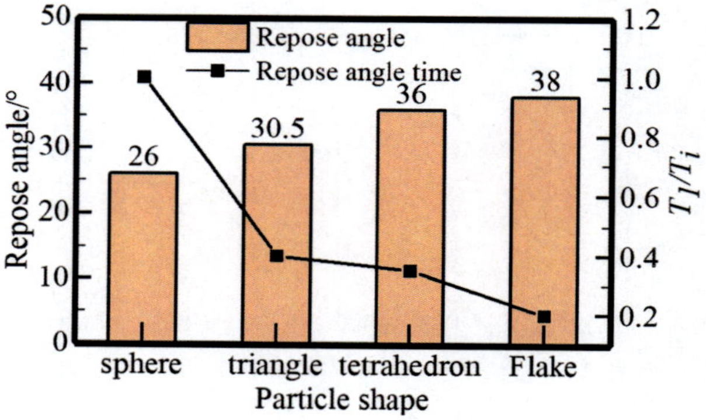

Fig. 2. Variation of repose angle with particle shape

2.3 Modeling and Simulation

The shovel plate part of the roadheader is located below the cutting part, connected with the first conveyor, and in front of the walking part and the body part. The shovel plate part is composed of a shovel plate, a driving device and a driven wheel device. The shovel plate part drives the star wheel to rotate through the left and right hydraulic motors, to realize the function of loading coal and rock.

Discrete element software can establish the geometric model exactly the same as the structure of roadheader loader, but it will cause too much computation. In this paper, the simplified model of roadheader loader is used for simulation analysis. The simplified roadheader loader is mainly composed of shovel plate and star wheel, as shown in Fig. 3. The shovel plate of the roadheader consists of three parts: the middle shovel plate, the left shovel plate and the right shovel plate. The simplified model of the shovel plate used in the performance simulation of the loading mechanism in this paper is to combine the main shovel plate, the left shovel plate and the right shovel plate into a whole, omit the position where the main shovel plate is installed on the first conveyor, and simplify the driving device of the star wheel. In order to study the relationship between material loading performance and the number of star gear teeth, two teeth, three teeth, four teeth and five teeth star gears are used for discrete element simulation.

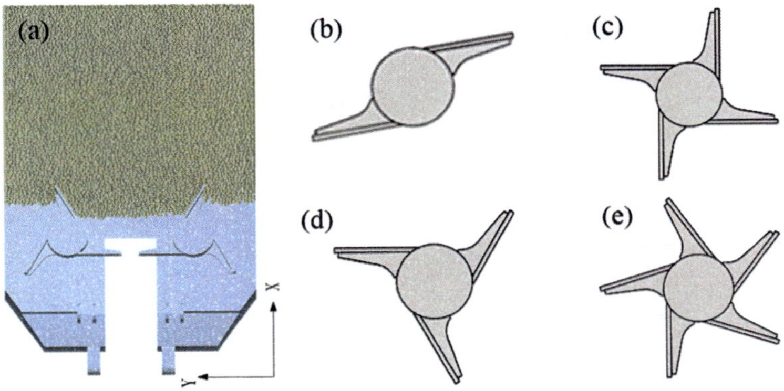

Fig. 3. Loader mechanism analysis model: (a) shovel plate; (b) the star wheel with two teeth; (c) the star wheel with three teeth; (d) the star wheel with four teeth; (e) the star wheel with five teeth.

3 Results Analysis and Discussion

3.1 Effect of Rotating Speed on Particle Loading

In order to explore the influence of different number of teeth on the material particle loading process at different speeds, the propulsion speeds of two teeth, three teeth, four teeth and five teeth star wheels are set to 0.05 m/s, and the speeds are set to 20 rpm, 25 rpm, 30 rpm, 35 rpm and 40 rpm respectively. The influence characteristics of rotating speed on materials loaded by star wheels with different teeth numbers are obtained, and the results are shown in Fig. 4.

The results show that the loading quantity curve of material particles gradually presents a linear inclined trend during the advancement of the shovel plate. During the half to the end of the discrete element simulation, that is, the loading process of the star wheel on the material particles tends to be stable, and the loading quantity curve of the material particles can be approximately regarded as a linear change. Under the same rotating speed of the star wheel, the loading quantity curve of material particles with different number of star wheel teeth changes obviously. With the increase of the number of star gear teeth, the inclined slope of the material loading quantity curve increases gradually. It should be noted that under the condition of the same number of starwheel teeth, changing the rotating speed of the starwheel will affect the inclination of the material particle loading quantity curve. The slope of loading quantity curve will increase with the increase of star wheel speed. This situation is more obvious on the loading quantity curve of the star wheel with a small number of teeth. On the loading quantity curve with a large number of star gear teeth, the slopes tend to be close to each other.

At the same speed, with the increase of the number of star gear teeth, the material loading speed increases, and the material particle loading quantity curve becomes smoother and smoother. The conclusion shows that with the increase of the number of star gear teeth, the randomness of loading decreases and the reliability increases. Therefore, the application of multi teeth star wheel can be properly considered to improve the loading reliability and efficiency.

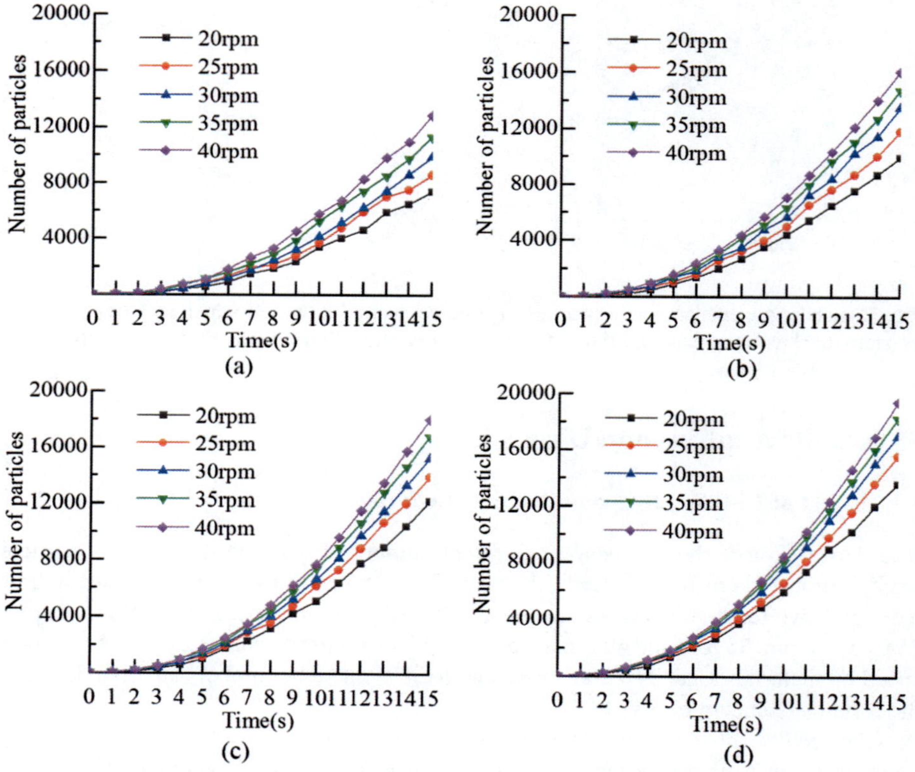

Fig. 4. The influence characteristics of rotating speed on materials loaded by star wheels with different teeth numbers

3.2 Effect of Number of Teeth of Star Wheel on Particle Distribution

To study the influence of the number of star wheel teeth on the particle distribution during loading, the particle loading effects of different star wheels at the same time are obtained under the condition of star wheel speed of 30 rpm, the results are shown in Fig. 5.

The results show that at the same time and speed, the number of particles attached in front of the teeth of two teeth, three teeth, four teeth and five teeth star gears decreases in turn. At the same time, it can be seen that the number of particles staying on the back and both sides of the shovel plate of the two teeth, three teeth, four teeth and five teeth star wheel is also decreasing. Two teeth, three teeth, four teeth and five teeth star wheels also expand the speed range of unloaded particles in front of the shovel plate. This is because under the premise of the same time and speed, the number of loading processes of the star wheel is directly related to the number of teeth of the star wheel, that is, in one rotation cycle of the star wheel, the two teeth star wheel can be loaded twice, the three teeth star wheel can be loaded three times, and so on, the five teeth star wheel can be loaded five times. The star wheel with a large number of teeth has a higher loading capacity, and each teeth has less load particles in the loading process. Correspondingly, the wear degree of its teeth is light and its service life is long. The number of teeth has a great impact on the star loader. The more the number of teeth of the star wheel, the more the loading capacity of the loader per unit time.

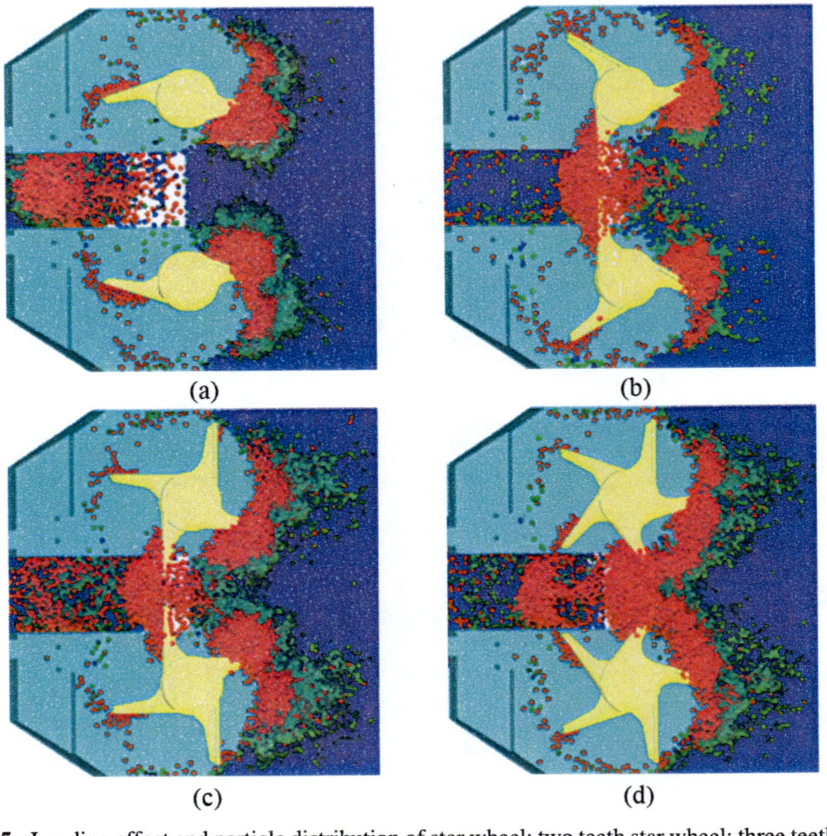

Fig. 5. Loading effect and particle distribution of star wheel: two teeth star wheel; three teeth star wheel; four teeth star wheel; five teeth star wheel.

3.3 Influence of Rotating Speed on Material Distribution and Movement

To study the influence of star wheel speed on particle motion during loading, simulations were carried out at speeds of 20 rpm, 25 rpm, 30 rpm, 35 rpm and 40 rpm respectively. Taking the four teeth star wheel as an example, the particle distribution state at different speeds is obtained, as shown in Fig. 6. As shown in Fig. 7, the influence characteristics of rotating speed on the moving speed of loaded particles are obtained.

As shown in Fig. 6, for a star wheel with the same number of teeth, with the increase of rotating speed, the amount of particles adhering to the front of the teeth is gradually decreasing, and the amount of particles staying on both sides and rear side of the shovel plate is also decreasing. This is because during the loading process, the rotation of the gear teeth transfers the kinetic energy to the particles, and the particles obtain the speed to separate from the gear teeth in the loading area of the shovel plate, so as to complete the loading process. The star wheel with high speed can make the particles obtain higher speed, and it is easier to separate the particles from the gear teeth to complete the loading. Properly increasing the rotating speed of the star wheel can reduce the number of particles

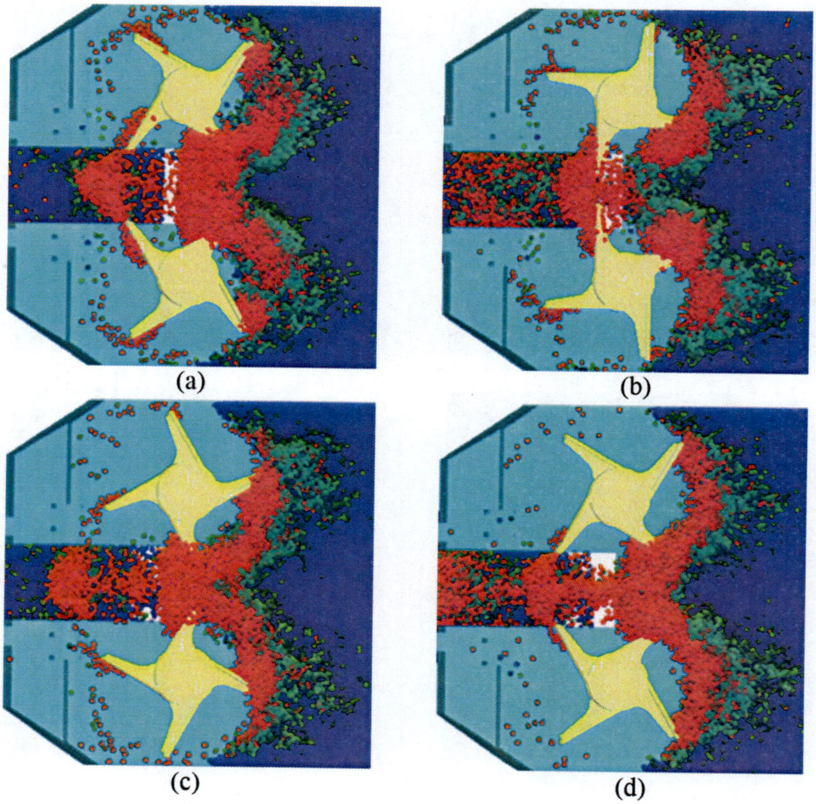

Fig. 6. The particle distribution state of the four teeth star wheel at different speeds: (a) 25 rpm; (b) 30 rpm; (c) 35 rpm; (d) 40 rpm.

staying on the shovel plate, reduce the wear of the shovel plate and improve the service life of the loader.

As shown in Fig. 7, when the star wheel speed is the same, the fluctuation range and fluctuation period of particle speed also show an obvious decreasing trend with the increase of the number of teeth; The fluctuation period of particle velocity decreases with the increase of rotating speed. It is worth noting that the fluctuation range of particle velocity loaded by two teeth star wheel changes obviously with the speed, while the fluctuation range of particle velocity loaded by four teeth and five teeth star wheel changes less. On the whole, the more the number of star gear teeth, the more stable the particle velocity is. When the loader works, you can choose higher speed and more star gear teeth to ensure the stability of the working process.

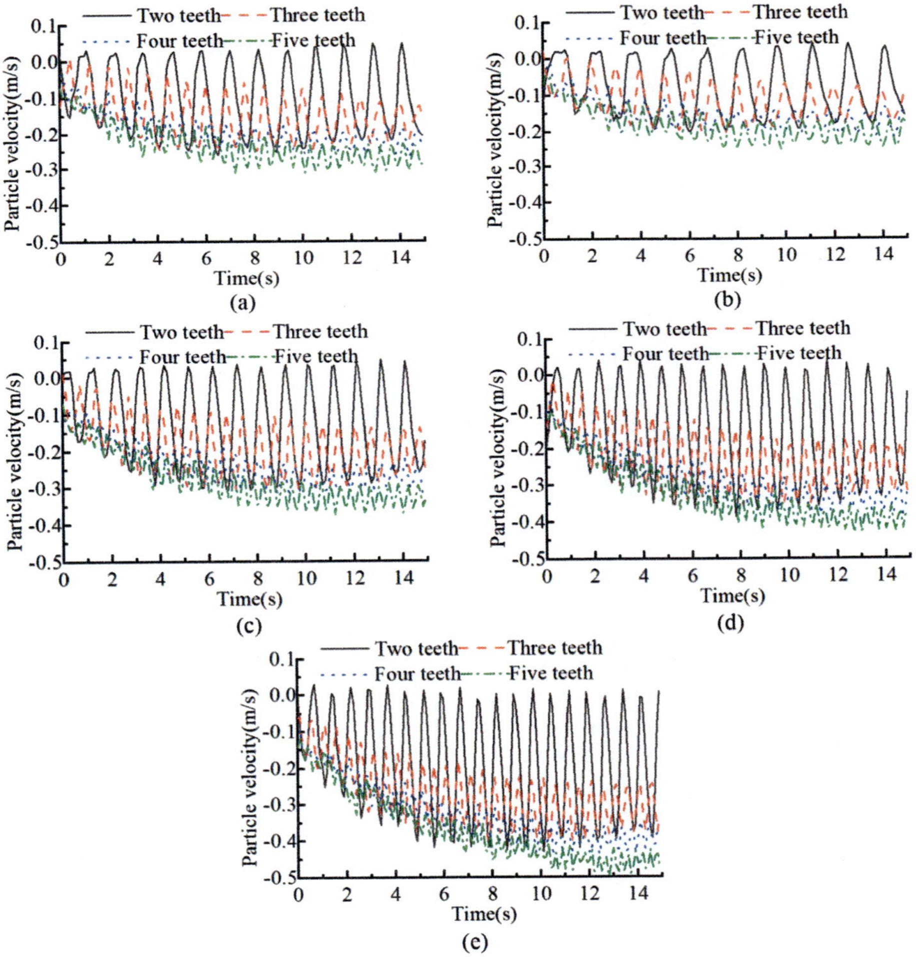

Fig. 7. The influence of rotating speed of star wheel on particle velocity: (a) 25 rpm; (b) 30 rpm; (c) 35 rpm; (d) 40 rpm.

3.4 Average Speed and Loading Efficiency of the Particles

In order to study the relationship between particle motion and loading capacity, the influence characteristics of star wheel speed and number of star wheel teeth on particle average speed and loading efficiency are obtained, as shown in Fig. 8.

As shown in Fig. 8(a), for the star wheel with the same number of teeth, the average speed of particles gradually increases with the increase of the star wheel speed, and the average speed difference between the two adjacent speeds is basically equal; On the other hand, under the same rotating speed, the average particle velocity increases with the increase of the number of star gear teeth. In general, the difference between the two adjacent speeds of the star wheel with more teeth is greater than that of the star wheel with less teeth, and the absolute value of the slope of the high speed curve is greater than that of the low speed curve, that is, the increase of the speed will expand the gap of the average particle speed between the star wheels with different teeth. In this way, when the number of teeth of the star wheel is small, the rotating speed of the star wheel can be changed to increase the particle speed.

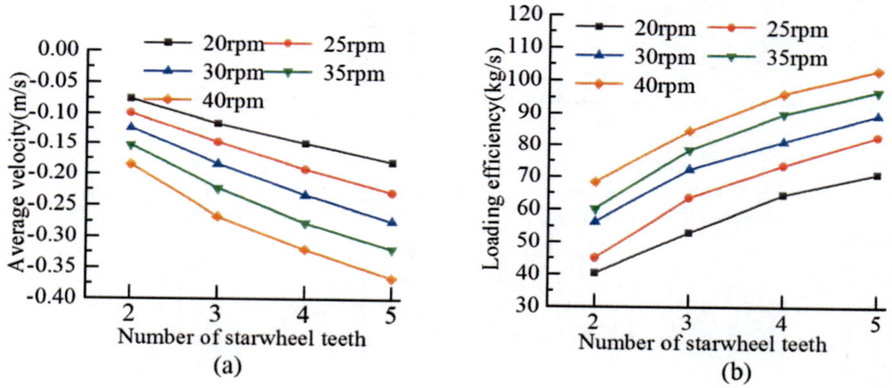

Fig. 8. Average speed and loading efficiency of the particles: (a) Average speed; (b) loading efficiency.

As shown in Fig. 8(b), the loading efficiency increases with the increase of star wheel speed. In particular, the influence of the speed of the five teeth star wheel on the loading efficiency is more significant than that of the two teeth star wheel. Combined with the analysis of Fig. 8(a), this is related to the x-direction movement of material particles. Increasing the speed in x-direction can effectively improve the loading probability of materials. However, with the increase of material movement speed, the collision between star wheel and material intensifies, and the contact slip between particles and shovel plate increases. The wear life of shovel plate and star wheel should be considered in practical use. In addition, with the increase of the number of star gear teeth, the material loading efficiency is significantly improved. The five teeth star wheel has high loading efficiency at 40 rpm and basically realizes full loading. However, with the increase of the number of teeth of the star wheel, the influence of the number of teeth on the loading efficiency decreases, especially when the speed is 20 rpm. Moreover, the increase of rotating speed can also increase the difference of loading efficiency of star wheels with different teeth numbers.

4 Conclusion

In order to study the performance of star wheel loading mechanism of roadheader, the loading model of star wheel is established by discrete element software. The particle loading process of star wheel under different number of teeth and different speed is simulated. The particle distribution, particle velocity and star wheel loading efficiency are studied. The main conclusions of this paper are as follows:

The number of teeth and rotating speed of the star wheel play an important role in particle loading. Increasing the number of teeth and rotating speed can improve the particle loading capacity. Increasing the number of star gear teeth can effectively improve the stability of particle movement speed and the loading reliability of roadheader, and can increase the service life of loading mechanism. In practical application, star wheels with different number of teeth and rotating speed can be selected to obtain a suitable loading capacity. The research provides a reference for the design and speed matching of roadheader star wheel.

Acknowledgment. This work was supported by the National Natural Science Foundation of China (52174146) and the Project of Shandong Province Higher Educational Young Innovative Talent Introduction and Cultivation Team (Performance enhancement of deep coal mining equipment).

References

1. Deshmukh, S., et al.: Roadheader-a comprehensive review. Tunnell. Underground Space Technol. **95**, 103–148 (2021)
2. Zong, K., Fu, S., Li, X.: Modelling and response analysis of multibody large-scale displacement of boom-type roadheader. Proc. Inst. Mech. Eng. Part K-J. Multi-Body Dyn. **235**(3), 326–337 (2021)
3. Kahraman, S., Sercan Aloglu, A., Aydin, B., Saygin, E.: The needle penetration index to estimate the performance of an axial type roadheader used in a coal mine. Geomech. Geophys. Geo-Energy Geo-Resour. **5**(1), 37–45 (2018). https://doi.org/10.1007/s40948-018-0097-3
4. Shaowu, W.: Test and research on parameters of planetary gear loading mechanism for mine roadheader. Coal Sci. Technol. **33**(9), 46–48 (2005)
5. Xiaodong, L., Xiaohuo, L.: Determination of star wheel loading mechanism's lowest rotating speed of road-headers which don't jam. J. Liaoning Techn. Univ. **25**(s2), 226–227 (2006)
6. Guodong, Z.: Confirmation of star wheel loading machine technical parameters. Coal Mine Mach. **27**(3), 379–381 (2006)
7. Gao, S., Mu, S., Qi, Z.: Design and research on star wheel loading mechanism for mine roadheader. Coal Mine Mach. **34**(8), 13–14 (2013)
8. Zhuo, X., Jun, M.: Research and confirmation structure size for star wheel loading machanism of roadheader. Coal Mine Mach. **30**(10), 9–10 (2009)

9. Xiaohuo, L., Chunhua, L., Jiquan, Y., Di, W.: Analyses of harmonic response for three helical-teeth star-wheel loading mechanism. World Sci-Tech R&D **32**(3), 339–341 (2010)
10. Linhui, Z.: Analysis about the influence factors of EBZ220 roadheader's loading device structural design. Colliery Mech. Elect. Technol. **6**, 74–77 (2012)
11. Fernandez, J.W., Cleary, P.W., Mcbride, W.: Effect of screw design on hopper drawdown of spherical particles in a horizontal screw feeder. Chem. Eng. Sci. **66**(22), 5585–5601 (2011)
12. Shimizu, B.Y., Cundall, P.A.: Three-dimensional DEM simulations of bulk handling by screw conveyors. J. Eng. Mech. **127**(9), 864–872 (2001)
13. Moysey, P.A., Thompson, M.R.: Modelling the solids inflow and solids conveying of single-screw extruders using the discrete element method. Powder Technol. **153**(2), 95–107 (2005)
14. Mcbridge, W., Cleary, P.W.: An investigation and optimization of the 'OLDS' elevator using discrete element modeling. Powder Technol. **193**(3), 216–234 (2009)

Enabling Sustainable Manufacturing in the Fashion Retail Industry Through the Demployment of Industry 4.0 Concept

Olivia Martinez and Yi Wang(✉)

Department of International Shipping, Logistics, and Operations Management, Plymouth Business School, Plymouth University, Plymouth 4 8AA, UK
{O.Martinez,yi.wang}@plymouth.ac.uk

Abstract. This paper identifies that Industry 4.0 (I4.0) paradigm does result in increased cyber-security risks for companies and industries that employ it. However, cyber-threats existed prior to that of I4.0 and although there is greater vulnerability under this paradigm there is also greater ability to protect against such attacks as technology evolves. For companies to maintain as high a possible level of cyber-security they must adapt their systems with the times staying as up to date as possible. In conclusion, after conducting a literature review to understand every aspect of the problem and solution and assessing their compatibility through the completion of the critical analysis; I4.0 is a suitable solution to the sustainability problems of manufacturing in the fashion retail industry. There are gaps in the research identified and resulting limitations to the ability of I4.0 to be the solution to all the sustainability problems identified. However, with technological and digital development at the centre of this paradigm and the resulting growth in abilities and knowledge offers the ability for further solution to be identified and these issues overcome.

Keywords: Industry 4.0 · Supply chain · Fashion supply chain · Sustainability

1 Introduction

This highly volatile business models applied in the fashion retail industry (FRI) are based around catering for instantaneously changing consumer preferences, short product lifecycles and product value determined by the ability to meet ever changing trends. This paved the way for brands such as 'H&M, Topshop and Zara' that were massive successes, receiving higher growth and profits than traditional brands [1].

The lifecycle of products has reduced 'from months and years to weeks and months' [2]. The combination of increased production, population, consumption, 'impulse' behaviours, and decreased product usage and lifecycle, has increased waste. This is not only of the final product, but the waste created during the manufacturing process also [3].

The FRI is one of the most polluting industries in the world [4]. Polluting in the areas of water, waste, air, chemical, and consumption of 'water… fossil fuels and energy' [5].

Y. Wang et al. (Eds.): IWAMA 2021, LNEE 880, pp. 759–765, 2022.
https://doi.org/10.1007/978-981-19-0572-8_97

Statistics show that annually 4–5 billion tonnes of CO2 is produced by the fashion industry, that also consumes '79 trillion litres' of water and pollutes another '190,000 tonnes' annually. The CO2 production is from unsustainable energy sources used by the industry [6]. The material waste from manufacturing accounts for 10% of each ton of clothing produced [7].

As well as this waste production there is the issues of social sustainability as brands chase the cost minimising routes of manufacturing in 'low-wage' developing countries To cut costs ever further fashion brands have disregarded the safety of these workers leading to 'sweat shops' [8] and thousands of deaths under these poor conditions.

This paper aims to identify and understand the growing sustainability concerns and issues associated with manufacturing in the fashion retail industry and illustrate some of the impacts these are having on the environment and society. Industry 4.0 (I4.0) is the new industrial revolution in manufacturing [9] and is suggested as the solution to the sustainability issues of the fashion retail industry. This paper will also identify and understand the concept of this paradigm and its operations and the application of this paradigm to achieving sustainability in manufacturing. Following on from this the information gathered and new information will provide an argument for the benefits and limitation of I4.0 and the application of I4.0 to the issues of the fashion retail industry. To conclude, an overall judgment will be made as to whether I4.0 is a justifiable and effective solution to the issue identified, remarks will also be made on the future of this paradigm and the limitations identified.

2 Literature Review

This revolution is proposed to increase global and domestic competitive advantages [10] and enable increased sustainability [11]. It aims to join physical 'resources, services, and humans' simultaneously during production. Along with the aid of 'sensors, actuators, and embedded software' for processing and communicating information that enables automation, simplification and flexibility [12]. This is a new system of manufacturing centred on value creation [13] and the 'adoption of artificial intelligence' [14].

The key technologies of this paradigm are: 'smart manufacturing... cyber-physical systems (CPS), cyber-physical production systems (CPPS)... internet of things (IoT)' [15, 16] 'cloud services' [11], and 'big data analytics (BDA)' [17].

The I4.0 paradigm is a German concept focussed on increasing efficiency, automation, and sustainability in production by creating 'self-aware and self-learning' machines that manage products entire life cycle. Enabled through CPS, IoT, cloud services and BDA these machines analyse and communicate available information directly to manufacturing so to instantaneously meet individual customer needs [11, 16, 17].

'Smart factory') 'smart manufacturing, smart products, smart supply chain and smart working' [18] are all dimensions of I4.0 that contribute to improved production, quality, innovation, speed, and product development [11, 18]. Smart products merge the information of customers and corporations to develop products and solutions for customers [19].

There are three forms of integration associated with I4.0, as briefly discussed. These enable value creation at each stage of the product lifecycle through increased customization of products [12, 13].

To make something sustainable is described as ensuring 'the needs of the present' are met without 'compromising the ability of future generations' in meeting their needs [20]. 'The triple bottom line' (TBL) is a measurement of sustainability in businesses. The TBL consists of three performance aspects; 'social, environmental and financial'/economic [21].

I4.0 presents the ability to make manufacturing sustainable through the use of 'digital solutions' [18]. These solutions enable more sustainable resource allocation of 'products, materials, energy and water' within and between factories through 'industrial symbiosis' [22]. As a result of this 'costs and pollution, raw materials and CO2 emissions', will decrease [23].

The smart dimensions discussed earlier are proposed to be 'key requirements' in firms attaining sustainability [18] through the new technological structure of manufacturing throughout 'value creation networks' [22].

The functioning of smart manufacturing and factories works together to identify the newest trends [18] for the customization of products. This results in better and more sustainable resource consumption [11] in efficiently delivering what consumers want. Smart factories work on decentralizing decision making to allow for the instantaneous manufacturing of products. This works to reduce waste that would be produced through planning and decision-making inefficiencies [12]. Smart manufacturing means designing products to be reusable, re-manufacturable, recyclable etc. at the end of their life. Putting sustainability at the centre of the design means ensuring an extended product lifecycle [24] again working towards closing the loop and avoiding waste.

3 Discussion

I4.0 enables more sustainable resource allocation of 'products, materials, energy and water' [22]. As a result of this 'costs and pollution, raw materials and CO2 emissions', will decrease [23]. These are the areas the FRI is known for polluting [25] as well as negative consumption of 'water… fossil fuels and energy' [5].

The issue of the tonnes of CO2 produced [3] using unsustainable energy sources [6], and the trillions of litres of water consumed and polluted annually during the material manufacturing processes [3]. The use of renewable smart energy provided by the smart grid is a solution. Allowing smart factories to become ever 'self-sufficient'. They are supplying and consuming their own energy through I4.0 [22].

Another solution provided by I4.0 to address these problems is the decentralized approach to decision making. In machines being able to make their own intelligent decisions their self-aware operations perform on the best efficiency and optimizing their functions reducing their consumption of energy and production of gases. The 'water 4.0' concept operates on the same basis however aiming for 'resource-efficient and flexible water management' [12].

Smart manufacturing is a solution applicable to the problem of the increased waste creation from the growing production, population, consumption, 'impulse' behaviours, and decreased product usage and lifecycle [3]. In designing products to be reusable, re-manufacturable, recyclable etc. at the end of their life. Sustainability is at the centre of the design ensuring an extended product lifecycle [24] and working towards closing the loop and avoiding waste. Another function of smart manufacturing and factories to avoid fashion textile waste is working together to identify the newest trends [18] for the customization of products. This results in better and more sustainable resource consumption [11] in efficiently delivering what fashion consumers want.

I4.0 do offer forms of solutions to the issues of sustainability in the manufacturing of the fashion retail industry. However, there is a gap in the literature in providing an element of I4.0 to apply to and solve the major sustainability issue of the use and pollution of chemicals in the FRI [26].

Smart working works to boost the productivity of and provide aid to human workers through the support of technology (Kagermann, et al. 2013, cited in Frank, et al. 2019). In working with these machines humans can gain skills and competencies that further increase workers efficiency (Zhong, et al. 2017). This is one way in which efficiency can be improved through I4.0 regarding the employment of human resources.

Using horizontal integration smart supply chains operate to improve sourcing of raw materials, improve delivery time and costs of operations [11]. The intelligence of machines means the immediate sharing of information and data enables the most educated decisions to be made and the most effective use of materials and resources as a result [17, 22]. This better allocation of resources results in a cost-reduction as waste product decreases [27].

However, this interconnection and intelligence makes manufacturing systems vulnerable to cyber or internet-based attacks. Such attacks can target programming and functioning of machines, causing mass disruption. Therefore, the efficiency of I4.0 would be compromised [28].

Closing the loop by retrofitting existing equipment with modern digital components of I4.0 uses CPS [22] extends the lifecycle of manufacturing products to save costs on replacement machinery and equipment [23]. Closing the loop also in the use of smart manufacturing and factories leads to manufacturing for an extended lifecycle [24] and against the most recent and desired trends identified [18]. This improvement in the function of operations increases manufacturing efficiency in working to avoid waste of the final product.

The use of renewable smart energy provided by the smart grid is a solution. Allowing smart factories to become ever 'self-sufficient'. They are supplying and consuming their own energy through I4.0 [22]. This is a form of energy efficiency that sees environmental costs decrease [27].

The decentralizing of decision making enables machinery and manufacturing to be more efficient. Removing the inefficiencies of long planning and decision-making processes allows machines to make instantaneous decisions removing the waste of time [12] and decision inefficiencies.

There are two perspectives on the outcome of this on employees. The increased automation of machines and implementation of smart systems creates high risks of job losses for the human aspect of manufacturing. This decrease in workers will leave only a few skill and knowledge-based jobs to monitor the machines [29].

Considering the location of most manufacturing operations, such as the fashion industries, being in developing countries; this is where there is a lack of education and high level of poverty [30].

An analysis of the employment effects of I4.0 in Germany shows that it does not have any significant or determinable effects. Although there were job losses there followed job creation, this resulted in a structural change in employment. Manufacturing positions declined however was counteracted by the creation of technological positions [29].

However, contradicting these claims is another study on the impacts of I4.0 on employment in the UK. This research claims that the workforce of the future needs to be educated in meeting the changing demands of future employment under I4.0. Underperformance in the areas of education is predicted to cause a shortage in the skilled labour required [31].

Along with the increased digitalization and automation of I4.0 come severe concerns over the increasing 'cyber-security risks' accompanying it [28]. The commitment of companies to I4.0 presents the highly complex job of navigating the cyber-security threats [32].

The existence of IoT and growth in data accumulation and interconnectivity has created increased vulnerability to and adaptation of cyber-threats. Cyber-attacks can occur for such reasons as 'financial and strategic' gain, and as technology evolves so does the sophistication of these attacks [33].

There are many precautions industries can take to reduce the threat of cyber-attacks – firewalls, remote accessing, employees have cyber-safety awareness, strengthening passwords, etc. However, as technology develops so do threats too technology; therefore no cyber-security that exists can be solely relied on currently and into the future [32, 33].

4 Conclusions

The FRI's sustainability issues in the manufacturing process identified many social and environmental impacts. The review of sustainable manufacturing enabled through I4.0 identified the many solutions directly applicable to the issues in the FRI. These opportunities are the reason the I4.0 paradigm has been suggested as the solution. It can be determined that the whole paradigm of I4.0 focusses on the improved efficiency and allocation of all resources. Enabling this through the implementation of smart dimensions and automated equipment in turn reduces costs from reducing the amount of waste. To solve social issues through I4.0 several approaches are suggested that focus on intensive training and motivation of employees. This literature will be conducted to gain an in-depth understanding of the sustainability issues faced by manufacturing in the fashion retail industry. Also a detailed description of a decision-making paradigm suggested as the solution to these issues and explanation of the ability of this solution to achieve sustainability.

The use and impact of chemicals in the manufacturing process is another unsustainable practice of the fashion industry. The production of 'synthetic and semi-synthetic fibres' used in fashion clothing items are comprised of a mass of 'highly toxic' chemicals with adverse effects on humans and the environment. Naturally sourced fibres were the base for most clothing fabrics in the past; these were derived from sustainable sources such as animals or vegetables. However, again relating back to the fashion industry manufacturing in developing countries already lacking education and finances; this adaptation of training and education of the workforce may not be possible. I4.0 is not seen to create or destroy jobs directly. Indirectly however this is not the case. Jobs under this paradigm would become more skilled and knowledge based requiring an adaptation of education and training, which developed countries can cater for. However, in the developing countries where manufacturing jobs exist this may not be an option and therefore leave the workforce unemployed. Therefore, there is a combination of the need for training, the ability to make educated decisions, implementation of more machines to aid humans and the resulting risk of job loss.In consideration of these factors, I4.0 does not appear to be a solution to the social issues in manufacturing. But instead contribute to them; in developing countries of rife poverty the creation of job losses does not offer solution or aid.

References

1. Sull, D., Turconi, S.: Fast fashion lessons. Bus. Strateg. Rev. **19**(2), 5 (2008)
2. Fernie, J., Sparks, L.: Logistics and Retail Management, 3rd edn., p. 32. Kogan Page, London (2009)
3. Niinimäki, K., Peters, G., Dahlbo, H., Perry, P., Rissanen, T., Gwilt, A.: Author correction: the environmental price of fast fashion. Nat. Rev. Earth Environ. **1**(5), 189–192 (2020)
4. Shen, B., Li, Q., Dong, C., Perry, P.: Sustainability issues in textile and apparel supply chains. Sustainability **9**(9), 1592 (2017)
5. Kozlowski, A., Bardecki, M., Searcy, C.: Environmental Impacts in the fashion industry. J. Corp. Citizsh. **2012**(45), 16–36 (2012)
6. Perry, P.: Exploring the influence of national cultural context on CSR implementation. J. Fashion Mark. Manag. Int. J. **16**(2), 141–160 (2012)
7. Mendes, F., Dos Santos, M.: Fashion garment manufacturing – FGM and cyclability theory. IOP Conf. Ser. Mater. Sci. Eng. **254**, 1–2 (2017)
8. Khan, Z., Rodrigues, G.: Human before the garment: Bangladesh tragedy revisited. Ethical manufacturing or lack there of in garment manufacturing industry. World J. Soc. Sci. **5**(1), 22–24 (2015)
9. Lasi, H., Fettke, P., Kemper, H., Feld, T., Hoffmann, M.: Industry 4.0. Bus. Inf. Syst. Eng. **6**(4), 239–242 (2014)
10. Castelo-Branco, I., Cruz-Jesus, F., Oliveira, T.: Assessing Industry 4.0 readiness in manufacturing: evidence for the European Union. Comput. Ind. **107**, 22–32 (2019)
11. Frank, A., Dalenogare, L., Ayala, N.: Industry 4.0 technologies: implementation patterns in manufacturing companies. Int. J. Prod. Econ. **210**, 15–26 (2019)
12. Stock, T., Obenaus, M., Kunz, S., Kohl, H.: Industry 4.0 as enabler for a sustainable development: A qualitative assessment of its ecological and social potential. Process. Saf. Environ. Prot. **118**, 254–257 (2018)
13. Wang, S., Wan, J., Li, D., Zhang, C.: Implementing smart factory of Industrie 4.0: an outlook. Int. J. Distrib. Sens. Netw. **12**(1), 1–3 (2016)
14. Fedorov, A., Goloschchapov, E., Ipatov, O., Potekhin, V., Shkodyrev, V., Zobnin, S.: Aspects of smart manufacturing via agent-based approach. Proc. Eng. **100**, 1572 (2015)
15. Rossit, D., Tohmé, F., Frutos, M.: Industry 4.0: smart scheduling. Int. J. Prod. Res. **57**(12), 3802–3813 (2018)
16. Vaidya, S., Ambad, P., Bhosle, S.: Industry 4.0 – a glimpse. Proc. Manuf. **20**, 233–238 (2018)
17. Zhong, R., Xu, X., Klotz, E., Newman, S.: Intelligent manufacturing in the context of Industry 4.0: a review. Engineering **3**, 616–617 (2017)

18. Longo, F., Nicoletti, L., Padovano, A.: Smart operators in industry 4.0: a human-centered approach to enhance operators' capabilities and competencies within the new smart factory context. Comput. Ind. Eng. **113**, 144–159 (2017)
19. Dalenogare, L., Benitez, G., Ayala, N., Frank, A.: The expected contribution of Industry 4.0 technologies for industrial performance. Int. J. Prod. Econ. **204**, 385 (2018)
20. Brundtland, G.: World Commission on Environment and Development: Our Common Future. Oxford University Press, Oxford (1987)
21. Slaper, T., Hall, T.: The triple bottom line: what is it and how does it work? Indiana Bus. Rev. 1 (2011)
22. Stock, T., Seliger, G.: Opportunities of sustainable manufacturing in Industry 4.0. Proc. CIRP **40**, 536–541 (2016)
23. Carvalho, N., Chaim, O., Cazarini, E., Gerolamo, M.: Manufacturing in the fourth industrial revolution: a positive prospect in sustainable manufacturing. Proc. Manuf. **21**, 671–678 (2018)
24. Sharma, R., Jabbour, C., Jabbour, A.: Sustainable manufacturing and industry 4.0: what we know and what we don't. J. Enterp. Inf. Manag. **34**(1), 230–266 (2020)
25. Bhardwaj, V., Fairhurst, A.: Fast fashion: response to changes in the fashion industry. Int. Rev. Retail Distrib. Consum. Res. **20**(1), 165 (2010)
26. Bhalla, S., Singh, Z.: Toxicity of synthetic fibres & health. Adv. Res. Text. Eng. **2**(1), 1–2 (2017)
27. Machado, C., Winroth, M., Ribeiro da Silva, E.: Sustainable manufacturing in Industry 4.0: an emerging research agenda. Int. J. Prod. Res. **58**(5), 1462–1484 (2019)
28. Prinsloo, J., Sinha, S., von Solms, B.: A review of industry 4.0 manufacturing process security risks. Appl. Sci. **9**(23), 5105–5133 (2019)
29. Enzo, W.: Industry 4.0: job-producer or employment-destroyer? Inst. Employ. Res. **2**, 2–5 (2016)
30. Tilak, J.: Education and poverty. J. Hum. Dev. **3**(2), 192–193 (2002)
31. Gao, J., Souri, M., Keates, S.: Advances in Manufacturing Technology XXXI. IOS Press, Netherlands (2017)
32. Lezzi, M., Lazoi, M., Corallo, A.: Cybersecurity for Industry 4.0 in the current literature: a reference framework. Comput. Ind. **103**, 97–105 (2018)
33. Ustundag, A., Cevickan, E.: Industry 4.0: Managing the Digital Transformation. SSAM, Springer, Cham (2018). https://doi.org/10.1007/978-3-319-57870-5

Blockchain Technology and the Efficiency of Supply Chains in the Marine Freight Industry

Alissa Schwab and Yi Wang(✉)

Department of International Shipping, Logistics, and Operations Management, Plymouth Business School, Plymouth University, Plymouth PL4 8AA, UK
{alissa.schwab,yi.wang}@plymouth.ac.uk

Abstract. This paper deals with the influence of blockchain technology on the efficiency of supply-chains in the marine freight industry. Blockchain is described and the issues of the shipping industry investigated. Following this, the implementation of blockchain technology in marine freight is critically evaluated in regard to the described problems of this industry. Furthermore, this paper points out the advantages and disadvantages of blockchain followed by a conclusion which shows that this new technology can resolve certain problems if a number of obstacles have been overcome.

Keywords: Blockchain · Supply chain efficiency · Marine freight

1 Introduction

The technology behind the blockchain ha been used in many industries such as the marine freight. Some companies like IBM and Maersk have already started to test this technology in logistics. Shipping is currently the main global transportation mode and the demand for cargo shipping is continuously increasing. Forecasts indicate that the value of the global marine freight industry will grow to $298.8 billion until 2024 which would be an increase of 21% since 2019 [1]. Container shipping is the most cost-effective way for transporting goods overseas, which is why it is particularly important for international supply-chains [2]. In spite of and more explicitly because of this growth some of the procedures and strategies need to be modernised and improved. Blockchain brings a huge potential for the shipping industry but also some hurdles. In this paper the advantages and disadvantages of blockchains are evaluated and it is shown how this could be a way to improve the efficiency of supply-chains in the marine freight industry. This paper starts with a brief introduction of the background, then it reviews the blockchain applying in the maritime industry, afterwards a critical evalution of blockchain is discussed. The paper ends with a conclusion.

2 Literature Review

Currently shipping is the main global transportation mode for products in the globe [3]. Nevertheless, problems like 'delays, damaged goods, […] theft and unsuitable terminal

Y. Wang et al. (Eds.): IWAMA 2021, LNEE 880, pp. 766–771, 2022.
https://doi.org/10.1007/978-981-19-0572-8_98

loading' is daily fare [4]. Hereafter, the main reasons for this inefficiency as well as the lack of effectiveness is shown.

One of the biggest issues in the shipping industry is the inefficiency that occurs especially when transporting goods across multiple borders. This lack of efficiency arises when intermediaries and interactions with a third party becomes necessary to ensure trust. [5] Participants in the marine freight industry are amongst others; importers, exporters, carriers, ship operators and governments [6]. Many stakeholders are included which means that the communication between the different parties is necessary to guarantee a continuous flow of goods and information. Especially the communication 'between private companies in and around ports, also including local authorities' [4] needs so be ensured. This lack of cooperation and communication which can be observed at the moment needs to be solved.

Despite the digitization paper-documentation is still used which is unpractical and error-prone [5]. For instance, bills of loading are very often written on paper. In that way, the information is out of date very quickly and the accuracy which is needed for a successful supply-chain is suffering [7]. One can imagine how much effort it takes to communicate between all those involved. It is also depicted how a port community system improves the efficiency of information flows [4]. With a port community system, the participants can access the same date but this means that if a shipping line works with many ports which they do most of the time, they have to connect to several systems. It will be explained further down how blockchain can help to improve these communication processes even more.

In the shipping industry many complicated regulations and laws need to be obeyed which leads to slower processes [8]. Most of the time codes of conduct of more than one country has to be followed. Therefore, lawyers and experts for different region are needed. It is also very time-consuming to collect all the information for a marine-insurance-contract. If a dispute occurs, it is often very time-consuming and difficult to settle this, as many parties are involved who are working with different information [7].

The chains represent the cryptographical connection of the blocks. A central authority, managing the database is not needed. Through cryptography the blocks are connected and the participants who want to access the data are verified. Once the data is entered into the system it cannot be changed. If modifications seem necessary one have to enter a new record [6].

One part of the blockchain technology are smart contracts. This is a method with which several parties can agree on a topic with up-to-date information and the content of the contract can be executed through the blockchain technology [7]. However, smart contracts are not only a digital version of a paper contract, but a piece of code [9]. In that way they can be 'self-verifying, self-executing, and self-response agreements' [4].

3 Discussion

3.1 Positive Aspects of Applying Blockchain

In real-time, information about 'location, temperature, capacity, movement and damage' can be shared. This information is more up-to-date and accurate than paper documents [4]. Data can not only be shared with business partners but also with tax authorities and governments. In the shipping industry, information like the ship details, the exact journey as well as certificates of insurance can be provided to permissioned stakeholders. Through that transparency and availability of information one can adapt quicker according to changing circumstance. For example, if a container gets lost, it is recognized immediately and a solution as well as the responsible party can be found more quickly [7]. This transparency is also guaranteed as the information stored in blockchains cannot be deleted. One can access the history of transactions at any time and miscommunication can be eliminated as every business partner has access to the same data. Tracing products and knowing the origin is especially important if perishable freight like in the food sector is shipped [6]. Maersk, the biggest ocean carrier and IBM, an international software company, established a joint venture to develop a cloud-based platform on blockchain technology. As the Chief Executive of this new company Micheal J. Whites says: 'opportunities coming from streamlining and standardizing information flows using digital solution' play a huge role [10]. All in all, it is clear, that benefits through an improved communication along the supply chain can be found while using blockchains.

[10] shows that there is a huge potential to optimize processes in the shipping industry. The implementation of blockchain technology can optimize the supply chain processes by saving time and costs. Thus, companies gain competitive advantages when working with blockchains [11]. Through the visibility which was explained above, ports and shipping companies can plan ahead and therefore operate more efficiently. Forecasting can be done more precisely [6]. Time can be saved, because disputes can be solved way quicker in processes in which blockchain was integrated as confusion gets eliminated. If an incident is content of a smart contract it is very clear who is responsible at what time for the cargo. Smart contracts are being executed in real-time which is why the whole process is much quicker compared to traditional contracts [5]. If for example goods are received, the payment can made automatically and quickly, without any intermediaries. Furthermore, risks for marine-insurance-contracts can be calculated automatically and therefore fairness is guaranteed. Every party can reconstruct the insurance price [7]. By cutting down paperwork and administrative tasks accuracy increases, and fewer errors caused by humans are made. No more filling of many paper documents also saves time [6]. In addition to that, costs can be reduced as less intermediaries are necessary. This disintermediation includes for example lawyers and accountants who would not be needed anymore when smart contracts are installed [12]. One can conclude that the increased cooperation of all the parties included, improves the supply-chain efficiency when using blockchain technology.

Space of improvement can also be found in the security of cargo while being at sea, as still many containers get lost every year. In the period of 2017 to 2019 the average annual loss was 779 containers [13]. If a container falls overboard, sensors could recognize this and put this information into a blockchain. The affected party could then react quickly to resolve the issue. Through the implementation of the blockchain technology the security of goods improves enormously. For instance, only freight forwards who can show the digital approval which they get from the blockchain are allowed to pick up the goods [4]. In addition to that, security can get improved as the data is encrypted and therefore protected from cybercrime. Smart contracts are secure as they are permanently stored in the blockchain and cannot be changed without creating a new record [6].

3.2 Negative Aspects of Applying Blockchain

To Information must be ensured that the input is correct, otherwise the output will be false [14]. It is important that this awareness of validation is created between the users. There is no easy solution yet for 'linking physical objects with digital identities'. RFID-tags which are used for creating this connection are expensive to implement and know-how in this area is necessary [15]. Furthermore, the implementation requires know-how and financial resources. Especially public blockchains are expensive to set up [16]. Blockchain requires adaption of many if not all parties included in a supply chain to profit from this technology. This is due to the fact that value is created to more people are participating [9].

Even if the technology is used from all parties of the supply-chain, many regulations need to be fulfilled. Meaning that it is not as easy as "one solution firs all", especially if several companies from different countries are included [15]. Johnson states that it is not likely that the whole industry is joining one solution. Instead many software firms are developing several different private blockchains [17]. These permissions blockchains are characterised by an authority level which is responsible for controlling the access to the data only to verified users [18].

The adaptation of values which is necessary to trust a public blockchain with thousands of participants won't happen in only a few years [17]. Figure 4 displays the results of a survey in which employees of logistics companies in the UK have been asked about, how they rate the influence of blockchain technology for their business [19]. One can see that only 5% of the respondents see a high relevance to their company. 44% evaluate the impact with "little relevance" or even with "very little relevance" which leads to the conclusion that the acceptance and the awareness of this innovation is still quite low

Another problem when applying blockchain in the shipping industry is that there is not a lot of expertise about this technology in this field. One the one hand, there is not much public knowledge, however, on the other hand also the experts are still figuring things out [6]. For a successful implementation professional in IT as well as in operations are required. It definitely can be said that more research regarding the actual implementation is needed. It is not a project that will be successful in the short term. The complexity of blockchain technology requires 'individuals with interdisciplinary talents' [20]. These talents need to be educated, but at the moment there are very little institutions which are teaching about this topic. Achieving staff to be more trained in this field could also be support by the government. For the marine freight industry this means that the willingness for improving the efficiency of the supply-chains must be created.

4 Conclusions

Many experts believe that the implementation of blockchains in shipping can be one of the biggest transformations of the last few year. There are several weighty advantages which cannot be neglected by doubters. Especially the improved visibility and the optimization through smart contracts impact the efficiency. Additionally, the data-security aspect is getting more and more important. In that connection it is important to note that not everyone is convinced yet that the blockchains are eliminating the impact of hacker attacks.

As stated before, many organisations are experimenting with blockchain but most of them are big companies. For this concept to be successful it is important that as many as possible parties participate. Meaning, that in the future also small companies should start dealing with it in order to find out if the costs of moving goods actually decreases. To overcome the disadvantages, it is also necessary to educate workers and managers about this topic as well as to create the awareness that change is needed. The negative aspect of dependency could not be avoided completely as this is the premise which creates trust within the participants.

To summarize, one can say that blockchain technology has a huge potential to improve the processes in the shipping industry but there is still a lot of space to develop a widespread implementation. Saying so, detailed and concrete implementation possibilities should be outlined to bridge the gap between theory and reality. In that context, a solution to link physical objects with digital information is one of the most important preconditions. Ascertainment regards regulation and legislation needs to be discussed and provided. That is why governments should deal with this innovation and eventually even create an institution which is developing international guidelines. All in all, blockchain technology is a concept that has a huge potential to revolutionize the freight industry if the disadvantages are dealt with.

References

1. WHO. MarketLine, "Global - Marine Freight". https://advantage-marketline-com.plymouth.idm.oclc.org/Analysis/ViewasPDF/global-marine-freight-101415 Accessed 1 Feb 2021
2. Statista (2020). Container shipping worldwide. https://www-statista-com.plymouth.idm.oclc.org/study/13992/container-shipping-statista-dossier/ Accessed 1 April 2021
3. Yang, C.-S.: Maritime shipping digitalization: Blockchain-based technology applications, future improvements, and intention to use. Transp. Res. Part E: Logistics Transp. Rev. **131**, 108–117 (2019)
4. Tsiulin, S., Reinau, K.H., Hilmola, O.-P., Goryaev, N., Karam, A.: Blockchain-based applications in shipping and port management: a literature review towards defining key conceptual frameworks. Rev. Int. Bus. Strategy **30**(2), 201–224 (2020)
5. Field, A.M.: Blockchain For Freight? J. Commer. **18**(5), 88–92 (2017)
6. Jugović, A., Bukša, J., Dragoslavić, A., Sopta, D.: The possibilities of applying blockchain technology in shipping. Pomorstvo **33**(2), 274–279 (2019)
7. Tuften, S.: Getting smart about marine. J. Australian New Zealand Inst. Ins. Financ. **41**(2), 1–4 (2018)
8. Riley, S.: How blockchain is poise to impact supply chain. Supply Demand Chain Exe. **18**(4), 24–25 (2017)
9. Johnson, E.: "Smart contracts for shipping offer shortcut to blockchain adoption". JoC Online, pp. 1–5 (2018)
10. van MARLE, G.: "Maersk and IBM launch blockchain joint-venture to 'reshape global supply chains.", Canadian Sailings, pp. 72–78 (2018)

11. Elrom, E.: The Blockchain Developer A Practical Guide For Designing, Implementing, Publishing, Testing, And Securing Distributed Blockchain-Based Projects/Elad Elrom. Apress, Springer eBook Collection (2019)
12. Iansiti, M., Lakhani, K.R.: The truth about blockchain. Harv. Bus. Rev. **95**(1), 118–127 (2017)
13. World Shipping Council (2020). "Containers Lost at Sea - 2020 Update" https://www.google.com/url?sa=t&rct=j&q=&esrc=s&source=web&cd=&ved=2ahUKEwj90MGF3PXtAhW08uAKHW6pCcEQFjADegQIARAC&url=https%3A%2F%2Fwww.worldshipping.org%2FContainers_Lost_at_Sea_-_2020_Update_FINAL_.pdf&usg=AOvVaw3O10ysIHoIwvvC88z2VHHV. Accessed 30 Dec 2020
14. Szewczyk, P.: "Application of blockchain technology in supply chain management", scientific papers of silesian university of technology. organization and management / Zeszyty Naukowe Politechniki Slaskiej. Seria Organizacji i Zarzadzanie **136**, 591–600 (2019)
15. Katsikouli, P., Wilde, A.S., Dragoni, N., Høgh-Jensen, H.: On the benefits and challenges of blockchains for managing food supply chains. J. Sci. Food Agri. **2**, 34–40 (2020)
16. Pirrong, C.: Will blockchain be a big deal? reasons for caution. J. Appl. Corp. Financ. **31**(4), 98–104 (2019)
17. Johnson, E.: Building blockchains: questions abound in the long, slow march to adoption in shipping and logistics. J. Commerce **19**(16), 10–14 (2018)
18. Gai, K., Wu, Y., Zhu, L., Xu, L., Zhang, Y.: Permissioned blockchain and edge computing empowered privacy-preserving smart grid networks. IEEE Int. Things J. **6**(5), 7992–8004 (2019)
19. Statista (2019). "Relevance of blockchain for logistics companies in the UK 2019"https://www-statista-com.plymouth.idm.oclc.org/forecasts/1015512/relevance-of-blockchain-for-logistics-companies-in-the-uk. Accessed 1 Feb 2021
20. Junyan, H., Siming, R., Weichao, L.: Thinking about development of blockchain. China Econ. Trans. (CET) **3**(2), 45–59 (2020)

Author Index

Y. Wang et al. (Eds.): IWAMA 2021, LNEE 880, pp. 773–775, 2022.
https://doi.org/10.1007/978-981-19-0572-8